D0164330

ANATOMY OF THE CHORDATES

ANATOMY
OF THE
CHORDATES

FOURTH EDITION

CHARLES K. WEICHERT

Late Professor of Zoology
McMicken College of Arts and Sciences
University of Cincinnati

McGraw-Hill Book Company

New York   St. Louis   San Francisco   Düsseldorf
London   Mexico   Panama   Sydney   Toronto

ANATOMY OF THE CHORDATES

Library of Congress Catalog Card Number 70-121668

ISBN 07-069007-3

4567890 HDBP 76543

PREFACE

The reception accorded the first three editions of "Anatomy of the Chordates" has been most gratifying to the author. Communications from students, teachers, and researchers from many parts of the world have emphasized the need for gathering, organizing, and presenting the vast amount of material included in this volume.

With the great upsurge in scientific research during the past decade, supported largely by the federal government, state governments, and various foundations, a bewildering array of new discoveries has necessitated a revision of many previously held concepts. One might say that anatomy is anatomy and it does not change. A knowledge of anatomy, however, without an understanding of functional or physiological implications, has little meaning, nor is it of much interest to students seeking an understanding of morphological phenomena. In this volume, unlike most textbooks on anatomy, every effort has been made to include discussions of the functions of body structures along with anatomical descriptions. Older concepts delineating various scientific disciplines are falling by the wayside. New discoveries in chemistry, as applied to certain biological phenomena, indicate evolutionary trends previously unsuspected. Discoveries made possible by the development of the electron microscope have revolutionized our concepts of the ultrastruc-

ture of many anatomical units and revealed details which have greatly clarified our knowledge of how parts function. Such information has been included wherever appropriate and feasible.

A companion volume, "Representative Chordates," also published by the McGraw-Hill Book Company, is available for those students who wish to supplement their reading by making direct observations of anatomical structures in the laboratory. It is essentially a laboratory manual, including directions for dissection as well as anatomical descriptions of the lamprey, the spiny dogfish, the mud puppy *Necturus,* and the cat. Depending upon the selection of material and attention to detail, it may be used for one-quarter, one-semester, or even one-year laboratory courses.

The present book begins with a review of the phylum Chordata. Much of this portion is presented according to the most generally used scheme of classification of chordates. Natural history, evolutionary trends, and nontechnical aspects of the animals included in the phylum are discussed briefly. Emphasis is placed, for the most part, upon more familiar living chordates. References to fossil forms have been included where necessary and desirable. The material is meant to parallel certain aspects of vertebrate paleontology but not to include it. The general review of the phylum is followed by a short chapter on development. An understanding of the significance of germ layers and tissues is basic, in many cases, to an understanding of the comparative anatomy of the organ systems of chordates.

Each organ system is covered in a separate chapter. Contrary to the convictions of certain biologists, the endocrine glands are treated as a separate system because of their profound effect upon the development and function of so many body structures. The nervous system and sense organs are, for convenience, treated independently despite their obviously close connection. At the end of each chapter is a summary of the main points emphasized in that chapter. Some critics have objected to the sequential arrangement of the chapters on the organ systems. The subject matter, however, in most cases is presented in such a manner that the information discussed in one chapter does not presuppose a knowledge of that contained within a previous chapter. The instructor should feel free to follow a sequence of his choice.

A summary chapter, giving characteristics and advances of members of the phylum, has been placed at the end of the book. Although this chapter is organized in a manner similar to that of Chap. 2, there is little duplication of subject matter.

Certain correspondents have deplored the paucity of reference material in former editions. The sheer volume of subject matter is so great that inclusion of detailed reference data would make the book entirely too voluminous. References are given in many places in the text where new concepts or new discoveries are introduced.

The author wishes to express his appreciation to his wife, Kathryn H. Weichert, for her understanding and forbearance during the many months when he was occupied with the preparation of the manuscript.

Every effort has been made to present the subject matter clearly and accurately. The author will appreciate it if readers will call to his attention any inaccuracies and inconsistencies which may have escaped his notice.

Charles K. Weichert

CONTENTS

| 1 |

INTRODUCTION

The science of biology is the study of living things in all their aspects. The scope of biology is so vast that no single person can hope to master all its ramifications. It is realized today that no scientific discipline can be sharply delineated. The highly complicated ultrastructure of cells of various kinds as revealed by the electron microscope, the ever-increasing knowledge of molecular interactions, the elaborately complex influence of enzymes on all phases of life processes, the vital role played by deoxyribonucleic acid (DNA) and ribonucleic acid (RNA) in forming proteins (including enzymes), the way cells react to each other, the manner in which cell products may influence cells, tissues, and organs far removed from them, etc., are all of basic concern to the modern biologist and must be considered by him in acquiring an understanding of living things in the past, present, and future. Since most familiar living things come under two general groups, plants and animals, biology has been conventionally subdivided into two great areas: botany, which deals with plants, and zoology, which treats of animals. With the tremendous growth of scientific knowledge in recent decades, the distinctions between these areas of biology, at least at molecular and ultrastructural levels, are less and less apparent. The close interrelationship of biology, chemistry, and physics must be emphasized. Many biological phenomena can be explained only in physical and chemical terms. Without a knowledge of chemistry and physics the biologist cannot go far in his study of living things. In this volume we are concerned primarily with a restricted, but large, group of animals, the *Chordates.*

METHODS OF APPROACH TO THE STUDY OF BIOLOGY

There are several general methods of approach to the study of biology. The more important ones are as follows:

Study of the Physical and Chemical Properties of Protoplasm, the only known form of matter in which life is manifested. Such study today involves many scientific disciplines.

Morphology, the science of form and structure of living things. It may be subdivided in turn into a number of more specialized fields of study which include the following:

Gross anatomy, dealing with structures which can be seen with the unaided eye without the aid of optical instruments. Dissection is the method of approach that is used in gross anatomy.

Histology, or microscopic anatomy, which is the study of the more minute details of the relations of normal tissues. Preparations which can be examined under the microscope are used in histological studies.

Cytology, or cellular anatomy, which today not only includes study of the detailed structures of cells but is also concerned with correlating structural features with cell function as well as with the interactions of cells.

Embryology, or developmental anatomy, which treats of the form and structure of the developing embryo and which utilizes gross, histological, and cytological methods of approach.

In recent years the scope of cytological and embryological studies has widened. Emphasis has been focused upon the biochemical mechanisms controlling cellular and developmental activity.

Physiology, the branch of biology which deals with the functions of the living organism and its parts. These activities are so complex and the various structures of the body are so closely interrelated that the functions of organs must always be studied in relation to those of other organs.

Ecology, the science of organisms as they are affected by their environment. Both morphological and physiological aspects of animals in relation to their environment are considered in the study of ecology.

The Study of Evolution, the fundamental biological principle that existing organisms are the result of descent, with modification, from those of past times. In relation to natural changes which have occurred on the earth, the higher forms of life as we know them have evolved from lower forms. Although the principle of evolution has been accepted by biologists since the time of Darwin (1859), the mechanism by which it could have taken place has only recently been elucidated. Genetic and biochemical studies of bacteria, viruses, and higher forms, have shown that *genes,* the ultimate units of hereditary material, ultramicroscopic particles capable of self-reproduction and occupying definite loci on chromosomes of the cell, are responsible for the specific chemical activities which modify development and provide the means by which evolution could occur. The origin of life itself probably goes back about 2 billion years to the chance assembly from nonliving ingredients of molecules having the capacity of self-duplication even of their chemical variations. It is chemical differences which basically determine variations in form and function of cells. These are, in each case, the result of unique action of enzymes which are proteins whose synthesis is directed by genes. Genes are composed of deoxyribonucleic acid (DNA), a large organic molecule (Fig. 1.1) consisting of two long strands of nucleotides twisted in the form of a helix. Nucleotides are made up of small units: a pentose sugar (deoxyribose), phosphoric acid, and a nitrogenous base. The latter, in the form of a side chain, may be a purine (adenine or guanine) or a pyramidine (cytosine or thymine). The two sugar phosphate strands are connected to each other by hydrogen bonds at a thousand or more sites along their course by purine-pyramidine linkages. Adenine and thymine together form a hydrogen-bonded structure; guanine and cytosine form similar bonds. Adenine on one strand always is bonded with thymine on the other; cytosine on one strand fits with guanine on the other. There are thus four possible kinds of bondings between any two nucleotides: adenine-thymine, thymine-adenine, cytosine-guanine, guanine-cytosine. Innumerable variations are thus possible in the sequence in which the linkages occur at the very numerous positions along these chains. It is probable that during mitosis the two strands separate from each other and each then serves as a template for a new strand which is synthesized from available chemicals and is a mate, not a model, of the strand. The two new double-stranded molecules are thus duplicates of each other and exact counterparts of the original. Any variation in sequence of the nitrogenous bases, whatever its cause, would be perpetuated and thus furnish a tangible difference from the original. Natural selection as elucidated by Darwin could act

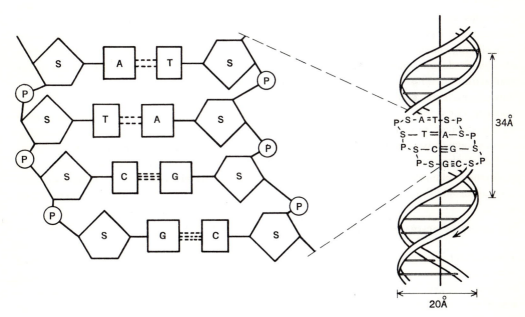

Fig. 1.1 The Watson-Crick model of DNA. Å, angstrom units; A, adenine; C, cytosine; G, guanine; P, phosphoric acid; S, pentose sugar (deoxyribose); T, thymine.

Introduction 3

upon the new variation in bringing about evolutionary changes. How the genes actually function is not yet definitely known. It is thought, however, that the DNA of which the gene is composed directs the synthesis of a complementary substance, ribonucleic acid (RNA), which becomes associated with ribosomes, minute cytoplasmic entities. Here amino acids are assembled into proteins, forming the enzymes so important in growth and function. The variations in sequence of the units of the RNA molecule, and the DNA from which it came, may thus account for the variations in enzyme structure which the nucleotides control.

Genetics, the branch of biology dealing with heredity and variation among related organisms, largely in their evolutionary aspects.

Taxonomy, or systematic biology, the science which has to do with the classification of organisms according to their natural relationships. It also includes the laws and principles of such classification. Although taxonomy is based primarily upon morphological data, changes which take place during embryonic development are also taken into consideration. Moreover, taxonomic affinities are often substantiated by certain physiological likenesses and differences. Serological studies, in which reactions of certain blood groups have been observed, have established the accuracy or inaccuracy of taxonomic relationships based on structural similarities and differences.

THE STUDY OF COMPARATIVE ANATOMY

The student of chordate anatomy should realize at the beginning that all the methods of approach are based in part upon comparative anatomical relationships. Comparative anatomy may also be used to explain some of the phenomena that appear during embryonic development. A knowledge of structure alone, however, has very little meaning unless it is interpreted in terms of function. Morphology and physiology complement each other. So far as is possible, it is desirable that both structure and function be considered together in studying animals. Nevertheless, a preliminary knowledge of morphology is a great aid to the study of physiology.

Most students of biology are interested in man, but a study of man alone would lead to an extremely narrow conception of his place on the earth. In order properly to understand the position of man in the world of life, a knowledge of his relations to other living things is almost essential.

Man is classified as a member of a large group of animals called the *vertebrates*. They are forms which have a backbone, or vertebral column, and include such diverse creatures as snakes, fishes, birds, frogs, elephants, and mice. The ancestors of these animals are the ancestors of man; the story of their origin is similar to his. A thorough understanding of the body, mind, and activities of man involves a study of his vertebrate ancestry.

Method of approach to comparative anatomical study. When similarities and differences in structural organization of the bodies of different animals are made objects of study, and when the facts obtained are compared, then from the study of structure general principles are derived from which deductive conclusions may be drawn. This is the method of the comparative anatomist, who seeks to explain the variations in structure that are found in the bodies of animals with a view to tracing biological relationships.

Although the method of comparative anatomy has been used in studying the relationships of almost all forms in the animal kingdom, the scope of the subject is too vast to be considered in a single volume. We shall confine our attention here to a comparative study of the vertebrates and a few of their closest relatives, all of which are included in the phylum *Chordata*.

The student, when in the laboratory, studies and dissects representative chordate animals and gains an intimate knowledge of the detailed structure of each. This knowledge does not constitute a science unless the facts learned are properly correlated and blended into a harmonious whole. It is the aim of this volume to effect such a correlation.

At the beginning of a course of study it is valuable for the student to have an idea of its objectives so that time may be most profitably utilized in putting the emphasis upon the facts of greatest importance. The comparative anatomist recognizes that the vertebrates and some of their close relatives are built upon the same fundamental plan and that there is often a close correspondence even in detail. Within the general vertebrate plan there may be many variations which are, for the most part, adaptive in nature. By this is meant that they are modifications which meet particular needs in relation to the environment in which the animal lives. The object of a course in comparative anatomy of vertebrates, therefore, is to acquaint the student with the plan of vertebrate structure. Such knowledge serves as a background for the better appreciation of the studies of embryology, human anatomy, and medicine.

Homology. The term *homology* refers to the correspondence in type of structure between parts or organs of different animals. Because of evolutionary differentiation, these may have diverged considerably from the same or corresponding part or organ of some remote ancestral form, so that superficially they bear but little resemblance to one another. For example, observation of the skeleton of a cat shows that its forelimb is composed of a single bone in the upper leg, two in the lower leg, and a number of intermediate bones situated between the lower leg and the toes. The wing of a bird, which is utilized in so different a manner, would at first sight seem to be a thing apart and in no way like the forelimb of a cat. The skeletons, however, reveal basic similarities in structure. The arm of man, the flipper of the whale, the hoofed leg of the horse, and the wing of a bat are all built upon the same fundamental plan and are, therefore, homologous (Fig. 1.2). In some cases differences in structure are so great that their homologies might be unsuspected. Observation of a series of intermediate stages between the two extremes may often clearly indicate their basic similarities. The study of homology of parts in various forms is one of the most interesting features of comparative animal studies. Observations on the details of embryonic development have in many instances rendered great service to the comparative anatomist in revealing such homologies. An embryological approach to comparative studies will, so far as is possible, be utilized in this volume.

The biogenetic law. It has long been realized by biologists that early embryos of all vertebrates show remarkable similarities in many respects. During the development of higher forms, certain anatomical features, apparently of little significance, make their appearance. In most cases these become modified, or they may degenerate and disappear. Many speculations have been made as to the significance of the presence of such structures. For example, at a certain stage of development in all vertebrates a series of pouches pushes out from the pharynx to

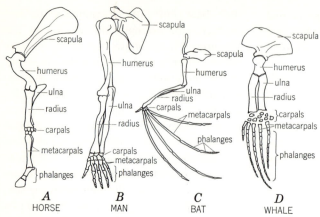

Fig. 1.2 Skeleton of forelimb, showing modifications of same fundamental type of structure. (*A, B, and C, after LeConte,* "Evolution," *Appleton-Century-Crofts, Inc. By permission.*)

establish connection with the outer covering of the body. In such lower forms as fishes perforations occur at the points of contact, becoming gill slits which persist throughout life. In the embryos of higher vertebrates, however, the pharyngeal pouches usually fail to open to the outside. They persist for a time but then become modified in various ways, retaining little or none of their original character. Why should pharyngeal pouches appear at all in higher forms? Another illustration of the principle under discussion is to be found in the appearance, in the region of the pharynx in all vertebrates during embryonic development, of certain blood vessels called *aortic arches*. The structural plan of the aortic arches in the *embryo* of a bird, for example, is somewhat similar to that found in the gill region of *adult* fishes. It is *almost identical* with that found in the *embryos* of such diverse forms as fishes, snakes, frogs, and man. In the bird, as well as in other higher forms, the aortic arches soon undergo considerable modification, ultimately forming an arrangement of vessels in the posterior part of the neck

region or in the anterior part of the thoracic region which is characteristic of each particular group.

Several ideas have been advanced to explain why such structures as pharyngeal pouches and aortic arches should make their appearance in the embryos of higher vertebrates. Perhaps the best known of these is the *recapitulation theory,* or *law of biogenesis,* which, when originally formulated, stated that during development an individual passes through certain ancestral stages through which the entire race passed in its evolution and that the embryos of higher forms in certain ways resemble the adults of lower forms. In other words, *ontogeny* (individual life history) recapitulates *phylogeny* (racial history). Although this concept has, in the past, furnished a good working hypothesis for embryologists in their attempts to work out homologies, there has been much debate as to its truth. Biologists today realize that ontogeny does *not* actually recapitulate phylogeny; that at no time does a mammalian embryo, for example, closely resemble an *adult* fish, amphibian, or any other lower ver-

Anatomy of the Chordates

tebrate. The embryos of all vertebrates, however, show many remarkable similarities. Those of higher vertebrates undergo fundamental changes which are like those of lower forms and occur in the same sequence. From such basically similar embryonic stages, which are believed to be characteristic of ancestral forms, each group of vertebrates diverges in one direction or another, so that when fully developed they differ widely in structure and habits.

It is not to be assumed that during development an animal goes through *every* stage that its ancestors did. Some stages are omitted or are passed through so rapidly that they can scarcely be discerned. Some characteristics by which they differ have arisen as a result of genetic variation, so that the ancestral conditions do not always appear in a typical form. Other structures of significance in the adaptation of the embryo to its environment are obviously unrelated to ancestral conditions or resemblances. In addition, the principle fails in many cases to explain the appearance of new, or *cenogenetic,* structures in certain forms. The theory does, however, account for the fact that the early embryonic stages of all vertebrates are remarkably similar (Fig. 1.3) up to a point at which each group diverges in its own characteristic manner. Despite its shortcomings, the concept has been useful to biologists in establishing homologies, in making clear certain phylogenetic relationships, and in helping to give a picture of the probable structure and appearance of ancestral forms and stages in our evolutionary history.

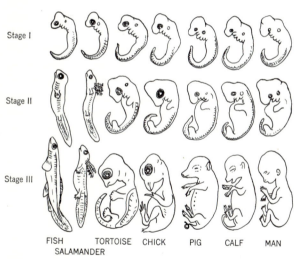

Stage I

Stage II

Stage III

FISH   TORTOISE   CHICK   PIG   CALF   MAN
  SALAMANDER

Fig. 1.3 Three comparable stages in the development of embryos of several vertebrates. Such membranous structures as yolk sac, amnion, chorion, and allantois, which appear in some embryos but not in others, have been omitted in order to emphasize similarities and to minimize differences. (*After Haeckel, modified from Daugherty, "Principles of Economic Zoology," W. B. Saunders Company. By permission.*)

CHARACTERISTICS OF CHORDATES

Members of the phylum Chordata are commonly referred to as *chordates*. There are three characters of prime diagnostic importance which are possessed by all chordates, as follows:

A notochord is present sometime during life. A more or less typical invertebrate animal, such as a crayfish, possesses a hard, nonliving, external covering, the *exoskeleton,* which is deposited on the surface of the body by the secretory activity of the underlying cells. It protects the animal, serves for muscle attachment, and is provided with joints which permit movement and locomotion (Fig. 1.4). The vertebrate skeleton, on the other hand, is essentially a living *endoskeleton,* which provides protection for the delicate internal organs, gives support to the body, and is jointed so that movement is possible. Although all chordates have an endoskeleton, turtles, armadillos, and a few others possess external skeletal structures as well (dermal skeleton), but these are primarily composed of living material and are not at all comparable to the exoskeleton that is found in the invertebrate.

In all chordates a notochord, which is a primitive endoskeletal structure, is present during embryonic life. It is pliant, rodlike, and made up of vesicular connective tissue (Fig. 1.5). The notochord is located along the middorsal line, where it forms the axis of support for the body. In some animals it persists as such throughout life, but in most chordates it serves as a foundation about which the vertebral column, or backbone, is built. A salamander larva or a frog tadpole, for example, at first possesses a notochord (Figs. 1.6 and 1.7), but it is gradually replaced, first by a cartilaginous and then by a bony vertebral column. By the time a tadpole has metamorphosed into an adult frog, the notochord has for the most part disappeared.

A hollow dorsal nerve tube is present sometime during life. The type of central nervous system found in arthropods, as well as in some lower phyla, is located on the ventral side of the body except for a dorsal ganglionic mass, known as the *brain,* which connects to the ventral portion by esophageal connectives (Fig. 1.4). This type of nervous system is sometimes called the "ladder type" because it is made up of a double chain of enlargements, called *ganglia,* united by

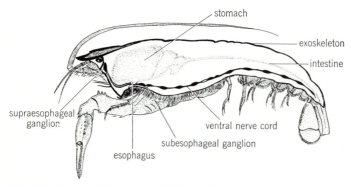

Fig. 1.4 Diagram showing exoskeleton, digestive tract, and position of ventral ganglionated nerve cord of crayfish.

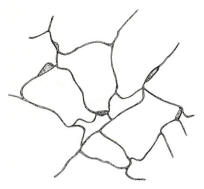

Fig. 1.5 Vesicular connective tissue of which notochord is composed, as seen under the microscope. Note the large, closely packed cells distended with fluid and the eccentric position of the nuclei.

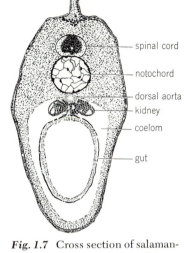

spinal cord
notochord
dorsal aorta
kidney
coelom
gut

Fig. 1.7 Cross section of salamander larva showing notochord lying beneath spinal cord.

nervous connections in both longitudinal and transverse directions and so has a ladderlike appearance. In the crayfish there is a considerable degree of fusion between right and left halves. The ventrally located nerve cord is solid and lies in the coelom.

The central nervous system of chordates, on the other hand, is located in a dorsal position. It lies just above the notochord and entirely outside the coelom (Figs. 1.6 and 1.7). Moreover, it is a tubular structure, having a small, hollow canal running from one end to the other. The dorsal hollow nerve tube per-

sists throughout adult life in almost all chordates, but in a few it degenerates before maturity has been attained.

Gill slits connecting to the pharynx or traces of them are present sometime during life. Many types of respiratory organs are found among animals belonging to phyla below the chordates in the evolutionary scale. Some respire through moist integuments, some have tracheal tubes, others have book lungs, and still others have gills of various kinds (Fig. 1.8). In none of these types,

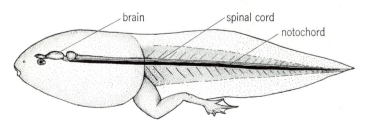

brain
spinal cord
notochord

Fig. 1.6 Diagram showing position of notochord in relation to brain and spinal cord in frog tadpole.

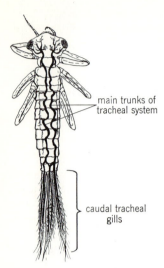

main trunks of
tracheal system

caudal tracheal
gills

Fig. 1.8 Damsel fly larva
with caudal tracheal gills
used in respiration.

however, do the respiratory organs have any connection with the digestive tract.

A great many aquatic chordates respire by means of gills made up of vascular lamellae or filaments lining the borders of gill slits which connect to the pharynx and open directly or indirectly to the outside (Fig. 1.9). Even terrestrial chordates, which never breathe by means of gills, nevertheless have traces of gill slits present as transient structures during early development. There are no vascular lamellae lining these temporary structures, nor do they open to the outside. However, the fact that they are present in all chordates is of primary importance in denoting close relationship.

The three characteristics listed above are possessed only by members of the phylum Chordata. In addition, chordates have certain features which are common to members of other phyla as well.

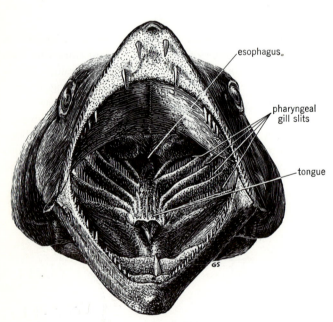

esophagus

pharyngeal
gill slits

tongue

Fig. 1.9 The open mouth of a fish, the barracuda, showing the typical relation of gill slits to the pharynx.

Bilateral symmetry. All chordates are bilaterally symmetrical, at least in the embryonic stage. Bilateral symmetry is the symmetry of the right and left sides of an organism in which any structure on one side of the median plane is a mirror image of the corresponding organ on the other side.

Cephalization. In bilaterally symmetrical animals there is a concentration of nervous tissue and sense organs in or toward the head. This is known as *cephalization.*

Metamerism. When certain structures, primarily like each other, are repeated one after another, they are said to be metameric. Among metameric structures are included certain nerves and blood vessels as well as vertebrae, ribs, muscles, etc.

A true body cavity, or coelom, is present. Such animals as ascarid worms have a body cavity separating the alimentary tract from the muscles of the body wall. This cavity is not lined completely with tissue derived from mesoderm (one of the primitive embryonic germ layers). It is therefore not considered to be a true coelom. All chordates, as well as annelids and some others, have a true coelom lined entirely with mesoderm.

Direction of blood flow. In typical invertebrates a pulsating organ, or heart, is situated on the *dorsal* side of the body. In general, blood is forced anteriorly, then ventrally and posteriorly, returning finally to the dorsal heart. The chordate heart, on the other hand, is located on the *ventral* side of the body. Blood is pumped anteriorly and forced to the dorsal side. It then courses posteriorly and returns to the heart by veins. The larger veins are ventral in position.

| 2 |

CLASSIFICATION
AND EVOLUTION
OF THE
CHORDATES

Before undertaking a study of the anatomy of the chordates it is desirable to have a working knowledge of the animals included in the phylum. Zoologists have grouped these animals in an orderly fashion according to their natural and anatomical affinities. This grouping serves as a basis for classifying all members of the phylum. It is important that the student of comparative anatomy have some knowledge of the classification of the chordates; otherwise a study of their anatomy from a comparative viewpoint will have little meaning. In a book of this kind it is not feasible, nor is it necessary, to discuss all the animals concerned. In the following pages, therefore, emphasis has been placed on those living forms of which some knowledge is almost essential to an understanding of the principles underlying the subject of comparative anatomy.

The chief principle used in classifying animals involves the division of larger groups into smaller groups according to certain distinctive traits which they possess in common. Thus phyla are divided into classes, classes into orders, orders into families, families into genera, and genera into species. The characteristics which separate the lower divisions are of lesser taxonomic importance than those used as a basis for separating the higher, or larger, groups. Often intermediate groupings are used, particularly when the number of animals included in a certain category is large. Thus there may be subphyla, superclasses, subclasses, superorders, etc.

The scientific name of an animal is made up of its generic and specific names, the genus always being written with a capital letter and the species with a small letter. Thus the common leopard frog is *Rana pipiens*. This system was first devised by the great Swedish naturalist Linnaeus and is known as the system of *binomial nomenclature*.

Authorities differ somewhat in the weight they give to certain categories in the general scheme of classification. These are usually of a minor nature but are, nevertheless, confusing to the student. An attempt has been made, therefore, in the present volume to conform in general to the schemes used by certain biologists and paleontologists who are pre-eminent in their respective fields and who have made a real effort to bring about some uniformity in an otherwise rather perplexing area of science.

PHYLUM CHORDATA

The phylum Chordata is subdivided into four main groups known as *subphyla*. The first three of these include a few relatively simple animals which lack a brain and a cranium. For this reason the term *Acrania* is used by some authors in referring to them collectively. The animals included in this category are believed to show similarities to the ancestors of the chordates and hence are frequently designated as the *protochordates*. The fourth subphylum, *Vertebrata,* includes those chordates possessing a brain and an endoskeleton. It is with the members of this subphylum that the present volume is primarily concerned.

Subphylum I. Hemichordata

Many zoologists consider the hemichordates to be the lowest form of chordate life,

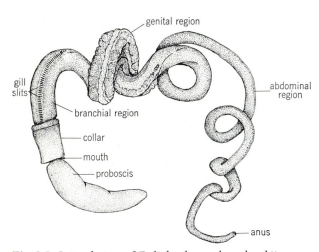

genital region

gill slits

branchial region

abdominal region

collar

mouth

proboscis

anus

Fig. 2.1 Lateral view of *Dolichoglossus kowalevskii.*

although there is some dispute about their classification. There has been a recent tendency among taxonomists to place them in a separate phylum. They are of special interest because they furnish a clue to the link which bridges the gap between chordates and some of the lower phyla. A free-swimming larva, the *tornaria,* occurs in the development of certain members of the subphylum. The tornaria is similar in appearance to the larval stages of some echinoderms.

A typical example is *Dolichoglossus kowalevskii* (*Saccoglossus kowalevskii*) (Fig. 2.1), a fragile, burrowing, wormlike marine animal of the Atlantic Coast. Three body regions are present: (1) a proboscis, (2) a collar, and (3) an elongated trunk. The trunk, in turn, is composed of an anterior branchial region with a row of transverse gill slits on either side, a genital region of irregular outline, and a posterior, abdominal region. The only obvious chordate character possessed by this animal is its pharyngeal gill slits. A peculiar forward extension of the gut into the proboscis (Fig. 2.2) has in the past been homologized with the notochord of higher forms. Most zoologists today doubt whether such a homology exists. *Dolichoglossus* possesses both dorsal and ventral nerve strands. The dorsal strand, which is the larger, is tubular only in the collar region. Nevertheless it is possibly homologous with the dorsal hollow nerve tube of higher forms.

Subphylum II. Cephalochordata

There are but two genera and about 20 species of animals included in the subphylum *Cephalochordata.* The common name *amphioxus* is applied to the more abundant genus, *Branchiostoma.* The primitive structure of amphioxus is of special interest to the zoologist. Furthermore, its early development (Figs. 3.1, 3.4, 3.6, and 3.7) illustrates with almost diagrammatic simplicity certain features which occur during embryonic stages of higher chordates.

Amphioxus (Figs. 2.3 and 2.4) is a small marine animal about 2 in. long. It is rather sedentary in habit and generally lies partly buried in the sand of the ocean floor with its anterior end protruding. The body of amphioxus is pointed at either end and during life is semitransparent. At the anterior end an integumentary fold, the *oral hood,* surrounds a cavity, or *vestibule,* which leads to the mouth. The oral hood is encircled by 22 papillalike projections, the *buccal cirri.*

A median dorsal fin extends almost the entire length of the body and is continuous with the caudal fin posteriorly. On the ventral side the caudal fin continues anteriorly as far as an opening, the *atriopore,* and then divides, extending forward as a pair of *metapleural folds* (Fig. 2.5). These were at one time believed to represent the beginning of the paired appendages of vertebrates, a

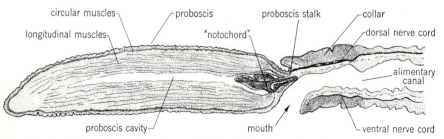

Fig. 2.2 Sagittal section of anterior end of *Dolichoglossus kowalevskii.*

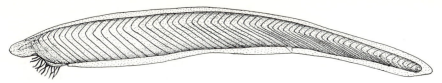

Fig. 2.3 Lateral view of amphioxus.

theory which is no longer tenable. The notochord extends practically the entire length of the body, projecting farther forward than the anterior end of the dorsal hollow nerve cord. The latter lies directly above the notochord. A slight enlargement, the *cerebral vesicle,* at the anterior end can scarcely be referred to as a brain. Metamerism is clearly indicated by the arrangement of the <-shaped muscle segments, or *myotomes,* of which there are between 50 and 65, depending on the species. They make up the greater part of the body wall. Those on one side are arranged alternately with those on the other.

The digestive system is simple in structure. The vestibule leads to the mouth opening located in a membranous *velum.* In front of this is a system of ciliated bands, the *wheel organ,* or *organ of Müller,* derived from the preoral pit of the larva (p. 394). The mouth leads to the pharynx. Projecting backward from the velum into the pharynx are 12 short *velar tentacles* which, in addition to the buccal cirri, serve as a sort of sieve, permitting only fine particles to enter the digestive system. Mucus from *Hatschek's pit* in the roof of the oral hood is discharged on the wheel organ, trapping food particles which are swept toward the mouth. The large *pharynx* with its numerous gill slits contains a system of ciliated and glandular grooves which direct food particles to the short esophagus and intestine. A small outpocketing of the floor of the pharynx is the *endostyle.* It consists of longitudinal glandular tracts, the mucuslike secretions of which serve to enmesh food particles. The possible homology of the endostyle of amphioxus with the thyroid glands of vertebrates is referred to on page 338. A pouch, the *hepatic caecum,* projects forward from the ventral side of the intestine on the right side of the pharynx. The intestine terminates at the anus located on the left side at some distance back of the atriopore.

Each of the numerous gill slits is secondarily divided by a *tongue bar.* The gill slits of the adult do not lead directly to the outside but communicate with an *atrium,* or *peribranchial chamber,* which surrounds the

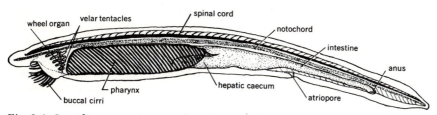

Fig. 2.4 Semidiagrammatic view of amphioxus, showing parts of the digestive tract.

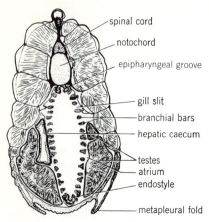

spinal cord

notochord

epipharyngeal groove

gill slit

branchial bars

hepatic caecum

testes

atrium

endostyle

metapleural fold

Fig. 2.5 Cross section through posterior pharyngeal region of amphioxus (anterior view).

greater part of the digestive tract except on the dorsal side. The atrium, which opens to the outside through the atriopore, should not be confused with the coelom.

A pulsating tube, homologous with the hearts of higher chordates, lies ventral to the pharynx (Fig. 12.17). Blood is pumped anteriorly and then dorsally through the gill region.

The anatomy of amphioxus indicates in every way that the animal is truly a chordate. Further details concerning its anatomy are discussed under appropriate headings in later pages.

Subphylum III. Urochordata

The urochordates, frequently referred to as the *tunicates,* include a peculiar group of widely distributed marine animals commonly known as sea squirts, or ascidians. They are of no economic importance. Many of them become sessile after a free-swimming larval period and attach themselves to a wharf pile, rock, or similar object. *Molgula manhattensis* (Fig. 2.6) is a fairly typical example of the subphylum. In its adult state *Molgula*

bears little resemblance to a typical chordate. It lacks a notochord, and its nervous system has been reduced to a small ganglionic mass. Only the pharynx with its numerous gill slits gives evidence of its chordate relationships. An adult specimen is about an inch long, round or oval in shape, with two openings, or *siphons,* at its free end. A stream of water enters the larger *incurrent siphon* and leaves through the smaller *excurrent siphon.* The animal is invested by a rather tough capsule, or *tunic.* This is composed of a substance called *tunicin* secreted by the cells of the *mantle* beneath. Tunicin has the same chemical formula as cellulose, which makes up the walls of plant cells. The greater part of the cavity enclosed by the mantle is occupied by the pharynx. The nerve ganglion lies embedded in the mantle in the region between the siphons. This is the dorsal side of the animal. Almost completely surrounding the ganglion is an *adneural gland* which opens by means of a duct into the pharynx. Some zoologists have homologized the adneural gland with the pituitary gland of vertebrates, but there is a difference of opinion concerning this. A ciliated groove, the *endostyle,* extends along the midventral line of the pharynx to the esophagus. Food particles that enter the pharynx with the incurrent stream of water are gathered by the endostyle and swept on to the esophagus. Some biologists believe that the endostyle may be the forerunner of the thyroid gland of higher forms. Certain other structures of *Molgula* are indicated in Fig. 2.6.

Sea squirts were formerly classified as mollusks because of their peculiar structure in the adult stage. It was not until 1866 that the proper classification was determined when Kowalevski, a Russian biologist, found that the ascidian larva possesses typical chordate characteristics. During its larval existence a sea squirt somewhat resembles

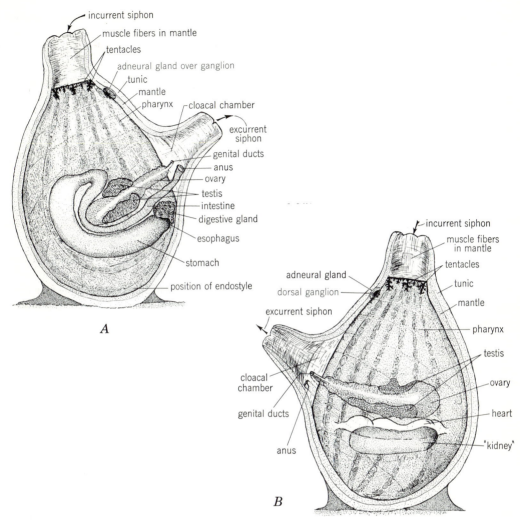

Fig. 2.6 *A*, internal organs of *Molgula manhattensis* as viewed from the left side; *B*, internal organs of *Molgula manhattensis* as viewed from the right side.

a tadpole in appearance. It has a notochord which is confined to the tail, a hollow dorsal nerve cord, and a pharynx perforated with gill slits. After a free-swimming period it finally attaches itself to some object and becomes sessile. The animal then undergoes a metamorphosis during which the notochord disappears and the nerve cord is re-duced to a simple ganglion. The only one of the three essential chordate characters which persists is the pharynx with its numerous gill slits communicating indirectly with the exterior.

Sessile urochordates may be either soli-tary or colonial. Not all urochordates are sessile, however, for some are free-swimming, transparent pelagic forms throughout life.

Subphylum IV. Vertebrata (Craniata)

All chordates, with the exception of the protochordates, are included among the vertebrates. In these the anterior end of the dorsal hollow nerve cord is enlarged to form a brain, and a protective and supporting endoskeleton is present.

With the appearance of the brain a skeletal structure called the *neurocranium,* or simply the *cranium,* makes its appearance, serving to protect that delicate organ. The cranium forms part of the skull. Openings, or *foramina,* are present for the passage of cranial nerves, blood vessels, and the spinal cord. The latter is likewise protected by a series of metamerically arranged skeletal elements, the *vertebrae,* associated with the notochord in their development. Openings for the passage of spinal nerves are present between adjacent vertebrae.

Vertebrates are most frequently divided into two large groups or *superclasses.* The first of these, the superclass *Pisces,* includes those lower forms commonly known as "fish." The second superclass, *Tetrapoda,* embraces the remaining vertebrates which are basically four-footed animals, although in some the limbs have been lost or modified in one way or another.

Superclass I. Pisces
Fishes are entirely aquatic and, in most cases, respire by means of gills lining pharyngeal gill slits. Included in this large superclass are some jawless forms exemplified by living lampreys and hagfishes as well as numerous fossil relatives collectively referred to as *ostracoderms.* Some authorities do not believe that these jawless animals should be included among the fishes and, therefore, place them in a separate subphylum, the *Agnatha.* The remaining vertebrates, all of which have jaws, are then included in another subphylum, the *Gnathostomata.* We shall consider the jawless, fishlike forms to be members of the lowest *class* of fishes, the *Agnatha.*

Class I. Agnatha
In both living and fossil agnathans there is an absence of jaws. Living species and most fossil forms also lack paired appendages but some ostracoderms possessed paired spiny structures, lateral body lobes, or flaplike flippers posterior to the head region (Fig. 2.7) which have been interpreted as possible forerunners of paired pectoral appendages. The primitive ostracoderms, which have been grouped into three or four separate orders, are the earliest vertebrates on record. Some possessed a single, dorsal nostril between the eyes. Others lacked the dorsal nostril. We shall not, in this volume, consider the detailed differences which are used in placing the various kinds of ostracoderms in separate orders. Unlike living agnathans, the bodies of ostracoderms were covered with a heavy armor of bony, dermal plates. It was formerly believed that the earliest vertebrates possessed cartilaginous skeletons much like those of the lowest vertebrates living today and that the appearance of bone was a later development. The presence of bony plates (and even of a bony endoskeleton, of a sort, in a few forms) in the fossil remains of the oldest vertebrates has changed the ideas of zoologists and paleontologists in this regard. The modern concept is that the presence of bone was a primitive character and that living cyclostomes and elasmobranch fishes with cartilaginous skeletons represent degraded descendants of bony ancestors which have lost their bony skeletal elements and armor secondarily.

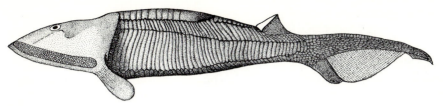

Fig. 2.7 Lateral view of the ostracoderm, *Hemicyclaspis*.

Order I. Cyclostomata. Living agnathans, the lampreys and hagfishes, are included in this order. Although primitive in many respects, they are believed to be modified and degenerate offshoots of the primitive vertebrate stock. In some respects they are so highly specialized that homologies with other vertebrates are difficult to ascertain.

Cyclostomes have rounded, eellike bodies with laterally compressed tails. Median fins are present supported by cartilaginous fin rays. Lacking jaws, there is a round, suctorial mouth (Fig. 5.5). The skin is soft and devoid of scales.

Suborder I. Petromyzontia. Lampreys are not to be confused with eels, which belong to a different class. Both fresh-water and marine forms exist. Some fresh-water forms are nonparasitic, but most lampreys are parasitic, the adults preying upon fishes. The marine lamprey, *Petromyzon marinus* (Fig. 2.8), is a parasitic form often used for dissection in courses in comparative anatomy. The animal, which attains a length of about 3 ft, lives in the sea but ascends rivers to spawn. The larval form, known as the *ammocoetes* (Fig. 2.9), lives in burrows in the sand for 3 years or more before undergoing metamorphosis. The young lampreys migrate downstream and out to sea where they remain 3 or 4 years before becoming sexually mature.

Seven pairs of gill slits are present, each opening separately to the outside. The suctorial funnel at the anterior end is ventral in position and beset with horny teeth. When feeding, the lamprey fastens its funnel to the side of a fish, usually just in back of the pectoral fin. Horny teeth on the tongue are used in rasping, and the animal feeds upon blood and mucus obtained in this manner.

The harmless ammocoetes larva is of special interest to the comparative anatomist since it is a very primitive and generalized vertebrate.

The marine lamprey has, over the last 45 to 50 years, gradually invaded the Great Lakes, from Lake Erie to Lake Superior, where it has become firmly established. It first appeared in Lake Erie in 1921, having worked its way from the St. Lawrence River through the Welland Canal, thus bypassing Niagara Falls. The fishing industry, particularly of

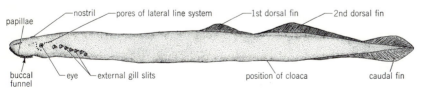

Fig. 2.8 Lateral view of lamprey, *Petromyzon marinus*.

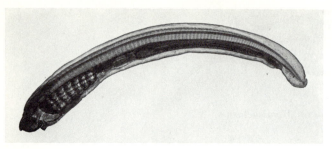

Fig. 2.9 Lateral view of ammocoetes larva of lamprey. (*Courtesy of Ward's Natural Science Establishment, Inc.*)

lake trout and burbot, which was formerly of considerable magnitude, was virtually destroyed by the ravages of the lamprey. In this connection it is of interest that with the disappearance of these large fishes, a small fish, the alewife, which also originally gained access from the sea to the Great Lakes via the Welland Canal, has multiplied in excess. Too small to be attacked by the lamprey, the alewife numbers around 175 billion in Lake Michigan alone, it is estimated. Scientists have attempted by various means to eliminate the lamprey population but, until lately, have met with little success because measures taken to destroy them have also had deleterious effects upon other fishes. Rather recently it has been discovered that a chemical called 3-trifluormethyl-4-nitrophenol (TFM), added to the water of lamprey-spawning streams in proper concentration, will effectively destroy the ammocoetes larvae, yet not harm young fishes. TFM was first used in 1958 on streams flowing into Lake Superior. By 1962 the first signs of success became evident. As adult animals die they are not being replaced. It is to be hoped that by using this method to control lampreys the fish population of the Great Lakes will eventually be restored. Chemical stimuli in the lamprey are perceived by olfaction. It is of interest, however, that the final localization of prey is accomplished by means of an electrical field produced by the animal in the water surrounding the head. This field becomes effective at a distance of 5 to 10 cm from the prey (H. Kleerekoper, 1958, *Anat. Record,* **132**:464).

Suborder II. Myxinoidea. *Myxine glutinosa,* the hagfish (Fig. 2.10), is the most common species included in the order. It is a marine form somewhat like the lamprey in appearance. Its mouth is terminal and has four pairs of tentacles about its margin. Only a single pair of external gill openings is present, located some distance behind the head. The animal feeds by rasping a hole in the side of a large fish, crawling inside the body cavity, and eating the viscera.

In members of the genus *Bdellostoma* there are 6 to 14 pairs of external gill openings, depending upon the species.

Gnathostomes

Members of the remaining groups of vertebrates, frequently referred to collectively as the *gnathostomes,* have paired pectoral and pelvic appendages, true upper and lower jaws, paired nostrils, and a well-developed endoskeleton. The cranium is better developed than in the agnathans.

Thus the other classes of *fishes* have

paired appendages. These are fins, of which there are usually two pairs. Median fins are also present. Dermal scales of various types are to be found in the skin with few exceptions.

Class II. Placodermi

Placoderms are known only through fossil remains. During Devonian times, approximately 350 million years ago, ostracoderms gradually became extinct and placoderms replaced them. The placoderms developed primitive jaws and paired fins but retained, to varying degrees, the bony armor of ostracoderms. This consisted of large bony plates arranged in different ways, which probably gave rise to the several types of dermal scales found in modern fishes. There was great diversity among these extinct, primitive fishes, which have been grouped into five or six separate orders. Some bore numerous spiny fins (Fig. 10.69); others had jointed necks; still others possessed true and typical paired fins. Placoderms were not successful on the earth from a long-range point of view. They gradually disappeared as two new types of fishes evolved, the cartilaginous and the bony fishes.

Class III. Chondrichthyes

Sharks, dogfishes, skates, rays, and chimaeras are examples of cartilaginous fishes.

Most are marine forms with cartilaginous skeletons, placoid scales, and ventral, subterminal mouths. All have tails of the heterocercal type (Fig. 10.67) in which the dorsal flange is larger than the ventral. Members of this class lack swim bladders, thus differing markedly from the bony fishes. The cartilaginous fishes are descendants of those which lived in Devonian times and which evolved from placoderms or placodermlike ancestors.

The skeleton in the *Chondrichthyes* is primitive in many respects but not in the fact that it lacks bone and is entirely cartilaginous. The lack of bone is believed to represent a retrograde condition. The bone in the basal plates of the placoid scales in the skin is all that remains of the ancient bony armor.

Subclass I. Elasmobranchii

The elasmobranch fishes, as members of this order are generally called, include those cartilaginous fishes in which the gill slits open separately to the outside. There are five to seven pairs of gill slits in addition to a modified pair, the *spiracles,* opening on top of the skull, just posterior to the eyes.

Order I. Selachii. Sharks and dogfishes living today, as well as numerous extinct forms, belong to the order *Selachii*. They differ from skates and rays in that their pectoral

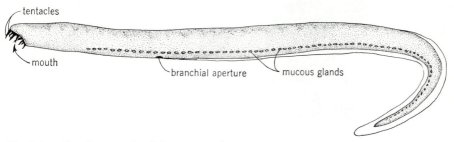

Fig. 2.10 The glutinous hagfish, *Myxine glutinosa.*

Fig. 2.11 Great hammerhead shark, *Sphyrna diplana*. (*Courtesy of the Chicago Natural History Museum.*)

Fig. 2.12 Dorsal view of clear skate. (*Courtesy of the Chicago Natural History Museum.*)

fins are distinctly marked off from cylindrical bodies, and in that they have laterally placed gill slits (Fig. 2.11). Since many structures in elasmobranchs are primitive or generalized in character, the shark or dogfish is an ideal animal for laboratory study as an introduction to the anatomy of higher vertebrates. Dogfish, which are among the smaller species, are plentiful. Among the better-known sharks are the hammerhead, white shark, sand shark, and whale shark, the latter being the largest living fish.

Order II. Batoidea. Skates and rays are included in this order. They might be regarded as modified sharks which have become flattened in a dorsoventral direction. Head and trunk are widened considerably, and the pectoral fins are not distinctly marked off from the body (Fig. 2.12). There is a distinct demarcation between body and tail, however. The gill slits are located ventrally. Skates are egg-laying forms, whereas rays give birth to living young. Among batoideans are included such forms as the gigantic devilfish, or Manta ray; the sting ray, or stingaree; the torpedo, or electric ray; and the sawfish. The body of the

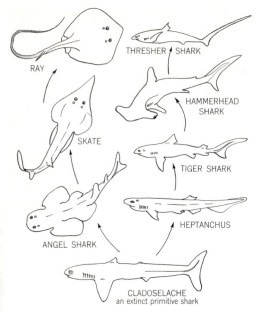

Fig. 2.13 Adaptive radiations among elasmobranchs.

last is not flattened to the same extent as in most members of the suborder. The varied body forms of skates and rays are regarded as adaptive radiations from primitive shark-like ancestors (Fig. 2.13).

Subclass II. Holocephali

This order is made up of a small group of fishes called chimaeras (Fig. 2.14). They are of special interest because they occupy a position between elasmobranchs and higher fishes. Among primitive features are a persistent notochord and poorly developed vertebrae. Advance is indicated by the presence of an *operculum* which covers the gill chamber on each side. Scales are practically absent in the adult.

Class IV. Osteichthyes

This class includes the bony fishes, the skeletons of which are bony to some degree. Several types of dermal scales are to be found within the group. The mouth is usually terminal, and an operculum covers each gill chamber. These fishes are divided into two subclasses, the distinctions between them being based primarily on fin structure. Members of the first subclass, *Actinopterygii*, are commonly known as ray-fins; those of the second, or *Sarcopterygii*, as the lobe-fins. The latter were formerly referred to as the *Choanichthyes* because certain members of the subclass possessed internal nares, or *choanae* (hence the name). Since living sarcopterygians do not possess choanae (a fact

Fig. 2.14 Lateral view of *Chimaera monstrosa*.

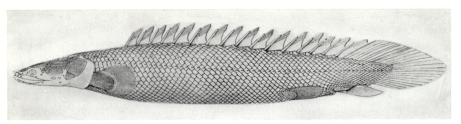

Fig. 2.15 *Polypterus bichir. (Courtesy of the American Museum of Natural History.)*

that is a fairly recent discovery), the name *Choanichthyes* scarcely seems appropriate and has generally been discarded. Members of the class *Osteichthyes* are to be found in both fresh and salt waters all over the earth. They range in size from gigantic creatures to very small forms considerably less than an inch in length. An extraordinary degree of diversity in body form is to be observed in this class, correlated with adaptations to countless environmental conditions with which the animals have had to cope in order to survive since the Devonian Period when, according to the fossil record, they first appeared on earth.

Subclass I. Actinopterygii

In the members of the subclass *Actinopterygii* all the fins, paired and unpaired, are supported by skeletal elements, the dermal fin rays.

Superorder I. Chondrostei. Only a few representatives of this ancient superorder are living today. Two genera, *Polypterus* (Fig. 2.15) and *Calamoichthys,* are found only in Africa. It is a question whether these fishes should more properly be included in the subclass *Sarcopterygii* since their pectoral fins have a fleshy basal portion. The structure, however, is somewhat different from the other lobe-finned fishes. The caudal fin, unlike the heterocercal structure of the *Chondrichthyes,* is symmetrical but not

typically diphycercal (Fig. 10.67). These fishes also were known as the "fringe-finned ganoids" because of the fringe-like appearance of the eight or more dorsal fin elements and the presence of ganoid scales (page 126). In some ways the young of *Polypterus* resemble amphibian tadpoles.

Other members of the superorder *Chondrostei* include the spoonbill, or paddlefish, and sturgeons. They too are believed to be survivors of primitive ray-finned fishes which have lost the greater part of their bony skeletons. The skeleton is almost completely cartilaginous. Their tails are of the heterocercal type.

Polyodon, the spoonbill (Fig. 2.16), inhabits the Mississippi River and its larger tributaries. A close relative, *Psephurus,* is found in China. Large specimens may weigh from 80 to 150 lb. A unique feature is the large "paddle" extending from the anterior end of the snout. A few small, rudimentary, placoid scales lie in back of the operculum, indicating a probable elasmobranch relationship. Small teeth are present on the jaws.

Adult sturgeons are large animals with five rows of keeled, bony, modified ganoid scales extending the length of the body. The jaws of the adult are toothless but the palate occasionally bears rudimentary teeth. The snout is prominent in the sturgeons and the mouth is located ventrally. Caviar is prepared from eggs taken from the body of the female.

Fig. 2.16 The spoonbill or paddlefish, *Polyodon folium.* (*Courtesy of the Chicago Natural History Museum.*)

Superorder II. Holostei. Only two genera of living fishes belong to the superorder *Holostei.* Both are fresh-water forms and survivors of ancient groups of fishes which formerly inhabited the ocean. *Amia,* the mudfish, fresh-water "dogfish," or bowfin, is commonly found in the Great Lakes region and in the Mississippi basin. The garpikes (Fig. 2.17) include several species, the most common of which is the long-nosed gar, *Lepisosteus osseus,* found in the fresh waters of North America east of the Rocky Mountains, for the most part. The body, which may be from 5 to 6 ft long, is covered with an extremely hard layer of rhomboid, ganoid scales (Fig. 4.14). Members of this genus range into parts of Central America and Cuba. The term "ganoid fishes" was formerly applied to a group of fishes which included *Lepisosteus, Polypterus, Calamoichthys,* as well as numerous fossil forms. The term is no longer in general use.

Superorder III. Teleostei. All the remaining ray-finned fishes are included in the superorder *Teleostei;* 95 per cent of all the fishes of the world come under this category. They are by far the most numerous vertebrates and approximately 20,000 species have been identified. In all the posterior, or occipital, region of the skull is bony. The tail is usually of the homocercal type (Fig. 10.67) with equally well-developed dorsal and ventral flanges. In some species it is diphycercal, lacking dorsal or ventral flanges. It is never heterocercal.

Ichthyology, the study of fishes, is a most fascinating subject. Many fishes, because of their economic importance, unusual body structure, or peculiar breeding habits, are of special zoological interest.

The division of the subclass *Actinopterygii* into three superorders, as described above, is not agreed upon by all authorities. Differences of opinion are based upon the fact that certain characters are not limited entirely to one group or another. It is impossible, for example, to draw a sharp boundary between all chondrosteans and all holosteans. Likewise, no clear-cut distinction

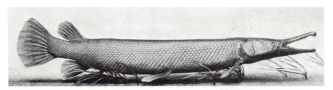

Fig. 2.17 Lateral view of the short-nosed gar, *Lepisosteus tristoechus.* (*Courtesy of the Chicago Natural History Museum.*)

can be drawn between holosteans and teleosts if all forms, including fossils, are considered. Nevertheless, the classification as presented here forms a good working basis for the student of comparative anatomy and is in accord with that used by a number of other contemporary zoologists and paleontologists.

Subclass II. Sarcopterygii

The lobe-finned fishes, of which this subclass is composed, are known chiefly from ancient fossil remains. The living forms, *Polypterus* and *Calamoichthys,* previously mentioned (see page 24) possibly should be included in this category. Within the subclass *Sarcopterygii* are to be found the first vertebrates in which nasal passages connect the mouth cavity with the outside. This characteristic does not appear in all, however. Although it is possible that some of the extinct members of the subclass used their nasal passages in connection with lung respiration, there is no clear-cut evidence that species living today actually employ them in such a manner. Nevertheless, the presence of choanae is of great importance from an evolutionary point of view, for it suggests that amphibians, the first air-breathing terrestrial vertebrates, may have evolved from similar fishes. The subclass is divided into two orders, which include the typical lobe-fins and the true lungfishes, respectively.

Order I. Crossopterygii. These are the typical lobe-finned fishes which make up a small but ancient group known mainly from fossil forms. They appeared on earth at the beginning of the Devonian Period. The fins of crossopterygians differ from those of all other fishes. Each is borne on a fleshy, lobelike, scaly stalk extending from the body. Pectoral and pelvic fins have articulations resembling those of tetrapod limbs. Crossopterygians apparently took two different lines of descent and are, accordingly, separated into two groups or suborders, the rhipidistians and the coelacanths.

Suborder I. Rhipidistia. The rhipidistian crossopterygians became extinct near the end of the Devonian Period. There is considerable evidence that it was through members of this suborder that evolutionary progress was made. The skeletal elements in the paired fins of the fossil *Eusthenopteron* (Fig. 10.76), one of the better-known rhipidistians, have been rather clearly homologized with those in the limbs of tetrapods (page 485). Fossils of primitive amphibians have been found together with those of rhipidistians in the same geological deposits.

Suborder II. Coelacanthini. The coelacanth crossopterygians were a more stable and conservative group which showed little change for millions of years. Until 1938 it was believed that coelacanths had become extinct about 90 million years ago. In that year a strange fish was caught off the east coast of South Africa. To the surprise of zoologists and paleontologists, this animal proved to be a coelacanth. It was named *Latimeria chalumnae* (Fig. 2.18). The specimen was poorly preserved and little of its anatomy could be ascertained. In more recent years several additional specimens have been obtained, all in the vicinity of the Comoro Islands in the Indian Ocean and at a depth of about 140 fathoms. These are being studied in detail by French scientists. Gradually a general picture of the anatomy of this strange fish is emerging. There is no doubt that *Latimeria* is a survivor of the oldest stock represented among living verte-

Anatomy of the Chordates

Fig. 2.18 *Latimeria chalumnae,* a surviving coelacanth.

brates. It would seem that the ancient rhipidistians did not persist long as fishes, but gave rise to coelacanths on one hand and land vertebrates on the other. *Latimeria,* therefore, cannot be considered as a "missing link" in the popular sense. In all the time that has elapsed since coelacanths were plentiful on the earth, *Latimeria* has remained relatively unchanged. An adult is 4 or 5 ft long, steel-blue in color, with a large head and a moderately laterally compressed body. Its diphycercal tail is extended into a short, supplementary, median fin, or lobe, the whole tail terminating in a fringelike fin. In addition to its rather complex paired pectoral and pelvic fins, it has two unpaired dorsal fins and a single, ventral anal fin. All except the first dorsal fin are typical lobe fins. The paired fins have a general resemblance to short, scaled, arms and legs with a terminal fringe. The animal is not a powerful swimmer, using its pectoral, second dorsal, and anal fins in a curious rotatory manner. Internal nares are lacking in *Latimeria.* It is even uncertain whether extinct crossopterygians possessed them. *Latimeria* may be an exception in this respect. Other details of the anatomy of this strange animal are described at appropriate places in the following pages.

Order II. Dipnoi. Three different genera of so-called lungfishes are included in the order *Dipnoi.* They include the Australian lungfish, *Epiceratodus* (Fig. 2.19), found in the rivers of Queensland, Australia; the South American lungfish, *Lepidosiren* (Fig. 2.20), which lives in the Amazon River and its tributaries; and three species of *Protopterus* (Fig. 2.21), living in the swamps and rivers of Africa. *Epiceratodus* is the most primitive of the living lungfishes. The presence of a pair of internal nares, or nostrils (choanae), is characteristic of all.

Most members of the class *Osteichthyes* possess an internal saclike structure filled with gas and called the swim bladder, or air bladder. This arises as a single or paired diverticulum of the digestive tract in the region of the pharynx. It may retain its connection with the pharynx (*physostomous* condition) or lose it and become a blind sac (*physoclistous* condition). In the true lungfishes the swim bladder is of the physostomous type. It is well developed, highly vascularized, and used as a lung in respiration. Such swim bladders are more efficient respiratory organs than the lungs of many amphibians. Nevertheless, because of certain other characteristics, notably the structure of the fins, it is believed that rhipi-

Fig. 2.19 The Australian lungfish, *Epiceratodus*. (*Courtesy of the Chicago Natural History Museum.*)

distian crossopterygians, rather than dipnoans, were ancestors of the amphibians.

Epiceratodus is found at the present time only in the limited region of the Mary and Burnett rivers in Queensland. It lives in water holes and stagnant pools which generally contain some water at all times of the year. The respiratory activity of the swim bladder supplements that of the gills. At times the fish rises to the surface to expel old air and to take in a new supply. When the water becomes so foul that other fish die, this animal seems to thrive. It does not undergo estivation, as do the other lungfishes, but lives in water throughout the year. The gills are better developed than those of the other lungfishes. When removed from the water, *Epiceratodus* will live only a short time.

The South American lungfish, *Lepidosiren,* is more eellike in appearance than its close relatives. It also rises to the surface

at times in order to obtain air. As the dry season approaches and the water level becomes lower and lower, *Lepidosiren* buries itself in the mud and secretes a sort of cocoon or capsule about its body. The animal then becomes dormant. The cocoon hardens, preventing loss of moisture from the body. Breathing is accomplished through a tube which leads to the surface. The entrance to the tube is closed by a perforated plug of clay which somewhat resembles a sieve. The fish escapes from its burrow when the rainy season returns.

The African lungfish, *Protopterus,* can creep over the bottom of marshes and lowlands by means of its long filamentous fins. Like *Lepidosiren* it forms a cocoon about itself at the approach of the dry season. So much does it depend upon its lungs for respiration that if prevented from reaching the surface it will soon drown. The gills are greatly reduced in size, being almost vestig-

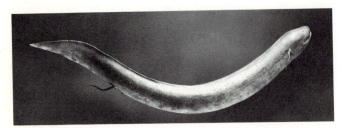

Fig. 2.20 The South American lungfish, *Lepidosiren.* (*Courtesy of the American Museum of Natural History.*)

Anatomy of the Chordates

Fig. 2.21 The African lungfish, *Protopterus*. (*Courtesy of the American Natural History Museum.*)

ial. In the water the fish rises to the surface at about 15-minute intervals in order to fill its lungs with air.

Superclass II. Tetrapoda

Tetrapods are those members of the subphylum *Vertebrata* having paired appendages in the form of limbs rather than fins. In some the limbs have been lost or modified. The basic plan from which all tetrapod limbs have evolved is the five-toed, or pentadactyl, appendage. Among other characteristics which distinguish tetrapods from fishes are: a cornified, or horny, outer layer of skin; nasal passages which communicate with the mouth cavity and which transport air; lungs used in respiration; and a bony skeleton. There has also been a reduction in the number of skull bones.

The first tetrapods are believed to have evolved from rhipidistian crossopterygian fishes which they rather closely resembled. The fossil remains of primitive tetrapods have been found in the eastern part of Greenland in deposits which date back to the end of the Devonian Period. These specimens possess characteristics which place them in a category intermediate between late crossopterygians and early amphibians. They were far different from any tetrapods living today.

The superclass *Tetrapoda* is divided into four classes made up of amphibians, reptiles, birds, and mammals, respectively.

Class I. Amphibia

The living representatives of the class include salamanders, newts, frogs, toads, and some less familiar burrowing, legless forms, the *caecilians*. Frogs and toads, often thought of as primitive tetrapods, are actually far removed from the original amphibian stock. Some skeletal structures of certain amphibians show a higher degree of specialization than do corresponding parts of lizards, which belong to the class *Reptilia*. Not all amphibians, therefore, are at a lower level in the evolutionary scale than reptiles. It is probable that both groups arose from a common ancestral stock but that amphibians, in most cases, failed to attain the degree of advancement and specialization generally found among reptiles.

The transition from aquatic to terrestrial life is clearly indicated in the class *Amphibia*. These are the first vertebrates to live on land, although they lay their eggs in water or in moist situations. Larvae with integumentary gills develop from the eggs. After a varying period of time, depending upon the species, metamorphosis usually occurs, after which the animal may spend the greater part of its life on land, although generally in a moist situation. Upon metamorphosis, among other changes, the gills usually disappear and lungs develop, supplementing the vascular skin as organs of respiration. Some salamanders never develop lungs even though they lose their gills. A few salamanders like the mud puppy, *Necturus* (Fig. 2.25), never undergo complete metamorphosis. They develop lungs but retain their gills throughout life. They are known as *perennibranchiates*.

The chief differences between amphibians and the truly terrestrial vertebrates lie in their aquatic reproductive habits, the lack of certain embryonic membranes, and the rather poorly developed corneal layer of the epidermis of the skin. The ends of the digits lack claws. The failure of amphibians to rise to a more dominant position lies chiefly in their mode of development. They are, so to speak, "chained" to the water. Although some are fairly well fitted for life on land, their conquest of the land has been only partially successful.

The most primitive amphibians, known from fossil remains, are referred to as the

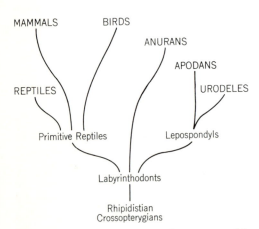

Fig. 2.22 Simplified diagram showing possible direction of evolution of modern amphibians, as well as other tetrapods, from primitive labyrinthodont and rhipidistian crossopterygian stock.

labyrinthodonts, the name being based upon the complex folding of the enamel layer of the teeth. These animals are sometimes called *stegocephalians* because of the solid roofing of the skull. In certain features they were similar to rhipidistian crossopterygians. It is probable that from such an ancestral stock there evolved three main groups, or lines (Fig. 2.22), the members of which were distinguished chiefly by differences in the detailed structure of their vertebral columns, or backbones. One of these lines was apparently an early offshoot of the primitive labyrinthodont stock, made up of a group of fairly small animals, the *lepospondyls,* which ultimately gave rise to living salamanders, or urodeles, and some other less well-known forms, the apodans, or caecilians. With the demise of the labyrinthodonts proper, two other lines are believed to have evolved. One gave rise to primitive reptiles and thence to higher forms of vertebrate life. From the other and larger group, in addition to numerous forms which became extinct, arose the tailless amphibians, or anurans, represented today by frogs and toads. In evolutionary terms, those amphibians living today are relatively unimportant forms which represent "dead ends" insofar as evolutionary progress is concerned.

The ancient labyrinthodonts are generally included in a special superorder of the class *Amphibia* called the *Labyrinthodontia,* made up of several orders. The three types of living species are divided into three orders.

Order I. Anura. Frogs and toads, which come under this category, lack a tail in the adult stage. Head and trunk are fused and there is no neck region (Fig. 2.23). Two pairs of well-developed limbs are present, the hind pair being particularly fitted for leaping. The feet are webbed and adapted for swimming. Frogs and toads are the first vertebrates

Fig. 2.23 Female green frog, *Rana clamitans.* (*Courtesy of the American Museum of Natural History.*)

to have vocal cords for sound production and used primarily in species identification. The anuran larva, or tadpole, does not resemble the parent. Head and body are fused into a single, egg-shaped mass, and the long tail is equipped with median fins. Horny jaws are used in lieu of teeth in feeding. Metamorphosis is clearly defined. At this time not only are the gills lost as lungs develop, but legs appear and the tail is resorbed. Anurans as a group are better fitted for terrestrial existence than are other amphibians. There are no superficial characteristics which distinguish frogs from toads. Frogs usually have a smoother and more moist body than do toads with their warty skin.

The order *Anura* and two others, made up of extinct ancestral forms, are often grouped together in a separate superorder, *Salientia.*

Order II. Urodela (Caudata). The urodele, or caudate, amphibians include salamanders (Fig. 2.24) and newts, the latter being small, semiaquatic forms. Urodeles are found, for the most part, in temperate and subtropical climates in the Northern Hemisphere. The elongated body consists of head, trunk, and tail, the tail being retained throughout life. Two pairs of weak limbs are present in most species. Larvae closely resemble adults and, like adults, possess teeth in both upper and lower jaws. Salamanders are common in the United States and often may be found under rotten logs in wooded areas and in moist situations.

The mud puppy, or water dog (Fig. 2.25), is a urodele amphibian frequently studied in courses in comparative anatomy as an example of a primitive terrestrial vertebrate. Members of the genus *Necturus* are found throughout North America east of the 100th meridian. *Necturus maculosus* is the most widely distributed species. This large salamander, adult specimens of which vary in length from 11 to 18 in., is commonly found in the Mississippi River system from the Arkansas River and northern Alabama northward into Canada and eastward into the mountainous regions of North Carolina, Virginia, West Virginia, and Pennsylvania. Its range includes the Great Lakes and St. Lawrence River and their tributaries in Manitoba, Ontario, and Quebec; the Hudson River drainage system; and the Susquehanna and Delaware rivers. *Necturus* spends its entire life in the water and, unlike most amphibians, does not metamorphose, retain-

Fig. 2.24 Female marbled salamander, *Ambystoma opacum,* guarding eggs in nest.

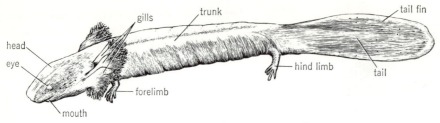

Fig. 2.25 Dorsolateral view of mud puppy, *Necturus maculosus.*

ing its external gills throughout life. This condition, in which larval characteristics persist during adult life, is referred to as *neoteny.* The neotenic condition is considered to be degenerate rather than primitive. It has already been mentioned that amphibians which retain their gills throughout life and fail to undergo metamorphosis, despite the development of lungs, are called *perennibranchiates.*

Necturus is frequently caught with hook and line by fishermen, many of whom believe the animal to be poisonous. Actually it is harmless. Even a bite by *Necturus* is of little consequence since the teeth are small and are not apt to cause skin abrasions.

Necturus feeds on crayfish, worms, aquatic insects, and their larvae. The animal is of little or no economic value although occasionally used as food for human beings. These amphibians are sluggish creatures which usually forage at night.

Order III. Apoda (Gymnophiona). The members of the order *Apoda,* as well as the urodeles, are believed to be descendants of the lepospondyls (page 30). Living members of the order are found only in India, Africa, and tropical America. They include the caecilians, which are burrowing forms with snakelike bodies, lacking limbs. The tail is very short, and the anus is almost terminal. Unlike other amphibians, some caecilians have dermal, fishlike scales in the skin. Adults

lack gills and gill slits. The very small eyes are buried beneath the skin or under the skull bones. Large caecilians may attain a length of 26 in. or more.

Amniota and anamniota

The three remaining classes of tetrapods, reptiles, birds, and mammals, are referred to collectively as the *Amniota.* The lower classes together form a group called the *Anamniota.* The essential difference between the two lies in the presence of certain membranes associated with the embryos of terrestrial forms during development. These membranes are the amnion, chorion, and allantois. The name *Amniota* is, of course, taken from the first of these. The amnion is a fluid-filled sac which comes to surround the embryo completely soon after development has begun. It first appears in reptiles and is of definite advantage to the embryo in preventing desiccation and otherwise offering protection during developmental stages. The remaining membranes are concerned with respiration and other processes essential to the developing amniote embryo.

Class II. Reptilia

Reptiles are represented today by four orders of living forms which include turtles and tortoises, alligators and crocodiles, lizards and snakes, and a peculiar, aberrant, lizardlike animal, the tuatara, *Sphenodon punctatum,* of New Zealand. Reptiles are believed to have made their first appearance on earth in the coal-swamp faunas existing about 250 million years ago (D. Baird, 1958, *Anat. Record,* **132**:407). Most reptiles live in tropical and subtropical regions. The reptiles of today are small and insignificant compared with the tremendous monsters of prehistoric times. During the Age of Reptiles, which began about 190 million years ago, reptiles dominated the earth. The extinct forms, known only by their fossil remains, have been grouped into numerous orders. Some lived on land, others took to the air, and still others were adapted to an aquatic habitat.

The most ancient reptiles were the *cotylosaurs,* some times called the "stem" reptiles. The best known of these, *Seymouria,* so closely resembled labyrinthodont amphibians that it has not been easy for paleontologists to set the two groups apart. Several groups of reptiles seem to have evolved from the cotylosaurs. Some were important for a time and then lost ground and disappeared; others have persisted to the present day with little change; still others evolved further and not only gave rise to most of our modern reptiles but to birds and mammals as well (Fig. 2.26).

Among early forms which went up a blind alley, so to speak, were such large aquatic reptiles as the short-necked ichthyosaurs and long-necked plesiosaurs. Modern turtles also are believed to have stemmed directly from an early side shoot of the cotylosaurs. They have shown little advance. Another group, tracing its origin to an early branch of the cotylosaurs, was that which gave rise to mammals. The earliest reptiles antecedent to the mammalian line are known as *pelycosaurs.* Certain features of their skulls indicate that progress toward mammalian heights probably began with a group such as this. Successive forms, over a great period of time, ultimately gave rise to the *cynodonts,* ad-

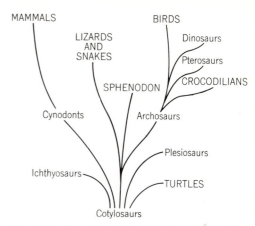

Fig. 2.26 Simplified diagram showing possible direction of evolution of reptiles, birds, and mammals from primitive cotylosaurs. The larger lettering indicates forms surviving today.

vanced mammal-like reptiles, the skulls and teeth of which more nearly resembled those of mammals than those of other reptiles. Still another stem from the early base is believed to have branched in two directions. One branch gave rise to *Sphenodon* and to the lizards and snakes of today. The other was made up of a great group of reptiles, the *archosaurs,* sometimes called the "ruling reptiles." To this group belonged the many and varied types of *dinosaurs* and the flying *pterosaurs* (Fig. 2.27). The only reptilian survivors of archosaurian origin are crocodiles and alligators. Even they have become considerably modified from the ancestral stock. Birds probably evolved from another branch of archosaurs. Like pterosaurs they developed the ability to fly, but the structure of the wings in the two groups is very different (Fig. 10.86).

At the close of the Age of Reptiles great climatic changes occurred affecting both animal and plant life. Thousands of years elapsed before the great dinosaurs, their smaller relatives, and the pterosaurs disap-

Fig. 2.27 Restoration of pterosaur by E. M. Fulda. (*Courtesy of the American Museum of Natural History.*)

peared and mammals, heretofore an insignificant group, began to flourish.

Although reptiles and amphibians are believed to have arisen from a common ancestral stock, the gap between the two groups is large. This is most strikingly brought out by their manner of development. Eggs of reptiles are always laid on land, and a resistant shell covers the large-yolked eggs. Of even greater significance is the presence of the embryonic membranes mentioned above.

As in lower forms, the body temperature of reptiles varies with the environment. In this respect they are said to be *poikilothermous*. The skeletons of reptiles are more completely bony than those of amphibians; their bodies are covered with dry, horny scales, and the skin is practically devoid of glands. Except in snakes and in certain lizards which lack appendages, the limbs are characteristically pentadactyl. Gill pouches are present only in the embryo, and respiration is entirely by means of lungs except in certain aquatic turtles in which cloacal respiration is employed as well. The young resemble their parents, and metamorphosis does not occur. Only the four orders of living reptiles are discussed below.

Order I. Chelonia. Chelonians are not closely related to other modern reptiles since they represent a very old group which has persisted with but little change for some 175 million years. Terrestrial forms are usually called tortoises, whereas turtles are generally semiaquatic. Terrapins are edible freshwater species.

The bodies of chelonians are relatively short and wide. In certain marine forms the pentadactyl limbs have been modified to form flippers. Teeth are lacking, the jaws being covered with sharp horny beaks. The anal opening is in the form of an elongated, longitudinal slit. Most noteworthy is the presence of a shell which encases the body. The more rounded dorsal portion is the *carapace;* the flatter ventral region is the *plastron*. The shell has an underlying layer of bone over which horny scales are arranged in a rather similar manner in most species.

Variations in size are extreme among turtles, some large marine forms attaining a weight of 1,500 to 2,000 lb.

Order II. Rhynchocephalia. The tuatara of New Zealand and about 20 surrounding islands, *Sphenodon punctatum* (Fig. 2.28), is the only living member of the order. *Sphenodon* has been called a living "fossil" because of its primitive and generalized structure. It is one of the oldest types of reptiles known,

Fig. 2.28 *Sphenodon punctatum. (Courtesy of the American Museum of Natural History.)*

even when extinct forms are taken into consideration. Adults attain a length of about 21 in. A significant feature is the presence of a well-developed, median, parietal, or parapineal, third eye in the middle of its forehead. Some lizards have a similar median eye (see page 622), but it is never as well developed as that of the tuatara. The anal opening is a transverse slit.

Despite the lizardlike appearance of *Sphenodon,* certain features of its skull indicate a lack of close relationship to modern lizards. The animal is rapidly approaching extinction.

Order III. Squamata. The order includes lizards and snakes. They are placed in separate suborders. These reptiles apparently came from the same ancestral stock which gave rise to *Sphenodon* but have flourished and evolved further instead of remaining at a stationary level. In evolutionary terms they represent the most recent of reptiles. Snakes are believed to have sprung from lizard stock.

The horny scales of snakes and lizards are derived from the epidermis of the skin. Sometimes bony dermal scales lie under the horny scales. As in *Sphenodon,* the anal opening is in the form of a transverse slit.

Suborder I. Sauria (Lacertilia). Lizards usually have two pairs of pentadactyl limbs,

but some, known as *glass snakes,* as well as certain others, do not have legs, so this feature cannot be used to characterize lizards as a group. The eardrum, or tympanic membrane, is not flush with the surface, and external ear pits are visible. Contrary to the condition in snakes, lizards have movable upper and lower eyelids as well as a nictitating membrane. The two halves of the lower jaw are united, and the animals lack the ability to open their mouths in the manner of snakes.

Lizards and salamanders are commonly confused by the layman. The shape of the body in these two groups is essentially similar, but the smooth moist skin of the salamander and its toes without claws are quite in contrast to the dry scaly skin of the lizard, the toes of which bear claws.

One characteristic for which certain species of lizard are famous is their ability to change color rapidly (*metachrosis*) so as to blend in with the background of their environment. Not all lizards have this ability, however. As an adaptation for self-protection many lizards can leave their tails behind them when escaping from enemies. A new tail regenerates under the influence of ependymal cells from the adjacent spinal cord (p. 605).

The largest species of lizard extant is the dragon lizard, *Varanus komodoensis,* of the East Indies. It may reach a length of 12 ft.

Fig. 2.29 The Gila monster, *Heloderma suspectum.* (*Courtesy of the American Museum of Natural History.*)

The Gila monster, *Heloderma* (Fig. 2.29), is the only poisonous lizard known. There are two species.

Suborder II. Serpentes (Ophidia). Snakes have lost their limbs in the course of evolution, but some, like pythons and boa constrictors, still retain skeletal vestiges of the pelvic girdle and limbs. Locomotion is accomplished in two ways, by sideward muscular undulations of the body and by movements of the transverse ventral scales. Snakes lack external ear pits and tympanic membranes. The eyelids are immovably fused, and are transparent. The loose ligamentous attachment of the jawbones to each other and to the cranium enables the snake to stretch its mouth to a remarkable degree. Snakes vary greatly in size, ranging from species only 5 or 6 in. long to 30-ft pythons.

Poisonous snakes fall into two groups, the vipers and the cobras. Some of the more familiar vipers, such as the rattlesnake and copperhead, are called *pit vipers* because of the presence of a pitlike depression between the eye and the nostril. The pit is a sense organ, sensitive to heat, enabling the animal to become aware of a warm object some distance away.

Order IV. Crocodilia (Loricata). Crocodilians are the only reptilian survivors of the ancient terrestrial archosaurs. The hind legs of many archosaurs were elongated and fitted for bipedal locomotion. With the demise of the archosaurs apparently one branch survived. The animals gradually changed back to the quadruped method of locomotion and assumed an amphibious life. Crocodilians are modified descendants of these archosaurs and are quite unlike other reptiles living today.

Crocodilians have laterally compressed tails and two pairs of short legs. There are five toes on the forefeet and four on the hind feet. The toes are webbed. The tympanic membrane is exposed but protected by a fold of skin. Eyes, nostrils, and ears are in a straight line on top of the head. This enables the animal to use its major sense organs when only a small part of the body is exposed above water. The anal opening, like that of chelonians, is a longitudinal slit. The skin is thick, with bony plates underlying the horny scales on the back and ventral sides.

With the exception of certain giant marine turtles, crocodiles and alligators (Fig. 2.30) are the largest reptiles living today. Caymans, of South America, and gavials, of India, are also included in the order. The feature that distinguishes crocodiles from alligators is that in alligators the fourth tooth on each side of the lower jaw fits into a pit in the upper jaw when the mouth is closed. In most crocodiles the fourth tooth of the lower jaw fits into a notch on the *outer* side of the upper jaw and is exposed when the mouth is closed. The snout of the American crocodile is much narrower than that of the American alligator.

Class III. Aves

Birds are the only animals which possess feathers. Feathers are modifications of the reptilian type of epidermal scale. Together with the scales on feet and legs of birds, they indicate a close relationship between birds and reptiles. This is also borne out in many other features. We have already alluded to the fact that birds did not evolve from pterosaurs. They probably arose from another type of archosaur with bipedal locomotion, in which the scales covering the body had become modified into feathers. Feathers on the forelimbs, rather than membranous wings, are used to resist the air in flight (Fig. 10.86).

Birds are the most highly specialized of vertebrates. As a group they have become adapted to aerial life, although there are a number of exceptions. Adaptive features include the light, hollow bones; loss of right ovary and oviduct in females of most species; the exceptionally well-developed eyes; the highly specialized lung and air-sac system; the modifications of the forelimbs to form wings; the presence of feathers which although light in weight offer an effective

resistance to air when the bird is in flight. There is less deviation within the entire class than is found within a single order in some of the other classes.

At the time of hatching, young birds are of either of two types: *altricial* or *precocial*. Altricial birds are those born in a helpless, more or less undeveloped condition. They must be cared for by the parents for a variable period before being able to fend for themselves. Examples of altricial birds are robins and sparrows. Precocial birds, on the other hand, have their eyes open at the time of hatching and are covered with a fluffy down. They soon run about in search of food. Chickens, ducks, and quail are examples of precocial birds. They may even be reared in brooders without any parental care. The average number of eggs in the nest is usually much less in the case of altricial birds than in precocial ones.

Birds have a high and constant body temperature independent of the environment, and are said to be *homoiothermous*. The young of altricial birds, however, at the time of hatching, are virtually poikilothermous and in this respect are similar to animals belonging to the lower classes. The homoiothermous condition becomes established shortly before the young leave the nest. In precocial birds the condition appears relatively much earlier.

Although the forelimbs of most birds have been modified into wings, they are not always adapted for flight. The hind limbs of various

Fig. 2.30 *Alligator mississippiensis* and nest. (*Courtesy of the Chicago Natural History Museum.*)

Fig. 2.31 Fossil remains of *Archaeopteryx*. (*Courtesy of the American Museum of Natural History.*)

types of birds show much variation. The feet usually have four toes terminating in claws (Fig. 2.35).

Birds are divided into two subclasses, one of which contains one or possibly two species known only by fossil remains. The second includes all other birds, extinct and modern.

Subclass I. Archaeornithes

The famous *Archaeopteryx lithographica* (Fig. 2.31) represents the subclass. It is known from two specimens and from the impression of a single feather found in the lithographic limestone in a quarry at Solenhofen in Bavaria, Germany. This bird was about the size of a crow. Its long jointed tail bore feathers along each side. Each wing had three clawed digits. The hind feet had four toes as in modern birds. Teeth, set in

sockets, were present in both jaws. Riblike dermal bones, called *gastralia,* found in numerous reptiles but not in other birds, were present in *Archaeopteryx*. In many respects, then, this fossil bird showed distinct reptilian characteristics. The presence of feathers, however, makes it imperative to classify it with birds, for in this feature it differs from any known reptile. *Archaeopteryx* is regarded as an almost ideal connecting link between these two great classes of tetrapods.

Subclass II. Neornithes

In members of the *Neornithes* there are 13 or fewer compressed vertebrae in the tail. The wing bones have been reduced in number, and, with rare exceptions, there are no free, clawed digits on the wings. A few primitive fossil forms, some extinct flightless

Fig. 2.32 *Hesperornis regalis. (Courtesy of the Chicago Natural History Museum.)*

species, and all modern forms are included in the subclass.

Superorder I. Odontognathae. The fossil *Neornithes* include *Hesperornis regalis* (Fig. 2.32), a flightless, swimming bird, and *Ichthyornis victor,* with well-developed wings. Both birds had true teeth.

Superorder II. Paleognathae (Ratitae). To this group belong six or seven orders of flightless, toothless, running birds. A few, such as the giant moas and elephant birds, have become extinct within memory of man. Living examples include the ostrich, rhea, emu, cassowary, and kiwi (Fig. 2.33). Lacking the ability to fly, they have failed to become widely distributed. Their wings are rudimentary or are too small and weak to be used in flight. The tinamous of Central and South America are exceptions.

Superorder III. Neognathae. All remaining birds belong to this group, in which 22 to 24 orders are recognized. All are flying birds, except the penguins, in which the forelimbs have been modified into paddlelike swimming organs (Fig. 2.34). No teeth are present.

Numerous adaptations of birds are of interest to the comparative anatomist. Those which fit them for aerial life as well as for life on land or water, the extraordinary migrations of many species, their methods of courtship and nest building have all been carefully studied. The feet and beaks of birds show many curious adaptive variations (Figs. 2.35 and 2.36).

Class IV. Mammalia

Mammals are homoiothermous tetrapods which have hair and mammary glands. All mammals have some hair, and only mam-

Fig. 2.33 The kiwi, *Apteryx australis,* with egg. *(Courtesy of the Chicago Natural History Museum.)*

Fig. 2.34 Jackass penguin. (*Courtesy of the Chicago Natural History Museum.*)

mals have hair. Mammary glands, found only in mammals are specialized skin glands which secrete milk used to nourish the young.

It is believed that mammals may have evolved from the mammal-like cynodont reptiles previously mentioned. It is possible, however, that several groups of mammal-like reptiles may have contributed to the pedigree of early mammals. Many such groups are known through fossil remains of skulls, jaws, teeth, etc. Only forms living today are described in the following account. Probably the intricate mechanism which ensures a constant body temperature independent of the environment has had much to do with the present wide distribution and supremacy of mammals. Birds, the only other homoiothermous animals, are also widely distributed. They have specialized along lines which enable them to fly. Mammals have specialized in another direction, i.e., development of the nervous system, particularly the cerebral portion of the brain.

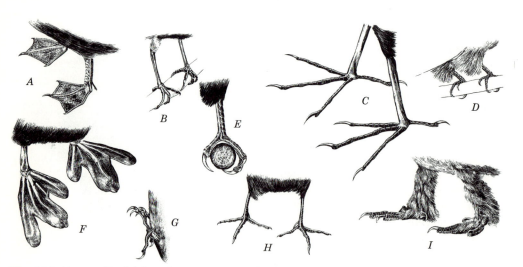

Fig. 2.35 Types of birds' feet, showing various adaptations to different modes of life: *A*, duck (swimming); *B*, robin (perching); *C*, heron (wading); *D*, cuckoo (climbing); *E*, hawk (seizing); *F*, grebe (swimming); *G*, woodpecker (clinging); *H*, rail (walking and running); *I*, ruffed grouse (scratching).

Fig. 2.36 Types of beaks in birds, showing various adaptations to different kinds of food: *A*, red-tailed hawk (flesh); *B*, robin (worms, insects, berries); *C*, cardinal (seeds); *D*, flycatcher (insects); *E*, red-breasted merganser (fish); *F*, wood duck (aquatic vegetation, fish); *G*, American bittern (fish, crustaceans); *H*, warbler (insects); *I*, woodpecker (grubs); *J*, crossbill (seeds in fir and pine cones).

In size, mammals vary from a small shrew, the body of which is less than 2 in. long, to the blue whale, which may reach a length of 103 ft or more. Mammals have become adapted to a diversity of habitats. Some, like moles, live almost entirely underground; whales are at home only in the water; bats rival birds in their ability to fly; most monkeys are arboreal; but the large majority of mammals walk or run on the ground. Many mammals, entire orders in some cases, have become extinct.

A few mammals (three species) lay eggs, but the rest give birth to living young. Accordingly, the class has been divided into two subclasses, of which the first contains only the egg-laying forms.

Subclass I. Prototheria

Eggs of prototherians are incubated outside the body, but the young are nourished by milk from mammary glands. No nipples are present, the glands opening directly onto the surface of the skin. The three living species are included in a single order.

Order I. Monotremata.

To this order belong the duckbill platypus, *Ornithorhynchus anatinus,* and two genera of spiny anteaters, or echidnas: *Tachyglossus* of Australia, Tasmania, and New Guinea; and *Zaglossus* of New Guinea and Papua.

The platypus of Australia (Fig. 2.37) is semiaquatic in habit, its beaverlike body, tail, and webbed feet adapting it to life in the water. Its ducklike bill is soft, flexible, and moist, and richly supplied with sense organs. It bears only a superficial resemblance to the bills of certain birds. No biologist believes that the platypus is a connecting link between birds and mammals.

In the spiny anteaters the skin is covered

Fig. 2.37 Duckbill platypus, *Ornithorhynchus anatinus*. (*Courtesy of the American Museum of Natural History.*)

with spines and coarse hair (Fig. 2.38). A short tail and external ears are present. The protrusible tongue is used for feeding on ants and termites.

Monotremes do not have a highly developed mechanism for controlling body temperature, which is, therefore, rather inconstant. The average temperature of the echidna is 31°C. Its oxygen consumption is only half that expected for a placental mammal. This group of mammals apparently has no close relatives. Several features indicate that they have retained certain reptilian characters. They are basically primitive forms, despite the fact that they possess certain specialized features, and should be con-

sidered as occupying an intermediate position between the mammal-like reptiles and higher mammals.

Subclass II. Theria

Therians bring forth their young alive and have mammary glands provided with nipples or teats. The subclass is divided into two infraclasses.

Infraclass I. Metatheria

The young of metatherians are usually born in an extremely immature condition and undergo further development in a marsupial pouch on the ventral side of the mother (Fig. 2.39). The term *mammary fetus* is sometimes applied to the pouch young. In some metatherians the marsupial pouch is reduced or wanting.

Order I. Marsupialia. The marsupials of today are confined almost entirely to Australia and the neighboring islands. Exceptions include the opossums of North, Central, and South America and the small opossum rat, *Caenolestes*, of Ecuador and Peru. The latter lacks a marsupial pouch. The Australian group embraces such forms as

Fig. 2.38 Spiny anteater, *Tachyglossus*. (*Courtesy of the Chicago Natural History Museum.*)

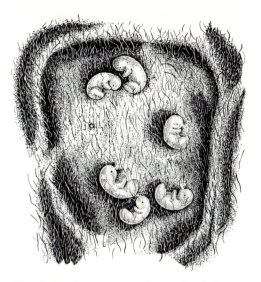

Fig. 2.39 Open marsupial pouch of Virginia opossum, *Didelphis marsupialis,* showing young at about the third day after birth.

kangaroos; the thylacine, or marsupial wolf; the wombat (Fig. 2.40), a small bearlike animal; the bandicoot, which resembles a rabbit; the marsupial mole; the marsupial flying squirrel; and a host of other forms which show likenesses to rodents, carnivores, and insectivores of other climes—but only in a superficial way.

The common Virginia opossum, *Didelphis marsupialis,* of the eastern and southern United States, is the most familiar form. The young at birth (Fig. 2.39) are more like the embryos of placental mammals. They make their way to the marsupial pouch without aid on the part of the mother. Here they undergo the rest of their development. After having grown still further, they leave the pouch, at first temporarily and then permanently. The Virginia opossum might be considered to be a "living fossil," having undergone little change in 120 million years. Probably because of its small size and generalized structure it has been able to survive in competition with more highly developed and specialized placental mammals.

In both sexes of most marsupials a pair of long, flat, *epipubic* bones extends forward from the lower part of the pelvic girdle. They are embedded in the muscles of the lower abdominal wall which they help support. Similar bones are found in monotremes. The fact that certain primitive reptiles also possessed a similar support of the ventral abdominal wall offers a possible clue to the ancestral relations of metatherians.

The peculiar geographical distribution of marsupials is difficult to explain and has

Fig. 2.40 The wombat, or marsupial bear. (*Courtesy of the Chicago Natural History Museum.*)

aroused many interesting speculations. Fossil remains of numerous marsupials have been found in South America, indicating that they once were a thriving race on that continent. A rather widely accepted explanation suggests that when mammals arose to a dominant position they had already split into two basic groups, the marsupials and the placentals. At this time Europe, Africa, Asia, and North America were united into a continuous land mass, or World Continent, whereas Australia and South America were probably island continents. It is known from the fossil record that primitive marsupials of little relative importance were present, at least in the European and North American regions of the World Continent. Placental mammals, however, were predominant. In Australia and South America the mammalian populations differed from those of the World Continent. The Australian forms were probably all marsupials, whatever their origin may have been. In South America herbivorous placental mammals and carnivorous marsupials seem to have coexisted for some time. Later on, when South America and North America became reconnected, there was apparently an invasion of competitively superior carnivorous placental mammals from the north, and the

marsupials in South America gradually became extinct except for a few surviving forms. Those in Australia, on the other hand, not in competition with placentals, flourished and slowly evolved into the many types which are now indigenous to that continent.

Many biologists today are of the opinion that both marsupial and placental mammals originally sprang from the same basic mammalian stock which underwent an early dichotomy. Neither type was more primitive or more advanced than the other, and both were equally progressive in their evolution and ability to adapt themselves efficiently to changing environmental conditions. The extinction of South American marsupials in competition with placental carnivores does not necessarily imply that marsupial mammals were inferior or more primitive than placentals, but rather that the types of placental mammals which invaded South America from the World Continent had already become competitively superior.

Infraclass II. Eutheria (Placentalia)
Eutherians are often referred to as placental mammals since their developing young are nourished by means of a placenta attached

Fig. 2.41 Common mole, *Scalopus aquaticus*. (*Courtesy of the American Museum of Natural History.*)

Anatomy of the Chordates

Fig. 2.42 European shrew, *Sorex oraneus*. (*Courtesy of the American Museum of Natural History.*)

to the lining of the uterus of the mother. An exchange of gases, nutritional substances, and excretory products between mother and young takes place at the placenta. Marsupial pouches and epipubic bones are never found in eutherians. To this infraclass belong most of the familiar mammals. Although as many as 28 separate orders, including fossil forms, have been recognized, we shall list only 16, which include the more familiar forms living today.

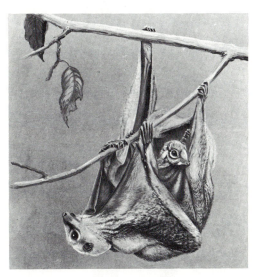

Fig. 2.43 Flying lemur, *Galeopithecus*. (*From a painting of a live animal by W. A. Weber. Courtesy of the Chicago Natural History Museum.*)

Order I. Insectivora. To this rather primitive order belong the moles, shrews, and hedgehogs. They are rather small mammals, usually with elongated snouts.

Moles (Fig. 2.41) live underground in burrows. Their eyes are extremely small and hidden in the fur. The heavily clawed, paddle-shaped forelimbs are used in digging; the hind limbs are small and weak.

Shrews (Fig. 2.42) include the smallest of all mammals and are frequently confused with mice. They are voracious and pugnacious.

In America the name *hedgehog* is often used to refer to the porcupine, which belongs to the order *Rodentia*. The European hedgehog, *Erinaceus,* is a true insectivore, the body of which is covered with short spines intermingled with hair.

Order II. Dermoptera. Only one genus, consisting of two species, is included in the order. *Galeopithecus,* the flying lemur, or colugo, of the Malay region (Fig. 2.43), is representative. It is larger than any of the insectivores and approximately the size of a cat. A well-developed fold of skin, the *patagium,* extends along either side of the body from neck to tail, enclosing the limbs. The feet are webbed. The presence of the patagium enables the animal to soar for some distance.

Order III. Chiroptera. Bats are highly specialized mammals which show unmistakable relations to insectivores. They are the only mammals that can actually fly. The wing of the bat (Fig. 10.86) is constructed entirely differently from the wing of the bird. The fingers are greatly elongated and connected by a web which offers resistance to the air. The web is attached to the body, hind limbs, and tail, if a tail is present. Whereas in the prehistoric flying reptiles (pterosaurs) (Figs. 2.27 and 10.86) the fourth finger of the hand was greatly elongated, in the bats fingers 3, 4, and 5 are very long and the first two are short. The first finger is free, but the rest are fastened to the web. The sternum is carinate (keeled) like that of the flying birds and serves for the attachment of the strong pectoral muscles used in flight.

Order IV. Primates. It may seem strange to the student that the order of mammals to which man belongs should be placed among the lowest orders. Although the nervous system, especially the cerebral portion of the brain, is best developed of all in primates, nevertheless in other respects they show many generalized characteristics. The development of the human brain far exceeds that of other primates, but, aside from this, man is clearly related, even in minor details, to the rest of the animals making up the order. The evolution of man is still in its ascendency but has reached the point at which his cultural evolution is preponderant over his biological evolution. He may ultimately even be able to direct his own evolution via scientific discoveries which are opening up new vistas.

In primates the limbs are unusually long and the pentadactyl hands and feet are relatively large. In many forms both thumb and big toe are opposable, although in man this applies only to the thumb. The limbs, in general, are adapted for arboreal existence. The eyes are directed forward. The order is divided into three suborders.

Suborder I. Lemuroidea. The lemurs (Fig. 2.44) of today are confined to Madagascar, Ethiopia, and parts of Asia. Fossil remains indicate, however, that they once were widely distributed. Lemurs are arboreal animals of moderate size, and nocturnal or crepuscular as to habit. The head is foxlike with a pointed muzzle; the tail is long but not prehensile. The second digit on each hind foot bears a sharp claw, whereas the other digits on both forefeet and hind feet are provided with flattened nails.

Fig. 2.44 Sportive lemur. (*Courtesy of the Chicago Natural History Museum.*)

Suborder II. Tarsioidea. Tarsiers (Fig. 2.45) are small lemurlike creatures with long ears, large and protruding eyes, and elongated heels. The second and third digits of the hind feet bear claws; those of the forefeet and the remainder on the hind feet have nails. These animals occur in the Philippines and in the islands between India and Australia. They are arboreal and nocturnal.

Suborder III. Anthropoidea. Monkeys, apes, and man, all of which are included in this order, bear digits with flattened or slightly rounded nails. They are arboreal or terrestrial, and diurnal as to habit. Within the group there is a tendency to walk upright, but only man has actually attained the upright posture. The increased specialization of the nervous system, with its epitome in man, is the most prominent feature of the group.

Fig. 2.45 Tarsier, *Tarsius spectrum* (*Courtesy of the American Museum of Natural History.*)

The suborder is made up of two large divisions, the *Platyrrhinii,* or South American monkeys, and the *Catarrhini,* or Old World forms. The chief anatomical feature which distinguishes the two groups lies in the wide distance between the outwardly directed nostrils in the former and the narrow septum between the downwardly directed nostrils of the latter. Although some of the *Platyrrhinii* have prehensile tails, this is never true of the *Catarrhini.* No connecting links have been found which give a clue to the relations of the two groups.

Among the *Platyrrhinii* are included marmosets (Fig. 2.46), and capuchin, spider, howling, and squirrel monkeys. The *Catarrhini* include macaques, Barbary apes, baboons, gibbons, orangutans, gorillas, chimpanzees, and man. The term *ape* may be used to designate any monkey, but it is generally applied to the larger Old World forms. The anthropoid apes are those which most closely resemble man.

The well-developed brain, the generalized condition of parts of the skeleton and other body structures, together with the fact that he is homoiothermous and can adapt himself to almost any climatic condition, all account for the success of *Homo sapiens* (man) in attaining his present leading position in the animal world.

Biologists do not believe that man arose from the apes in evolution but rather that man and the apes sprang from a common ancestral stock, each having diverged in a different direction. No single fossil has yet been found which might be considered as the ancestral form. Discovery of the remains of manlike apes in South Africa within recent years indicates that in this region paleontologists may ultimately find the crucial missing links in the human pedigree.

It has recently been suggested, on the basis of certain blood tests, that the chimpanzee and gorilla, now included in the

Fig. 2.46 Golden marmoset. (*Courtesy of the Chicago Natural History Museum.*)

family *Pongidae*, be placed in the family *Hominidae*, which heretofore has been reserved exclusively for man (*Lancet*, **1964-I** (7334):651).

Order V. Edentata. The lack of teeth or the presence only of poorly developed molar teeth characterizes the edentates. Some forms that are toothless as adults have teeth when young, but they are lost as the animals grow older. Among the edentates are included the South American anteaters, armadillos (Fig. 2.47), and sloths. Only one species, the nine-banded armadillo, is native to the United States. It is a nocturnal animal found in the southwestern part of the country but is spreading eastward. Armadillos are the only mammals which possess dermal skeletal structures, the body being covered with a bony case of armor. There are large, bony plates on the head, shoulders, and hindquarters. The shoulder and hindquarter plates are connected by a series of bony rings, the number of which is characteristic of each species.

The great South American anteater (Fig. 2.48) has a narrow, elongated head and a plumelike tail covered with coarse hair. Its forefeet are furnished with long, heavy claws which are used to advantage in tearing down large anthills and termite nests. When the disturbed insects scurry about, the animal sweeps them into its mouth with its long, extensile, sticky tongue. Locomotion is carried on with some difficulty, for the animal folds up its claws against a rough pad on its palm and makes contact with the ground with the back of its claws.

The peculiar sloths (Fig. 2.49) are arboreal animals which spend the greater part of their lives hanging upside down in trees. When placed in the water these animals swim an overarm backstroke. There are two types: the two-toed and the three-toed sloths. These terms refer to the number of toes on the forelimbs only, since both species have three toes on the hind feet.

Order VI. Pholidota. Only a single genus, *Manis,* the pangolin, or scaly anteater (Fig. 2.50), is included in the order. Its native habitat is in eastern Asia and Africa. The animal is covered with large, horny, over-

Fig. 2.47 Nine-banded armadillo. (*Courtesy of the Chicago Natural History Museum.*)

lapping scales among which a few hairs are interspersed. Teeth are lacking.

Order VII. Rodentia. The gnawing teeth of rodents are their most outstanding feature. These are the two chisel-shaped incisor teeth in front of both upper and lower jaws. They continue to grow throughout life. Canine teeth are lacking, but grinding, posteriorly located premolar and molar teeth are present.

Rodents are by far the most numerous of mammals and are distributed all over the world. Among the more familiar forms are rats, mice, squirrels, beavers, chipmunks, porcupines, and guinea pigs. The largest rodent of all is the capybara, *Hydrochoerus*, a South American form (Fig. 2.51).

Order VIII. Lagomorpha. Rabbits, hares, and pikas are included among the lagomorphs. Like rodents they lack canine teeth and their incisors grow continually. They differ from true rodents, however, in other respects. They have four incisor teeth in the upper jaw rather than two. The second pair

Fig. 2.48 Great South American anteater. (*Courtesy of the Chicago Natural History Museum.*)

Fig. 2.51 Part of capybara habitat group. (*Courtesy of the Chicago Natural History Museum.*)

Fig. 2.49 Two-toed sloth. (*Courtesy of the Chicago Natural History Museum.*)

lies behind the first and is smaller. Their tails are short and stubby.

Order IX. Carnivora. Carnivores are flesh eaters, for the most part, although some are omnivorous. All have three pairs of small incisor teeth in both upper and lower jaws and large, well-developed canine teeth. Their limbs are typically pentadactyl. There are two suborders, the distinctive features of

Fig. 2.50 Asiatic pangolin, or scaly anteater. (*Courtesy of the Chicago Natural History Museum.*)

which have to do with the condition of the toes. In the suborder *Fissipedia* the toes are separated, whereas in the suborder *Pinnipedia* the toes are webbed, forming flippers.

The fissiped carnivores include such familiar forms as bears, doglike mammals, cats, raccoons, weasels, skunks, and others. Among more unfamiliar species are civets, mongooses, genets, langsangs, and murkats. Seals, sea lions, and walruses are placed in the suborder *Pinnipedia*. They are well adapted for aquatic life but move about on land with difficulty. Most are fish eaters.

Order X. Cetacea. Whales, porpoises, and dolphins are members of the order *Cetacea*. Although porpoises and dolphins are relatively small in size, in this group are found the largest animals inhabiting the earth. Cetaceans have undergone profound modifications in adapting themselves to marine conditions. Nevertheless, they are true mammals. They are homoiothermous, breathe air by means of lungs, bring forth their young alive, and nurse them with mammary glands. Only traces of a hairy covering remain,

Fig. 2.52 Sperm whale. (*Courtesy of the Chicago Natural History Museum.*)

shown by a few scattered hairs on the snout. The webbed pectoral appendages are flippers. Pelvic appendages are lacking, except for rudiments of the pelvic girdle, embedded in the flesh. The tail is flattened horizontally and is referred to as the "flukes." Since whales are homoiothermous and have a thick, insulating layer of fat, or blubber, beneath the skin, they are at home even in the Arctic and Antarctic seas.

Whales are separated into two large groups, the toothed whales and the whalebone whales. The toothed whales are primarily fish eaters. They are usually not of exceptional size and include the dolphin, porpoise, and grampus. The sperm whale (Fig. 2.52) and the bottle-nosed whale are among the larger species. The whalebone whales (Fig. 2.53) have no teeth and are characterized by the great sheets of baleen, or whalebone, which hang from the roof of the mouth. The baleen is used in straining from the water the great numbers of microscopic plants and animals which form the chief food supply of these huge animals. The giant whales are members of this group. They include the gray, blue, sulfur-bottomed, hump-backed, and right whales.

The so-called "spout" of the whale is not a column of water, nor is it merely a condensation of moisture from the warm, exhaled air coming from the lungs. It is composed largely of mucus, gas, and emulsified oil that gather in the lungs. The mixture is expelled in the form of foam through the nostrils located on top of the head.

Order XI. Tubulidentata. The aardvark, *Orycteropus* (Fig. 2.54), is the only representative of the order. It is a rather large, African, termite-eating, burrowing mammal with a thickset body, large pointed ears, and

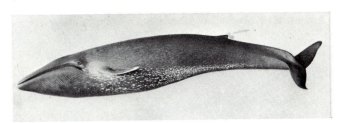

Fig. 2.53 Blue whale. (*Courtesy of the Chicago Natural History Museum.*)

Fig. 2.54 Aardvark. (*Courtesy of the Chicago Natural History Museum.*)

a long snout. There are a few poorly developed permanent teeth. Incisors and canines are lacking. It is not related to other anteaters previously discussed.

Order XII. Proboscidea. Elephants are characterized by their large size, thick skin, and scanty hair coat. The nose and upper lip have been extended to an extreme degree to form a long, prehensile proboscis, called the trunk, with nostrils at its free end. There are two large genera of elephants living today, the African and Indian forms. Indian elephants have small ears, tusks in the male only, a single pointed fingerlike tip at the end of the proboscis, five toes on the forefeet and four on the hind feet. These animals are rather easily tamed. African elephants have large, fanlike ears, tusks in both sexes but larger in males, two opposable fingerlike tips at the end of the proboscis, four toes on the forefeet and three on the hind feet. They are savage and not easily tamed. The pygmy elephants of the Congo are small animals of the African type. Extinct members of the order include the mastodon and the mammoth. Mastodons were older and more primitive. The proboscideans of today seem to be on their way to extinction. Their ancestors were at one time present in great numbers and in great variety on all the continents of the earth except Australia. Pale-

ontologists have gathered a great deal of information on the evolutionary history of the proboscideans, which has proved to be rather complex.

Order XIII. Hyracoidea. Only one genus, *Hyrax* (*Procavia*), belongs to this order. These small herbivorous creatures, commonly known as conies (Fig. 2.55), inhabit certain regions of Arabia, Syria, and Africa. They somewhat resemble guinea pigs in shape and size. Four digits are present on the forefeet and three on the hind feet. All digits are provided with hooflike nails, with the exception of the second digit on the hind foot, which bears a claw. The upper incisor teeth, like those of rodents, grow continually.

Order XIV. Sirenia. The manatees (Fig. 2.56) and dugongs, commonly called sea cows, are the only representatives of the order. They are herbivorous, and whalelike in external form, but are actually more closely related to members of the orders *Hyracoidea* and *Proboscidea*. Their bones are heavy and dense, a feature of importance

Fig. 2.55 Group of conies. (*Courtesy of the Chicago Natural History Museum.*)

Fig. 2.56 Florida manatees. (*Courtesy of the Chicago Natural History Museum.*)

in connection with their bottom-feeding habits. Manatees are about 9 or 10 ft long. They inhabit the rivers along the Atlantic coasts of South America and Africa. Some go as far north as the Everglades of Florida. Dugongs are oriental and Australian forms. Steller's sea cow, extinct since 1768, reached a length of 24 ft.

Order XV. Perissodactyla. This order of odd-toed, hoofed mammals includes horses, donkeys, zebras, tapirs, and rhinoceri. All are herbivorous. They walk on their nails (hoofs) and usually only on the middle finger or toe. The functional axis of the leg passes through the middle toe. The toes of the tapir have not been reduced to the extent of those of the horse family, there being four toes (one of them small) on the forefeet and three on the hind feet. The nose and upper lip have been drawn out into a relatively short proboscis.

There are one-horned and two-horned species (Fig. 2.57) of rhinoceri. The horns are located on the median line of the snout. Here, too, there are four toes on the forefeet (one of them small), and three on the hind feet.

Fig. 2.57 Rhinoceros. (*Courtesy of the Chicago Natural History Museum.*)

Order XVI. Artiodactyla. These are the even-toed hoofed mammals. All members of the order walk on the nails (hoofs) of the third and fourth toes. The other digits are greatly reduced or absent. The functional axis of the leg passes between the third and fourth toes. Members of the orders *Perissodactyla* and *Artiodactyla* are often called ungulates because of their unguligrade foot posture in which only the hoof is in contact with the ground.

In the *Artiodactyla* there is a division into two groups: the cud chewers, or *ruminants*, and the *nonruminants*. The first includes cattle, sheep, goats, camels, llamas, antelopes, deer, and giraffes. Among nonruminants are pigs, hippopotami, and peccaries. Cud chewers first swallow their food which is later regurgitated into the mouth for thorough mastication. Many members of the order bear horns or antlers projecting from the frontal bones of the skull.

SUMMARY OF CLASSIFICATION

    Phylum Chordata
        Subphylum I. Hemichordata. *Dolichoglossus*
        Subphylum II. Cephalochordata. Amphioxus
        Subphylum III. Urochordata. *Molgula*
        Subphylum IV. Vertebrata (Craniata)
           Superclass I. Pisces
             Class I. Agnatha
               Order I. Cyclostomata
                 Suborder I. Petromyzontia. Lampreys
                 Suborder II. Myxinoidea. Hagfishes
             Class II. Placodermi. Fossil placoderms
             Class III. Chondrichthyes
               Subclass I. Elasmobranchii
                 Order I. Selachii. Sharks; dogfishes
                 Order II. Batoidea. Skates; rays
               Subclass II. Holocephali. Chimaeras
             Class IV. Osteichthyes
               Subclass I. Actinopterygii. Ray-finned fishes
                 Superorder I. Chondrostei. *Polypterus; Calamoichthys;*
                      Polyodon; Acipenser
                 Superorder II. Holostei. *Amia; Lepisosteus*
                 Superorder III. Teleostei. Teleost fishes
               Subclass II. Sarcopterygii. Lobe-finned fishes
                 Order I. Crossopterygii
                   Suborder I. Rhipidistia. *Eusthenopteron*
                   Suborder II. Coelacanthini. *Latimeria*
                 Order II. Dipnoi. Lungfishes: *Epiceratodus; Protopterus;*
                    Lepidosiren

Superclass II. Tetrapoda
  Class I. Amphibia
    Order I. Anura. Frogs and toads
    Order II. Urodela (Caudata). Salamanders; newts
    Order III. Apoda (Gymnophiona). Caecilians
  Class II. Reptilia
    Order I. Chelonia. Turtles; tortoises
    Order II. Rhynchocephalia. *Sphenodon*
    Order III. Squamata
      Suborder I. Sauria (Lacertilia). Lizards
      Suborder II. Serpentes (Ophidia). Snakes
    Order IV. Crocodilia (Loricata). Alligators; crocodiles
  Class III. Aves
    Subclass I. Archaeornithes. Fossil *Archaeopteryx*
    Subclass II. Neornithes
      Superorder I. Odontognathae. Fossils: *Hesperornis; Ichthyornis*
      Superorder II. Paleognathae (Ratitae). Ostrich; kiwi
      Superorder III. Neognathae. Most familiar birds
  Class IV. Mammalia
    Subclass I. Prototheria
      Order I. Monotremata. Platypus; echidnas
    Subclass II. Theria
      Infraclass I. Metatheria
        Order I. Marsupialia. Opossum; kangaroo
      Infraclass II. Eutheria (Placentalia)
        Order I. Insectivora. Moles; shrews
        Order II. Dermoptera. Flying lemur
        Order III. Chiroptera. Bats
        Order IV. Primates
          Suborder I. Lemuroidea. Lemurs
          Suborder II. Tarsioidea. Tarsiers
          Suborder III. Anthropoidea. Monkeys; man
        Order V. Edentata. Sloths; South American anteaters
        Order VI. Pholidota. Scaly anteater
        Order VII. Rodentia. Rats; mice; squirrels
        Order VIII. Lagomorpha. Rabbits; hares; pikas
        Order IX. Carnivora. Dogs; cats; seals; walrus
        Order X. Cetacea. Whales; porpoise
        Order XI. Tubulidentata. Aardvark
        Order XII. Proboscidea. Elephants
        Order XIII. Hyracoidea. Cony
        Order XIV. Sirenia. Manatee; dugong
        Order XV. Perissodactyla. Horse; zebra
        Order XVI. Artiodactyla. Cattle; deer; sheep

| 3 |

EARLY
DEVELOPMENT
AND
HISTOGENESIS

In both plants and animals there are two general types of reproduction, asexual and sexual. The asexual, or agamic, method of reproduction in animals is confined to members of some of the lower phyla. Budding and fission are examples of this method.

Sexual reproduction is the rule in the phylum Chordata. Parthenogenesis, the development of an egg without fertilization, is considered to be a form of sexual reproduction since a sexual element, the egg, is involved. It has long been known that under experimental conditions the eggs of such vertebrates as frogs and rabbits (G. Pincus, 1939, *J. Exp. Zool.,* **82**:85) may be induced to develop parthenogenetically. For example, pricking an unfertilized frog's egg with a sharp needle will induce parthenogenetic development, but only if a bit of foreign protein is transferred to the egg in the process. In other cases, parthenogenesis may occur spontaneously. Although natural parthenogenesis is common among invertebrates, only recently have authenticated cases of natural parthenogenesis among vertebrates been described. These include the turkey (M. W. Olsen, 1960, *Proc. Soc. Exp. Biol. Med.,* **105**:279), the Caucasian rock lizard (I. S. Darevsky, 1966. *J. Ohio Herp. Soc.,* **5**:115), and some others.

EARLY EMBRYOLOGY

In the usual type of sexual reproduction the union of egg and spermatozoön results in a fertilized egg, or *zygote.* Fertilization of the egg brings about changes which result in the development of an embryo. The embryo is built up by a series of cell divisions in which the resulting cellular units do not dissociate but remain attached to each other and become differentiated or specialized to form the tissues and organs of the adult. Although it might appear that the stages in embryonic development differ in the several chordate groups, nevertheless the significant phases are basically similar in all. Such variations as do occur are primarily related to differences in the sizes of egg cells. The size of the egg depends, for the most part, upon the quantity of yolk that is present.

Yolk content of eggs. Eggs may be classified according to the amount of yolk which they contain. *Microlecithal,* or *alecithal,* eggs are those with a very small amount of yolk or without yolk. Many mammals and numerous invertebrates have eggs of the microlecithal type. Some embryologists are of the opinion that there are no eggs totally lacking in yolk content. *Meiolecithal* eggs, such as those of amphioxus, contain a small amount of yolk. *Mesolecithal* ova, typical of amphibians (Fig. 8.12), have a medium amount of yolk and are quite in contrast to the large, yolk-laden *polylecithal,* or *macro-lecithal,* eggs of reptiles and birds.

Distribution of yolk. Eggs are also classified according to the manner in which yolk is distributed throughout the ovum. In *isolecithal,* or *homolecithal,* eggs the yolk is rather evenly distributed throughout the cytoplasm. Microscopic mammalian eggs are of this type. *Telolecithal* ova, on the other hand, are those in which the yolk is concentrated more on one side than the other. This condition is encountered in the eggs of fishes, amphibians, reptiles, and birds. In insects and other arthropods the ova are usually *centrolecithal,* since the yolk is confined to the central region of the egg. No chordate has centrolecithal ova.

Polarity. In most forms, even in the single-cell stage and during subsequent development, it is possible to recognize a definite axis in the embryo. The extremes of the axis consist of unlike *poles,* which are called the *animal* and *vegetal poles,* respectively. The existence of such an axis is termed *polarity.* It is first expressed by the arrangement of egg substances (yolk, nucleus, etc.) and later by differential rates of cleavage. In most eggs there is a tendency for the nucleus and the clear cytoplasm to be located near the animal pole and for the yolk to concentrate at the vegetal pole.

Cleavage. The term *cleavage* is applied to the early mitotic divisions of the fertilized egg, the resulting cells being called *blastomeres.* During cleavage in microlecithal, meiolecithal, and mesolecithal ova, the zygote first divides into two smaller cells of equal size. These in turn divide into smaller and smaller units with no perceptible increase in size of the original mass. In this respect cleavage differs from the mitotic divisions of ordinary cells which usually grow to a size equal to that of the parent cell before dividing. Recent research indicates that the adhesion rather than the dissociation of cells of the embryo during early developmental stages is dependent upon the presence of calcium in the form of some calcium-macromolecular substance such as ribonucleic acid-protein-calcium complex (L. T. Stableford, 1967, *Develop. Biol.,*

16:303). When growth begins and differentiation of parts becomes apparent, the cleavage stage of development has been completed. Cleavage in large polylecithal ova follows a different pattern (page 60). The rate of cleavage is determined to a great extent by the amount of yolk present in the egg. Cleavage stages of amphioxus, frog, and bird eggs are shown in Figs. 3.1, 3.2, and 3.3.

Types of cleavage. The type of cleavage which takes place in an ovum is also governed largely by the amount of yolk present. *Complete,* or *holoblastic, cleavage* is typical of microlecithal, meiolecithal, and mesolecithal ova. The first cleavage plane divides the egg into two complete halves, passing through both animal and vegetal poles. The second cleavage is at right angles to the

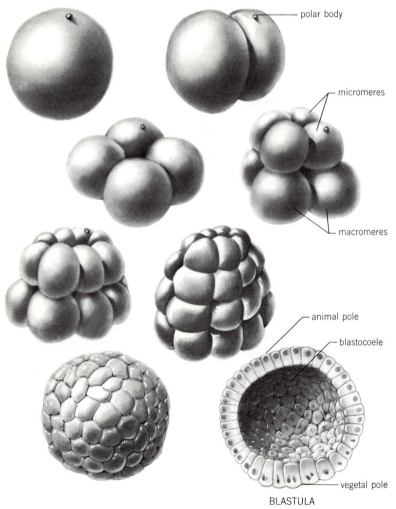

polar body

micromeres

macromeres

animal pole

blastocoele

vegetal pole

BLASTULA

Fig. 3.1 Cleavage stages of amphioxus egg. The figure to the lower right is that of a blastual shown as though it were cut in half. (*Drawn by G. Schwenk.*)

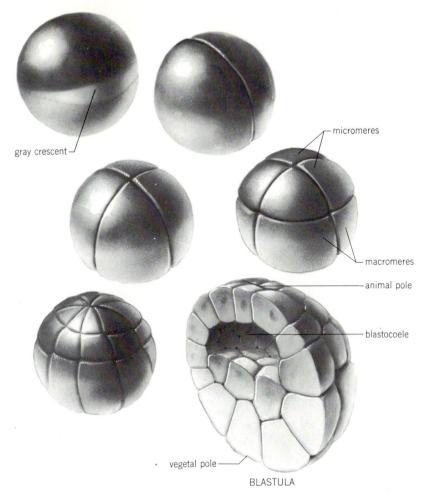

gray crescent

micromeres

macromeres

animal pole

blastocoele

vegetal pole

BLASTULA

Fig. 3.2 Cleavage stages of frog egg. The enlarged figure to the lower right is that of an early blastula, shown as though it were cut in half. (*Drawn by G. Schwenk.*)

first and is also meridional since it passes through the poles. There is, however, a difference in various eggs in regard to the third cleavage plane. This is latitudinal in relation to the two poles, so that after the third cleavage the embryo is in the eight-cell stage. In alecithal eggs the eight cells are approximately equal in size. This type of cleavage is called *holoblastic equal cleavage*. On the other hand, in meiolecithal and mesolecithal ova the third cleavage, though latitudinal, is nearer the animal pole. As a result, the four cells near the animal pole are smaller than the four yolk-laden cells near the vegetal pole. They are referred to as *micromeres* and *macromeres,* respectively. In this case the type of cleavage is termed *holoblastic unequal cleavage.*

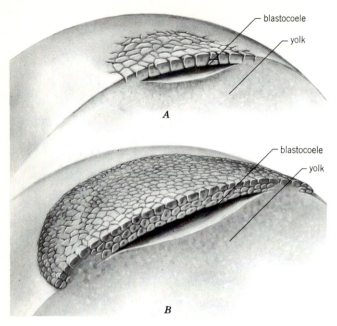

blastocoele
yolk

A

blastocoele
yolk

B

Fig. 3.3 Meroblastic discoidal cleavage of the polylecithal bird ovum: *A*, early blastula stage; *B*, later blastula stage. (*Drawn by G. Schwenk.*)

In large polylecithal ova complete cleavage does not take place. Instead, the nucleus and the clear cytoplasm undergo a series of incomplete divisions which are confined to a small area at the animal pole, the *blastodisc.* A small disclike area composed of cleaving cells, therefore, appears at the animal pole. Cleavage of this type is referred to as *incomplete,* or *meroblastic discoidal cleavage.* It is typical of the eggs of fishes, reptiles, and birds (Fig. 3.3).

Blastula. As cleavage progresses and the original ovum becomes subdivided to form the multicellular embryo, the cells usually become arranged so as to form a sphere with a single layer of cells surrounding a central cavity (Figs. 3.1 and 3.2). At this stage the embryo is referred to as a *blastula* and the

central cavity is spoken of as the *blastocoele,* or *segmentation cavity.* The blastula stage of polylecithal ova (Fig. 3.3) differs in that it is not spherical but consists of a disc of cells lying on the yolk. The structure, now referred to as the *blastoderm,* is several cells thick. Those in the central area are free from the yolk beneath, the blastocoele, a fluid-filled space, separating the two. The area over the blastocoele is the *area pellucida;* that around the margins and in contact with the yolk is the *area opaca.* Polarity in microlecithal, meiolecithal, and mesolecithal ova is still evident in the blastula stage. The cells at the vegetal pole are generally larger than those at the animal pole because of the greater amount of yolk which they contain. The blastula of amphioxus never consists of more than a single layer of cells. In the frog, however, the cells of the blastula multiply in such a manner that several layers of cells ultimately surround the blastocoele.

Gastrula: epiblast and hypoblast. The blastula stage is succeeded by the *gastrula,* the process bringing about the change being called *gastrulation.* In amphioxus, as a result of gastrulation, a double-walled, cuplike structure is formed (Fig. 3.4). The larger, yolk-laden cells from the original vegetal pole form the *hypoblast,* or inner layer of the cup, and the smaller, animal-pole cells make up the *epiblast,* or outer wall. The terms *endoderm* and *ectoderm* are frequently used for hypoblast and epiblast, respectively. Since, however, in later development the hypoblast of amphioxus gives rise to endoderm, mesoderm, and notochord, it is preferable to apply the terms ectoderm and endoderm at a later stage when all three *germ layers* as well as the notochord have become established. The original blastocoele is obliterated during gastrulation, and a new cavity, the *gastrocoele,* or *archenteron,* is formed. This is lined entirely with hypoblast and opens to the outside through the *blastopore.* The endoderm derived from the hypoblast is destined to form the lining of the alimentary, or digestive, tract and its derivatives. In mesolecithal ova the larger amount of yolk in the vegetal-pole cells is responsible for the fact that in the gastrula stage a plug of yolk-filled cells usually protrudes through the blastopore (Figs. 3.5 and 3.9). Moreover, in such gastrulae the cells of the hypoblast lining the archenteron are not of equal size. In mesolecithal amphibian gastrulae at this stage of development the hypoblast forming the roof of the archenteron is often referred to as *chorda-mesoderm* since it is destined to give rise to the notochord and some prechordal mesoderm in the midline, with somites and lateral mesoderm on either side. The latter proliferates and spreads laterally and ventrally. The cells of the hypoblast making up the floor and side walls of the archenteron may now be considered to be endodermal. As the mesodermal cells are spreading laterally and ventrally between epiblast and endoderm, the cells along the marginal borders of the endoderm proliferate and grow upward, ultimately to meet dorsally. The archenteron thus becomes completely lined with endoderm (Fig. 3.9).

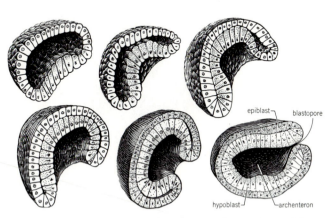

Fig. 3.4 Successive stages during gastrulation in amphioxus. (*Drawn by G. Schwenk.*)

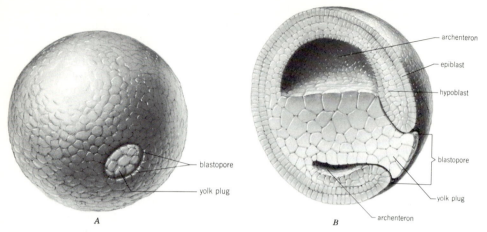

Fig. 3.5 Gastrula stage in development of frog: *A*, external view showing yolk plug filling the blastopore; *B*, median sagittal section of same, showing internal arrangement of cells making up epiblast and hypoblast. (*Drawn by G. Schwenk.*)

The way that blastula and gastrula form in polylecithal ova differs considerably from the above description, the differences being related to the enormous amount of yolk present. It is not necessary for our purposes here to discuss the details of these processes about which there is still some contention. Suffice it to say that the gastrula stage consists of a two-layered disc lying on the yolk with which it is in contact only at the periphery. The upper layer comprises the epiblast; the lower, the hypoblast. The small space between epiblast and hypoblast represents the blastocoele; that between hypoblast and yolk is the future archenteron. Soon after the egg is laid and incubation has begun, the area of the blastoderm expands. Concomitantly it may be observed that a thickened area appears in one quadrant. It is destined to become the caudal end of the embryo. The thickened area, where epiblast and hypoblast are in contact, soon begins to show an anterior-posterior elongation, extending along the midline of the developing structure to a point somewhat anterior to the center.

This longitudinal structure is called the *primitive streak*. It establishes the anterior-posterior axis of the embryo. The entire blastoderm has now assumed an elliptical shape. Epiblast and hypoblast are in contact at the primitive streak and at the peripheral margins of the blastoderm.

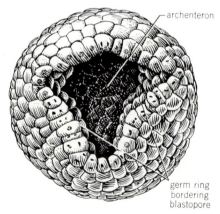

Fig. 3.6 Posterior view of gastrula of amphioxus, showing germ ring bordering the open blastopore. (*Drawn by G. Schwenk.*)

Closure of the blastopore. In the usual type of gastrula an elongation in the direction of the longitudinal axis passing through the blastopore next takes place. It is chiefly due to rapid cell divisions in the area about the margins of the blastopore, particularly on the dorsal side where epiblast and hypoblast are in continuity. This area is known as the *germ ring* (Fig. 3.6). As development progresses, the blastopore becomes smaller and smaller. In most species the edges of the germ ring, or margin of the blastopore, finally coalesce. In polylecithal ova the primitive streak is considered by most biologists to represent the fused margins of an elongated blastopore which is never open to the outside, the enormous amount of yolk precluding a development comparable to that of lower forms. In the embryos of certain invertebrates the blastopore remains open to

become the anus, but in chordates the blastopore closes and an anal aperture is established at a later stage.

Mesoderm. Usually toward the end of gastrulation the third germ layer, designated as *mesoderm,* appears and comes to lie between ectoderm and endoderm. In amphioxus the mesoderm first arises as a series of pouches which pinch off from the dorsolateral angles of the hypoblast lining the archenteron (Fig. 3.7).

Mesoderm formation in amphibians has been described above. In higher forms, except those with polylecithal ova, it usually originates by a proliferation of cells in the region of the germ ring. In polylecithal eggs, mesodermal cells grow out from either side of the primitive streak, extending laterally between epiblast and hypoblast (Fig. 3.8).

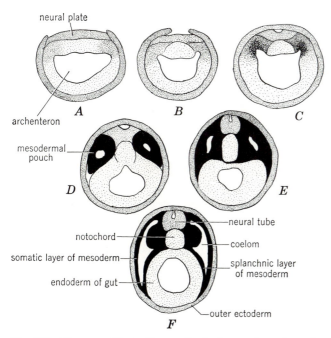

Fig. 3.7 Diagrams illustrating method of mesoderm formation in amphioxus.

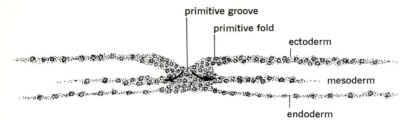

primitive fold

ectoderm

mesoderm

endoderm

Fig. 3.8 Cross section through primitive streak of bird embryo, showing proliferation of mesoderm.

The mesodermal cells push in between epiblast and hypoblast and fill the crevice which represents the remnant of the original segmentation cavity. Usually the mesoderm is at first a solid mass, but it soon splits peripherally into two layers with a space, the body cavity, or *coelom,* lying between. In amphioxus the cavities within the original mesodermal pouches become the coelom. With the establishment of the mesoderm and notochord (page 65) the terms ectoderm and endoderm, rather than epiblast and hypoblast, are appropriately used. The outer layer of mesoderm, called the *somatic,* or *parietal, layer,* is closely applied to the ectoderm, the two together, known as the *somatopleure,* being destined to form the definitive body wall. The inner, *splanchnic,* or *visceral,* layer of mesoderm is closely applied to the endoderm to form the *splanchnopleure,* which gives rise to the digestive tract and its derivatives. The coelom, which is thus lined entirely with mesoderm, represents the future peritoneal, pericardial, and pleural cavities of higher forms, as the case may be.

Differentiation. The cells making up the germ layers of an embryo are said to be undifferentiated; i.e., they do not possess distinctive or individual characteristics. Further development of an embryo entails, among other things, a differentiation or specialization of various groups of cells to form the several types of tissues and organs which make up the body of the individual.

Notochord. Usually around the time that the mesoderm is differentiating into somatic and splanchnic layers, the notochord begins to form. In amphibians, as previously noted, this event takes place somewhat earlier. In amphioxus the notochord arises as a mid-dorsal thickening of the wall of the archenteron (Fig. 3.9). In many chordates, however, it originates from cellular proliferation in the region of the dorsal lip of the blastopore between epiblast and hypoblast. In birds, a thickening, known as *Hensen's node,*

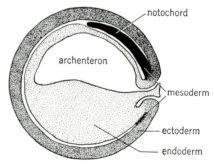

Fig. 3.9 Diagram of median sagittal section of frog gastrula in late yolk-plug, or neural-plate, stage, showing germ-layer relationships.

Anatomy of the Chordates

appears at the anterior end of the primitive streak. Accompanied by streaming movements of surrounding cells, Hensen's node gradually recedes in a posterior direction, leaving in its wake the notochord (Fig. 11.1). The latter forms an elongated, rodlike structure which comes to lie in the middorsal region of the body, between ectoderm and endoderm, separating the dorsal mesoderm of the two sides (Fig. 3.9). In most chordates the vertebral column later forms about the notochord, which is gradually replaced.

Neural tube. In the details of its development the nervous system of amphioxus differs somewhat from that of higher chordates and need not be considered here. In vertebrates, as the above-mentioned changes are going on, the flattened layer of ectoderm along the middorsal side of the gastrula becomes thickened. It is known as the *neural,* or *medullary, plate.* Proliferative changes in the cells of this region result in the formation of a longitudinal depression, the *neural groove,* along the middorsal line, flanked on either side by an elevated *neural fold.* The neural folds gradually approach each other dorsally and fuse in such a manner as to form a hollow *neural tube* (Fig. 11.1). During its formation the neural tube gradually sinks down to a deeper position in the embryo. The

neural tube is the forerunner of the brain and spinal cord.

The differentiation of the neural tube seems to be influenced by chemical factors (by macromolecules in the form of nucleotides) presumably elaborated by the surrounding cell systems (notochord and somites). An opening, the *neuropore,* at the anterior end of the neural tube persists for a time but ultimately closes. The outer ectodermal fold on each side meets its partner along the midline above the neural tube, forming a continuous layer from which the neural tube soon becomes completely separated. Thus the neural tube lies under the outer surface of the body. Small dorsolateral masses of ectodermal cells grow out on each side in the crevices formed between neural tube and outer ectoderm (Figs. 3.10 and 11.1). They are termed *neural crests.* Neural-crest formation seems to be dependent upon the release of a metabolic precursor, phenylalanine, from the roof of the archenteron (C. E. Wilde, Jr., 1956, *J. Exp. Zool.,* 133:409). The neural crests are of significance in the development of numerous nerve ganglia, the adrenal medulla, and in the formation of certain pigment cells, known as *chromatophores,* found in many lower vertebrates. Neural crests, at least in amphibians and birds, also appear to contribute to the

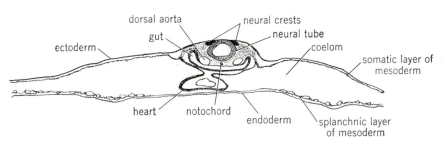

Fig. 3.10 Cross section of 33-hour chick embryo, showing neural tube and neural crests in relation to other embryonic structures.

mesenchyme from which a number of cartilages in the branchial region are derived. The mesenchyme forming the *dental papilla* of the developing tooth also seems to be of neural-crest origin.

Further differentiation of mesoderm. Mesoderm exists in the body either in the form of sheets of cells lining cavities of one kind or another, or as masses of loose, branching cells called *mesenchyme,* sometimes referred to as *embryonic connective tissue.* Sheets of mesodermal cells lining the body cavities are known as *mesothelia;* those lining the cavities of blood vessels, lymphatic vessels, and the heart are *endothelia.* Both epiblast and hypoblast may originally contribute to mesenchyme formation. Mesenchymal cells migrate from their place of origin and fill in spaces between structures developing from other sources. Mesenchyme is a rather diffuse tissue consisting both of cells and intercellular substance in which the cells lie. The intercellular substance is of such a nature as to permit diffusion of nutrients, gases, wastes, and water over a considerable distance at a time when blood vessels are developing and tissues are not and cannot be supplied directly with blood.

During the time that the above changes have been taking place in vertebrate embryos, the mesoderm not only has pushed in between ectoderm and endoderm but has grown dorsally on either side of the notochord and neural tube. Three different regions or levels of mesoderm may then be recognized: (1) an upper, or dorsal, *epimere,* (2) an intermediate *mesomere,* or *nephrotome,* and (3) a lower *hypomere,* or *lateral plate* (Fig. 3.11). The epimeric mesoderm becomes marked off into a longitudinal, metameric series of blocklike masses, the *mesodermal,* or *mesoblastic, somites,* which form in succession, beginning toward the anterior end of the embryo (Fig. 3.12). The somites later become separated from the remainder of the mesoderm. Although the mesoderm of the mesomere shows evidence of metamerism in some lower forms, this is not evident in embryos of higher vertebrates except, perhaps, toward the anterior end of the body. The hypomeric mesoderm does not become segmented. It is this which is composed of somatic and splanchnic layers separated by coelom.

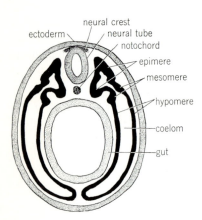

Fig. 3.11 Diagram of cross section of hypothetical vertebrate embryo, illustrating the three different regions, or levels, of mesoderm.

DERIVATIVES OF THE PRIMARY GERM LAYERS

All structures of the body trace their origin to one or more of the three primary germ layers. Further development is the result of the operation of a number of processes which are effective in shaping the body and in forming the various tissues and organs. Multiplication of cells goes on at a rapid rate, and the newly formed cells increase in size. Some clumps of cells form localized thickenings or swellings; others undergo a thinning

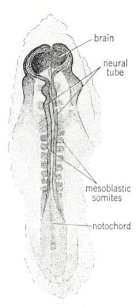

Fig. 3.12 Rabbit embryo, 190 hours after copulation, showing series of paired mesoblastic somites in process of forming.

brain

neural tube

mesoblastic somites

notochord

or spreading out. Certain masses of cells may evaginate (push outward) or invaginate (push inward) from a surface. Some groups of cells migrate to different parts of the body and become rearranged. Diverse masses of cells may fuse or else may split apart from each other; structures or even regions may form temporarily and then undergo resorption and disappear. Various cell aggregates differentiate to form tissues of different types.

Growth of the embryo and differentiation of organs from germ layers begin before the germ layers are completely formed. The entire process of development is gradual, and although the various phases occur in a logical, step-by-step order, they are not in themselves distinct but one merges imperceptibly into the next. Experimental work has shown that the determination of cells, germ layers, tissues, and organs is not invariably the same but is directed, in the main, by relationships to other cells and by the influence of outside factors of various kinds. The term *induction* is used to describe the directive influence which one developing part, or group of cells, has upon the differentiation and development of other tissues and organs. It is believed that induction is brought about by organizers, or evocators, the nature of which is presently unknown. The dorsal lip of the blastopore of the amphibian embryo, for example, is the most active organizing center at the critical stage of development when invagination occurs and the main axis of the embryo is established. The developing optic vesicle of the brain induces the lens of the eye to form from adjacent superficial ectoderm (page 670). The different germ layers and the structures which arise from them are considered to be homologous in all vertebrates. The germ-layer concept is of importance chiefly because it furnishes a convenient method of classifying organs according to their embryonic derivation and of tracing their homologies.

Abnormal embryonic development is not unusual and is presently the subject of intense research. The term *teratology* refers to the study of serious malformations or deviations from normal structure. The drug thalidomide (N-pthalylglutamic acid imide), taken as a tranquilizer before its complete effects were known, induced abnormal limb development in a large number of human embryos. Protein-zinc-insulin administered to white rats during pregnancy brings about abnormal development of offspring. Numerous other drugs, such as *colcemid* and *actinomycin D,* administered to pregnant mice, induce various kinds of teratogenesis. Recently a number of *Hyla* tree frogs with five legs have been found in North Carolina. Whether this is a genetic mutation has not as yet been determined. It is possible that some chemical pollutant in the pond water in which these frogs developed could have modified the normal DNA pattern so as to induce the limb duplication which has been genetically perpetuated.

Ectoderm. The entire outer covering of the body of the developing vertebrate embryo consists of a layer of ectodermal cells referred to as the *superficial ectoderm.* During development, when the somites, which have formed from the epimeric region of mesoderm, are undergoing differentiation, the outer portion of each somite (dermatome) becomes mesenchymatous. These mesenchymal cells proliferate and migrate in such a manner as to underlie the outer ectodermal covering of the body, at least in the dorsal and dorsolateral regions. The two layers together will form the future skin, or integument. The ectoderm gives rise to the *epidermis* of the skin; the underlying mesenchyme forms the *dermis.* It has come to be recognized that in some forms the somatic layer of mesoderm contributes to the dermis of the skin in the flank and ventral surfaces of the body. Many structures associated with the skin are derived from the epidermis and are thus of ectodermal origin, including various kinds of skin glands, epidermal scales, hair, feathers, nails, claws, hoofs, and horns. Certain specialized smooth-muscle fibers of ectodermal origin, called *myoepithelial cells,* are associated with salivary, lacrimal, sweat, and mammary glands (Fig. 9.15). They aid in emptying the glands of their secretory products. The origin of these cells from ectoderm is unlike that of most smooth muscle, which, in general, is of mesodermal origin.

The archenteron, which is destined to form the digestive tract, is lined with endoderm. At first, except in embryos of polylecithal ova, it opens to the outside through the blastopore, which soon closes, leaving the archenteron as a blind tube without mouth or anus. At the anterior end of the embryo an ectodermal invagination occurs, pushing inward to meet the endoderm of the archenteron. The ectodermal invagination is referred to as the *stomodaeum,* or primitive mouth. A two-layered membrane, the *oral plate,* or *oral membrane,* at first separates the two cavities. This persists but for a short time and then ruptures, so that the stomodaeum, or mouth, becomes continuous with the remainder of the digestive tract (Fig. 14.23). A similar ectodermal invagination forms at the posterior end of the embryo. This is the *proctodaeum.* An *anal plate,* or *anal membrane,* at first separates archenteron and proctodaeum. With the perforation of the anal plate a complete tube is established which terminates anteriorly with the mouth and posteriorly with the anus. Derivatives of the ectoderm lining the stomodaeum include the lining of the lips and mouth, enamel of the teeth, glands of the oral cavity, covering of the tongue, and the anterior and intermediate lobes of the pituitary gland. The proctodaeal ectoderm gives rise to the lining of the anal canal or, in most forms, to a portion of the cloacal lining. It is very difficult to ascertain the exact point of transition between endoderm and proctodaeal ectoderm. For this reason there is some question as to the precise germ layer from which some cloacal derivatives arise. Certain anal and cloacal glands are believed to be derived from ectoderm of the proctodaeum.

The origin of the neural tube and neural crests has already been reviewed. The neural tube is the primordium of the brain and spinal cord, which make up the *central nervous system.* Early evidence of metamerism in the neural tube is shown by the temporary appearance of segments known as *neuromeres.* The anterior end of the neural tube enlarges rapidly to form the brain, the remainder becoming the spinal cord. The brain has three primary divisions referred to, respectively, as *forebrain (prosencephalon), midbrain (mesencephalon),* and *hindbrain (rhombencephalon).* From the posterior part of the forebrain a pair of lateral evaginations, the *optic vesicles,* grows out to come in contact with the superficial ectoderm of the head. Each optic vesicle soon becomes flattened and then indented in such a manner that an incompletely double-walled *optic cup* is formed (Fig. 14.2). This gives rise to the *retina* and to part of the *iris* of the eye. The *lens* of the eye develops from the outer, or superficial, layer of ectoderm opposite the optic cup. With the exception of the conjunctiva and a portion of the cornea, both of which are derived from superficial ectoderm, and the muscles of the iris which come from the outer layer of the optic cup, the other structures of the eye are of mesodermal origin.

The sensory parts of the remaining sense organs also come from ectoderm. The organs for the sense of smell first make their appearance as a pair of thickenings, the *olfactory placodes,* situated on either side of the head. These thickenings invaginate to become *olfactory pits.* In those vertebrates which breathe by means of lungs, and in a few others, the olfactory pits establish connection with the stomodaeum. A membranous *olfactory plate* at first separates the olfactory pit from the stomodaeum, but this soon ruptures and the two cavities become continuous. Some of the cells lining the olfactory pit become sensory and send out fibers which grow back to the forebrain with which they make connection. These fibers form the *olfactory nerve.*

On each side of the head, opposite the hindbrain, a thickened mass of superficial ectodermal cells, the *auditory placode,* is destined to form the inner ear. An invagina-

tion, the *auditory pit,* appears which soon pinches off from the outer ectodermal layer. It is then called the *auditory vesicle* (Fig. 14.15). The auditory vesicle undergoes rather complex changes, ultimately forming the semicircular ducts and other portions of the inner ear which in their aggregate make up the *membranous labyrinth.* Groups of cells in various portions of the membranous labyrinth develop sensory functions. In lower aquatic vertebrates similar placodes on the head and along the sides of the body give rise to the sensory elements of the *lateral-line system.*

The organs for the sense of taste arise, for the most part, from stomodaeal ectoderm, but some are of endodermal origin. They consist of clusters of *taste cells* and supporting nonsensory cells. Fibers of the seventh and ninth cranial nerves supply the gustatory receptors.

The various *cranial* and *spinal nerves,* which form connections with the brain and spinal cord, as the case may be, make up the *peripheral nervous system.* The fibers of the peripheral nerves, depending upon the nerve in question, arise from neural-crest cells, cells from the ventrolateral parts of the neural tube, or from thickened placodes of superficial ectoderm. The *autonomic portions* of the peripheral nervous sytem, regulating those activities of the body under involuntary control, are derived, for the most part, from neural-crest cells, but in some forms the ventrolateral portions of the neural tube contribute certain components. The peripheral nervous system is entirely ectodermal in origin, since the neural crests, neural tube, and placodes from which it arises are all ectodermal structures.

Some neural-crest cells migrate to a position near the kidneys, where they become differentiated as suprarenal tissue or, in mammals, as the medulla, or central portion, of the adrenal glands. These tissues are endocrine in nature and secrete substances of great physiological importance into the blood stream.

Associated with tissues of the central nervous system and giving them support and protection are some other components known as *neuroglia*, or *glia cells*. Several types of glia cells, all of ectodermal origin, are recognized. They are derived originally from the neural tube.

Two or three other structures of ectodermal origin should also be mentioned here. They are derivatives of the posterior portion of the forebrain. A ventral evagination of the floor of the brain in this region is called the *infundibulum*. It grows down to meet an evagination from the roof of the stomodaeum called *Rathke's pocket*. A portion of the infundibulum together with Rathke's pocket give rise to the pituitary body. The portion derived from the infundibulum becomes the posterior lobe. The anterior and intermediate lobes of the pituitary gland come from Rathke's pocket which, except in a few lower forms, loses its connection with the stomodaeum. One or two dorsal evaginations grow out from the posterior part of the forebrain. In some forms the more anterior of the two, the *parietal*, or *parapineal*, *body*, becomes a median, eyelike organ. The posterior evagination differentiates into a structure called the *pineal body*. The former is lacking in higher vertebrates.

Endoderm. The archenteron, or primitive digestive tube, is lined entirely with endoderm. Its communication with the stomodaeum anteriorly and with the proctodaeum posteriorly has already been alluded to. All endodermal structures of the body are derived from the archenteron. In polylecithal ova, because of the enormous amount of yolk present, the archenteron is formed somewhat differently by a complicated manner of folding of the endoderm. The end result, however, is a tubular structure much like those of other forms. The archenteron elongates as the embryo grows. With progressive development it gradually differentiates into various parts which, beginning from the anterior end, include the pharynx, esophagus, stomach, intestine, and the greater part of the cloaca when such a structure is present. The inner portion of the splanchnic, or visceral, layer of mesoderm surrounding the archenteron becomes differentiated into mesenchyme. From this come the connective tissues and smooth-muscle coats which, together with the endoderm, make up the wall of the digestive tract and of the various structures derived from it.

Two important digestive glands, liver and pancreas, arise as evaginations, or diverticula, of the digestive tract. The endodermal cells push out into masses of mesenchyme where they branch profusely. These organs are thus actually made up of tissues coming from two germ layers, the epithelial cells alone being of endodermal origin. The ducts of the liver and pancreas usually open into the digestive tract at the points where the original evaginations occurred.

The pharyngeal region is of particular interest, since many important organs are derived from it. Several pairs (usually four or five in higher vertebrates) of lateral pharyngeal pouches push out from the walls of the pharynx through the mesoderm until they come in contact with the outer, superficial ectodermal covering of the body with which they fuse. In cyclostomes, fishes, and amphibians the pouches break through to the outside and thus form a series of gill slits. In cyclostomes and fishes, but not in amphibians, gill lamellae, richly supplied with blood vessels, arise from the walls of the pharyngeal pouches. Larval amphibians develop external gills of a different type. Although in higher vertebrates (amniotes) the pouches form in the characteristic manner, they rarely break through to the outside. Their existence is temporary, except for the first pouch which persists in modified form as the middle ear and Eustachian tube. From the others are budded off groups of cells which go to form portions of such structures as the palatine tonsils, thymus, and parathyroid glands, and the small, irregular ultimobranchial bodies frequently found in the neck region. The appearance of pharyngeal pouches in higher forms of vertebrates is clearly indicative of their ancestry from gill-breathing, aquatic predecessors.

The thyroid gland originates as a small, midventral evagination of the pharynx. In some groups certain pharyngeal pouches contribute to the formation of the thyroid gland.

Another midventral outgrowth of the pharynx in air-breathing vertebrates develops into larynx, trachea, and lungs. It soon divides into two branches, each of which may divide many times to form the lobes of the lungs.

The yolk sac is an additional structure which is attached to the midventral part of the archenteron somewhat near its middle portion. It is composed of splanchnopleure (endoderm covered with splanchnic mesoderm). In many embryos in which abundant yolk is present, it grows out to surround the yolk. In certain teleost fishes, as in the muskellunge, a different arrangement exists, the inner layer of the so-called yolk sac consisting of a syncitial protoplasmic envelope containing giant nuclei and usually referred to as the *periblast*. The cells overlying the syncitial envelope are mesodermal and ectodermal in origin (W. Bachop, 1965, *Trans. Amer. Microscop. Soc.*, **84**:80). The yolk sac is a typically embryonic structure. It becomes smaller and smaller as the yolk

which it contains is utilized in nourishing the developing embryo. Eventually it disappears except in unusual and abnormal cases. It is of interest that in such forms as mammals, the eggs of which contain practically no yolk, a yolk sac appears, nevertheless, as a transitory structure.

Still another endoderm-lined structure of great significance in the development of reptiles, birds, and mammals is an embryonic, membranous sac, the allantois, which consists of a more or less extensive outpocketing of the posterior end of the archenteron. The allantois is lost at birth. A portion of the allantois may be utilized in the development of the urinary bladder, although the greater part of this organ is derived from cloacal endoderm, as is the urethra, which carries urine from the bladder to the outside.

Mesoderm. The origin and differentiation of mesoderm into epimere, mesomere, and hypomere has already been referred to. The segmentally arranged somites, derived from the epimeric mesoderm, become mesenchymatous in their dorsolateral and midventral portions. As previously mentioned, the cells of the dorsolateral mesenchyme (*dermatome*) proliferate and migrate so as to lie under the layer of ectoderm covering the body. They give rise to the dermis of the skin. The mesenchyme from the midventral portion of the somite (*sclerotome*) grows in and fills the spaces about the notochord and neural tube. In most forms it is destined to form the various elements of the vertebral column. The remaining portion of the somite is called the *myotome,* or muscle segment. Adjacent myotomes are separated from each other by partitions of connective tissue, the *myocommata.* During development each myotome grows down in the body wall between the superficial ectoderm and the somatic layer of mesoderm to meet its partner from the other side at the midventral line of the embryo (Fig. 11.3). With some exceptions the myotomes give rise to the greater part of the voluntary musculature of the body. The precise origin of the voluntary muscles of the limbs in most vertebrates has been debated. Whether or not they are derived from myotomes, they are, at any rate, of mesodermal origin.

The mesomere which, except in lower forms, is for the most part unsegmented is concerned with the development of the urogenital organs and their ducts. The terminal parts of these ducts may, however, be lined with epithelia of ectodermal or, in some cases, of endodermal origin.

The mesothelial splanchnic layer of the hypomere surrounding the archenteron, or primitive gut, becomes mesenchymatous on the side adjacent to the endoderm. From this mesenchyme arise the involuntary muscles and connective tissues of the gut and other structures which ultimately come to surround the endodermal lining of the digestive tract and its derivatives. The heart itself is derived from splanchnic mesoderm. The remainder of the splanchnic layer, together with the somatic layer, forms the mesothelium lining the coelom and contributes to the pericardium, pleurae, or peritoneum, as the case may be. Mesenteries and omenta are also derived from the splanchnic mesoderm.

The parts of the skeleton other than the vertebral column, whether made up of cartilage, bone, or other connective tissues, are all derived from mesenchyme in different parts of the body and of various origins. Mesenchyme also gives rise to blood and lymphatic vessels, blood corpuscles, lymph glands, and other blood-forming tissues. The heart, however, comes from the splanchnic mesoderm of the hypomere, as mentioned above. Other structures derived from mesen-

chyme include various parts of the eye, dentine of the teeth, steroidogenic tissue (page 369), or, in mammals, the cortices of the adrenal glands.

It is known that all mesenchymal cells do not differentiate into the various structures mentioned. Some remain undifferentiated even in adult life. They form a reserve from which connective tissues of various kinds, to be mentioned later, may be formed and which may be called upon for repair of injured structures.

TISSUES

It will be recalled that the branch of biology which deals with the minute structure of tissues is known as *histology*. Comparatively recent developments in electron and other types of microscopy have almost revolutionized the subject of histology within the last few years. Many details of structure which were formerly unknown or were vague have now been demonstrated with exactitude and clarity. Moreover, new discoveries in the fields of physics and biochemistry, when correlated with these detailed anatomical findings, have given an insight into functional relationships not previously conceived.

The use of the electron microscope in minute anatomical studies has depended in turn upon the development of techniques by means of which tissues may be properly prepared for examination. Today it is possible to cut tissue sections about 0.000001 in. in thickness for study under the electron microscope.

Formation of Tissues

When tissues form from the undifferentiated cells of which the germ layers are composed, two general processes are concerned: (1) the cells multiply by numerous mitotic divisions; (2) they then undergo differentiation.

Undifferentiated cells are characterized by having relatively large nuclei and very little cytoplasm. They also have exceptional ability to divide mitotically. With a subsequent increase in cytoplasm the rate of cell division is retarded. Differentiation consists in the cell taking on a new form and changing its arrangement in relation to surrounding cells. Changes in the cytoplasm are most characteristic of differentiating cells, although the nuclei are frequently involved but to a lesser degree. Secretions or products given off by the cells may in some cases contribute to tissue formation. In other cases, accumulations of cell products within the cells themselves may characterize certain kinds of tissues. The physical and chemical properties of the cells actually undergo changes which fit the tissue to carry on the specialized function that it performs in the body.

Types of Tissues

Four general types of tissues are recognized: (1) epithelial, (2) connective, (3) muscular, and (4) nervous.

Epithelial tissue. An epithelium may be defined as a tissue composed of one or more layers of cells covering an external or internal surface of the body or embryonically derived from such a surface. Epithelia generally consist of closely connected cells which have very little intercellular substance between them. They cover the outside of the body, give rise to most glandular structures, form parts of all the sense organs, and line all internal cavities of the body which communicate with the outside. The term *endothelium,* instead of epithelium, is used to designate the simple layer of cells lining the inner surfaces of the walls of blood vessels

and lymphatic vessels and of the heart. Similarly, the word *mesothelium* is used for the layer covering the surface of the membrane lining the various parts of the coelom. All canals and cavities of the body which communicate with the exterior are lined with *mucous membrane* which is composed of a superficial layer of epithelium, the cells of which produce mucus, together with a layer of connective tissue called the *lamina propria.* The alimentary canal and its diverticula, the respiratory tract and its connections, and the urogenital tract are all lined with mucous membrane. The term *serous membrane* is applied to the thin layer of loose connective tissue covered with mesothelium which lines the body cavities. Peritoneum, pleurae, and pericardium are serous membranes. Mesenteries and omenta are covered with serous membrane on both surfaces. Such membranes produce a watery fluid which serves to lubricate the surfaces of organs as they move over one another when carrying on their activities. Cells which are present in the serous fluid are derived from the serous membrane.

Epithelia play an important role in the metabolism of the body and are concerned with protection, secretion, absorption, respiration, assimilation of nutritive substances, and elimination of waste products. Various types of epithelia are present in the body. They are usually classified according to the form and arrangement of the cells of which they are composed. Two main types are recognized: (1) *simple epithelia,* composed of a single layer of cells, and (2) *stratified epithelia,* made up of several cellular layers. So-called *basement membranes,* often referred to in connection with epithelia and upon which the epithelial cells rest, are actually composed of modified intercellular substance (page 78) produced by connective tissue beneath.

In general, three varieties of epithelial cells, differing primarily as to shape, may be distinguished. *Squamous cells* are broad and flat. Their height is insignificant in comparison with their length and breadth. They may be so thin that the nucleus, which is generally in the center, causes a slight bulge. *Cuboidal cells,* as their name implies, are about as tall as they are long or wide. In cross section they appear to be square. *Columnar cells,* on the other hand, are much taller than they are wide. The shapes of these three types of cells are subject to considerable variation depending upon whether the surface they form is stretched or contracted. The more they are stretched the flatter they become. Conversely, contraction of the surface tends to increase their height.

Adjacent cells in epithelia adhere to one another with considerable tenacity. Many investigators in the past have attempted to determine the nature of the cohesive force or substance. It was formerly held that the cells making up epithelia were fastened together by a cement substance, forming

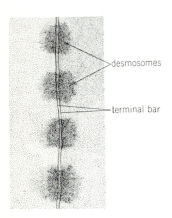

desmosomes

terminal bar

Fig. 3.13 Desmosomes, as seen under the electron microscope, arranged along a terminal bar between portions of two adjacent epithelial cells.

terminal bars. Electron microscopy has now revealed that small bodies, called *desmosomes,* often present along cell membranes of adjacent cells, hold the cells tightly together at the sites where they exist (Fig. 3.13). The cell membranes of adjacent cells are somewhat thickened where desmosomes are located, and the intercellular space at such locations is diminished. It is still not clear, even at a magnification of 70,000, how desmosomes actually hold the cells together. Minute interdigitating ridges and grooves of contiguous cell surfaces may also aid in this process. So-called intercellular bridges (Fig. 3.14A) connecting adjacent epidermal cells were formerly thought to pass from one cell to another. Each such "bridge" is now known to consist of two arms in close contact, one extending from one cell and one from the other (Fig. 3.14B). When, in some specialized tissues, no cell walls can be detected between adjacent "cells" and the nuclei seem to be lying in the same cytoplasmic mass, the term *syncytium* is used. Syncytia are rare among epithelia. In a developing embryo, the outer layer of one of the embryonic membranes (plasmoditrophoblast of the chorion) is in the form of a syncytium.

SIMPLE EPITHELIA. Seven types of simple epithelia are usually recognized. They are distinguished basically by the type of cell of which they are composed.

Simple squamous epithelium is made up of flat, platelike, squamous cells which fit together in the manner of tiles on a floor (Fig. 3.15A). This type of epithelium is found lining the membranous labyrinth of the inner ear, in portions of the kidney tubules, and in the minute ducts of numerous glands. Mesothelium and endothelium are also composed of squamous cells but are not usually considered to be epithelia.

Simple cuboidal epithelium (Fig. 3.15B)

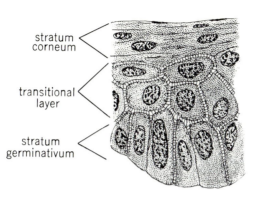
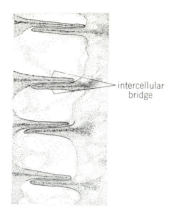

stratum corneum

transitional layer

stratum germinativum

intercellular bridge

Fig. 3.14 A, section through epidermis of frog, showing so-called intercellular bridges. B, semidiagrammatic representation of so-called intercellular bridges between two epidermal cells, as seen under the electron microscope.

appears in cross section as a sheet of cells which are approximately square in shape. The free surface may appear, however, as a mosaic of irregularly hexagonal units. Cuboidal epithelium is found in certain glands and lining numerous ducts. The thyroid gland, covering of the ovary, and portions of the kidney tubules are typically composed of epithelium of this type. In some lower vertebrates the mesothelium over the peritoneum and mesenteries may be of the cuboidal rather than of the squamous type.

Simple cuboidial ciliated epithelium may frequently be found in such places as portions of the kidney tubules. The free borders of the cuboidal cells bear cilia.

Simple columnar epithelium is composed of a sheet of more or less hexagonal columnar cells which adhere to each other along their lateral or longitudinal surfaces. The nuclei of the columnar cells are sometimes located in the center of the cell but are more commonly found in the basal portion (Fig. 3.15C). The lining of the alimentary canal, from stomach to anal region, is of the simple columnar type of epithelium. Certain cells

seem specialized to secrete mucus and are referred to as *goblet cells* because when such a cell becomes distended by its accumulated secretion it has a gobletlike appearance. Goblet cells are also abundant in the lining of the trachea. Other columnar cells lining the alimentary canal are specialized for the absorption of water and the end products of digestion. It was formerly believed that the free surfaces of such cells were covered with a thin, protective cuticle which showed minute striations. The electron microscope has revealed that the latter are really large numbers of tiny, fingerlike, cytoplasmic projections which have been termed *microvilli* (Fig. 3.16). It has been estimated that a single cell may bear as many as 3,000 microvilli and that in 1 sq mm of the lining of the intestine there may be 200 million microvilli. Their role in increasing the surface for absorption or other functions should be obvious.

Simple columnar ciliated epithelium is similar to the regular columnar type except for the presence of cilia on the free surface of each cell. Oviducts, small bronchi, and the central canal of the spinal cord of embryos

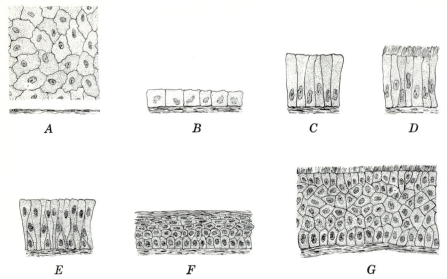

Fig. 3.15 Various types of simple and stratified epithelia: *A*, two views of simple squamous epithelium, top view above and side view below; *B*, simple cuboidal epithelium; *C*, simple columnar epithelium; *D*, simple columnar ciliated epithelium; *E*, pseudostratified columnar epithelium; *F*, stratified squamous epithelium; *G*, stratified columnar ciliated epithelium.

are lined with this type of columnar epithelium (Fig. 3.15*D*).

Pseudostratified columnar epithelium, when viewed in cross section, gives the impression of being made up of several layers of cells. Close examination, however, shows that there is actually only a single layer of columnar cells but that their nuclei are at different levels, thus giving the illusory effect (Fig. 3.15*E*). The fact that certain cells taper at either or both ends and appear to be squeezed in between other and wider cells helps to give the impression of stratification. Pseudostratified epithelium occurs in the urethra of the male human being and in the large ducts of a number of glands such as that of the parotid salivary gland.

Pseudostratified columnar ciliated epithelium is similar to ordinary pseudostratified columnar epithelium, but the free borders of

the cells bear either motile *kinocilia* or nonmotile *stereocilia*. Examples of this type of epithelium are to be found in the lining of the respiratory passages, Eustachian tubes and middle ears, and the ducts of the male reproductive system (epididymis and ductus deferens).

STRATIFIED EPITHELIA. Stratified epithelia are generally classified into five different types. Their names are based primarily on the shape of the cells forming the outermost layers. The type of cell making up this layer varies greatly, depending primarily upon whether the surface is normally wet or dry.

Stratified squamous epithelium consists of several layers of cells (Fig. 3.15*F*) and is found in areas in which various degrees of protection are required. The primary function is never that of absorption or secretion.

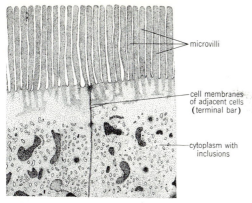

Fig. 3.16 Portions of two adjacent columnar epithelial cells from the intestine as viewed under the electron microscope. Note the numerous microvilli on the free surface of each.

The number of layers shows much variation in different regions of the body. The basal layer appears to be of the low columnar or cuboidal type. The cells of this layer, by mitotic proliferation, bring about the formation of other layers which are nearer the surface. A gradual change in shape of the cells occurs as they approach the surface which is composed of flat, platelike, squamous cells. These squamous cells are either sloughed or wear away and are constantly being replaced by new cells from beneath. Stratified squamous epithelium makes up the outer part (epidermis) of the skin, the lining of the mouth and esophagus, the outer portion of the cornea of the eye, the lining of the vagina, and is present in other regions as well.

The character of the superficial cells of stratified squamous epithelium on dry surfaces is somewhat different from that on wet surfaces. In the former, as in the epidermis of the skin of terrestrial vertebrates, the outer cells form a layer of tough, nonliving *keratin* which is in close union with the underlying living cells of the epithelium. In this connec-

tion it is interesting to note that in individuals subjected to prolonged vitamin A deficiency, keratinization of the superficial layer of cells of the stratified squamous epithelium may occur in areas which are normally kept moist and which otherwise would never show this modification.

Stratified cuboidal epithelium is similar to stratified squamous epithelium in the basal layers. The superficial layer, however, is composed of cuboidal cells. This type of epithelium is found in the epidermis of many tailed amphibians (urodeles). Secretory activity of certain epidermal cells may account for their cuboidal shape when they are distended with accumulated secretory products.

Stratified columnar epithelium is also similar to the aforementioned types in the lower layers, but the outer layer is made up of columnar cells. This type of epithelium is not found in abundance and is usually confined to restricted areas such as on the epiglottis, on a part of the urethra, in the folds of the conjunctiva, and in other scattered places.

Stratified columnar ciliated epithelium is like that just described, but the superficial cells are ciliated (Fig. 3.15*G*). The larynx and upper surface of the soft palate are lined with epithelium of this type.

Transitional epithelium was formerly thought to represent a transition between stratified squamous and columnar epithelia such as occurs at the point where the esophagus joins the stomach. However, since it occurs in the lining of the pelvis of the kidney, ureters, urinary bladder, and part of the mammalian urethra, all of which are nontransitory zones, the original name does not seem particularly apt. Transitional epithelium is found in areas which frequently undergo distention and contraction. Its appearance, therefore, varies considerably. The deeper portion is like that of other stratified epithelia, but the cells of the outer layer differ. They are convex on the free edge but are concave below. They are so arranged as to be able to slide past each other when the epithelium is stretched and yet return to their original position when contraction occurs. When the epithelium is stretched, only two layers of cells may appear to be present, an underlying cuboidal layer and an outer squamous layer. The stratified appearance is assumed during the contracted state.

FUNCTIONAL CLASSIFICATION OF EPITHE-LIA. Although structural characteristics have been used here as a basis for classifying epithelia, functional properties are sometimes utilized in describing them. The term *glandular epithelium,* for example, is frequently used in referring to columnar epithelium, which not only lines glandular organs but the cells of which actually produce the secretion characteristic of the gland. *Sensory epithelia* form the sensory portions of sense organs, such as the retina of the eye and the olfactory region of the nasal passages. *Germinal epithelium* is that part of the sex glands from which the germinal elements (eggs and spermatozoa) arise. *Cuticular epithelium* secretes a substance, the *cuticle,* which covers the cells and forms a tough, resistant, noncellular, protective layer. The chitinous exoskeleton of certain arthropods represents such a secretion.

Connective tissue. The principal functions of connective tissues are to bind other tissues together and to give support to various structures of the body. Connective tissues are distinguished by the presence, in most cases, of considerable quantities of *intercellular substance,* or *matrix.* The nature of the intercellular material is characteristic of each type.

Connective-tissue cells themselves usually contribute only an inconspicuous portion to the tissue. It is the intercellular substance, produced by the cells, which makes up the main bulk of the connective-tissue mass. Some types of connective tissue have very little intercellular substance, consisting, for the most part, of cells which may have additional functions. The production of intercellular substance is almost exclusively confined to connective-tissue cells. It is this nonliving material which connects and binds other tissues together and provides support for them.

Connective tissues are of mesodermal origin and develop from mesenchyme. Mesenchyme is often referred to as *embryonic connective tissue* and consists both of cells and of intercellular substance. When mesenchymal cells first appear in an embryro they have the potential, or capacity, to differentiate along any one of a number of lines leading to the formation of the several types of connective-tissue cells. Even in an adult there remain large numbers of undifferentiated mesenchymal cells with the capacity of developing into one kind of connective tissue or another.

There are, in general, four types of connective tissue: (1) connective tissue proper, (2) blood and lymph, (3) cartilage, and (4) bone. It is not always possible to make a sharp distinction between the various types of connective tissue since there are several intergrading forms.

CONNECTIVE TISSUE PROPER consists of several types showing rather wide differences in the nature of the intercellular substance. It is composed, in large part, of fibers, among which various types of cells are scattered. It is now fairly clear that the fibers are not formed within cells, but rather at cell surfaces or even at some distance from cells. It is possible that a precursor substance is formed in the cytoplasm of the cells. This is believed to be secreted by the cells and passes outside of the cell wall where it becomes polymerized to form fibers. The term "polymerization" refers to the formation of a compound from several single molecules of the same substance. Mechanical forces undoubtedly are important in influencing the manner and direction in which the fibers are arranged.

An important constituent of connective tissue proper is referred to as the *amorphous ground substance*. This is an intercellular substance, present even in mesenchyme, in which cells, fibers, etc., are embedded. The amorphous ground substance is difficult to demonstrate except when very special staining techniques are employed. It has the capacity to permit diffusion of nutrient substances, water, gases, and wastes over considerable distances. This is of importance in areas where small blood vessels are absent, particularly during developmental stages.

Loosely organized connective tissue is the most widely distributed type. The subcutaneous tissue between skin and underlying muscles is the most familiar example. It is composed of loose fibers which ramify in all directions, and of scattered cells of various types. The term *areolar tissue* is often used for this kind of connective tissue because under certain conditions bubblelike spaces appear among the fibers. These are actually small air bubbles which are drawn into the tissue when it is stretched in preparation for study. Since they are not present in intact tissues, the name "areolar" is not really quite appropriate.

Two types of fibers, collagenous and elastic, are usually recognized in loosely organized connective tissue (Fig. 3.17). *Collagenous fibers* course in all directions and have the appearance of ribbons composed of several parallel strands or fibrils. The fibrils themselves do not branch, but groups of them may separate so as to give a branched appearance to the fiber as a whole. It is not

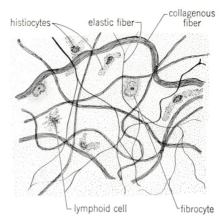

Fig. 3.17 Loosely organized connective tissue.

possible to locate the ends of collagenous fibers. They are very tough, resistant to stretching, and do not have elastic properties. Studies, using the electron microscope, reveal that collagen fibrils are composed of microfibrils which show a uniform cross banding of regularly alternating light and dark areas, each area in turn being made up of smaller cross bands. Adjacent microfibrils are in phase with each other so far as arrangement of these cross bands is concerned. It seems that the microfibrils themselves are composed of extremely fine filaments which, in chemical terms, probably represent polypeptide chains. *Elastic fibers* are fairly long. They show no striations and are not composed of fibrils. They branch and may unite with similar fibers so as to form a rather loose network. Except in the very young, numbers of elastic fibers, when viewed together, have a yellowish appearance. They may be pulled or stretched rather easily but spring back to normal position when the tensile force is released. In contrast to the highly organized structure of collagenous fibers, the fine structure of elastic fibers is essentially amorphous.

The cellular elements in loosely organized

connective tissue are of several types. Some apparently *undifferentiated,* or perhaps *partially differentiated, mesenchymal cells* persist, usually in close proximity to blood vessels. These may at various times give rise to various types of connective-tissue cells. *Fibroblasts,* or *fibrocytes,* are large, somewhat flattened cells which are usually seen in close association with collagenous fibers and in some cases may also be observed along the surface of elastic fibers. These cells are believed to be concerned with the formation of collagen fibers, as previously described, and with the production of the amorphous components of loosely organized connective tissue. *Macrophages,* or *histiocytes,* are almost as numerous as fibroblasts. Their nuclei, which have a characteristic shape, are smaller than those of fibroblasts, are indented on one side, and stain more deeply. Under normal conditions they do not move about. They are able to ingest and store certain kinds of microscopic particles. Under inflammatory conditions they may become wandering cells which migrate about the tissue spaces by ameboid movement, engulfing bacteria and other small particles. The shape of these cells varies, their cytoplasm extending in the form of blunt, irregular processes. *Multinucleated giant cells* are sometimes encountered. These apparently represent a fusion of a number of macrophages which have united to form a cellular mass large enough to surround a foreign body too big to be enclosed within a single macrophage. *Lymphoid cells,* commonly found in loosely organized connective tissue, appear to be identical with lymphocytes which are included among the various types of leukocytes, or white corpuscles, of the blood. They are irregular in appearance, have large nuclei with but little cytoplasm, and have a phagocytic action. *Mast cells,* which show distinctive staining properties, are also present in loosely organized connective tissue. In amphibians their cytoplasm may be extended into fairly long processes, whereas in mammals they are usually oval in shape. The cytoplasm of these cells is packed with rather large granules. It has been demonstrated that mast cells contain considerable amounts of histamine, a substance which is involved in allergic and inflammatory reactions. They also contain serotonin and heparin. Serotonin (see page 599) acts locally as a constrictor of blood vessels and generally to reduce blood pressure; heparin inhibits the clotting of blood. Whether mast cells actually produce these substances, which are also found elsewhere, or merely segregate and store them, has not been determined. *Fat cells* may occur singly or in groups, usually along blood vessels in loosely organized connective tissue. Before they store fat they closely resemble fibroblasts. Fatty droplets accumulate within the cytoplasm. The droplets within a cell gradually coalesce until the cell is filled with fat. The cytoplasm of the distended cell is stretched out to a thin layer and the nucleus is pushed to the outer edge. *Chromatophores,* or *pigment cells,* of lower forms are derived from neural crests and are therefore of ectodermal origin. In such animals as fishes, amphibians, and reptiles, they are present in connective tissue, particularly in the outer part of the dermis of the skin. They are irregularly branched cells in which the pigment granules they contain show, at different times, varying degrees of dispersion or concentration within the cells (Fig. 4.5). The pigment is in the form of small granules which are manufactured by the cells and incorporated within the cytoplasm. Different names are applied to chromatophores, depending upon the color of the pigment (page 109). In mammals such pigment cells are rare. *Melanocytes,* that also manufacture the pigment (melanin) which they contain, are confined almost exclusively to the epidermis. An exception seems to occur in newborn babies of the Mongol race in which a dark area in the skin over the sacral region, called the *Mongol spot,* owes its color to *dermal* melanocytes which contain pigment formed by the cells themselves. Other cells, present in both epidermis and dermis, may engulf pigment particles instead of manufacturing them, thus behaving much as do macrophages. They are frequently present in the outer part of the dermis of certain mammals. The term chromatophore is often applied to such cells, which should not be confused with the chromatophores of lower forms.

Loosely organized connective tissue is so constructed as to be strong as well as elastic. It forms an ideal groundwork for supporting and holding together other tissues and fills in spaces between certain structures. In parts which undergo movement and those which are concerned with such processes as secretion, absorption, and excretion, the loose connective tissue plays an important role. It is present in the walls of blood vessels and supports the intricate capillary network of the lungs. It has been mentioned previously that gases, water, nutritive substances, and excretory products must pass through the amorphous ground substance on their way to and from capillaries. The part played by the ground substance in metabolism should be apparent. This type of connective tissue is of great importance in reconstruction or repair after injury and also in combating infection.

Densely organized connective tissue is composed essentially of the same materials as are found in loosely organized connective tissue, but the collagenous fibers are thicker and more numerous. They are very closely interwoven so as to form a dense matting. Cellular elements are relatively few in number. The bundles of collagenous fibers are woven together in such a way, and enough elastic fibers are provided, that the tissue is permitted a certain degree of stretch. The dermis of the skin, portions of the walls of the digestive tract and blood vessels, as well as certain parts of the urinary excretory ducts are composed of densely organized connective tissue.

Regularly arranged connective tissue is sometimes considered to be a variety of densely organized connective tissue. It is made up, for the most part, of densely organized collagenous fibers which show a definite arrangement. The fibers run parallel to each other, and fibrocytes are the only cellular elements present. *Tendons,* by means of which muscles are attached to bones or other structures, are composed of regularly arranged connective tissue. They are very strong and inelastic. The fibers are arranged in bundles which are fastened together with loosely organized connective tissue. *Ligaments,* which connect bones or support internal organs, are somewhat similar in structure to tendons. However, the collagenous fibers are not so regularly arranged as those of tendons, and elastic fibers are present in addition. The *nuchal ligament* at the nape of the neck of certain grazing mammals has an abundance of yellow elastic fibers. In these animals it serves to sustain the weight of the head and thus helps to prevent fatigue.

Specialized connective tissues are present in addition to those already referred to. MUCOUS CONNECTIVE TISSUE is a type found in embryos. It is composed of scattered cells which seem to be connected with each other by means of long, fine, branching processes. The intercellular substance, or matrix, is abundant and gives the staining reaction for mucin. In the umbilical cord the

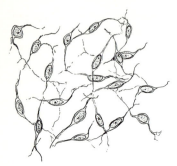

Fig. 3.18 Mucous connective tissue, showing a few cells from Wharton's jelly in the umbilical cord of a pig embryo.

so-called *Wharton's jelly* is perhaps the best example of mucous connective tissue (Fig. 3.18). It lies between blood vessels and certain ducts which are connected to the body of the embryo (yolk stalk and allantois). The comb of the cock is largely made up of this type of tissue, the development of which, in this case, is regulated by the male hormone secreted by the testes.

ELASTIC TISSUE, in which parallel yellow elastic fibers predominate, is found in various parts of the body. The fact that it can be stretched and then spring back to its normal position when released is of great importance in the region where this tissue occurs. The true vocal cords and the flaval ligaments between adjacent vertebrae which serve to maintain the upright posture of man are examples of elastic tissue.

ADIPOSE, OR FATTY, TISSUE is really loosely organized connective tissue in which fat cells are particularly abundant. They take the place of other connective-tissue cells which are, therefore, far less numerous than otherwise. Adipose tissue serves to store neutral fat which is then available as a source of heat and other forms of energy for the body. In histological preparations, unless special techniques are followed, the fat is dis-

solved out of the cells and only a skeletal framework of adipose tissue remains (Fig. 3.19).

In certain mammals a special type of adipose tissue of brownish color and called *brown fat* is present in such regions as the neck, between the scapulae, etc. The nuclei of the cells composing brown fat are rather centrally located in the cell and the fatty droplets do not tend to coalesce. Pigment incorporated in the fat is responsible for its color. In some species, particularly among rodents, this tissue is present throughout life. In others, including man, it occurs in certain areas during embryonic life but disappears shortly after birth, being transformed into common fatty tissue. Histologically, brown fat somewhat resembles some of the lipid-storing endocrine glands. This material is frequently referred to as the *hibernating gland* since in hibernating animals it may, during the active season, store glycogen and lipids which are then made available as nutrient materials during hibernation. Under conditions of stress there is rapid discharge and depletion of stored lipids and glycogen from brown fat. Experiments have shown that if the nerve supply to brown fat bodies is interrupted, depletion of these substances does not occur under stress. Some

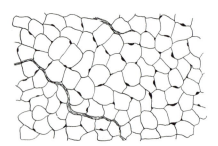

Fig. 3.19 Adipose, or fatty, tissue. In this preparation the fat has been dissolved and removed from the cells so that only a skeletal framework remains.

authorities are of the opinion that brown fat tissue should be looked upon as adipose tissue that has not developed beyond its embryonic state.

RETICULAR TISSUE consists of fibers which have a special affinity for silver salts and turn black when impregnated with them. Except for this staining reaction, they rather closely resemble fine collagenous fibers even to the extent of showing similar light and dark areas when examined under the electron microscope. Their different staining properties may be due to a relatively greater concentration of certain sugars contained within the fibers.

PIGMENTED TISSUE is characterized by the presence of numerous cells, the cytoplasm of which is laden with pigment (melanin) granules *formed within the cells.* Loosely organized connective tissue in certain regions may contain an abundance of such cells. The outer layer of the choroid coat of the eye is made up of pigment tissue composed largely of elastic fibers and large pigmented cells which are believed to manufacture their own melanin and store it *in situ.* When choroid and sclera are separated, portions of the pigment tissue may adhere to the sclera.

VESICULAR CONNECTIVE TISSUE is present in the notochords of vertebrates. It consists of large cells which are distended with fluid and closely packed together (Fig. 1.5). The intercellular substance is reduced in amount. This type of tissue, which persists only in some lower vertebrates, is usually considered to be of mesodermal origin, but it is really derived from the undifferentiated region at the dorsal lip of the blastopore.

BLOOD AND LYMPH are classified among the connective tissues even though the cellular elements are free and the intercellular substance is fluid. Both cells and fluid course through the body in vessels which are lined with endothelium. Certain of the cellular elements, the white cells of the blood, or leukocytes, are constantly passing back and forth between connective tissue proper and the blood or lymph, migrating en route through the walls of capillaries and small venules. It is, therefore, somewhat difficult to separate these two types of tissue categorically. The behavior and even the appearance of the leukocytes may differ significantly in these different kinds of environments.

The fluid part of the blood is known as *plasma.* The role which it plays in the metabolism of the body is of great importance. It is composed mostly of water but really is a solution of both colloids and crystalloids. Plasma is the medium which carries nutritive substances, waste products, hormones, and even gases (to a minor extent) to all parts of the body where they are utilized or eliminated, as the case may be. The constancy in the percentage of the various constituents of plasma is one of the marvels of nature and is brought about by the exchange of substances between plasma and the tissue juices, or fluid.

Tissue fluid is actually derived from that part of the blood plasma capable of diffusing through the endothelial walls of capillaries. The walls are permeable to solutions of crystalloids, which then can pass out of the blood and make up the tissue fluid. The cellular elements of the blood and the colloids in the plasma do not normally pass through the endothelium and are retained within the blood vessels. Tissue fluid, therefore, differs considerably from plasma in that, for the most part, colloid is lacking. The similarity in the composition of tissue fluid and sea water, in which living things are believed to have originated, is of special interest.

The cellular, or formed, elements of the

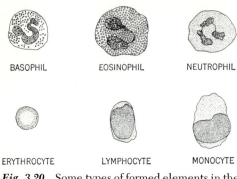

BASOPHIL   EOSINOPHIL   NEUTROPHIL

ERYTHROCYTE   LYMPHOCYTE   MONOCYTE

Fig. 3.20   Some types of formed elements in the blood of mammals.

blood include red corpuscles, or *erythrocytes;* white corpuscles, or *leukocytes;* and *blood platelets,* or *thrombocytes,* which are found only in mammals but which have a function similar to that of *spindle cells* found in the blood of certain lower vertebrates.

Although erythrocytes in various vertebrates differ in structure and appearance, their function in all is to carry oxygen to the tissues and to aid in the removal of carbon dioxide. It is the pigment hemoglobin, protein in nature and contained within the erythrocytes, which is responsible for the greater part of the oxygen-carrying property of blood. Erythrocytes are confined to blood vessels under normal conditions. They are absent in lymph. In mammals they lose their nuclei before entering the blood stream. In man the normal shape of an erythrocyte is that of a biconcave disk. It is of interest that in certain diseased conditions erythrocytes of altered appearance may be found in the blood stream. The peculiar shape of the erythrocytes under various conditions of this sort is often of diagnostic value.

Leukocytes are of several types (Fig. 3.20) but may be separated into two main groups on the basis of whether they possess granules in the cytoplasm or lack them. *Granular leukocytes* are of three main types, although intergrading forms are abundant. They are all rather similar in size and are distinguished chiefly by the staining properties of the granules which they contain. *Basophils,* which stain with basic dyes, contain relatively large granules which are less abundant than in the other two types of granular leukocytes. The nuclei are twisted and show one or more constrictions. These corpuscles, which are quite distinct from mast cells of loosely organized connective tissue, are seen infrequently in human blood, making up only about 0.5 per cent of all the blood leukocytes. In some vertebrates they may be absent entirely; in others they are more numerous and show considerable variation in the size and staining properties of their granules. *Eosinophils,* the granules of which have an affinity for acid dyes, possess nuclei which consist of two lobes connected by a fine strand of nuclear material. These leukocytes are almost twice the diameter of erythrocytes. Eosinophilic leukocytes are not abundant but are present in all vertebrates except certain species of fishes. *Neutrophils,* often referred to as *polymorphonuclear leukocytes,* are by far the most numerous type. In man they make up between 60 and 70 per cent of the entire leukocyte count. Their granules are very fine and stain with neutral dyes, but in other vertebrates they may react to acid and basic dyes as well. The nuclei show great variation in shape but usually appear in the form of three or more lobelike regions connected by a delicate strand. It appears that neutrophils are lacking in turtles (H. A. Charipper and D. Davis, 1932, *Quart. J. Exp. Physiol.,* 21:372).

Nongranular, or *agranular, leukocytes* are of two kinds, *lymphocytes* and *monocytes.* Lymphocytes are small for the most part and in human beings are approximately similar in size to red corpuscles. Small, medium, and large lymphocytes are recognized, however. The cells are more or less rounded and contain large nuclei which have an indentation on one side. A thin layer of cytoplasm surrounds the nucleus. Monocytes appear to be somewhat like large lymphocytes but have more abundant cytoplasm. The nucleus, which does not stain very deeply, is sometimes markedly constricted.

Certain leukocytes engulf bacteria and other foreign particles. They are known as *phagocytes.* Neutrophils are the principal phagocytic agents. Occasionally monocytes exhibit phagocytic activity. It will be recalled that fixed and free histiocytes, or macrophages, present in loosely organized connective tissue as well as in other regions, are of great importance because of their ability to ingest various types of particulate matter. Leukocytes are able to migrate throughout the body by ameboid movement as long as they have some solid foundation on which they can move. They may enter and leave capillaries by squeezing between the squamous endothelial cells. This process is referred to as *diapedesis.*

Tissue which gives rise to various types of blood corpuscles is called *hemopoietic tissue.* There are actually two varieties of this. In the adults of higher vertebrates, erythrocytes, granular leukocytes, and thrombocytes trace their origin to red bone marrow to which the name *myeloid tissue* has been applied. Lymphocytes and monocytes, the nongranular types of leukocytes, are formed, for the most part, in *lymphoid tissue.* It is in anuran amphibians that red bone marrow first becomes hemopoietic tissue. In these forms lymphocytes originate, in part, in lymphoid tissue in the intestinal wall. Lymph nodes, which also first appear in amphibians, produce great numbers of lymphocytes, monocytes, and plasma cells, the latter being of special importance in antibody forma-

tion. During embryonic development in higher forms, before spleen and bone marrow develop, mesenchyme and liver are the chief tissues in which red corpuscles are formed. Later the spleen and bone marrow take over this function but after birth the red bone marrow is the chief site of erythrocyte formation. The intestinal wall and lymph nodes retain their role of producing lymphocytes. In many fishes and in urodele amphibians granular leukocyte formation takes place in the opisthonephros and even in portions of the gonads. All these structures should, therefore, be included among the hemopoietic tissues.

In vertebrates below mammals the blood contains *spindle cells,* or *thrombocytes,* which are believed to play an important role in the clotting of blood. These cells are of approximately the same size as small lymphocytes and are biconvex disks. They have a spindle-shaped appearance when viewed edgewise. The cytoplasm is clear and the nucleus, which stains deeply, is oval or rounded. A nucleolus is absent. In mammals there are numerous small bodies in the circulating blood which are not cells but which appear to be fragments of cytoplasm. These are called *blood platelets.* Unfortunately they are also referred to as *thrombocytes,* thus causing some confusion with the spindle cells, mentioned above, which are entirely different in structure even though their function may be similar. Platelets are even smaller in size than red corpuscles. No nucleus is present. It is difficult to study them because they clump together readily as soon as blood leaves the vessels and stick to any surface with which they happen to come in contact. They disintegrate rapidly. In man 1 cu mm of blood contains between 250,000 and 500,-000 blood platelets. They are believed to be fragments of cytoplasm pinched off certain giant cells called *megakaryocytes* present

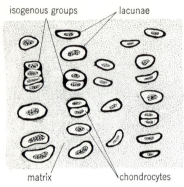

isogenous groups   lacunae

matrix   chondrocytes

Fig. 3.21   Microscopic structure of hyaline cartilage.

in the red bone marrow. Experiments using radioactive substances reveal that platelets live in the blood stream only from 3 to 7 days.

CARTILAGE is a form of connective tissue in which the intercellular matrix is abundant and firm. The cellular elements, or *chondrocytes,* which are mesenchymal derivatives, give rise to the intercellular substance and are embedded within it (Fig. 3.21). During life they lie in and completely fill small spaces in the intercellular substance called *cartilage lacunae.* In early embryonic stages and at the borders of cartilages the chondrocytes closely resemble fibrocytes in appearance. When a chondrocyte divides in a young or developing animal, each daughter cell soon forms a matrix about itself, so that usually only one or two cells are seen within a single lacuna. Cell division occurs rapidly, so that occasionally three or four cells, but seldom more, are present in a lacuna. In any case the daughter cells appear to lie in groups within the matrix. Such a group is said to be *isogenous.* Cartilage, which is sometimes called gristle, forms an important part of the skeleton. It is also present in certain organs and in other parts of the body as well. In cyclostomes, elasmobranchs, and certain other fishes, the entire endoskeleton is carti-

laginous. In other vertebrates, although varying amounts of bone are present in the endoskeleton, some cartilage is always present. The skeleton goes through a cartilaginous stage in the embryos of most vertebrates. In some elasmobranchs the cartilage in certain regions becomes calcified, or impregnated with calcium salts, and is hard and rigid. Upon superficial examination calcified cartilage may be mistaken for bone, but it has an entirely different histological structure. Cartilage is surrounded with a layer of densely arranged connective tissue called the *perichondrium.* The transition zone between these two types of connective tissue shows that they are closely related. It is to the perichondrium that tendons and muscles are attached.

Cartilage exists in three main forms: hyaline, elastic, and fibrous. In *hyaline cartilage* the intercellular substance appears to be homogeneous, translucent, and of a bluish-green cast. Actually it contains considerable numbers of collagenous fibers which are embedded in amorphous intercellular material and which are not ordinarily discernible since both have approximately the same refractive index. Hyaline cartilage possesses some degree of flexibility. In man it is found in such places as the lower parts of the ribs (costal cartilages) and at the ends of bones in joint cavities. *Elastic cartilage,* in addition to containing collagenous fibers, has large numbers of elastic fibers embedded in its intercellular substance. This type of cartilage is more flexible than others and possesses a considerable degree of elasticity. In the fresh state it is opaque and of a yellowish color. The external ear of mammals, epiglottis, walls of the Eustachian tube, and external auditory canal are composed in part of elastic cartilage. *Fibrous cartilage* has little intercellular substance, but it contains large collagenous fibers which run parallel to each other. It is really an intermediate type of connective tissue and represents a transition between true cartilage and densely arranged connective tissue. It is found in the discs between vertebrae and also in regions where tendons and ligaments are attached to cartilage of the hyaline or elastic types.

Since cartilage contains no blood vessels, the chondrocytes receive their nourishment by diffusion of materials through the intercellular substance which surrounds them. Transplants of cartilage from one part of the body to another may survive for long periods of time since they need not receive a blood supply in order for their cells to live. Transplants in which the cells have been killed for one reason or another do not survive as well as those with living chondrocytes.

BONE, which forms the greater part of the skeleton in most vertebrates, is the last type of connective tissue to make its appearance during embryonic development. One might think that calcification of cartilage, in producing rigid, inflexible, skeletal material, might provide needed stiffness and serve the same function as bone. Deposition of calcium salts in the intercellular matrix of cartilage, however, interferes with the diffusion of gases and other substances from which the chondrocytes obtain their nourishment. Hence the cells die and the intercellular material is gradually resorbed. The requirement of higher vertebrates for a strong skeletal framework to support the weight of the body could not, therefore, be satisfied by mere calcification of cartilage. The evolution of bone is believed to have come about because of the inability of calcified cartilage to provide the needed requirements of higher vertebrates for survival, at least as far as their skeletal system is concerned. The intercellular substance of bone is very dense, and the bone cells, or *osteocytes,* lie in spaces called *lacunae,* scattered throughout the matrix. During the developmental stages of bone, osteocytes have numerous, irregularly branching processes which join similar processes of other bone cells. The dense intercellular substance is formed around them. As the developing bone matures, the cytoplasmic processes gradually are withdrawn, leaving in their places innumerable small canals, called *canaliculi,* which ramify throughout the intercellular substance. There is thus provided a system whereby tissue fluid may be conducted to the osteocytes. Nevertheless, the latter cannot be too far removed from capillaries if they are to survive, since the canalicular system at best is not too efficient a method of providing for the nourishment of cells and removal of the waste products of metabolism. A fundamental difference, therefore, between cartilage and bone is that whereas cartilage is an avascular tissue, bone is highly vascular, having a rich capillary supply. Another primary difference lies in the fact that the matrix in bone becomes calcified very soon after it is formed and becomes rigid. It is not capable of expansion. During growth new bony tissue is formed and added to one of the surfaces by apposition. Collagenous fibers, apparently identical with those of loosely organized connective tissue, are embedded in the matrix of bone but are difficult to demonstrate unless very special techniques are used. The fibers are so arranged as to provide for maximum strength. Studies using the electron microscope have revealed many fine points of difference in structure of the matrices of calcified cartilage and bone. The hard, rigid consistency of bone is actually the result of impregnation of the matrix with certain inorganic salts, which make up about 30 per cent of the weight of bony tissue. Calcium phosphate, calcium carbonate, sodium chloride, and magnesium phosphate are the principal compounds concerned. Of these, calcium phosphate is the most abundant, comprising about 85 per cent of the total.

Bone generally exists either in spongy or compact form. *Spongy,* or *cancellous, bone* consists of numerous small bony plates and bars joined together in an intricate manner. The small spaces between the plates and bars contain *bone marrow. Compact,* or *periosteal, bone* appears as a hard solid mass without any spaces except those of microscopic size. The two types of bone are not basically different but merely represent a different arrangement of the elements of which they are composed. The expanded extremities of long bones of the body are composed of spongy bone underlying an outer compact layer which is of variable thickness. In the bones of the skull a layer of spongy bone (diploë) lies between two layers of compact bone, the *inner* and *outer tables.*

Spongy bone ensures considerable strength with a minimum of weight. Compact bone generally encloses a variable amount of spongy bone. It forms the hard, tubular shaft of the typical long bone and encloses the marrow cavity. The outer surfaces of bones are made up of compact bone except on their articular portions. Compact bone is covered with a dense layer of connective tissue, the *periosteum.* It is by means of the periosteum that muscles and tendons are attached to bones. The cavities of bones which are filled with bone marrow are lined with a thin layer of connective tissue called the *endosteum.*

A properly prepared cross section of compact, or periosteal, bone, when examined under the microscope (Fig. 3.22), gives a typical picture of bony connective tissue. Rounded, relatively large spaces, the *Haversian canals,* are numerous. In living bone they serve for the passage of blood vessels. Arranged in fairly regular concentric circles about each Haversian canal are the bony *lamellae.* Between adjacent lamellae are located the lacunae in which the osteocytes are situated. Canaliculi may be seen to extend from one lacuna to another. The blood vessels in Haversian canals supply nourishment to the osteocytes by transfer through tissue fluid, filling the canaliculi and whatever space may be present in the lacunae in which the cells lie. A

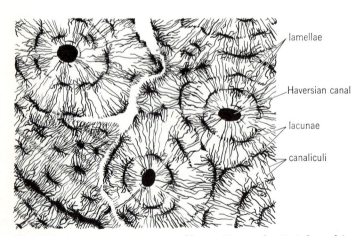

lamellae

Haversian canal

lacunae

canaliculi

Fig. 3.22 Microscopic structure of bone. (*Drawn by G. Schwenk.*)

Anatomy of the Chordates

Haversian system consists of a Haversian canal together with the surrounding lamellae, lacunae, and canaliculi. Between adjacent Haversian systems are irregularly disposed *interstitial lamellae,* lacking Haversian canals. They represent the remains of former Haversian systems which persist while new systems are being formed during growth processes, when reconstruction of bone is taking place. *Circumferential lamellae,* also lacking Haversian systems, are present just adjacent to periosteum and endosteum of compact bone. They run parallel to the outer and inner surfaces. The bone making up the irregular plates and bars of spongy bone lacks Haversian systems, but the osseous tissue usually shows a lamellar arrangement. It is an interesting fact that the Haversian systems of different mammals, although basically similar in structure, show enough species variation to make it possible to identify certain animals by the idiosyncrasies in arrangement of their Haversian systems (D. J. Gray, 1941, *Anat. Record,* **81:**163). Haversian systems are not always present in compact bone of vertebrates. They are lacking in most amphibians, certain reptiles, and small mammals. In such forms, nevertheless, an adequate blood supply is provided via numerous vascular channels.

Muscular tissue. Although the property of contractility is possessed by all living protoplasm, it is particularly well developed in muscular tissue. In such tissue the direction of contraction is along definite lines which correspond to the long axes of the muscle cells. Although animals in some of the lower phyla move by means of contraction of certain *epitheliomuscular cells,* in most metazoan animals all movement is brought about by the contraction of muscle cells. Such contractions are responsible, not only for locomotion, but also for movements of the vari-ous internal organs, for the beat of the heart, for the propulsion of blood and lymph through vessels, for the passage of food through the digestive tract, and for the passage of glandular secretions and excretory products through ducts leading from one part of the body to another.

With the exception of the muscles of the iris of the eye and the myoepithelial cells of sweat, lacrimal, salivary, and mammary glands (Fig. 9.15), all ectodermal in origin, muscular tissue is derived from mesoderm (mesenchyme).

Three main types of muscular tissue are recognized. The distinctions among them are based both on structural and functional differences. *Smooth,* or *involuntary, muscle* makes up the contractile tissue of hollow visceral organs, ducts, and blood vessels. The action of this type of muscular tissue is not under control of the will. *Striated, voluntary,* or *skeletal muscle* forms the greater part of the body musculature and is under control of the will. A third type of muscular tissue, called *cardiac* muscle, represents an intermediate form, since it is striated in the manner of skeletal muscle but is, nevertheless, free of voluntary control.

Smooth-muscle tissue (Fig. 3.23) is composed of long, narrow, spindle-shaped cells each of which bears an elongated nucleus in its central portion. Threadlike *myofibrils,* which are demonstrated with difficulty, run in a longitudinal direction through the cytoplasm, or *sarcoplasm* as it is called in muscle cells. They are thought to be the actual contractile elements. Smooth-muscle cells, or fibers, may occur individually or in the form of bundles or sheets. Their average length is about 0.2 mm. Individual fibers in close association with collagenous and elastic connective-tissue fibers are found in such regions as the skin about the nipples, the wall of the scrotum, and the villi of the small intestine.

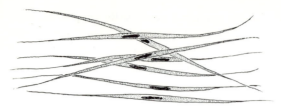

Fig. 3.23 Smooth-muscle cells teased apart.

Bundles of smooth-muscle fibers surrounded by connective tissue are represented by the *arrector pili* muscles which are attached to the bases of hairs and cause hair to stand erect at certain times. The most usual arrangement of smooth-muscle fibers, however, is in the form of sheets composed of several layers of cells. In the walls of hollow viscera they are usually arranged in two main layers called *circular* or *longitudinal,* depending on the direction in which the fibers extend. These layers vary in thickness according to their state of contractility. The longitudinal layer is external to the circular layer. This arrangement is characteristic of the muscular wall of the intestine and certain other organs. In some regions the fibers may not show such a distinct stratification. Bundles of fibers run in different directions, but the fibers within a bundle are parallel to each other.

The close association of smooth muscle and connective tissue is everywhere apparent. Muscle cells are often arranged in bundles surrounded by loosely organized connective-tissue fibers which may even penetrate between individual cells. The cellular elements of connective tissue do not, however, appear between individual cells of the smooth-muscle type.

Smooth muscle undergoes slow and rhythmic contractions controlled by the autonomic nervous system which functions independently of the will. Peristaltic and segmenting contractions of the digestive tract,

responsible for the propulsion of food down the digestive tract and for mixing it thoroughly with digestive juices and enzymes, are brought about by contraction of smooth muscle. Contraction of the circular layer causes a narrowing of the lumen and a lengthening of the organ; constriction of the longitudinally arranged fibers increases the diameter of the lumen and causes the digestive tract to shorten. In this manner ingested food is propelled down the alimentary tract, in the various parts of which different phases of digestion occur.

It has been demonstrated (J. C. Thaemert, 1959, *J. Biophys. Biochem. Cytol.,* **6**:67) in electron microscopic studies that protoplasmic bridges occur between smooth-muscle cells in the digestive tract. It is believed that by this means impulses for contraction may be conducted or conveyed from one muscle cell to another.

Striated, or *skeletal, muscle tissue,* of which all voluntary muscles are composed, consists of long fibers which are, in reality, multinucleated cells. A *muscle* consists of bundles of striated-muscle fibers bound together in a sheath of connective tissue. The length of the fibers varies somewhat. In short muscles they may extend the entire length of the muscle, but in longer muscles they are attached to tendons or to other muscle fibers in most cases. Individual fibers (Fig. 3.24) are independent, show no branching, and course parallel to each other. Each fiber is bounded by a membrane, the *sarco-*

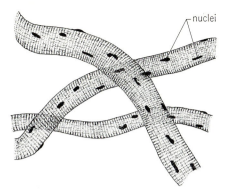

nuclei

Fig. 3.24 Three striated-muscle fibers teased apart.

lemma. Numerous *myofibrils,* which are present within the fiber, are surrounded by sarcoplasm, or a sarcoplasmic reticulum, which may be discontinuous. The myofibrils consist of minute, alternating dark and light plates, or discs. These are so arranged within a fiber that all the dark discs lie at the same level, giving the appearance of dark bands within the boundaries of the sarcolemma. The light discs, lying side by side, form light bands in a similar manner (see page 506).

The close association between muscular tissue and connective tissue is again brought out in a study of striated muscle, particularly in the region where the muscle is attached to some connective-tissue structure on which it exerts its pull when contracted. Such a structure may be a tendon, an aponeurosis, a raphe, perichondrium, or any structure in the body composed of densely organized connective tissue. A sheath of connective tissue, the *epimysium,* surrounds an entire muscle. Bundles of muscle fibers are intimately interdigitated with the collagenous fibers of a tendon or other structure. Tendons are actually continuations of the *perimysium,* the regularly arranged connective-tissue fibers of which lie between bundles of muscle

fibers, which they help to bind together and with which they form a close union. A thin network of connective tissue, the *endo-mysium,* surrounds the individual muscle fibers. It is continuous with the perimysium and contains connective-tissue cells in addition to the fibrous elements. Elastic fibers may be found in more than ordinary numbers in certain muscles, such as those which cause eye movements and those of the tongue and face.

The growth of striated muscles and the enlargement which occurs as a result of considerable exercise is due to an increase in the size of individual fibers rather than to an increase in number.

Attention should be called to the red, pink, or white color of muscle in different vertebrates and even in different regions of an individual animal. They differ in their fine structures, functions, and capillary patterns. Red muscles have more abundant sarcoplasm and are more vascular; white muscles have less sarcoplasm and are less vascular. In pigeons both pectoral and leg muscles are red, whereas in chickens the pectoral muscles are white and the leg muscles are red. In chickens the pectoral muscles are used relatively little but in the pigeon they are of great importance in flight.

Cardiac-muscle tissue, confined to the muscular wall of the heart and to the roots of the large blood vessels joining the heart, shows some characteristics of both smooth and skeletal muscles. It contracts rhythmically and involuntarily in the manner of smooth muscle. In structure, however, it more closely resembles skeletal muscle since it is striated. Alternating dark and light bands or discs are present in the myofibrils. The nuclei of the fibers, however, are spaced a fair distance apart and lie deep within the fiber in its axial portion. They are quite in

Fig. *3.25* Cardiac-muscle fibers, showing intercalated discs.

contrast to those of skeletal muscle which are peripheral in position, lying close to the sarcolemma. Morever, cardiac-muscle fibers branch so as to form a network, but each fiber tends to course in the same direction as its neighbors. As a result, spaces between the fibers have a slitlike appearance. Endomysium is located in these spaces. An abundant supply of capillaries accompanies the endomysium, making close contact with the cardiac-muscle fibers. Lymph capillaries are present in addition. Nerve fibers, which terminate on the cardiac-muscle fibers, also accompany the endomysium.

Peculiar, transverse, *intercalated discs* are present at intervals in cardiac-muscle fibers (Fig. 3.25). Their nature and significance were not understood until recently. It was formerly believed that the fibers of cardiac muscle were not composed of cells but, instead, formed a complicated syncytial network. Studies by Muir (A. R. Muir, 1957, *J. Biophys. Biochem. Cytol.,* 3:193) have changed this concept. The intercalated discs are now known to represent the cell membranes of adjacent cardiac-muscle cells. Each cell, therefore, is a distinct entity and contains its own nucleus.

In various conditions in which the heart may become enlarged for one reason or another, increase in size, as in the case of skeletal muscle, is the result of an increase in size of the fibers rather than an increase in their number.

Nervous tissue. All protoplasm is irritable. It has the power of responding to stimuli or disturbing influences such as may be produced by mechanical forces, sound, heat, cold, light, chemical reaction, and electricity. Protoplasm also has the property of conductivity, by means of which impulses set up by stimuli are transmitted from one portion of the protoplasmic mass to another. In the cells which make up *nervous tissue,* the properties of irritability and conductivity are more highly developed than in any other tissue within the body. The greater part of the *nervous system* is composed of nervous tissue. Its function is to receive stimuli and to send impulses from one part of the body to another. In this manner the functions of many organs and parts of the body are coordinated and integrated. It shares these functions with the endocrine system. The functional relations of these two systems are so striking that the term *neuroendocrine system* is appearing with increasing frequency in scientific literature. Nervous tissue is also the seat of all conscious experience. The nervous system is a dominant system. It must function properly if the integrity of the body is to be maintained.

The structural units of nervous tissue are *neurons,* or *nerve cells,* together with their processes. The part of the cell in which the nucleus lies is called the *cell body.* Masses of nerve-cell bodies, if located in the brain or spinal cord, are referred to as *nuclei.* Such nuclei should not be confused with nuclei of individual cells. A mass of nerve-cell bodies located *outside* the central nervous system is called a *ganglion.* A ganglion may contain only a few nerve-cell bodies or as many as 50,000.

The processes of a neuron are of two kinds: (1) *dendrites,* usually numerous, which are short processes showing a high degree of branching close to the nerve-cell body; (2) a single, slender, *axon,* or *axis cylinder,* which is a long process with branches, the *axon end-ending,* or *terminal arborization,* at its end (Fig. 3.26). Collateral branches may be given off by the axon along its course, always at a node and at right angles. These branches are frequently lacking. Axons of certain neurons may be very long, as in those which conduct impulses from the spinal cord to the extremities of the appendages. The number of cytoplasmic processes extending from a nerve-cell body varies and forms the basis for classifying different types of neurons.

The nucleus of a nerve cell is large and spherical and contains a single, distinct, densely staining nucleolus. Special staining techniques reveal the presence of numerous basophilic bodies in the cytoplasm. They are called *Nissl's granules* or *bodies.* Under the electron microscope they are seen to be composed of rough-surfaced endoplasmic reticulum and ribosomes. The significance of Nissl's granules is not completely understood despite the vast amount of study directed toward them. Most significant findings suggest their involvement in the formation of new cytoplasm in the nerve-cell bodies. This then passes into the axons, the semifluid cytoplasm of which is constantly being replenished. The axon of a very long nerve cell may contain several hundred times the amount of cytoplasm that is present in the cell body. It seems that Nissl's granules are in some way related to protein synthesis within the nerve cell. Their importance in connection with the metabolism of neurons is indicated by the very definite changes which they undergo under varying physiological conditions. Nissl's granules are present in the cytoplasm of dendrites, particularly in the region adjacent to the cell body. They are not present in the axon. Very fine *neurofibrils,* in the form of an intricate network of threads, are also present in the cytoplasm. They extend even into the smallest branches of both dendrites and axons. The neurofibrils probably play an important part in the transmission of impulses.

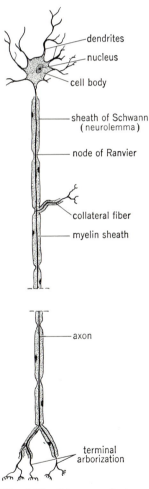

Fig. 3.26 Diagram of a typical medullated neuron.

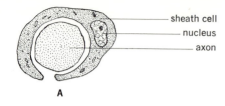

— sheath cell
— nucleus
— axon

A

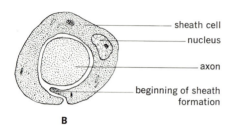

— sheath cell
— nucleus

— axon

— beginning of sheath formation

B

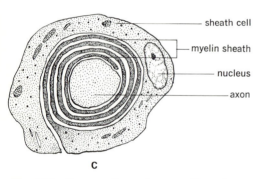

— sheath cell

— myelin sheath

— nucleus

— axon

C

Fig. 3.27 Cross sections of a nerve fiber, showing stages *A*, *B*, and *C* in the development of the myelin sheath from a sheath-of-Schwann cell.

The axon with its enveloping sheath or sheaths is spoken of as a *nerve fiber.* Those nerve fibers located outside the brain and spinal cord are covered by a thin, continuous, encapsulating membrane, the *neurolemma* (*sheath of Schwann*). The neurolemma is composed of *sheath cells* (*Schwann's cells*) which are present in series beginning at the brain or spinal cord, or at a ganglion, and ending almost at the terminal arborization. Each sheath cell with its cytoplasm and flattened nucleus surrounds a section of the axon. The sheath cells, like neurons, are of ectodermal origin. It is believed that they represent neuroglia cells (page 70) from the central nervous system which have moved out to a peripheral position. They are not connective-tissue cells. Their presence seems to be essential for the proper functioning of the axons of peripheral nerves.

Nerve fibers are usually said to be either *medullated* (myelinated) or *nonmedullated* (nonmyelinated), depending upon whether or not they are surrounded by a distinct sheath of glistening white material called *myelin* which lies between the axon and the neurolemma. However, unlike the neurolemma, which is continuous, the myelin sheath is interrupted at rather regular intervals by circular constrictions, the *nodes of Ranvier.* Sections between adjacent nodes are called *internodes.* A single sheath cell of neurolemma covers an internode. It has been shown that the myelin surrounding an axon originates as a result of the peculiar manner in which the sheath cells, comprising the neurolemma, grow and encircle the axon, rotating about the latter in a manner comparable to a jelly roll (Fig. 3.27). As the series of rings continues to wind around the axon, the cytoplasm within them is squeezed out or, at any rate, disappears. The resulting myelin sheath, formed largely by adjacent cell membranes which fuse, has been shown to be composed of alternating concentric layers of lipid (fatty) and protein materials. Schwann's cells are essential for the formation of myelin, at least in peripheral nerves.

The function of the myelin sheath has been the subject of much research. One idea is that it serves as an insulating substance, much in the manner of the coating of an electric wire, preventing loss of energy of the nerve impulse during its passage along the fiber. There is clear-cut evidence that the myelin sheath surrounding an axon is related to the velocity of conduction of the impulse. The nerve fibers of vertebrates which conduct impulses at the fastest rates are of medium size but have thick myelin sheaths comprising as much as 50 per cent of the total diameter of the fiber.

Medullated fibers in the brain and spinal cord, making up the so-called white matter, are not surrounded by a neurolemma. Instead, certain neuroglial (page 98) "satellite" cells (astrocytes and oligodendroglia) form very incomplete sheaths about the myelin but are probably responsible for its formation, nevertheless. They make contact with nerve fibers at the nodes of Ranvier. The presence of the neurolemma on peripheral nerves is believed to account for the fact that such fibers may regenerate after injury. Fibers in the brain and spinal cord, lacking a neurolemma, do not regenerate after they are severed or destroyed.

Until rather recently, so-called nonmedullated fibers were believed to lack myelin sheaths, and the neurolemma was reported to surround the axon directly. Electron microscopy has demonstrated that even in such fibers a trace of myelin may be present. At any rate, the term "nonmedullated fiber" is often retained in current literature. A dozen or more of such nonmedullated, or sparsely medullated, fibers may be surrounded by the same sheath cell. Such fibers appear gray, in contrast to the white appearance of medullated fibers.

The neurons in amphioxus and cyclostomes are reported to lack myelin sheaths. Even certain giant Müller's fibers which course in the spinal cord of the lamprey from the brain to the posterior end of the cord are nonmedullated. As might be expected, such fibers have been shown to conduct impulses at a comparatively slow rate.

A *nerve,* or *nerve trunk,* is made up of numerous fibers outside the central nervous system, bound together in parallel bundles by connective tissue. Most peripheral nerves are composed of medullated fibers. Sparsely medullated fibers are found chiefly as components of nerves of which the sympathetic nervous system is composed.

In the animal body, impulses always travel in one direction in nerve cells, passing from dendrites to cell body and then down the axon. Thus one end has a receptive function, whereas impulses are discharged at the other. This is spoken of as the *polarity* of the neuron. It has been shown, however, that under experimental conditions it is possible for impulses to travel in the reverse direction. This does not occur in the body because of the manner in which the various neurons are arranged and because of the fact that they are polarized. The nervous system is mostly made up of neurons arranged in chains in such a manner that impulses can be transmitted from one to another in series. The terminal arborization of one cell comes in close, and even intimate, contact with the dendrites of one or more other neurons. The cells do not fuse, however, and the substance of one cell is separated from that of the next by a slight gap (page 96). Such a point of "almost" junction is known as a *synapse.* An impulse does not travel across the synapse from the terminal arborization of one cell to the dendrites of the next. Rather, a fresh impulse is set up on the other side of the synapse. In this manner impulses are conducted from one part of the body to another. One neuron in most cases forms synapses with many other neurons. There have been many speculations as to the nature of the synapse and the manner in which impulses are transmitted. The two sides of the synapse are not alike structurally or functionally. Often the terminal branches of the axon bear small, clublike enlargements which are closely applied to the dendrites or even to the cell body of another neuron. In some cells they form small networks which appear to be in close contact with dendrites of another cell. It is now fairly well established that a minute amount of a chemical substance called *acetylocholine* is given off when a nervous impulse reaches the terminal arborization of an axon, except in most of those postganglionic fibers belonging to the sympathetic nervous system which, instead, give off a chemical substance, formerly known as *sympathin,* but now known to be a catecholamine called *norepinephrine* (page 358). It seems plausible that acetylcholine and norepinephrine are formed in nerve cells and stored in tiny neurovesicles, which electron microscopy has demonstrated to be present as cytoplasmic processes on the clublike enlargements of the axon endings. Liberation of acetylcholine is believed to cause a change in the permeability of the membranes separating the terminal arborization, or axon ending, from the next neuron or neurons. This changes the electrical potential of the membranes, with the result that an impulse is set up in the cell or cells with which the terminal arborization connects.

In some parts of the body *neurosecretory cells* (page 655) are present. These are nerve cells in which a secretion formed within the cell body passes down the axon and is discharged directly into the blood stream.

It is not definitely known whether synapses in the central nervous system involve chemical mediation like those of peripheral synapses, although it appears probable that they operate in the same manner.

Nervous regulation, which involves the release of neurosecretions, is essentially a chemical phenomenon. Neurohumors, released at synapses and at points of junction between terminal processes and effector organs, indicate that secretion is as basic a functional characteristic of neurons as is the rapid transmission of impulses. The part played by neurosecretions in coordinating body functions antedates that of the endocrine glands (page 335), which appeared later in evolution and which, for the most part, are regulated by neurosecretory cells (page 655). If a stimulus is too weak to initiate a nerve impulse and another stimulus of the same magnitude is applied immediately afterward, a nerve impulse will then start up The first seems to sensitize the nerve substance in such a manner that the second stimulus may then act. The combined activity of two such stimuli applied close together is known as *summation.* The presence of the synapse may possibly be of importance in connection with this phenomenon.

The terminal processes of neurons in contact with epithelial, muscle, glandular, and even certain connective-tissue cells show many interesting variations. Some are *receptor,* or *afferent, dendrites,* capable of receiving sensory stimuli; others are *effector,* or *efferent, axons,* either causing muscle cells to contract or causing glandular cells to secrete. Receptors are homologous with dendrites, modifications of dendrites, or are comparable to dendrites. Effectors are homologous with the terminal arborizations of axons. The various modifications of receptor and effector types of nerve endings are discussed in Chap. 14.

TYPES OF NEURONS. Nervous tissue passes to practically every part of the body. Nerve cells, therefore, are very numerous. It has been estimated that there are between 9 and 10 billion neurons in the cortex of the human brain alone, plus around 100 billion neuroglia cells.

A great number of types of neurons may be observed in various parts of the nervous system. *Unipolar neurons* are those in

Anatomy of the Chordates

which only a single process, the axon, passes from the cell body and no dendrites are present. Such neurons are rarely found in the central nervous system of adults, but they are present in abundance during developmental stages in early embryonic life.

Typical *bipolar neurons* have a single dendrite and a single axon projecting from opposite ends of the cell. Such neurons are present in the retina of the eye, in the olfactory epithelium, and in the ganglia of the auditory nerve. In certain regions a further differentiation occurs, so that both axon and dendrite appear to arise from the nerve cell by a single process. Intermediate stages are shown in Fig. 3.28. Neurons of this type are found in the ganglia of cranial and spinal nerves. They are sometimes spoken of as *pseudounipolar,* or *unipolar-ganglion, neurons.* The dendrite of such a cell is highly specialized. It may be very long, and its microscopic structure similar to that of the axon. The chief difference between an axon and a dendrite of this type is that they conduct impulses in opposite directions in reference to the cell body. Sometimes such dendrites are even referred to as peripheral sensory axons.

The shape of *multipolar neurons,* which are most numerous of all, is variable and depends upon the number, arrangement, and length of the processes. They may, in general, be resolved into two main types. In *Golgi Type I* there is a single long axon and numerous short dendrites; in *Golgi Type II* the axon is very short. The cell bodies of multipolar neurons are located in the gray matter of the brain and spinal cord. They are often referred to as *interneurons,* or *association neurons,* and are neither sensory nor motor. In Golgi Type II the cells are confined to the central nervous system, but the axons of many neurons of Golgi Type I form the motor components of spinal nerves and of certain cranial nerves. The latter type neurons also constitute the ganglia of the autonomic nervous system.

GRAY MATTER AND WHITE MATTER. Groups of nerve-cell bodies, their dendrites, and the proximal ends of medullated axons have a grayish appearance. The term *gray matter* is applied to those parts of the brain, spinal cord, and peripheral ganglia in which nervous tissue of this type is present. Postganglionic fibers of the sympathetic nervous system are sparsely medullated and appear gray in color. *White matter,* on the other hand, is present in the brain, spinal cord, and most peripheral cranial and spinal nerves. It is composed of large bundles of medullated nerve fibers with their glistening white myelin sheaths. Generally speaking, white matter serves to conduct impulses from one part of the body to another. Gray matter, on the other hand, functions in integrating impulses. It should be recalled that supporting, ectodermal *neuroglia cells* are present in the brain and spinal cord. Special histological techniques have demonstrated that there are several types of neuroglia cells and it is believed that in addition

Fig. 3.28 Intermediate stages in the transition of a bipolar neuron to one of the pseudounipolar type.

to providing support they may play an important part in the proper functioning and metabolic activity of neurons. It has already been mentioned that astrocytes and oligodendroglia, neuroglial derivatives, are suspected of being involved in the formation of myelin sheaths of neurons in brain and spinal cord which lack a neurolemma. Neuroglia cells show definite reactions in response to certain pathological processes. Their role in the body economy should not be underestimated.

ORGANS AND ORGAN SYSTEMS

As observed in an earlier part of the chapter, any part of the body which carries on some special function is referred to as an *organ*. Organs are composed of several types of tissues which work in harmony in the normal functioning of the body. Organ systems are composed of several organs which unite in a common function. The organ systems of the body include the following:

1. Integumentary system
2. Digestive system
3. Respiratory system
4. Excretory system
5. Reproductive system
6. Endocrine system
7. Skeletal system
8. Muscular system
9. Circulatory system
10. Nervous system and receptor organs

In the following pages a comparative study of the organ systems of vertebrates is presented.

SUMMARY

The sexual method of reproduction is the rule in members of the phylum Chordata. A fertilized egg or zygote undergoes a series of cell divisions resulting in the development of an embryo. The manner of development is influenced greatly by the amount and distribution of yolk within the egg cell. An embryo during early development passes through blastula and gastrula stages during which two layers, epiblast and hypoblast, are formed. A third layer soon appears which comes to lie between the other two. Three germ layers are thus established, ectoderm, mesoderm, and endoderm. All structures in the body trace their origin to one or more of these three primary germ layers. The following outline summarizes the derivation of the various structures of the body from the three primary layers.

I. ECTODERM
   A. Skin. Epidermis, skin glands, hair, feathers, nails, claws, hoofs, horns, epidermal scales, covering of external gills
   B. Lining of mouth. Enamel of teeth, glands of the mouth, covering of tongue and lips, anterior and intermediate lobes of the pituitary gland
   C. Nervous system. Brain and spinal cord, cranial and spinal nerves, autonomic portion of peripheral nervous system, sensory parts of all sense organs, medulla of adrenal gland, infundibulum, and posterior lobe of the pituitary gland
   D. Miscellaneous. Lens of eye, intrinsic eye muscles, neuroglia, pineal and parapineal bodies, lining of anal canal and derivatives, lining of a portion of cloaca, myoepithelial cells of sweat glands, lacrimal, salivary, and mammary glands.

II. Endoderm*

A. Alimentary canal. Pharynx, esophagus, stomach, intestine, liver, pancreas, lining of most of cloaca

B. Pharyngeal derivatives. Larynx, trachea, lungs, gills of the internal type, middle ear, Eustachian tube, tonsils, thyroid, parathyroids, thymus, ultimobranchial bodies

C. Miscellaneous. Allantois, urinary bladder, urethra, yolk sac

III. Mesoderm

A. Muscles. Smooth, striated, cardiac

B. Skeleton. Cartilage, bone, other connective tissues

C. Excretory organs. Kidneys and their ducts

D. Reproductive organs. Gonads, ducts, accessory structures

E. Circulatory system. Heart, blood vessels, blood, spleen, lymphatics, blood-forming tissues

F. Miscellaneous. Dentine of teeth, dermis of skin, cortex of adrenal glands, lining of body cavities, mesenteries and omenta, portions of the eye

The cells making up the germ layers do not possess distinctive differences and are said to be undifferentiated. Further development of the embryo entails, among other things, a differentiation or specialization of groups of cells to form the several types of tissues which make up the body of an individual. A tissue is an aggregation of similarly specialized cells united in the performance of a particular function such as secretion, support, protection, or contraction. A complex of several tissues may together form an organ. Organs have definite form and perform special functions. Groups of organs, working together in performing one or more closely allied functions, constitute an organ system.

Tissues are classified as follows:

I. Epithelial Tissue

Composed of one or more layers of cells covering an external or internal surface of the body or embryonically derived from such a surface

A. Simple epithelia. Consisting of a single layer of cells

1. Simple squamous epithelium. Lining of blood vessels and body cavities
2. Simple cuboidal epithelium. Lining of many glands and their ducts
3. Simple cuboidal ciliated epithelium. Portions of lining of kidney tubules
4. Simple columnar epithelium. Lining of alimentary canal
5. Simple columnar ciliated epithelium. Lining of oviducts and small bronchi
6. Pseudostratified columnar epithelium. Lining of urethra of human male and of the large ducts of certain glands (parotid)

---

* The listing refers only to the epithelial lining of the organs. All are surrounded or supported by tissues of mesodermal origin.

7. Pseudostratified columnar ciliated epithelium. Lining of respiratory passages and of ducts of the male reproductive system

    B. Stratified epithelia. Composed of several layers of cells
        1. Stratified squamous epithelium. Epidermis of skin, lining of mouth
        2. Stratified cuboidal epithelium. Epidermis of many tailed amphibians
        3. Stratified columnar epithelium. Epiglottis, lining of part of urethra
        4. Stratified columnar ciliated epithelium. Lining of larynx, upper surface of soft palate
        5. Transitional epithelium. Lining of urinary bladder, pelvis of kidney

II. CONNECTIVE TISSUE

Binds other tissues together and gives support to various structures in the body. It is distinguished by the presence of intercellular material, or matrix

    A. Connective tissue proper
        1. Loosely organized connective tissue. Under skin, walls of lungs, most visceral organs
        2. Densely organized connective tissue. Dermis of skin, part of walls of urinary ducts
        3. Regularly arranged connective tissue. Tendons, ligaments
        4. Specialized connective tissues
            a. Mucous. Umbilical cord, cock's comb
            b. Elastic. True vocal cords, flaval ligaments
            c. Adipose. Fat
            d. Reticular. Groundwork of most glandular organs
            e. Pigmented. Choroid coat of eye
            f. Vesicular. Notochord
    B. Blood and lymph. Fluid connective tissue
    C. Cartilage. Firm and abundant intercellular matrix
        1. Hyaline. Costal cartilages
        2. Elastic. External ear, epiglottis
        3. Fibrous. Intervertebral discs
    D. Bone. Dense, hard, intercellular matrix

III. MUSCULAR TISSUE

Composed of cells in which the property of contractility is particularly well developed

    A. Smooth or involuntary. Arrector pili muscles, walls of intestine and blood vessels
    B. Striated, skeletal, or voluntary. All voluntary muscles
    C. Cardiac. In wall of heart

IV. NERVOUS TISSUE

Composed of cells in which the properties of irritability and conductivity are extremely well developed. The greater part of the nervous system of the body is composed of nervous tissue

| 4 |

INTEGUMENTARY
SYSTEM

The integument, or outer covering of the body, is commonly referred to as the skin. Together with its derivatives it makes up an important organ system of the body, the *integumentary system*. It is continuous with the mucous membrane lining the eyelids, mouth, nostrils, and the openings of the rectum and urogenital organs. The primary function of the skin is to cover and protect the tissues lying beneath it, since this is the part of the body which comes in contact with the environment. It is abundantly supplied with sensory nerve endings which are affected by environmental stimuli. The role played by the integument in the general metabolism of the body is of vital importance. The character of the skin and its derivatives shows much variation in different regions of the body, in different individuals, and even in the same individual as age advances. Differences in the integuments in animals making up the various groups of vertebrates are particularly striking. The type of environment, whether aquatic or terrestrial, is an important influence in determining the character of the variations to be found. Nevertheless, basic similarities exist in the integuments

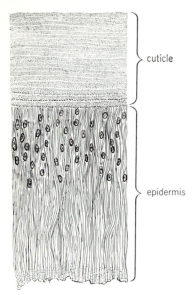

Fig. 4.1 Section through a portion of the integument of a crayfish, a typical invertebrate.

of all vertebrates. The relationships of function to structure are very well brought out in a study of the integumentary system.

INTEGUMENT PROPER

In such invertebrates as arthropods, annelids, mollusks, and some others, the integument consists of a single layer of cells, the *epidermis,* together with a noncellular *cuticle* (Fig. 4.1) secreted by the cells. The cuticle may be very thin, as in annelids, or a heavy layer composed of chitin, calcareous material, or other substances. In arthropods the rigid cuticle makes up the *exoskeleton.* The term *ecdysis* refers to the periodic shedding of this outer layer. The integument of vertebrates (Fig. 4.2) consists of an outer layer, the *epidermis,* composed of cells which are derived from ectoderm, and an underlying mesodermal layer known as the *dermis,* or

corium, derived from dermatome and the somatic layer of mesoderm (page 64). In tetrapods, ecdysis consists of a shedding, or sloughing, of the outer layer of the epidermis. In some forms this is shed as a whole, but in others it is given off in fragments of various sizes. Under the dermis lies a loose layer of connective tissue, the *subcutaneous tissue.* In skinning an animal the integument is readily detached from the rest of the body. In many regions fat may be deposited in the subcutaneous tissue. The fatty layer is known as the *panniculus adiposus.* In obese individuals, the panniculus may be extensively developed and several inches in thickness. In such places as the palms of the hands and soles of the feet the fibers of the dermis are more or less tightly interwoven with those of the subcutaneous layer so that the skin in these regions is more firmly attached. A layer of striated muscle, the *panniculus carnosus,* lies beneath the sub-

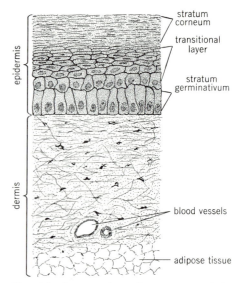

Fig. 4.2 Section through a portion of the integument of a typical tetrapod (semidiagrammatic).

cutaneous tissue. It is usually well developed in lower vertebrates but in man is poorly represented. The platysma muscle in the superficial region of the neck is its chief remnant.

Skin of man. In order to understand more readily the modifications of the integument found in various vertebrates, it is convenient first to discuss that of the human being, since it is most familiar.

The integument in man varies in thickness from less than 0.5 mm to 3 or 4 mm in different parts of the body. It is generally thicker on the dorsal surface than on the ventral surface. It is also thicker on extensor surfaces of the appendages than on flexor surfaces. In males the skin is thicker than in females although in the latter the subcutaneous tissue tends to be thicker. In considering the thickness of the skin in different areas it is important to consider whether epidermis, dermis, or both are involved.

The skin of infants is practically free of creases except for congenital *flexure lines*. These consist of fixed creases, sometimes referred to as skin joints, which are present in areas where flexion and extension of appendages or parts of appendages occur. They are to be found on the palms, soles, fingers, and toes. In the young child creases soon appear in certain regions where muscles exert a direct contractile action on the integument. The circumanal area and the scrotum are examples. In older individuals, with loss of elasticity and diminution of fat deposits, the skin tends to sag, and wrinkles of a permanent nature appear.

The integument proper under normal conditions is germproof and serves as an effective barrier against disease organisms which might otherwise gain access to the body. It is, to a great extent, impervious to water, helps to regulate body temperature, and

upon exposure to ultraviolet light can manufacture Vitamin D, the vitamin which prevents the development of rickets.

EPIDERMIS. The outer epithelium, or epidermis, is made up entirely of cells which are arranged in more or less distinct layers. It is closely applied to the dermis beneath. The living cells of the epidermis constitute the *Malpighian layer*. In the deepest layer, one cell in thickness, the cells are columnar in shape and arranged perpendicular to the dermis. These cells, which are continually dividing mitotically, make up the *stratum germinativum*. As new cells are formed they gradually approach the surface, becoming flattened as they do so. The region in which this occurs is called the *transitional layer*. On the outer surface the cells are scaly and dead and have lost their nuclei. The outer layer of flattened cells is the *stratum corneum*, or *horny layer*, of the skin. Its chief constituent is *keratin*, a very hard, tough, insoluble protein. The stratum corneum is known as the horny, or cornified, layer because in some vertebrates it becomes very hard in certain regions. The epidermis of man is an excellent example of stratified squamous epithelium (Fig. 3.15*F*).

Certain properties of the skin depend upon the presence of keratin in the stratum corneum. Since keratin is waterproof, it is this which prevents the living cells lying beneath it from absorbing water when the body is immersed in it; it also prevents evaporation of water from these cells. The tough consistency of keratin prevents damage to underlying cells otherwise susceptible to injury from the ordinary vicissitudes of exposure. Whereas the layer of keratin in most regions of the skin is rather flaky, that on the palms and soles is particularly thick, thus enabling the individual to resist the extraordinary wear and tear to which these surfaces are subjected. Keratin also is resistant to bacteria and serves as a defense against infection.

The cells of the stratum germinativum lie in direct contact with the dermis. A basement membrane, separating epidermis and dermis, has been described, but there is little agreement among investigators concerning its composition and nature. From the dermis with its rich supply of blood vessels these cells derive their nourishment and are thus able to grow and divide. As new cells are constantly forming, the outer, flattened cells of the stratum corneum would ultimately make up a very thick layer were it not for the fact that they are continually being sloughed or worn off from the surface and replaced by new cells. The rate of proliferation of the cells of the stratum germinativum is approximately the same as that at which the outer corneal cells are being desquamated, so that the thickness of the epidermis is relatively constant. An increased amount

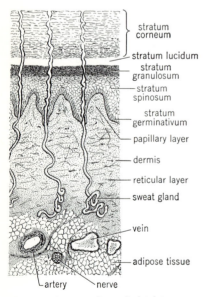

Fig. 4.3 Section through thick integument on the palm of a human being (semidiagrammatic).

of friction in a given area of the epidermis seems to stimulate the cells to divide more rapidly and results in the formation of calluses. Such thickened areas afford a greater protection to the delicate tissues lying beneath.

In many tetrapods the dead corneal layer is shed at intervals as a continuous sheet, but in man small fragments are continually being given off, and shedding is ordinarily not noticeable. After a severe sunburn the corneal layer of the human skin may, however, be sloughed in fairly large shreds. It is a familiar experience that after perspiring, portions of the skin may easily be rubbed off in small fragments. Shedding of the corneal layer seems to be at least partly controlled by the secretion of the thyroid gland.

The epidermis in certain parts of the body in man shows modifications from the conditions described above. For example, in the thick skin on the soles of the feet and palms of the hands (Fig. 4.3), transition from the columnar cells of the stratum germinativum to the flattened corneal cells is not so abrupt. Three rather distinct layers represent the so-called transitional layer. Just above the stratum germinativum is a layer of variable thickness, the *stratum spinosum,* or prickle-cell layer. It was formerly believed that the cells in this layer were held together by "intercellular bridges" which, under the light microscope, gave the cells a prickly appearance. Under the electron microscope each "bridge" has been shown actually to consist of two arms, one from each cell, in close apposition (Fig. 3.14*B*). Peripheral to the stratum spinosum lies the *stratum granulosum* in which an accumulation of keratohyalin granules occurs. The outer cells contain an increased number of these granules. It is in this layer that the epidermal cells die. Another layer, the *stratum lucidum,* lies peripheral to the stratum granulosum in these thickened regions of the integument and forms the remainder of the transitional layer. The stratum lucidum is rather transparent and resists ordinary stains in sections prepared for histological study. It is reported to consist mostly of a chemical substance called *eleidin,* which is supposedly an intermediate product in the transformation of keratohyalin granules to keratin which is the main constituent of the outer stratum corneum. These tough layers serve as a protection to the cells of the stratum germinativum in regions most likely to come in contact with external objects.

It is interesting that the epidermis in various parts of the body shows differences of such magnitude that one has difficulty in appreciating that all types of epidermis have a similar origin. The hairy integument of the scalp, for example, is quite in contrast with the skin on the palms and the soles. Once differentiation of the skin occurs, it is maintained. Transplanting one type of integument to another region will not bring about a change in its character. On the other hand, in the healing of superficial wounds, surface epithelium may form from cells derived from the ducts of sweat glands and those forming the outer sheaths of hair follicles. Differentiation of various regions of the skin occurs at different times. That of the lips, nose, and eyebrows develops earlier than that in other regions.

DERMIS. The dermis, sometimes called the true skin, is best developed in mammals. In contrast to the epidermis it is not composed entirely of cells but consists largely of connective-tissue fibers extending in all directions and forming a fairly elastic covering which makes up the greater portion of the skin. Cells appear among these fibers. In addition to connective tissue, the dermis contains nerves, smooth muscle fibers, blood vessels, and certain glands. In preparing leather, the epidermis is first removed by maceration and the connective-tissue fibers of the dermis are thickened and toughened by the action of tannin, alum, chromium salts, or other so-called tanning agents.

The dermis is actually made up of two layers which are not distinctly separated from each other. The outer or *papillary layer* is thin and lies immediately beneath the stratum germinativum of the epidermis. The *reticular layer* forms the remainder of the dermis. Collagenous fibers in the papillary layer are of finer texture than those in the reticular layer.

Patterns. On the surface of the human skin are many small grooves and ridges which intersect so as to bound small triangular and quadrangular areas. On the palms and soles the grooves and ridges generally run parallel to each other for some distance. The side of the dermis in contact with the epidermis in these areas is thrown into prominent rows of *papillae.* This makes up the papillary layer in which the collagenous and elastic fibers are delicate and rather widely separated. The papillae are quite definitely arranged so that a double row lies beneath one of the external ridges (Fig. 4.4). The grooves between the external

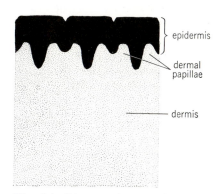

Fig. 4.4 Diagram showing the relation of dermal papillae to epidermis in friction area of skin.

ridges are situated above slight extensions of the lower part of the epidermis into the dermis. In other regions the papillae show variations in size and arrangement. The ridges on the palms and soles are arranged in such a manner as to provide the greatest possible friction with a surface. They tend to prevent slipping should the surface be very smooth.

The finely detailed patterns formed by depressions and friction ridges are very useful in making fingerprints and footprints. The patterns do not change from birth to death. There is no longer any question as to the constancy of fingerprints, and this fact has been used extensively in preparing systems of identification now so widely used. Use of fingerprints as a means of identification goes back to ancient times when the thumb impression of a monarch was used as a signature. The patterns on the various fingers and toes of a single individual show as much variation as do those of separate persons. With millions of fingerprints now on record, a system of cataloguing them had to be devised in order to identify quickly any print that was being traced. There are four main types of patterns: arches, whorls, loops, and composites. In each type distinct subdivisions occur, and these in turn show detailed variations. The number of possible patterns is almost limitless. The Federal Bureau of Investigation in Washington maintains the largest collection of fingerprints in the world. It cooperates with 77 foreign countries and four territorial possessions in the international exchange of fingerprint-identification data. So well catalogued are the prints that one can be traced almost in a matter of minutes.

By use of various laboratory techniques, the epidermis and dermis can be cleanly separated. If the underside of the epidermis is then examined under a microscope, depending upon the region of skin in question, it may readily be noted that numerous types of arrangements of interlocking of epidermis and dermis exist. Much depends upon the arrangement of hairs and sweat glands in various regions.

PIGMENT. Pigment in vertebrates occurs in certain cells in the form of small granules of *melanin,* the actual chemical structure of which is not known with certainty. The human integument does not contain the kinds of dermal *chromatophores** (pigment-bearing cells) that are found in cyclostomes, fishes, amphibians, and reptiles; these are ectodermal in origin, being derived from neural crests, manufacture their own pigment, and respond to stimulation by changing the distribution of the pigment granules within the cells (Fig. 4.5). Instead, the color of the skin of man depends upon the presence of certain cells called *melanocytes* in the lower layers of the epidermis. Melanocytes are differentiated melanoblasts, cells also of neural-crest origin. They manufacture their own pigment, but pigment distribution within the cell is not variable. Pigment seems to be fairly evenly distributed, however, among all the deep epidermal cells. It is believed that they may receive pigment from the melanocytes which lie among them. Certain cells in the dermis (also, unfortunately, often referred to as dermal chromatophores) appear to be pigment cells but they are actually phagocytes, engulfing pigment granules rather than manufacturing them. It might be supposed

* The terminology used for cells containing the pigment *melanin* is often confusing. The nomenclature used here has been officially adopted by the National Research Council's Committee on Pathology (T. B. Fitzpatrick and A. B. Lerner, 1953, *Science,* 117:640). *Melanoblasts,* of neural-crest origin, are immature cells which give rise to *melanocytes; melanocytes* are mature melanin-forming cells; *melanophores,* also of neural-crest origin, are *chromatophores* in which distribution of the pigment melanin varies under different conditions; *melanophages* are actually macrophages of mesenchymal origin, located in the dermis, which engulf melanin pigment granules.

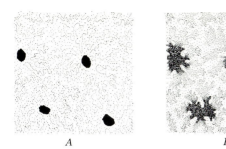

A   B   C

Fig. 4.5 Three stages in concentration and dispersion of pigment in chromatophores in the skin of certain vertebrates: *A*, pigment concentrated in center of cells; *B*, intermediate condition; *C*, pigment dispersed throughout the cytoplasm of chromatophores.

that all the cells of the epidermis would contain pigment since they are derived from the stratum germinativum. This, however, is not the case, for much of the pigment is lost as the cells approach the surface. When a water blister forms in the skin, the fluid collects between the outer and inner layers of the epidermis. The skin covering a blister in a Negro as well as in a white man is practically devoid of pigment. With the exception of certain white animals and albinos, the integument of all vertebrates bears pigment granules. In true albinos there is a total lack of pigment in all parts of the body.

Wide variations in skin color exist among races of man and even among individuals of the same race. The color of the human skin actually depends upon three varying factors: its basic color is yellowish; the blood vessels in the skin give it a reddish hue; the pigment melanin in different concentrations causes various shades of brown. The actual amount of pigment present is relatively small. It has been estimated that there is about 1 g of pigment in the entire skin of an average Negro.

A contested point but one of considerable interest is that of the relationship between the degree of human pigmentation and the degree of intensity of the sun's ultraviolet rays in different regions of the earth. In general, the Negroid races nearest the equator are the darkest. In the area about the Mediterranean Sea the dark Spanish and Italian races are to be found, while the regions farther north are inhabited by the lighter Germanic and Scandinavian races with the least degree of pigmentation. There are many exceptions, however, to the general rule. The Eskimo, for example, with dark skin and hair inhabits the most northern regions. The discrepancy may perhaps be attributed to migration from a more southerly clime. At any rate, the degree of pigmentation seems to be associated with protection from the sun's rays. Formation of pigment by melanocytes is stimulated by mild exposure to ultraviolet light. In the lighter-colored races this is commonly known as "tanning." A dense pigmentation gives its possessor a certain advantage in his power of enduring the heat of the sun. It is probable that a heavy layer of pigment in the stratum germinativum of the epidermis prevents the actinic rays from reaching the capillaries nearest the surface. These may then expand naturally so that a relatively large amount of blood is brought to the surface of the body from which excess heat is given off by radiation.

In cases of sunburn the ultraviolet rays apparently kill the outer cells of the skin. The cells are believed to release a metabolic substance which has the effect of dilating the the skin capillaries. Hence the sunburned area becomes flushed. Evidence that the local reaction spreads beyond the limits of skin exposed to light gives conclusive support to the view that a vasodilator substance is released. In cases of severe sunburn, blisters form in regions where the capillaries are excessively dilated. The fluid contained within the blister is exuded by the capillaries. Leukocytes may gather and give the affected area a yellow color. The dead skin is finally sloughed.

The difference between becoming sunburned and tanned should be apparent. Two entirely different processes are involved. In sunburn a severe, rapid, and harmful action occurs; whereas in tanning, which is at best a slow process, there is but mild or short exposure and the skin is stimulated to form pigment which acts as a protection from too great exposure to the sun's rays.

The degree of pigmentation in various parts of the body of an individual also shows variation. In the darker races the back and dorsal regions of the arms and legs are darker than the ventral aspects. The palms and soles in particular are lighter in color. Such a distribution of pigment might be correlated with exposure to light if man were to assume the position taken by most tetrapods. On the other hand, regions such as the axillae, nipples, external genitalia, and groin, as well as the circumanal area, all show a greater degree of pigmentation than do other parts of the body. A pigmented area appears in the skin over the sacral region in young babies of the Mongol race (page 81). No satisfactory explanation has been offered which throws any light on the function of pigment in these regions. It scarcely seems plausible that it may be correlated with exposure to the sun's rays.

In human beings the skin usually tans rather evenly. In some individuals, however, pigment is stimulated to form in small patches called *freckles*. A circumscribed pigmented spot or area in the skin of man is often referred to as a *mole* or *pigmented nevus*. Pigmented tumors of the skin are very common. They are usually harmless but may occasionally become malignant and very dangerous.

In certain apes the skin appears to be blue in one area or another. The blue coloration is related to melanin pigment lying deep in the dermis. The fact that the color is observed through a greater thickness of overlying tissue accounts for the bluish appearance. A certain type of skin tumor has a bluish color for the same reason.

Generally speaking, the skin is pigmented in those vertebrates which lack accessory integumentary structures such as hair or feathers. In those in which the body is thickly covered, the accessory structures receive the pigment and the integument proper has a scant supply.

Among vertebrates of all classes innumerable types of pigmentation patterns are evident. These are determined by (1) the genetic endowment of the neural-crest cells which give rise to melanoblasts; (2) the migration of melanoblasts to various parts of the body; (3) the area, or environment, in which these cells ultimately become located; (4) the responses to endocrine secretions of the gonads and thyroid gland; (5) the effects exerted by melanoblasts upon each other.

Comparative anatomy of the integument proper. AMPHIOXUS. The skin of amphioxus represents chordate integument reduced to its simplest form. The epidermis consists only of one layer of columnar epithelial cells which secrete a thin, noncellular cuticle perforated by minute pores. The dermis is thin and made up of soft connective tissue. Pigment is lacking.

CYCLOSTOMES. Cyclostome integument shows a greater degree of development than that of amphioxus. The epidermis is made up of several layers of cells. There is no dead stratum corneum, since the outer cells have nuclei, are living, and are active enough to secrete a thin cuticular covering. The dermis is even thinner than the epidermis and consists merely of a meshwork of connective-tissue fibers intermingled with blood vessels, nerves, and smooth-muscle fibers. Horizontal bands of fibrous tissue are interrupted at definite intervals by vertical strands, the *myocommata,* which lie between successive myotomes. The color of the skin is due to the presence of chromatophores which, as in fishes, are located in the dermis.

FISHES. In fishes the integument shows little if any change over the condition in cyclostomes. There is no dead corneal layer. However, in certain teleosts during the breeding season, so-called "pearl organs" appear on the skin. These are cornified areas of the epidermis. The dermis consists of connective tissue, nerves, blood vessels, and smooth-muscle fibers and shows some degree of stratification. Scales are embedded in the dermis.

The color of fishes is due primarily to the presence of chromatophores in the dermis. These pigment cells sometimes wander into the epidermis. Crystals of guanin located in cells called *iridiophores,* or *guanophores,* associated with scales, are responsible for the iridescence and reflection so characteristic of the integument of many fishes. The chromatophores of lower forms, being derived from neural crests, are of ectodermal origin. They contain pigments of various colors. *Melanophores* contain black or brown pigment; *erythrophores,* red pigment; *xanthophores,* yellow pigment. Other chromatophores may contain granules of an orange color. The red, yellow, and orange types are sometimes referred to collectively as *lipophores.* Chromatophores have many irregular, branching processes. When the pigment granules are dispersed throughout the cell it displays the greatest amount of color (Fig. 4.5C). When the granules become concentrated about the nucleus only a small spot of color is visible. Various colors may be produced by combinations and blending of chromatophores bearing different kinds of pigment granules and by various degrees of dispersion of the granules. The background color of pigment in the cells of the stratum germinativum is also of importance in producing any effect. The ability to change color is known as *metachrosis.* Changes are probably mediated, for the most part, by impulses reaching the brain from the eyes. Two methods of control have been discovered, nervous and endocrine. In elasmobranch fishes, amphibians, and most reptiles, epinephrine from the adrenal gland brings about a concentration of pigment, and intermedin from the intermediate lobe of the pituitary gland brings about its dispersion. In experiments on melanophores on the scales of the killifish, *Fundulus,* NaCl induces dispersion of pigment, whereas KCl causes it to aggregate. Many teleost fishes, and chameleons, among lizards, evidence a nervous control of chromatophore-pigment dispersion and concentration via the autonomic nervous system. The nerve endings presumably give off certain chemicals such as epinephrine or norepinephrine which actually bring about the reaction.

AMPHIBIANS. The epidermis of amphibians (Fig. 4.6) is composed of several layers of cells and is the first to have a dead stratum corneum. The stratum corneum is best developed in those amphibians which customarily spend considerable time on land.

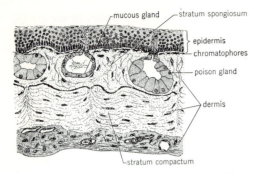

Fig. 4.6 Section through integument of frog, *Rana pipiens*. Note upper loose layer (stratum spongiosum) and lower compact layer (stratum compactum) of dermis.

The dead corneal layer is an adaptation to terrestrial life. It not only aids in protecting the body but helps prevent loss of excessive amounts of moisture. When ecdysis occurs, the corneal layer may be shed as a whole or come off in fragments. In the larval or tadpole stage, the epidermis is unusual in that even prior to hatching it is made up of two layers of cells. In the earliest stages the epidermis is ciliated.

A so-called "basement membrane" separating epidermis and dermis is very evident in amphibians and has been the subject of numerous electron microscopic studies. It is a fairly complicated structure which has no counterpart in the mammalian skin.

The dermis is relatively thin in amphibians. It is composed of two layers: an outer, looser layer, the *stratum spongiosum,* and an inner, more compact layer, the *stratum compactum.* The outer part of the stratum spongiosum is often more compact than the remainder of this layer. Blood vessels, lymph spaces, nerves, and glands are abundant in the stratum spongiosum. The amphibian skin is an important organ of respiration, and the dermis is supplied with an unusually large number of blood vessels.

The ability of the integument to change color so as to blend in with the environment is related to the chromatophores which are present, for the most part, between epidermis and dermis.

The dorsal integument of *Rana pipiens* larvae contains two types of light-reflecting iridiophores located in the two regions of the dermis. One type lies above the dermal melanophores, the other below. Reflecting plates within the cells are composed of crystals of guanine, adenine, and hypoxanthine. Unlike fishes, there seems to be a lack of nervous control of chromatophores in amphibians. Such factors as changes in temperature and light, affecting the adrenal glands and intermediate lobe of the pituitary gland, seem to be responsible for color changes.

REPTILES. The epidermis of reptiles is characterized by the extremely well-developed corneal layer and is admirably adapted to life on land. The scales of reptiles are derived from the horny stratum corneum. Ecdysis occurs at intervals and results in the removal of dead outer layers.

The dermis consists of superficial and deep layers. In many snakes and lizards the superficial layer has an abundance of chromatophores, but these are not particularly conspicuous in other reptiles. In chameleons, as previously mentioned, the chromatophores are under nervous control; otherwise, the slower-acting endocrine mechanism seems to prevail in snakes and lizards. The deep layer of the dermis is composed mostly of bundles of connective tissue. Leather can be prepared from the skins of certain reptiles. Some reptilian leathers have a high commercial value.

BIRDS. The epidermis of birds is thin and delicate except in such exposed regions as the legs and feet, where the thick, corneal layer forms protective scales. Feathers, of course, cover and protect the bodies of birds.

They, too, are modifications of the stratum corneum.

The dermis is also thin and made up mostly of interlacing connective-tissue fibers. Muscle fibers are particularly abundant, being used in raising and lowering the feathers. In a few birds, such as the ostrich, the integument is so thick that it can be prepared as leather.

Pigment in the skin of birds is generally confined to the beak or bill, feathers, and scales. No chromatophores of the type found in lower forms are present.

MAMMALS. The human skin, previously discussed, is typical of the mammalian integument, but detailed differences among various species are often striking. A major distinction is that the dermis of mammals is much thicker than the overlying epidermis.

STRUCTURES DERIVED FROM THE INTEGUMENT

Glands

One of the most important functions of epithelial tissues is the part played by them in the metabolism of the body. Absorption of certain external substances and the liberation of others may take place through the epithelium. Modifications of the epithelium in the form of glands aid in carrying on its secretory function. The many and varied types of glands found in the skin are not essentially different from those found in other parts of the body, but the stratum germinativum alone is concerned with their formation. All of them are *exocrine*, or externally secreting, glands which pour their secretions onto an epithelial surface either directly or through ducts of varying degrees of complexity. Although *endocrine*, or internally secreting, glands are usually derived from epithelial surfaces, none is associated with the integument except as it may be derived from ectodermal epithelium in the embryo.

Structure

According to structure, the exocrine glands of the skin are of two general types, *unicellular* or *multicellular*, depending upon whether they consist of isolated units or numbers of similar cells joined together to form the glandular element.

Unicellular glands. The simplest of all glands found in the skin are unicellular glands, which may be scattered as modified single cells among other epithelial cells covering the body in amphioxus, cyclostomes, fishes, and amphibian larvae. These glandular cells are, for the most part, known as *mucous*, or *goblet, cells*. They secrete a substance known as *mucin*, which is protein in nature. Mucin together with water forms a slimy, viscid material called *mucus*. When a unicellular gland of the goblet type is secreting, it contains a mass of mucin somewhat resembling a goblet in shape, with its wide end near the surface and a more or less narrow stalk at its base. The nucleus is generally located at the basal end of the cell. Mucin protrudes through an opening on the free surface. In some cases the elimination of mucin is gradual, and it retains its goblet-like shape. In others, the mass of mucin is liberated and the cell collapses, only to fill up gradually again with secretion, after which the process is repeated. Goblet cells may pass through a number of such phases of secretory activity but finally die and are shed. In the meantime, new goblet cells have been formed. The slippery mucus serves to lubricate the surface of the body, thus lessening the degree of friction with surrounding water. Other types of unicellular glands include *granular gland cells*, *thread cells*, and large *beaker cells* of cyclostomes and fishes. A beaker cell may extend from the lowest layer of epidermis to the surface. The secretion of the unicellular glands of an amphibian larva digests the egg capsule and frees the embryo.

Unicellular glands do not occur in the integuments of reptiles, birds, or mammals. They are numerous, however, in such places as the lining of the intestine and the tracheal lining. It has been demonstrated that the free surfaces of such goblet cells are covered with microvilli (Fig. 3.16) but that these disappear during the time the cell is giving off its secretion.

Multicellular glands. The multicellular glands of the skin are formed by ingrowths of the stratum germinativum into the dermis. In some cases the ingrowth may be a hollow structure from the beginning, but in others it is a solid structure in which a lumen appears later. As development progresses, the mass of cells may give off side shoots, resulting in the formation of compound glands. The branches in most cases, therefore, connect to a single duct which opens onto the outer surface. The glandular portions push deeper down into the dermis where their actively secreting cells are nourished by the blood vessels so abundant in that region. Multicellular glands are classified according to shape.

TUBULAR GLANDS. The tubular gland, as one would infer from its name, is a tube of practically uniform diameter without any bulblike expansion at its end.

Simple tubular glands. Glands of the simple tubular type are short, blind tubes lying partly in the dermis but extending to the outer epithelial surface (Fig. 4.7A). The skin of *Mormyrus,* a fresh-water fish inhabiting some of the larger streams of northern and central Africa, contains such glands. Certain specialized glandular areas in amphibians, such as the swollen thumb pads of male anu-

rans and the *mental glands* of male plethodontid salamanders, appear to be masses of simple tubular glands. The *glands of Moll,* which are found on the margin of the human eyelid, might be considered to be of this type. They apparently are modified sweat glands as are the ceruminous, or wax-producing, glands of the external ear passage.

Simple coiled tubular glands. Sweat glands are typical examples of the simple coiled tubular type of integumentary gland and are present only in the mammalian skin. Each gland consists of a long narrow tube, the distal end of which is coiled into a small ball that lies in the dermis (Fig. 4.7B). The proximal end serves as a duct which courses through the epidermis and opens onto the surface of the skin by a slightly widened aperture. The openings are sometimes referred to as the "pores" of the skin. The glandular epithelium itself is composed of a single layer of cuboidal or columnar cells.

Simple branched tubular glands. The duct of the simple branched tubular gland, which leads to the surface, divides at its lower end into two or more branches (Fig. 4.7C). They may or may not be coiled in their terminal portions. Some of the large sweat glands in the axillae, or armpits, are of this type.

Compound tubular glands. Compound tubular glands consist of a varying number of tubules, the excretory ducts of which unite to form tubules of a higher order. These tubules in turn combine to form others of a still higher order, etc. (Fig. 4.7D). The unit structure of the compound tubular gland is like a simple tubular gland. Mammary glands of monotremes are compound tubular glands. Except for these it is doubtful if any other integumentary glands of this type exist in vertebrates, although several glands associated with internal organs are classified as compound tubular glands.

Anatomy of the Chordates

SACCULAR GLANDS. The saccular, aci-
nous, or alveolar gland differs from the tubular
glands in that there is a spherical expansion
at the terminal portion of a tubular duct. The
actual secretory cells are in the expanded
portion, which lies in the dermis. The duct
serves as a passageway by means of which
the secretion is discharged onto the epithelial surface. There are three general types
of saccular glands.

Simple saccular glands. If only one expanded bulb, or acinus, is at the end of a duct,
the gland is of the simple saccular type (Fig.
4.7E). Numerous simple saccular mucous
and poison glands are found in the skin of
amphibians.

Simple branched saccular glands. If several acini are arranged along a single excretory duct, as in the *tarsal,* or *Meibomian,*
glands of the eyelid, or if a single acinus is
divided by partitions into several smaller
acini (Fig. 4.7F and G), as in the sebaceous
or oil glands of the skin, the gland is said to be
of the simple branched saccular type.

Compound saccular glands. Compound
saccular glands consist of several portions
called *lobules.* The smallest unit of the lobule
corresponds to a simple saccular gland. Several of these unit structures unite to form a
common duct. This in turn unites with similar
ducts of other lobules to discharge the secretion through a main duct which opens on the
surface (Fig. 4.7H). The mammary glands of
metatherians and eutherians are examples of
such glands. Each lobe of the mammary
gland is a compound saccular gland.

Method of secretion

In addition to the classification of glands
based on structural characteristics, glands
are sometimes classified according to their
manner of secreting.

Merocrine glands include those in which
the glandular cell bodies are not injured or

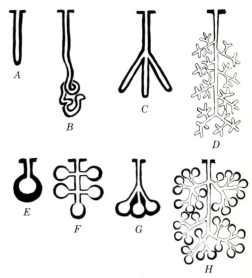

Fig. 4.7 Diagrams representing various types
of integumentary glands, all of epidermal
origin: *A,* simple tubular gland; *B,* simple
coiled tubular gland; *C,* simple branched tubular
gland; *D,* compound tubular gland; *E,* simple
saccular gland; *F* and *G,* two types of simple
branched saccular glands; *H,* compound saccular
gland.

destroyed during elaboration of the secretion. Unicellular integumentary glands are
merocrine glands as are sweat glands of the
eccrine type (see page 119). Among nonintegumentary glands which secrete in this
manner are the salivary glands and the
pancreas.

Holocrine glands are those in which there
is an accumulation of secretion within the
cell bodies. The cells die and are discharged
with their contained secretion. New cells
are constantly being produced, so that the
process is continuous. Sebaceous glands of
the skin are of the holocrine type. The term
necrobiotic is frequently applied to holocrine
glands.

Apocrine glands are those in which the
secretion gathers at the outer ends of the

glandular cells. The accumulated secretion is pinched off with a portion of the cytoplasm. Most of the cytoplasm, however, as well as the nucleus, remains unchanged, and after a time the process is repeated. Certain sweat glands, to be described later (page 120), are reported to be of the apocrine type, as are mammary glands. Observations with the electron microscope seem to indicate that secretion droplets of so-called apocrine glands during their discharge become surrounded by a membrane that is in continuity with the cell membrane. Further detailed studies may indicate that certain glands, long regarded as apocrine glands, may belong to an entirely different category.

Type of secretion

Glands are also sometimes classified according to the type of secretion which they produce.

Mucous glands secrete mucin. The unicellular glands of cyclostomes and the simple saccular glands in the integuments of fishes and amphibians are examples.

Serous glands form a thin, watery secretion. Sweat glands are examples.

Mixed glands secrete a mixture of mucous and serous fluids. The submaxillary salivary glands are nonintegumentary mixed glands. It is doubtful whether any skin glands are of this type.

Fat, or oil, glands are exemplified by sebaceous glands of the skin and Meibomian glands of the eyelids, which are considered to be modified sebaceous glands.

Comparative anatomy of integumentary glands

Amphioxus. Only unicellular mucous glands of the merocrine type are present in the integument of amphioxus. They are mostly goblet cells and are scattered among the other columnar epithelial cells which cover the body.

Cyclostomes. Goblet cells, granular gland cells, and beaker cells, all of which are unicellular merocrine glands, are the only integumentary glands found in cyclostomes. The mucus secreted by them renders the surface of the body extremely slippery. In the hagfish the skin may give off an almost incredible amount of mucus. The hagfish integument also possesses pockets of so-called *thread cells*. The protoplasm in each of these cells secretes a spirally coiled thread which, under the proper stimulus, is shot out and unwound, often to a considerable distance. The threads are added to the other mucous secretion so that a protective layer is formed about the animal.

Fishes. The skin of fishes is particularly abundant in unicellular and multicellular mucous glands. The unicellular glands are like those of cyclostomes, whereas the multicellular glands are of the simple saccular type.

An unusual adaptation in which mucous integumentary glands play an important part is to be found in the South American and African lungfishes *Lepidosiren* and *Protopterus*. During the dry season these fishes bury themselves in the mud of dried-up river beds. The abundant secretion of mucous glands of the skin forms a sort of "cocoon," or capsule, about the fish in which it lies during estivation.

A few elasmobranch and teleost fishes have multicellular poison glands derived from the epidermis, which are used in a protective manner. In one species, *Synancia*, from the Indian Ocean, the poison organ is well developed. The terminal half of each dorsal spine is provided with a deep groove on either side. A sac containing the poison lies at the lower end of each groove. A duct from the sac extends to the top of the spine and runs along the groove. Fishermen accustomed to these fish are adept at handling them without injury. The fish generally lies hidden in the sand. Not infrequently a person wading with bare feet will step upon the spines and by causing pressure on the poison sac will inject the poisonous secretion into the wound. The common catfish has a poison gland opening at the base of a spine on each pectoral fin. The spine, when erected, may inflict painful wounds. Other fishes have similar glands associated with spines on the operculum. The spines are often hollow tubes with a poison sac opening at the base of each. They function in a manner analogous to a hypodermic needle. Erection of the spine brings pressure to bear on the gland, causing the poison to be ejected. This is, perhaps, the first example where the vertebrate skin is modified as a special device for protection from enemies.

Some of the deep-sea elasmobranchs and teleosts, living in almost total darkness, have luminous phosphorescent organs in the skin. The arrangement of these organs varies considerably, but most frequently they lie in longitudinal lines near the ventral surface of the body. The luminous organs, or *photophores,* are modifications of integumentary glands. Some are quite complicated in structure. In one particular species, *Porichthys,* in which development of photophores has been studied, certain cells bud off the stratum germinativum and come to lie in the dermis. Here they become differentiated into a lower glandular layer and an upper layer which forms a lenslike structure. A reflecting layer surrounded by pigment forms below the glandular layer. The glandular cells give off phosphorescent light, and the other structures serve to transmit it to the outside (Fig. 4.8). In many cases the "cold" light is produced by oxidation of a protein, *luciferin,* by

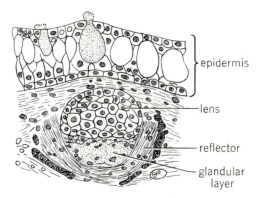

Fig. 4.8 Section through luminous photophore of the fish, *Porichthys. (Redrawn from Kingsley, after Greene.)*

an enzyme, *luciferase.* In others, phosphorescent bacteria, living symbiotically in the organ, furnish the course of light. It is possible that the light emitted by luminous organs of this sort may help certain deep-sea fishes to attract their prey. It is more probable, however, that photophores are of survival value in that nearsighted predators, looking upward from the waters below, may not see a sharply silhouetted figure emitting bright points of light. Instead, the light points may appear to be blended together with the result that the image is blurred or fuzzy and, therefore, the object seen is less likely to be regarded as a possible source of food (W. B. Clarke, 1963, *Nature,* **198**:1244). In a certain fish, *Astronethes,* a spherical, luminous organ in the skin posterior to the eye is provided with a muscle which enables it to rotate, and at the same time the light is dimmed.

Still another type of integumentary gland found in fishes is the *pterygopodial gland* located at the bases of the clasping organs of certain species of skates and rays. Pterygopodial glands are multicellular mucous glands, the function of which is little understood.

Amphibians. Since the skin of many amphibians is an important respiratory organ, it must be kept moist so that exchange of gases may take place between the dermal capillaries and the environment. In these forms the large numbers of integumentary glands that are present aid in keeping the skin moist. The glands are usually simple saccular structures, the bulblike expansions of which lie in the stratum spongiosum of the dermis. Because of differences in structure and secretion, two varieties are recognized: mucous and poison glands (Fig. 4.6).

Mucous glands are generally more numerous and smaller than poison glands. They are found all over the surface of the body but are more abundant in some parts than in others. In one species of frog, *Rana fusca,* there may be as many as 60 to each square millimeter of surface. Around each gland is a coat of smooth-muscle fibers which aids in expelling mucin from the cavity of the gland.

Poison glands are larger but not so numerous as mucous glands and are not so evenly distributed over the surface of the skin. In many species they are more abundant on the dorsal side of the body and hind legs. The so-called "warts" of toads and the *parotoid glands* which lie on the neck are actually masses of poison glands. The secretion of these poison glands appears as a whitish fluid which is said to have a burning taste. It may be secured by compressing the glands in the skin of a toad or by exciting the animal and thus causing a natural exudation of poison. When given to a dog the poison brings about trembling, weakness of legs, and nausea. A large dose may cause death, acting upon the heart and nervous system. A small quantity placed on the tongue causes a great stimulation of the salivary glands and induces nausea. A young dog that catches his first toad quickly learns to leave toads alone. A tree frog, *Hyla vasta,* which is found in Haiti,

gives off a poisonous secretion so strong that when the animal is handled the poison causes inflammation of the skin. The poison glands in the skin are thus an adaptation for protection. Many amphibian collectors have had the experience of finding all their specimens dead after a day of collecting, with the exception of perhaps one species. The poison of the one species has a lethal effect on the others, but apparently members of its own kind are immune. Stories of getting warts from handling toads are fictitious. Poison glands, in general, are more abundant in toads than in frogs.

Some specialized integumentary glands in amphibians have a tubular structure. They include those on the feet of certain tree-dwelling frogs and toads. Suctorial discs on the toes aid in climbing. The swollen glandular thumb pads of male frogs and toads during the breeding season aid them in clasping the females in amplexus. The males of some plethodontid salamanders develop a glandular area, the *mental gland,* on the chin as the breeding season approaches (Fig. 4.9). Its secretion seems to attract the female. Unicellular glands also occur in amphibians, being present on the snouts of tadpoles and urodele larvae. Their secretion is believed to have digestive properties, thus aiding in freeing the larva from the egg capsule during early development. Large *glands of Leydig* are unicellular glands of uncertain function in the epidermis of some larval urodeles.

Reptiles. Reptiles, contrary to amphibians, have rough and scaly skins which are primarily adapted for life on land. Correlated with this is an almost total lack of integumentary glands. This paucity of glands in the skin is an adaptation which prevents any unnecessary evaporation of water from the body and accounts for the fact that reptiles

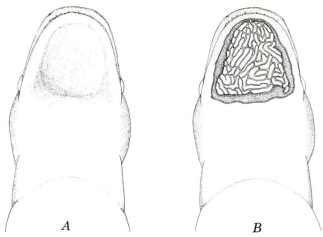

Fig. 4.9 Mental gland of male salamander, *Eurycea bislineata*, as it appears prior to and during the breeding season: *A*, ventral view of head showing position of glandular area; *B*, the same as *A*, with the skin below the gland removed to expose the tubular nature of the gland.

can generally live for a considerable time without water. In many species the water contained in the food is enough to give them an adequate supply. Many snakes, however, contrary to popular opinion, do drink water when it is available. The integumentary glands of reptiles, therefore, are fewer in number, and those which are present are scent glands, for the most part associated with sexual activity.

Crocodilians have two pairs of skin glands which secrete musk. One pair is located on the throat on the inner halves of the lower jaw. A slitlike external opening leads to a sac which, in large specimens, is about the size of a walnut. The sac contains a pale, brownish substance which is concentrated essence of musk, much prized by the natives of certain localities. These glands secrete most actively during the breeding season. A second pair of glands of the same type lies just within the cloacal aperture. They are possessed by both sexes and are probably

hedonic in function. In crocodilians a row of glands runs down either side of the back between the first and second rows of dermal plates. If any function is attached to these glands, it is unknown.

The skin of lizards is practically devoid of glands. So-called *femoral glands* are found in males on the undersurfaces of the thighs. Each opens through a short duct which passes through a conical projection of the skin. During the breeding season they give off an abundant yellowish secretion. Their function is undoubtedly hedonic. The spiny projections apparently serve as an aid in copulation.

Some snakes have glands in the cloacal region which seem to be used in defense. They secrete a milky fluid which has a nauseating odor. The fluid cannot be projected but has a high protective value, as anyone who has handled snakes will agree.

In certain turtles musk glands are present beneath the lower jaw and also along the

Fig. 4.10 Section through uropygial gland of bird. (*After Schuhmacher.*)

line of junction of plastron and carapace. They are probably used in sexual allurement.

Birds. As we shall learn later, birds have many characteristics resembling those of reptiles. Like the reptilian integument, the skin of birds is practically devoid of glands. Usually the only gland that is present in the skin of birds is the *uropygial gland,* a simple branched saccular oil gland located on the dorsal side of the body at the base of the *uropygium,* or tail rudiment. In the chicken it is made up of two lobes which in most cases have separate openings (Fig. 4.10). A septum divides the gland into two halves. The oil collects in a central cavity. The bird, by squeezing out a quantity of oil on its beak, or bill, oils its feathers as it preens them, making them quite impervious to water. The uropygial gland is best developed in aquatic birds. In some, several ducts lead from the gland. It is said that if a duck is washed thoroughly with soap and water so as to remove the oil, the animal will sink when it takes to water.

Not all birds possess a uropygial gland,

it being notably absent in the Paleognathae and in a few other species such as parrots and some varieties of pigeons.

The only other skin glands found in birds are certain modified oil glands in the region of the external ear opening. These are found in only a few gallinaceous birds such as the American turkey.

Mammals. The skin of mammals is particularly abundant in glands, and a considerable variety is found. There are two essential types from which all are probably derived, sebaceous glands and sweat glands.

SEBACEOUS GLANDS. Sebaceous, or oil, glands are distributed over the greater part of the surface of the skin, being absent, however, from the palms and soles. With few exceptions, the duct of the gland opens into a hair follicle. Often several glands are associated with a single hair follicle. The oily secretion given off by these glands serves to keep the hair and skin smooth and soft and also imparts to the animal an individual scent, or odor. Sebaceous glands are present, without being connected to hairs, on the corners of the mouth and lips, the glans penis, internal surface of the prepuce, labia minora, and mammary papillae. These glands, although associated with sexual activity, should not be confused with primary sex glands. It has been reported that the activity of sebaceous glands in man is regulated by androgenic hormones (see page 368). The Meibomian, or tarsal, glands of the eyelids also have no connection with hairs, although it seems probable that at one time they were associated with them. They are sebaceous glands with fairly long, straight ducts into which the separate alveoli open. There are about 30 Meibomian glands in the upper eyelid of man and about 20 in the lower. Their oily secretion forms a film over the layer of lacrimal fluid which is thus held evenly over the surface of the eyeball. The film also prevents tears from overflowing onto the cheeks under normal conditions. Small *glands of Zeis* on the eyelids are sebaceous glands which open into the hair follicles of the eyelashes. Infection of one of the glands of Zeis results in a *sty,* or *hordeolum.* Secretion of sebaceous glands during fetal life contributes to the *vernix caseosa* (page 139), the white, cheesy material found on the surface of the skin of older human fetuses. Sebaceous glands are lacking in whales and porpoises.

SWEAT GLANDS. Sweat glands, or sudoriparous glands, are found only in mammals. According to structure they are either simple coiled tubular or simple branched tubular glands. Several mammals have no sweat glands. Among these are the spiny anteater, moles, sirenians, cetaceans, the scaly anteater, and certain edentates such as the two-toed sloth.

Sweat glands in man are abundantly distributed all over the surface of the body except on the borders of the lips, the eardrum, the glans penis, the glans clitoridis, and the nail bed. They have two very important functions: getting rid of metabolic wastes of various kinds, and helping to maintain a constant body temperature. In hot weather or after strenuous exercise when the body temperature tends to be higher than normal, the sweat glands pour out their secretion onto the surface of the skin. The cooling effect of evaporation of sweat is important in maintaining a constant body temperature. As much as 2 or 3 liters of sweat and even more may be given off in a 24-hour period.

Deficiency or absence of sweat glands occurs occasionally in man. Several individuals lacking sweat glands have been reported in the state of Mississippi. Frequent plunges in cool water during hot weather enable them to keep their body temperature within normal range.

According to their manner of secreting, sweat glands have been shown to be of two general types: *eccrine* and *apocrine.* Eccrine sweat glands are actually merocrine glands in which the cells involved in secreting are not injured or destroyed during the process of secretion. In apocrine sweat glands portions of the cytoplasm containing secretory products protrude from the cell surface. They are believed to separate from the rest of the cell and form part of the secretion.

Eccrine sweat glands are best developed in primates, especially in man, in whom they are almost unique. Apocrine sweat glands of one type or another are found in most mammals, but in man are restricted to certain areas and are relatively few in number as compared with the eccrine variety.

Eccrine sweat glands. Eccrine sweat glands are simple coiled tubular glands. Although restricted in general to catarrhine primates, in other mammals a few may be present on those surfaces of the feet or paws which make contact with the ground. Among primates the chimpanzee and gorilla are the only forms other than man in which eccrine sweat glands exceed the apocrine type in number. Lemurs and platyrrhine primates are reported to possess only apocrine sweat glands.

It has been estimated that in an average human being there are about 2.5 million sweat glands in the skin. Accurate counts indicate an average distribution of 143 to 339 glands per square centimeter of the surface. They are most numerous in regions devoid of hair or where the hairy covering is scant. On the palms and soles they are particularly abundant, their ducts opening onto the surface of the skin along the tops of the small ridges rather than through the furrows between ridges (Fig. 4.11). According to Krause, there are about 370 sweat glands per square centimeter on the palms of the hands and a slightly

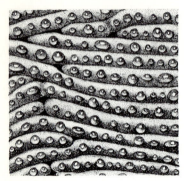

Fig. 4.11 Droplets of sweat accumulating at the openings of sweat glands on friction ridges of the palm of the hand (highly magnified).

The coiled portion of an eccrine sweat gland, which averages about 0.3 to 0.4 mm in diameter, lies deep in the dermis. The duct is about 3 mm long and 0.06 mm wide. Some spindle-shaped *myoepithelial cells* are located around the coiled secretory portion. They are supposedly modified smooth-muscle cells of ectodermal origin which may aid in the expulsion of the secretion (see page 68).

The activity of eccrine sweat glands varies considerably. Those in various areas respond differently to thermal stimulation. The sweat glands on the palms and soles respond least to temperature variations but are the first to exhibit activity when sensory and psychic stimuli are applied. The term "cold sweat" is sometimes used to characterize the sweating on the palms and soles when no thermal stimuli are involved.

Apocrine sweat glands. Apocrine sweat glands are of many types and are to be found in most mammals. In man they are confined to certain areas, such as the axillae or armpits, the pubic region, circumanal area, vicinity of the nipples, inner surface of the prepuce or foreskin of the penis, and on the labia minora of the female external genitalia. A few scattered apocrine sweat glands may be found in other areas. Most apocrine sweat glands are simple coiled tubular glands, but those in the axillae may show some degree of branching and their coiling is less pronounced. These, as well as those in the circumanal area, are unusually large and may measure as much as 3 to 5 mm in diameter. They are reddish in color and lie rather deep in the subcutaneous tissue. The myoepithelial cells associated with them are larger and more numerous than those of eccrine sweat glands. Apocrine sweat glands in man usually open into hair follicles above the opening of the duct of the sebaceous gland. In lower primates they open directly onto the surface. It has been reported that these glands do not se-

smaller number on a corresponding area of the soles of the feet. Krause has made the following estimates for the number of sweat glands in various regions of the body: 200 per square centimeter on the back of the hand; 175 per square centimeter on the forehead, front and sides of the neck; 155 per square centimeter on the chest, abdomen, and forearm; 60 to 80 on the legs, back of the neck, and trunk. Other investigators report larger numbers than does Krause in the various regions mentioned above. Generally speaking, the extensor surfaces of the appendages have fewer sweat glands than do flexor surfaces; the arms have many more than the legs; the chest and abdominal regions have larger numbers than the back. The glans penis, the inner surface of the prepuce, the clitoris, and the labia minora, lack eccrine sweat glands. It is doubtful whether the various races of man differ significantly in the numbers of sweat glands present in the skin or in their distribution. Eccrine glands are formed during fetal life and new ones do not develop after birth. The skin of an infant, therefore, has many more glands per comparable unit area than does an adult.

crete continuously even upon stimulation, there being an inactive period of 24 hours or more between active periods. The secretion of apocrine sweat glands of this type is relatively thick and contains fat droplets as well as pigment granules. This explains why sweat in certain parts of the body such as the axillary region may appear stained. Bacterial action upon fatty and other material secreted by these glands imparts a characteristic odor to the axilla when cleanliness is not practiced.

There is some evidence that the apocrine sweat glands of the axilla in women undergo periodic changes in correlation with the menstrual cycle. An enlargement of the cells and lumina of the glands is reported to occur during the premenstrual period. During the menstrual period itself regressive changes occur. Changes are also reported to take place during pregnancy. Recent investigations cast some doubt on the authenticity of these reports.

Pigment granules incorporated with the secretion accounts for the fact that the sweat of a certain South African antelope, *Cephalophus,* is of a bluish cast. That of the hippopotamus, in the head and back regions, has a reddish color and is rather albuminous in consistency. It has been reported that the gorgeous color of the fur of the male red kangaroo, *Macropus rufus,* is caused by red pigment granules in the sweat which dries on the hair.

The horse is supplied with an abundance of exceptionally large apocrine sweat glands. In the bear and hippopotamus, also, apocrine sweat glands are present all over the body. Many textbooks and articles state that sweat glands are lacking in the hairy skin of the dog. On the contrary, large numbers of apocrine sweat glands are present. Experiments have shown that the sweat glands in the hairy part of the dog's skin do not play an important part in regulating body temperature but serve

chiefly in protecting the skin from an excessive rise in temperature (T. Aoki and M. Wada, 1951, *Science,* **114**:123). Many mammals have a limited number of sweat glands confined to certain localities. Sweat glands in the platypus are present in the snout region. In deer they are arranged about the base of the tail. In mice, rats, and cats they are present on the undersides of the paws. Rabbits have them around the lips. The moisture on the muzzles of sheep, goats, cattle, pigs, dogs, and others is due to the secretion of sweat glands. It is improbable that the presence of sweat glands in such restricted areas is related to the temperature-regulating mechanism of the body.

The *ceruminous,* or wax-producing, glands in the external ear passages are modified apocrine sweat glands having a peculiar arrangement and function. Their secretion, together with that of the sebaceous glands in the external ear passages, is known as *cerumen,* or *ear wax.* The glands are unusually large, their secretory portions branch, and their ducts, which also may branch, either open directly into the external ear passage or may join those of large sebaceous glands which open into hair follicles. Cerumen is brown in color, with a waxy consistency and a bitter taste. It protects the tympanic membrane and the skin of the ear passage from excessive drying and prevents the entry of insects.

Circumanal glands and the glands of Moll in the margins of the eyelids are also considered to be modified sweat glands which are larger than those of the ordinary type. In the latter, the terminal portions are not coiled but may show a slight degree of twisting. The ducts, which have rather wide lumina, may open directly on the skin surface or into the hair follicles of the eyelashes.

SCENT GLANDS. Scent glands are found in many mammals. Their structure indicates that they are either modified sweat glands or sebaceous glands, but the nature of their secretion is known in only a few cases. Scent glands of various types serve several functions. In some animals they are useful in attracting members of the same species. In other forms they are used in attracting members of the opposite sex and may be present only in the male. In skunks and some others, scent glands have a definite function in serving as a protection from enemies. It is possible that in certain other cases the scent may act as a lure or even to frighten foes away.

The location of scent glands varies greatly. In the deer family they are located on the head in the region of the eyes. Many carnivores have sebaceous scent glands near the anal opening. During the rutting season the male, by rubbing the anal glands on trees, stones, etc., marks out a territory in which he will permit no other males to encroach. Skunks and weasels have saclike scent glands which open into the rectum just inside the anus. They may be everted, and, since they are surrounded with muscle fibers, their foul-smelling secretion may be expressed. Scent glands may be located at the openings of the reproductive organs as in many rodents; on the face, as in bats; between the hoofs, as in pigs; at the base of the tail in dogs; over the temporal bone in the elephant; and in still other animals on the arms, legs, and other parts of the body. The callosities on the legs of horses and closely related forms are believed to be vestiges of former glandular areas.

MAMMARY GLANDS. The mammary, or milk-producing, glands which are present in all mammals and only in mammals are really modified sweat glands of the apocrine type, as indicated by their manner of development and their structure, particularly in some of the lower forms. The main portion of each gland lies in the subcutaneous tissue. The ducts open directly or indirectly onto the surface of the skin. The evolutionary origin of mammary glands as organs used to nourish the young is obscure.

Milk is essentially of the same composition in all forms of mammals. The percentages of the different constituents, however, vary in different species. Mammary glands are active only at certain times: immediately after the young are born and generally as long as active sucking continues. The development and functioning of the mammary glands are largely controlled by hormones secreted by the ovaries, the anterior and posterior lobes of the pituitary gland, and the adrenal cortex.

Although the details of development show variations in different species, nevertheless most mammary glands are built on the same general plan. The actively secreting mammary gland is made up of many small masses called *lobules*. Each lobule, in turn, consists of large numbers of *alveoli* which contain the actual secretory cells. The small ducts leading from the alveoli in a lobule gradually converge to form a larger duct. This generally unites with similar ducts from other lobules, and the common duct or ducts thus formed lead to the outside. In the inactive mammary gland the alveoli and lobules are reduced or wanting and the glandular tissue consists mainly of branching ducts. Since abundant adipose, or fatty, tissue usually surrounds the ducts and alveoli, this contributes much to the size of the *mammae,* or breasts.

In the Prototheria the mammary glands are of the compound tubular type. No nipples are present, and the glands open directly onto the surface of the skin. The mammary area, however, is depressed, and the milk is exuded onto this depressed area. The young animals, in lieu of nipples, grasp tufts of hair, which project from the depression, and thus obtain their nourishment

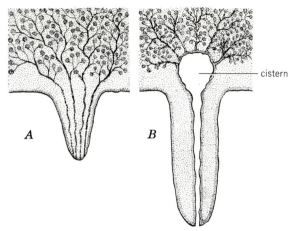

cistern

Fig. 4.12 Diagrammatic sections: *A*, through nipple; *B*, through teat.

by lapping, licking, or sucking. In the Metatheria and Eutheria the mammary ducts lead to nipples or teats which are grasped by the mouths of the young when sucking. Nevertheless, the mammary glands of prototherians and metatherians are structurally more similar than are those of metatherians and eutherians.

The distinction between a nipple and a teat is not always apparent on superficial examination (Fig. 4.12). A nipple is a raised area on the breast through which the mammary duct or ducts open directly to the outside. In some mammals, a single duct leads to the surface, as in certain rodents, marsupials, and insectivores. In other forms, several ducts may open on the nipple, as is the case in some carnivores and in man, in which as many as 20 separate ducts may be present. In the false nipple or teat, which is present in horses, cattle, and others, the skin of the mammary area grows outward to form a large projection. The mammary ducts open into a "cistern" at the base of the teat, and the milk is then carried by the secondary duct, or tube, to the surface.

Many laymen believe that the udder of the cow is a simple sac which fills with milk and must be drained periodically by milking. Nothing is further from the truth. The udder is made up of mammary and other tissues through which the mammary ducts ramify and converge at the bases of the teats (Fig. 4.13). If a milch cow which had been giving several quarts of milk per day were slaughtered, it is doubtful if more than a pint of milk could be obtained from her udder. It is the secretory activity of the cells in the living alveoli which forms milk from substances brought to them by the blood stream.

Although nipples, teats, and mammary tissue may be present in both sexes, functional mammary glands occur normally only in lactating females. In exceptional cases, functional mammary glands have been known to occur in males. Such a condition is referred to as *gynecomastia*. Although there have been reports that in certain species the male may actually aid in nursing the young, these have not been well authenticated. It is more probable that such cases as have been observed have been abnormal, inter-

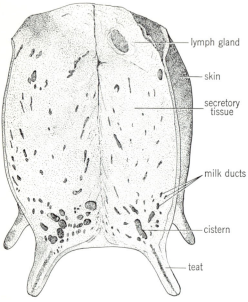

lymph gland

skin

secretory tissue

milk ducts

cistern

teat

Fig. 4.13 Cross section through the rear quarters of a cow's udder. (*From C. W. Turner, Univ. Missouri Agr. Exp. Sta. Research Bull. 211. By permission.*)

to normal. Nipples and teats are usually paired, their number being roughly proportional to the number of young delivered at birth. They vary from 1 pair in man and some others to 11 pairs in certain insectivores. Occasionally, in a particular species, extra nipples or teats may be present, a variation which harks back to primitive ancestral forms or to abnormal embryonic development in which more than the usual number of glandular areas develop in the mammary ridge. If more than the usual number of mammary glands or breasts occurs, the condition is referred to as *hypermastia,* or *polymastia.* Perhaps the most interesting case on record is that of a Polish woman who possessed eight functional breasts. If only supernumerary nipples are present, the term *hyperthelia* is applied. Hyperthelia is not uncommon in men, in whom extra nipples appear as small pigmented areas on the thoracic or abdominal walls.

The distribution of mammary glands varies greatly in different groups of mammals and is related to the habit of the mother when nursing her young. Animals such as dogs, cats, and pigs, which lie on their sides when the young are feeding, have mammary glands in two ventrolateral rows extending from axilla to groin. Horses and cows, which stand up when nursing, have their teats in a protected location between the hind legs. In elephants a single pair is located between the front legs, whereas in whales they are confined to the inguinal region. Primates, with their arboreal habits, have pectoral mammary glands, as do the manatees and dugongs of the order Sirenia. In human beings mammary glands occasionally are found in such unusual situations as the axillae, hips, thighs, or back. This has no real morphological significance and is due to dis-

sexual individuals. It is, however, not uncommon for milk to be present in the mammary glands of human males as well as females at birth and at puberty. The milk which is present in the mammary glands of the human infant at birth is sometimes called "witches' milk."

During early embryonic development in mammals an elevated band of ectoderm, called the *mammary ridge,* or *milk line,* appears on either side of the body, extending from the base of the pectoral limb bud to the pelvic limb bud. Depending upon the species, thickenings appear at various points along the mammary ridge, each of which develops into a separate mammary gland with its nipple or teat, as the case may be. The tissue between adjacent glands reverts placement rather than reversion to ancestral conditions.

Scales

In many vertebrates the body has a covering of scales, which gives it protection. Scales, in general, are of two types: epidermal and dermal. The structure and mode of development of these two types are very different, and it is important not to confuse them.

Epidermal scales

These are cornified derivatives of the stratum germinativum of the epidermis and are found primarily in terrestrial animals. A well-developed stratum corneum is characteristic of vertebrates living on land. Thus, few examples of epidermal scales are to be found in amphibians, but in reptiles, birds, and certain mammals they are very well developed. Epidermal scales, with few exceptions, are usually shed and replaced. Large epidermal scales, such as those on the shell of the turtle or the head of a snake, are often called *scutes*. The same term is frequently applied to large dermal plates or scales. The student should not infer from this that epidermal and dermal scales are in any way homologous.

Dermal scales

As their name implies, these scales are located in the dermis of the skin and are mesenchymal in origin. They are found, for the most part, in fishes, and take the form of small bony or calcareous plates which fit closely together or overlap. Both epidermal and dermal scales are present in certain reptiles and a few other forms. It is important to distinguish between the two. Dermal scales are remnants of the dermal skeleton (page 403).

Comparative anatomy of scales

Amphioxus. No scales of any kind are present in the skin of amphioxus.

Cyclostomes. The ancient ostracoderms were covered with bony dermal scales in the trunk and tail regions where freedom of movement for swimming was important. Heavy, bony dermal plates covered the head region. Integumentary scales are absent in modern cyclostomes. Nevertheless, the epidermal teeth found in the buccal funnel and on the tongue (Figs. 5.5 and 5.18) are really modified epidermal scales.

Fishes. Epidermal scales are lacking in fishes, but dermal scales, making up part of the *dermal skeleton,* are abundant and of several types in this class of vertebrates. Not all fish have scales. The common catfish, or bullhead, and *Torpedo,* the electric ray, are examples of fishes which lack integumentary scales. It is commonly believed that eels have no scales, but they are present, nevertheless. Scales of eels are very small and are deeply embedded in the dermis. In certain other fishes, as in chimaeras, scales may occur only in localized regions. The scales of fishes are colorless. Numerous pigment cells (chromatophores), which give the fish its color, are located in the outer part of the dermis both above and below the scale. The presence of water on a descaled fish will frequently reveal the pattern of the scales. In the sea horse and a few other fishes, bony plates form a veritable armor which covers the entire body.

Fish scales as well as the flat membrane bones in the skulls of fishes and higher vertebrates are believed to be remnants of the bony armor present in ostracoderms and found also in a number of placoderms. The armor of these extinct forms consisted of four rather distinct layers. The deepest layer (*isopedine*) resembled compact bone; the next layer was composed of spongy bone containing numerous vascular spaces; the third layer was of more compact consistency and made up of a hard dentine (*cosmine*) often in the form of separate elevations or ridges; the outer layer was thin and of a hard, enamel-like consistency.

Two types of scales have been identified among the earliest primitive bony fishes. One type, the *cosmoid scale,* not found in any form living today with the possible exception of *Latimeria,* was constructed essentially on the plan of ostracoderm or placoderm armor plates. The scales, however, were in the form of small, separate elements. Cosmoid scales were present in primitive members of the subclass Sarcopterygii, i.e., crossopterygians and dipnoans. *Latimeria* has cosmoid scales, but they are somewhat simpler in structure than the usual type in that the cosmine layer is present only in the part of the scale that is exposed. The scales

are large and overlapping, with rough, tuberculate surfaces. The other type of scale, the *ganoid scale,* found in primitive ray-finned fishes, existed in two forms. One, the *paleoniscoid ganoid scale,* persists today in *Polypterus.* Here the basal layer is of compact bone (isopedine); the spongy-bone layer has been lost; the cosmine layer has been reduced; the outer portion of the scale is composed of layers of a hard, shiny, translucent material of mesodermal origin and called *ganoin.* The other, the *lepisosteoid ganoid scale,* has eliminated both the spongy-bone and cosmoid layers so that the ganoin lies directly upon the layer of compact bone. This type of scale persists today in the garpike.

Dermal fish scales of four types, ganoid, placoid, ctenoid, and cycloid, are recognized in species living today. They are all fundamentally similar in origin, but their form varies considerably.

GANOID SCALES. The structure of the two types of ganoid scales as found in *Polypterus* and *Lepisosteus* has already been described. The scales fit closely together like tiles on a

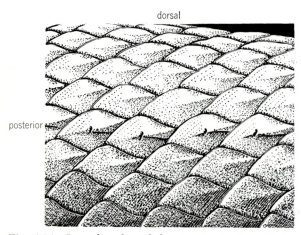

dorsal

posterior

Fig. 4.14 Ganoid scales of the gar, *Lepisosteus.* Note perforations for openings of the lateral-line canal.

Anatomy of the Chordates

anterior

posterior

Fig. 4.15 Enlarged photograph of skin of shark, *Squalus acanthias*, showing numerous placoid scales.

floor and are arranged in diagonal rows (Fig. 4.14). *Amia* has modified ganoid scales on the head. Sturgeons have similar scales in the tail region. In both these fishes the layer of ganoin is lacking.

PLACOID SCALES. With few exceptions, placoid scales are found only in the elasmobranch fishes. Each scale consists of a basal bony plate which is rounded or rhomboid in shape and embedded in the dermis. From the plate a spine projects outward through the epidermis and points posteriorly (Fig. 4.15). The spine is made up of dentine covered with a hard layer of *vitrodentine,* both of mesodermal origin. A pulp cavity lies within the spine and opens through the basal plate. This offers a point through which blood vessels enter the pulp cavity. Many authorities have stated in the past that the placoid scale is covered by a layer of enamel. Researchers indicate, however, that although the scale may be formed under the organizing influence of an enamel organ (page 167), no

actual enamel is formed and the spine is composed of dentine alone. The general similarity in structure of placoid scales to teeth of higher forms should be apparent. Both are considered to be modified remnants of the bony armor of such primitive vertebrates as ostracoderms and certain placoderms.

The scales of dogfishes and sharks are numerous and set closely together, but in other forms they are usually large and scattered about on different parts of the body. They are frequently particularly well developed along the middorsal line. The teeth on the rostrum, or saw, of the sawfish (Fig. 4.16) are actually the extremely large spines of modified placoid scales embedded in sockets at their bases. Placoid scales may develop during embryonic life in chimaeras but soon disappear except in certain scattered areas.

CTENOID SCALES. The ctenoid scale derives its name from the fact that its free edge bears numerous comblike projections (Gr. *ctenos*, comb). It is a rather common

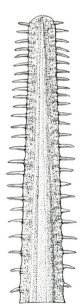

Fig. 4.16 Saw of sawfish, *Pristis pectinatus.* The teeth are modified placoid scales.

type and is present in most teleost fishes. Ctenoid scales are thin, translucent plates composed of an underlying layer of fibrous material covered by a layer which somewhat resembles bone. These scales might be compared with ganoid scales from which the layer of ganoin has disappeared and the underlying layer of bone modified. Each scale is embedded in a small pocket in the dermis. The scales are obliquely arranged so that the posterior end of one scale overlaps the anterior edge of the scale behind it. When a fish is scaled preparatory to cooking, the scales are readily flipped out of their pockets by scraping with a knife in a tail-to-head direction. The basal end of the ctenoid scale is usually scalloped (Fig. 4.17). Lines of growth are present which, if examined under a microscope, will give an exact indication of the age of the fish. The number of lines of growth varies according to the species in question and the part of the scale examined. When several lines of growth are widely separated, it indicates that that portion of the scale was formed during the warm season when food was plentiful and rapid growth took place. When several lines of growth are placed close together, the cold season is denoted. Thus, an area on a scale which covers a series of wide lines and a series of close lines would correspond to a year in the life of the fish.

Fig. 4.18 Cycloid scale of white sucker. From a fish that has lived through eight winters. (*From a photograph by Dr. W. A. Spoor.*)

CYCLOID SCALES. The cycloid scale seems rather primitive in structure as compared with a ctenoid scale. It is roughly circular in outline (Fig. 4.18) with concentric lines of growth which may be used as a means of determining the age of the animal. Cycloid scales are also located in pockets in the dermis but are somewhat more loosely attached than ctenoid scales.

Both ctenoid and cycloid scales overlap, much in the manner of shingles on a roof. In certain fishes, such as flounders, scales of both types may be present, those on the underside being cycloid and those on the upper side being ctenoid. It is probable that complicated ctenoid scales were developed from the more simple cycloid type. Intermediate types of scales, some of which bear spines, are present in a variety of fishes. It is interesting to note that the scales covering the lateral line (page 723) are frequently perforated and permit the passage of the small connectives of the lateral-line canal to the outside (Figs. 4.14 and 14.27).

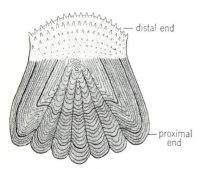

Fig. 4.17 Ctenoid scale.

Amphibians. The skin of modern amphibians lacks scales except in a few toads and in some of the burrowing, limbless caecilians. The skin is usually smooth and moist.

EPIDERMAL SCALES. In the spade-foot toads, *Scaphiopus* (American) and *Pelobates* (European), a highly cornified area of epidermis on the inner side of the hind foot, which is used in digging, might be considered to be an epidermal scale. In the South African clawed toad, *Xenopus laevis,* the dark cornified epidermis at the ends of the first three digits on the hind feet has a clawlike appearance. The salamander, *Hynobius,* also has similar structures at the ends of the digits.

DERMAL SCALES. In a few toads bony plates are embedded in the skin of the head or back. Extinct labyrinthodonts, which are usually considered to be the ancestors of modern amphibians, possessed large bony plates in the skin. True dermal scales are found in the integument of certain caecilians. They lie between ringlike folds in the skin of these animals and alternate with areas which are typically glandular in nature. The scales of caecilians resemble those of fishes in that they are embedded in pockets in the dermis, but instead of a single scale being confined to a pocket, several (four to six) may be present.

Reptiles. Both dermal and epidermal scales exist in the class Reptilia, and the two types are frequently associated with each other.

EPIDERMAL SCALES. Epidermal scales are extremely well developed and are characteristic of the class as a whole. The scales are of two general types: those present in snakes and lizards and those found in turtles, crocodiles, and alligators. In the former type each scale projects backward and overlaps the scales behind (Fig. 4.19). The formation of this type of scale is of interest. Raised and depressed areas appear in the integument

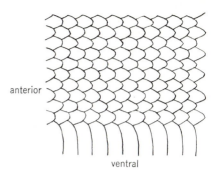

Fig. 4.19 Overlapping epidermal scales of snake (diagrammatic).

(Fig. 4.20). Each thickened area, which consists of a covering of epidermis together with the dermis beneath, is known as a *papilla.* The flattened papillae grow upward and posteriorly and become lopsided, since the cells on the upper or outer surface multiply faster than do those beneath. The rapidly dividing epidermal cells are nourished by materials brought to them by the blood vessels in the dermis. Dermis and stratum germinativum of the epidermis gradually retract, leaving the dead, cornified scale on the surface. The hollow scale soon collapses, and all that is left is a flat, thin structure. Thus, although the fully formed scale is entirely epidermal, it is formed under the influence of the dermis. The scales are continuous with each other at their bases. Snakes and lizards periodically undergo ecdysis. Before ecdysis occurs a new set of scales has been formed beneath the old. The old corneal layer is shed as a whole and in snakes is turned inside out. Several days before the layer is to be cast off, the colors of the snake seem to fade and the scales over the eyes become whitish. This is because the colorless, semitransparent old corneal layer is separating from the new one beneath. It first loosens on the top of the head and margins of the mouth. The snake then literally crawls out of its old skin, aided

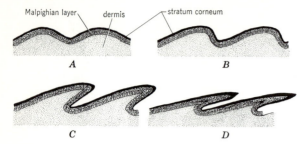

Fig. 4.20 Stages in development of epidermal reptilian scales of the type found in snakes and lizards (diagrammatic).

partially by squeezing its body through a narrow place. Lizards might be said to creep out of their castoff scaly covering, for it is not turned inside out during ecdysis. The frequency of shedding depends on such factors as the amount of food eaten and the activity of the thyroid gland and anterior lobe of the pituitary gland. A healthy snake that is well fed may shed its old corneal layer at intervals of 2 months or less.

The scales on the ventral side of the body in snakes differ from those on the dorsal and lateral sides in being transversely arranged. Their importance in locomotion is discussed on page 528. The scales on the dorsal surface of the head are considerably modified, forming a close-fitting layer over the bones. Their pattern, which is often used as a means of identifying species, does not correspond to that of the bones over which they lie. These scales form a continuous sheet with the other scales and are shed along with them during ecdysis.

Some special modifications of epidermal scales are to be observed in certain lizards and snakes. The so-called horns of the horned lizard ("toad") are really sharp, spiny, scales, each covering a bony projection of the skull. The rattle of the rattlesnake is made up of a series of old, dried scales loosely attached to

each other in sequence. At the time of ecdysis the scale at the very tip of the tail is not shed with the others because of a peculiar bump which causes it to adhere to the newly formed scale (Fig. 4.21). Thus a series of scales forms the rattle, each new scale being slightly larger than its predecessor. The exact age of a rattlesnake cannot be determined by counting the number of rattles, since ecdysis may occur several times a year. Old rattles at the tip may gradually wear off or be broken off. There are usually less than 10 rattles, although more than this number are frequently observed. The so-called "warning" of the rattlesnake is not a warning in the real sense of the word. The vibration of the tail, which produces the buzzing sound of the rattler, is, however, of protective value to the snake,

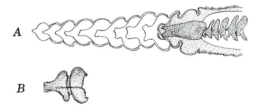

Fig. 4.21 Rattle of rattlesnake: A, median sagittal section of end of tail; B, a single horny element of the rattle. (*After Czermak.*)

Anatomy of the Chordates

since the sound is enough to frighten even the most formidable of enemies.

In most turtles large epidermal scales cover the carapace and plastron, but the pattern of the scales does not conform to that of the bony plates beneath. Each scale develops separately. Periodically the stratum germinativum under each scale grows peripherally, and thus the area of each scale is increased. A new cornified layer is then formed, which pushes the older scales away from the shell. This results in a piling up of the older scales, which then appear from above as a series of irregular concentric rings (Fig. 4.22). The number of rings indicates the number of growth periods, and hence the number of ecdyses. In temperate zones each ring represents a year's growth, since during hibernation there is an almost complete cessation of growth. The old, smaller, outer scales frequently break off or are worn off, so that the number of rings does not necessarily represent the exact age of the animal. In such forms as the painted turtle the old scales characteristically peel off and only the smooth, most recently formed scale is present at any one time. Scales also cover the softer parts such as the neck and legs. Soft-shelled turtles and the so-called leatherback sea turtle possess a rather soft, leathery "shell" lacking epidermal scales as well as underlying bony dermal plates.

Pigment in the turtle's integument is located in the stratum germinativum. In some cases it diffuses into the horny layers above. The beautiful tortoise shell of commerce owes its color to the various combinations of pigment which have diffused into the corneal layer.

Epidermal scales of crocodilians (Fig. 4.23) cover the entire body. On the lateral and ventral sides, as well as on the tail, each scale bears a pitlike depression. A small sen-

Fig. 4.22 Arrangement of scales on carapace of wood turtle.

sory capsule is located at this point, and the scale is not cornified in the region of the capsule. Crocodilian scales do not undergo periodic ecdyses as in other reptiles. Instead there is a wearing away and a gradual replacement. Patches, made up of a few scales, are occasionally sloughed. Alligator skins that are used commercially consist of the tanned dermis with the epidermal scales removed.

DERMAL SCALES. Perhaps the best example of highly developed dermal scales exists in the turtles. Here the bony plates in the dermis form a rigid dermal skeleton which becomes intimately connected with the endoskeleton. The carapace, or dorsal part of the turtle's shell, beneath the epidermal scales is composed of bony costal plates fused to the neural arches of vertebrae in the midline, and laterally to the ribs beneath (Fig. 10.62). The plastron, or ventral part of the shell, is usually made up of nine large dermal plates covered with epidermal scales.

Fig. 4.23 Arrangement of scales on neck and shoulder region of alligator.

In crocodilians the dermis is thick and soft except on the dorsal side of the body and occasionally on the throat, where bony plates lie beneath the epidermal scales. In caymans, thin bony scutes also occur under the ventral scales. The dermal plates are much reduced in size and are not fused to one another. Also present in these reptiles are dermal "ribs," called *gastralia,* located in the ventral abdominal region. The rhynchocephalian *Sphenodon* also possesses gastralia. They should not be confused with true ribs.

Small dermal scales are present in certain lizards and snakes, but most of these animals lack such bony elements. The membrane bones of the skull, however, represent dermal scales.

Birds. With the exception of the membrane bones of the skull, dermal skeletal structures, or scales, are usually lacking in this class of vertebrates. Epidermal scales and their derivatives, however, are exceptionally well developed.

EPIDERMAL SCALES. Epidermal scales of reptilian type are confined to the lower part of the legs, the feet, and the base of the beak. They generally overlap and are formed in the same manner as those of lizards and snakes.

A bony projection of the tarsometatarsus in the males of certain species of birds is known as the *spur.* It is covered with a horny, scalelike epidermal sheath which may be very sharp and pointed. The spur is best developed in some gallinaceous birds in which it is used in fighting. Spurs may also occur on the wings (metacarpus) in certain forms.

The webs on the feet of aquatic birds, such as geese, ducks, and swans, are modified regions of the integument. The skin forming the web is characteristically scaled.

FEATHERS. The outstanding feature of birds is the presence of feathers, which are found in no other group of animals. Feathers are actually modified reptilian scales. This is very well shown in certain varieties of chickens such as cochins, brahmas, and langshans in which feathers appear on the legs and toes in close relation to the scales. In feathers the stratum corneum has reached the height of modification and specialization, as witness the elaborate covert feathers making up the "tail" of the (male) peacock, and the beautiful plumage of certain birds of paradise.

The varied and sometimes gorgeous coloring of feathers is based upon two main factors: (1) pigmentary, or chemical, coloration and (2) structural, or physical, coloration. Presence of pigment in the feather substance, deposited during development, is the most common basis for coloration. The color that is observed is due to the absorption of certain wavelengths of light by the pigment. The light that is not absorbed is that which affects the eye of the observer. Black, red, and yellow pigments in various combinations have been identified. Perhaps green should be added to the list, although most greens are due to physical factors mentioned below. White color is not due to white pigment but rather to an altogether different principle. Whiteness is caused by reflection of light without the absorption of any of its component rays. Reflection from a polished surface like that of a mirror does not cause whiteness. Irregular reflection which scatters or reflects light in all directions is responsible for the white color. The presence of innumerable air spaces in a white feather causes irregular reflection. This embodies the same principle as that involved when a transparent piece of ice is crushed into many smaller particles which then give the mass a white appearance.

Structural, or physical, coloration is brought about by such physical phenomena as refraction and interference of light rays as they are reflected from minute, irregular surfaces of parts of the feather. Iridescent hues, metallic colors, grays, and certain shades of blue are due to such structural irregularities. When physical coloration is superimposed upon the effect of the basic pigment, it may result in a color which differs considerably from that of the pigment itself.

Endocrine factors have been shown to play an important part in deposition of feather pigments during development.

In general, there are three main types of feathers, known as *filoplumes,* or hair feathers; *plumulae,* or down feathers; and *plumae,* or contour feathers.

Filoplumes. Hair feathers, as their name implies, appear superficially like hairs but have an entirely different structure and origin. They can usually be observed on a chicken after it has been plucked preparatory to cooking and are commonly singed or burned off. The structure of a hair feather is simple, since it consists only of a long, slender shaft which may bear a few barbs at its distal end (Fig. 4.24). The shaft is embedded in the skin and surrounded by the *feather follicle* at its base. Hair feathers are usually scattered over the surface of the body, but in some birds, such as flycatchers, they are concentrated about the mouth and serve as an aid in catching insects. In peacocks they are of unusual length.

Plumulae. The down feather is more complex than the hair feather (Fig. 4.25). It is composed of a basal, short, hollow *quill* which is embedded in the integument. Numerous *barbs* arise from the free end of the

Fig. 4.24 A hair feather.

Fig. 4.25 A down feather.

quill, and these bear tiny *barbules* along their edges. The transient nestling down which may be observed on the young bird as its first feather covering is shed when the later-appearing contour feathers emerge. Nestling downs often adhere to the tips of the newly erupting contour feathers. The type of down feather seen in an adult bird is sometimes called "powder" down. Such feathers lie under the larger contour feathers and in many cases form a warm insulating layer which not only aids in keeping the body temperature of the bird constant but assists in warming the eggs during incubation. The down of the eider duck, a large sea duck of northern regions, is used for stuffing pillows and has rather high commercial value.

Plumae. Except in penguins, ostriches, and toucans, plumae, or contour feathers, arise from certain areas of skin called feather tracts, or *pterylae* (Fig. 4.26). Large areas between the pterylae are devoid of contour feathers and are known as *apteria*. Hair feathers and down feathers may occur in the apteria, however. Under experimental conditions, supernumerary pterylae may be induced to develop, generally in the midventral apterium, by implanting various kinds of

living or inanimate objects in the ventral body wall of chick embryos. Neural-tube implants are most active in this respect, but agar, containing brain extract, is also highly effective (S. and M. Kieny, 1967, *Develop. Biol.,* **16**:532). Contour feathers are so named because they give the body its outline, or contour. If they are removed by plucking, the body configuration is decidedly altered. Special contour feathers located on the wings are referred to as *remiges,* or flight feathers. Those on the tail are *rectrices,* or tail feathers.

The typical contour feather (Fig. 4.27) consists of a long *shaft* and a broad, flat portion called the *vane.* The shaft is made up of two parts: the hollow *quill,* or *calamus,* which is embedded in the skin, and a solid *rachis* which bears the vane. At the lower end of the quill is a small opening, the *inferior umbilicus.* At the junction of rachis and quill is another opening, the *superior umbilicus.* On the underside of the rachis and extending from the superior umbilicus to the tip is a groove, the *umbilical groove.*

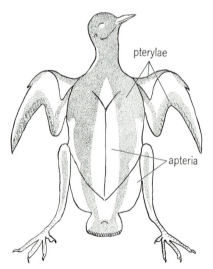

pterylae

apteria

Fig. 4.26 Distribution of feather tracts in the pigeon. (*After Nitzsch.*)

Fig. 4.27 A contour feather from the wing of a chicken.

The vane is composed of a number of *barbs* which arise from the rachis. Each barb in turn bears small *barbules* on both distal and proximal sides. The lower part of each distal barbule bears small *barbicels* which terminate in tiny *hooklets,* or *hamuli,* which fasten onto the proximal barbules of the next adjacent barb. Hooklets are lacking from the proximal barbules (Fig. 4.28). As a result of this arrangement the barbs are hooked together via their barbules, and the vane offers a flat, wide, firm, and unbroken surface except at the base where the distal barbules of a variable number of barbs lack hooklets. This part of the feather may then have a rather fluffy appearance. Should certain barbs become separated, the bird can easily put them in place again by preening, or drawing the barbs together with its beak or bill. In some of the smaller contour feathers of the wings, a considerable portion of the lower part of the vane may consist of loose barbs since hooklets are lacking from their barbules.

In many birds at the junction of rachis and quill is located what appears to be another feather. This is called the *aftershaft,* or *hyporachis.* It also bears barbs and barbules, but hooklets are lacking. The aftershaft is generally smaller than the main shaft and may consist only of a small downy tuft. In some birds, such as the emu and cassowary, the aftershaft with its accessory parts may be as long as the main shaft and the feather appears to be double (Fig. 4.29).

Feathers are shed periodically in a process known as *molting.* During molting not all the feathers are shed at once. If such were the case, the bird would be greatly handicapped by being temporarily unable to fly. The common barnyard chicken characteristically molts in the latter part of the summer in temperate zones. New feathers are in the process of formation at the time that the old ones are shed.

Development. A study of the development of a feather emphasizes its homologies with the lizard and snake type of scale,

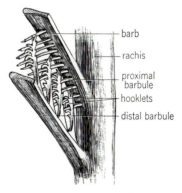

Fig 4.28 Arrangement by means of which barbs of a feather are held together.

Fig. 4.29 Feather from a cassowary. The main shaft and aftershaft are of similar dimensions.

since each is formed from a dermal papilla covered with epidermis.

In the development of the typical down feather the first indication of its formation is seen in the appearance of the dermal papilla with its epidermal covering. Instead of being flat and platelike, as in the reptile, the papilla is elongated and round. The structure is, however, essentially the same. An *annular groove,* which is the beginning of the feather follicle, appears around the base of the papilla. During development the blood vessels of the dermal pulp supply nourishment to the growing structure. The outer, thin, cornified layer of the epidermis forms a sheath known as the *periderm.* This is ultimately sloughed. First, however, the epidermis beneath the periderm gradually forms a series of longitudinal folds or ridges arranged on the surface of the dermal pulp into which they extend (Fig. 4.30). Actually these ridges arise from a collar of cells, the stratum germinativum, at the base of the

papilla. The entire structure is now called the *feather germ.* The longitudinal epidermal ridges are destined to give rise to the barbs of the feather. It is at this time that pigmentation of the feather cells may occur. The feather germ with its peridermal sheath grows rather rapidly and soon projects from the feather follicle above the surface of the skin as a "pinfeather." The dermal pulp and stratum germinativum gradually retract. Soon the periderm at the apex splits, and the tips of the cornified epidermal ridges dry and crack. The distal ends of the barbs can then be observed projecting from the apex of the sheath. As the feather continues to grow, the proximal parts of the elongated ridges become cornified and separate. The periderm is shed in the form of small, dandruff-like scales, and the barbs spread out. They remain attached at their bases to the quill, which lies beneath the surface of the skin and does not split. Small, pithy partitions inside the quill represent dried remnants of the dermal pulp. A small papilla remains at the base of the feather. This later gives rise

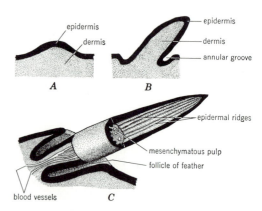

Fig. 4.30 Stages in the development of a down feather. (*Modified from Kingsley, "Comparative Anatomy of Vertebrates," The Blakiston Division, McGraw-Hill Book Company. By permission.*)

to the feather which will replace its predecessor when the latter is shed.

The development of the contour feather is similar to that of the down feather up to a point, after which it becomes considerably more complex. Observations (F. R. Lillie and M. Juhn, 1938, *Physiol. Zoöl.*, 11:434) have shown that during development a differential growth occurs on one side of a collar of stratum germinativum at the base of the feather germ. This is considered to be the dorsal surface of the developing feather, and the region opposite is the ventral surface. The middorsal portion of the collar starts to grow

outward, carrying the barbs on either side along with it (Fig. 4.31). This becomes the rachis. The umbilical groove on its undersurface represents the line of fusion of the two sides of the collar as it grows outward. As the series of barbs moves dorsally with the concrescence of the two sides of the collar, new barbs appear in the space thus provided on the ventral side. As in the case of the down feather, the contour feather is at first enclosed by a sheath of periderm. When the periderm dries and sloughs, the developing feather splits along the midventral line and flattens out. The underside of a feather,

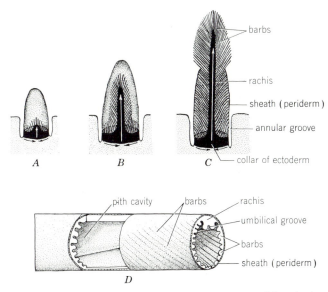

Fig. 4.31 *A*, *B*, and *C*, schematic representation of developing contour feather from a collar of ectodermal cells which shows differential rates of growth. In *C*, the periderm at the distal end has ruptured. The barbs of the two sides have separated along the midventral line and have flattened out. *D*, a stereogram of part of a developing contour feather. (*A*, *B*, and *C*, based on figures by Lillie and Juhn, 1938, Physiol. Zoöl., 11:442. By permission of the University of Chicago Press. D, after Kingsley, "Comparative Anatomy of Vertebrates," The Blakiston Division, McGraw-Hill Book Company. By permission.)

therefore, represents that portion which was originally adjacent to the feather pulp. Again, as in the down feather, the quill is that portion which lies beneath the skin and has failed to split. Transverse, pithy partitions inside the quill represent the remainder of the dermal pulp. The superior umbilicus and inferior umbilicus are merely openings through which the dermal pulp was originally in continuity with the dermal papilla. The development of the aftershaft is like that of the main shaft except that differential growth occurs along the midventral line as well.

Mammals. Except for the membrane bones of the skull, most mammals lack dermal skeletal structures. Armadillos have the best-developed dermal skeletons among living mammals. Bony plates, or scales, lie under the epidermal scales which they reinforce. The extinct glyptodons, belonging to the order Edentata, possessed a rigid, bony, dermal armor. In certain whales bony plates may be present on the back and in the dorsal fins. In mammals in general, as in birds, there has been a tendency toward elaboration of epidermal structures and a loss or diminution of dermal derivatives.

EPIDERMAL SCALES are present in many mammals. With the exception of the scaly anteaters and armadillos, these are generally confined to the tails and paws. When scales do occur they are usually associated with hairs. This, among other reasons, has given rise to the belief that mammalian hair was originally derived from scales. The question of the degree to which scales, feathers, and hairs are comparable is of interest, but it is clear that feathers are more closely related to scales than are hairs.

In the scaly anteaters the body, except on the ventral side, is covered with large, overlapping, horny epidermal scales (Fig. 2.50). They are typically reptilian in structure, but

ecdysis occurs singly. In the armadillos the large scales fuse to form plates over the head, shoulders, and hindquarters as well as the ringlike bands which surround the midbody region except along the midventral line. As mentioned above, dermal bones underlie the epidermal scales. These hark back to the fishlike ancestors of prehistoric times. There is no true ecdysis but instead a gradual wearing away from the surface and replacement from beneath. Transient epidermal scales are present on the body of the fetus of the brown bear and the European hedgehog. They are numerous and interspersed with hairs.

On the tails of many rodents, such as the rat, mouse, muskrat, and beaver, imbricated epidermal scales are present. They are truly reptilian in structure and development, but the degree of cornification is not pronounced, nor is there a period of ecdysis.

Of great interest is the arrangement of hairs in relation to scales. The hairs project from beneath the scales. On the tail of the rat, for example, three hairs project from beneath each scale. The one in the center is larger and coarser than the two lateral hairs (Fig. 4.32). The astonishing fact is that the hairs on other parts of the body have a similar arrangement. The skin of the pig is devoid of scales, but

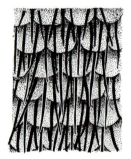

Fig. 4.32 Arrangement of hairs in relation to scales on the tail of the rat.

the grouping of the hairs, or bristles, gives evidence of origin from scaled ancestors. Coarse hairs are usually arranged in clusters of three and interspersed with finer hairs. Examination of a pair of pigskin gloves will clearly indicate this relationship. In other mammals, including man, similar arrangements are found, hairs being grouped in twos, threes, fours, and fives. Often this relationship, although clear during development, is obscured in the adult.

In most mammals the undersurfaces of the hands and feet are scaled or bear evidence of the former presence of scales. The friction ridges (page 106) familiar to everyone and which in man are used in making fingerprints really represent such scale rudiments. In certain animals definite elevated pads on the undersurfaces of the feet bear friction ridges. Such pads are known as *tori*. On the foot of a rat, for example (Fig. 4.33), there are 11 elevated tori. They include 5 *digital tori* at the ends of the digits, 4 *interdigital tori* below the spaces between the digits, 1 *thenar torus* at the base of the first digit, and 1 *hypothenar torus* at the base of the fifth. In man, in which no elevated tori are present in the adult, the friction ridges have become flattened and spread over a considerable surface. However, groupings in the regions where one would expect tori to occur indicate a relationship similar to that described above.

In the development of the integument in embryos of reptiles, birds, and mammals the first layer of cells given off by the stratum germinativum forms a continuous sheet called the *periderm*. The cells of the periderm possess characteristic staining properties. This layer is later shed during development and in mammals contributes to the *vernix caseosa,* a cheesy material present on the skin of the newborn. That the periderm may represent the scaly covering of ancestral forms is indicated by the fact that the only

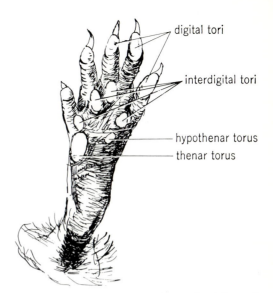

Fig. 4.33 Arrangement of tori on hind foot of rat.

place where it is not shed is on the palms and soles, and in these regions it is reported to contribute to the formation of friction ridges. The term *epitrichium* is often applied to the periderm of mammals since developing hairs push it up, causing it to be shed. Of course no hairs develop on the palms and soles of mammals where the periderm persists.

Hair. True hair is found in all mammals, and only mammals have hair. In most forms practically the entire integument is covered with hair, but in some the hairy coat has disappeared for the most part and only traces remain. In the adults of some of the larger whales, for example, only a few coarse, stubby hairs are present in what might be called the snout region. During fetal development in all mammals the body is at one stage covered with a coating of fine hair called the *lanugo.* This is absent only on the ventral surfaces of the hands and feet. The lanugo hair is transient and is usually shed some time before birth. In the human fetus the lanugo is fully developed by the seventh month. Shedding begins during the eighth month and most of it has been lost by the time of birth or shortly after, except on the margins of the eyelids, the region of the eyebrows, and on the scalp. In these regions the hairs persist and become somewhat coarser and stronger. Later on these hairs are also shed and are replaced by still coarser ones. A new growth of hair takes place over most of the rest of the body, forming a fine downy coat referred to as the *vellus*.

In cases where hairs have failed to develop, the condition is known as *atrichosis*. There are occasional human beings whose bodies are entirely devoid of hair. In such circumstances the nails and even the teeth may be lacking. When the hairy coat is unusually scanty, the term *hypotrichosis* is applied. Abnormally excessive hairiness is called *hypertrichosis*. This condition is frequently observed in men but rarely in women, in whom it is often associated with abnormal functioning of the cortical portion of the adrenal glands. In extreme cases the entire body of a human being may be covered with a thick coating of hair like that of a dog or cat. The celebrated "Jo-Jo the dog-faced man" of circus fame (Fig. 4.34) is an example of this condition which is alluded to as *pseudohypertrichosis*. It is possibly the result of overdevelopment of the lanugo or vellous coat, but this is not certain. No adequate explanation has been offered.

STRUCTURE, DEVELOPMENT, AND GROWTH OF HAIR. Hair is entirely epidermal in origin. During development a small thickening in the epidermis, which is to become the *hair follicle* from which the hair will arise, pushes down into the dermis and finally becomes cupped at its lower end (Fig. 4.35). Connective tissue from the dermis extends into the cuplike depression forming a *dermal papilla*. The blood vessels which extend into the

Fig. 4.34 Jo-Jo, the dog-faced man, an example of pseudohypertrichosis.

papilla bring nourishment to the stratum germinativum of epidermal cells. The follicle, which is at first a solid cord of cells, grows down in an oblique direction to the level of the panniculus adiposus. The lower part of the follicle enlarges, this portion being called the *bulb*. Hair follicles usually are formed singly. In man an additional follicle customarily develops on either side of the original so that there is a tendency to form groups of three. Two thickenings, one proximal and one distal, appear along the side of a developing follicle. The proximal thickening is destined to form a *sebaceous gland* which, therefore, is of ectodermal origin; the distal thickening will provide the point of attachment for one of the *arrector pili* muscles, previously mentioned, which develop from surrounding mesenchyme (page 142).

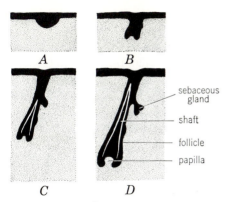

A   B

sebaceous gland

shaft

follicle

papilla

C   D

Fig. 4.35 Semidiagrammatic representation of stages in the development of a hair.

As this growth process is going on, the epithelial, epidermal ingrowth undergoes a change so as to form a central *shaft* surrounded by a space between it and the follicular wall. This is actually the result of the gradual keratinization, or cornification, of cells derived from the stratum germinativum. Keratinization of the shaft becomes complete about half the distance from the surface. As new cells continue to be formed and become keratinized, the central shaft increases in length and literally pushes its way through the solid cord of cells between it and the surface. Thus the hair emerges through a slightly depressed area which marks the place where the original invagination of epidermal cells occurred. In the meantime the proximal thickening, mentioned above, is proliferating and developing into a sebaceous gland. This opens directly into the hair follicle. Within the follicle (Fig. 4.36) the hair shaft is surrounded by two layers of cells which do not extend beyond the limits of the follicle. *Huxley's layer* is the inner layer of the sheath, and *Henle's layer* lies outside this. The hair shaft is entirely cellular and is composed of a central core, or *medulla;* a middle *cortex;* and an outer covering, the

cuticle, consisting of a single layer of scaly, cornified cells which have lost their nuclei. The scales overlap, much like shingles on a roof, but the free edges point away from the skin, rather than toward it (Fig. 4.37). Fine, downy, vellous hairs lack a medulla. At the base of the follicle the hair is expanded slightly to form a bulblike enlargement, which is sometimes called the "root" of the hair. All growth takes place in the root, where the cells of the stratum germinativum divide actively. Beyond this point the cells gradually die and the shaft of the hair is thus composed of dead, cornified cells.

It is of interest that cutting hair and shaving have no influence on the rate at which hairs grow or on their texture. The average weekly growth of hair in man varies from 1.5 to 2.7 mm in different parts of the body.

The character of the surface of the hair varies from smooth to decidedly rough and scaly. The hair of sheep, commonly known as wool, is an example of the latter type. The small scales of adjacent hairs interlock in such a manner that the hairs cling together. Wool, therefore, is admirably adapted for spinning and gives a high quality of cloth. The smooth type of hair naturally gives a glossier appearance. The oily secretion of the sebaceous glands which open into the hair

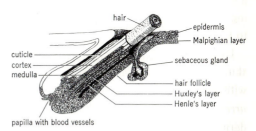

hair

epidermis

Malpighian layer

cuticle
cortex
medulla

sebaceous gland

hair follicle
Huxley's layer
Henle's layer

papilla with blood vessels

Fig. 4.36 Diagram showing structure of a hair and its follicle. (*After Kingsley, "Comparative Anatomy of Vertebrates," The Blakiston Division, McGraw-Hill Book Company. By permission.*)

follicles serves to keep the hair smooth and glossy and aids in shedding water.

The nature and distribution of hair in man shows great variation in different parts of the body, as witness the relatively coarse, long hairs on the scalp, eyebrows, and eyelashes, and in the axillary and pubic regions.

The manner in which hairs of various types grow is said to be either *definitive* or *angora.* Angora hairs grow to a considerable length before they loosen and are shed. The head hair of man is of this type. Definitive hairs, however, grow to a certain length and then stop. They are then shed and quickly replaced. Eyelashes, eyebrows, and body hairs are of the definitive type. Angora hairs may persist for several years, but definitive hairs usually last only for a few months. The length of hairs in different parts of the human body varies from less than 1 mm to extreme conditions in which scalp hair may be 4 or 5 ft long.

A general belief exists that hair may continue to grow after death. Since hair is made up of cornified cells produced by the proliferation of preexisting cells, it is obvious that after death, when oxygen and nutriment are no longer supplied by the blood stream, cell divisions must cease. After death, especially when the body is embalmed, the skin shrinks, thus exposing hair shafts which had hitherto been concealed. A preparator not infrequently may observe what appears to be a growth of beard several days after embalming a corpse. Possibly this has given origin to the fantastic story mentioned.

Hairs do not emerge vertically from the skin but project at an acute angle. Associated with each hair group is a small, involuntary *arrector pili* muscle. This develops in the dermis and, for a short distance, consists of a single bundle of fibers. The muscle then subdivides, with a branch going to each hair of a hair group where it inserts on the lower part of the follicle on the side toward which the hair slopes, or the side forming an obtuse angle with the epidermis (Fig. 4.37). These muscles vary in width from 0.05 to 0.2 mm. Contraction of the arrector pili muscles, by pulling on the bases of hairs, tends to cause them to stand erect. In man this brings about a condition called *cutis anserina,* or gooseflesh (duck bumps). Such contraction squeezes the sebaceous gland, which lies in the area between the follicle and arrector pili muscle, causing accumulated *sebum,* the secretion of the gland, to be expressed. In dogs and cats the raising of the hair, particularly along the back of the neck, may be observed when the animal is confronted with danger. It is supposed that this is a protective adaptation which gives a false contour, or outline, to the body and might cause an enemy to "miss his mark" when attacking. Arrector pili muscles are not associated with the eyelashes or with nasal hairs.

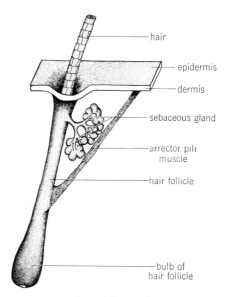

Fig. 4.37 A hair follicle, showing its relationship to the skin, a hair, a sebaceous gland, and an arrector pili muscle.

Hairs range in thickness from approximately 0.005 mm, as in lanugo or vellous hairs, to as much as 0.2 mm in coarse beard hairs.

Variations in the shape of hairs in cross section are believed to be responsible for the degree of straightness or curliness. A straight hair, which is typical of the Mongoloid races, is round; that of the generally kinky-haired Negro is flat. Oval and elliptical hairs show different degrees of curliness or waviness. Follicles producing curly hairs are flatter than those from which straight hairs emerge. Keratin, which is the chief chemical constituent of hair, is rather resistant to chemical change. Its molecules consist of long, parallel, polypeptide chains connected to each other by side links. Normally these molecules are arranged in a zigzag manner, but when moist they may be readily bent or stretched. It is, therefore, much easier to "set" wet hair to bring about a curly or wavy effect. For the same reason atmospheric humidity may have a bearing on the curliness of hair. In certain individuals, on wet or damp days there is a tendency for the hair to curl. Since the hair is somewhat porous, it will absorb a certain amount of moisture and swell along the line of least resistance, which is the diameter. If the diameter increases, the length will decrease, so that if the hair is short its end will curl in the manner of a figure 9, Long hairs do not exhibit this phenomenon to the same degree. In the modern beauty parlor curly or wavy hair is produced by chemicals and instruments which actually flatten hair by causing longer-lasting changes in alignment of the polypeptide linkages.

Hairs in different parts of the body show some variation in the direction in which they grow or slant. There are usually different patterns, or hair currents, along which the hairs grow. These are best seen when the hair is short or clipped but are more obscure when the hair is long. Where hair currents diverge, *parts* occur. When they converge, *crests* or *tufts* may be present. In certain regions *whorls,* or *vortices,* indicate a spiral arrangement. The familiar cowlick on the human forehead is an example of a whorl. Although in most mammals the hair parts on the middorsal line and slopes to the ventral side, the sloths show the opposite condition. Here, associated with the animals' habit of hanging upside down in trees, parting occurs on the midventral line and the hair actually slopes dorsally. This enhances the shedding of water in the tropical rain forests in which the sloths are at home.

COLOR OF HAIR. The color of hair in various mammals actually shows very little range, and the vivid colors seen in other vertebrate classes are lacking. To be sure, in certain apes like the mandrills, highly colored skin areas are present, but in these it is the skin itself rather than the hair which bears the pigment. There are very few mammals that can be said to be highly colored.

Four factors are responsible for the color and luster of the hair: (1) the color of pigment which is present in the cortex, (2) the amount of pigment, (3) the character of the hair surface, whether smooth or rough, and (4) the amount of air contained within the intercellular spaces of the medulla.

Pigment in hair is present in the intercellular spaces of the cortex. The amount that is present determines the shade. Browns seem to predominate in most forms, but in man the color seems to have black and red pigments as its foundation. Absence of one or both or different combinations or dilutions of the two apparently bring about the many variations in color that may be observed.

As in the case of the white feather, white hair is not due to the presence of white pigment. Instead there is a lack of pigment, and light is reflected in all directions from the air spaces between the cells, particularly those of the medulla. Gray hair may be the result of reduction of pigment and a reflection from an increased number of air spaces. Usually, however, the hair is called gray when white hairs are intermingled with pigmented hairs.

Pigment deposition in a developing hair follows the same pattern as that found in epidermal cells in general. Although pigment deposition usually does not cease until the latter part of life, this is not true of all mammals. The varying hare, arctic fox, stoat, and certain weasels are remarkable in that, as winter approaches, the old brown pelage is replaced by a white coat which provides protective or aggressive coloration, as the case may be. In these mammals, pigment deposition is seasonal.

In man the deposit of pigment in hair may sometimes be interrupted or permanently stopped after sickness or extreme nervous shock. It is, of course, impossible for hair to turn white overnight, since pigment already present in the hair cannot be retracted. In mammals with white-spotting patterns, or with a piebald hairy coat, the areas free of pigment have been shown to be the result of a failure of melanoblasts to differentiate into melanocytes in those areas, not of a defect in the migration of melanoblasts from their site of origin from neural crests during embryonic development (T. C. Mayer, 1966, *Amer. Zoologist*, 6:548).

SHEDDING OF HAIR. In mammals which show decided differences in their summer and winter coats, the hair is shed periodically. Horses and cattle shed their long, thick coats in the spring. The buffalo has a shaggy, moth-eaten appearance when large patches of hair are being shed. In many mammals, however, there is constant shedding and re-

newal. If a hair is pulled out, the place where it "gives" is usually just above the stratum germinativum at the base of the follicle. A new hair may then grow in its place. When there is complete destruction of the stratum germinativum and papilla, regeneration will not occur. Under normal conditions when a hair is nearing the termination of a growth period, cell divisions in the stratum germinativum slow down and finally cease. The dermal papilla atrophies. The base of the hair shaft and the root become thinner and the layers of cells covering the dermal papilla change into a club-shaped mass. This is now referred to as a *club hair*. It loses its connection with the dermal papilla and moves outward, ultimately being shed or pulled out. The follicle then remains quiescent for a variable length of time. Later it usually becomes active again; the base of the follicle elongates and becomes thicker, a new bulb is formed, and finally a new hair emerges. It is not certain whether the remnant of the old dermal papilla becomes reactivated or whether a new papilla forms. In man, unlike the condition in many mammals in which there is wholesale shedding of hair, each follicle has its own, independent cycle.

Rather interesting studies have recently been conducted on the causes of baldness in the human being. This condition is far more prevalent in men than in women. Obviously, hereditary factors are concerned but, in addition, certain endocrine secretions of the anterior lobe of the pituitary gland and of the testes are involved. According to an older report (D. Osborn, 1916, *J. Heredity*, 7:347) baldness seems to be dominant in men and recessive in women, but it is not a sex-linked character. Men may become bald if they are heterozygous. Now it is believed that in addition to carrying the hereditary factor, the male sex hormone, testosterone (page 307), must be present in the circulatory system of the individual concerned if the condition is to manifest itself. Castration, with the resulting lack of testosterone production, in men usually prevents the hereditary factor from becoming operative. Although a high degree of virility is often compatible with baldness, it should not be inferred that individuals with excellent heads of hair may be less virile. It is only when the hereditary factor for baldness is present that the male hormone manifests its presence in this manner. It is of interest that the same pattern of baldness is usually present in a family strain.

DENSITY OF THE HAIRY COAT. The density of hair on the body shows much variation in different mammals. Those living in cold climates, either far north or south or at high altitudes, possess the heaviest hair coats of all. Tropical forms are generally sparsely covered. A permanent abode in the water is associated with an almost total loss of hair in sirenians and cetaceans. Even among the races of man, rather marked variations exist. Certain members of the white race, the aborigines of Australia and the hairy Ainus of the isle of Yesso, are at one extreme. On the other hand, the natives of the Malay Peninsula with their smooth, practically beardless skin furnish quite a contrast.

Hair in man is of little or no importance in keeping the body warm. The follicles, however, are significant in that they provide points of origin for regeneration of epidermal tissue in the repair of epidermis destroyed by burns or abrasions of one kind or another. It is they which are primarily responsible for formation of new epithelium in damaged areas when split-skin grafting is practiced.

The quality of a fur, or pelt, is determined primarily by the degree of development of the underfur. This consists of large numbers of short, densely arranged hairs interspersed among the larger and coarser ones. Animals inhabiting cold climates have the heaviest underfurs. As the cold season approaches, the underfur begins to grow; and when it is fully developed the fur is said to be *prime*. At this time its commercial value is greatest. The reason that furs from the far north have a much greater value than those from subtropical or tropical regions should be evident.

In certain mammals, such as the fox, the long hairs are the ones which make the pelt attractive. In others the coarse, long hairs are removed in the preparation of the pelt for market. Hudson seal is really dyed muskrat with the long hairs removed. True seal and beaver pelts are also treated in the same manner.

SPECIAL TYPES OF HAIR. The type of hair in different parts of the body of the same animal varies. Manes like those of the horse or lion, crests and tufts in certain regions, long tail hairs, eyelashes, eyebrows, dust-arresting hairs in the nose, all are hairs of special types which are, in the main, adaptive in function. They serve as an aid in protection from enemies, in defense against insects, and in guarding delicate membranes from foreign particles which might otherwise be injurious. In man the head hair clearly serves the function of protection from sun and rain. Pubic and axillary hair aids in decreasing friction between the limbs or between the limbs and body during locomotion. According to one theory, the persistence of pubic and axillary hair goes back in time to arboreal ancestors in which the longer hair in these parts furnished a means by which the young could cling to the parent, thus permitting a greater freedom of limb movement.

In the category of special hairs are the *vibrissae* (whiskers or feelers) found on the snouts of most mammals. These hairs are unusually sensitive. They are best developed in nocturnal mammals and their follicles have an abundant nerve and blood supply at the base. The blood vessels are in the form of cavernous sinuses, closely resembling those of erectile tissue (page 543). It seems that an increase in pressure of the blood in the cavernous sinuses increases the sensitivity of the nerves at the base of the follicle, thus making the hair more sensitive to tactile stimuli and pressure changes. Other hair follicles, even though possessing a good nerve supply, are less sensitive than vibrissae, presumably because there is no cavernous tissue associated with them.

The porcupine is of special interest because practically all the main types of hair are to be found in this animal. The spines, or quills, of the porcupine, which are really modified hairs, have barblike scales at their tips and are loosely attached at the base. They cannot be projected but are easily pulled out when the barbed ends become embedded in the flesh of an enemy.

The fur of the chinchilla, a small South American rodent, is very soft, of a pearly gray color, and very expensive. It is peculiar in that the hair follicles are compound. As many as 75 hairs, each derived from its own follicle, emerge through a common opening (H. H. Wilcox, 1950, *Anat. Record,* **108**:385). There is but one arrector pili muscle to such a cluster. This condition should not be confused with multiple hairs in which two or possibly more hairs are derived from the same follicle under the influence of a lobed dermal papilla.

Differences in quality and distribution of hair in different regions are often quite marked in the sexes and furnish some of the outstanding secondary sex characters that distinguish male from female. The beards of men and the hairy covering of the chest, arms, and legs are in marked contrast to the relatively smooth condition of the skin in women. The heavy mane of the lion clearly marks him from the lioness. Even in rats, where secondary sex characters are not very clear cut, the coarser texture of the fur of the male makes it easy to distinguish the sexes for anyone who has become accustomed to handling these animals in the laboratory.

Beaks and Bills

In turtles and tortoises and in all modern birds, teeth are lacking. Each jawbone is covered with a modified epidermal scale which forms the beak or bill. In turtles and tortoises the beak is hard, and the bite of a large animal may be serious.

Among birds, great variation is to be seen in the shape of the beak (Fig. 2.36), correlated with its use in procuring food. Seed-eating birds usually have short, rather blunt beaks; insect eaters possess long and narrow beaks which are not so strong as those of the seed-eaters. The long, strong, hooked beaks of birds of prey are well fitted for their methods of obtaining food.

As applied to birds, the words *beak* and *bill* are used interchangeably. The use of one or the other depends on the preference of the user. The term bill is more commonly used by ornithologists (personal communication from Dr. Harry C. Oberholser, Cleveland, Ohio).

The bill of the egg-laying mammal, the duckbill platypus, is soft and pliant. It is not covered with a modified epidermal scale and should not be confused with the type of bill found in birds.

Claws, Nails, and Hoofs

The hard structures which are present at the distal ends of the digits are derived from the horny layer of the integument. They differ from other epidermal structures in that they grow parallel to the surface of the skin. Wearing away occurs at the tip. Nails, claws, and hoofs are built upon the same plan, and even superficial observation reveals their homologies. The stratum lucidum of the epidermis is exceptionally well developed at the base of these structures. True claws are present only in reptiles, birds, and mammals. They have also been described in certain fossil amphibians. In the larval stage of the salamander, *Onychodactylus,* horny epidermal caps at the ends of the digits are elongated so as to from sharp "claws" which are much like those of lizards in appearance. The South African "clawed toads," *Xenopus, Hymenochirus,* and *Pseudohymenochirus,* are amphibians which have clawlike tips on the first three digits of the hind feet. The salamander, *Hynobius,* also possesses similar structures. These are actually conelike, cornified, black epidermal caps. Such structures may possibly foreshadow the appearance of claws in higher classes of vertebrates but they are not true claws. Mammals are sometimes grouped into those which bear claws, nails, or both, and those possessing hoofs. The hoofed animals are usually referred to as *ungulates.*

Claws. A claw is composed of a dorsal scalelike plate called the *unguis,* and a ventral plate, the *subunguis,* or *solehorn.* The unguis is the better developed of the two and is of greater importance. The claw covers the terminal bony phalanx of the digit and is thus reinforced.

In the typical reptilian claw (Fig. 4.38) the unguis is curved both longitudinally and transversely and encloses the subunguis between its lower edges. The claw forms a sort of cap at the end of the digit. The outer layers of reptilian claws are shed and renewed periodically.

Claws of birds are typically reptilian in structure, but many variations are to be found, associated with the mode of life of the

Fig. 4.38 Structure of a typical reptilian claw. (*After Bütschli, from Walter and Sayles, "Biology of the Vertebrates," copyright 1949 by The Macmillan Company and used with their permission.*)

bird (Fig. 2.35). The strong, curved talons of the birds of prey; the sturdy, stubby claws of the gallinaceous birds; and the sharp, rather slender structures of birds that perch or cling to surfaces are representative of the many types that may be observed. In birds, shedding of the entire claw is unusual, and growth and wearing away occur at a more or less constant rate. Although claws are generally present only on the feet of birds, the young hoatzin, *Opisthocomus cristatus* (Fig. 4.39), of British Guiana and the valley of the Amazon bears claws on the first two digits of the wings, and these are used as an aid in climbing. In this respect the young hoatzin resembles the fossil *Archaeopteryx* which had three clawed digits on each wing. As the bird grows older the claws disappear.

In the claws of mammals the subunguis is reduced in size and is continuous with the torus, or pad, at the end of the digit, which bears the friction ridges. Members of the cat family possess retractile claws which, when not in use or not extended, are withdrawn into a sheath. They are thus protected and

Fig. 4.39 Wing of young hoatzin, showing claws on first two digits. (*After Pycraft.*)

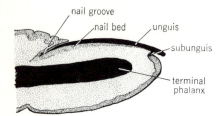

Fig. 4.40 Longitudinal section through end of human finger, showing arrangement of nail.

kept sharp for the purposes for which the animals use them. In some mammals, as in certain lemurs, nails are present on some digits and claws on others (page 46). The tarsier has claws on the second and third digits of the hind feet, the others being supplied with nails.

Nails. In nails the dorsal unguis is broad and flattened, and the subunguis is reduced to a small remnant which lies under the tip of the nail (Fig. 4.40). The so-called "root" of the nail, or region where growth of the unguis takes place, lies embedded in a pocket under the skin called the *nail groove,* or *sulcus unguis.* The *nail bed* lies beneath both the

nail and its root. It is made up of proximal, middle, and distal regions, which show differences in structure. The proximal part, or *matrix,* is the most important. It is this portion which is concerned in the formation of the nail. Its anterior portion in man may be seen through the base of the transparent thumbnail and forms the whitish *lunula,* a crescent-shaped area. The lunula is not so conspicuous on the other digits. The stratum germinativum above the base of the nail forms a rather rough margin where the nail emerges from the sulcus unguis. This is called the *eponychium,* or *cuticle.*

Hoofs. In the hoofs of ungulates the unguis curves all the way around the end of the digit and encloses the subunguis within it (Fig. 4.41). The torus, or pad, lies just behind the hoof and is called the *frog.* Since the unguis is of a harder consistency than the subunguis, it wears away more slowly and a rather sharp edge is thus maintained. The sure-footedness of many ungulates is dependent upon this factor. Work horses are customarily shod with steel horseshoes to prevent the unguis from wearing away too rapidly. Shoes are nailed onto this portion of the hoof. Since

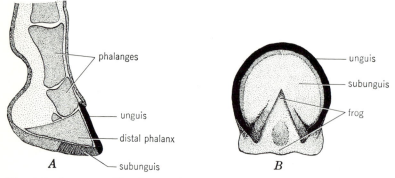

Fig. 4.41 Structure of hoof: A, median sagittal section of distal end of leg of horse; B, ground surface of hoof. (*After Ellenberger in Leisering's Atlas, based on Sisson, "The Anatomy of the Domesticated Animals," W. B. Saunders Company. By permission.*)

the hoof grows continuously, the shoes must be removed at intervals. The unguis is then pared down, and the shoe nailed on again.

Horns and Antlers

Except for the hornlike structures in a few lizards, horns are found only in mammals. Certain dinosaurs possessed horns on the head, consisting of bony projections, each of which was probably covered with a horny epidermal cap. The so-called "horned toads" of the genus *Phrynosoma*, indigenous to the southwestern United States, are lizards which have numbers of such structures on their heads. Each bony projection of the skull is covered with a horny epidermal scale, forming a rather sharp spine. They are of high protective value.

Among mammals, horns are found only in certain members of the order Artiodactyla and in the rhinoceros of the order Perissodactyla. Four types are recognized: the keratin-fiber horn, the hollow horn, the pronghorn, and the antler.

Keratin-fiber horns. The keratin-fiber horn is found only in the rhinoceros. It is a conical structure composed of a mass of hardened, keratinized cells growing from the epidermis covering a cluster of long, dermal papillae. A fiber, somewhat resembling a very thick hair, grows from each papilla. Cells which develop in the spaces between the papillae serve as a cement substance, binding the whole together. The fibers are not true hairs since their bases are not located in follicles extending into the dermis. These unusual and formidable weapons are median in their position on the head. In the Indian rhinoceros only one horn is present, but in the African species there are two, the larger one being in front. A roughened, bony excrescence of the fused nasal bones supports the great horn in the Indian species. Another roughened area,

posterior to the first but less elevated, serves for attachment of the second and smaller horn of the African species of rhinoceros. It is located on the frontonasal region of the skull.

Hollow horns. The hollow horn (Fig. 4.42) is the type found in cattle, sheep, goats, buffaloes, and others. In certain species they are found only in males, but this is not true of all. The hollow horn consists of a projection of the frontal bone of the cranium covered by a cornified layer of epidermis. A cavity, which is continuous with the frontal sinus, extends into the bony projection. The horny layer is not shed. The powder horns of early settlers of America were prepared by loosening the horny covering from the bone beneath.

In a very young animal which will bear horns of this type, a loose buttonlike mass of bone, the *os cornu,* can be felt under the skin which lies over the frontal bone. If the os cornu is destroyed by caustic or is otherwise removed, the horn will not develop. If it is left intact, the horn will form and the os cornu ultimately becomes fused with the frontal bone. In certain domestic animals it is often desirable either that the horns be prevented from developing or that they be removed (polled). The former method, involving destruction of the os cornu, is more de-

Fig. 4.42 Hollow horn of domestic cow with bony core removed.

sirable, since polling cattle with well-developed horns is much more difficult and may cause harm to the animal which is treated.

Pronghorns. A unique type of horn, the pronghorn, is found only in the antelope, *Antilocapra americana,* of the western United States. It consists of a projection of the frontal bone covered with a horny epidermal sheath. The sheath usually bears one prong (Fig. 4.43), although as many as three have been observed. The unusual feature of this type of horn is that the horny covering is shed with annual periodicity and a new horn forms from the epidermis which persists over the bony projection.

Antlers. The male members of the deer tribe possess branched antlers which project from the frontal bones. Only in the reindeer and caribou do both sexes bear antlers. In its fully developed state an antler is composed of solid bone. It is therefore entirely mesodermal in origin and properly should not be called a horn at all. Antlers are to be con-

sidered as parts of the dermal skeleton. Since, however, the antler is formed under the influence of the integument, it is expedient to include it under the discussion of integumentary derivatives.

In the young male an outgrowth of the frontal bone develops on each side (Fig. 4.44). This is covered with soft skin, and the blood vessels in the overlying dermis bring minerals and other elements necessary for the growth of the structure. The antler is then said to be "in velvet." When it has reached the full extent of its development, a wreathlike burr grows out around the base. This cuts off the blood supply to the skin covering the antler. The skin then dries, cracks, and is rubbed off in shreds by the deer. This usually occurs as autumn approaches. At the end of the first season the antler appears as a single "spike." A young buck with a single spike on each side is referred to as a "spike buck." In the spring, because of certain degenerative processes which occur between the frontal bone and the burr, the antlers become loosened and are shed. The skin then closes over the area. Soon a new antler begins to grow, but this time it is branched. In fact, each year more branches appear, and the structure becomes more complicated. The developing antlers remain in velvet until autumn approaches. During this time they are sensitive and warm to the touch. Again the burr cuts off circulation, and the skin covering the antler is sloughed. The entire bony antler is shed the following spring.

In the giraffe both sexes possess antlers. These, however, are small and rather inconspicuous. They are never shed and remain permanently in velvet. Antlers are secondary sex characters in other species with the exception of the caribou and reindeer. They reach the height of their development just before the mating season, when it is cus-

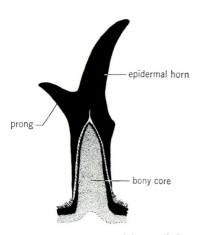

Fig. 4.43 Structure of horn of the pronghorn antelope.

- epidermal horn
- prong
- bony core

Anatomy of the Chordates

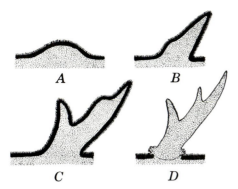

A B

C D

Fig. 4.44 Stages in the development of an antler. D shows the final stage after the covering skin has been sloughed.

Starvation and death of both combatants usually follows.

Antler growth is apparently regulated by hormones secreted both by the testes and the anterior lobe of the pituitary gland (G. B. Wislocki, J. C. Aub, and C. M. Waldo, 1947, *Endocrinology,* **40**:202). In the Virginia deer, castration before the eighth month of life completely suppresses antler growth. If the antlers have already developed and if they are in velvet at the time when castration occurs, then they are not shed but remain permanently in velvet. However, castration of males with antlers which have lost their velvet results in a prompt shedding. Subsequently, new antlers grow, and these remain permanently in velvet. Antler growth has been induced in castrated (ovariectomized) female Virginia deer by administration of the male hormone testosterone.

tomary for the males to fight for domination of the herd. Occasionally the antlers of two bucks become interlocked in combat in such a manner that they cannot be extricated.

SUMMARY

1. The vertebrate integument consists of an outer, ectodermal, epithelial layer called the epidermis and an underlying mesodermal layer, the dermis, composed largely of connective tissue, blood and lymphatic vessels, nerves, and smooth-muscle fibers.

2. Pigment in the integument occurs either in the form of small granules in the cells of the lowest layers of the epidermis or else in special branched cells, called chromatophores, in which the pigment may become dispersed throughout the cell or concentrated in its center around the nucleus. Chromatophores of this type are usually located between the epidermal and dermal layers. They are not present in birds or mammals.

3. Both epidermis and dermis give rise to various integumentary structures which are for the most part, protective and adaptive in character.

4. From the epidermis are derived: (*a*) skin glands of the following types: mucous, poison, sweat, sebaceous, ceruminous, mammary, and scent (photophores are light-producing modifications of skin glands); (*b*) scales, formed from the hardened, horny, corneal layer. They include epidermal scales of reptiles, birds, and mammals; feathers; hair; claws; nails; hoofs; spurs; true horns; beaks and bills.

5. From the dermis are derived: (*a*) bony scales of the ganoid, placoid, ctenoid, and cycloid types; (*b*) bony plates in the skin of certain reptiles and mammals; and (*c*) membrane bones of the skull.

6. The dermis, through its blood vessels, supplies nourishment to the developing epidermal structures but is not otherwise involved. Dermal structures arise within the dermis itself. In certain cases bony dermal plates are closely associated with epidermal scales which lie above them.

7. The evolution of the integument is fundamentally correlated with the transition from aquatic to terrestrial life. The highly developed epidermis of terrestrial forms, with its many modifications fitting the animal for life on land, is indicative of the trend. Dermal derivatives are basically more primitive.

| 5 |

DIGESTIVE
SYSTEM

In order to produce heat and other forms of energy, the living organism is constantly using up materials which have been stored within it in one form or another. It is clear that the body would soon be depleted of such energy-producing substances if they were not replenished. The primary concern of every living thing is to obtain certain necessary substances so that it may continue to live. Oxygen, food, and water are necessary for the continuance of life. Oxygen, present in the air (or dissolved in water), enters the body through the agency of the respiratory system, to be discussed later; but food and water are, in most animals, first taken into the digestive system. Water can be absorbed unchanged through the walls of the alimentary tract, but few kinds of food are available to a vertebrate in such a form that they can be used or absorbed directly.

The chief function of the digestive system is to prepare foods so that they may be made available to the body for use in growth, structural maintenance, and the production of various forms of energy. The main principle involved in digestion is the breaking down of complex molecules of certain foods, by a series of chemical changes, into molecules of simpler structure. These can then either be absorbed in soluble form through the walls of the digestive tract and enter the blood vessels (in the case of proteins and carbohydrates) or be otherwise prepared so that they may enter the small vessels or lacteals of the lymphatic system (if the food is in the form of oil or fat). It seems, nevertheless, that some fat enters the hepatic portal vein.

Food that is present in the digestive tract is not strictly inside the body. Not until it has been digested and enters the blood-vascular or lymphatic system, or is otherwise incorporated within the body tissues, can it be spoken of as actually being within or part of the body. Ingested material that cannot be digested thus never really enters the body proper. It merely passes down the digestive tract and is eliminated unchanged.

GENERAL STRUCTURE

In all vertebrates the digestive, or alimentary, tract is *complete,* having a *mouth* at one end and an *anus* at the other. The greater part of the lining of the alimentary tract (mesodaeum) is of endodermal origin, but the regions of ingress (mouth, or stomodaeum) and egress (anal canal, or proctodaeum) are lined with epithelium of ectodermal derivation. In the completely developed alimentary tract there is no exact point which would indicate the original line of junction of stomodaeum or proctodaeum with the mesodaeum. Within rather rough limits it may be stated that the original stomodaeal region ends just posterior to the teeth and that the original proctodaeum extends inward only to the point where the urogenital ducts enter the cloaca.

During embryonic development in many vertebrates the endodermal portion of the digestive tract contains relatively large amounts of yolk. In embryos which develop

from polylecithal ova there is so much yolk that it cannot be contained within the limits of the body wall. Most of it is located in a *yolk sac* which is attached to the alimentary canal by means of a *yolk stalk* (Fig. 5.1). As the embryo develops, the blood vessels surrounding the yolk sac gradually absorb yolk and carry it to the developing tissues which utilize it in growth processes. The yolk never enters the digestive tract directly. The yolk sac becomes smaller and smaller until finally nothing remains save a small, scarlike remnant on the intestine which indicates the original point of attachment of the yolk stalk.

The original body cavity, or coelom, in vertebrates becomes partitioned off to varying degrees in members of the several classes, forming such cavities as the *pericardial cavity,* in which the heart lies, the *pleural cavities,* containing the lungs, and the *abdominal,* or *peritoneal, cavity* in which the various abdominal viscera are located. In the lower classes of vertebrates only pericardial and *pleuroperitoneal* cavities are present, since the pleural cavities, which in mammals become completely separated from the abdominal cavity, have not as yet become specific entities. In some forms even the pericardial and pleuroperitoneal cavities are not completely separated, being connected with

Fig. 5.1 Embryo of dogfish shark, *Squalus acanthias,* showing yolk sac attached to body by means of yolk stalk. (*Drawn by G. Schwenk.*)

each other by a *pericardioperitoneal canal.*

Inside the large pleuroperitoneal or peritoneal cavity, as the case may be, the body wall is lined with a smooth, shiny membrane, the *parietal peritoneum.* Reflected over the surface of the alimentary tract in this region is the *visceral peritoneum.* The parietal peritoneum and the visceral peritoneum are connected to each other along the middorsal line of the body cavity by the *dorsal mesentery.* This is actually a two-layered sheet continuous dorsally with the two halves of the parietal peritoneum. The relationships of these layers are indicated in Fig. 5.2. With the differentiation of the alimentary tract into various regions, appropriate names are applied to certain portions of the dorsal mesentery. Thus, the term *mesogastrium (great omentum)* is used to denote the part of the dorsal mesentery which supports the stomach, that associated with the colon is the *mesocolon,* etc. In reptiles and mammals the dorsal mesentery is continuous, extending from one end of the alimentary tract to the other; in amphibians and birds, however, portions may disappear so that the structure becomes discontinuous. A complex degree of folding of the dorsal mesentery is evident in those forms in which the intestine is elongated and coiled. A *ventral mesentery,* connecting the gut with the ventral body wall, is usually present during developmental stages. Thus for a time the body cavity is divided into right and left halves. The ventral mesentery disappears before long in most cases. In some regions a part of the alimentary tract is connected to an adjacent organ or part by a persistent remnant of the ventral mesentery. These are generally referred to as *omenta* or *ligaments.* The names of the various omenta and ligaments are indicative of the structures which they connect.

The wall of the alimentary tract (Fig. 5.3) is, in general, made up of four main layers, or coats: serous, muscular, submucous, and mucous. The outer *serous layer* is the *visceral peritoneum* which almost entirely surrounds those parts of the alimentary tract that lie within the coelom. This layer, then, is lacking in the region of the esophagus. The cells of the serous layer rest on loose connective tissue. Beneath this lies the external *muscular coat,* which in turn is composed of two layers often referred to as the *outer longitudinal* and *inner circular* layers. This is because in cross section the smooth-

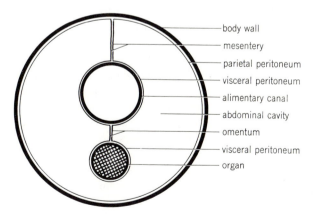

body wall

mesentery

parietal peritoneum

visceral peritoneum

alimentary canal

abdominal cavity

omentum

visceral peritoneum

organ

Fig. 5.2 Diagram showing relationship of a mesentery and an omentum to parietal and visceral peritoneum.

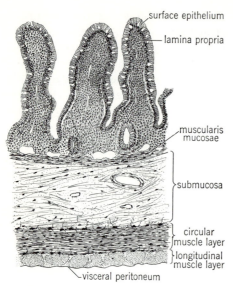

surface epithelium

lamina propria

muscularis mucosae

submucosa

circular muscle layer

longitudinal muscle layer

visceral peritoneum

Fig. 5.3 Part of transverse section of ileum showing three villi and the layers of the intestine.

muscle fibers of the inner layer appear to be arranged in a circular manner, whereas those of the outer layer are cut transversely. It has been demonstrated that both these layers are actually arranged in the form of spirals. The inner layer is a very close spiral with frequent turns; the outer layer is in the form of an open spiral, the turns of which are rather far apart. Some fibers from each layer cross over into the other. Between the two muscle layers lies a nervous network composed of nerve cells and bundles of nerve fibers, the parasympathetic *myenteric plexus,* or *plexus of Auerbach.*

Next to the circular muscle layer is the *submucosa,* composed of dense connective tissue and elastic fibers with a small island of fatty, or adipose, tissue here and there. Intestinal glands may lie in this layer, particularly in the duodenal portion of the intestine. Blood and lymphatic vessels are present in the submucosa in addition to a plexus of

nerve cells and fibers, the submucous *plexus of Meissner.* The nerve fibers associated with the plexus of Meissner are postganglionic fibers of the sympathetic nervous system originating in the superior mesenteric ganglion (page 652). Some ganglionic cells of the parasympathetic division of the autonomic nervous system are also present. The submucosa is the strongest layer of the alimentary tract. The *mucous coat (mucosa),* or innermost layer, is in turn composed of three portions. Next to the submucosa is the *muscularis mucosae,* a thin layer of smooth-muscle fibers and elastic connective tissue consisting of inner circular and outer longitudinal layers. A third layer of oblique fibers is sometimes present in part of the stomach wall. Internal to the muscularis mucosae is a layer of peculiar connective tissue, the *lamina propria* of the mucosa. Numerous small nodules of lymphatic tissue lie within the lamina propria, as do blood vessels, lymphatic vessels, and nerves. The lamina propria is covered with a basement membrane which supports a single layer of simple, columnar, epithelial cells, constituting the *surface epithelium.*

The muscular layers are under the involuntary control of the autonomic nervous system of which the plexuses of Auerbach and Meissner are parts. Contraction of the smooth-muscle fibers is responsible for peristalsis, which propels food materials along the intestine, and for segmenting and pendular movements which do not propel food but aid in kneading it and mixing it thoroughly with digestive juices and enzymes. Striated, or voluntary, muscle fibers are found only at the two ends of the canal.

The digestive tract is admirably suited for its functions. Its wall contains muscles which thoroughly mix the contents and propel them along; it is provided with glandular elements which secrete fluids and ferments (enzymes) necessary for proper chemical reactions to take place. The thin cell walls of the lining epithelium are well suited for absorption, and the rich supply of blood and lymphatic vessels not only ensures the proper nourishment of the tissues of the digestive tract itself but serves as the pathway by means of which digested substances are transported to the liver, where assimilation occurs, and thence to all parts of the body.

In vertebrates the alimentary tract is typically divided into several sections. These are, in order, the *mouth, pharynx, esophagus, stomach, small intestine, large intestine,* and *cloaca* (Fig. 5.4). The term cloaca has been adopted from the Latin word meaning sewer. It is a common chamber into which the intestine, urinary ducts, and reproductive canals discharge. A cloaca is present in the adult stage of birds, reptiles, amphibians, and many fishes, being notably absent in teleosts, chimaeras, *Polypterus,* sturgeons, and others. Among adult mammals, too, a cloaca is lacking except in monotremes and the pika. However, even in higher mammals a cloaca is present for a time during embryonic development.

Such digestive glands as the *salivary glands, liver,* and *pancreas* are derived from the alimentary canal with which they are connected by means of ducts.

Other structures derived from the endoderm of the primitive digestive tract in the region of the pharynx, but which are not concerned with digestion, include the *thyroid, thymus,* and *parathyroid glands;* the *tonsils; middle ear; Eustachian tubes;* and the *respiratory organs* (Chap. 6, pages 201 to 249).

MOUTH AND STRUCTURES ASSOCIATED WITH IT

It will be recalled that the primitive mouth, or stomodaeum, which develops early during

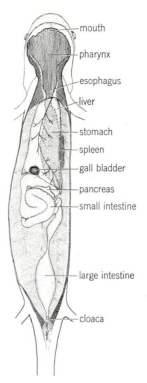

Fig. 5.4 Digestive organs of salamander, *Eurycea bislineata.* The other organs have been removed.

embryonic life, is lined with ectoderm. It is, for a time, separated from the endodermal portion of the digestive tract by an oral, or pharyngeal, membrane. This soon ruptures and the stomodaeum and gut become continuous. The stomodaeum develops into part of the mouth, the lining of which is ectodermal. The exact point where ectoderm and endoderm merge is often impossible to discern, and there is much variation. It is important to remember that the various structures derived from the stomodaeum are, for the most part, ectodermal in origin or are at least covered with material derived from ectoderm.

Amphioxus

At the anterior end of amphioxus is a funnel-shaped structure called the *oral hood*. This is bordered by a number of tentaclelike *cirri*. The cavity within the oral hood is spoken of as the *vestibule*. This is not comparable to the mouth cavity of higher forms. The true mouth opening is located at the apex of the vestibule. On the walls of the oral hood are several lobelike structures covered with cilia. These form the so-called *wheel organ*. Posterior to this is a membranous *velum* which contains the true mouth opening in its center and which also bears several *velar tentacles*.

Cyclostomes

In cyclostomes a *buccal funnel* is located at the anterior end. This is bordered by many small *papillae*. In lampreys the lining of the buccal funnel is beset with numerous *horny teeth* (Figs. 5.5 and 5.18), not in any way

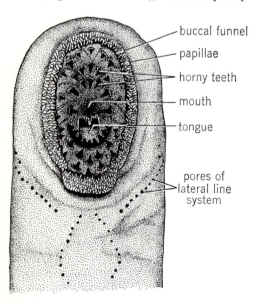

Fig. 5.5 Ventral view of anterior end of lamprey showing buccal funnel and tongue beset with horny, epidermal teeth.

comparable to the true teeth of higher forms. At the apex of the funnel is the mouth opening through which the tongue projects. The tongue also bears horny teeth. No jaws are present, and the mouth is permanently open. The name Cyclostomata (round mouth) has been given to the lowest group of vertebrates because of this characteristic.

In cyclostomes the buccal funnel actually represents only the posterior portion of the original stomodaeum. The single, median nostril on the dorsal surface of the head, together with the olfactory sac and nasopharyngeal pouch (Fig. 5.18) are really derivatives of the stomodaeum which have become separated from the main portion. This is due to a shifting of these structures during developmental stages from their original ventral position (Fig. 14.23). The front part of the head itself in the adult is thus covered with stomodaeal ectoderm.

Gnathostomes

All the remaining vertebrates possess both upper and lower jaws and are generally referred to as *gnathostomes*. The size and shape of the mouth opening are varied, being generally correlated with the type of food eaten and the manner in which it is obtained. *Cheeks* may mark the boundary of the opening of the mouth on either side. Many gnathostomes, however, have no cheeks and may open their mouths to an astonishing degree.

Lips

The upper and lower boundaries of the mouth opening are known as *lips*. In all forms below mammals they are immovable and consist of epithelial folds or pads or connective tissue which meet when the mouth is closed. In turtles and birds the jawbones are covered with hard, horny beaks. Truly movable lips occur only in mammals. The presence of muscles in the lips is associated with the development of dermal facial muscles, which reach their most complex development in man. Notable exceptions to the movable condition of the lips in mammals include the platypus and whales. The bill of the platypus is not horny but is soft, pliant, and extremely sensitive. Whales merely have soft pads of connective tissue which serve to keep the mouth tightly closed.

Mammalian lips are covered with skin on the outside and lined with mucous membrane on the inside. A modified area lies between the two. This is most conspicuous in man and is richly supplied with sensory receptors. Among other functions, the lips of mammals serve to keep the food between the teeth while eating. They may be used as prehensile organs in grasping or picking up food and other objects; they aid in the process of drinking in many species and are particularly well adapted for sucking. In the human being they are also used in modifying the voice and in osculation.

Vestibule

The space between the lips and jaws is the *vestibule*. Lateral to the mouth opening the vestibule may be bounded on the outside by the cheeks and on the inside by the gums. In mammals a *labial frenum,* composed of a fold of mucous membrane, connects the middle of each lip to the gum, or *gingiva*.

Cheeks, when present, are covered with skin but are lined with a rather thick, non-keratinizing, stratified squamous epithelium of a type typically present on wet surfaces subjected to wear and tear. Absorption does not take place through this lining epithelium.

In certain mammals cheek pouches of considerable size are present but not all are homologous. In some species, as in the pocket gopher, they are actually *external* pouches, lined with fur and used as tem- porary storage places for food when these rodents are foraging. Squirrels have the same habit but to a lesser degree. Certain Old World monkeys also have well-developed cheek pouches. According to H. Burrell (1927, "The Platypus," Angus and Robertson, Ltd.) the cheek pouches of the platypus are used chiefly for holding fine gravel, which is used as an aid in masticating chitinous or other hard food.

A variety of glands opens into the vestibule. These are, for the most part, mucous glands. *Labial mucous glands* are small structures opening on the inner surface of the lips. In man they may be identified as small lumps when one pushes the tongue against the lips. In some mammals *molar mucous glands* open near the molar teeth. The *parotid duct* (*Stensen's duct*), which drains the parotid salivary gland, opens into the vestibule opposite the first or second upper molar teeth in man and at a corresponding point in other mammals.

Oral cavity

In fishes, the mouth, in addition to serving as a passageway for food, is concerned with the passage of water which contains the dissolved oxygen used in respiration. In most fishes the nasal cavities are quite independent of the oral cavity, but in the Sarcopterygii they first communicate with the mouth by means of a pair of *internal nares,* or *choanae.* In amphibians the lining of the oral cavity is ciliated. The tissue underlying the epithelium is unusually vascular in amphibians, serving as an important aid to respiration.

The roof of the mouth and pharynx is often referred to as the *palate*. In most reptiles and birds a pair of *palatal folds* grows medially for a variable distance on either side. The palatal folds do not meet in the median line, thus forming a *palatal cleft* through which the nasal and mouth cavities are in communication. The upper portion provides a region through which air can pass more freely than otherwise. In both crocodilians and mammals horizontal shelflike processes of certain bones of the skull within the palatal folds fuse with their partners of the opposite side, thus forming a *secondary palate* which effectively separates the nasal passages from the mouth cavity. These passages thus communicate with each other more posteriorly than they would otherwise. The anterior part of the secondary palate is called the *hard palate* since it is reinforced by portions of the premaxillary, maxillary, palatine, and, in some forms, of the pterygoid bones. The posterior part in mammals is the *soft palate* since it has no bony foundation. Abnormal development of the palate in mammals may result in a condition spoken of as *cleft palate*. The nasal passages communicate with the pharynx posterior to the soft palate. In man, the soft palate ends in a fleshy, pointed elongation, the *uvula,* which hangs downward and backward. This serves to close off the nasal passageway from the mouth during the act of swallowing. In many forms the palate is thrown into transverse ridges, the *palatine rugae.* These are not so prominent in man as in such carnivores as dogs and cats in which they undoubtedly serve as an aid in securing prey. Both epidermal and true teeth are found with some frequency as palatal developments among various vertebrates (see pages 168 and 170).

It is convenient at this point to mention the baleen, or "whalebone," found in the mouths of the so-called whalebone whales. Baleen consists of large, often enormous, horny sheets, or plates, which hang in series from the palate and edges of the upper jaw. The lower ends of the plates are fringed and serve to strain from the water the microscopic organisms on which these huge animals feed. Baleen is formed from very large

cornified papillae which become fastened together.

Glands of the oral cavity. In cyclostomes a so-called *salivary gland* opens into the mouth cavity on either side just below the tongue. The gland secretes an anticoagulant, called *lamphedrin,* which facilitates the flow of blood from an animal when it is attacked by the lamprey.

Fishes, and amphibians that spend their entire life in the water, have no glands other than simple mucous cells opening into the mouth cavity. Glands first appear in the mouths of terrestrial vertebrates, their prime function being to moisten the food and render it slippery so as to facilitate swallowing. In higher forms specialization of the mouth glands occurs, and they may play an important part in capturing prey and in digestion. In poisonous snakes and lizards certain mouth glands serve as dangerous organs of defense.

AMPHIBIANS. In terrestrial amphibians a mucous gland called the *intermaxillary,* or *internasal, gland* lies in the nasal septum between the premaxillary bones and nasal capsule. It is larger in anurans than in urodeles, but in both cases the ducts open in the front of the mouth. The secretion of this gland helps to give the tongue its adhesive properties. The intermaxillary gland is lacking in caecilians. Another gland, referred to as the *pharyngeal gland,* is present in frogs and toads. It lies near the internal nares, its secretion passing into them. Mucous *lingual glands* are numerous on the protrusible tongues of frogs and toads as well as in certain salamanders. Their secretion aids in the capture of prey. In some frogs a digestive enzyme, *ptyalin,* is secreted by the mouth glands.

Reptiles. The oral glands of reptiles show a much more distinct grouping than do those of amphibians. A *palatine gland* is present which is homologous with the inter-maxillary gland of amphibians. In addition, *lingual, sublingual,* and *labial glands* are present.

In poisonous snakes the gland which secretes the poison, or venom, is apparently a modification of a labial gland of the upper jaw. It may be homologous with the parotid salivary gland of mammals. The labial poison gland is encapsulated by a strong sheath of connective tissues. Its duct opens into the cavity or groove of the poison fang (Fig. 5.6).

In the Gila monster, *Heloderma,* which is the only poisonous lizard extant, it is the sublingual gland which is modified to form the poisonous secretion. This passes through four ducts which penetrate the bone of the lower jaw to emerge in the vestibule in front of the grooved teeth.

In marine turtles and in crocodiles and alligators, oral glands are poorly developed.

BIRDS. Well-developed *anterior* and *posterior sublingual glands* opening in the floor of the mouth are present in birds. Another gland at the angle of the mouth, the so-called *angle gland,* may possibly be homologous with the labial glands of reptiles. In birds, labial and intermaxillary glands are missing. Numerous groups of small glands open separately on the roof of the mouth. A digestive

function has been ascribed to the mouth glands of many birds, the enzyme ptyalin being present in their secretions.

MAMMALS. Many small mucous glands are located on the palate and tongue in mammals. Here, however, for the first time, we find the large and distinctly grouped *salivary glands.* There are usually three sets of salivary glands, which are named according to their position. The *parotid gland* lies in the region of the ear, usually beneath and somewhat anterior to the external auditory meatus. Its position may vary within limits in different species. The *parotid duct (Stensen's duct)* courses over the masseter muscle of the cheek to enter the vestibule of the mouth opposite the upper molar teeth. The *submandibular,* or *submaxillary gland* lies in the posterior part of the lower jaw. Its duct, the *submandibular,* or *submaxillary duct (Wharton's duct),* opens in front of the tongue near the lower incisor teeth. Rabbits and horses lack the submandibular gland. The *sublingual gland* is smaller than the other salivary glands. It is composed of a major portion, the sublingual gland proper with its duct *(of Bartholin),* and numerous smaller elements each with its own duct *(of Rivinus).* The ducts open separately into the oral cavity along the jaw-tongue groove and lateral to

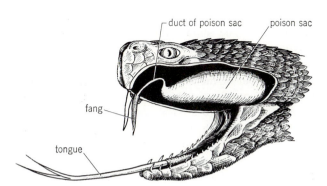

duct of poison sac   poison sac

fang

tongue

Fig. 5.6 Head of rattlesnake dissected to show relation of poison gland to hollow fang.

the submandibular duct, with which they frequently unite. The sublingual gland is not present in the mouse, mole, or shrew. Salivary glands are secondarily reduced or wanting in whales and sirenians.

The secretions of the three sets of salivary glands vary. In man the submandibular and sublingual glands produce most of the mucin in the saliva. This gives saliva its mucilaginous character. The secretion of the parotid gland is more watery and contains the enzyme ptyalin in greater amounts than do the other two glands. Ptyalin plays a role in digestion, since in its presence and in a neutral or slightly alkaline medium, starches are converted into maltose (malt sugar) in a step-by-step process. Maltose is of a simpler chemical structure than starch. Ptyalin is not secreted by the salivary glands in all mammals. Another enzyme, *maltase,* is also present in the saliva of some forms.

The disease known as *mumps,* which affect human beings, is technically known as *contagious parotitis.* It is marked by an inflammation and swelling of the parotid gland. The causative organism is a virus. Mumps may occur on both sides or on one side only. Sometimes the submandibular and sublingual glands may be involved, and occasionally the testes, mammary glands, and labia majora become swollen.

Mucous *molar glands* are especially well developed in such herbivorous forms as the Artiodactyla. Their secretion aids greatly in swallowing the coarse vegetation on which these animals feed. The large *orbital glands* of the dog family are mucous glands, the ducts of which open into the mouth in the region of the last molar teeth.

Tongue. The tongues of vertebrates show much diversity, and not all are homologous with the mammalian organ with which we are most familiar.

CYCLOSTOMES. The cyclostome "tongue" is a very highly specialized structure which is not homologous with the tongues of higher vertebrates. Cyclostomes attach themselves to fishes by means of the buccal funnel, which acts as a sort of suction cup. They then rasp the skin and muscles of their prey by a back-and-forth movement of the muscular tongue which bears horny teeth on its surface (Fig. 5.5).

FISHES. The tongue, or rather the *primary tongue,* of fishes should scarcely be called a tongue. It is merely a fleshy fold which develops from the floor of the mouth between the mandibular and hyoid arches. The hyoid arch frequently extends into this fold and supports it. The tongue is devoid of muscles and can be moved only within narrow limits by varying the position of the hyoid arch. Sensory receptors may be present. In some fishes the tongue bears small papillae, and in a few teleosts, such as the salmon, teeth may even be present on the tongue. These are supported by a *glossohyal bone* which unites the hyoid arches of the two sides.

AMPHIBIANS. Four different conditions pertaining to the tongue prevail in amphibians. In a certain group, called the *aglossal toads,* no tongue of any sort is present. This represents a degenerate condition. In some urodeles, such as *Necturus,* in which the entire life is spent in the water, the tongue shows very little change over the condition found in fishes. Other amphibians have movable tongues which may be thrust out of the mouth and used in capturing the animals on which they feed. These tongues consist of a basal portion which is homologous with the primary tongue of fishes and an expanded, anterior, glandular portion which is well supplied with protractor and retractor muscles. These together make up the so-called *definitive tongue.* In frogs and toads the base of the tongue is attached anteriorly at the margin of the jaw. Its free end is folded back on the floor of the mouth when at rest and is flipped out of the mouth during activity. In urodeles the tongue has a more extensive attachment to the supporting hyoid arch. Some salamanders, such as *Eurycea* (Fig. 5.7), can thrust out their tongues (boletoid type) directly and pull them back again with lightninglike rapidity.

REPTILES. In turtles, crocodiles, and alligators the tongue is not protrusible and lies on the floor of the mouth. Snakes and lizards, on the other hand, have well-developed tongues which can be extended and retracted. Many possess a sheath into which the greater part of the tongue can be withdrawn. Because of the presence of a small notch at the tip of the lower jaw, snakes can extend their bifurcated tongues out of their mouths even when the latter are tightly closed. Among the lizards, the tongue of the chameleon is best developed and is used as a prehensile organ in capturing prey. It has been reported that the bifurcated tongues of certain lizards and of snakes, when retracted, are placed in the blind pockets of Jacobson's organ (see page 209) which opens on either side into the mouth cavity near the choanae. This seems to be an accessory olfactory device. Volatile chemical substances, particles of which cling to the surface of the tongue, may thus be detected by the animal.

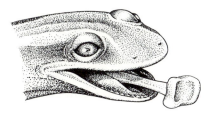

Fig. 5.7 Head of salamander, *Eurycea bislineata,* with tongue extended.

The reptilian tongue represents a higher degree of development than that of amphibians. It includes a fold over the hyoid arch and a homologue of the glandular tongue, or *tuberculum impar,* seen in amphibians. This arises between the basihyal bone and the lower jaw. A pair of *lateral lingual folds* above the mandibular arch also contributes to tongue formation. The tongue is thus composed of four portions which have fused together.

BIRDS. The tongue of birds is practically lacking in intrinsic muscles. It is usually covered with horny material and may be provided with pointed, thornlike projections. In some birds it is bifurcated, and in others it is split at the end, forming a brushlike structure. Since the tongue lacks intrinsic muscles, the only way it may be moved is by altering the position of the hyoid apparatus which supports it at its base. Although most birds can move the tongue only slightly, the woodpeckers can extend it for a considerable distance. The hyoid apparatus of the woodpecker is long, slender, and coiled when at rest (page 463). When it is extended the coils straighten out and the bird is thus able to secure the insect larvae which make up the bulk of its food.

In birds the tongue has lost the lateral lingual folds derived from the mandibular arch and which are possessed by reptiles. It is largest in the birds of prey and in the parrots. The familiar thick, soft tongue of the parrot is large, not because of muscular development but rather on account of the fat, the blood vessels, and the glandular elements which it contains.

MAMMALS. The mammalian tongue is best developed among vertebrates and shows many modifications not only in form but in function. In all groups except the whales it is movable because of the well-developed intrinsic muscles which are contained within it. These are striated muscles, the fibers of which are arranged in bundles which course among each other in three planes, thus accounting for the high degree of mobility of which the mammalian tongue is uniquely capable. In the anteaters the tongue is best developed of all, and its muscles may extend back as far as the sternum.

The mammalian tongue is derived from five separate portions which have united. These include the unpaired tuberculum impar, paired lateral lingual swellings from the mandibular arch which make up the *body* of the tongue, and paired fleshy ridges, contributed by the hyoid arch and the third and fourth branchial arches, which form the pharyngeal portion, or *root,* of the tongue. The hyoid apparatus supports the tongue at its base. The thyroid gland originally arises from the midline at the point where the anterior and posterior portions of the tongue come together. A small depression, or pit, the *foramen caecum,* at the base of the tongue (Fig. 5.8), represents the point of origin of the thyroid gland.

The mucous membrane under the front of the mammalian tongue forms a median fold, the *lingual frenum,* which is attached

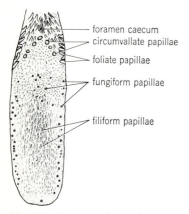

foramen caecum
circumvallate papillae
foliate papillae
fungiform papillae
filiform papillae

Fig. 5.8 Tongue of cat, showing arrangement of papillae.

to the floor of the mouth. Occasionally this frenum in human beings is too short and thus restricts the freedom of movement of the tongue. A person with an abnormally short lingual frenum is said to be tongue-tied. Speech may be difficult, at least when the production of certain sounds is attempted. A simple operation which involves cutting the frenum usually will remedy this condition.

The tongue bears papillae of various kinds (Fig. 5.8) which may or may not be associated with taste buds. Among those most commonly found in mammals are *filiform, fungiform, foliate,* and *vallate* or *circumvallate papillae. Filiform papillae* are small, conical projections making up the greater part of the plushy surface of the tongue. In the cat family (Felidae) the epithelium covering the filiform papillae is highly cornified. These modified papillae are used in cleaning the fur as well as in rasping the flesh off bones, which may be cleaned completely. *Fungiform papillae,* as their name implies, are toadstool-shaped structures scattered over the surface of the tongue. *Foliate* papillae are broad and leaflike. Their numbers vary in different species, but they are generally situated near the base of the tongue. They are lacking in man. *Circumvallate papillae* are few in number but large. Each is surrounded by a trench. These papillae are arranged in the form of an inverted V with the apex pointing toward the base of the tongue. The foramen caecum lies just beyond the apex of the V. For a further discussion of the lingual papillae and the taste buds associated with them, the reader is referred to Chap. 14, pages 715 to 719.

The tongue of the dog is highly important in keeping the body temperature constant. Although dogs have sweat glands in the skin (page 121), they probably play little, if any, part in temperature regulation. Some

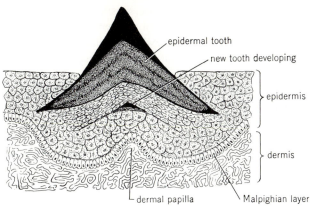

epidermal tooth

new tooth developing

epidermis

dermis

dermal papilla

Malpighian layer

Fig. 5.9 Section through epidermal tooth of lamprey. (*After Warren.*)

mechanism other than evaporation of sweat is necessary for the elimination of excess heat. In the dog this is accomplished by panting. The moist, highly vascular tongue swells and hangs from the mouth. The rapidly expired and inspired air, passing over the surface of the tongue, evaporates the moisture and thus produces a cooling effect.

In ruminating animals the tongue is used as a prehensile organ and is important in browsing. In the giraffe the extremely long tongue may be twisted about a twig to strip off the leaves in a remarkable manner. Other mammals use their tongues in lapping up fluids and in manipulating food that is within the mouth. In man, articulate speech is made possible largely by movements of the tongue.

Teeth. Teeth are primarily employed by animals in manipulating food, but in some species they serve an additional function as structures used in attack or defense. In most vertebrates teeth are used only to hold captured prey. In others they may be modified and are of use in cutting, crushing, or grinding food. Among vertebrates, two types of teeth are found: *epidermal* and *true* teeth.

EPIDERMAL TEETH. These consist of hard, pointed, cornified, epithelial projections. They are present in but few vertebrates and are in no way homologous with true teeth discussed below. Perhaps the best-developed epidermal teeth are those of cyclostomes. They are conical, horny structures which in the lamprey are arranged in a definite pattern on the inner walls of the buccal funnel and on the highly specialized rasping tongue (Figs. 5.5 and 5.18). Each tooth is derived from the epidermis overlying a papilla of dermis, or corium (Fig. 5.9). In the hagfish a single median tooth lies above the mouth aperture and two rows of small teeth are located on the tongue.

Among amphibians, epidermal teeth are found on the edges of the jaws of most larval anurans. They are in reality cornified papillae which are used in scraping off algae on which the larvae feed.

Epidermal teeth are also found in the duck-bill platypus. True teeth are present in the animal during developmental stages, but they are lost even before birth. When the true teeth are erupting, the surrounding ectoderm becomes cornified and grows under them. Thus, when the true teeth are shed, a horny plate remains which serves as an aid in holding and perhaps in crushing food. Horny crushing plates are also present in members of the mammalian order Sirenia, but these animals possess true teeth as well.

TRUE TEETH. All other teeth found in vertebrates are built upon the same general plan although variations exist in the composition of the hard, shiny, translucent covering of the exposed surface. In mammals this outer layer is made up of *enamel,* derived from ectoderm; in most other forms possessing true teeth the hard covering layer is composed of *vitrodentine* of mesodermal origin. The structure of the latter type of tooth is, in general, rather similar to that of the placoid type of scale found in elasmobranch fishes (page 127).

An opinion has been held in the past that true teeth have been derived phylogenetically from the placoid type of scale. The modern view is that both are modified remnants of bony dermal plates which formed the armor of such ancestral forms as ostracoderms and placoderms and are, therefore, homologous. The type of true tooth covered with vitrodentine, accordingly, deviates less from the ancestral type than does the mammalian tooth with its covering layer of enamel.

Structure and development. The portion of the true tooth that is exposed is called the *crown.* It is the crown that is covered with vitrodentine or enamel, as the case may be. Under this lies the somewhat softer *dentine* (ivory) composed of a peculiar type of calcified connective tissue, the structure and chemical composition of which differ somewhat from bone (see page 167). The dentine, in turn, surrounds a *pulp cavity* filled with soft connective tissue similar to mesenchyme except for the portion adjacent to the dentine which is composed of a layer of *odontoblasts,* or dentine-forming cells. The pulp is supplied with small nerves and blood vessels, thus accounting for its sensitivity. Nerve fibers are not present in the dentine which, nevertheless, is somewhat sensitive because of its close proximity to the nerve fibers in the pulp.

The embryonic origin of the mammalian type of tooth is twofold. Enamel is derived from stomodaeal ectoderm; the rest of the tooth from mesoderm. The vitrodentine-covered type of tooth is entirely of mesodermal origin but is formed under the organizing influence of an enamel organ which comes from stomodaeal ectodermal epithelium.

When teeth develop, the first indication of their formation is to be seen in a thickening in certain regions of the mouth or, in some forms, even of the pharynx. This ectodermal thickening, the *dental primordium,* by a process of infolding, or invagination, pushes into the surrounding mesenchyme to become the shelflike *dental lamina.* This may be continuous or discontinuous but most often is in the form of a continuous longitudinal strand of ectodermal tissue surrounded by mesenchyme. From this, at intervals, small epithelial buds proliferate, each of which develops into a *tooth germ* by multiplication of the ectodermal cells. The mesenchymal cells under the ectoderm also increase in number and push against the latter so that an inverted, cup-shaped structure, the *enamel organ,* takes form (Fig. 5.10). Secondary enamel organs usually develop along with the primary ones being discussed. These will give rise to the teeth that appear later on in life. In those forms in which a succession of teeth occurs throughout life, new tooth germs develop at intervals from persistent portions of the dental lamina. The mesenchyme within the enamel organ is known as the *dental papilla.* It is of neural-crest origin. Those cells of the dental papilla next to the enamel organ become differentiated into *odontoblasts,* which deposit a layer of dentine just under the ectodermal cells of the enamel organ. Dentine is harder than bone but resembles it in certain ways. Its organic and collagen content is less than that of bone. Whereas growth of bone takes place by the addition of new material applied to its various surfaces, dentine grows only where odontoblasts are located, i.e., on the side adjacent to the pulp, of which they form a part. In bone the bone-forming cells become trapped within the matrix they secrete (page 87) and are, therefore, included within the bony tissue. Dentine-forming cells, on the other hand, are in the form of a layer. They are provided with small processes which extend outward toward the enamel organ in the form of parallel fibers (*Tomes's dentinal fibers*) which secrete dentine and which ultimately and persistently lie in small canals or *dentinal tubules*. The dentinal fibers and tubules thus become longer and longer as the layer of dentine thickens. In older individuals the dentinal tubules may become filled with deposits of calcium salts and the dentine then becomes less sensitive.

The cells of the enamel organ become differentiated into two layers, the *outer* and *inner enamel layers*. Looser cells, making up the *enamel pulp,* lie between them. In mammals and a few others the cells of the inner enamel layer become *ameloblasts,* or enamel-forming cells. It has been reported that although the enamel organ induces the differentiation of cells making up the dental papilla, including the odontoblasts, the odontoblasts in turn induce the cells of the inner enamel layer to form enamel. It is not certain whether the ameloblasts secrete enamel or whether their cytoplasm becomes converted into enamel. At any rate, the layer of enamel covers the dentine like a cap. Enamel is the hardest substance in the body and will give off sparks when hit against steel. It is translucent and of a bluish-white cast. Vitro-

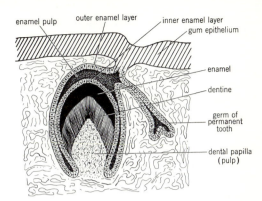

Fig. 5.10 Diagrammatic section showing structure of primary and secondary enamel organs.

dentine, when present, is actually only an extremely hard outer layer of dentine. Continuous addition of new dentine from below forces the tooth to the surface and it is then said to *erupt.* The part of the dental papilla that remains after the dentine is formed makes up the pulp of the tooth, part of which, as mentioned before, is composed of odontoblasts.

In most cases after a tooth has completely formed it ceases to grow since the odontoblasts fail to function any longer. In certain special cases, as in the incisor teeth of rodents and lagomorphs and the tusks of elephants, the odontoblasts function throughout life and the teeth continue to grow.

In mammals a layer of peculiar histological structure, the *cementum,* forms around the *root* of the tooth, or that part below the crown. The cementum covering the upper, or coronal, half of the root is noncellular and represents a calcified form of connective tissue; that covering the remainder of the root contains some collagenous fibers and is cellular, with *cementocytes,* similar to bone cells, or osteocytes, lying within lacunae. Canaliculi are present but Haversian systems are lacking. Electron microscopic studies indicate an occasional canalicular communication between the cementum and the dentine (C. H. Tonge and E. H. Boult, 1960, *Anat. Record,* **136**:292). At the tip, or apex, of the root is a small opening through which nerves and blood vessels enter the pulp cavity. It is referred to as the *apical foramen*.

Replacement of teeth. Except in mammals and a few others, teeth may be replaced an indefinite number of times (*polyphyodont*) and tooth replacement continues throughout life. This is made possible by the formation of a succession of enamel organs from persistent portions of the dental lamina. These bring about the development of new teeth which gradually grow and erupt. At the same time the bases of the old teeth are undergoing resorption or are losing their firm anchorage on the skeletal elements to which they are attached. Replacement of old teeth with new ones is not a haphazard process, as might be supposed from casual observation of the jaws of an animal in which this is occurring. It generally takes the form of waves of replacement (A. G. Edmund, 1958, *Anat. Record,* **132**:431) which actually begin in the front part of the jaw and proceed posteriorly in such a manner that the odd- and even-numbered teeth in a row alternate with each other to the extent that the odd-numbered teeth may be old ones undergoing degeneration, whereas the even-numbered teeth have but recently erupted or are undergoing development. The next wave stimulus would have just the opposite effect, bringing about the appearance of new teeth in the odd-numbered positions, and thus replacing the old ones, at the same time that degeneration and loss of those in the even-numbered positions is occurring.

In most mammals only two sets of teeth develop (*diphyodont*), but in a few there is only one set (*monophyodont*). In certain mammalian species, as in man, not all the teeth are replaced. The persistent molar teeth (the last three teeth at the rear of both upper and lower jaws on each side) have no successors and actually belong to the first set of teeth.

Location of teeth. Teeth may be located almost anywhere in the oral or pharyngeal walls where stomodaeal ectoderm is present and where there is cartilage or bone to support them. In addition to their usual location on premaxillaries, maxillaries, and mandible, in certain forms they may be found on the vomer, palatine, pterygoid, splenial, and parasphenoid bones as well. In some fishes, teeth are located on the tongue and attached to the hyoid arch. In others they may even be found on the posterior visceral arches.

Shape of teeth. When all the teeth are of a similar form, or shape, the dentition is said to be *homodont*. When teeth are arranged in groups which differ in shape and function, the term *heterodont* is applied to the dentition. Most vertebrates below mammals that possess teeth have homodont dentition. There are several kinds of fish, however, which have heterodont dentition. In the Port Jackson shark, *Heterodontus* (*Cestracion*), for example, the anterior teeth in each jaw are pointed and used in a prehensile manner, whereas the posterior teeth are scroll-like and used for crushing. Although a few mammals, such as the toothed whales, have homodont dentition, most are heterodont, the teeth being differentiated into *incisors, canines, premolars,* and *molars.*

Attachment of teeth. The manner in which teeth are attached at their bases also varies. One type of arrangement found among vertebrates is the *thecodont* condition in which the teeth are set in sockets in the jawbones (Fig. 5.11*C*) to which they are firmly attached. This is accomplished by means of a *peridontal membrane* consisting of numerous collagenous *fibers of Sharpey*. The ends of

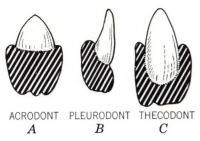

ACRODONT PLEURODONT THECODONT
A   B   C

Fig. 5.11 Diagram showing three methods of tooth attachment.

these fibers are embedded, respectively, in the bone surrounding the socket or alveolus, and in the cementum covering the root of the tooth. Deposition of bone or cementum about the ends of the fibers during development accounts for the firm attachment of teeth to the surrounding bone. Some fishes possess the thecodont method of tooth attachment, but otherwise only crocodilians and mammals, among living forms, have utilized this method. Numerous fossil reptiles had thecodont dentition, as did *Archaeopteryx*, among birds. The roots of mammalian teeth are longer and more deeply embedded than those of lower forms with thecodont dentition. There are two kinds of roots, open and closed (Fig. 5.12). In the first type the pulp cavity, or root canal, is wide open, and by addition of new layers of dentine from beneath, such teeth may continue to grow throughout life. The tusks of elephants and the incisors of rodents and lagomorphs, mentioned previously, are examples of this type of tooth. In the closed type, which is the usual form, the root canal is very small and serves only for the passage of nerves and blood vessels. These teeth fail to grow after they have reached their definitive size. Any increase in dentine through activity of odontoblasts would serve only to reduce the size of the pulp cavity.

In other vertebrates the method of tooth

attachment is more simple. In sharks and rays the teeth are not directly attached to the cartilaginous jaws but are embedded in the fibrous and tough mucous membrane which covers the jaws. In most vertebrates, however, the teeth are fused to the underlying bone (*acrodont*) (Fig. 5.11*A*). Such teeth are apt to break off easily but may be replaced. In some fishes and snakes certain of the teeth may be *hinged*. Hinged teeth yield to pressure and bend backward, only to snap back to the upright position when the pressure is released. Thus, objects are permitted to pass into the mouth, but egress is prevented when the teeth assume their nor-

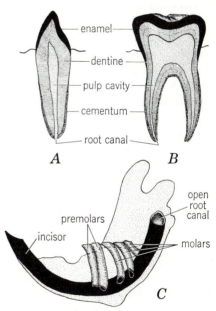

Fig. 5.12 A, section through incisor tooth; B, section through premolar tooth; both with closed pulp cavities. C, lower jaw of rodent, *Geomys*, showing large incisor tooth with open pulp cavity. (*C, after Bailey, modified from Wiedersheim, "Comparative Anatomy of Vertebrates," copyright 1907 by The Macmillan Company and used with their permission.*)

mal vertical position. Hinged teeth are held in position by dense, fibrous ligaments which radiate from the posterior sides of their bases to the bone supporting them. The fronts of the teeth are free at their bases and lift away from the bone when the teeth are pressed backward. Hinged teeth are found in the pike, hake, angler, and some other fishes, which also possess numerous teeth firmly fastened to the jawbones by the acrodont method of attachment. Another method of attachment of teeth is the *pleurodont* type. Here the teeth are united not only to the bone below but also along a shelflike indentation on the inside of the jawbone (Fig. 5.11*B*). In teeth of the acrodont or pleurodont type, where there are no roots, nerves and blood vessels enter the pulp cavity at the base.

Comparative anatomy of teeth. CYCLOSTOMES. Only epidermal teeth are present in cyclostomes (page 165).

FISHES. A few species of fishes are naturally toothless, at least in the adult condition. Among these are sturgeons, sea horses, and pipefish. Teeth in fishes are all of the true type. Dentition in most cases is polyphyodont, acrodont, and homodont. However, many teleosts and some elasmobranchs have heterodont dentition, and in certain forms such as the haddock, garpike, barracuda, and several others the thecodont method of attachment is to be found.

A primitive type of tooth, referred to as *labyrinthine,* occurs in fossil crossopterygians and in the living coelacanth *Latimeria.* Such teeth are of a simple, conical shape, bearing longitudinal ridges and grooves which represent a rather complex folding of the covering layer of enamel. Fish teeth usually consist of little more than conical projections, but in some, as in sharks, they are vertically flattened and triangular in shape. In chimaeras and in the Dipnoi several teeth appear to have fused to form platelike structures used for crushing. The number of teeth is highly variable, ranging from a few flattened plates up to several thousand. In different species teeth may be located in various parts of the mouth or pharynx wherever bony or hard parts are situated.

AMPHIBIANS. Save for the epidermal teeth found in larval anurans, the teeth of amphibians are all true teeth. Teeth are lacking in toads of the genera *Bufo* and *Pipa.* In several accounts the salamander *Siren* is described as being toothless. Actually it bears a patch of numerous splenial teeth on either side of the lower jaw and two elongated patches, each containing many vomerine teeth (Fig. 5.13) in the anterior portion of the roof of the mouth (C. J. Goin, 1942, *Ann. Carnegie Museum,* **24:** 211). In frogs, teeth are lacking in the lower jaw except in members of the genus *Amphignathodon.*

Some of the ancient fossil amphibians possessed conical teeth of the labyrinthine type, but in modern forms they are simple, peglike structures. Amphibian teeth are more restricted as to location than are the teeth of fishes, being confined to the jawbones, palatines, and vomers. In a few forms (*Spelerpes, Plethodon, Batrachoseps*) they are even attached to the parasphenoid bone. Amphibian teeth, for the most part, are polyphyodont, homodont, and acrodont.

REPTILES. Turtles are the toothless representatives of the class Reptilia. The horny beaks of these animals are sharp and strong and serve them in lieu of teeth. Some of the prehistoric pterosaurs also lacked teeth.

In other reptiles the teeth are situated on the jawbones, but in snakes and lizards they may also occur on the pterygoids and palatines. *Sphenodon* has vomerine teeth, an unusual feature among the reptiles. *Sphenodon* is also monophyodont since its single

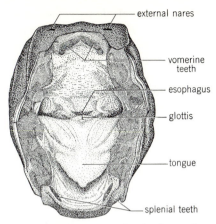

external nares

vomerine teeth

esophagus

glottis

tongue

splenial teeth

Fig. 5.13 Open mouth of salamander, *Siren*, showing numerous vomerine and splenial teeth. The angle of the mouth has been cut in order to depress the lower jaw. (*Drawn by R. Speigle.*)

set of teeth is not replaced. Otherwise most reptiles have homodont and polyphyodont dentition, but variations occur in the method of tooth attachment. In snakes and lizards teeth are either acrodont or pleurodont, but in several ancestral fossil reptiles as well as living crocodilians thecodont dentition occurs. Certain extinct reptiles, the theriomorphs, or cynodonts, had heterodont teeth very similar to those of mammals. A few modern reptiles approach this condition.

The poison fangs of certain snakes are specialized teeth concerning which particular mention should be made. The fangs are attached to the maxillary bones. Some fangs, like those of the cobras, are permanently erect. Those of the vipers lie close to the roof of the mouth and are covered with folds of mucous membrane when not in use. When the mouth is opened to strike, however, the fangs are erected. This is made possible by the shifting in position of a chain of loosely articulated bones in the skull (page 458). There are two types of poison

fangs, but both are basically similar. The first type, found in the cobras, bears a groove or poison duct on its anterior surface. In the vipers a hollow canal lined with enamel runs through the tooth and opens at the tip. In the latter type it is as though the edges of the former had come together and fused to form a tube (Fig. 5.14). In both cases the duct of the poison gland enters at the base of the fang. When the snake is about to strike, the muscles around the poison glands contract and the venom is forced through the fangs much in the manner of liquid passing through a hypodermic needle. It is doubtful whether a snake can strike for a distance over half the length of its body. The head of the snake lunges forward at the same time that the venom is being ejected, and when the snake strikes some soft substance, such as flesh, the poison is actually injected. It is of interest that the young of poisonous species are equipped at birth with a functional poison apparatus which, although but a miniature of the adult, is just as effective for its size. The coral snake of the South Atlantic and Gulf states is the only native representative of the cobra group in the United States. Its bite is dangerous and often fatal. In the cobras the poison fangs are either the most anterior or most posterior of several maxillary teeth, but in the vipers there is but one fang on each maxilla with a few reserve fangs posterior

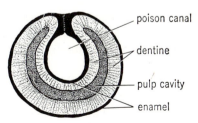

poison canal

dentine

pulp cavity

enamel

Fig. 5.14 Section through poison fang of rattlesnake.

to it. The fangs of poisonous snakes exhibited in circuses and sideshows are customarily pulled out, so that there is little danger from snake bite. The operation must be performed periodically; otherwise replacement teeth will soon take over the function of the original fang.

The lower front teeth of the Gila monster are strong and slightly curved. They bear grooves on their anterior surfaces which serve for the passage of the secretion of poisonous sublingual glands. There are no true fangs, and the animal must hold tightly to its prey in order to make its virulent venom effective.

What is sometimes called an "egg tooth" is present on the tip of the upper jaw in the embryos of lizards and certain snakes. It projects beyond the other teeth and serves as an aid to the young in emerging from the leathery eggshell. It is lost soon after hatching. In *Sphenodon,* turtles, and crocodilians a *horny, epithelial* egg tooth at the tip of the snout of the embryo serves the same purpose. This is also a transitory structure and, although analogous to the egg tooth of snakes and lizards, is not homologous in any way.

BIRDS. No modern birds possess teeth. *Archaeopteryx* had thecodont teeth embedded in sockets like those of crocodilians and mammals. In *Ichthyornis* the sharp, pointed, conical teeth were set in shallow sockets; in *Hesperornis,* they were set in a continuous groove.

In many birds a transitory, horny egg tooth is present in the embryo and aids in breaking the shell at the time of hatching. It soon dries and falls off in the form of a small scale.

MAMMALS. Only a few toothless mammals are known. The adult platypus has epidermal teeth but no true teeth are present. During development a single set of teeth appears, but these are present only tempo-

rarily and are soon shed. In echidnas no teeth are present at any time. The whalebone whales lack teeth in the adult, as do certain anteaters. Although other anteaters, armadillos, and certain other forms possess teeth, they are imperfect structures which lack roots and enamel.

The thecodont condition is the rule in mammals. Most members of this class are diphyodont and have but two sets of teeth. The first set, known as the baby teeth, milk teeth, lacteal teeth, or deciduous teeth, is lost and then replaced by a permanent set. In certain forms, as in bats and guinea pigs, the milk teeth are lost even before birth. A few mammals are monophyodont, having but a single set of teeth. The platypus, toothed whales, sloths, and sirenians are examples. Marsupials are an intermediate group since they retain all their milk teeth except the last premolars.

In certain marsupials a prelacteal dentition has been observed, and occasionally in the human being a postpermanent dentition has been recorded. Thus in mammals it seems that there may originally have been four sets of teeth, a complication not so far removed from the polyphyodont condition of the lower vertebrates.

Among mammals only certain cetaceans are homodont. The teeth of the toothed whales, for example, are all similar in shape. These animals feed entirely under water, and their teeth are adapted for catching and holding prey but not for chewing, since most aquatic animals swallow their food whole. These homodont mammals have anywhere from 2 to 200 teeth.

With the above exception, the heterodont condition usually prevails in mammals. Pinnepedians show a tendency toward homodont dentition, however. Four types of teeth are commonly identified in the upper jaw. *Incisors* are in front, and their roots are embedded in sockets, or *alveoli,* in the pre-maxillary bones. On either side the incisors are followed by a single *canine,* which is the most anterior tooth in the maxillary bone. It usually differs from the remaining teeth in being larger and having a single, pointed crown and a single root. Following the canines are the *premolar,* or bicuspid, teeth which are more complex in form and have two roots. Lastly come the *molars* with several roots. There is but a single set of molars, which should really be considered as belonging to the milk dentition but differing from the others in not being replaced. The mandible, or lower jaw, bears teeth corresponding to those just described. A natural space between two types of teeth is referred to as a *diastema.* In rodents and lagomorphs, which lack canine teeth, a diastema is present between the incisor and cheek teeth. In the cat there is a diastema between the canine and first premolar on each side of the lower jaw.

Only in mammals is there a fixed number of teeth characteristic of each species. The maximum number of teeth in *placental* mammals with the heterodont dentition is 44. There are usually fewer than this because of reduction in number of one or more types. The number of teeth in any particular species is surprisingly constant. For purposes of classification, a so-called *dental formula* has been devised. Since the two halves of the jaw correspond, only the numbers on one side are recorded. Those of the upper and lower jaws are separated by a horizontal line, so that they appear in the formula as numerator and denominator. The kind of tooth is represented by the initial letters *i, c, p, m,* indicating incisor, canine, premolar, and molar, respectively. Thus, the dental formula of man is $i\frac{2}{2}, c\frac{1}{1}, p\frac{2}{2}, m\frac{3}{3}$. In order to simplify these dental formulas the initial letter is customarily

omitted, so that for man it appears as $\frac{2.1.2.3}{2.1.2.3}$. When a certain type of tooth is lacking, a zero is used to indicate this fact. Thus in a beaver, which lacks canines, the dental formula is $\frac{1.0.1.3}{1.0.1.3}$. The cow, which lacks upper incisors and canines, is represented by $\frac{0.0.3.3}{3.1.3.3}$.

From numerous studies it is believed that the hypothetical ancestral form from which existing mammalian dentitions have been derived had the formula $\frac{5.1.4.5}{5.1.4.5}$. This formula is almost attained by some marsupials, but in higher mammals there is always a reduction in the number of incisor and molar teeth. A few dental formulas of familiar mammals are given below.

Cat	$\frac{3.1.3.1}{3.1.2.1}$	Squirrel	$\frac{1.0.2.3}{1.0.1.3}$
Dog	$\frac{3.1.4.2}{3.1.4.3}$	Rat	$\frac{1.0.0.3}{1.0.0.3}$
Mole	$\frac{3.1.4.3}{3.1.4.3}$	Horse	$\frac{3.1.4.3}{3.1.4.3}$

Identification of an individual tooth may be indicated by following the initial letter with the exponent, if on the upper jaw, or with a subscript, if on the lower jaw. Thus p^2 and p_2 refer to the second premolar teeth on the upper and lower jaws, respectively.

In different mammals, the shapes of the molar and premolar teeth vary according to their feeding habits and the type of food eaten. The cheek teeth in the terrestrial carnivores have sharp, cutting crowns. This condition is known as *secondont.* A so-called *carnassial tooth* in each jaw, a premolar above and a molar below, is specially developed for shearing. In man and some other

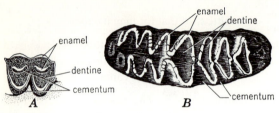

Fig. 5.15 A, grinding surface of selenodont tooth from lower jaw of a sheep; B, grinding surface of lophodont molar tooth of African elephant. (B, after Tomes.)

mammals, the flattened cheek teeth with their small tubercles are of the *bunodont* type and used for grinding. In ruminating mammals, the horse, and some others, the cheek teeth are characterized by the presence of vertical, crescent-shaped folds of hard enamel enclosing softer areas of dentine. Cementum may fill the interstices (Fig. 5.15A). This condition is known as *selenodont*. Because the softer dentine wears away more rapidly, a rasping surface, well adapted for grinding, is maintained. In elephants, enormous grinding teeth are present which may measure as much as 1 ft in length and 4 in. in width. There is an intricate folding of the enamel and dentine to form transverse ridges. Such teeth are of the *lophodont* type (Fig. 5.15B). In elephants as well as the manatee there is an almost constant succession of molar teeth from behind throughout life. The

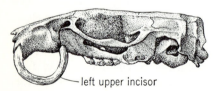

Fig. 5.16 Lateral view of skull of experimental rat. The lower incisor teeth were cut off at intervals. The unopposed upper incisors then grew abnormally in the curved manner shown in the figure.

front ones are pushed forward and are shed as new ones form behind. Only one fully developed molar tooth is present at a time on each side of each jaw in elephants.

Certain unusual teeth found in mammals deserve special mention. The tusks of elephants are incisor teeth with open root canals. They grow continuously throughout life. The gnawing incisor teeth of rodents and lagomorphs are also examples of teeth with open pulp cavities. These chisellike teeth also grow continuously. Only the anterior face of the tooth is covered with enamel. The teeth of the two jaws oppose each other and are kept sharp by whetting against each other since the softer dentine wears away more rapidly than the hard enamel. If the jaw is damaged so that one or the other incisor tooth has nothing to oppose it, it may grow in a spiral manner (Fig. 5.16) and become so large that it prevents the jaws from closing. Animals with such teeth usually die of starvation. The tusks of the male wild boar are greatly developed canine teeth which also have large, persistent pulp cavities. Both upper and lower canines curve upward and are kept sharp by working against one another. Tusks of both male and female walrus are also enlarged upper canine teeth.

In a number of mammals rather striking sexual differences are apparent in the dentition. In addition to those animals mentioned above, the apes show sexual dimorphism in this respect. The canines and first premolars are decidedly larger in the male than in the female.

Embryonic monotremes, being the only mammals to develop within an eggshell outside the body of the mother, are the only mammals which have a horny egg tooth at the end of the snout. This serves the young in escaping from the eggshell.

The origin of the rather complex cheek teeth of mammals from the simple conical type found in most other vertebrates has long been a subject for conjecture. The simple, single-rooted incisor and canine teeth show so little modification that they need not enter into the discussion. Two main theories have been advanced to explain the origin of the more complex premolars and molars. The first theory embodies the idea that the cheek teeth originated by fusion of two or more conical teeth. Indeed, in the dugong several enamel organs fuse to form the molar teeth, but this condition is exceptional. Embryological evidence which should support the soundness of this theory is fragmentary and indecisive. The other theory, which is more generally held and for which there is considerable paleontological evidence, holds that starting with a primitive conical tooth, two additional projections, or buds, developed, giving rise to the

so-called *triconodont shape*. Later these cones shifted in position so as to give rise to separate tubercles or cusps arranged in a triangle. This has been called the *trituber-cular position* (Fig. 5.17). Still later, other parts may have developed from these three original tubercles so as to form additional cusps, ridges, and folds, and thus are finally arrived at the many and varied types of mammalian cheek teeth that exist today. For further details and elaborations of this theory of tooth evolution, the reader is referred to the works of Cope and Osborn, who originally propounded it.

Anterior and intermediate lobes of the pituitary body

The pituitary body, or gland, can under no circumstances be considered as a part of the digestive system. However, since its anterior and intermediate lobes arise from the ectodermal epithelium of the primitive mouth cavity, they should be mentioned here as stomodaeal derivatives. For a complete discussion of this important endocrine organ, the reader is referred to pages 381 to 397.

PHARYNX

The first part of the endodermal portion of the alimentary tract, and the portion which is directly continuous with the ectoderm-

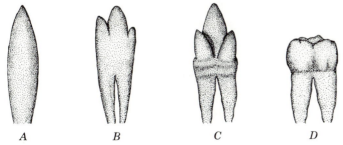

A   *B*   *C*   *D*

Fig. 5.17 Theoretical development of tooth cusps from a single pointed tooth. (*After Osborn.*)

lined mouth, is the pharynx. Its function as part of the digestive system is merely to serve as a passageway between the mouth and esophagus. It is in the walls of the pharynx that the muscles which initiate swallowing movements are located. The pharynx is also very important in connection with respiration. Gill clefts, lungs, and several other structures are derived from its walls. These are more properly described in their relation to the respiratory system and will be discussed in that connection. The presence of teeth in the pharynx is an indication that stomodaeal ectoderm has invaded this region.

ESOPHAGUS

The esophagus is that portion of the alimentary canal immediately following the pharynx. At its other end it joins the stomach. In some forms, particularly among the lower vertebrates, the point at which the esophagus ends and the stomach begins is difficult to distinguish. The length of the esophagus is related to the length of the neck region.

The histological structure of the esopha-

gus is similar, in general, to that of the remainder of the digestive tract except that, since it does not lie within the coelom, it is not surrounded by a layer of visceral peritoneum. Furthermore, the mucous membrane lining the esophagus is of the stratified squamous type of epithelium rather than being composed of a single layer of columnar epithelial cells. In some vertebrates the squamous cells become cornified, but in others, as in man, these cells do not undergo true cornification. In the upper part of the esophagus the muscle fibers, in most cases, change gradually from the striated, or voluntary, type to the smooth, or involuntary, type. There are exceptions to this, particularly in the ruminating, or cud-chewing, mammals, in which the striated fibers extend throughout the entire length of the esophagus. Contraction of the esophagus in these animals is under voluntary control.

Cyclostomes. In the adult lamprey a very specialized condition occurs for which there is no parallel in other vertebrates. From the mouth cavity two tubes extend posteriorly (Fig. 5.18): a dorsal esophagus, and a ventral pharynx which ends blindly

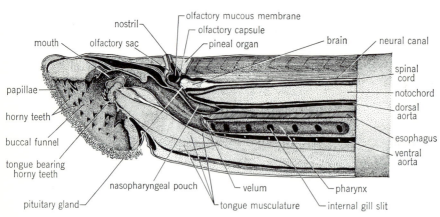

Fig. 5.18 Sagittal section of anterior end of lamprey.

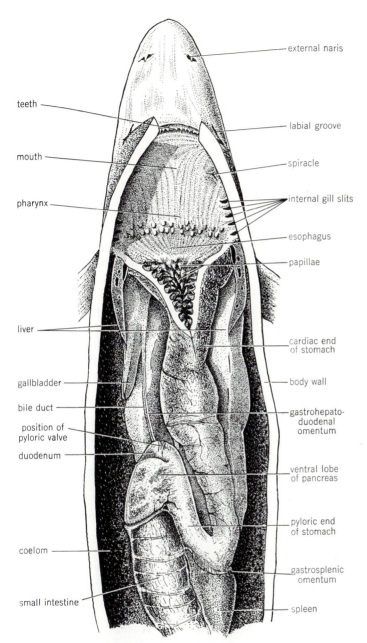

external naris

teeth

mouth

pharynx

liver

gallbladder

bile duct

position of
pyloric valve

duodenum

coelom

small intestine

labial groove

spiracle

internal gill slits

esophagus

papillae

cardiac end
of stomach

body wall

gastrohepato-
duodenal
omentum

ventral lobe
of pancreas

pyloric end
of stomach

gastrosplenic
omentum

spleen

Fig. 5.19 Ventral view of abdominal viscera of the elasmobranch
fish, *Squalus acanthias.*

and is concerned entirely with respiration. During the ammocoetes stage a single tube connects the mouth with the remainder of the digestive tract. Later on a peculiar change takes place in this region whereby pharynx and esophagus become separated in such a manner that each has its own connection with the mouth (Fig. 6.8). At the entrance to the pharynx is a valvelike *velum*. The esophagus is lined with numerous folds. In the hagfish the condition is much as in other vertebrates, and the esophagus merely extends posteriorly from the pharynx in back of the last internal gill aperture.

Fishes. In fishes the esophagus is very short and its junction with the stomach is almost imperceptible. In certain elasmobranchs, as in *Squalus acanthias,* numerous backward-projecting papillae line the esophagus as well as the first part of the stomach (Fig. 5.19). The esophagus commonly bears longitudinal folds which permit a considerable degree of distention. Some fishes are actually capable of swallowing others larger than themselves.

Amphibians. The esophagus in amphibians is extremely short and consists of little more than a constricted area of the alimentary tract. In most amphibians the lining of the esophagus as well as of the mouth is ciliated, and by this means small food particles are swept on to the stomach. Secretory cells in the esophageal epithelium of the frog are reported to have a digestive function. They secrete a substance called *propepsin,* which, however, is ineffective until it reaches the stomach. When acted upon by hydrochloric acid in the stomach, propepsin is converted to the enzyme *pepsin,* used in digestion.

Reptiles. In reptiles the esophagus is generally longer than in lower forms. Longi-

tudinal folds in the walls permit considerable expansion, which is of special use in snakes all of which are capable of swallowing large objects. The lining of the esophagus of certain marine turtles is covered with cornified papillae which point backward (Fig. 5.20).

Birds. In many birds the esophagus is likewise lined with horny papillae. Its junction with the stomach is abrupt. In grain-eating birds as well as birds of prey, the esophagus forms a large sac, or else a ventral, pouch-like outgrowth is present (Fig. 5.21). This is called the *crop,* or *ingluvies.* It is primarily useful in permitting the bird to secure an abundance of food in a short period, thus lessening the time during which it is in danger from enemies. It also enables the bird to compete with others for a limited amount of food. Food is moistened within the crop. Although it is doubtful whether any digestion occurs here, such food particles as grain may swell and are thus rendered more capable of being

Fig. 5.20 Cornified papillae lining esophagus of a marine turtle. (*Drawn by G. Schwenk.*)

Anatomy of the Chordates

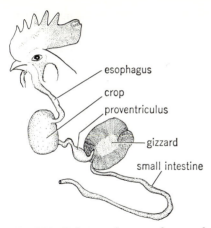

esophagus

crop

proventriculus

gizzard

small intestine

Fig. 5.21 Relation of crop and stomach regions of the domestic fowl.

digested later on. The crop releases portions of its contents at intervals. The food then passes on to the stomach. The *furcula,* or *wishbone,* supports the crop.

In pigeons a peculiar adaptation is found in the so-called *crop glands,* which are two modified areas in the walls of the crop in both males and females (Fig. 5.22). These are really not glands since they are cytogenic, or cell-forming, structures. Upon proper stimulation, the crop "glands" enlarge and very

rapid cellular proliferation occurs. A nutritious, cheesy material called *pigeon milk* is sloughed into the cavity of the crop. This is regurgitated into the throats of the young squabs and serves as their food. It has been found that the activity of the crop glands is controlled by the *lactogenic hormone* secreted by the anterior lobe of the pituitary gland (page 393). Development of the crop glands seems to be stimulated via the pituitary by the act of incubating the eggs, in which both male and female participate. After the young are reared, the crop glands return to their resting condition.

Mammals. In mammals there is a clear distinction between stomach and esophagus. The length of the esophagus varies with the length of the neck, the giraffe having the longest esophagus of all. In man it is about 9 or 10 in. long, on the average. On its way to the stomach it must pass through the diaphragm. That portion below the diaphragm is covered by serosa, or visceral peritoneum, which is lacking from the upper portion.

A number of older textbooks state that the first three parts of the stomach region of ruminants (page 182) are actually modified

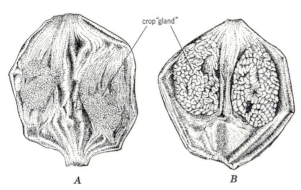

crop "gland"

A   B

Fig. 5.22 Crops of pigeons cut open to show crop "glands." A, during resting, or inactive, period; B, from a bird previously injected with lactogenic hormone.

Digestive System   179

regions of the esophagus. This is because these regions, like the esophagus, are lined with stratified squamous epithelium. The embryological studies of Lewis (F. T. Lewis, 1915, *Anat. Record,* **9**:102) indicate that those who hold this view are in error and that these structures are modified regions of the true stomach.

STOMACH

The stomach is basically a dilatation of the digestive tract for the temporary storage of food. Only when its lining epithelium contains gastric glands (with certain special exceptions) is it properly called a true stomach. Its digestive function is apparently a secondary acquisition. The shape of the stomach is related to the shape of the body. In such elongated animals as snakes, it extends longitudinally, but in those with wider bodies it occupies a more transverse position. The end of the stomach which connects with the esophagus is nearer the heart and, therefore, is called the *cardiac end.* The main portion is called the *body.* The *pyloric end* connects with the intestine and terminates at a valve, the *pylorus,* or *pyloric valve.* This consists of

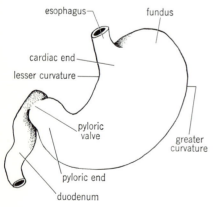

Fig. 5.23 Outline of human stomach, showing regions (ventral view).

a fold of the lining mucous membrane surrounded by a thick, involuntary, sphincter muscle which regulates the passage of the contents of the stomach into the intestine.

In those animals in which the stomach has assumed a transverse position, it is U-shaped or J-shaped. A twisting has occurred so that the cardiac end is on the left side of the body and the pylorus on the right. Two faces, the lesser and greater curvatures, are thus formed (Fig. 5.23). The lesser curvature is actually ventral, and the greater curvature dorsal. The expansion at the cardiac end of the stomach, formed by the greater curvature, is the sac-like *fundus.*

Longitudinal branching folds of the lining of the stomach may be observed when the organ is empty. They are called *gastric rugae.* When the stomach is full the rugae flatten out and disappear.

The muscular walls of the stomach above the middle of the "body" do not exhibit contractions of a peristaltic nature. Those of the pyloric end contract in such a manner as to mix or churn the contents of the stomach thoroughly. Peristaltic waves originate near the body of the stomach and pass through the pyloric end down to the intestine, forcing the contents along. The rate of peristalsis and the strength of the contractions vary with the type and amount of food eaten. In the stomach region there are usually three rather than two layers making up the external muscular coat. The additional layer is composed of obliquely arranged fibers lying between the circular layer and the submucosa.

Cyclostomes. The cyclostome stomach is very poorly developed and consists of nothing more than an almost imperceptible enlargement at the posterior end of the esophagus.

Fishes. There is practically no distinction between esophagus and stomach in fishes, and the longitudinal folds of the former may extend for some distance into the stomach. A considerable variety of stomach shapes may be observed. Some are simple, straight tubes without any digestive function, as in Dipnoi, chimaeras, and a number of teleosts. In others, as in *Polypterus,* the cardiac and pyloric limbs have fused along the line of the lesser curvature so that the stomach appears much as a blind pouch.

In elasmobranchs the J-shaped stomach is typical. The pyloric limb is smaller than the cardiac portion (Figs. 5.19 and 5.25*A*). Many other fishes have similarly shaped stomachs. Among teleosts there is greater variety of stomach shapes than in the other groups of fishes. In some teleosts a ciliated lining is present.

Amphibians. In frogs the cardiac end of the stomach is wide, there is no fundus, and the pyloric end is short and narrow. In certain salamanders the stomach is straight (Fig. 5.24). In some amphibians patches of ciliated epithelium may be found in the lining mucosa. It is probable that most, if not all, amphibian stomachs have a digestive function.

Reptiles. No striking deviations are to be observed in the stomachs of reptiles. Snakes and lizards have long, spindle-shaped stomachs in correlation with their elongated and narrow body shape. There is a clear-cut line of demarcation between stomach and esophagus, however. Of all the reptiles, the crocodilians have the most specialized gastric organs. Part of the stomach is like that of birds, being modified into a gizzardlike muscular region.

Birds. In accordance with the lack of teeth and the type of food eaten by birds, the stomach of most has been modified greatly for trituration. It has become differentiated into

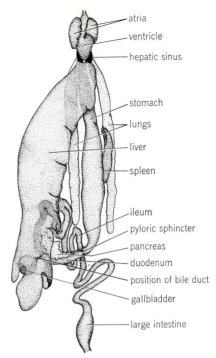

Fig. 5.24 Ventral view of digestive organs and lungs of *Necturus.*

two regions (Fig. 5.21). The first, which is continuous with the esophagus, has a glandular lining which secretes gastric juices. It is known as the *proventriculus.* Food, which is mixed with the digestive fluid in the proventriculus, passes into the much modified and highly muscular *gizzard,* which represents the pyloric portion of the stomach. The muscles form a pair of disclike areas with tendinous centers. The glandular cells lining the gizzard secrete a tough, horny layer which in some cases bears bumps, or tubercles, on its surface. These aid in the grinding process. In grain-eaters, pebbles are taken into the gizzard. They serve as an aid in reducing the contents to a pulp. In commercial poultry establishments it is the practice to supply grit or pebbles to the birds; otherwise the food cannot be properly utilized. The gizzard is best developed in grain-eating birds. It is less well developed in insectivorous forms and in birds of prey, in which little differentiation is noticeable.

Mammals. VARIATIONS IN STRUCTURE. Many modifications exist in the transversely arranged stomachs of mammals, although most of them are of the typical kind. In monotremes the lining epithelium lacks glands, and the pouchlike structure, which serves merely for the storage of food, is therefore not considered to be a *true* stomach. In the platypus the two limbs have fused along the lesser curvature, so that the organ appears as a wide sac (Fig. 5.25*B*). In some monkeys, rodents, and other mammals a constriction marks off pyloric and cardiac regions. Such a stomach is often referred to as an *hourglass* stomach. It is occasionally encountered in man.

The gastric organs of herbivorous mammals are larger than those of carnivores. They are frequently divided into two or more compartments. Ruminating mammals (those which chew their cud) have the most complex stomachs of all. The stomach in cattle consists of four separate chambers: the

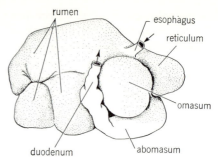

Fig. 5.26 Stomach of ox viewed from the right side, showing normal structural relationships. (*After Sisson, "The Anatomy of the Domesticated Animals," W. B. Saunders Company. By permission.*)

rumen (*paunch*), reticulum (*honeycomb*), omasum (*psalterium* or *manyplies*), and abomasum (*rennet*) (Figs. 5.26 and 5.27). Although the first three, as mentioned previously, are often considered to be modified portions of the esophagus, embryological studies show that they should properly be considered to be parts of the stomach. The rumen is a large, sacculated compartment which serves mainly for the storage of food. According to H. H. Dukes (1947, "The Physiology of Domestic Animals," Comstock Publishing Associates, Inc., pp. 307–325), the animal feeds fairly rapidly and fills the rumen with grain, grass, or other herbage. In cattle, fluid is added, and, by muscular contractions, the food in the rumen is churned about. The food also undergoes some bacterial fermentation during its stay in the rumen. It is believed that bacterial action on relatively simple nitrogenous compounds brings about a synthesis of proteins and vitamins of the B complex, which are important food constituents. While this is going on, the animal frequently lies down and chews its cud at leisure. Food in the rumen passes by degrees into the reticulum or directly to the esophagus and, together with a fair amount

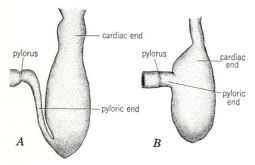

Fig. 5.25 A, J-shaped stomach of elasmobranch fish; *B*, stomach of platypus in which the two limbs have fused along the lesser curvature.

Anatomy of the Chordates

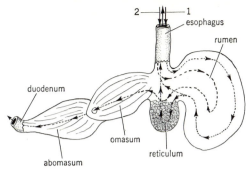

2——1
esophagus
rumen
duodenum
omasum
reticulum
abomasum

Fig. 5.27 Diagram of stomach of ruminant, showing functional relationships: 1 denotes initial swallowing; 2 indicates path taken by food upon reswallowing. (*Modified from Kingsley, "Comparative Anatomy of Vertebrates," The Blakiston Division, McGraw-Hill Book Company. By permission.*)

of fluid, is then regurgitated into the mouth. It is referred to as the cud, or bolus. The voluntary musculature of the esophagus makes this possible. When the animal begins to chew its cud it first swallows most of the fluid, which passes back into the rumen. When the cud has been well masticated and mixed with secretions of the salivary glands, it is swallowed once more and again passes into the rumen. A new bolus is then formed, regurgitated, and the entire process repeated. Most of the food which has been thoroughly masticated goes from the rumen into the reticulum and then passes directly to the omasum, a small chamber with many well-defined longitudinal folds in its walls, and thence to the abomasum. The latter is the only chamber of the stomach provided with gastric glands.

In camels the stomach is not quite so complicated, since an omasum is lacking. Pouchlike diverticula, called *water cells,* arise from both rumen and reticulum. Their openings are guarded by sphincter muscles. It is generally believed that the camel stores water in its water cells and, therefore, may go for long periods without drinking. This belief is not tenable (J. E. Hill, 1946, *Natural History,* **55**:387). The pouches can by no means hold as much water as the animal would require on a prolonged journey through the desert with no water available. Many camels have carried a standard load of 400 lb 25 to 30 miles per day for 8 days without water. The water cells, or pouches, sometimes do contain almost pure water, but this is generally metabolic water drawn from other parts of the body and is used to moisten food undergoing digestion. Some water is stored in the muscles and connective tissues, but most of it comes indirectly from the breakdown of glycogen stored in the muscles and of fat stored in the hump. Physiologists have determined that for every 100 g of fat oxidized in the body 110 g of water is produced. As much as 8 gal of water may be obtained by the animal as a result of processes involved in the breakdown of fat stored in the hump alone and which is used under certain conditions by the body as a source of energy.

The stomachs of whales and hippopotami are divided into several compartments. In the kangaroo the pyloric portion has many peculiar sacculated folds in its walls. The pyloric portion of the stomach of the blood-sucking bat, *Desmodus,* is elongated into a caecumlike structure which fills with blood when the animal is engaged in feeding.

SECRETIONS. In the typical mammalian stomach, secretory cells are arranged so as to form a great number of tubular gastric glands. It has been estimated that there are about 35 million such glands in the stomach of a human being. Occasionally glands will show a slight degree of branching, but usually they are simple in structure. Three kinds of gastric glands are present, named according to their location: *cardiac, fundic,* and *pyloric glands.* They differ both as to histological structure and type of secretion. The cardiac and pyloric glands secrete an alkaline mucus. It is the glands of the fundus which are of greatest importance in digestion. Three kinds of cells are recognized in the fundic glands: *mucous neck cells; chief,* or *zymogen, cells;* and *parietal,* or *acid-secreting, cells.*

The mucous neck cells, as their name implies, secrete the mucus which coats the lining of the stomach; the chief, or zymogen, cells are concerned with the manufacture and secretion of gastric proenzymes; the parietal cells secrete hydrochloric acid.

By use of the electron microscope it has been shown that each large parietal cell contains an *intracellular canaliculus* which opens into the lumen of the gastric gland. Great numbers of microvilli extend into the canaliculus, which is literally crammed full of these tiny projections. They provide a relatively large area of parietal-cell membrane through which hydrochloric acid is secreted.

In most forms two *proenzymes* are secreted by the chief cells of the fundic glands. These are called *propepsin* and *prorennin,* respectively. In the presence of hydrochloric acid, secreted by the parietal cells, these are converted into the active enzymes, *pepsin* and *rennin.* Another enzyme called *gastric lipase* is also secreted by the gastric glands, but it has not as yet been determined in which cells it is formed. *Gastric juice* is a mixture of all these secretions and thus contains mucus, hydrochloric acid, pepsin, rennin, and gastric lipase. The concentration of hydrochloric acid in human gastric juice varies between 0.05 and 0.3 per cent. In the dog it may reach a concentration as high as 0.6 per cent.

Control of the secretion of gastric juice by gastric glands is fairly complex. The sight, smell, or taste of food alone is sufficient to cause them to secrete. The stimulus causing this so-called *psychic secretion* is mediated through the vagus nerves. If the vagi are cut, the psychic phase of gastric secretion is abolished. Many food substances, when they enter the stomach, may serve as direct stimuli to secretion of the gastric glands; others, after having undergone partial digestion, apparently act indirectly, causing the cells of the pyloric end of the stomach to liberate a polypeptide known as *gastrin,* which is absorbed into the blood stream and, being carried by the blood stream back to the stomach, stimulates the gastric glands to secrete, particularly those of the fundus.

This reaction occurs whether or not the nerve supply to the stomach is intact. Still another method by which gastric glands are stimulated is often referred to as the *intestinal phase.* When digestion of some foods has reached a certain point in the intestine, absorption of some unknown substances (perhaps amino acids) into the blood stream takes place. When these reach the stomach via the blood stream they seem to act as a stimulus to further activity of the gastric glands.

Pepsin is said to be a proteolytic enzyme, since it is concerned with the digestion of *proteins.* It acts only in an acid medium and splits large protein molecules into smaller molecules of *peptones* and *proteoses.* Rennin, which is especially abundant in the gastric juice of young mammals, is a coagulating enzyme. It functions in a slightly acid medium and in the presence of calcium, to coagulate casein, the soluble protein in milk. This forms an insoluble compound called *calcium paracasein,* which is referred to as *curd.* A soluble, proteoselike substance, the *whey protein,* remains. The curd is then acted upon by pepsin in the usual manner. Gastric lipase, the third gastric enzyme, is weak and of little, if any, importance to an adult. Fats in a fine state of emulsion, such as the butterfat in milk and the fat of egg yolk, may possibly be acted upon by this enzyme. Mucus from the gastric glands has the power of combining with acid and, since it coats the lining of the stomach, may help protect the mucosa from the action of gastric juices.

The rapid passage of food through the mouth makes it almost impossible for the enzyme ptyalin to act to any great extent upon starches within the mouth by splitting them into molecules of maltose. It is a question as to how much of the starchy food that is mixed with saliva can be digested when it gets to the stomach. Ptyalin acts only in an alkaline, neutral, or faintly acid medium, and it would seem that the relatively high percentage of hydrochloric acid within the stomach would soon stop all activity. Investigations have shown that if the stomach of a rat which has been fed differently colored foods is removed, frozen, and sectioned, the food is observed to be arranged in concentric layers. Thus, before the contents of the stomach can be thoroughly churned, some time may elapse. During this interval and before the acidity increases to a point where reaction will cease, the ptyalin can act in the breaking down of starches.

It was formerly believed that the opening and closing of the pyloric valve, which regulates the passage of the contents of the stomach into the intestine, was controlled by the relative acidities of the materials in the stomach and intestine. It is now held that the pylorus is partly open most of the time and that emptying of the stomach is related to peristaltic muscular contractions, or waves, that pass from the stomach to intestine, and also to the degree to which the stomach contents have been rendered fluid or semifluid. The amount of material in the upper part of the intestine may also be a determining factor.

It has been shown that in a condition known as *pernicious anemia,* in which there is a deficiency of red corpuscles, the bone marrow is unable to form normal red blood cells in adequate numbers. Apparently this is caused, at least in part, by a lack of vitamin B_{12} (cyanocobalamin). The daily requirement of B_{12} in man is extremely small, and ample quantities are ordinarily present in the food. It seems that some factor secreted by the cells of the stomach is necessary for the absorption of B_{12} by the intestine. The substance appears to be a mucopolysaccharide but the exact source of its secretion is not clear. An individual who, for one reason or another, is unable to secrete the specific mucopolysaccharide required cannot absorb vitamin B_{12} from the intestine, and the cells of the bone marrow, without this factor, are unable to carry on their functions normally.

ABSORPTION. Although selective absorption of digested food substances is, for the most part, the function of the intestine, a few substances are known to be absorbed through the gastric mucosa. These include: water, simple sugars (monosaccharides), alcohol, salts, and certain drugs.

Among the mesenteries supporting the alimentary canal are the *mesentery proper,* suspending the small intestine, and the *mesocolon* and *mesorectum,* supporting the large intestine. The *dorsal mesogastrium,* or peritoneal fold which passes to the dorsal border or greater curvature of the stomach, takes a rather indirect course. During development the stomach undergoes a change from its original position so that the greater curvature swings downward and to the left. In doing so the dorsal mesogastrium necessarily is pulled along. It later becomes much elongated and forms a double fold which may extend posteriorly as far as the pelvis. This fold, which comes to lie between the viscera and ventral body wall, is called the *great omentum.* It actually consists of four sheets of serous membrane. The two layers of the descending part of the great omentum enclose the left half of the pancreas and the spleen. The portion of the great omentum between stomach and spleen is spoken of as the *gastrosplenic omentum.* The great omentum is usually a site for the deposition of adipose tissue. Fat may accumulate here in considerable quantity.

Between the two folds of the great omentum is a portion of the peritoneal cavity called the *omental bursa,* or *lesser peritoneal cavity.* This is in communication with the rest of the peritoneal cavity through an opening, the *formen of Winslow,* or *epiploic foramen,* which lies adjacent to the bile duct near the posterior end of the liver on the right side.

The *lesser omentum* is a smaller sheet which connects the liver with the duodenum and the lesser curvature of the stomach. It is sometimes referred to as the *gastrohepatic omentum* or *duodenohepatic omentum,* depending upon the region concerned.

INTESTINE

The portion of the alimentary tract following the stomach is the *intestine.* Here the acid *chyme* from the stomach is mixed with alkaline bile from the liver, pancreatic juice from the pancreas, and the secretions of great numbers of small glands in the walls of the intestine itself. These secretions finally convert food into such a form that it may be absorbed through the wall of the intestine into the blood or lymphatic systems, which then distribute it to all parts of the body. The intestine usually consists of two main parts: a long but narrow *small intestine* and a short, but wider terminal portion, the *large intestine,* or *colon.* The part of the small intestine which immediately follows the stomach is called the *duodenum.* It is here that the ducts from liver and pancreas open.

Many modifications are to be found in the intestines of vertebrates. They serve, for the most part, to increase the surface area of the intestinal epithelium for secretion of digestive fluids and for the absorption of digested foods. The intestine may become elongated and coiled; its diameter may be increased; saclike diverticula called *caeca* may be present; the lining is often thrown into longitudinal folds; circular folds of various types may develop; there may be great numbers of tiny projections, or *villi,* all over the lining surface. In some, a fold, called the *spiral valve,* extends throughout the length of the small intestine. Alternate contractions of circular and longitudinal muscles in the intestinal wall are responsible for three types of contractions that may be observed. These include peristalisis, segmenting movements, and pendular movements. Together they result in a churning and mixing of the contents and the propulsion of materials toward the posterior end.

The length of the intestine is related to the feeding habits of the animal. It is relatively short in carnivorous forms and long in herbivores. This is unusually well shown in the frog. During its existence as a larva, or tadpole, it feeds on vegetation, and the intestine is long and coiled. In the mature frog, which is carnivorous, the intestine is not only relatively but actually shorter than that of a tadpole half its size.

Cyclostomes. The cyclostome intestine is straight. At its posterior end it enlarges slightly to form a rectum which terminates in an anus. This opens into the anterior end of the cloacal depression (Fig. 7.7 and 8.2). A longitudinal fold, the *typhlosole,* which takes a somewhat spiral course, projects into the cavity of the intestine.

Fishes. In elasmobranchs the small intestine is shorter than the stomach. It is wide, but rather straight, and contains a well-developed

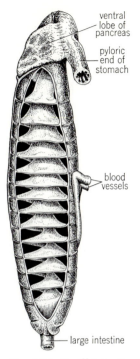

ventral
lobe of
pancreas

pyloric
end of
stomach

blood
vessels

large intestine

Fig. 5.28 Small intestine
of elasmobranch fish,
Squalus acanthias, cut
open to show spiral valve.

spiral valve which greatly increases the surface for digestion and absorption (Fig. 5.28). In *Squalus acanthias* the spiral valve takes 14½ turns. In some elasmobranchs the number of turns is far in excess of this. Spiral valves are also found in the small intestines of chimaeras, Dipnoi, *Latimeria,* members of the superorder Chondrostei, *Amia, Lepisosteus,* and others. Traces are to be found in certain teleosts. The presence of a spiral valve in some fossil placoderms is an indication of its primitive character. The spiral valve in certain elasmobranchs, as in *Carcharias,* has been modified to form a scroll-like structure extending the length of the small intestine. The elasmobranch large intestine is a short passageway extending from the small intestine to the anus. It bends

slightly before opening into the cloaca. A long, slender, fingerlike *rectal gland* connects to the intestine by means of a duct near the point where small and large intestines join. Some biologists in the past have homologized the rectal gland with the colic caecum (page 189) of higher forms. This is highly unlikely. Its function was long in doubt. It is now known (J. W. Burger and W. N. Hess, 1960, *Science,* **131**:670) that the rectal gland secretes a highly concentrated solution of NaCl, thus ridding the blood of excess salt.

In the Dipnoi a *cloacal caecum* is present. Many fishes have *pyloric caeca* coming off the part of the intestine which immediately follows the pylorus (Fig. 5.29). In *Polypterus* there is but one. In some teleosts, as in the mackerel, for example, approximately 200 separate pyloric caeca may be present. Sometimes the pyloric caeca are bound together in a compact mass by connective tissue.

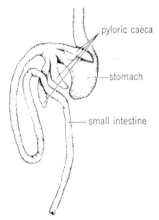

pyloric caeca

stomach

small intestine

Fig. 5.29 Digestive tract of perch showing pyloric caeca. (*After Wiedersheim, "Comparative Anatomy of Vertebrates," copyright 1907 by The Macmillan Company and used with their permission.*)

A circular valve often separates the small and large intestines. In many fishes the anus opens separately to the outside. This is particulary true of teleosts, chimaeras, sturgeons, *Polypterus,* and some others. In other fishes a true cloaca is usually present. When no cloaca is present, the posterior part of the large intestine is known as the *rectum.*

Amphibians. In caecilians the intestine is not differentiated into large and small regions and shows but a slight degree of coiling. In salamanders a greater degree of coiling is evident, and in anurans this tendency is much more marked. The large intestine in urodeles and anurans is short, straight, and plainly marked off from the small intestine. It opens into a cloaca. A ventral diverticulum of the amphibian cloaca gives rise to the *urinary bladder.* It is difficult to discern exactly from which germ layer, ectoderm or endoderm, the lining of the cloaca is derived in amphibians. Most biologists are of the opinion that it is endodermal and that the bladders of amphibians are homologous with those of higher forms. Frequently an *ileocolic valve* is present between small and large intestines. The modification of the lining of the small intestine to form villi is first apparent in certain members of this class although villi are lacking in the common leopard frog and many others. Circular folds, the *valvulae conniventes,* are also present in the small intestines of some amphibians.

Reptiles. The small intestine in reptiles is elongated, coiled, and of fairly uniform diameter. The large intestine is generally straight, is of greater diameter, and opens into a cloaca. An ileocolic valve lies at the junction of small and large intestines, and at this point, except in crocodilians, a *colic caecum* arises. This is the first place in the vertebrates where true colic caeca are found.

Birds. A tendency to greater length is evident in the small intestine in this group. The large intestine is straight and relatively short and terminates in a cloaca. A colic caecum is lacking in parrots, woodpeckers, and some others, but in most forms one or two such structures are present at the junction of small and large intestines. They are not always functional. In certain birds (ducks, geese, turkeys, ostriches, etc.) the colic caeca attain a very large size and the walls may even bear villi. The relatively enormous caecum of the ostrich contains a spiral fold not found in other birds. An important disease causing a great mortality in chickens and turkeys is known as *coccidiosis.* The causative organism is a sporozoan parasite which lodges primarily in the caecum and attacks its walls.

Mammals. The intestines of mammals are more elaborately developed than those of other vertebrates. The coiled small intestine is made up of three regions: *duodenum, jejunum,* and *ileum.* The name duodenum is derived from the Latin word *duodecimus,* which means *twelve.* Old anatomists, accustomed to measuring things according to the breadth of the fingers, gave it this name because the average human duodenum is about as long as the combined breadths of 12 fingers. The name jejunum, which is that part of the small intestine following the duodenum, is also taken from the Latin. The word means "empty." This part of the intestine is usually found empty soon after death. It comprises about two-fifths of the small intestine. The ileum extends from the jejunum to the large intestine and makes up the remaining three-fifths.

A pouchlike structure, known as *Meckel's diverticulum,* is sometimes found projecting from the lower part of the ileum. It represents a portion of the embryonic yolk stalk which has failed to degenerate in the normal

Anatomy of the Chordates

manner. A Meckel's diverticulum is found in about 2 per cent of all human adults. Its average length is about 2 in., but sometimes it is as long as 7 in. Usually the diverticulum is a blind pouch. It may be fastened by a cord of connective tissue to the abdominal wall at the umbilicus. In some cases there may even be an opening onto the outside of the abdomen at the navel through which partially digested food may escape. Although ordinarily unimportant, the presence of a Meckel's diverticulum may cause certain conditions which require surgical attention. In some cases it may push back into the cavity of the ileum and cause an intestinal obstruction. In other cases a loop of the small intestine may become strangulated if it should become twisted about the cord of connective tissue by which the diverticulum is fastened to the body wall.

It is generally stated that the average length of the small intestine in man is 22½ ft. Actually there is much variation. According to a report on 100 autopsies, the small intestines of men ranged from 16 ft to 25 ft 9 in.; of women from 11 ft to 23 ft 6 in. (B. M. L. Underhill, 1955, *Brit. Med. J.,* 2:1243). The large intestine, or *colon,* is much shorter than the small intestine but of considerably greater diameter. In man it averages about 4 to 5 ft in length. The colon terminates in a rectum, which opens to the outside through the anus. Among mammals only the monotremes and the pika, a lagomorph, possess a cloaca. The intestine of herbivorous mammals is very long and may be from twenty to twenty-eight times the length of the body. In the cow the average length is 165 ft, and in the horse approximately 95 ft. Carnivorous forms have an intestine only five or six times the length of the body.

At the junction of the ileum and colon lies the *ileocolic* (sometimes called *ileocaecolic*) *valve,* which regulates the passage of material from the small to the large intestine

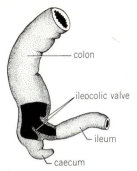

Fig. 5.30 Junction of ileum and colon of cat, showing caecum and ileocolic valve.

(Fig. 5.30). Coming off the colon at this point of juncture is a single colic caecum, found in almost all mammals. Some edentates have two, but this is exceptional. Great variations occur in the form and size of the caecum. It is small in bats, carnivores, some edentates, and certain whales. In man its distal end has degenerated and the remnant is represented by the *vermiform appendix,* the seat of so much trouble (Fig. 5.31). The human appendix under normal conditions lies deep to *McBurney's point* (Fig. 5.32). This is situated on the abdomen about 2 in. from the right anterior spine of the ilium on a line between the latter and the umbilicus, or navel. A vermiform appendix is also found in certain monkeys, civets, and a few rodents. In marsupials, herbivores, and some rodents the caecum is relatively enormous and may even exceed the length of the body. In *Hyrax* a peculiar pair of caecal diverticula arises from the colon at some distance from the usual caecum.

The intestinal wall contains myriads of intestinal glands which are of two main types. The first of these are the simple tubular *glands,* or *crypts, of Lieberkühn,* found throughout the entire length of the small and large intestines. Their secretory portions are

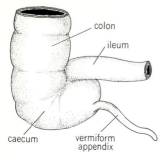

Fig. 5.31 Junction of ileum and colon of man, showing caecum and vermiform appendix.

confined to the lamina propria, dipping down almost to the muscularis mucosae. The glands of Lieberkühn in the small intestine open to the intestinal surface between the villi. Their cells are primarily responsible for secretion of intestinal enzymes but the presence of goblet cells indicates that they also produce mucus. Two peculiar types of cells have been identified at the bases of the crypts of Lieberkühn. So-called *Paneth cells* seem to have the ability to concentrate the element zinc, but the part they play in the general body economy is not clear. *Argentaffin cells,* the cytoplasmic granules of which stain with silver and chromium salts, may possibly be of ectodermal origin and bear some relation to the autonomic nervous system. They secrete a substance called *serotonin* which is transported by the blood platelets (see page 599). It acts as a local vasoconstrictor as well as being a blood-pressure-reducing agent which may help to arrest bleeding.

In the caecum and large intestine, in which villi are not ordinarily present after birth, the character of the glands of Lieberkühn changes somewhat. They are longer than those in the small intestine and goblet cells are more numerous.

The second of the two types of intestinal glands are the branched tubular mucous *glands of Brunner.* They are confined to the region of the duodenum. The secretory portions are located in the submucosa; their ducts pass through the muscularis mucosae to open into the crypts of Lieberkühn. It is unlikely that the glands of Brunner secrete any digestive enzymes.

Two hormonal substances, referred to as *secretin* and *pancreozymin,* respectively, are formed by cells in the duodenal mucosa when the acid secretion of the gastric glands comes in contact with them. These hormones, carried by the blood stream to the pancreas, stimulate the production of pancreatic juice. Pancreozymin is primarily concerned with pancreatic enzyme production; secretin stimulates the formation of bicarbonate and some other ingredients of pancreatic juice. Secretin has also been reported to stimulate the secretion of bile by the liver. It is a polypeptide.

Another hormonal substance known as *cholecystokinin* is believed to be formed by the cells of the mucosa of the upper intestine in response to stimuli by the products of fatty digestion. When carried by the bloodstream

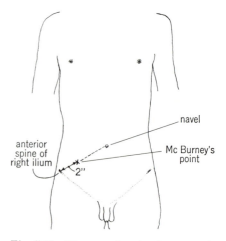

Fig. 5.32 Diagram showing location of appendix at McBurney's point.

to the gallbladder, it stimulates the latter to evacuate its contents.

The existence of still another hormone-like secretion of the duodenal mucosa has been postulated. When soaps, fat, fatty acids, and certain sugars come in contact with the mucosa of the upper small intestine, they cause the release of *enterogastrone*. This inhibits motor and secretory activity of the stomach. This is demonstrable in 1 to 7 minutes after fat has been introduced into the intestine. Enterogastrone has not as yet been chemically identified. It contains some amino acids.

The secretion of the intestinal glands is called the *succus entericus*. Several enzymes are secreted by the intestinal glands. What was formerly referred to as *erepsin* is now known to be composed of a whole battery of proteolytic enzymes called *peptidases* which bring about the final breakdown of peptids into amino acids; *lipase*, which together with lipase from the pancreas aids in the digestion of fats; various enzymes which break down complex sugars into glucose, *sucrase* for cane sugar, *maltase* for malt sugar, and *lactase* for milk sugar. Another enzyme, *enterokinase*, is secreted by the intestinal glands and functions to activate trypsinogen coming from the pancreas (page 197).

In addition to the minute digestive glands, the epithelium of the small intestine bears great numbers of villi of various shapes. Those in man vary from 0.2 to 1 mm in height. They give a velvety appearance to the lining of the small intestine. In a human embryo, villi are also present in the large intestine, but they gradually disappear, so that at birth the lining epithelium is relatively smooth. Villi contain blood vessels and lymphatics (lacteals) which collect the products of digestion after they have been absorbed. The free surfaces of the villi bear innumerable microvilli, as studies with the electron microscope have revealed. The absorptive surface of the mucosa of the small intestine is increased immeasurably by their presence.

Nodules of lymphoid tissue, called *Peyer's patches,* are large oval areas in the small intestine, most numerous in the region of the ileum. Occasionally some are present in the duodenum and jejunum. They lie on the side of the intestine opposite that to which the mesentery is attached. In the human intestine the average number of Peyer's patches ranges from 30 to 40, although as many as 60 have been observed. Villi are lacking from the epithelium which covers these structures. In old age the patches of Peyer undergo a retrogression. Lesions of these structures occur in typhoid fever.

Several types of folds in the intestinal lining are found in mammals. In man these include (1) *plicae circulares,* or *valvulae conniventes (valves of Kerckring),* which are circular folds found only in the jejunum and ileum; (2) *plicae semilunares,* or internal transverse folds of the colon, containing circular muscle fibers; and (3) *plicae transversales* in the lining of the rectum. Bulges of the colon between the plicae semilunares are known as *haustra.* In the region of the anal canal several longitudinal folds called *rectal columns* are present.

PROCTODAEAL DERIVATIVES

Derivates of the proctodaeum are few in number when compared with those of stomodaeal origin. It is very difficult to determine the point of transition between the endoderm lining the cloaca and the ectoderm lining the proctodaeum. Although some disagree, it is generally conceded that the urinary bladder of amphibians, which opens directly into the cloaca (Fig. 5.33), is lined with *endoderm* derived from the cloaca and *not* with procto-

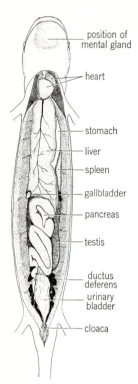

position of
mental gland

heart

stomach

liver

spleen

gallbladder

pancreas

testis

ductus
deferens

urinary
bladder

cloaca

Fig. 5.33 Internal or-
gans of male salaman-
der, *Eurycea bislineata,*
showing relation of uri-
nary bladder to cloaca.

daeal ectoderm. The *bursa Fabricii,* present
in most young birds, is another endodermal
cloacal derivative, often thought to be derived from the proctodaeum. It forms a secondary connection, however, with the proctodaeum during early development. The bursa
Fabricii comes to lie in the body cavity between the spinal column and posterior part
of the large intestine and enters the cloacal
orifice just beyond the openings of the urogenital ducts. It apparently functions as a
lymphoid organ during early life (see page
590) but degenerates as the animal approaches sexual maturity. In this way it

somewhat resembles the thymus gland, another lymphatic organ. The bursa Fabricii
seems to be basically responsible for the
formation of cells involved in antibody production and in reactions against invading
bacteria. Certain so-called *anal,* or *cloacal,
glands,* used primarily for defense or sexual
allurement, are probably derived from the
ectodermal, proctodaeal invagination.

LIVER

The liver is primarily a digestive gland, although it has many other functions. It is the
largest gland in the body. During embryonic
development the liver arises as an outgrowth
of the endodermal wall of the primitive gut,
or archenteron. The region from which it
arises becomes the duodenal portion of the
small intestine. The original outgrowth gives
rise to a hollow hepatic diverticulum, which
soon differentiates into two parts: an anterior, or cranial, portion which proliferates to
become the large, glandular mass of the liver
and its bile ducts; a posterior, or caudal, part
which gives rise to the gallbladder and cystic
duct. A fairly typical arrangement of these
various parts is to be found in man (Fig.
5.34). Two main ducts, one from the right
lobe and one from the left, drain the liver.
These unite to form a single *hepatic duct.* The
gallbladder, in which bile is stored, is an enlarged, pear-shaped, saccular structure, its
narrow neck connecting with a *cystic duct,*
which in turn joins the hepatic duct to form
the *bile duct,* or *ductus choledochus,* which
opens into the duodenum. The *pancreatic
duct* (ventral pancreatic duct, duct of Wirsung) joins the bile duct shortly before it
enters the duodenum. The common orifice of
the two ducts opens at the tip of a *duodenal
papilla.* The short segment common to the
bile and pancreatic ducts is dilated, forming
an ampulla (*ampulla of Vater*). It courses

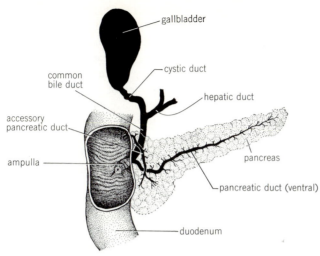

Fig. 5.34 Fairly typical arrangement of gallbladder, bile ducts, and pancreatic ducts in relation to duodenum as found in man (semidiagrammatic).

obliquely a short distance through the duodenal wall. The terminal, or preampullary, portion of the bile duct, just before its junction with the pancreatic duct, is surrounded with a strong *sphincter muscle of Boyden* (E. A. Boyden, 1957, *Anat. Record,* **127**:399). This prevents bile, coming from the liver via the hepatic duct, from entering the duodenum except at such times as the sphincter of Boyden is relaxed. Bile, therefore, passes under some pressure via the cystic duct to the gallbladder where it accumulates and is stored temporarily. Another, but weaker, sphincter muscle, the *sphincter of Oddi,* surrounds the ampulla as well as the distal portion of the ventral pancreatic duct.

The liver proper consists of great numbers of small units called *liver lobules.* There is considerable confusion in the literature as to exactly what constitutes a liver lobule. This is a matter for detailed consideration by histologists and will not be touched upon further in this volume. Suffice it to say that the cells composing each tiny liver unit are

in close proximity to a branch of the bile duct (bile capillary), a branch of the hepatic artery, a branch of the hepatic portal vein, and one or more small lymphatic vessels. Large numbers of branched columns of cells, called *liver trabeculae,* are present in each lobule. They are separated from one another by small blood spaces, or *sinusoids,* which empty into a *central vein* of the lobule. Blood in the sinusoids is contributed by both the hepatic portal vein and the hepatic artery. The liver receives its blood supply from two sources. The first is the usual type of oxygenated blood which supplies all organs. It reaches the liver through a hepatic artery, which is a branch of one of the main vessels coming off the aorta. The hepatic portal vein, which begins in capillaries in the digestive tract and spleen, brings the products of digestion, especially those of carbohydrates and proteins, to the liver. Here the vein breaks up into capillaries (sinusoids), which ramify throughout that structure. The two types of blood lose their identity in the liver.

Narrow *bile capillaries* course between the columns of liver cells and anastomose with each other. Larger ducts drain the bile capillaries, and these finally unite to form the hepatic duct.

There are many variations from the arrangement just described to be found in different vertebrates. Some of the variations are described below. The shape of the liver varies according to the shape of the animal and is typically divided into right and left lobes. It is generally larger in herbivorous forms than in carnivores. The liver is held firmly in place by ligamentous folds of peritoneum. Its main attachment is to the *septum transversum,* a structure which in mammals contributes to the formation of the diaphragm.

Comparative anatomy of the liver. AMPHIOXUS. Although no true liver is found in amphioxus, the presence of such a structure in higher chordates is foreshadowed in amphioxus by a hollow, forward-projecting, ventral *hepatic caecum* which comes off the intestine just posterior to the branchial region (Fig. 2.4). The lining of this pouch is ciliated, and it may have some digestive function. A system of veins coming from the intestine breaks up in capillaries on the hepatic caecum, thus presaging the appearance of the hepatic portal vein of higher forms.

CYCLOSTOMES. The liver in cyclostomes is unusally small. In the lamprey it appears to be single-lobed, but in the hagfish it consists of two halves, this representing the most typical condition of the liver. The ducts from these two parts in myxinoids open separately into the gallbladder. Neither gallbladder nor bile duct is present in an adult lamprey. It is of interest, however, that the ammocoetes larva of the lamprey possesses both these structures which degenerate during metamorphosis.

FISHES. The livers of fishes are lobed and relatively very large. Some are of considerable economic value because of the oil and vitamins which they contain. A gallbladder is almost uniformly present, being lacking only in a few species of sharks. The color of fish livers shows a great range, particularly among the teleosts.

AMPHIBIANS. Amphibians' livers are large in proportion to the size of the body. They are lobed, and a gallbladder is present.

REPTILES. No important deviations in liver structure occur in reptiles. In snakes, correlated with the long and narrow body, the liver consists only of one elongated lobe. All reptiles possess gallbladders.

BIRDS. Lobed livers are the rule in birds, but in many a gallbladder is lacking. In the pigeon, which lacks a gallbladder, there are two bile ducts which open separately into the duodenum. One comes from each main liver lobe, but the right one is the more slender of the two. It has been reported that a transitory gallbladder is present in the pigeon for a short time during embryonic development.

MAMMALS. More variations occur in the lobulation of the mammalian liver than in any other group. The two main lobes are usually subdivided into smaller units which are arranged in various ways. In some mammals there may be as many as six or seven lobes. In man there are five. Many mammals lack gallbladders. These include certain rodents, whales, *Hyrax,* some of the *Artiodactyla,* and all the *Perissodactyla.* Although the rat lacks a gallbladder, there is some evidence that a minute rudiment of a gallbladder is present for a day or two during early embryonic development.

The liver is supported by several ligaments. A *falciform ligament* is attached to the ventral part of the diaphragm and ventral abdominal wall. A fibrous cord running from the umbilicus to the undersurface of the liver is the *round ligament,* or *ligamentum teres.* It represents the remnant of the left umbilical vein (see page 587). The peritoneum is reflected over the ligamentum teres, which passes to the liver within the free margin of the falciform ligament. A single *coronary ligament* and two *lateral triangular ligaments* from the dorsal and dorsolateral portions of the diaphragm, respectively, also support the liver.

Functions of the liver. The functions of the liver are numerous. Bile, which is secreted by the liver cells, is a complex fluid. It has an alkaline reaction and contains, among other things, bile pigments, bile salts, lecithin, and cholesterol. No digestive enzymes are present in bile, but nevertheless it plays a very important part in digestion. Certain excretory products are also eliminated by the liver and discharged with the bile into the intestine, from which they are expelled with the feces. Other functions of the liver include formation of blood corpuscles in the embryo, storage of excess carbohydrate in the form of glycogen, liberation of glycogen in the form of glucose when the concentration of the latter in the blood stream falls below normal, formation of urea, storage of copper and iron, formation of vitamin A from carotene, enzyme formation, and the elaboration of fibrinogen which is very important in the clotting of blood. The liver is also of great importance in the metabolism of fats and proteins.

The liver plays an important but as yet undetermined role in connection with certain steroid hormones elaborated by some of the endocrine glands. Cholesterol and the steroid hormones have a basically similar chemical structure consisting of what is called the cyclopentanoperhydrophenanthrene nucleus. Apparently, along with cholesterol, the sex hormones as well as certain other steroid hormones liberated into the blood stream by the adrenal cortex are secreted by the liver into the bile. When the bile enters the intestine at least some of the steroid hormones are absorbed through the intestinal mucosa, once more entering the circulatory system. Whether some alteration in chemical structure of the hormones takes place during these processes is not quite clear. Much remains to be learned concerning steroid hormone metabolism.

The normal liver contains a factor which is specific in treating pernicious anemia, a disease characterized by a deficiency of red blood corpuscles. The substance is formed in the mucosa of the pyloric end of the stomach and is stored in the liver. This factor permits absorption from the intestine of vitamin B_{12} which is essential for normal production of red blood cells in the bone marrow. Extracts of the liver taken by mouth or by intramuscular injection are effective in alleviating this condition.

A gallbladder is not indispensable since it is lacking in many mammals and can be removed from human beings and others without serious effect. It functions mainly as a storage place for bile which then can be emptied into the duodenum in quantity at appropriate intervals. Bile is also concentrated by the gallbladder. The cells of the lining epithelium bear numerous microvilli, thus providing an extensive surface for absorption of water and inorganic salts. There is no evidence that these cells are secretory. When bile is secreted by the liver it cannot enter the duodenum unless the sphincters of Boyden and Oddi are relaxed. It, therefore, backs up into the gallbladder where it accumulates and from which it is discharged intermittently. The role of a hormonal substance, *cholecystokinin,* secreted by the upper intestinal mucosa, in bringing about contraction of the gallbladder and relaxation of the sphincters of Boyden and Oddi, has already been referred to (page 193).

If bile pigments are not properly eliminated from the body, they escape into the tissues and impart a yellow color to them. This shows up most noticeably in the "whites" of the eyes, but the entire skin may be affected. Such a condition is known as yellow jaundice and is caused by mechanical blocking of the hepatic duct or ductus choledochus. Tumors, gallstones, or parasites may be instrumental in causing such a blockage. In jaundice, normal digestion, particularly of fats, is interfered with.

Gallstones are usually composed of cholesterol, a polyatomic alcohol normally present in solution in bile. When this is thrown out of solution under certain abnormal conditions, it will deposit concretions about tiny fragments of tissue or pigment to form gallstones. In some types, calcium and bile pigments also contribute to their formation. There may be a single large stone, a few smaller ones, or even hundreds of tiny stones which appear much like gravel. The passage of gallstones is not only excruciatingly painful but may block the duct and cause jaundice. Bile may even back up into the pancreatic duct and cause acute pancreatitis, a very serious condition.

In so-called typhoid carriers, or persons who carry and distribute typhoid bacilli without necessarily having the disease themselves, the organisms usually grow in the gallbladder and are discharged from the body with the feces.

PANCREAS

The pancreas is the second largest of the digestive glands. It arises from the endoderm of the primitive gut in the same general region from which the liver develops. A single dorsal, and usually one or two ventral, pancreatic diverticula develop in most vertebrates. The proximal portions of the diverticula form the pancreatic ducts, and the distal parts, by an intricate process of budding, give rise to the main mass of pancreatic tissue. The ducts undergo various degrees of fusion or reduction, so that the number and arrangement show considerable diversity. Usually only one or two ducts persist. They may open independently into the duodenum or may join the bile duct. The chief *pancreatic duct* (*ventral pancreatic duct, duct of Wirsung*) arises in the embryo as a diverticulum of the bile duct. This duct and the common bile duct have a common opening into the duodenum. A *dorsal pancreatic duct* (*accessory pancreatic duct, duct of Santorini*) opens into the duodenum independently. The pancreas is a single organ of irregular shape and may consist of several lobes. It generally lies in a loop between the stomach and duodenum. In such animals as calves and sheep it is large enough to serve as food for human beings and is sold under the name "stomach sweetbreads."

The pancreas plays a dual role in the body, serving both as an exocrine and endocrine gland. The structure of the exocrine portion of the pancreas closely resembles that of the salivary glands. Only one type of cell is present. The exocrine secretion of the pancreas is called *pancreatic juice*. The manner in which secretin and pancreozymin, hormonal substances from the duodenal mucosa, stimulate the secretion of pancreatic juice has already been described (page 190). When the alkaline pancreatic juice is poured into the duodenum and neutralizes the acid chyme from the stomach, the secretion of pancreatic juice ceases. This is because liberation of secretin and pancreozymin is normally brought about by the action of acid coming in contact with the intestinal mucosa. That part of the pancreas which is responsible for the endocrine secretion consists of small groups of cells called the *islands (islets) of Langerhans,* which in most forms are scattered throughout the main mass of the gland. Since they are not concerned in the digestive processes, they are more properly discussed in Chap. 9, Endocrine System (pages 360 to 364).

Comparative anatomy of the pancreas.
AMPHIOXUS. No pancreas is present in amphioxus. The hepatic caecum is considered by some authorities to be a possible homologue of both liver and pancreas, but this is doubtful.

CYCLOSTOMES. In the adult lamprey no well-defined pancreas is evident. Recognizable pancreatic tissue is present, however, embedded within the liver and in the wall and typhlosole of the intestine. This probably represents only the endocrine portion of the pancreas. No traces of an exocrine part are to be found. The hagfish possesses a small pancreas located near the bile duct into which several pancreatic ducts open independently.

FISHES. In certain fishes (Dipnoi and many teleosts) the pancreas is so diffuse as to be almost unrecognizable. In some, the endocrine portion is separated from the remainder of the pancreas and is spoken of as the *principal island* (page 360). Elasmobranchs have a well-defined pancreas consisting of dorsal and ventral lobes connected, in many cases, by a narrow isthmus. A single duct enters the intestine. The pancreas of elasmobranchs arises entirely from a single dorsal diverticulum of the archenteron.

AMPHIBIANS, REPTILES, AND BIRDS. No noteworthy features are exhibited by the pancreas in members of these three classes of vertebrates. One duct or several may be present, and they may open directly into the duodenum or indirectly through the bile duct.

MAMMALS. Generally two pancreatic ducts are present in mammals. The ventral duct opens into the bile duct near the ampulla (Fig. 5.34). The dorsal duct enters the duodenum directly, although its position varies in different forms. Both ducts are functional in the horse and dog. In the cat, sheep, and man the ventral duct is the functional duct, although the dorsal accessory duct may persist in reduced form. In the pig and ox only the dorsal duct persists. A *pancreatic bladder* is sometimes found in the cat. Pancreatic juice may accumulate here and then be released into the intestine in much the same manner as bile is released from the gallbladder.

Digestive function of the pancreas. Pancreatic juice is alkaline in reaction and contains a number of enzymes. The most important ones are *amylase, lipase, maltase, pancreatic rennin,* and a proenzyme, *trypsinogen.** Amylase has an action similar to ptyalin from the salivary glands but is much stronger. Any starches that have escaped the action of ptyalin in the mouth or stomach are broken down to maltose, or malt sugar, by the action of amylase in the intestine. Only traces of the enzyme maltase are present in pancreatic juice, but this, together with the maltase secreted by the intestinal glands, effects the final breakdown of malt sugar to glucose. In the form in which trypsinogen is secreted by the pancreas, it is practically inactive. *Enterokinase* in the small intestine activates trypsinogen, converting it into the enzyme *trypsin.* Trypsin acts only in an alkaline medium. The action of trypsin, which is a proteolytic enzyme, carries the

* Other enzymes secreted by the pancreas include *chymotrypsin, ribonuclease, deoxyribonuclease, carboxypeptidase,* and *elastase.*

digestion of proteins further than does pepsin from the stomach. It serves to break down peptones and proteoses into smaller units called peptids, which are really small groups of amino acids. These are finally acted upon by erepsin secreted by the intestinal glands. Pancreatic lipase acts in the intestine to split fats into fatty acids and glycerin. Pancreatic rennin is similar in action to the rennin secreted by gastric glands.

It should be clear that the pancreatic enzymes serve both to initiate action on foods that have escaped the effect of salivary and gastric enzymes and to carry the breaking-down processes still further. It is the intestinal enzymes which in most cases complete the digestive activity.

DIGESTION OF FOOD (A SUMMARY)

The principles of digestion have been most thoroughly worked out in man. With minor variations, the same processes take place in other vertebrates. The following account embraces the main factors concerned in digestion.

Foods exist in three main forms: carbohydrates, fats, and proteins. In order to produce food substances, a great deal of energy is required. In the last analysis it is the energy from the sun which is used in food production. Plants, under the influence of sunlight, can synthesize carbohydrates from water and carbon dioxide. Carbohydrates under certain conditions may be converted into fats. Other inorganic elements may be taken from the soil or air to manufacture proteins. Herbivorous animals can make use of this stored-up energy in building up their own bodies. Carnivores are but one step further removed.

When food is taken into the mouth, mucus is added which renders it capable of being swallowed more easily. Ptyalin in the saliva may begin to break down starches into the smaller molecules of malt sugar. This process may even continue for a time after the food has reached the stomach, because of the process of *layering*. In the stomach, hydrochloric acid, propepsin, prorennin, and perhaps some gastric lipase are secreted by the glands in the fundic region. Additional mucus is added by gastric glands, particularly those in the cardiac and pyloric regions. The muscular walls of the stomach contract and churn the contents thoroughly. The acid in the stomach soon stops the action of ptyalin on starches but activates the proenzymes to form pepsin and rennin. Rennin, in the presence of calcium, acts upon the soluble protein, or casein, in milk to form the curd referred to as calcium paracasein. The proteoselike whey protein remains. If not all the milk is changed in this way, pancreatic rennin will later act upon the remainder in the intestine. Pepsin, in the acid medium of the stomach, acts upon proteins, including calcium paracasein, and breaks them down into peptones and proteoses. This reaction, however, may not be completed in the stomach.

When the food in the stomach has reached a fluid or semifluid state, it is known as *chyme*. Muscular peristaltic waves intermittently force the chyme through the pylorus into the duodenum. The waves of peristalsis continue throughout the entire length of the intestine. Bile, which has been stored in the gallbladder, is released into the duodenum, and to this is added the alkaline pancreatic juice which contains the enzymes amylase, maltase, lipase, and rennin and the proenzyme, trypsinogen. Amylase continues the breaking down of starches which have escaped the action of ptyalin and reduces them to malt sugar. Maltase from the pancreas, together with that formed in the intestine, splits the malt sugar into molecules of the more simple sugar, glucose. This passes through the wall of the intestine into small blood vessels which lead into the hepatic portal vein. Other sugars such as sucrose and lactose, which may be present in the food and cannot be absorbed as such, are acted upon by the intestinal enzymes sucrase and lactase, respectively, and are also reduced to glucose.

Proteins, peptones, and proteoses are at the same time acted upon by trypsin, when trypsinogen becomes activated by enterokinase from the small intestine. Trypsin carries the digestion of proteins past the peptone stage until only peptids remain. The final breakdown of peptids to amino acids, the "building stones" of proteins, is accomplished by a number of intestinal enzymes (peptidases) formerly referred to collectively as erepsin. Amino acids pass into the blood vessels of the intestinal walls, and thence to the liver via the hepatic portal vein.

Fats are not appreciably digested until they reach the intestine. Lipase from the pancreas and intestinal glands breaks the fat molecules down into fatty acids and glycerin. The alkali in the intestine forms soaps with some of the fatty acids, and this aids in the emulsification of the remaining fats. Bile salts which are present in the bile are even more important in this respect, since they are instrumental in lowering surface tension. A greater suface of fat particles is thus exposed to the action of lipase. Bile salts also assist in the passage of fatty acids through the intestinal wall. About 60 per cent of the digested fat passes into the lacteals, which are small lymphatic vessels, to enter ultimately the thoracic lymph duct which joins the left subclavian vein. The remainder of the digested fat is apparently absorbed into the capillaries which lead into the hepatic portal vein.

It has been determined through use of the electron microscope (S. L. Palay and L. Karlin, 1959, *J. Biophys. Biochem. Cytol.,* 5:363) that not all fat must be broken down to fatty acids and glycerin before it can be absorbed or taken into the cells of the intestinal mucosa. In the rat very small fat droplets have been observed between the microvilli within 20 minutes after ingestion of a fatty meal. It would seem that a process similar to phagocytosis or pinocytosis (see page 541) takes place, since each tiny droplet is surrounded by a membrane when in the cytoplasm. The droplets move to the sides of the cell and pass through the cell wall into the small adjacent lymphatic vessels.

In the ileum, the contents of the intestine are still fluid in consistency. When they reach the caecum and colon, the remaining products of digestion are absorbed. The absorption of water takes place, for the most part, in the first part of the colon, and the indigestible residue becomes firmer and is known as *feces.* Very little undigested food is present in feces. It is composed mostly of indigestible matter, old epithelial cells, leukocytes, and enormous quantities of bacteria.

In herbivorous animals, cellulose, which makes up the walls of plant cells and which is ordinarily not digested, is acted upon for the most part by bacteria in the colon and broken down into substances which can be utilized by the body.

The structure of the digestive tract is admirably suited and adapted for the functions it performs. When one considers the labors that a chemist would have to go through to bring about all the reactions that take place within a relatively short time in the digestive tract, one must marvel at the provisions made by nature whereby animals can utilize the foods that are available to them.

SUMMARY

The digestive tracts of vertebrates are built upon the same fundamental plan and function in a comparable manner. Many variations are to be found which may be considered as adaptations to meet particular needs. The signs of evolution are not so apparent in a study of the digestive system as in some of the other organ systems, but the adaptive radiations from a simple plan are strikingly brought out.

Among the more important advances that are to be noted within the group are:

1. Separation of mouth and nasal passages
2. Change to heterodont condition of teeth
3. Development of muscles in lips and tongue
4. Beginning of digestive function on part of mouth glands
5. Distinct separation of esophagus and stomach
6. Digestive function of stomach added to that of mere storage
7. Complicated stomachs in many forms providing for proper utilization of type of food eaten
8. Differentiation of intestine into distinct regions
9. Tendency toward lengthening of alimentary canal as a whole
10. Provisions for increasing surface for absorption and secretion
11. Differentiation of cloaca so that openings of digestive and urogenital systems are separate

| 6 |

RESPIRATORY SYSTEM AND RELATED STRUCTURES

Every living cell in an organism consumes oxygen. A supply of oxygen is essential to that phase of metabolism known as *catabolism*. Oxidation of substances within the cells results in the liberation of heat and other forms of energy and in the production of carbon dioxide. This end product of metabolism, unless removed from the cell (or body), acts as a poison which is harmful to protoplasm. It has been estimated that the average human adult exhales about 20 liters of carbon dioxide per hour. In a poorly ventilated room occupied by several people, the percentage of carbon dioxide is not poisonous unless the gas is present in large quantities; air containing more than 15 parts in 10,000 is not fit for respiration under any conditions.

RESPIRATION

Aerobic respiration. The term *aerobic respiration* is used to denote the exchange of oxygen and carbon dioxide between an organism and its environment. In this type of respiration oxygen is taken into the body and carbon dioxide is given off.

Anaerobic respiration. In certain organisms *anaerobic* respiration takes place. Carbon dioxide is given off, but no oxygen is taken in. The necessary oxygen is obtained, in such forms, by the incomplete breakdown of carbohydrates, and possibly of fats, in the body.

The role of blood in respiration. Although in members of certain lower phyla there is a direct exchange of gases between the cells and the environment, in vertebrates and some other forms the blood serves to bring oxygen from the environment to the cells and to carry carbon dioxide from the cells to the environment. It is the presence of the pigment hemoglobin that gives vertebrate blood its unusual capacity for carrying oxygen. Hemoglobin is present in the blood of all vertebrates and even in certain invertebrates. In vertebrate blood it is confined to the red blood corpuscles, or erythrocytes. It is interesting that the muscles of birds and mammals also contain a form of hemoglobin called *myoglobin*. There are minor differences in the hemoglobins of different forms and they vary somewhat in their properties. The importance of hemoglobin in connection with respiration lies in its ability to combine with oxygen through the medium of iron which it contains. Two atoms of oxygen, under favorable conditions, unite with each atom of iron in the hemoglobin complex. Thus $Hb_4 + 4O_2 = Hb_4O_8$. Hemoglobin which has combined with oxygen in this manner is spoken of as *oxyhemoglobin*. It is scarlet red in color and a very unstable compound, since it gives up its oxygen so readily. After having given off its oxygen it is known as *reduced hemoglobin*, or simply *hemoglobin*, and has a purplish-blue color.*

The red corpuscles with their contained hemoglobin also play a prominent role in transporting carbon dioxide. When carbon dioxide leaves the cells and tissues of the body it enters the bloodstream and diffuses into the red blood corpuscles. Here most of the carbon dioxide combines with water to form carbonic acid as follows: $CO_2 + H_2O = H_2CO_3$. An enzyme, *carbonic anhydrase*, present in the red cells, facilitates the reaction. The carbonic acid thus formed combines with potassium (formerly in combination with hemoglobin) to form potassium bicarbonate. Bicarbonate ions then diffuse out into the plasma in exchange for chloride ions which enter the cell. In the plasma the bicarbonate ions combine with sodium to form sodium bicarbonate. Thus, carbon dioxide is transported by the plasma as well as by the red corpuscles. At the respiratory surface oxygen enters the red blood corpuscles and the above processes are reversed. Carbon dioxide is then liberated as a free gas.

Phases of respiration. In animals in which the blood is used in carrying oxygen and carbon dioxide, two phases are involved in respiration. They are referred to as *external* and *internal respiration.* The term external respiration is used to denote the exchange of gases between the blood and the environment. This usually takes places in the capillaries of the gills or lungs, but in some cases other structures such as the skin are utilized. In gills the capillaries are almost in direct contact with the water in which oxygen is dissolved. In lungs they are practically in contact with air that has been taken into the cavities of the lungs. Even though lungs are situated within the body, nevertheless the term external respiration is used. During

---

* In the following pages the term "oxygenated" blood is used to refer to blood carrying a high percentage of oxyhemoglobin; "unoxygenated" blood refers to that in which *most* of the hemoglobin has been reduced. Actually there is always at least a small amount of oxygen present in blood since it is carried by the plasma to a minor extent. Furthermore, not every red corpuscle in so-called "unoxygenated" blood need of necessity have given off its oxygen.

external respiration the blood takes up oxygen and loses most of the carbon dioxide which it is carrying. Internal respiration refers to the gaseous exchange between the blood and the tissues or cells of the body. Oxygen is taken from the blood, which in turn gains carbon dioxide given off by the tissues. The principles involved in internal and external respiration follow the "laws of gases" and are physical rather than biological in nature.

In vertebrates, respiratory organs are present which serve to facilitate external respiration. Certain requirements are demanded of respiratory organs in order that they may function properly: (1) a large, vascular surface area must be provided so that an ample capillary network may be exposed to the environment; (2) the membrane surfaces through which gaseous exchange occurs must be moist at all times and be thin enough to permit the passage of gases; (3) provision must be made for renewing the supply of the oxygen-containing medium (air or water) which comes in contact with the respiratory surface and for removing the carbon dioxide which has been given off from that surface; and (4) blood in the capillary network must circulate freely.

With few exceptions, the organs of respiration in vertebrates are formed in connection with the pharynx. In some forms, notably amphibians, the skin itself is an important respiratory organ. In a fish, the loach, *Misgurnus fossilis,* a peculiar method of respiration is utilized. The animal has the habit of swallowing air, passing the air bubble the length of the intestine, and voiding it at the anus, oxygen being absorbed en route by blood vessels in the extremely vascular intestine.

The function of the pharynx as part of the digestive tract is merely to serve as a passageway from mouth to esophagus and to initiate swallowing movements. However, the part it plays in respiration is of much greater importance. The internal gills of aquatic vertebrates and the lungs of air breathers are structures, derived from the pharynx, which are particularly adapted for respiration. In some cases the vascular epithelium of the pharyngeal wall itself is important in respiration. Other structures not associated with respiratory activity are derived from the pharynx. Among these are the swim bladders of certain fishes, tonsils, middle ear and Eustachian tube, as well as such glands as the thyroid, thymus, parathyroids, and ultimobranchial bodies.

NASAL PASSAGES

In air-breathing vertebrates there is a close association between the olfactory organs and the organs of respiration. In lower aquatic forms, however, these structures are usually completely divorced from each other. The advantages of a close relationship between the organs of smell and those of respiration should be obvious. When air is drawn from the outside environment into the lungs, volatile substances carried by the air may stimulate the sensory endings of the olfactory nerves situated in the nasal passages. In most of the Sarcopterygii and in the larval forms of some of the more primitive salamanders, such as the newt, water enters the mouth through the nares and leaves via pharyngeal gill slits. Chemical substances in the water may be detected by the olfactory apparatus located in the nasal passages.

Comparative anatomy of nasal passages. CYCLOSTOMES. In lampreys the single, median, nasal aperture on the top of the head leads to a blind olfactory sac from which a long nasopharyngeal pouch extends ventrally (Fig. 5.18). The latter also ends blindly. This complex serves primarily as an olfactory apparatus and has no connection with the pharynx or with respiration. The hagfish possesses a rather similar apparatus but the lower end of the nasopharyngeal pouch forms a connection with the pharynx. In *Myxine,* therefore, a respiratory current of water may reach the pharynx and gills via the passage furnished by the olfactory apparatus and nasopharyngeal pouch.

FISHES. In most fishes there is no connection between nostrils and mouth cavity. In some elasmobranchs, however, an open *oronasal groove* on each side forms a channel connecting the olfactory pit to the mouth. There is some dispute as to whether the *labial grooves* at the lateral edges of the mouth in *Squalus acanthias* (Fig. 6.1) are homologous with the oronasal grooves of other elasmobranchs. Oronasal grooves foreshadow the appearance of a direct connection between the two in higher forms. It is in the fishes of the subclass Sarcopterygii that for the first time a direct connection exists between nasal and mouth cavities in the form of a pair of closed tubes. The openings to the outside are known as the *external nares;* those opening into the mouth are the

internal nares, or *choanae.* In these fishes the external nares open to the outside beneath the upper lip but outside the mouth. They cannot be seen when the mouth is shut. Internal nares apparently were secondarily lost in the coelacanths and are not present in *Latimeria,* which, nevertheless, and for other reasons, is classified with the Sarcopterygii. One would not expect to find internal nares in a deep-sea dweller like *Latimeria* in which even the swim bladder is degenerate (page 222). In a certain group of teleosts, the stargazers (*Astroscopus*), a well-developed passageway leads to the oral cavity and internal nares are present (J. W. Atz, 1952, *Anat. Record,* **113**: 105). These, however, are not homologous with the nasal passages of the Sarcopterygii or tetrapods even though they are used to some extent in respiration and probably have an olfactory function. None of the living fishes possessing internal nares is known to use them to breathe atmospheric air (J. W. Atz, 1953, *Quart. J. Biol.,* **27**: 366). Water is drawn into the mouth through both the mouth and nasal passages. In the Sarcopterygii olfactory epithelium is located in the dorsal part of each nasal passage, thus indicating that the nasal passages serve to increase the effectiveness of olfaction.

In most vertebrates during embryonic development two ectodermal thickenings appear in the surface epithelium at the ventro-lateral portions of the head. These are the *olfactory placodes* which soon invaginate to form *olfactory pits.* They form connections with the oral cavity by means of a groove similar to the oronasal groove of elasmobranchs. The edges of the olfactory pits and grooves fuse together, forming a pair of tubular connections between the external nares and the mouth.

AMPHIBIANS. External and internal nares are present in amphibians. The nasal passages connecting them are short, the internal

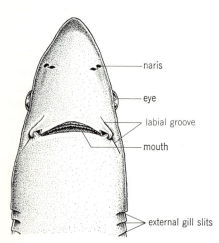

Fig. 6.1 Head of dogfish shark, *Squalus acanthias,* ventral view.

naris

eye

labial groove

mouth

external gill slits

nares being located just inside the upper jaw (Fig. 6.2). In larval urodeles the current of water passing through the nasal passages is produced either by ciliary action or by movements of the gill apparatus and muscles of the lower jaw. In many urodele larvae and in the tadpoles of anurans, valves around the internal nares control the direction of flow of water.

With the appearance of lungs at the time of metamorphosis in most urodeles, smooth-muscle fibers are generally found to be present around the external nares, thus providing a means for regulating the size of the aperture. This is the first instance in which atmospheric air is drawn through the nasal passages. Adult anurans use a peculiar device for closing their nostrils. A small, rounded, conical projection, the *tuberculum prelinguale,* at the tip of the lower jaw, is thrust forward and upward and pushes apart the two premaxillary bones which lie above it in the upper jaw. The movement of these bones, by changing the position of a portion of the nasal cartilages, effectively closes the nasal passages.

Each nasal passage in amphibians is surrounded by a cartilaginous *nasal capsule* which is bounded by nasal, premaxillary, maxillary, and vomer (prevomer) bones. In a few urodeles a projection into the nasal passage from the lateral wall on each side indicates the first appearance of the *conchae,* which become highly developed in certain mammalian forms. The nasal passages themselves consist of upper olfactory and lower respiratory regions.

REPTILES. Beginning with reptiles there is a tendency toward elongation of the passageway between the external and internal nares. This is brought about by the development of a pair of palatal folds. These are horizontal, shelflike projections of the premaxillary, maxillary, palatine, and even of

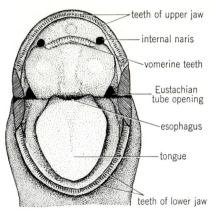

Fig. 6.2 Oral cavity of salamander, *Ambystoma texanum.* The angle of the mouth on each side has been cut in order to depress the lower jaw.

teeth of upper jaw
internal naris
vomerine teeth
Eustachian tube opening
esophagus
tongue
teeth of lower jaw

the pterygoid bones. Only in the crocodilians among reptiles, however, do the processes from the two sides fuse along the median line, thus forming a bony *secondary palate* which separates the nasal passages from the mouth cavity. The internal nares communicate with the mouth cavity posterior to the secondary palate. In crocodiles and alligators they are located rather far down the throat. In reptiles a single concha, supported by the maxillary bone, projects into the nasal passage on each side from its lateral wall. In crocodilians the concha divides anteriorly. Although salt-secreting glands have been described in marine turtles, one marine lizard, and many terrestrial lizards, it has only recently been discovered that marine sea snakes also possess such structures. A glandular mass, called the *nutrial gland,* which serves as an extrarenal organ for ridding the body of excess NaCl, lies in the interpalatine groove in the soft tissues under the premaxillary bone, anterior and medial to Jacobson's organ, and anterior to the internal choanae.

Paired ducts lead to the interpalatine space. The secretion, a concentrated salt solution, is apparently pushed out of the mouth by movements of the tongue (A. M. Taub and W. A. Dunson, 1966, *Amer. Zoologist*, 6:565). In marine turtles, orbital, or lacrimal, glands are primarily responsible for handling the salt load.

BIRDS. The greater part of the upper portion of the beak, or bill, of birds consists of the very much enlarged and fused premaxillary bones, together with the maxillaries at the base. The bones are covered with a hard, horny sheath. The external nares are usually located at the base of the beak, but those of the kiwi are placed almost at the tip. The secondary palate, like that of most reptiles, is incomplete, consisting of a pair of palatal folds which fail to fuse medially. The choanae lie above the palatal folds. In most birds the nasal passages are relatively short because of the basal position of the external nares and the cleft condition of the palate.

Three folds, or conchae, extend into the nasal passages of birds. They are supported by *turbinate bones*. Only the posterior concha is covered with olfactory epithelium. The mucous membrane covering the conchae aids in warming and moistening the air on its way to the lungs. The choanae connect with the pharynx posterior to the conchae.

A pair of *nasal glands* in birds has long been known to be considerably larger in marine forms than in terrestrial species. These have been shown to function by secreting salt and are referred to as salt glands (R. Fänge, K. Schmidt-Nielson, and H. Osaki, 1958, *Biol. Bull.*, 115:162). They supplement the kidneys in the excretion of sodium chloride and, in some forms, are even more important than the kidneys in eliminating excess salt from the body. Salt glands, the gross and microscopic structures of which are generally similar, have been studied in herring gulls, pelicans, cormorants, eider ducks, petrels, etc. They develop as outgrowths of the nasal cavity and are composed of lobes made up of branched tubular glands arranged in a radial manner about a central canal. In the herring gull the lobes are situated on top of the skull in the supraorbital grooves of the frontal bones. Two ducts lead from the glandular mass on each side to the anterior, or vestibular, region of the nasal cavity. At times when the nasal glands are secreting, the salty discharge passes through the external nares and drips from the end of the beak.

MAMMALS. The nasal passages of mammals have been greatly elongated and are larger and more complicated than in lower forms. Medial projections from the premaxillary, maxillary, and palatine bones fuse in the midline to form the *hard palate*. The pterygoids, which are reduced in mammals, do not contribute to palate formation. Extending posteriorly from the hard palate is the *soft palate* composed of muscle and connective tissue covered with mucous membrane and without any bony foundation. The presence of the soft palate carries the nasal passages posteriorly, so that the choanae open into the pharyngeal region.

The presence of a *nose* is characteristic of mammals. That of man is of a special nature and not comparable to those found in most other forms. In some mammals the nose is excessively developed and forms a *proboscis* at the end of which the external nares are located. A proboscis is more or less well developed in moles, shrews, tapirs, and elephants. The trunk of the elephant actually represents the very much drawnout nose and upper lip.

The nasal passages of mammals are divided into three general regions: vestibular, respiratory, and olfactory. The *vestibu-*

lar region, lined with skin, leads from the outside to the inner mucous membrane where the squamous epithelium is not keratinized. Dust-arresting hairs, sweat glands, and sebaceous glands may be abundant in the front part of the vestibule.

Into the *respiratory* and *olfactory regions,* which make up the greater part of the nasal cavity, extends a veritable labyrinth of conchae. These are thin, coiled, and scroll-like projections of the nasal, maxillary, and ethmoid bones covered with mucous membrane. The bony projections are referred to as *turbinate bones* and known as *naso-turbinates, maxilloturbinates,* and *ethmo-turbinates,* respectively. The maxilloturbinate supports the *inferior,* or *ventral, nasal concha* which is homologous with the single concha observed in reptiles. The remaining turbinates on each side support conchae which appear as a series of parallel ridges projecting into the nasal cavity from the lateral wall. The most dorsal, or *superior nasal concha,* is supported at least in part by a projection from the nasal bone which is, therefore, referred to as the *nasoturbinate.* Mammals may possess several *middle,* or *lateral, nasal conchae,* each supported by an *ethmoturbinate bone.* The *respiratory region* is lined with mucous membrane often spoken of as the *Schneiderian membrane,* covered with "respiratory" epithelium. This is a ciliated, pseudostratified, columnar epithelium, rich in goblet cells. Mucous and serous glands also open onto its surface. The middle and inferior nasal conchae are covered with respiratory epithelium which keeps the nasal passages moist. The mucous membrane lining the nasal cavities is highly vascular, having an abundant supply of arteries, capillaries, and veins. That covering the middle and inferior conchae has an abundant supply of venouslike spaces which closely resemble those of erectile tissue (see page 543). The venouslike spaces are ordinarily collapsed, but under certain conditions may become distended with blood, resulting in a thickening of the mucous membrane. This may, in certain individuals, be so pronounced as to obstruct the nasal passages and render breathing through the nose difficult. Whether this is true erectile tissue, comparable to that found in the penis and clitoris, has been questioned by some authorities because of the absence of septa containing smooth-muscle fibers. It is of interest, however, that in certain individuals it is affected by erotic stimuli.

The *olfactory region* occupies the innermost and upper recesses of the nasal passages. It is covered with "olfactory epithelium" which contains nerve endings for the sense of smell. The olfactory epithelium is, in general, confined to the roof of the nasal passages but covers much of the surface of the superior nasal conchae as well, thus increasing the sensory area for olfaction. Here the mucous membrane has a yellowish-brown appearance which is quite in contrast to the pink or reddish color of that of the respiratory region. In man, with his relatively poorly developed sense of smell, there has been a great reduction of the conchae as compared with many other mammals. For this reason they merely project into the nasal passages in the form of rather shallow ridges. The olfactory epithelium in man is confined on each side to an irregular area measuring roughly 250 sq mm.

Sinuses, or spaces, within certain bones communicate with the nasal passages. The chief sinuses found in mammals are those of the frontal, maxillary, ethmoid, and sphenoid bones. They are lined with ciliated mucous membrane like that of the respiratory region, but glandular cells are smaller and less numerous. The sinuses, which in man are so often the seat of infection and

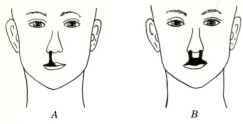

Fig. 6.3 A, unilateral harelip; B, bilateral harelip.

Jacobson's organ. In many tetrapods a peculiar structure, the *vomeronasal organ* of Jacobson, is present. It consists of a pair of blind diverticula usually extending from the ventromedial portion of the nasal cavity. In amphibians, reptiles, and mammals it receives branches from the terminal (O), olfactory (I), and trigeminal (V) cranial nerves. Although the function of Jacobson's organ is obscure, it is believed to serve as an accessory olfactory device and aids in the recognition of food, since it is best developed in animals which hold food in their mouths.

give so much trouble, have no known function. Their presence lightens the skull and gives resonance to the voice.

The nasal passages of mammals are formed in a manner somewhat different from that of lower forms. On each side of the head an olfactory pit develops which communicates with the mouth cavity by a groove similar to the oronasal groove of certain elasmobranchs. Here the similarity ends, for the nasal pits grow inward and establish secondary connections with the oral cavity. The original grooves close over and fuse. Failure of the grooves to close on one or both sides results in unilateral or bilateral *harelip,* as the case may be (Fig. 6.3).

The vomeronasal organ first appears in amphibians. In frogs it is situated in the anteromedial portion of each of the nasal cavities, but in urodeles it is usually lateral in position. Its connection with the nasal cavity is constricted, so that it appears in the nature of a duct. The organ is lacking in *Necturus* and in the cave salamander, *Proteus*.

In *Sphenodon*, snakes, and lizards (Fig. 6.4), Jacobson's organ is best developed of all. Instead of opening into the nasal cavity, in these forms the duct connects directly with the mouth cavity but near the choanae. In other reptiles and in birds the organ ap-

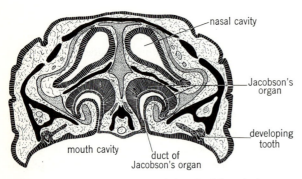

Fig. 6.4 Cross section through head of lizard showing Jacobson's organ. Black areas represent bone; stippled areas represent cartilage. (*After Schimkewitsch.*)

Anatomy of the Chordates

pears only as a rudiment during embryonic development. The tips of the forked tongues of snakes and of some lizards, when retracted, are placed in the small, blind pockets of the vomeronasal organs. These apparently are able to detect chemical substances that may have adhered to the surface of the tongue when it was thrust out of the mouth. In mammals there is much variation. In some, as in man, it appears only as an embryonic vestige in the lower medial side of each nasal cavity, but in others it persists throughout life as a definite structure. Jacobson's organ is best developed in monotremes in which it is surrounded by cartilage through which a branch of the olfactory nerve passes. A small concha even projects into its cavity, much in the manner of the conchae of the main nasal cavity referred to above. The organ is small but well-developed in marsupials, insectivores, and rodents. In some rodents the vomeronasal organ opens into the nasal passage, but in others its duct penetrates the secondary palate, opening into the mouth cavity in a manner similar to that of *Sphenodon*, snakes, and lizards.

PHARYNGEAL POUCHES

In the embryos of all chordates a series of pouches develops on either side of the pharnyx. These endodermal structures push through the mesenchyme until they come in contact with invaginated *visceral furrows* of the outer ectoderm to which they fuse, forming thin, platelike areas. The pharyngeal pouches arise in succession in an anterior-posterior direction. They develop in such a manner that they become successively smaller from first to last (Fig. 6.5), and the pharynx therefore is more or less funnel-shaped and tapers toward the esophagus. Except in the highest vertebrate classes, perforations usually occur where endoderm and

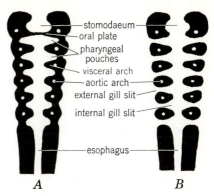

Fig. 6.5 Diagram showing arrangement of pharyngeal pouches: *A*, during development; *B*, after connection with the outside has been established and the oral plate has disappeared.

ectoderm come in contact. Even in the higher classes, however, temporary openings are occasionally established. The *visceral,* or *pharyngeal, pouches* are then called *clefts.* Primitively the cavity of the pharynx is thus connected by a series of clefts to the outside. The openings connecting the pharynx proper with the clefts are called *internal gill slits;* those connecting the clefts with the outside are *external gill slits.*

The number of pharyngeal pouches or clefts is greatest in the lowest groups of vertebrates and least in the higher classes. Thus, among cyclostomes, certain forms (*Bdellostoma*) possess as many as 14 pairs, although the lamprey (*Petromyzon*) has 7 pairs and the hagfish (*Myxine*) but 6. In fishes, amphibians, and reptiles 5 or 6 pairs commonly appear in the embryo, but birds and mammals possess the reduced number of 5 or 4 as the typical condition. The pharyngeal pouches are of greatest significance in the lower aquatic vertebrates since they bear gills and are directly concerned with respiration. Those of amniotes do not bear gills and generally disappear, except for the first which becomes the Eustachian tube and middle ear. Remnants of the others persist in the form of certain glandular structures, to be discussed later.

The visceral clefts are separated from one another by septa, which are mesodermal structures, covered with epithelium derived either from ectoderm or endoderm, depending upon whether it is toward the exterior or interior surface. Within each septum lies a cartilaginous or bony barlike structure, the *visceral arch*. This serves to support the septum. Blood vessels, called *aortic arches*, branch from the ventral aorta and course through the septa, which also receive branches of certain cranial nerves to be described later. The visceral arches, which make up the so-called *visceral skeleton*, are modified in higher vertebrates to form various portions of the skeleton in the head region (pages 434 to 436). The aortic arches also undergo marked changes in the different classes (pages 559 to 569). Metamerism, or segmentation, of the branchial region, as indicated by the arrangement of the visceral arches, aortic arches, nerves, and muscles in this area, is a reflection of the arrangement of the series of visceral pouches or clefts rather than being related to the segmental arrangement of the trunk myotomes. The latter is based upon the appearance of

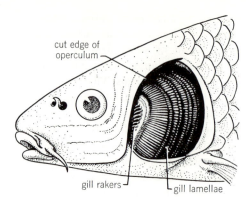

Fig. 6.7 Head of bony fish (carp). The operculum has been cut away in order to expose the internal gills. (*From Storer, "General Zoology," McGraw-Hill Book Company. By permission.*)

metameric mesodermal somites which appear during early development (page 66).

GILLS

Gills are composed of numerous *gill filaments* or *gill lamellae,* which are thin-walled extensions of epithelial surfaces. Each contains a vascular network. Blood is brought extremely close to the surface, thus facilitating the ready exchange of gases. In their aggregate, gills present a relatively large surface for respiratory exchange.

Types of gills. Gills are of two general types, external and internal. *External gills* (Fig. 6.6) develop from the integument covering the outer surfaces of visceral arches. They are usually branched, filamentous structures covered entirely with ectoderm and are not related to the visceral pouches. *Internal gills* (Fig. 6.7) are usually composed of a series of parallel *gill lamellae,* although in some forms they may be filamentous. They may be borne on both sides of the interbranchial septa but in some cases are

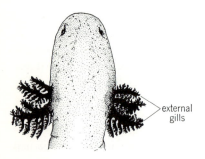

Fig. 6.6 Head of salamander larva with external gills.

present on one side alone. A series of lamellae on one side of an interbranchial septum is termed a *half gill,* or *hemibranch.* Two hemibranchs enclosing between them an interbranchial septum make up a *complete gill,* or *holobranch.* The two hemibranchs bounding a gill cleft thus belong to different holobranchs. It is generally assumed that internal gills are covered entirely with endoderm, but there is considerable controversy on this point and the question of exact origin has not been definitely settled. In some animals both external and internal gills are present.

The functioning of external gills poses no problems since the filaments are in direct contact with the water containing dissolved oxygen. When internal gills are used in respiration, in most cases, water, containing dissolved oxygen, is taken into the mouth and passes through the internal gill slits into the gill clefts. As the water passes over the gill lamellae, oxygen is taken from the water and carbon dioxide is released. The water then passes through the external gill slits to the outside.

In lampreys and certain marine teleosts it has been determined that internal gills, in addition to their respiratory function, play an important role in the excretion of salt from the blood. In teleosts, such nitrogenous wastes as ammonia have been shown to be excreted via glandular cellular elements present in the gill lamellae. The gills may thus supplement the kidneys as excretory organs (see page 270).

Comparative anatomy of gills. AMPHIOXUS. The mouth of amphioxus leads into the pharynx, the opening being guarded by a fold called the *velum,* bearing 12 *velar tentacles.* The walls of the pharynx are provided with large numbers of vertically elongated gill clefts. Over a hundred of these may be present. The gill clefts are separated by *primary gill bars* of a stiff, gelatinous consistency, somewhat resembling chitin which makes up the exoskeleton of insects. The primary gill bars bifurcate at their ventral ends. During development a *secondary bar,* or *tongue bar,* grows down from dorsal to ventral sides between two primary gill bars and thus divides each primary gill cleft in two. The secondary bars are *not* bifurcated at their lower ends. Still later, small *crossbars* appear, connecting one primary bar with the next so that the gill clefts become further subdivided. The crossbars pass *over* the secondary bars, but the two usually fuse after a time. There are no gill filaments or lamellae, and blood vessels course through the pharyngeal bars. A few of the anterior gill slits, for a time, open directly to the outside. With the development of the ectoderm-lined *atrium,* however, the pharynx no longer forms direct connections with the outside and all gill clefts come to open into the *atrial cavity,* or *peripharyngeal chamber.* A single opening to the outside, the *atriopore,* is located on the ventral side of the animal, about two-thirds of the way back.

Although much of the respiration of amphioxus takes place through the skin, a constantly renewed stream of water passes from mouth to pharynx, through the gill clefts, and into the peripharyngeal chamber. In passing over the gill bars, the blood there is oxygenated. Water leaves the peripharyngeal chamber through the atriopore.

CYCLOSTOMES. The respiratory system of an adult lamprey is highly specialized and on first inspection seems to be atypical. Study of its development in the ammocoetes larva, however, reveals that it is built upon the typical vertebrate plan. In the pharyngeal wall of the larva eight pairs of gill pouches begin to develop, but the first pair flattens out before long and disappears. Seven pairs of gill clefts remain, each with its internal and external gill slits. At this stage the pharynx connects with the mouth in front and with the esophagus behind. During further development, and with the formation of the round, sucking mouth, a peculiar change occurs in the pharnyx-esophagus relationship. These structures become split apart, or separated, in such a manner that each has its own connection with the mouth (Fig. 6.8). The esophagus then lies dorsal to the pharynx. The latter becomes a blind pouch, the opening of which is guarded by a *velum*. Seven pairs of internal gill slits open from the pharynx into seven pairs of gill clefts which are rather large and spherical in shape. The gill lamellae are arranged in a more or less circular fashion, but, nevertheless, each gill cleft is bordered by a hemibranch on both anterior and posterior walls. There are thus 14 hemibranchs on each side, but only 6 holobranchs, since the first and last hemibranchs are not parts of holobranchs (Figs. 6.9 and 12.18).

During respiration the lamprey normally

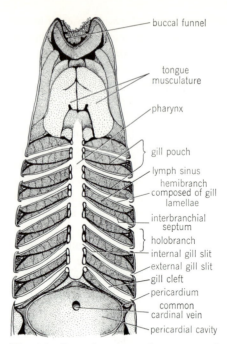

Fig. 6.9 Frontal section of anterior end of lamprey, showing arrangement of the gills, as seen from below.

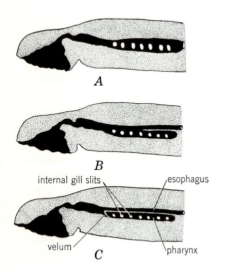

Fig. 6.8 Stages *A, B,* and *C* in the separation of esophagus and pharynx in the lamprey (semidiagrammatic).

takes water in and out of the gill clefts through the *external* gill slits. Thus the gill lamellae are bathed by a constantly replenished supply of water. This method of respiration is quite in contrast to that employed by true fishes in which water enters the mouth and passes over the gills on its way to the outside through the external gill openings. The method utilized by the lamprey is necessary because when the animal is attached to some object or engaged in feeding, the mouth opening is blocked. It is, of course, possible that when the animal is free, water may pass through the buccal funnel into the mouth cavity and to the outside via pharynx, internal gill slits, gill clefts, and external gill slits. It is doubtful, however, whether this method is employed, at least to any appreciable extent.

When blood, mucus, and tissue fragments,

on which the animal feeds, pass into the mouth cavity, they go down the esophagus to the remainder of the digestive tract. The velum at the anterior end of the pharnyx prevents the passage of the bloody food into the respiratory system and also prevents dilution of the blood with water which might enter the mouth from the pharynx. When the lamprey is getting ready to feed (T. E. Reynolds, 1931, *Univ. Calif. Publ. Zool.,* **37**:15), the buccal funnel is first collapsed or flattened and the water already in it is forced over the tongue into the mouth cavity and *hydrosinus* above. The hydrosinus is an anterodorsal extension of the mouth cavity. Next, the tongue is used to block the passage between the buccal funnel and the mouth cavity. The suctorial disc is then arched, or warped, outward, thereby increasing the volume and the degree of suction of the buccal funnel. Water in the hydrosinus and oral cavity is then forced past the velum into the pharynx, the velum serving as a one-way valve which prevents the return of water into the mouth cavity. The tongue, up to this point, is still blocking the passage between the buccal funnel and mouth cavity. The discharge of water from mouth cavity to pharynx creates a low-pressure area extending from the attached disc to the velum. The tongue is now free to move back and forth in rasping. Tissue fragments, and blood flowing from the wound, pass readily into the mouth cavity and thence to the esophagus and remainder of the digestive tract. In the meantime the animal is respiring by taking water in and forcing it out of the external gill slits as described above.

In the hagfish the pharynx forms a direct connection between mouth and esophagus. Six pairs of internal gill slits and gill pouches are present. Only a single pair of external openings exists, however. A series of long tubes coming from the gill pouches unites to form a common duct on each side, and

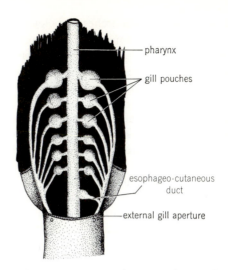

Fig. 6.10 Diagram showing relation of gill pouches of hagfish, *Myxine,* to the pharynx and to the single pair of external gill apertures, ventral view. (*After Müller.*)

this leads to the external aperture (Fig. 6.10), which is situated near the midventral line at some distance from the anterior end. This may be of advantage to the animal when it is feeding, or boring its way into the body of a fish in order to devour the soft internal organs. An *esophageo-cutaneous duct* connects the esophagus with the common duct on the left side in *Myxine.* It lies posterior to the last gill pouch on that side and is similar to a gill cleft but lacks gills.

In cyclostomes of the genus *Bdellostoma* the number of gill clefts varies from 6 or 7 up to 13 or 14 pairs. The gill clefts connect internally with the pharnyx which is similar to that of the hagfish and not a blind pouch of the type found in the lamprey. An esophageo-cutaneous duct is also found on the left side in *Bdellostoma.* The animal resembles lampreys, however, in that the external gill slits open separately to the outside through small apertures.

In cyclostomes, with their poorly developed skeletons, visceral arches homologous with those of higher forms are essentially lacking. A peculiar, irregular, cartilaginous branchial basket (page 444) provides some support but lies peripheral to the gill clefts, just beneath the skin.

FISHES. In fishes and tetrapods a series of skeletogenous *visceral arches* encircles the pharynx. In tetrapods these become greatly modified, but in fishes they serve primarily to support the gills. They are located between the gill clefts, one behind the other, at the bases of the interbranchial septa. The *first* is called the *mandibular arch;* the *second* is the *hyoid arch.* The remaining visceral arches are referred to by number (3, 4, 5, 6, etc.). The first gill pouch or cleft lies between the mandibular and hyoid arches and is often referred to as the hyomandibular cleft. In fishes it is either modified to form a *spiracle* or is closed altogether. The ancestral placoderms had a full-sized gill slit between the mandibular and hyoid arches.

Two main types of *internal* gills are to be found among fishes. The first and more primitive type is typical of elasmobranchs. In this group the interbranchial septa are exceptionally well developed and extend beyond the hemibranchs which are closely applied to them (Figs. 6.11*A* and 6.12*A*).

Each bends posteriorly at its distal end in such a manner that a row of separate *external* gill slits is formed. Thus the interbranchial septa, in addition to separating the gill clefts, serve to protect the gills themselves. In the frilled shark, *Chlamydoselachus,* the interbranchial septa are continued to an unusual degree and project *over* the external gill slits. In the dogfishes and sharks the external gill slits are lateral in position, but in adult skates and rays they are ventrally located even though they are in a lateral position during the early part of embryonic life. Small cartilaginous *gill rays,* or *branchial rays,* in the form of a single row project from each visceral arch into the interbranchial septum to which they give support. Rigid, comblike *gill rakers* usually project from the gill arches. Their function is to prevent food and foreign bodies from entering the gill clefts.

The second type of gill is found in the remaining fishes. In these the interbranchial septa are reduced to varying degrees (Fig. 6.11*B* and *C*) to form slender structures from which the hemibranchs protrude into a single *branchial,* or *extrabranchial, chamber* located on each side between the operculum and gills. The distinction between internal and external gill slits is less obvious in these. Two rows of gill rays are usually

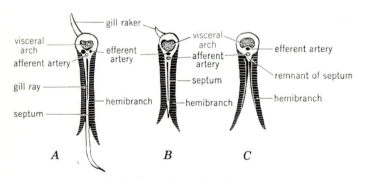

Fig. 6.11 Types of fish gills: A, elasmobranch; B, chimaera; C, teleost.

Anatomy of the Chordates

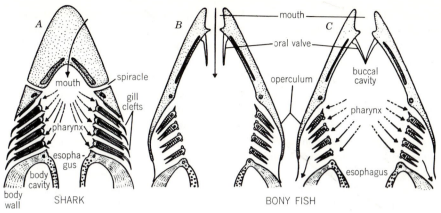

Fig. 6.12 The respiratory mechanism of fishes; diagrammatic frontal sections (lobes of oral valve actually are dorsal and ventral); arrows show paths of water currents. Shark: *A*, water enters ventrally placed mouth which then closes and floor of mouth region rises to force water over the gills and out the separate gill clefts. Bony fish: *B*, inhalant, opercula closed, oral valve open, cavity dilated, and water enters; *C*, exhalant, oral valve closes, buccal cavity contracts, water passes over gills in common cavities at sides of pharynx and out beneath opercula. (*From Storer, "General Zoology," McGraw-Hill Book Company. By permission.*)

present, rather than one as in the elasmobranchs. Each row supports a hemibranch. Beginning with the Holocephali, the septum of the hyoid arch is enormously developed and grows caudad as a cutaneous fold, covering the gill clefts posterior to it. The gill cover, or *operculum,* which in bony fishes is strengthened by a number of thin *opercular bones,* protects the gills in the branchial chamber which thus opens to the outside through one gill aperture. In *Latimeria* the operculum is poorly developed and partially ossified. In a few forms, as in the eel, the opening of the branchial chamber to the outside is small and round, but it is usually long and slitlike. In one group of fishes, the symbranchiates, there is a single, midventral opening. Although in elasmobranchs, chimaeras, sturgeons, and some others a true hemibranch receiving unoxygenated blood is located on the posterior

side of the hyoid arch, this is lacking in *Polypterus, Amia, Polyodon,* teleosts, and Dipnoi. Teleosts have four complete holobranchs on each side. Since in fishes no aortic arch courses through the tissue posterior to the last gill cleft, no hemibranch is found on the posterior wall of the last gill cleft except in *Protopterus* in which its homologies are doubtful.

In fishes having an operculum, a pliant, *branchiostegal membrane* supported by bony *branchiostegal rays* usually extends from the inner surface of the operculum to the body wall. It serves as a one-way valve which permits water to leave the branchial chamber through the opercular opening but does not allow it to enter. Other folds, from the maxillary and mandibular regions, form *oral valves* located just within the mouth opening. These permit water to enter the mouth but not to leave it. Gill respiration in these fishes (Fig. 6.12B and C) is brought about by raising the opercula and closing the branchiostegal folds. This causes a change in pressure within mouth and pharynx, so that the oral valves are opened and water enters the mouth. Lowering the operculum, with a resulting increase in pressure, causes the oral valves to close and the branchiostegal valves to open. Oral valves seem to be lacking in *Latimeria*. As water passes over the gills, emerging through the opercular slit, oxygen is taken up, and carbon dioxide liberated. Respiration in fishes thus involves a series of muscular contractions in the walls of mouth and pharynx, bringing about a flow of water. It is of interest that some fishes use the stream of water emerging from the branchial chamber as a propelling force in locomotion. Certain rapidly swimming fishes keep their mouths open while swimming, thus forcing a stream of water to enter the mouth, bathe the gills, and leave the branchial chamber without resorting to muscular contractions and opercular movements. In most, however, the act of inspiration causes water to enter the mouth and pharynx, and expiration results in its expulsion. According to J. F. Daniel (1934, "The Elasmobranch Fishes," University of California Press), respiration in elasmobranchs is brought about by the interaction of a complicated series of buccal and pharyngeal muscles which, in the absence of specific valves, in most cases, regulate the passage of water over the gills. When water enters the mouth the external gill slits close, and when the slits open, the mouth closes. In one species, *Heterodontus francisci*, the number of respirations has been observed to be approximately 35 per minute when the animal is at rest.

In most elasmobranchs and in a few other fishes (*Acipenser, Polyodon,* and *Polypterus*) the first gill pouch has become modified and opens to the outside by means of a *spiracle*. In other fishes the spiracle has been lost or is merely represented by a blind pocket in this region of the pharynx. Rudimentary gill lamellae may be located on the *anterior* wall of the spiracle. Since the blood supply to these lamellae consists of oxygenated blood, they do not perform a respiratory function, and the term *false gill,* or *pseudobranch,* is customarily applied to them. The blood in the pseudobranch, however, is actually exposed to oxygenation a second time, so that the brain and eyes, to which this blood is diverted, receive an unusually high percentage of oxygen. This is undoubtedly of some advantage to the animal. The spiracles generally open on the top of the head posterior to the eyes. Even in skates and rays with their ventral mouth and ventrally disposed gill slits, the relatively enormous spiracles are in a dorsal position. In most of the free-swimming sharks water passes *out* of the pharynx through the spiracles as well as through the external gill slits. In many species the spiracles are provided with valves. In the rays, which are bottom-dwelling forms, most of the water *enters* the pharynx through the spiracles and little comes in through the mouth. The spiracular valve closes by contraction of the first dorsal constrictor muscle (page 524), and water is forced out through the external gill slits. Other valves in rays prevent water from leaving through the mouth, which does not usually close entirely. Under certain conditions rays have been observed to reverse the usual procedure and to "spout" water *out* through the spiracles. This seems to be a protective device enabling the animal to eject foreign matter that has entered the pharynx with the incurrent stream of water. It is of interest that in one of the ovoviviparous elasmobranchs, *Petroplatea micrura,* certain villous projections from the uterine

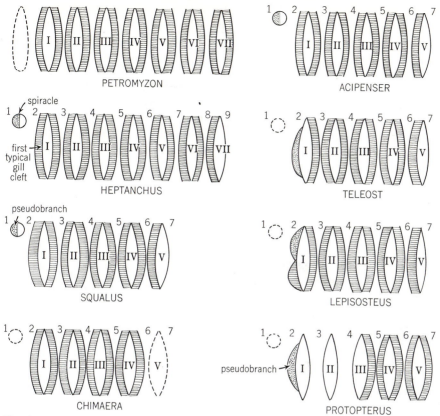

Fig. 6.13 Diagrams showing number of gill slits and arrangement of hemibranchs in lamprey (*Petromyzon*) and seven other representative fishes. Dotted line indicates gill slit or spiracle which has failed to open to the outside; solid, parallel lines indicate lamellae of true hemibranchs; stippled areas indicate pseudobranchs; 1–9 position of visceral arches; I–VII, typical gill slits.

wall of the mother enter the spiracles of the developing young, supplying nutriment directly to the walls of their digestive tracts.

Although in most fishes a *true* hemibranch, receiving unoxygenated blood (see footnote, p. 202), is lacking on the posterior side of the hyoid arch, a modified *opercular gill,* or *pseudobranch,* receiving oxygenated blood may be present. Opercular gills of this type are found in *Amia,* the Dipnoi, *Latimeria,* and many teleosts. Such oper-

cular gills probably represent the spiracular pseudobranch which, upon closure of the spiracle, has shifted its position. *Lepisosteus* is peculiar in having both dorsal and ventral gills on the inner surface of the operculum. The dorsal lamellae belong to a pseudobranch, whereas the ventral gill is a true hemibranch.

The number of gills and gill clefts varies among fishes (Fig. 6.13). There is some confusion as to just what each represents since different authors use different numbers in designating them. Throughout this volume we shall consider the mandibular arch to be the *first* visceral arch, the hyoid the *second,* and shall refer to the remaining arches as 3, 4, 5, 6, etc. The hyomandibular cleft between the mandibular and hyoid arches is actually the first cleft. However, since it is either modified to form a spiracle or is closed altogether, and because it never bears true gills, we shall refer to it only as the spiracle, spiracular cleft, or hyomandibular cleft. The cleft between the hyoid and third visceral arches is the first *typical* gill cleft and, therefore, will here be designated as gill cleft 1.

Most elasmobranchs have five pairs of clefts in addition to the spiracles. One form, however, *Hexanchus,* has six, and another, *Heptanchus,* has seven, exclusive of the spiracles. *Heptanchus* has the largest number of gill clefts of any gnathostome. Chimaeras have but four pairs of clefts, the spiracles being absent and the last cleft closed. *Polypterus* has only four pairs but *Acipenser* and *Polyodon,* like most elasmobranchs, have five pairs in addition to the spiracles. *Amia, Lepisosteus,* and teleosts, all of which lack spiracles, also have five pairs of clefts. The Dipnoi, which lack spiracles, show the greatest variations. *Epiceratodus* and *Protopterus* have five pairs of clefts, but *Lepidosiren* has only four. All the branchial arches of *Epiceratodus* bear functional gills. In *Lepidosiren* the first slit has been closed, and only three holobranchs are present. In *Protopterus* no gill lamellae border the first two clefts or the anterior wall of the third. A hyoidean pseudobranch lies anterior to the first gill cleft. The gill filaments of *Lepidosiren* and *Protopterus* are coarse and of little importance in oxygen uptake. They are important, however, for carbon dioxide elimination.

The presence of external gills is rare among fishes. In the larva of *Polypterus* a single pair of external integumentary gills is present in the region of the hyoid arch. Each gill consists of a narrow, main, central portion which bears a double row of filaments. Larval Dipnoi possess four pairs of external cutaneous gills located on the visceral arches. These external gills disappear in adult life, although vestiges may remain. In certain elasmobranchs during embryonic life the gill filaments are so long that they extend out the pharynx through the spiracles and function in the manner of external gill filaments. It is believed that they may even aid in the absorption of nutriment.

Most fishes die soon after being exposed to air, even though their gills are kept moist. Lack of water in the branchial chambers, as well as the accumulation of mucus, causes the gills to stick to each other with the result that the exposed respiratory surface is decreased to the point that exchange of gases is no longer adequate.

Some fishes, such as eels, which can migrate for some distance on the land, are able to retain water in the branchial chamber. The opercular aperture is small and rounded, thus more readily permitting the retention of water.

Although gills serve as the chief organs of respiration in most fishes, other respiratory mechanisms may be present in addition. The skin of the eel is also important in respiration, an unusual condition among fishes. Vascular areas in the skin, in some forms, may also aid in respiration. Some fishes possess pockets, projecting from the branchial chambers, in which water may be carried. Other accessory respiratory organs in fishes may enable the animals to utilize oxygen taken from the air. Certain species have posterior extensions of the gill chamber, filled with air, extending the entire length of the coelom, and lined with respiratory epithelium. Another very important structure which in some of the lower fishes serves as a lung is the *swim bladder,* or *air bladder,* to be discussed in detail further in this chapter.

AMPHIBIANS. Most members of the class Amphibia spend their larval life in water and, after a period of metamorphosis, go to the land. During the larval state gills of the external integumentary type are used as organs of respiration in addition to the highly vascular skin. In a few urodeles the gills are retained throughout life, but in most urodeles and all the tailless amphibians, they disappear at the time of metamorphosis. Newly developed lungs then usually take over the respiratory function. Cutaneous respiration is very important in this class both in larval and adult life. In adult salamanders of the family Plethodontidae, lungs fail to develop and, with the loss of the gills, the animals depend entirely upon cutaneous and buccopharyngeal respiration (page 240).

During embryonic development, with few exceptions, five pairs of pharyngeal pouches form in the characteristic manner. The first and usually the last do not become perforated, so that only numbers 2, 3, and 4 actually connect to the outside. The first gill pouch gives rise to the *Eustachian tube* and *middle ear,* at least in anurans. In most amphibians all the gill slits become closed over at the time of metamorphosis, but in certain urodeles some of them persist throughout adult life. In *Siren* all three clefts remain. *Necturus* and *Proteus* possess only two pairs of clefts, representing pharyngeal pouches 3 and 4, during the adult stage. In *Amphiuma* only one pair persists (pharyngeal pouch 4). In the hellbender, *Cryptobranchus alleganiensis,* at the time that the external gills disappear, the edges of the operculum (Fig. 6.14) fuse to the throat except on the dorsal side, leaving an opening on either side. These openings serve as outlets for water taken into the mouth during feeding. These animals depend upon their skin and lungs for respiration. Frequently one or the other of the openings fails to develop.

The external gills of amphibians consist of tufts of filaments with bases on the third, fourth, and fifth visceral arches. Arterial loops coming off the main aortic arches (Fig. 12.24) ramify through the external gills. In urodele amphibians and in early stages of development of anuran larvae, the aortic arches are not interrupted by gill capillaries as in fishes. Blood, therefore, may go either directly through the aortic arches or through the arterial loops in the external gill filaments. The gills are covered with ciliated epithelium. Waving movements of the gills themselves, together with movements of the cilia, ensure a constant change of the water with which they are in contact.

In the larvae of anurans an operculum develops shortly after the external gills appear. It arises from the hyoid region but

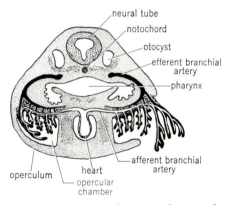

neural tube
notochord
otocyst
efferent branchial artery
pharynx
afferent branchial artery
operculum
heart
opercular chamber

Fig. 6.14 Diagram showing relations of external gills of anuran larva before (*right*) and after (*left*) the operculum has grown back to enclose the gills in an opercular chamber. (*After Maurer.*)

does not contain any skeletal elements. The operculum grows posteriorly and covers over the gill clefts, gills, and the region from which the forelimbs will later develop. It then fuses with the body, behind and below the gill region, in such a manner that the gills are confined to an *opercular chamber* (Fig. 6.14). The opercular aperture becomes greatly reduced in size and is sometimes referred to as the *spiracle*. It should not be confused with the spiracle present in certain fishes. There may be either a single midventral aperture or one that is more lateral in position. In a few forms paired apertures are present. The external gills soon degenerate, and a new set of gills develops from the tissue covering the same visceral arches. These are often referred to as *internal gills* because of their location *within* the opercular cavity. The entire opercular cavity is lined with ectoderm, a portion of which covers the internal gills. These, as well as the external gills, should be considered as integumentary derivatives, and it is indeed questionable whether they should be homologized with the internal gills of fishes. When the "internal" gills develop in anuran larvae, the continuity of the aortic arches is temporarily interrupted as new arterial loops appear. Upon metamorphosis, however, the aortic arches once more make direct connections with the dorsal aorta. Water passes from mouth to pharynx and then through the gill slits into the opercular chamber. It finally leaves through the opercular aperture.

An opercular fold, mentioned in connection with *Cryptobranchus*, also develops in other urodele amphibians, but it is usually much reduced in size and consists of a mere crease immediately anterior to the gill region. The gills of larval caecilians are, in general, similar to those of urodeles, a reduced opercular fold also being present in this group. In one form, *Caecilia compressi-cauda,* the gills are in the form of large, leaflike folds.

The gills of anurans are resorbed during metamorphosis. Among other changes which occur in this transformation from tadpole to adult, the developing forelimbs push out into the opercular chamber. In the leopard frog, *Rana pipiens,* the left limb passes through the opercular aperture, whereas the right one pushes through the wall of the operculum itself.

In urodeles several conditions obtain. Three pairs of gills and two pairs of gill clefts persist during adult life in *Necturus*. In *Amphiuma* the gills are resorbed, but a single pair of gill slits remains. In most salamanders, however, both gills and gill slits usually disappear during metamorphosis. A condition referred to as *neoteny* is encountered in certain salamanders. This term is used to denote the retention of certain larval characteristics beyond the normal period, or during adult life. For example, the tiger salamander, *Ambystoma tigrinum,* normally loses its gills at the time of metamorphosis and becomes a terrestrial animal. The Mexican *axolotl,* formerly believed to be another species, is an aquatic, gill-breathing animal even during its adult life. In 1876 Duméril demonstrated that under certain conditions the axolotl could be induced to lose its gills and become a typical lung-breathing salamander, in every way identical with *Ambystoma tigrinum*. No satisfactory explanation of the meaning of neoteny has been advanced. Not all urodeles which retain their gills may be induced to lose them under experimental conditions (page 341).

In some amphibians the gill stage is passed through very rapidly while the embryo is still enclosed within the jellylike egg envelopes. By the time of hatching, metamorphosis may already have occurred, and the animal is fitted for a terrestrial existence rather early in life. The salamander, *Pletho-don cinereus,* exemplifies this type of development.

REPTILES, BIRDS, AND MAMMALS. Five pairs of pharyngeal pouches are formed in reptiles during early embryonic life, but only four develop in birds and mammals. Although a fifth pair may form in these latter groups, it is rudimentary and attached to the fourth pair. The pouches do not usually break through to the outside, but they may do so occasionally. The first pharyngeal pouch persists to become the Eustachian tube and middle ear, but the other pouches become reduced and disappear except in snakes and burrowing lizards, leaving certain vestiges and derivatives to be discussed later in the chapter. The homologies of these pouches with those of lower forms should be apparent. Their development furnishes, in a limited way, an example illustrating the law of biogenesis. If the pharyngeal pouches fail to become obliterated in the normal manner, they may lead to the formation of branchial cysts and fistulae. In the latter condition there is an opening in the neck region which communicates with the pharynx. In such cases a perforated gill cleft has failed to disappear. Gills do not develop in association with the pharyngeal pouches of the three highest classes of vertebrates. Some transitory structures, which appear for a short time during development as outgrowths of the gill pouches in chick embryos and the embryos of certain turtles, may possibly be homologues of gills. It is certain that they have no respiratory function.

PHYLOGENETIC ORIGIN OF LUNGS AND SWIM BLADDERS

The swim bladder in various fishes shows a rather wide diversity in structure and function. Generally speaking, it is a gas-filled diverticulum which arises from the pharyngeal or esophageal region of the digestive tract. In certain primitive fishes its structure and function are such that it is difficult to make a distinction between swim bladder and lung. In fact, the swim bladder is a better lung in some fishes (Dipnoi) than are the lungs of most amphibians. The term *lung* is usually applied specifically to the respiratory organs of terrestrial vertebrates. The similarity of the lungs of tetrapods to the swim bladders of such fishes really makes such a distinction artificial. Some authors refer to the swim bladder as a lung if it has a ventral connection with the digestive tract. Nevertheless, in the following pages we shall refer to these structures in fishes as swim bladders since in most cases their primary function is not respiratory. A swim bladder may be single or double and may open into the digestive tract dorsally, ventrally, or not at all. It may be large and extend the entire length of the coelom or may be so small as to be practically indistinguishable. It is usually located dorsally and lies directly beneath the vertebral column, dorsal aorta, and opisthonephros, but outside the coelom. The peritoneum covers only its ventral surface. Swim bladders which retain the open *pneumatic duct,* or connection with the digestive tract, are said to be *physostomous,* whereas those which are completely closed are *physoclistous.*

Many authorities in the past have considered the swim bladder to be the forerunner of the lungs of higher forms. It was supposed to have arisen in fishes as an unpaired, dorsal evagination from the digestive tract and to have become modified gradually, in subsequent evolution, into the bilobed, ventrally located lungs of the higher vertebrates. The modern point of view, however, is that the presence of lungs was a primitive character and that the swim bladder of higher fishes is probably a specialized modification of the lung.

The origin of the swim bladder in the phylogenetic series has been obscure, and several theories, now only of historical interest, have been advanced to explain a possible mode of derivation. In 1941 (R. H. Denison, *J. Paleontol.,* **15**:553) it was discovered that lungs were present in the fossil remains of the Devonian placoderm, *Bothriolepis.* The placoderms, all of which are extinct, are considered to be the most primitive of the gnathostomes. It is now thought probable that the lungs of the lower fishes and of the tetrapods, together with the modified swim bladders of higher fishes, came from a common source among the placoderms. But few connecting links among the various groups have been discovered up to the present time.

Although in some of the lower fishes the swim bladder functions chiefly as an accessory respiratory organ, supplementing the gills in this respect, it may also serve as a hydrostatic organ and, in certain cases, as an organ for sound production and sound reception as well. The modified swim bladders of higher fishes seem to play little or no part in respiration and serve almost entirely as hydrostatic or equilibratory organs, enabling the fish to swim at different levels of water with but little effort.

Comparative anatomy of swim bladders.
PLACODERMS. The lungs (swim bladders) of placoderms, as evidenced by study of *Bothriolepis,* consisted of little more than a posterior pair of pharyngeal pouches extending ventrally which, in all probability, functioned as primitive pneumatic organs.

CHONDRICHTHYES. No swim bladder is present in the cartilaginous fishes, probably having been lost early in evolutionary history. In several species, however, a transitory rudiment appears during development. Opinions differ as to the significance of the rudiment. Nevertheless, it is entirely possible that it represents a lunglike development which was present in some remote progenitor.

LATIMERIA. The swim bladder of *Latimeria* is a small tube, only 2 to 3 in. long, prolonged as a filament to the end of the body cavity. Like the swim bladder of the Dipnoi and *Polypterus,* and the lungs of tetrapods, it connects with the ventral side of the gut. It is obviously degenerate as one might expect in an inhabitant of the deep sea.

POLYPTERUS. The most primitive of all the ray-finned fishes living at the present time is *Polypterus.* It is of particular interest because of the peculiar structure of its swim bladder, which is usually referred to as a lung. The smooth-walled swim bladder is bilobed and its pneumatic duct lies ventral to the digestive tract. It opens into the ventral part of the anterior end of the esophagus by means of a small, muscular *glottis.* The opening is slightly to the right of the midventral line. The two lobes of the swim bladder are of unequal size, the left being considerably smaller than the right. Where the right lobe extends beyond the left, it curves around the right side of the digestive tract and is dorsal to the intestine at its posterior end.

The arteries which supply the swim bladder of *Polypterus* arise from the last pair of efferent branchial arteries. Blood going to the structure has therefore already passed through the capillaries of the gills and under normal conditions would be oxygenated. The veins which drain the swim bladder enter the hepatic veins a short distance below the sinus venosus.

DIPNOI. The true lungfishes include but three surviving genera: *Protopterus, Lepidosiren,* and *Epiceratodus.* Study of the *sac-*

culated swim bladders of these three forms offers a clue to the origin of the specialized types of swim bladders found in the teleost fishes. During the dry season some of the Dipnoi estivate in burrows in the mud and utilize their swim bladders (lungs) in respiration. *Protopterus* and *Lepidosiren* are similar to *Polypterus* in that their swim bladders are bilobed and connect with the esophagus on its ventral side. The two lobes are, however, of practically the same size, although during development the left lobe is, for a time, smaller than the right. The blood supply of the swim bladder is derived from pulmonary arteries which arise directly from the dorsal aorta posterior to the last pair of aortic arches. In these fishes, nevertheless, at least partially oxygenated blood is carried to the swim bladder (page 549). Pulmonary veins carry the blood directly to the left atrium of the heart. Two atria, right and left, appear in the Dipnoi for the first time in the vertebrate series.

Epiceratodus is of particular interest because its swim bladder consists of a single lobe lying *dorsal* to the digestive tract and communicating with the esophagus by a tube, the *pneumatic duct*. The duct courses around the right side of the esophagus and opens through a glottis on its *ventral* side. A small diverticulum, present only during embryonic development, represents the left lobe, which is altogether lacking in the adult. A pulmonary artery arises from each posterior efferent branchial artery to supply the swim bladder. Two pulmonary veins unite to form a single vessel which opens into the left atrium of the heart. In *Epiceratodus* there is not so complete a separation of right and left atria as in *Protopterus* and *Lepidosiren*. Blood going to the swim bladder has, under normal conditions, been oxygenated in the gills.

The evolutionary step from the double,

ventrally connecting type of swim bladder, as found in *Protopterus*, to the lungs of tetrapods would seem to be comparatively simple and logical. The origin of the physoclistous organs of many teleosts from the dorsally located, single-lobed swim bladder of the type present in *Epiceratodus* would also appear to follow a logical sequence (Fig. 6.15).

TETRAPODS. In amphibians, reptiles, birds, and mammals the lungs arise during embryonic development as a diverticulum from the ventral portion of the pharynx. The diverticulum becomes bilobed. In most forms there is a tendency for the left lung to be smaller than the right, as is true of *Polypterus* and some other species of fishes. The walls of the lungs become subdivided into compartments of varying degrees of complexity. In mammals each lung is commonly divided into two or more large lobes. In structure and method of development, the tetrapod lung so closely resembles the swim bladder of *Protopterus* and certain other fishes that there is little doubt concerning homologies and the phylogenetic origin of the lungs of higher vertebrates.

HOLOSTEI. Included in the superorder Holostei are the bowfin, *Amia*, and the garpike, *Lepisosteus*. The swim bladder of *Amia* is unpaired, large, and highly vascular. It lies dorsal to the digestive tract and extends almost the entire length of the coelom. Contrary to the condition in *Epiceratodus*, however, the pneumatic duct opens through a muscular glottis on the *dorsal* side of the esophagus just behind the pharynx. In *Amia*, also, a rudiment of the left lobe appears during development, but it persists only for a short time. The arterial supply to the swim bladder is similar to that of *Polypterus* and *Epiceratodus*, coming from the last pair of efferent branchial arteries. The venous drainage is like that of *Polypterus*.

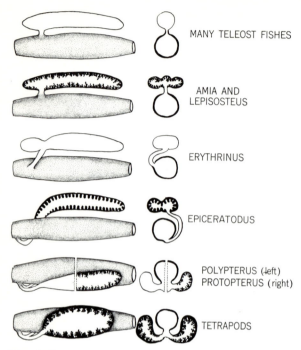

MANY TELEOST FISHES

AMIA AND
LEPISOSTEUS

ERYTHRINUS

EPICERATODUS

POLYPTERUS (left)
PROTOPTERUS (right)

TETRAPODS

Fig. 6.15 Various types of swim bladders (or lungs), shown in longitudinal and cross section (diagrammatic). (*After Dean.*)

The swim bladder of *Lepisosteus* is very much like that of *Amia*. It is, however, supplied with arteries which arise directly from the dorsal aorta, and the veins leading from the organ enter the hepatic portal and postcardinal veins. This condition is similar to that encountered in the teleosts.

TELEOSTEI.   Among the teleost fishes which make up the remainder of the subclass Actinopterygii are found the most specialized swim bladders of all. They bear the least resemblance to the primitive structure of *Polypterus*. There are several families of teleosts, however, in which a swim bladder is altogether lacking. Bottom-dwelling flat fishes, such as the flounder and halibut, lack a swim bladder in adult life. The typical teleostean swim bladder is a thin-walled, gas-filled sac which lies dorsal to the digestive tract and extends the entire length of the body cavity. Branches of the coeliac artery supply it with blood, and the venous return is through the hepatic portal or postcardinal veins. Teleostean swim bladders probably play little or no part in respiration, their primary function being concerned with hydrostasis and equilibration.

Among the major specializations to be found within the group is the loss of the pneumatic duct, which may atrophy and either disappear altogether or persist as a remnant in the form of a solid, fibrous cord.

The salmon, carp, pickerel, and eel are among commonly known teleosts which retain the open pneumatic duct. The swim bladder of the salmon is of simple construction. It consists of a single sac connected with the dorsal side of the esophagus. A sphincter muscle is present at the esophageal end of the pneumatic duct.

In the carp family there is a reduction in size of the swim bladder and a diverticulum grows forward into the head region. This is the *anterior chamber*. The *posterior chamber* is joined to the esophagus by a pneumatic duct. The anterior chamber is connected by a chain of four small bones, the *Weberian ossicles,* to the inner ear. Weberian ossicles, which are derived from portions of the anterior four vertebrae, are found in members of the order Cyprinoformes to which the carp, catfish, and sucker belong. In some other teleosts (herring, etc.), extensions of the swim bladder itself make direct contact with parts of the inner ear. They may thus be of importance in static or even auditory perception.

There is much variation in the degree of vascularity of the swim bladders of teleosts. In some the entire inner surface is uniformly vascular; in others, rather localized masses of interlacing, tightly packed capillaries called *retia mirabilia* are present. In certain teleosts with physostomous swim bladders, one or more retia mirabilia become further special-

ized to form peculiar structures known as *red bodies* which project into the cavity of the swim bladder near its anterior end. The lining epithelium of the swim bladder is continued over the red bodies, apparently without any modification. The common eel possesses several red bodies (Fig. 6.16), the largest of which lie near the opening of the pneumatic duct. Red bodies are able to remove oxygen, nitrogen, and carbon dioxide from the blood that is circulating through them and to pass these gases into the cavity of the swim bladder. The structures are unique in being able to remove free oxygen (O_2) from the oxyhemoglobin contained within the red corpuscles. How this is accomplished is little understood.

Completely closed physoclistous swim bladders are found in such teleosts as the perch, toadfish, cod, and haddock. With the disappearance of the pneumatic duct a modification of the red body occurs. The capillaries are still densely arranged, but the epithelium covering the projecting portion of these structures is of a glandular nature with numerous crypts and folds. The red body has now become the *red gland*. A single red gland is found in the cod and haddock, but in other teleosts several may be

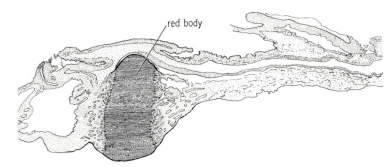

Fig. 6.16 Longitudinal section through red body dissected from an eel. The red body proper consists of numerous blood vessels of capillary dimensions, compactly arranged and running parallel to one another.

present, projecting from the dorsal or ventral walls of the anterior end of the swim bladder.

Physoclistous swim bladders show rather extreme variations in shape. Some have the configuration of a horseshoe. In others there may be distinct anterior and posterior chambers separated by a narrow opening or by a constriction which is provided with a sphincter muscle. In some there are two constrictions, so that it appears as though three sacs are present. Still other teleosts have simple or even complexly branched caecalike outgrowths, or diverticula, of the swim bladder.

Any excess gas secreted into the swim bladder in physostomous fishes may be liberated via the pneumatic duct. In physoclistous forms, gas which is secreted by the red glands at the anterior end of the swim bladder is, under certain conditions, absorbed by vessels which form a more diffuse rete mirabile at the posterior end. In species having two chambers in the swim bladder, the sphincter muscle separating them may regulate the passage of gas from anterior to posterior chambers and thus control the amount of gas which may be absorbed by the rete mirabile in the wall of the posterior chamber. In certain forms the posterior chamber with its rete mirabile is reduced to a small pocket, the *oval,* the entrance to which is guarded by a sphincter.

CHONDROSTEI. The swim bladders of the paddlefish and sturgeons, which are grouped together with *Polypterus* and *Calamoichthys* in the superorder Chondrostei, have not been discussed prior to this point since they do not fit into the general scheme of things as outlined here. Some sturgeons lack a swim bladder altogether. In others it is oval in shape, with a smooth inner surface lined with ciliated epithelium. It opens into the dorsal side of the esophagus by means of a rather wide aperture. During embryonic development the swim bladders of sturgeons seem to rise more posteriorly and nearer the stomach than in other fishes. In such forms gastric glands may be present in its walls for a time during development. This group of fishes is considered to be degenerate in other respects, and the loss or degenerate condition of the swim bladder may be nothing more than an association with bottom-dwelling habits.

Functions of the swim bladder. It has been mentioned that in some of the lower fishes the swim bladder serves as an important adjunct to the gills in respiration. The observed behavior of such fishes, and the structure of the swim bladder with its blood supply, support such an assumption. Its position in the body, and the fact that it contains gas within its cavity, indicate that the swim bladder may function also in hydrostasis, equilibration, and sound production. It is doubtful whether the physoclistous swim bladders of the higher teleost fishes are of any importance in respiration. They seem to serve chiefly as hydrostatic and equilibratory organs. Those teleosts having Weberian ossicles may possibly use their swim bladders in sound reception.

RESPIRATION. It is well known that *Lepisosteus,* the dipnoans, and a few other species of fishes are unable to live for any length of time if kept submerged and prevented from coming to the surface of the water to gulp air. Observations of *Lepisosteus* show that under normal conditions the animal comes to the surface and "breaks water" every 7 or 8 minutes. Such fishes normally inhabit rather shallow waters or stay rather close to the surface.

In most of these lower fishes the lining of the physostomous swim bladder is thrown into a complicated series of highly vascular folds and sacculations of an alveolar nature (Fig. 6.17), thus providing an excellent respiratory surface. Blood going to the swim bladder under normal conditions, however, has already been oxygenated in the gills since the arteries with which it is supplied arise from the posterior efferent branchial arteries or directly from the dorsal aorta. If the oxygen content of the water in which these fishes live is depleted and the gills cannot obtain enough oxygen, the ability of the swim bladder to take over the respiratory function entirely is of great adaptive value to the animal. On the other hand, it would seem that perhaps the respiratory surface of the gills is inadequate to secure the necessary amount of oxygen, and a supplemental supply may be obtained at the swim bladder. This seems to be particularly true of *Protopterus,* in which gills are lacking on the borders of the first two gill slits and anterior portion of the third. The anterior aortic arches in *Protopterus* thus form direct channels from ventral to dorsal aortae, and the pulmonary arteries, which arise from the dorsal aorta posterior to the gill region, supply the swim bladder with blood which has been only partially oxygenated.

The exact mechanism by means of which

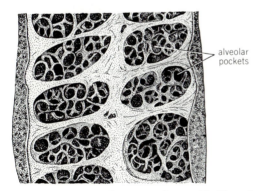

alveolar pockets

Fig. 6.17 Lining of portion of swim bladder of *Lepisosteus,* showing alveolar pockets.

air can be taken into the swim bladder is not well understood. Muscles are present in its walls in certain forms, and these may possibly play a part in inhalation and exhalation of air. It has been reported that in certain fishes muscles at the entrance of the pneumatic duct enable the animal literally to swallow air which then enters the swim bladder.

Some investigators believe that the physoclistous swim bladder may be used for the storage of oxygen which has been separated from the blood by the red glands. This oxygen then may be utilized when the oxygen content of the surrounding water is low. If this is true, then in such a sense the swim bladder may be considered to be an accessory organ of respiration. However, it has been demonstrated, under experimental conditions, that certain fishes may be asphyxiated even though the swim bladder contains a considerable amount of oxygen. The amount of carbon dioxide in the swim bladder has a very definite influence upon the amount of oxygen that may be absorbed by this organ. A more satisfactory interpretation of the role of the physoclistous swim bladder in respiration is indicated by experiments which show that if the amount of carbon dioxide in the water increases, the red gland secretes more gas into the swim bladder, thus causing the fish to rise almost automatically to the surface, where oxygen is available in greater quantities.

HYDROSTASIS. The hydrostatic function of the swim bladder has been amply demonstrated by experimental work. The organ serves to equilibrate the body of the fish with the surrounding water at any level. The animal literally "floats" *in* the water. Adjustment of the body in changing from one level in the water to another is not usually accomplished quickly. Change in the volume of the contained gas is one method by which this may possibly be effected. Muscular contractions, at least in part, may be used to change the volume. Increasing or decreasing the amount of gas contained within the swim bladder, however, is the more probable explanation, pressure within the bladder being maintained by muscles in the wall of the organ. In physostomous fishes passage of gases in or out of the swim bladder seems to be associated with the pneumatic duct. However, in physostomous teleosts, gulping of air has been shown to be but one means of increasing the volume of gas. Gases may also pass in a free state directly, from the blood in the retia mirabilia, or red bodies, into the swim bladder. The latter method of increasing the amount of gas requires a greater length of time than the former and is probably used primarily when the fish does not have ready access to surface air.

In physoclistous fishes, inasmuch as no pneumatic duct is present, any changes in the gaseous content of the swim bladder must be brought about solely by exchange of gases between the swim bladder and bloodstream. The peculiar concentration of parallel capillaries of the red glands is believed to be the chief center of gas secretion or diffusion. At the time when a red gland is giving off free oxygen, there is an increased flow of blood through the gland, the capillaries dilate, and free oxygen dissociates from the oxyhemoglobin in the red corpuscles. In fishes with closed swim bladders and lacking red glands, there is a physical diffusion of gases between the swim bladder and the blood vessels in its walls. In such cases diffusion does not take place so rapidly as in those in which red glands are present.

Gases pass into and out of the swim bladder depending upon environmental conditions, particularly in relation to pressure changes. The composition of the contained gas also shows variations in relation to pressure. The chief gases in the swim bladder, as previously stated, are oxygen, carbon dioxide, and nitrogen, with traces of other inert gases that are normally present in the atmosphere. When a fish is swimming near the surface the gases are in approximately the same concentration as in the air. But as the depth increases, the per cent of oxygen in the swim bladder increases. In some deep-sea species oxygen may make up as much as 87 per cent of the gaseous content of the swim bladder.

When fishes are suddenly brought to the surface from depths where the pressure of water on the body is great, various results are noted. Physostomous fishes release bubbles of gas from their swim bladders, thus adapting themselves to the change in pressure. Physoclistous fishes, however, do not have any such means of gas release. In some cases the swim bladder may burst, but more frequently the expansion of gas, which naturally follows when pressure is released, causes the swim bladder to become abnormally distended, and the fish is then incapable of swimming. The stomach may be pushed out through the mouth, or the intestine everted through the anus. Pressure of the inflated organ upon neighboring viscera is probably the major cause of their extrusion. Under normal conditions ascent and descent between surface water and depths take place rather gradually.

SOUND PRODUCTION. Among other functions of the swim bladder, the part which it may play in sound production is important. Sound production in fishes is by no means so rare as is generally supposed. Many are able to produce rasping or strident sounds by rubbing such structures as teeth, bones, and spines against one another. Sounds

Anatomy of the Chordates

comparable to those made by higher vertebrates, in which passage of air forms the basis of sound production, occur but rarely among fishes. In some cases the sounds originate by expulsion of air from the swim bladder through the pneumatic duct and mouth. In the loach, which employs intestinal respiration, voiding of air bubbles through the anus causes sound production of a sort.

In those fishes in which gas in the swim bladder is used in making sounds, some mechanism is generally present which is capable of compressing the contained gas. Muscles attached to the swim bladder, on the one hand, and to bony structures, on the other, may serve alternately to compress and expand the swim bladder. In some species, in which the swim bladder lacks muscles of its own, the organ may be at least partially invested with tendinous or muscular extensions of the body wall which can be utilized in producing pressure upon the gas contained within. In cases in which the organ is divided by constrictions into two or three chambers, the passage of gas from one chamber to another over the edges of loose septa may cause sounds to be produced.

Cyclostomes and elasmobranchs are not known to be capable of emitting sounds. Sound production in fishes is confined mostly to the teleosts, although there is evidence that *Polypterus, Protopterus, Calamoichthys,* and *Lepidosiren* are capable of sound production. The mechanism by means of which this is accomplished is not understood. In none of the fishes are skeletal supports to be found in the region of the glottis, which might be considered as homologues of the laryngeal structures of higher vertebrates.

Since water is a much better conductor of sound waves than is air, sounds produced by fishes may be carried for considerable distances. In certain cases the sounds may have some relation to mating activities, species recognition, or defense from enemies.

LUNGS AND AIR DUCTS

The diverticulum which in the embryo gives rise to lungs grows out ventrally from the floor of the pharynx posterior to the last gill pouch. It soon divides into two halves, the *lung buds,* which are destined to give rise to the bronchi and the lungs proper. The lung buds grow posteriorly, invested by an envelope of mesoderm, until they reach their final destination in the body. They may branch to varying degrees, depending upon the species. The original unpaired duct which connects the lungs to the pharynx serves to carry air back and forth and is known, in most forms, as the windpipe, or *trachea.* In most anuran amphibians the duct is so short as to be practically nonexistent. In many tetrapods the anterior end of the trachea becomes modified to form a voice box, or *larynx,* which opens into the pharynx by means of a slitlike *glottis,* the walls of which are supported by cartilages. At its lower end the trachea usually divides into two bronchi which lead directly to the lungs.

Larynx. The skeletal elements supporting the walls of the larynx are derivatives of certain visceral arches which, along with the loss of gills, have become modified during development. No such structural modifications have been observed to be associated with the pneumatic duct in fishes, the visceral arches of which are retained as such.

AMPHIBIANS. The simplest condition of the larynx is to be found in certain urodele amphibians, such as *Necturus,* in which a pair of *lateral cartilages* bounds the slitlike glottis. These cartilages are derived from the last pair of visceral arches. Other amphibians show further modification in that the lateral cartilages are supplanted by an upper pair of *arytenoid cartilages,* bounding the glottis, and a lower pair of *cricoid cartilages* (Fig. 6.18). The two cricoids may fuse to form a cartilaginous ring located within the walls of the larynx. The arytenoid cartilages are moved by muscles which work in antagonistic pairs and which are also derived from the gill region. They serve to open and close the glottis. In anurans two thickened ridges of tissue, composed of elastic fibers for the most part, project into the cavity of the larynx. These are the *vocal cords* (Fig. 6.19). They are arranged so as to parallel the glottis. The inner, or lower, rim of each vocal cord is set into vibration as air is forced from the lungs, thus bringing about the production of sound. Alterations of pitch are influenced by tightening or relaxing the vocal cords to different degrees. Although a few urodeles can produce sounds, they rarely amount to much more than a slight squeak or hiss. In anurans,

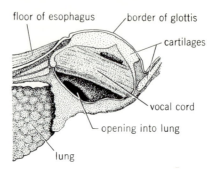

Fig. 6.19 View of interior of left half of larynx of bullfrog, *Rana catesbeiana,* showing position of vocal cord, attached at its ends and along its length to the wall of the larynx.

sound production is often well developed, particularly in males. Vocalization, in the form of mating calls, serves in species identification. Not only is the larynx usually much larger in males, but they also possess *vocal sacs* which, when dilated, act as resonators. The vocal sacs are diverticula of the mouth cavity which extend ventrally and laterally so as to lie under the outer skin and muscles of the throat region (Fig. 6.20). Air is forced back and forth between the inflated vocal sacs and lungs, the nostrils generally remaining closed. Sounds, therefore, may be produced when the animal is under water as well as when on land. Many curious variations of the vocal sacs are to be found within the group. They are usually paired. However, in some forms they are unpaired, median structures. In certain species the vocal sacs are very large and extend posteriorly into the large lymphatic spaces. Sounds produced by female anurans, with their smaller larynges and in the absence of vocal sacs, are of much less magnitude than those of males.

REPTILES. The larynx in reptiles is generally no better developed than in amphibians. It is closely associated with the hyoid

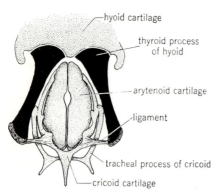

Fig. 6.18 Dorsal view of laryngeal cartilages of frog, *Rana esculenta.* Black represents bone; stippled areas, cartilage. (*After Gaupp.*)

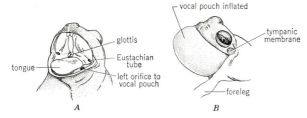

Fig. 6.20 Vocal pouch of American spadefoot toad, *Scaphiopus holbrookii.* A, showing location of openings to vocal pouch; B, the vocal pouch inflated. (*From Noble, "Biology of the Amphibia," McGraw-Hill Book Company. By permission.*)

apparatus, and its skeletal support consists of a well-developed pair of arytenoid cartilages and an incomplete cricoid ring. In crocodilians another cartilaginous element, the *thyroid cartilage,* is present in the larynx. Otherwise a thyroid cartilage appears only in mammals. A fold of mucous membrane of the mouth, which in certain lizards and turtles lies anterior to the glottis, is thought to represent the beginning of an *epiglottis* such as is found in mammals. In chameleons the lining of the larynx between the cricoid cartilage and the first tracheal cartilage extends outward, forming an inflatable *gular pouch.*

Most reptiles lack the ability to produce sounds except for a sort of hissing noise which may result from the rapid expulsion of air from the lungs. Vocal cords are present in the larynges of certain lizards (geckos and chameleons), and these are able to produce guttural noises. Male alligators make loud, bellowing sounds, particularly during the mating season when they may be heard over distances of a mile or more.

BIRDS. One would expect that in the highly vocal birds a well-developed larynx would be present. This is not the case, however. Another structure, the *syrinx,* located at the lower end of the trachea, is respon-

sible for sound production (page 234). The larynx in birds is poorly developed and incapable of producing sounds. Arytenoid cartilages, which are sometimes ossified, guard the glottis. In some birds the upper portions of the cricoids have separated into additional elements called the *procricoids.*

MAMMALS. The mammalian larynx is more complicated than that of other groups of vertebrates. In addition to the arytenoid and cricoid cartilages, other skeletal elements are present as well as extrinsic and intrinsic muscles. A *thyroid cartilage,* derived from portions of the fourth and fifth visceral arches, is located on the ventral side of the larynx. It articulates with the hyoid apparatus at its anterodorsal angles. The thyroid cartilage usually consists of two parts which, except in monotremes, are fused along the midventral line. In man this portion forms a prominent protuberance called the *pomum Adami (Adam's apple).* The arytenoid cartilages, covered with mucous membrane, form the lateral boundaries of the posterior portion of the larynx; the cricoid makes up the posterodorsal portion. Its middorsal part is frequently separated from the rest of the cricoid and forms the *procricoid.* The arytenoid and thyroid cartilages articulate with the cricoid. In some mammals small *cornic-*

ulate and *cuneiform* cartilages are arranged in close association with the arytenoid cartilages.

An epiglottis is characteristically present in mammals. It is composed of elastic cartilage and extends dorsally from the cephalic median portion of the thyroid cartilage in front of the glottis. It was formerly thought that the epiglottis served as a sort of cover, or lid, for the glottis over which food passed on its way to the esophagus. This view is no longer held. During the act of swallowing, however, the epiglottis stands erect and the larynx is raised by muscular contraction in such a manner that the glottis is sheltered by the epiglottis and the root of the tongue. Even if the epiglottis is removed, little trouble is encountered from food entering the larynx. Elevation of the larynx and closure of the glottis itself during the act of swallowing are the important factors in preventing food from entering the respiratory passages. If the larynx cannot be raised, as in certain diseased conditions, swallowing is virtually impossible. The homologies of the epiglottis are not clear.

The larynx extends from the root of the tongue to the caudal end of the cricoid cartilage where it is continuous with the trachea. It is divided into two regions by the presence of vocal folds, or *true vocal cords* (Fig. 6.25). Each of these consists of a band-like fold of yellow elastic tissue stretched between the cephalic portion of the thyroid cartilage and the caudal part of the arytenoid. The vocal cords are covered by a thin layer of mucous membrane with a stratified squamous epithelium of the nonkeratinizing type. The glottis lies between the vocal cords and is bounded by the arytenoid cartilages behind. Air going through the narrow slit-like passage between the true vocal cords, when they are contracted, causes them to vibrate and produce sounds. Pitch is regu-

lated by the degree to which the cords are contracted or relaxed by laryngeal muscles. The portion of the larynx in front of the true vocal cords is called the *vestibule*. A pair of crescentic *false vocal cords* (*ventricular folds*) extends into the vestibule for a short distance on either side. The portion of the vestibule between the true and false vocal cords is called the *ventricle* of the larynx. It is usually small but in some forms assumes considerable proportions. In the howling monkeys, for example, a large, saccular diverticulum of the ventricular portion of the larynx forms a resonating chamber (Fig. 6.21), supported by a greatly enlarged basihyal, which is responsible for the howling noises made by these forms. Such noises may be audible for considerable distances. Although both true and false vocal cords are present in most mammals, there are some exceptions. Elephants lack false vocal cords, and the hippopotamus lacks true ones.

In man the larynges of the two sexes differ but little in size until the age of puberty is reached. At that time the larynx of the male

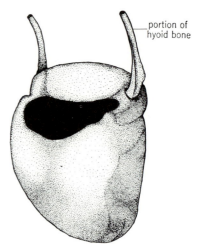

portion of hyoid bone

Fig. 6.21 Resonating chamber of larynx of howling monkey.

increases considerably in size and becomes much larger than that of the female. All the cartilages enlarge, and the thyroid cartilage forms a rather marked protuberance in the midline of the neck. The enlarged vocal cords are responsible for the lower pitch of the masculine voice. Growth of the male larynx at the time of puberty is a response of that structure to a hormone secreted by the testes. Removal of the testes before the age of puberty results in a failure of the larynx to grow to normal proportions, and the high-pitched voice of the boy is retained. After the larynx has grown to adult size, castration has little, if any, effect upon the pitch of the voice.

An additional function of the larynx is to prevent anything but air from gaining access to the lungs. A laryngeal spasm, in the form of a cough, results if any irritating substance enters the larynx. Occasionally an individual, believed to have drowned, is found, upon autopsy, to have little or no water in the lungs. In such cases it is probable that a laryngeal spasm, induced by water entering the larynx, causes asphyxiation.

Trachea and bronchi. The trachea is the main trunk of a system of tubes through which air passes between larynx and lungs. It is a cartilaginous and membranous tube the walls of which are composed of fibrous and muscular tissue stiffened by cartilages which prevent it from collapsing. The trachea is lined with mucous membrane, the epithelium of which is composed of columnar ciliated and mucus-secreting cells.

AMPHIBIANS. The trachea in most of the tailless amphibians is so short that it can scarcely be said to exist. However, in one group, the Pipidae, in which the lungs function as hydrostatic organs, a definite trachea is present. It divides into two bronchi at its lower end. In urodele amphibians the trachea is short but nevertheless exists as a definite structure which in some forms (*Amphiuma* and *Siren*) may be 4 or 5 cm long. The tracheal cartilages are small and irregular. They show a tendency to form bands. In caecilians the cartilages are in the form of half rings. The trachea in either case divides at its lower end into two bronchi which lead directly to the lungs.

REPTILES. In amniotes the trachea is elongated to varying degrees, depending upon the length of the neck. Among reptiles, the trachea of lizards is relatively shorter than in other groups. In crocodilians and turtles it is considerably elongated and in some turtles is even convoluted. The tracheal cartilages have become better developed and more complete. In *Sphenodon,* lizards, and some snakes, certain anterior cartilages are present in the form of complete rings. The remainder appear as imperfect bands located on the ventral and lateral parts of the trachea. These bands are deficient on the dorsal side. In snakes successive tracheal cartilages are frequently united. Supporting cartilages are also present in the walls of the bronchi in many reptiles. In some snakes only one bronchus and one lung (the right) are present.

BIRDS. In correlation with the long necks of many birds, and the location of lungs in the thoracic region, the trachea may be of unusual length. In many birds the length of the trachea even exceeds that of the neck. In such cases it is convoluted, the loops lying under the skin of the thoracic or abdominal region or between the muscles and the sternum. In swans and cranes a loop of the trachea lies within a peculiar cavity located in the modified keel of the sternum, which thus serves as a resonator. An unusual condition is encountered in the penguins, in which the trachea is double (Fig. 6.22). The duplication begins about 1 cm from the larynx. The tracheal cartilages in birds are

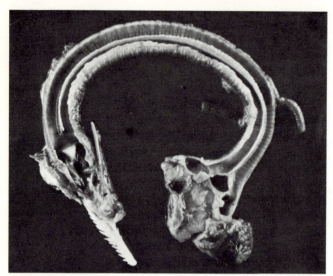

Fig. 6.22 Trachea of jackass penguin, dissected to show duplication which begins about 1 cm past the larynx.

usually in the form of complete rings and are often calcified or ossified. The bronchi are also provided with supporting cartilaginous rings.

It has already been emphasized that the larynx is poorly developed in birds and is not concerned with sound production. The *syrinx,* found only in birds, is the true organ for production of sound. It is located at the lower end of the trachea where that structure divides into the two bronchi.

The syringes of birds show much variation in structure, but in general three different types are recognized: *tracheal, bronchotracheal,* and *bronchial.* In the *tracheal syrinx* the lateral parts of the posterior tracheal rings have disappeared and a vibratile membrane closes over the gap thus formed. The *bronchotracheal syrinx* is the more usual type. Here the posterior tracheal rings are modified to form an enlarged resonating chamber, the *tympanum,* which often becomes ossified. Membranous folds, the *in-*

ternal and *external tympanic membranes,* project into the cavity from the medial and lateral walls of each bronchus. Often there is a skeletal structure, the *pessulus,* crossing the trachea in a dorsoventral direction at the point where it bifurcates. The lower portion of the pessulus bears a *semilunar membrane.* In certain aquatic birds, such as the male duck, the tympanum is very large and asymmetrical. In the *bronchial syrinx,* membranes between two successive cartilaginous rings of each bronchus form vibratile folds when the bronchi are contracted and the cartilages pulled closer together. In these various types of syringes the positions of the different membranes and skeletal elements are altered by the action of complicated sets of muscles so that production of a variety of sounds is made possible.

MAMMALS. Although the trachea is relatively short in the aquatic cetaceans and sirenians, in other mammals it is elongated and straight, its length varying with the

Anatomy of the Chordates

length of the neck. The trachea in a human adult measures about 4 to 5 in. in length. Tracheal cartilages in mammals are usually incomplete on the dorsal side, the esophagus pressing against the gap thus formed. In whales and sirenians they are arranged in the form of a spiral. In man the number of tracheal rings varies from 16 to 20, whereas in the long-necked giraffe over 100 may be present.

In most mammals the trachea divides into two main bronchi; but in pigs, whales, and some ruminants a third, the *apical,* or *eparterial, bronchus* arises independently from the right side of the trachea craniad the main point of bifurcation. In some other forms the apical bronchus is a main branch of the right bronchus. Cartilaginous rings support the walls of the bronchi in mammals.

Lungs. The lung buds of terrestrial vertebrates give rise to two saclike structures, the walls of which become subdivided in most cases. The divisions become more and more numerous and complex as the vertebrate scale is ascended, reaching the highest degree of branching in mammals. In the latter group the lungs form a spongy mass composed of minute chambers called *infundibula,* or *alveolar sacs,* lined with *alveoli.* The alveoli make up the respiratory surface of the lung. The greater the degree of branching, the greater is the area of respiratory surface.

AMPHIBIANS. The lungs of amphibians are relatively simple in structure. They connect directly with the larynx in most forms, although a short trachea may be present. In urodele amphibians the lungs consist of a pair of elongated sacs, the left often being longer than the right. In some the lining is smooth, whereas in others alveoli may be present. In many cases the alveoli are confined to the basal portion. The left lung is

very short in caecilians, and alveoli are present over the entire surface lining the right lung. In frogs and toads the lining of the lung sacs is somewhat more complex since the wall is thrown into numerous infundibular folds, which are, in turn, lined with alveoli. The more terrestrial amphibians generally possess a larger alveolar respiratory surface than do those which remain close to the water or which retain the aquatic mode of life.

Although it is usual for amphibians to develop lungs at the time of metamorphosis, this is not always the case. In some salamanders and frogs they develop during larval life. In salamanders of the family Plethodontidae no larynx, trachea, or lungs develop at the time that the gills are resorbed. The animals depend upon cutaneous and buccopharyngeal respiration during adult life. In certain forms, such as *Necturus,* which retains its gills, lungs are present in addition. It is doubtful whether they are used very effectively since they are supplied with blood which has already been oxygenated (page 565). However, they may be utilized in times of emergency when the oxygen supply of the water in which the animal lives becomes depleted.

Surgical removal or ligation of the lungs in certain newts has demonstrated that lungs are of little or no importance under normal conditions in satisfying the respiratory requirements of the body. Such operations, however, are followed by certain changes in the circulatory system, skin, and digestive tract (C. Philippi, L. Hausler, H. Bialy, and S. Jakowska, 1956, *Anat. Record,* **125**:656).

REPTILES. The lungs of reptiles are more complex structures than those of amphibians. Snakes and lizards are characterized by a lack of symmetry between the two sides. In some lizards the right lung is the larger, whereas in others the condition is exactly the opposite. In snakes the left lobe is smaller than the right or is altogether absent.

Sphenodon illustrates the most primitive reptilian condition, in which the lungs are rather simple sacs lined uniformly with infundibula. In snakes the infundibula are usually confined to the basal portion of the lung. The next advancement is seen in the lizards. In this group septa, or partitions, divide the cavity of the lung into several chambers. In some cases the chambers communicate with each other at the proximal end of the lung, and in others at the distal end. In some of the higher lizards, turtles, and crocodilians the bronchi divide into smaller and smaller branches which finally terminate in infundibula lined with alveoli. This arrangement gives the lungs a spongy consistency. The crocodilian lung most nearly approaches the condition found in mammals. In chameleons several thin-walled, saclike diverticula, which ramify among the viscera, come off the distal portion of the lung. Only the proximal part of the lungs is spongy. The projecting diverticula seem to foreshadow the appearance of air sacs, which reach their climax of development in birds. Throughout the reptilian class there is a tendency toward an increase in the alveolar respiratory surface.

BIRDS. The lungs of birds show many peculiarities not found in other groups. They are perhaps the most efficient of all from a physiological point of view. The lungs are small, very vascular, and capable of but little expansion. They are firmly attached to the ribs and thoracic vertebrae. The lower surface of each lung is covered by a membrane into which are inserted several muscles that arise from the ribs.

The main bronchus on each side enters the lung on its medial ventral surface and passes to the distal end. The portion within the lung proper is referred to as the *mesobronchus*. As it passes through the lung its cartilaginous rings disappear and a number of *secondary bronchi* arise from it. The secondary bronchi branch into numerous small tubes of rather uniform diameter called *parabronchi,* which form loops connecting with other secondary bronchi (Fig. 6.23). Surrounding each parabronchus are large numbers of minute tubules, the *air capillaries*, which form anastomosing networks and loops with each other and which open into the parabronchi. The arrangement is not that of a branching system as in the

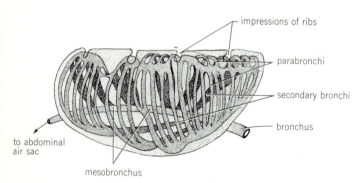

Fig. 6.23 Diagrammatic medial view of left lung of chicken, illustrating relations of bronchus, mesobronchus, secondary bronchi, and parabronchi. (*After Locy and Larsell.*)

lungs of mammals and some other vertebrates, but is, rather, a system of intercommunicating tubes within the lung itself. The actual respiratory surface consists of the very vascular lining of the air capillaries. The mesobronchus and several (usually four) secondary bronchi continue on through the walls of the lung and expand into large *air sacs* which ramify among the viscera. Branches even enter the cavities of several of the bones. Only in the primitive kiwi are the air sacs so poorly developed that they fail to enter the bones or to pass into the abdominal cavity. The air sacs do not furnish an increased respiratory surface since their walls are smooth and have a poor blood supply. Moreover, they are furnished with oxygenated blood. In addition to the one large tube which enters each air sac, there are several small tubes, called *recurrent bronchi,* which connect the sac with the adjacent portion of the lung.

During inhalation, at times when the bird is at rest, air is drawn from the outside through the trachea, bronchi, mesobronchi, and secondary bronchi into the air sacs. This is accomplished by the action of intercostal muscles which raise and lower the ribs and sternum, thereby varying the volume of the body cavity in the thoracic region. Air, however, is not simply drawn *into* the lungs; it actually passes *through* them in a much larger volume than these small structures could possibly accommodate. Upon exhalation, it is believed, a large quantity of air in the air sacs, on its way to the outside, instead of returning directly to the mesobronchus, passes through the recurrent bronchi to the lungs. The air is then forced, under pressure, into the intercommunicating system of parabronchi and air capillaries, the latter furnishing the surface where actual respiratory exchange takes place. The air continues its passage into one or another

secondary bronchus and thence to the mesobronchus. From here it may possibly go back again to an air sac and be recirculated. Finally, however, it passes through the bronchus and trachea to the outside. Since there are no blind alveolar pockets in the lungs of the bird and a constantly renewed supply of air passes over the respiratory surface, there is no stagnation of air as in the lungs of other vertebrates (residual air). During flight, movement of air in the air sacs is effected by pressure from surrounding viscera and by movements of the flight muscles and hollow wing bones which contain diverticula of the air sacs. At such times the skeleton must be kept rigid in order to brace the wings, and the ordinary method of inspiring and expiring air by the use of intercostal muscles must be abandoned temporarily. The more rapidly a bird flies, the more rapid is the circulation of air back and forth through the respiratory channels. The efficiency of this type of respiration should be obvious. The best fliers among birds are those having the most highly developed air sacs.

In addition to furnishing the lungs with oxygen, the air sacs may be of importance in reducing the specific gravity of the bird when in flight by reason of the warmer air which they contain. They may also play a significant role in temperature regulation by serving as an internal cooling device. The skin is of little help in this respect. Unlike that of mammals, it is practically devoid of glands and is covered with insulating feathers.

The air sacs in different species of birds show considerable variations in their detailed arrangement (A. M. Lucas, R. J. Keeran, and C. F. Coussens, 1959, *Anat. Record,* 133:452).

MAMMALS. In the development of mammalian lungs the original lung buds, which arise from the midventral wall of the pharynx, divide many times to form primary, secondary, tertiary, etc., bronchi. The smaller bronchi, in turn, give rise to *bronchioles* which are of several orders of magnitude. The terminal bronchioles usually divide into two or more *respiratory bronchioles* which then branch into several *alveolar ducts*. From these arise *alveolar sacs* into which numerous terminal *alveoli* open (Fig. 6.24). Alveoli are usually completely separated from each other by interalveolar septa but are sometimes connected by minute *alveolar pores*. The alveoli are supplied with a rich capillary meshwork; and since the alveolar walls are very thin, it is they which make up the true respiratory surface. The respiratory bronchioles derive their name from the fact that a few alveoli may be found projecting from their walls.

The various branches of a single bronchiole, together with the bronchiole itself,

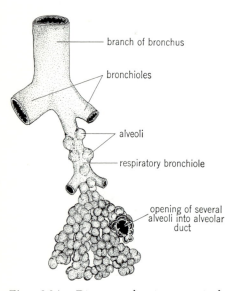

Fig. 6.24 Diagram showing terminal branching of a small bronchus in a typical mammalian lung.

branch of bronchus

bronchioles

alveoli

respiratory bronchiole

opening of several alveoli into alveolar duct

form units, or masses, within the lung which are referred to as *lobules*. In the pig, the lobules are separated from one another by interlobular connective-tissue septa. In rabbits such septa are lacking. Man represents a more or less intermediate condition.

The entire respiratory surface is believed to be of endodermal origin, since it is a derivative of the pharyngeal epithelium. The remainder of the lung is mesodermal and is composed of reticular and collagenous connective tissues, a large number of elastic fibers, and numerous blood vessels. The epithelial lining of the greater part of the "respiratory tree" is of the ciliated pseudocolumnar type but the cilia end a short distance down the respiratory bronchioles. Here the epithelium changes to a low cuboidal type. There has been some dispute about the actual nature of the epithelium lining the alveolar ducts, alveolar sacs, and alveoli. No cilia are present, and the lining cells are unusually thin. The cilia in the upper regions beat toward the trachea, serving to remove dust and other particles from the lungs. Recent studies, using the electron microscope, have shown details of structure of the interalveolar wall, or septum, which with usual techniques cannot be seen under the light microscope. Despite the thinness of the wall, it has been demonstrated that between the air in an alveolus and the blood in an underlying capillary there are present, as might be expected: the cytoplasm of the epithelial cells lining the alveolus, the basement membrane of the epithelial lining, the basement membrane of the capillary endothelium, and the cytoplasm of the endothelial cells of the capillary.

Cartilaginous rings, similar in structure to those of the trachea, are present in the portion of the bronchus between trachea and lung. Within the lungs they become smaller and smaller, and in their place are found small, irregular plates of cartilage, some of which may encircle the bronchi completely. A layer of smooth muscle is present between the cartilages and the mucous membrane lining the intrapulmonary bronchi. When the bronchioles are no more than 1 mm in diameter, the cartilaginous supports disappear altogether. Collapse is prevented by the presence of elastic connective-tissue fibers.

The entire outer surface of the lung down to the root, where bronchi and pulmonary vessels enter or leave, is covered with serous membrane, the *visceral pleura*. The lungs of mammals lie entirely within a pair of pleural cavities which are separated from each other as well as from the rest of the coelom.

In most mammals the lungs are subdivided externally into lobes, the number varying somewhat in different species. The lobulation depends to some extent on the manner in which the bronchi branch or divide during development. In some forms the lobes are completely separated from each other except for their connection by the bronchi and connective tissues. In others the separation is not so complete. The number of lobes on the right side generally exceeds that on the left. Thus in man there are three lobes on the right and two on the left. The cat has four on the right and three on the left (Fig. 6.25). In cattle four or five lobes may be present on the right side but only three on the left. The left lung of the rat is not lobulated even though the right side has three large lobes and one small one.

In many mammals the most cephalic lobe on the right side of the lung is called the *superior*, or *apical*, lobe. It is supplied by the *apical*, or *eparterial*, *bronchus* which is a cranially directed branch of the right bronchus. The name *eparterial* is applied because it lies upon, or is dorsal to, the right

pulmonary artery. The lungs of some mammals show little, if any, indication of lobulation. Whales, sirenians, elephants, *Hyrax*, and many members of the order Perissodactyla have lungs of this type.

MECHANISM OF LUNG RESPIRATION

The exact mechanism by means of which air is taken in and forced out of the physostomous swim bladder of the lungfishes is not clear (page 227). In higher forms inhalation and exhalation are accomplished in various ways, some of which are fairly well understood.

Amphibians. The lungs of amphibians lie in that portion of the coelom referred to

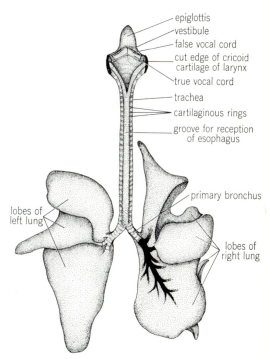

Fig. 6.25 Dorsal view of respiratory organs of cat. The larynx has been split open to show the vocal cords.

as the *pleuroperitoneal cavity.* A partition, the *septum transversum,* has developed posterior to the heart, cutting off a part of the coelom, the *pericardial cavity,* in which the heart lies, from the rest of the body cavity. The lungs of amphibians are located in the anterior lateral portions of the pleuroperitoneal cavity.

In several amphibians the lungs develop before metamorphosis when the larvae are still living in an aquatic environment. Such larvae frequently come to the surface and seem to snap for air. The air bubble thus obtained may be forced through the glottis and into the lungs by elevating the floor of the mouth. This method of obtaining air is sometimes employed by the adult *Necturus* in which the rather poorly developed lungs serve only as accessory respiratory organs. In such cases expiration is brought about by the pressure exerted by the elastic walls of the lungs when the glottis is opened. A means is thus provided for the escape of air which may or may not be depleted of its oxygen content.

When breathing air, both urodeles and anurans keep their mouths tightly closed. Air is sucked into the mouth and forced out again through the nares by alternately lowering and raising the muscular floor of the mouth. Smooth-muscle fibers around the external nares of urodeles help to control the respiratory current. Anurans, as has been alluded to previously, utilize the tuberculum prelinguale in moving the premaxillary bones and thus regulate the opening and closing of the nasal passages. This method of breathing is known as *buccopharyngeal respiration.* Some oxygen is absorbed by blood vessels in the lining of mouth and pharynx. Filling the lungs with air is accomplished in a somewhat different manner. The throat muscles are lowered to a greater degree than in buccopharyngeal respiration, and the nostrils are then closed.

The glottis is opened, and air from the lungs enters the mouth cavity, mixing with fresh air just obtained from the outside. Raising the floor of the mouth and keeping the nostrils closed forces the mixed air back through the glottis into the lungs. This process may be repeated one or more times. Sometimes when frogs are making strong respiratory movements, their eyes may be observed to sink down into the orbit for a moment. Since mouth and orbit are separated only by a thin layer composed of connective tissue and mucous membrane, retraction of the eyes aids in compressing the air in the mouth cavity and helps to force air into the lungs. After the lungs are filled, buccopharyngeal respiration may again be employed, and a fresh supply of air enters the mouth and pharynx. Repetition of these processes ensures a slow but adequate supply of oxygen to the lungs. The elasticity of the lungs themselves, possibly aided by pressure exerted upon them by contraction of the muscles of the body wall, is responsible for the escape of air from the lungs into the buccopharyngeal cavity. Even though many amphibians employ the above-described mechanisms in respiration, cutaneous respiration, in most forms, continues to be of utmost importance. In anurans the degree of pulmonary and buccopharyngeal respiration versus cutaneous respiration varies according to temperature, the former being used more extensively at higher temperatures (W. G. Whitford, 1965, *Amer. Zoologist,* **5**:705).

Reptiles. As in amphibians, the lungs of reptiles lie in that part of the coelom referred to as the pleuroperitoneal cavity, specifically in the anterolateral portions. In many reptiles, respiration is accomplished in the manner of amphibians in which throat movements and opening and closing of valves in the nostrils are utilized. However, the presence of ribs and rib muscles in reptiles aids these animals in carrying on respiration more effectively than is possible in the ribless amphibians. Raising the ribs and increasing the size of the pleuroperitoneal cavity reduces the pressure within the cavity. This, in turn, brings about an expansion of the lungs. Conversely, lowering the ribs results in an increased pressure upon the lungs, and expiration is the result. In certain reptiles, as in crocodilians, the pleuroperitoneal cavity is divided by coelomic folds into anterior and posterior portions. The lungs are shut off from the remainder of the coelom by these folds which are at least functionally comparable to the diaphragm in mammals.

The presence in turtles of a hard, rigid shell, formed in part by fusion of the ribs with the dermal plates of the carapace, makes general expansion and contraction of the body cavity by the usual method impossible. Turtle respiration, therefore, presents some unusual physiological problems. It was formerly thought that these animals employed throat movements, similar to those of amphibians, in respiration. However, it has been demonstrated (F. H. McCutcheon, 1943, *Physiol. Zool.,* **16:**255) that this is not the case and that turtles utilize a unique method of breathing. Expiration is the result of contraction of paired muscular membranes which enclose the viscera. Inspiration is brought about by the contraction of other paired muscular membranes enclosing the flank cavities. This brings about an increase in volume of the pleuroperitoneal cavity. The glottis remains closed except when movements of expiration and inspiration are in progress. When the glottis is closed, pressure on the air within the lungs, brought about by various degrees of contraction of the paired muscular membranes, varies independently of the atmospheric pressure. Alteration in pressure brings about changes in the diffusion gradient across the lung epithelium and results in an

exchange of gases from lung cavities to blood, or vice versa. The term *poikilobaric* has been proposed for air breathers of this type. *Homeobaric respiration* is the common type in which the respiratory organ is in continuous equilibrium with the atmosphere.

Birds. A membrane on each side, known as the *oblique septum* and sparsely supplied with muscle fibers, extends from the dorsal body wall to the septum transversum. This separates the lungs from the other viscera so that they are enclosed in separate *pleural cavities.* The remainder of the coelom is the *peritoneal,* or *abdominal, cavity.* The lungs are small and closely adherent to the thoracic vertebrae and ribs. *Costopulmonary muscles,* inserted in the membrane covering the lower surface of each lung, bring about expansion and contraction of these respiratory organs under normal conditions when the bird is not in flight. During flight, however, the body must be kept as rigid as possible so that the well-developed flight muscles are firmly anchored. At such times contraction and expansion of the air sacs, brought about by movements of viscera, muscles, and bones, cause air to circulate back and forth through the intercommunicating parabronchi and secondary bronchi. It has been shown under experimental conditions that if the trachea of a bird is occluded and an air sac, such as that in the humerus of the wing, is opened, the bird can continue to breathe without any difficulty. The fact that air may pass over the respiratory epithelium of the air capillaries several times in birds indicates that their lungs function more efficiently than those of most other forms.

Mammals. A musculotendinous diaphragm, present only in mammals, separates the pleuroperitoneal cavity into two parts. That portion anterior to the diaphragm contains the lungs. Each lung is enclosed in a separate *pleural cavity* which is lined with a fibro-elastic membrane called the *pleura.* The part of the pleura lining the wall of the pleural cavity is referred to as the *parietal pleura.* The lung itself is invested by a layer of *visceral pleura* reflected over its surface (Fig. 6.26). The outermost layer of each pleura, or that facing the pleural cavity, consists of squamous mesothelial cells. Visceral and parietal pleurae are continuous with each other at the root of the lung, or hilus, where the bronchi and blood vessels enter. Pleurisy is a disease marked by inflammation of the pleura. Toward the median line the parietal pleurae of the two sides come close together so as to form a two-layered septum, the *mediastinum,* separating the two pleural cavities. The space between the two layers of the mediastinum is called the *mediastinal space.* It widens as it nears the dia-

phragm and contains the aorta, esophagus, and posterior vena cava. In the region of the heart the mediastinal membrane passes over the parietal layer of the pericardium. The pleural membrane is rich in capillaries and lymphatic vessels. A serous fluid, present in the pleural cavities, enhances the movement of the lungs within the cavities. Under normal conditions there are practially no pleural cavities in existence, since the spaces are completely filled by the lungs except for the small amount of fluid which serves as a lubricant.

The pleural cavities are airtight. Increasing their size results in an expansion of the lungs, and air from the outside is drawn inside, passively. Additional blood is concomitantly drawn into the blood vessels of the lung and those of the thorax proper. Reduction in the size of the pleural cavities forces air out of the lungs. The mechanism of respiration in

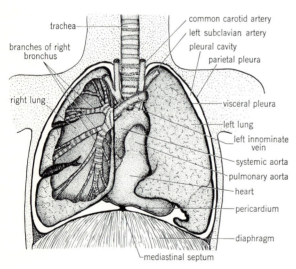

Fig. 6.26 Relations of lungs, pleural cavities, and heart in man, ventral view (semidiagrammatic). The pleural cavities have been exaggerated for the sake of clarity, since in life the visceral and parietal pleurae are in apposition.

Anatomy of the Chordates

mammals, therefore, consists of alternately increasing and decreasing the size of the pleural cavities. This is generally accomplished in two ways. Contraction of the muscles of the arched diaphragm flattens or lowers that structure. This increases the size of the pleural cavities, causing inspiration, and brings about an increased pressure on the abdominal viscera. Contraction of the abdominal muscles causes the abdominal viscera to push against the diaphragm, bringing about a reduction in size of the pleural cavities and a resulting expiration of air from the lungs. This method of breathing is known as *abdominal respiration*. The other method of expanding the pleural cavities is brought about by contraction of the external intercostal muscles. These pass from the lower border of one rib to the upper border of the rib below. Contraction brings about an elevation of the ribs. The method by which expiration is brought about in this type of breathing is not so well understood. In some mammals, such as the cat, contraction of the internal intercostal muscles reduces the size of the thoracic cavity. The elasticity of the expanded lung walls, together with the tension of the stretched body wall and ribs, undoubtedly accounts for the greater part of expiratory movement.

Just as the stomach, intestine, etc., actually lie outside the peritoneal cavity, since they are covered with visceral peritoneum, so the lungs, being covered with visceral pleura, lie outside the pleural cavities which should, therefore, be considered to be more in the nature of *potential* spaces rather than as actual spaces. If air enters the pleural cavity through a wound, or because of rupture of the lung wall in certain diseased conditions, the potential pleural cavity becomes a real cavity and the lung on that side will collapse since its tissues are under constant tension because of the elastic fibers present in them.

Such a condition is known as a *pneumothorax*. As a therapeutic measure, collapse of the lung is induced artificially in certain pulmonary lesions (tuberculosis) in which the disease is confined to one lung. The theory behind such treatment is to permit the lung to rest for a time by making it functionless. A pneumothorax is usually induced by injecting air into the pleural cavity. The air is gradually absorbed, and the procedure must be repeated at intervals. The term *hydrothorax* refers to the presence of an abnormal amount of fluid in the pleural cavity. This occurs in certain diseased conditions and here, too, the potential pleural cavity becomes a real cavity. Although the two pleural cavities are completely separated by the mediastinum, that membrane is so pliant that a pneumothorax or hydrothorax on one side affects the other side to some degree.

In the condition known as asthma, characterized by difficulty in breathing, usually in response to allergic conditions of one kind or another, there is a contraction of the smooth muscles of the smaller bronchioles. This is commonly accompanied by a swelling of the mucous membrane lining these small tubes. Epinephrine (adrenalin) or some other drug with similar properties (page 359) is often administered because of its effect in dilating the respiratory passages so that breathing can be more easily accomplished. The patient suffering from asthma usually has more difficulty in expelling air than in drawing it into the lungs.

MISCELLANEOUS RESPIRATORY ORGANS

Although gills are the chief respiratory organs of aquatic vertebrates, and lungs serve terrestrial forms in a similiar capacity, other structures are present in certain vertebrates which may be considered to belong in the category of respiratory organs.

Yolk sac. The blood vessels (vitelline arteries and veins) which supply the yolk sac, a structure present during embryonic stages of many vertebrates, not only function in absorbing yolk, which is used as food, but may serve in a respiratory capacity as well. In ovoviviparous sharks, for example, the blood vessels of the yolk sac (Fig. 5.1) lie in close apposition to highly vascular, villus-like projections of the lining of the uterus. Here the exchange of gases takes place between embryo and mother. In the bird embryo the yolk sac is the respiratory organ for the first few days of development or until another embryonic membranous sac, the allantois, pushes between yolk sac and porous shell and takes over the respiratory function. Even in the embryos of certain lower mammals (some marsupials) a *yolk-sac placenta* is present through which gaseous exchange occurs. The yolk sacs of most mammals, on the other hand, are but rudimentary vestiges.

Allantois. The allantoic sac, present in the embryonic stages of reptiles, birds, and mammals, is an important, if temporary, respiratory organ. The allantoic, or umbilical, blood vessels, with which it is furnished, are the ones through which oxygen is obtained and carbon dioxide eliminated during embryonic life. Even though the allantois is not well developed in many mammals, the allantoic circulation is of great importance in the exchange of gases in the placenta. The allantois is lost at the time of birth.

Skin. In some fishes the skin, in addition to the gills, may function in respiration. For example, the walking goby, *Periophthalmus,* on the shore of the island of Celebes, has been observed to emerge from the water, its head and trunk completely exposed but its very vascular caudal fin remaining submerged and undoubtedly functioning as a breathing organ. Delicate outgrowths of the pelvic fins of the South American lungfish, *Lepidosiren,* are believed to have a respiratory function.

A highly vascular integument is very important in amphibian respiration. In some species of frogs, over 70 per cent of the carbon dioxide may leave the body through the skin. Only in forms in which the skin is kept moist can it be used in respiration, since the presence of a moist membrane is one of the requirements for respiratory exchange. In the lungless salamanders of the family Plethodontidae, which also lack gills in the adult stage, the integument is highly important in respiration. Buccopharyngeal breathing is also employed by these amphibians. In the "hairy" frog, *Astylosternus robustus,* the sides of the body and thighs are covered with many filamentous, cutaneous outgrowths which give the animal a hairy appearance. These furnish an additional respiratory surface. Barbels on the head of the African clawed toad, *Xenopus laevis,* are thought to be respiratory in function. Even the skin of man might in one sense be considered as a respiratory organ of a sort, since a small amount of carbon dioxide is eliminated in the process of sweating.

Rectum and cloaca. A highly vascular rectum serves some fishes as an accessory organ of respiration. Water is alternately taken in and squirted out of the rectal chamber. Some aquatic turtles have a pair of thin-walled saclike extensions of the cloaca which are sometimes referred to as accessory urinary bladders. These sacs, which are abundantly supplied with blood vessels, are frequently filled and emptied of water through the anus and, therefore, may act as important respiratory organs, particularly when in emergencies the animal may remain submerged for an undue length of time.

OTHER DERIVATIVES OF THE PHARYNX

A number of structures derived from the pharynx, but not associated in any way with respiration, may be conveniently discussed at this point.

Except for the thyroid gland, the structures derived from the pharynx arise as modifications of pharyngeal pouches. During development, small dorsal and ventral recesses appear at the outer ends of the pouches. Cellular proliferation about these recesses brings about the formation of small, budlike masses of endodermal origin which represent the primordia of certain glands confined to the neck region. The thymus, tonsils (palatine), parathyroids, and ultimobranchial bodies are derived from these primordia. The history of the various structures arising from different pharyngeal pouches varies somewhat in the different vertebrate classes, and there is much confusion, particularly in the lower forms, as to what the various structures in the neck region actually represent. They are often referred to collectively as *epithelial bodies*.

Thyroid gland. The thyroid gland in lower vertebrates arises as a midventral, endodermal diverticulum of the pharynx in the region between the second and fourth visceral pouches. In higher forms it usually develops between the first and second pouches. In some it is paired, but in others it is a single organ. In the human embryo it has been reported that small, paired primordia from the fourth pair of visceral pouches are added to the median thyroid diverticulum and that all three parts combine to form the thyroid gland. Since the thyroid is an endocrine organ, the reader is referred to Chap. 9, Endocrine System, for a more complete discussion.

Thymus gland. For many years the thymus was included in the category of endocrine glands. However, its activity as an endocrine organ is now considered very doubtful. Evidence indicates that it belongs to the lymphatic system rather than to the endocrine system and may ultimately be responsible for some of, if not all, the functions of the body having to do with immunological reactions. In lower vertebrates it is difficult to ascertain whether certain tissues, ordinarily referred to as thymus tissue, are actually homologues of the thymus gland as we recognize it in higher forms.

COMPARATIVE ANATOMY OF THE THYMUS. *Cyclostomes.* In the lampreys all seven pairs of pharyngeal pouches are reported to be involved in the development of a number of separate elements which may possibly represent thymus tissue. Small primordia arise as buds from the dorsal angles of the pouches. These later become separated from the pharynx, come to lie dorsal to the gill region, and persist throughout life. The ventral portions of the pouches have also been reported to contribute to the thymus. If these structures actually represent thymus tissue, the lampreys exhibit the most generalized condition of all vertebrates in respect to thymus development, since all pharyngeal pouches contribute to its formation. Several lobulelike structures located posterior to the gill region in the hagfish are regarded by some authorities as thymus tissue. This is questioned by others.

Fishes. In fishes, in correlation with the reduced number of pharyngeal pouches, there is a reduction in the number of thymus primordia as compared with cyclostomes, and considerable variation is to be found within the superclass. Primordia arise from the dorsal angles of all the pharyngeal pouches except the first, and in a few species even the first pair of pouches is involved. In teleosts there is usually a fusion of the separate components on each side to form a single lobulated mass. It lies dorsal to the gill region beneath the dorsal musculature. In certain forms (*Heptanchus*), connections with the pharyngeal epithelium are retained.

Amphibians. Except in caecilians in which all six pouches seem to be concerned in contributing to the thymus gland, amphibians exhibit a reduction in number of thymus primordia. In urodeles the dorsal angles of pouches 1 to 5 are concerned with thymus formation but the portions from pouches 1 and 2 degenerate. In anurans the second pair of pouches alone furnishes permanent contributions. The thymus gland in urodeles usually consists of a lobulated mass of tissue on each side of the neck region representing the fused elements of that side. In the adult frog the thymus lies behind the tympanic membrane and under the depressor mandibulae muscle. It is a compact structure, lymphoid in character, and is reported to be considerably smaller in the adult stage than during larval life.

Reptiles. In lizards the second and third pharyngeal pouches apparently give rise to thymus primordia from their dorsal angles, whereas in snakes thymus buds have been described as arising from pouches 4 and 5. The elements may form separate compact masses, but there is a tendency toward fusion into lobular strands of tissue. This is best shown in crocodilians.

Birds. Only pharyngeal pouches 3 and 4 are responsible for thymus development in birds. The dorsal primordia arising from these pouches fuse to form an elongated thymus body on each side of the throat. In a young chicken the thymus gland is characteristically composed of 14 lobes, seven of which extend in a linear series on each side of the neck.

Mammals. Mammals are exceptional in that the thymus arises from the *ventral* portions of the pharyngeal pouches rather than from the dorsal. Only pouches 3 and 4 are involved, but the fourth contributes little and the main mass of the thymus is therefore derived from the ventral angles of the third pair of pouches. When fully developed, the portions from the two sides, although not actually fused, are bound together with connective tissue so that the thymus appears to be a single structure. It is located along the ventral surface of the trachea and extends as far back as the base of the heart. The thymus has commercial value as food and is sold on the market under the name of neck, or throat, "sweetbreads."

It has been reported that in some mammals the part of the thymus lying in the cervical region is derived from skin ectoderm. Since it is sometimes difficult to distinguish the point of transition between ectoderm and endoderm on the borders of a pharyngeal pouch, variations such as this may only be apparent. It is, therefore, unnecessary to assume that differences in homology exist.

Since the thymus gland is considered to belong to the lymphatic system, discussion of its histological structure and function is included in Chap. 12, Circulatory System (see pages 591 to 593).

Parathyroid glands. Definite parathyroid glands are lacking in cyclostomes and fishes. They first make their appearance in members of the class Amphibia. The parathyroids arise from the ventral angles of certain visceral pouches in amphibians, reptiles, and birds. In mammals, however, they come from the dorsal angles. These derivatives of the pharynx are endocrine glands related in function to the calcium and phosphorus metabolism of the body. Detailed discussion is reserved for Chap. 9, Endocrine System.

Tonsils. Lymphoid masses derived from the ends of the second pair of visceral pouches give rise to the *palatine tonsils,* which are present only in mammals and are located on either side where mouth and pharynx join. The pouch itself on each side becomes very much reduced and forms the fossa and covering epithelium of the tonsil. *Lingual tonsils* are also present only in mammals. They are situated at the base of the tongue and consist of lymphoid masses made up of numerous lymphatic nodules located in the lamina propria lying beneath the epithelium in this region. They contain lymphocytes and large numbers of plasma cells. Lingual tonsils are closely associated with lingual glands. In reptiles and birds, and occasionally in mammals, *pharyngeal tonsils* are present. These appear in the lamina propria of the mucous membrane on the roof of the pharynx and come to lie behind the choanae. Their general structure is similar to that of the palatine tonsils. Certain lymphoid masses in the roof of the pharynx in amphibians may be homologous with pharyngeal tonsils. In man, enlargements of these structures are known as *adenoids.* Adenoid tissue may occlude the nasal passages, necessitating breathing through the mouth. If adenoids are not removed in the young, malformations of the face may result. Thus in mammals a partial ring of lymphoid tissue composed of palatine, lingual, and pharyngeal tonsils surrounds the opening into the pharynx.

Whether the tonsils have any function aside from being part of the lymphatic system is not known. In the rabbit fetus, small basophilic cells appear in the mesenchyme of the developing tonsils. These can be identified as lymphocytes a few days before birth. The palatine tonsils are frequently the site of infection and often must be removed. Tonsillectomy is a relatively simple operation. However, since the ascending branch of the external carotid artery lies in close proximity to the tonsil, there is danger of severe and sometimes fatal hemorrhage.

Middle ear and Eustachian tube. The first, or hyomandibular, pharyngeal pouch in elasmobranchs and some other fishes forms the spiracle, which has already been discussed. In most fishes as well as in amphibians, reptiles, birds, and mammals, the first pouch fails to break through to the outside or becomes secondarily closed. In anurans, for the first time, the tissue between the pouch and the exterior forms the *tympanic membrane,* or eardrum. The pouch itself becomes somewhat enlarged at its outer end to form the *middle ear.* The narrowed portion connecting the middle ear to the pharnyx is called the *Eustachian tube.* Middle ear and Eustachian tube function primarily as a pressure-regulating mechanism which keeps the pressure on the two sides of the tympanic membrane equal. Description of the middle ear and Eustachian tube is more properly included under a discussion of the ear (Chap. 14, pages 703 to 707). In caecilians, urodele amphibians, a few anurans, snakes, and certain other reptiles, middle ears and Eustachian tubes have degenerated and are lacking.

Other structures. Among the derivatives of the pharyngeal pouches are some small, glandular-appearing structures variously known as postbranchial or ultimobranchial bodies. They arise from, or close behind, the fifth pair of pharyngeal pouches and come to lie in the region of the thyroid gland.

Experimentalists have been unable to demonstrate that they have any particular function. They are frequently confused with the parathyroid glands and may possibly represent accessory parathyroid tissue. In some forms, as in the golden hamster, the ultimobranchial bodies contribute to thyroid-gland formation.

The so-called *carotid body*, closely associated with the carotid artery at the point where it bifurcates into internal and external branches, is conspicuous in frogs and toads and is also found in the higher vertebrates. It is composed of masses of epithelial-like cells richly supplied with nerve endings and capillaries of a sinusoidal nature. The carotid body may or may not play a role in controlling circulation in the carotid vessels. In amphibians the carotid bodies arise near the ventral portions of the second pair of pharyngeal pouches, whereas in lizards they appear in the vicinity of the third pair. Their origin in birds and mammals is questionable. It is somewhat doubtful whether the carotid bodies actually contain true epithelial cellular elements. If not, the pharyngeal epithelium may not in any way be concerned with their formation. Confusion in regard to the carotid body has arisen from the fact that various investigators appear to have assigned the name *carotid body* to different structures in the neck region which vary from species to species. The epithelial-like cells of the carotid body, with their abundant nerve supply, seem to respond to variations in the oxygen and carbon dioxide content of the blood. Nerve endings, or chemoreceptors, in the carotid body send impulses, via a branch of the glossopharyngeal nerve, to brain centers controlling the heart, arterial vessels, and respiration. Surgical removal of even one of the carotid bodies has been shown to have very beneficial effects in a large percentage of cases when applied to human beings suffering from asthma.

SUMMARY

1. With few exceptions the organs of respiration in vertebrates are formed in connection with that part of the digestive tract known as the pharynx. In some forms the allantois and cloaca may also be utilized. Respiratory organs are usually covered or lined with epithelia derived from endoderm.

2. In all chordate embryos a series of visceral pouches develops on either side of the pharynx. The number of visceral pouches is greatest in the lowest groups of vertebrates and least in the higher classes. The pouches push through the mesenchyme to fuse with the outer ectoderm, forming thin, platelike areas. In fishes and larval amphibians perforations occur, forming gill slits which connect the pharynx to the outside. The pouches are then called gill clefts. They are separated from each other by connective-tissue septa within which lie visceral arches composed of skeletal material.

Internal gills, supported by visceral septa and arches, are vascular, lamellar, or even filamentous extensions of the epithelial surface of the gill pouches. External gills, generally covered with ectoderm, may project as filamentous outgrowths from the integument covering the visceral arches.

3. In lampreys and elasmobranch fishes each gill cleft opens separately to the

outside. In higher fishes an operculum, which is an extension of the hyoid arch, covers the gill chamber. Although most fishes possess internal gills, some have external gills in addition, at least during larval life.

4. Only external gills occur in amphibians. With few exceptions they disappear upon metamorphosis and the gill slits close over. In reptiles, birds, and mammals the pharyngeal pouches usually fail to break through to the outside and gills are lacking.

5. In anurans, most reptiles, birds, and mammals, the first gill pouch gives rise to the Eustachian tube and middle ear. The original lining of the remaining pouches gives rise to such structures as tonsils, parathyroid and thymus glands, and ultimo-branchial bodies. The thyroid gland arises as a midventral outgrowth of the pharynx.

6. Lungs develop as a bilobed diverticulum from the floor of the pharynx posterior to the last gill pouch. In higher forms the connection between lungs and pharynx lengthens and is called the trachea. Lungs show unmistakable homologies to the swim bladders of fishes, which in certain cases are used as lungs but in others have become specialized as hydrostatic organs. Many modifications of the swim bladder occur.

7. The upper part of the trachea becomes modified as a larynx, or voice box, the walls of which are supported by skeletal elements derived from the visceral arches. In birds, a syrinx which develops at the lower end of the trachea is the organ of sound production.

8. The lungs of lower forms are rather simple, vascular sacs, but in higher vertebrates the walls become subdivided into numerous pocketlike air spaces. The divisions become more and more complex as the vertebrate scale is ascended and reach the highest degree of branching in mammals. The lungs of birds are complicated in that they give rise to air sacs which penetrate among the viscera and even enter the hollow bones. Similar structures in certain lizards foreshadow their appearance in birds. The thin-walled, moist, highly vascular lining of the lungs forms an ideal respiratory surface.

9. Various methods are employed by vertebrates in inflating and deflating the lungs. The most complex condition is encountered in mammals, in which the lungs lie in separate pleural cavities partitioned from the abdominal cavity by a muscular diaphragm. In amphibians and reptiles they lie in the anterior part of the coelom which is known as the pleuroperitoneal cavity. The lungs of birds lie in separate pleural cavities, even though no diaphragm is present.

10. Nasal passages develop in connection with the olfactory apparatus. Blind nasal pits are found in most fishes, but in the Sarcopterygii they form connections between the oral cavity and the outside. This condition is retained in amphibians which, for the first time, employ the nasal passages for intake and outgo of air. In reptiles a secondary palate begins to form, partially separating the nasal and mouth cavities. In crocodilians and mammals the secondary palate becomes complete so that the two passageways are in communication only in the region of the pharynx. In these forms the internal nasal openings are located relatively far back and are situated near the opening of the trachea.

| 7 |

EXCRETORY
SYSTEM

The word *metabolism* is used to designate the different chemical changes that take place in the tissues of the body upon which heat production, muscular energy, growth, and maintenance of vital life processes depend. The part played by the end products of digestion, after they have passed into the circulatory system, in building up protoplasm and in providing materials to be used in essential body processes is but one phase of metabolism. The term *anabolism* is used to designate this positive, or building-up, phase. *Catabolism* is the expression which is used to signify the reactions involved in the breaking down of chemical substances in the body to simpler compounds by oxidation, with the release of energy in its various forms.

Some of the products of metabolism cannot be used as a source of energy. They are called the end products of metabolism, or *wastes*. The waste products formed within the cells pass slowly out of the cells into the tissue fluid and hence into the lymphatic system and bloodstream. Since they are of no further use to the body, they must be eliminated. Among the waste products of metabolism are carbon dioxide, urea, ammonia, uric acid, creatinine, various pigments, and inorganic salts. Carbon dioxide is eliminated, for the most part, through the gills in aquatic forms, through the skin and lungs in amphibians, and through the lungs in terrestrial vertebrates. The remaining substances, most of which are products of the metabolism of proteins, are excreted almost entirely through what are called the *excretory* or *urinary organs*. It has been demonstrated, however, that in certain fishes nitrogenous and other wastes may be excreted in considerable quantity by way of the gills. In man, small amounts may be eliminated through the sweat glands of the skin. It should be recalled at this point that although feces may contain substances excreted through the bile or by the cells of the intestine, the greater bulk of it is made up not of true excreta but of indigestible material which has never entered the body proper.

Most excretory substances are in solution in water. Water itself is not considered to be a waste product, but any excess is eliminated along with substances dissolved in it. The final product is known as *urine*. The organs which eliminate the urine are called *kidneys*.

Although the excretory system, particularly in the male, is usually closely associated with the reproductive system, nevertheless so far as is possible the two systems will be discussed separately. Frequently the excretory and reproductive systems are grouped together and spoken of as the *urogenital*, or *urinogenital*, *system*. All parts of the excretory system in vertebrates, except for a few structures near the terminal duct openings, are mesodermal structures which are derived from the embryonic mesomere (page 66).

TYPES OF KIDNEYS

In some of the invertebrates, excretory organs called *nephridia*, or *nephridial tubules*, are present. These are frequently segmentally arranged, with a pair in practically every segment. In the earthworm each nephridium occupies parts of two adjacent segments. In this animal the tubules are independent structures connecting with the coelom at one end by a ciliated funnel, or *nephrostome*, and at the other end opening to the exterior of the body by a pore, the *nephridopore*. Tissue fluid from the coelom passes along the tubule, which is relatively long and coiled into three loops. Water and certain other substances of importance to the body are resorbed back into the body tissues en route. The cells of the wall of the nephridial tubule excrete additional substances into the fluid as it passes along the tubule. The material which is discharged onto the surface of the body, therefore, is quite unlike the tissue fluid which originally entered the tubule. It no longer contains substances of importance to the body and is much more highly concentrated than it was formerly. There is little or no evidence that the nephridial tubules of invertebrates bear any relationship to the kidney tubules of vertebrates and no evolutionary trends are indicated even though the method of functioning of the two is basically similar.

Even the excretory tubules of amphioxus (Fig. 7.1) are of an entirely different type from those of vertebrates and seem to be more closely related to invertebrate structures. In amphioxus there is a series of such tubules opening into the atrium, or peribranchial space. Each lies on the outer dorsal side of a secondary gill bar. They are apparently of ectodermal origin, have no connection with the coelom, and are composed of numerous flame cells called *solenocytes*

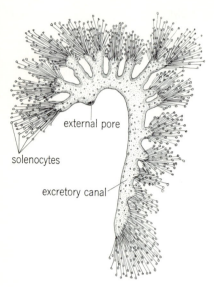

Fig. 7.1 One of the numerous excretory tubules of amphioxus. (*After Goodrich.*)

which collect wastes. The solenocytes are attached to the walls of blood vessels and are also bathed by coelomic fluid. Those belonging to a given tubule enter a common excretory canal, which in turn opens into the atrium through a small excretory pore (nephridiopore).

It is rather difficult to work out homologies among the many and varied types of excretory organs found among vertebrates. The variations that are encountered are, in all probability, correlated with problems with which vertebrates have had to cope in adapting themselves to the different environmental conditions under which they have lived. Whether animals remained aquatic or took to terrestrial existence posed important problems in connection with excretion. Whether aquatic forms lived in fresh water or salt water made a great difference in their adapting themselves to changes in osmotic pressures, variations in salt con-

centration, and the like. Comparative anatomists have long speculated concerning the evolution of the various types of vertebrate kidneys. Gradually a fairly logical sequence of events has been postulated. There has been, and still is, much confusion in this field, particularly in regard to terminology. Investigators who have studied kidney development in embryos of birds and mammals in the past have attempted to homologize these structures with the excretory organs of lower forms. It now appears that many of these speculations were erroneous. The student, therefore, should be wary in accepting certain interpretations of homologies that appear in the older literature on the subject. Often the ideas advanced have been correct, yet the terminology used has confused the issue.

Archinephros

The belief is now generally held that the primitive vertebrate ancestor possessed an excretory organ which is referred to as an *archinephros*, or *holonephros* (Fig. 7.2). This is supposed to have consisted of a pair of *archinephric ducts* located on the dorsal side of the body cavity and extending the length of the coelom. Each duct was joined by a series of segmentally arranged tubules, one pair of tubules to a segment. At its other end, the tubule opened into the coelom by a ciliated, funnel-shaped, peritoneal aperture called the *nephrostome*. Also formed in connection with each tubule was a so-called *external glomerulus*, a small knot or cluster of capillaries interposed within the course of an arteriole (a small branch of an artery) and located in close proximity to the nephrostome. A thin layer of peritoneal epithelium was reflected over the projecting surface of the external glomerulus. Tissue fluid, exuded at the glomeruli, passed into the coelom and thence through the nephro-

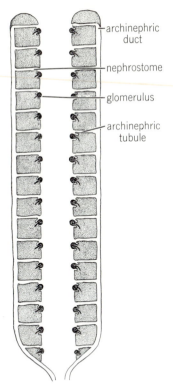

archinephric duct

nephrostome

glomerulus

archinephric tubule

Fig. 7.2 Diagram showing hypothetical structure of archinephros.

stomes into the kidney tubules and finally through the archinephric ducts to the outside. It is believed that the various kidneys of present-day forms may originally have been derived from a primitive type similar to the archinephros. It is interesting to note that even today the larval form of *Myxine,* the hagfish, and the larvae of some of the caecilians, or apodan amphibians, possess kidneys of the archinephric type.

The Anamniote Kidney

The anterior part of the archinephric kidney has persisted in only a few vertebrates, and even in the adult hagfish it has been somewhat modified. It appears in the em-

bryos of most vertebrates as a transitory structure, usually referred to as the *pronephros,* which degenerates soon after it has formed. In the few anamniote vertebrates in which the pronephros persists in the adult stage it is called the *head kidney.* The remainder of the kidney, posterior to the pronephric region, is known as the *opisthonephros.*

There is much confusion as to the correct term to be applied to the kidney duct among vertebrates. It has been variously called the pronephric duct, mesonephric duct, Wolffian duct, and ureter. When appropriated by the reproductive system for the transport of spermatozoa (page 316), parts of this same duct are called the ductus epididymidis and the ductus deferens. It is well to keep these terms in mind since different authors use them all rather freely and students of anatomy and embryology may become rather confused by the variations in nomenclature. The term *ureter,* as applied to this duct, should be avoided, however, as it is most properly applied only to the kidney duct in reptiles, birds, and mammals (*Amniota*). We shall use the term *archinephric duct* in referring to the primitive kidney duct as it appears in cyclostomes, fishes, and amphibians (*Anamniota*).

Pronephros. The pronephros in anamniotes actually consists of a varying number of anteriorly located pronephric tubules, together with a pair of archinephric ducts. The tubules and ducts lie in the dorsolateral mesoderm on either side of the mesentery that supports the gut. The ducts extend posteriorly and usually open into the cloaca. One end of each of the segmentally arranged tubules connects with the archinephric duct near its anterior end; in fact the duct is actually formed by successive tubules bending posteriorly and fusing with adjacent tubules. The other end of the tubule opens into the coelom by means of a nephrostome. The nephrostome and the part of the tubule near the nephrostome are ciliated. In a few forms, rounded external glomeruli (Fig. 7.4), arterial vessels of capillary dimensions covered with peritoneal epithelium, coming from segmental branches of the dorsal aorta, project into the coelom near the nephrostome. Most forms, however, possess *internal glomeruli.* These are small knots of interarterial capillaries, each surrounded by a double-walled structure called *Bowman's capsule,* the two together being known as a *renal,* or *Malpighian, corpuscle* (Figs. 7.3 and 7.5). The outer wall of Bowman's capsule is the *parietal layer;* the inner wall, everywhere in contact with the blood vessels of the glomerulus, is the *visceral layer.* Blood is brought to the internal glomerulus by an *afferent arteriole* and leaves through an *efferent arteriole.* The latter breaks up into *true* capillaries along the course of a pronephric tubule, and the blood is ultimately returned to the heart through one of the postcardinal veins. Even though pronephric tubules have internal glomeruli associated

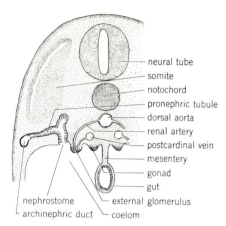

neural tube
somite
notochord
pronephric tubule
dorsal aorta
renal artery
postcardinal vein
mesentery
gonad
gut
nephrostome
external glomerulus
archinephric duct
coelom

Fig. 7.4 Diagrammatic section through part of a vertebrate embryo, showing relation of external glomerulus to pronephric tubule and nephrostome.

with them, a connection with the coelom via the nephrostome is retained (Fig. 7.5). Whether glomeruli are of the external or internal type is of no particular significance in their relation to the pronephros.

From the glomerulus a protein-free filtrate of blood plasma (tissue fluid) passes into the coelom or into the cavity of Bowman's capsule, depending upon whether the glomerulus is of the external or internal type. Cilia sweep this fluid into the tubule by way of the nephrostome or down the tubule from the capsule. Certain cells in the wall of the tubule probably excrete wastes into the lumen, where they are added to those in the fluid filtered from the glomerulus. A selective reabsorption of water and other constituents occurs as the fluid passes along the tubule so that a much smaller amount of fluid with a higher concentration of wastes in solution finally enters the archinephric duct. Electron microscopic observations on the functional pronephric tubules of *Rana pipiens* tadpoles indicate that their fine structure is similar to the tubules of adult kidneys with micro-

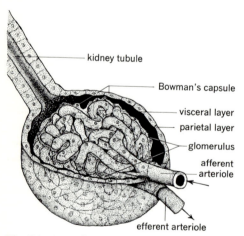

kidney tubule

Bowman's capsule

visceral layer
parietal layer
glomerulus
afferent arteriole

efferent arteriole

Fig. 7.3 A renal corpuscle (semidiagrammatic).

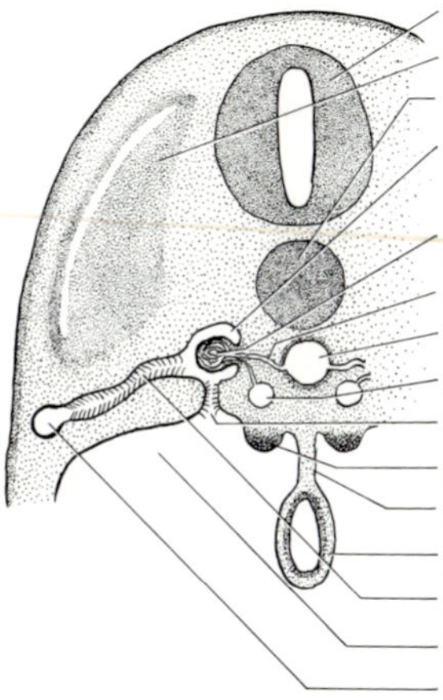
neural tube
somite
notochord
Bowman's capsule
internal glomerulus
renal artery
dorsal aorta
postcardinal vein
nephrostome
gonad
mesentery
gut
kidney tubule
coelom
archinephric duct

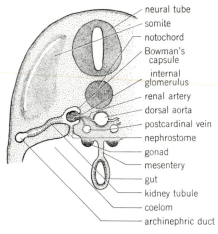

Fig. 7.5 Diagrammatic section through part of a vertebrate embryo, showing relation of internal glomerulus to kidney tubule and archinephric duct.

villi, pinocytotic vesicles, etc. (page 541). (C. W. Gibley, Jr. and J. P. Chang, 1966, *Amer. Zoologist,* **6**:610).

Sometimes several glomeruli unite to form a larger *glomus.* In some cases pronephric tubules expand so as to form *pronephric chambers,* or else several tubules may fuse to form one large chamber. The importance of the pronephros consists chiefly in the part it plays in forming the archinephric duct, which persists even though the pronephros itself disappears. In members of the class Chondrichthyes, the pronephros degenerates soon after it is formed and no vestiges remain. In some larval forms the pronephros functions in getting rid of wastes at a time when the opisthonephros is in the process of formation. In many fishes and larval amphibians it becomes modified and functions for a time as a true excretory organ. It is only in the hagfish and certain teleost fishes, such as *Zoarces, Fierasfer,* and *Lepadogaster,* that the head kidney with its coelomic connections persists in adult life. In myxinoids, the pronephros and opisthonephros become completely separated by degeneration of the portion between them. The nephrostomes of the pronephros connect with the pericardial cavity where a single large glomus is present. The fluid drained by the tubules atypically passes into a nearby vein instead of entering a duct.

Opisthonephros. Since the pronephros in most cases is but a transient structure, the opisthonephros is actually the more important part. It serves as the adult kidney in lampreys, most fishes, and amphibians.

In many accounts the term *mesonephros* is used in describing what we are here calling the opisthonephros. Biologists have come to realize that the opisthonephros of cyclostomes, fishes, and amphibians is not quite comparable to the mesonephros of embryonic amniotes even though the two are structurally similar in many ways. We shall, therefore, in the following account, use the term mesonephros in referring to the structure which appears only *during embryonic development* in reptiles, birds, and mammals.

The reason for the distinction lies in the fact that the three types of kidneys which appear in the embryos of amniotes—pronephros, mesonephros, and metanephros—represent developments from different levels of the primitive archinephros which appear in succession in an anterior-posterior direction. The transition from one to the next is almost imperceptible. The opisthonephros of cyclostomes, fishes, and amphibians actually extends over a region which in amniotes will form the mesonephros and metanephros. In forms possessing an opisthonephros there is a general tendency toward a concentration of kidney tubules toward the posterior end of the organ. The anterior portion frequently loses its significance as an excretory organ and, in the male, may become part of the reproductive system.

The opisthonephros differs from the pronephros in several respects. As we shall see, the persistent archinephric duct may be taken over almost entirely in males by the reproductive system, in which case additional, or accessory, urinary ducts are formed to carry away waste materials. The chief difference between the pronephros and opisthonephros lies in the fact that the segmental arrangement of the kidney tubules no longer exists in the latter and numerous tubules may lie within the confines of a single segment, this being particularly evident in the posterior portion of the opisthonephros. Furthermore, the connection of the kidney tubules with the coelom is lost in most cases, the presence of renal corpuscles with internal glomeruli being typical.

The kidney tubules, of which the bulk of the opisthonephros is composed, develop from the intermediate cell mass (mesomere, nephrotome, intermediate mesoderm) which lies between the somite and the lateral-plate mesoderm of the embryo or larva (Fig. 3.10). The archinephric duct, formed during development of the pronephros, is already present in this region. In some forms the anterior tubules of the opisthonephros appear in the same segments as the posterior pronephric tubules, so that there is some degree of overlapping.

At first the opisthonephric tubules are solid structures, but they soon develop lumina and begin to grow. One end of each tubule extends out to establish a connection with the archinephric duct. The tubule grows and becomes S-shaped. Its other end enlarges. This is soon invaginated by a tuft of tortuous interarterial capillaries formed by a small, segmental, renal artery coming from the dorsal aorta. The end of the tubule thus becomes a double-walled Bowman's capsule

with its contained internal glomerulus, the two together forming a renal (Malpighian) corpuscle. The tubules themselves vary somewhat in structure in different forms. Generally speaking, however, each tubule differentiates into a narrow neck at the end near the renal corpuscle, followed in turn by secretory and collecting portions. The collecting portion connects with the archinephric duct. The secretory part of the tubule forms two loops named the *proximal* and *distal convoluted segments,* or *tubules,* in relation to their proximity to the renal corpuscle (Fig. 7.6).

At first the opisthonephric tubules have a metameric arrangement, but secondary, tertiary, etc., tubules soon arise so that the segmental character of the opisthonephros becomes obscured. Each tubule has its own renal corpuscle, but the collecting ends of several tubules may unite to form a ureterlike duct which either opens into the archinepric duct or establishes an independent

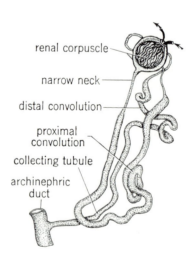

renal corpuscle

narrow neck

distal convolution

proximal convolution

collecting tubule

archinephric duct

Fig. 7.6 An amphibian kidney tubule of the opisthonephros, showing the renal corpuscle and secretory and collecting portions.

opening into the cloaca. In some forms several such urinary ducts may be present.

An afferent arteriole brings blood to the glomerulus which, it will be recalled, is really a tuft of tortuous small vessels of capillary dimensions which intervene within the course of an arteriole. Such an arrangement is sometimes called a *rete mirabile*. The afferent arteriole divides into a few main branches, each supplying a subdivision or "lobule" of the glomerulus. Anastomoses between capillaries within a lobule are common, but not between those of separate lobules. Main, or preferred, channels within the glomerulus are indicated. In any event, the small vessels finally converge and leave the glomerulus through an efferent arteriole. As this small vessel leaves the glomerulus, it is distributed along the proximal part of the kidney tubule where it breaks up into true capillaries. The vessels that collect the blood enter the postcardinal veins, which also receive blood from the renal portal system supplying the greater part of the kidney tubules proper.

There is not complete agreement as to the method of functioning of opisthonephric tubules since there is considerable variation in their detailed structure among the various groups comprising the Anamniota. Animals inhabiting salt water have different problems with which to contend than do fresh-water and land-dwelling species. In some, the renal corpuscles are unusually large; in others, they may be altogether absent. In some, portions of the kidney tubules may be ciliated; in others, cilia are lacking. In the frog, for example, long cilia project into the lumen from the cells of the neck of the tubule. Shorter cilia are present for a slight distance in the part of the tubule just beyond the neck. The beat of the cilia is directed toward the proximal convoluted tubule. Experiments on amphibians indicate

that at the glomerulus a protein-free filtrate of blood plasma is eliminated, probably because of pressure differences. A higher pressure (both blood and osmotic) exists in the glomerular capillaries than on the other side of the visceral layer of Bowman's capsule which closely invests the glomerulus. Although the afferent arteriole leading to the glomerulus is about twice the diameter of the efferent arteriole which leaves it, their lumina are of approximately the same diameter. The wall of the afferent vessel is thicker, thus accounting for the discrepancy in size. If the efferent vessel were a vein, the diameter of its lumen would be larger than that of its afferent companion. At any rate, the fact that the lumina of the two vessels are similar in size accounts for the fact that there is some resistance to the passage of blood through the glomerular capillaries, resulting in a building up of pressure within the glomerulus. The cells of the secretory portion of the tubule excrete some urinary wastes into the lumen of the tubule. The secretory tubule, particularly in its distal portion, is known to reabsorb water, chlorides, bicarbonates, and reducing substances from the contents of its lumen. The rate of reabsorption of each of these substances differs. Thus the fluid (urine) which reaches the collecting tubule is in a more concentrated form. It then passes into the archinephric duct or into an accessory, ureterlike duct which leads to the cloaca. According to this concept, then, glomerular filtration, excretion by certain cells of the tubule, and selective reabsorption by the secretory tubules are the important factors in kidney function.

Only internal glomeruli are usually present in the opisthonephros. Peritoneal funnels connect the tubules with the coelom in certain forms, but this is unusual. In most species such connections are lost, and the coelom is no longer continuous with the outside through the excretory system.

COMPARATIVE ANATOMY OF THE OPISTHONEPHROS. *Cyclostomes.* In the embryo hagfish the kidney is of the archinephric, or holonephric, type. In the adult, however, as has been pointed out, the anterior end has become modified to form a persistent pronephros, or head kidney. The remainder of the kidney, which is separated from the pronephros, becomes an opisthonephros which, in this species, differs but little in structure from the original archinephros except that the posterior tubules lose their peritoneal connections.

The opisthonephros of the lamprey consists on each side of a long, strap-shaped body without any peritoneal connections in the adult. The kidneys lie on either side of the middorsal line, from which they are suspended by mesenterylike membranes. The archinephric duct runs along the free edge of the kidney (Figs. 7.7 and 8.27). In *Petromyzon* a vestige of the part of the archinephric duct, which was associated with the degenerated pronephros, extends forward from the opisthonephros (Fig. 8.2). The ducts from the two sides unite posteriorly to open into a *urogenital sinus* which leads to the outside through an aperture at the tip of a small *urogenital papilla.* Two slitlike openings, the *genital pores,* connect the urogenital sinus with the coelom (Fig. 7.7). Reproductive cells (eggs or spermatozoa) escape from the body cavity through the genital pores and leave through the urogenital aperture. It is only here that the reproductive and excretory systems of the lamprey are associated. The condition is similar in both sexes.

Fishes. The opisthonephros of fishes shows great variation in shape but there are fundamental similarities in structure. It is dorsal in position in all species. In some fishes, as in elasmobranchs, it consists of long, narrow, strap-shaped bodies, lying against the dorsal body wall above the parietal peritoneum, on either side of the median line. The dorsal aorta and postcardinal veins lie

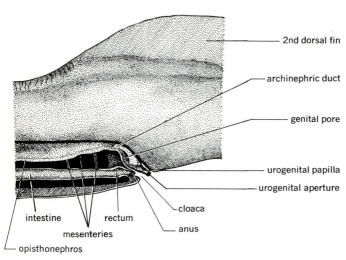

Fig. 7.7 Position and arrangement of left genital pore of lamprey. It is through the genital pores that the reproductive elements leave the coelom to enter the urogenital sinus.

Anatomy of the Chordates

between them. The kidneys extend almost the entire length of the body cavity. In other fishes they may be more voluminous, the two sides showing various degrees of fusion. In still others they are short and confined to the posterior part of the body cavity. Peritoneal funnels are retained in only a few forms, notably *Amia,* sturgeons, and certain elasmobranchs, such as *Squalus acanthias.* In the latter they lose their connections with the kidney tubules and end blindly. In certain marine teleosts no external or internal glomeruli are present, and their kidneys therefore are said to be of the *aglomerular type.* In the absence of glomeruli and Bowman's capsules, it is assumed that the proximal segments of the convoluted tubules play a more important role than otherwise in the excretion of waste materials.

Generally speaking, the opisthonephros of male fishes is longer than that of females, the anterior end in males having been appropriated by the reproductive system. In the males of some groups, small modified kidney tubules, now called *efferent ductules,* connect the testes with the archinephric duct. The archinephric duct then becomes the *ductus deferens,* serving primarily for sperm transport. It may, in addition, continue to carry wastes. However, there is a marked tendency in such cases for the posterior portion of the opisthonephros to assume the greater part of the excretory function, with separate ureterlike ducts developing to carry wastes directly to the cloaca or to the outside. Usually the connection of the testis and archinephric duct occurs at the anterior end of the opisthonephros (selachians, chondrosteans, and some others). In *Polypterus* the connection is confined to the posterior portion (Fig. 8.33*A*). In teleosts there is no connection between the testes and the opisthonephros, and the ducts from the testes may either join the archinephric ducts near their posterior ends or open independently to the exterior. Accessory ureterlike ducts may be present in some species in addition to the archinephric ducts. The term ductus deferens is properly used only in reference to an archinephric duct when it is used for sperm transport. Thus in teleosts and some other fishes, the sperm duct which transports spermatozoa to the outside, and which has no connection with the archinephros or archinephric duct (having an entirely different origin), should not be called a ductus deferens.

In female fishes the posterior ends of the archinephric ducts may enter a common cavity, the *urinary sinus,* inside a small *urinary papilla.* In Dipnoi and elasmobranchs this enters the cloaca, but in most other fishes it opens directly to the outside, a cloaca being absent. A similar union occurs in male elasmobranchs, but because of the association of excretory and reproductive systems in that sex, the terms *urogenital sinus* and *urogenital papilla* are used.

Dilatations of the archinephric ducts may form bladderlike enlargements for the temporary storage of urine, or, in forms in which the archinephric duct becomes a ductus deferens, seminal vesicles and sperm sacs may develop and serve for the temporary storage of spermatozoa. The anterior part of the archinephric duct may become highly convoluted, in which case it is sometimes called the *ductus epididymidis,* or simply the *epididymis.*

Amphibians. The primitive archinephric type of kidney, found in the larval stage of the hagfish, also occurs in larval caecilians in which there is a distinct metameric arrangement of kidney tubules, renal corpuscles, and nephrostomes. In an adult, the opisthonephros extends the greater part of the length of the coelom and is lobulated.

Although a small head kidney with peri-

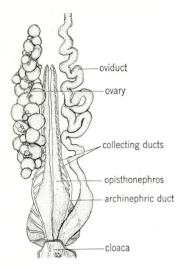

Fig. 7.8 Urogenital organs of female salamander, ventral view. The ovary on the left side and the oviduct on the right side have been removed for the sake of clarity.

up the main part or body of the opisthonephros. The archinephric duct courses along the lateral edge of the kidney a short distance from the kidney proper. Numerous collecting ducts or tubules leave the opisthonephros at intervals to join the archinephric duct. These are considerably shorter in females than in males (Figs. 7.8 and 7.9). In males, the collecting tubules, particularly those from the posterior portion, are much larger, longer, and more prominent than in females. They run a parallel course around the ventrolateral part of the kidney, converging before joining the archinephric duct proper at a point quite close to its entry into the cloaca. Although renal corpuscles may be present in both portions of the male opisthonephros, they are far more numerous in the posterior portion. The archinephric duct in the male serves primarily as a ductus defer-

toneal connections is present in many larval amphibians, it does not persist in the adult stage. It is of interest, though, that in adult frogs ciliated nephrostomes are found on the ventral surfaces of the kidneys. These are not, in most cases, connected with kidney tubules but have become secondarily associated with the renal veins. Coelomic fluid may enter the circulatory system directly by this means. In *Necturus,* peritoneal connections with some of the kidney tubules persist throughout life.

Urodele amphibians have an opisthonephros much like that of elasmobranchs. There are two regions: an anterior narrow portion which in males is referred to as the *epididymis* (C. L. Baker and W. W. Taylor, Jr., 1964, *J. Tenn. Acad. Sci.,* **24:**1) and which is primarily concerned with genital rather than urinary functions; and a posterior expanded region, which makes

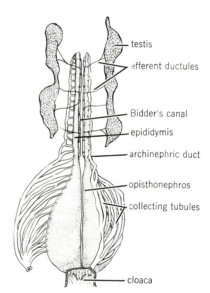

Fig. 7.9 Urogenital organs of male salamander, ventral view. The collecting ducts on the right side are shown detached from the cloaca and spread out for the sake of clarity.

Anatomy of the Chordates

ens but in the female is concerned only with transporting wastes. The archinephric ducts in both sexes open into the cloaca on either side through a small papilla.

The opisthonephros of anurans shows a more posterior concentration of tubules which are confined to the posterior part of the abdominal cavity. It is dorsally located, retroperitoneal, and flattened in a dorsoventral direction. There is no clear-cut distinction between anterior and posterior portions as in urodeles. An adrenal gland of a yellowish-orange color is located on the ventral side of the kidney to which it is closely attached (Fig. 7.10). The kidneys of female frogs and toads show no relation to the reproductive system, but in males an intimate connection exists. Certain anterior kidney tubules become modified as efferent ductules connecting the testis with the kidney and archinephric duct which serves, therefore, as a ductus deferens in addition

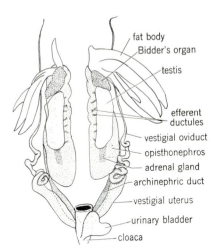

Fig. 7.10 Urogenital organs of male toad, *Bufo americanus*, ventral view. (*Modified from Witschi, in Allen, "Sex and Internal Secretions," 2d ed., The Williams and Wilkins Company. By permission.*)

Labels in figure:
fat body
Bidder's organ
testis
efferent ductules
vestigial oviduct
opisthonephros
adrenal gland
archinephric duct
vestigial uterus
urinary bladder
cloaca

to transporting urinary wastes. The archinephric duct, contrary to the condition in urodeles, is located *within* the kidney along its lateral margin. It leaves the opisthonephros near its posterior end and passes to the cloaca.

A thin-walled urinary bladder arises as a diverticulum from the ventral wall of the amphibian cloaca a short distance beyond the openings of the archinephric ducts. It is bilobed. There is no direct connection of the ducts with the bladder, so that the thin, watery urine first passes directly to the cloaca. Since the terminal cloacal opening is constricted most of the time, fluid is forced into the bladder, which may become distended to a considerable degree. The animal expels urine intermittently by opening the cloacal orifice and contracting the muscular wall of the bladder. It is possible that there may be some resorption of water through the bladder wall into the bloodstream. As stated previously, the amphibian bladder originates from the cloaca. It is extremely difficult to discern, in this region, the exact point of transition between the endoderm lining the cloaca and the ectoderm lining the proctodaeum. There is some difference of opinion, therefore, regarding the germ layer from which the urinary bladder of amphibians is derived. Most authorities are of the opinion, however, that it is of endodermal origin and homologous with the urinary bladders of higher forms.

The Amniote Kidney

In reptiles, birds, and mammals, three types of kidneys are usually recognized: pronephros, mesonephros, and metanephros. These appear in succession during embryonic development, but only one, the metanephros, persists to become the functional adult kidney. Mesonephros and metanephros actually represent different levels of the opisthonephros of the anamniota, the metanephros being the equivalent of the posterior portion. In all forms, an anteriorly located pronephros is present during very early stages of development, but it soon degenerates and the more posterior mesonephros then develops. The duct of the pronephros, however, persists to become the duct of the mesonephros. This is actually the same as the archinephric duct, to which considerable reference has been made. Since most accounts refer to this duct as the *mesonephric,* or *Wolffian, duct,* to avoid further confusion on the part of the student, we shall use the term Wolffian duct here in referring to the archinephric duct as it appears in amniotes.

The mesonephros persists for a time and then degenerates. In the meantime the metanephros has begun to develop from the region posterior to the mesonephros. Portions of the mesonephros may persist to contribute to the reproductive system in the male or to remain as mere vestigial structures without any apparent function.

The concept that the various kinds of vertebrate kidneys have evolved in a cranio-caudal sequence from an ancient archinephros gains some support from the fact that even in primitive forms the structure of the kidney shows a gradient of complexity from cranial to caudal ends. This may involve nothing more than better development of kidney tubules and glomeruli, but differences are evident, nevertheless.

Pronephros. The pronephros in amniotes forms in a manner similar to that of anamniotes. Segmentally arranged pronephric tubules appear in the intermediate cell mass in some of the anterior segments of the body. These are at first solid structures, but they soon hollow out, one end establishing a connection with the coelom. The

tubules appear in succession in a craniocaudal direction. The distal end of each tubule bends backward and fuses with the adjacent tubule. A variable number of such tubules forms in different species. In the chick, for example, 10 or 11 pronephric tubules form on each side from the fifth to the fifteenth or sixteenth segment. The last tubule grows caudally and establishes a connection with the cloaca. Thus, a long pronephric duct is formed, the anterior end of which is connected to a series of tubules with coelomic connections. The tubules soon disappear; in fact the anterior tubules may degenerate before the posterior ones even form. External glomeruli may or may not form. If they do, the sequence of their appearance may not even coincide with the time when the tubules are forming. In mammals, pronephric tubules appear only as the merest of vestiges. Hence, the pronephric duct can scarcely be said to be formed as the result of fusion of tubules. Nevertheless, it appears in the nephrotome region, first as a solid cord which grows back to the cloaca, hollowing out to become a typical pronephric duct. Thus the amniote pronephros is little more than a transient structure which undoubtedly has no value at all as an excretory organ. To the comparative anatomist it is of interest because its presence indicates that the amniote kidney, like that of the anamniotes, has evolved from the ancestral archinephric, or holonephric, type. When the duct of the pronephros becomes the duct of the mesonephros it is called the Wolffian duct.

Mesonephros. The mesonephros is sometimes called the *Wolffian body.* It extends over a greater number of segments than does the pronephros and develops after the pronephros has formed. Mesonephric tubules appear in the intermediate cell mass

Anatomy of the Chordates

in the region where the pronephric duct is located, but posterior to the pronephric tubules. In some forms the anterior mesonephric tubules appear in the same segments as the posterior pronephric tubules, so that there is a slight degree of overlapping. The mesonephric tubules differentiate in the same manner as the opisthonephric tubules described earlier (page 256). Some of the anterior tubules may form peritoneal connections, but this is unusual. Contrary to the condition in the pronephros, several tubules may form in a single segment. In the human embryo from 2 to 9 mesonephric tubules may develop in this manner, making a total of between 30 and 40 for each mesonephros. In other forms the number may be even greater. As a result, the mesonephros, at least in some forms, as in the pig, becomes rather voluminous, (Fig. 7.11) occupying an extensive portion of the body cavity.

In reptiles, birds, and mammals the mesonephros exists only temporarily and is followed in development by the metanephros,

which becomes the functional adult kidney. In the embryonic stages of the mouse, rat, and guinea pig, the mesonephros is so poorly developed and appears so early that it can scarcely be regarded as functional at any time. Certain embryonic membranes in these animals may possibly help to rid the body of excretory products in the interval before the metanephros develops. On the other hand, in reptiles and in such mammals as echidnas and certain marsupials, the mesonephros may persist for a time after birth. Well-developed peritoneal funnels, associated with the mesonephros, have been described for the monotremes.

As long as the mesonephros functions, the Wolffian duct is the urinary passage; but as soon as the metanephros becomes functional this duct degenerates in the female although it persists in the male. The remnants of the mesonephros, which are left after the main part has degenerated, include portions which become associated with the reproductive system. The male structures derived from mesonephric components include the *epididymis, ductus deferens, seminal vesicles, paradidymis,* and *ductus aberrans.* In the female such rudimentary structures as the *paraoöphoron, epoöphoron,* and *canal of Gärtner* represent remains of the mesonephros (page 303).

Metanephros. The metanephros, found only in amniotes, arises posterior to the mesonephros on each side and is more compact than the latter organ. It comes from a level which corresponds to the most posterior portion of the opisthonephros of the anamniota. The metanephros is made up of essentially the same parts as the mesonephros and consists of renal corpuscles, secretory tubules, and collecting tubules. No nephrostomes are present. Each kidney has a twofold origin. A diverticulum, the *ureteric bud,* or *metanephric diverticulum,* from

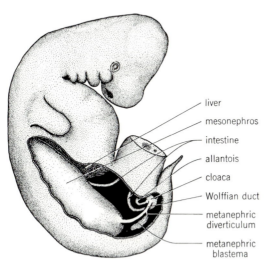

liver
mesonephros
intestine
allantois
cloaca
Wolffian duct
metanephric diverticulum
metanephric blastema

Fig. 7.11 Pig embryo of about 10 mm dissected to show position and relative size of mesonephros.

the Wolffian duct near its posterior end grows forward into the so-called *metanephrogenous tissue,* or *metanephric blastema,* which lies posterior to the mesonephros. The metanephric blastema is continuous with the nephrogenous tissue which gave rise to the mesonephric tubules. The diverticulum of the Wolffian duct, which is destined to form the *ureter,* branches and rebranches a varying number of times and ultimately forms large numbers of very fine *collecting tubules.* An expansion at the point where the ureter undergoes its primary divisions, at least in mammals, becomes the pelvis of the kidney (Fig. 7.12). Condensations in the mesenchyme of the adjacent metanephric blastema soon give rise to hollow *secretory tubules* which grow longer and become S-shaped. One end of each secretory tubule establishes a connection with a collecting tubule. The other end expands and is soon invaded by a glomerular tuft from a branch of the renal artery so that a typical renal corpuscle is formed. All blood going to the metanephros is arterial, since the renal portal system of veins does not exist in these forms. As devel-

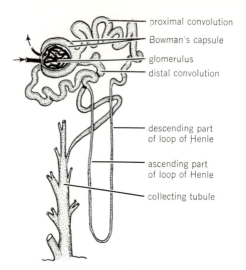

Fig. 7.13 A mammalian metanephric tubule, showing the renal corpuscle as well as secretory and collecting portions.

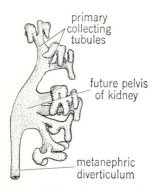

Fig. 7.12 Manner of branching of diverticulum of Wolffian duct during early development of metanephrios. (*After Huber.*)

opment progresses, there is a great increase in the number of secretory tubules.

The differentiation of a secretory tubule of the metanephros in mammals and, to a lesser extent, in birds is somewhat more elaborate than in the case of the mesonephros (Fig. 7.13). Each tubule, as it leaves Bowman's capsule, consists of a *proximal convoluted tubule,* a long *loop of Henle* with its *descending* and *ascending portions,* and a *distal convoluted tubule.* Secretory tubules with their Bowman's capsules are very numerous. It has been estimated that in the average human being there are between 1.3 and 4 million renal corpuscles and nephrons, or secretory tubules, in each kidney. A single nephron is approximately 50 to 55 mm long, their combined length being about 75 miles. Approximately one-fifth of the entire amount of blood in the body is said to circulate through the kidneys each minute so that there is a continuous elimination of waste materials from the blood. The metanephros functions, in general, in a manner similar to that of the opisthonephros (page 257). Wastes are carried from the kidneys by the ureters, which enter the cloaca or urinary bladder, as the case may be. Reabsorption of water in reptiles and birds also occurs in the cloaca, into which the ureters open. This results in a partial solidification of urine as well as of feces.

Comparative Anatomy of Metanephros. *Reptiles.* The kidneys of reptiles are restricted to the posterior half of the abdominal cavity and are usually confined to the pelvic region. They are generally small and compact, but the surface is lobulated (Fig. 7.14). The posterior portion narrows down on each side, and in some lizards the hind parts may even fuse. The degree of symmetry may vary and is most divergent in snakes and limbless lizards, which have excessively lobulated, long, narrow kidneys in correlation with the shape of the body. One kidney may lie entirely posterior to the other. The ureters in these reptiles are extremely

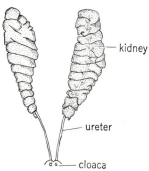

Fig. 7.14 Lobulated meta-nephros of lizard, *Monitor indicus.* (*After Wiedersheim, "Comparative Anatomy of Vertebrates," copyright* 1907 *by The Macmillan Company and used with their permission.*)

long. Turtles and crocodilians have very short ureters.

A urinary bladder is lacking in snakes, crocodiles, and alligators. Most lizards and turtles, however, have well-developed and usually bilobed bladders which open into the cloaca. The ureters from the metanephros open separately into the cloaca and, except in turtles, are not connected with the bladder. In reptilians the bladder is derived partly from the cloaca and partly from the base of the *allantois,* an endodermal embryonic structure important in fetal respiration. In some turtles a pair of accessory urinary bladders is also connected to the cloaca. They have been shown to function as accessory organs of respiration. In females they may be filled with water, which is used to soften the ground when a nest is being prepared.

Birds. In all birds the kidneys are situated in the pelvic region of the body cavity, and the two frequently unite at their posterior ends. They are lobed structures, deep fissures between the lobes serving for the passage of branches of the renal veins. The ureters of birds are short and open independently into the cloaca.

Except for the ostrich, birds have no urinary bladders. Urinary wastes, chiefly in the form of uric acid, are eliminated in a semisolid form along with the feces. The lack of a urinary bladder facilitates flight in birds since unnecessary ballast in the form of liquid urine need not be carried about.

Mammals. The typical mammalian metanephros (Fig. 7.15) is a compact, bean-shaped organ attached to the dorsal body wall. It is retroperitoneal. The ureter leaves the medial side at a depression called the *hilum,* or *hilus.* At this point a *renal vein* also leaves the kidney and a *renal artery* and nerves enter it. The metanephros is surrounded by a *capsule* of connective tissue

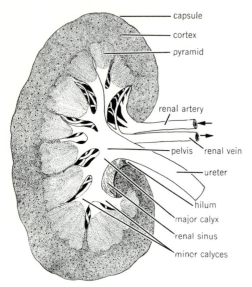

capsule
cortex
pyramid
renal artery
pelvis
renal vein
ureter
hilum
major calyx
renal sinus
minor calyces

Fig. 7.15 Sagittal section of metanephros of man (semidiagrammatic).

with which the distal convoluted portions of numerous secretory tubules connect and into which they drain. The connection of the secretory tubule with the branched collecting tubule is often referred to as the *arched collecting tubule.*

The renal pyramids themselves present a striated appearance. Here are located the straight collecting tubules and the loops of Henle. The inner border, or tip, of each pyramid, in the form of a blunt papilla, points toward the pelvis and actually projects into a branch of the pelvis known as a *minor calyx.* Several minor calyces join together to form a *major calyx.* The major calyces in turn open directly into the pelvis, of which they are actually primary branches (Fig. 7.15). About 16 to 20 straight collecting tubules, or papillary ducts, enter each minor calyx.

In the center of the kidney is the *sinus* of the metanephros. It contains blood vessels, nerves, and fatty tissue, in addition to the pelvis and calyces.

The glomerular filtrate, or tissue fluid, which has been removed from the blood in the renal corpuscles, first passes into the proximal convoluted tubules. The electron microscope has revealed that the surfaces of the cells of the proximal convoluted tubule facing the lumen are covered with exceedingly numerous *microvilli,* each about 1 micron (μ) in length. They appear to be rather similar to those lining the small intestine (Fig. 3.16). A matrix of some sort apparently fills the spaces between them. The first portion of the descending loop of Henle presents a similar appearance and really seems to be an extension of the proximal convoluted tubule, even if it is straight. It has been estimated that the total surface of the microvilli in all the proximal convoluted tubules in the kidneys of man represents an area of 50 to 60 sq m. It is highly probable

under which lies the *cortex.* The renal corpuscles and convoluted portions of the secretory tubules are confined to the cortical region. Immediately beneath the cortex is the *medulla,* with which it is intimately connected. It is made up of large areas known as *renal pyramids,* between which extend the *renal columns of Bertin* made up of cortical tissue. Each pyramid of medullary tissue is covered with a cap of cortical substance, the two constituting a *lobe* of the kidney. The lobes are further subdivided into *lobules,* which are most evident at the outer borders of the kidney. The kidneys of such mammals as rabbits, rats, and others are called *unilobar kidneys* since only a single pyramid with its cortical cap is present. The multilobar kidneys of man are composed of from 6 to 18 lobes, each actually equivalent to a unilobar kidney. In the cortex are numerous *medullary rays,* each making up the core of a lobule. In the center of a medullary ray is a *branched collecting tubule* that as much as two-thirds of the water of the glomerular filtrate is resorbed here. In addition, these cells exercise some selective resorption of other constituents of the glomerular filtrate. Practically all the sugar as well as a good portion of the sodium chloride, phosphates, and other materials seem to be resorbed through the cells of the proximal convoluted tubules. These cells are also believed to perform an active excretory role. Nevertheless, the resorptive function of the proximal convoluted tubules is by far the more important of the two.

The loops of Henle and the cells of which they are composed are so thin that it is unlikely that selective resorption and excretion of waste substances occur in this part of the kidney tubule. It has been claimed that only those animals in which the nephron contains a loop of Henle can eliminate urine which is hypertonic to the blood. This is of great importance to terrestrial animals in which water conservation is of primary advantage.

As the ascending portion of the loop of Henle approaches the distal convoluted tubule, its wall thickens. It then makes a direct contact with the renal corpuscle. At this point there is a small cluster of heavily nucleated cells known as the *macula densa*, the significance of which is not understood. The distal convoluted tubule with its thickened wall then pursues a mildly tortuous course and finally joins an arched collecting tubule which in turn empties into a branched collecting tubule. Unlike the cells of the proximal convoluted tubule, those of the distal convoluted segment of the nephron have only occasional microvillous projections. Here and there an isolated cilium may be present. Estimates indicate that the rest of the sodium chloride, other chlorides, and approximately 14 per cent of the water of the glomerular filtrate are resorbed in the distal convoluted tubule. There is some evidence

that it is here that *vasopressin,* the antidiuretic hormone of the posterior lobe of the pituitary gland (page 384), exerts its effect.

The function of the collecting tubules is primarily to conduct the waste materials eliminated by the nephrons to the pelvis of the kidney. It is possible, however, that a small amount of water may be resorbed through the walls of the collecting tubules. The largest portions of the pyramidal collecting tubules, which open into the minor calyces through the papillae of the pyramids, are called the *papillary ducts,* or *ducts of Bellini.*

It has been calculated that in human kidneys, within a 24-hour period, between 150 and 200 qt of fluid are filtered through the glomeruli into the secretory tubules. Approximately 99 per cent of the glomerular filtrate is resorbed through the walls of the secretory tubules and returned to the bloodstream. The concentrated remainder is *urine,* which passes down the collecting tubules in the renal pyramids into the minor calyces, then into the major calyces and pelvis. The wall of the kidney pelvis contains smooth-muscle fibers. It is possible that their contraction may play a part in helping to evacuate urine from the ducts of Bellini by a suctionlike milking action. From the pelvis, urine passes to the ureter and thence to the bladder (cloaca in monotremes). After being stored temporarily in the bladder, it passes through the urethra to the outside.

Recent studies, using the electron microscope, have brought to light some new and interesting facts concerning glomerular microstructure. The wall of the afferent arteriole leading to the glomerulus is thicker than that of the efferent arteriole even though their lumina are of practically the same size. The thicker wall is due to the presence of peculiar muscle cells of the tunica media (page 538), often referred to as *muscular media cells.* These actually resemble epithelial cells more than smooth-muscle cells and are sometimes called *juxtaglomerular cells.* They are in close contact with the cells of the macula densa, mentioned above, where the basement membrane of the tubular epithelium is lacking. It has been known for many years that in certain diseases of the kidney there is an elevation of blood pressure. This is called *hypertension.* Hypertension can be induced experimentally by injection of kidney extracts or by restricting the blood supply to the kidneys by means of clamps, etc. Under such circumstances an enzyme, *renin* (pronounced ree'-nin), passes from the kidney into the bloodstream. This acts upon a protein known as *hypertensinogen,* or *reninsubstrate,* derived from the liver, to form a substance called *hypertensin* or *angiotensin I* (*angiotonin*). Another enzyme acts upon the latter to produce *angiotensin II,* which has a profound effect in elevating blood pressure by constricting arterioles as well as causing an increase in heart rate and in the contractile force of the heart. These effects seem to be mediated via ganglia of the sympathetic nervous system. In the dog angiotensin II acts upon the inferior, or caudal, cervical ganglion in bringing about functional changes in the heart; in the cat it seems that the superior cervical ganglion mediates the changes. Evidence points to the juxtaglomerular cells as the site of origin of renin. Renin extracted from the kidneys of hogs exists in at least four different forms rather than in only one, as had been previously believed. It is ultimately inactivated by the liver. Angiotensin I is a peptide chain of 10 amino acids; angiotensin II is identical except that the two terminal amino acids, histidine and leucine, are lacking. The kidney is also suspected of producing another factor referred to as *erythropoietin,* which stimulates the production of erythrocytes, or red blood corpuscles. Little is known about its chemical nature.

The structure of the renal corpuscle itself is much more complicated than was formerly believed. The visceral layer of Bowman's capsule is folded and interdigitated with the glomerular capillaries to such a degree of complexity that its basement membrane is everywhere in contact with glomerular capillaries. The capillary cell walls are extremely thin except where the nuclei of the endothelial cells are located. Furthermore, and unlike capillaries in other parts of the body, the cells are fenestrated, containing numerous tiny pores which have a diameter of approximately 0.0001 mm. The cells of the visceral layer of Bowman's capsule are not typically squamous cells but are modified in a peculiar way to form *podocytes.* The main part of each podocytic cell is separated from its basement membrane by a *subpodocytic space,* but numerous tiny *feet* extend to the basement membrane (Fig. 7.16). The basement membrane, therefore, is all that separates the blood in the fenestrated capillaries from the podocytic spaces. It is this which serves as the dialyzing membrane through which the glomerular filtrate passes. The capillaries surrounding the proximal convoluted tubules are also fenestrated in much the same manner as those of the glomeruli.

The metanephros of many mammals with multilobar kidneys, in accord with its manner of development, shows marked lobation in the embryo, and in many forms this condition is retained throughout life (Proboscidea, Pinnipedia, Cetacea, Artiodactyla, some Carnivora, and some Primates). In other mammals the lobation is not superficially apparent in the adult and the kidney surface is relatively smooth. Occasionally the external lobation persists in man, but this is rather rare. Such anomalies as the presence of an extra kidney, fusion of the kidneys at their lower ends (horseshoe kidney), double ureters, and double pelves may

Anatomy of the Chordates

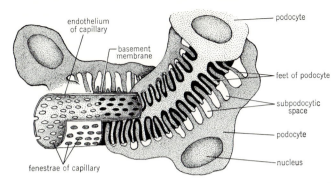

endothelium
of capillary

basement
membrane

podocyte

feet of podocyte

subpodocytic
space

podocyte

nucleus

fenestrae of capillary

Fig. 7.16 Diagrammatic representation showing relation of fenestrated glomerular capillary to podocyte and podocytic space.

be accounted for by abnormal embryonic development. It is interesting that the kidneys of a newborn child are approximately three times as large in proportion to body weight as they are in an adult. About one-third of the renal corpuscles of an adult are present at birth. New ones continue to be formed during the first two or three months. The increase in size of the kidney is due primarily to enlargement of tubules already present at birth. If one kidney is extirpated, the other undergoes a compensatory hypertrophy. This is the result of an increase in the size of the kidney tubules, not to an increase in their number.

Urinary bladders, which are present in all mammals, are muscular sacs which, like those of reptiles, are derived from a part of the ventral cloacal wall and possibly from a portion of the allantoic stalk. The bladder narrows down and opens to the outside through a urethra. The lower ends of the ureters, except in the monotremes, open directly into the bladder on its dorsal posterior surface and usually near the urethral end. In the monotremes the ureters open into the urethra through small papillae opposite the base of the bladder.

It was formerly believed that a sphincter

muscle at the junction of bladder and urethra regulated the passage of urine from the bladder and that this was contracted except during the act of urination. It has been found (R. T. Woodburne, 1961, *Anat. Record,* **139**:287) that no internal sphincter actually exists at the neck of the bladder. Here there is an accumulation of elastic tissue. A good portion of the bladder musculature, in the form of bundles of muscle fibers, or fascicles, continues down into the urethra. Urine flow ceases when the urethra is lengthened and the diameter of its lumen is reduced. It is really the entire urethra of the female and the prostatic portion of the urethra of the male that controls micturition. In males the urethra passes through the penis and opens at the tip of that organ through the *external urethral orifice,* or *meatus.* In females the condition varies; in some, as in the rat and mouse, the urethra opens independently to the outside, passing through the clitoris; in others it enters the urogenital sinus, or vestibule, which is the terminal part of the genital tract.

Even in monotremes, which, with the exception of the pika, are the only mammals having a cloaca, a urogenital portion is partially separated by means of a fold from the region into which the intestine opens. In Eutheria, the embryonic cloaca becomes separated into two distinct parts, each of which has an independent opening to the outside. The dorsal portion becomes the rectum with its anal opening; the ventral part becomes the urogenital sinus which receives the ducts of the reproductive system as well as the urethra. In certain marsupials the division of the cloaca is not complete and a slightly depressed area receives the urogenital ducts and rectum. In females of certain species (rats, mice, and most primates) further separation of the urogenital sinus brings about a condition in which urethral and reproductive openings are independent.

EFFECT OF ENVIRONMENT ON KIDNEY STRUCTURE AND FUNCTION

The importance of the kidney is not confined to its role in eliminating wastes. It is of great significance in maintaining a fluid environment in the body which provides the proper conditions for keeping the body cells in satisfactory condition. This is accomplished (1) by regulating the amount of water given off by the body and maintaining a normal fluid balance; and (2) by controlling the amounts of NaCl and other electrolytes eliminated through the urine and thus keeping their concentration in the blood within normal limits.

Some interesting speculations have been advanced in attempting to correlate various details in the evolution of kidney structure and function with environmental factors. The speculations are based upon the assumption that vertebrates evolved from ancestors inhabiting continental fresh waters. There is some disagreement concerning this theory; nevertheless, the concept is supported by considerable paleontological evidence.

The fresh-water ancestor was undoubtedly an organism with a relatively high concentration of salts in the blood. The osmotic pressure of the body fluids would thus be considerably higher than that of the water in which the animal lived. Under such conditions water would be constantly entering the body because of osmotic inflow. The necessity of ridding the body of this water was essential in maintaining a constant concentration of salts and other constituents of the blood. The glomerulus is believed to have arisen in evolution as a mechanism for ridding the body of excess water.

Those fishes which took to life in the sea had problems to contend with associated with the relatively high salt concentration and osmotic pressure of sea water. Osmotic extraction would tend to deplete the body of its water. But water must be retained so that the kidneys can form dilute urine for the elimination of wastes. One way in which this can be accomplished is by taking sea water into the digestive tract and absorbing most of the water and salts. The salts may be excreted through the gills and rectal gland, leaving the water free for the formation of urine. It has been discovered (J. W. Burger and W. N. Hess, 1960, *Science,* 131:670) that the rectal gland (page 187) of the elasmobranch fish, *Squalus acanthias,* gives off a secretion which is virtually a solution of sodium chloride. The salt concentration is even greater than that of sea water and about twice that of blood plasma. Another method of absorbing water is to be found in the elasmobranchs and coelacanths, in which, because of a high concentration of urea in the blood under normal conditions, the osmotic pressure is higher than that of the saltwater environment. Water can then be absorbed directly through gills and other membranes; it need not go through the digestive route. The permeability of the gills and membranes of the mouth to urea is

Anatomy of the Chordates

diminished in these fishes, a factor of importance in retaining a high concentration of urea in the blood.

In certain marine teleosts, there is no need to remove water from the body, since the osmotic pressure of the blood is lower than that of sea water. Nitrogenous wastes, as well as others, have been shown to be eliminated via the gills. In these marine teleosts, glomeruli appear during developmental stages but they later degenerate, the adults having aglomerular kidneys. Such kidneys are of little functional importance, and only a very small volume of urine is excreted in these fishes. The elasmobranch fishes, on the other hand, have well-developed kidneys with numerous glomeruli. In this group the osmotic pressure of the blood is high and, as in freshwater fishes, water must be constantly removed by glomerular filtration.

Since the filtered fluid, like the blood, has a high concentration of chemicals which are essential for life, it is important that they be retained by the body. The kidney tubules with their ability to reabsorb these chemicals selectively have undoubtedly arisen in evolution as a means of effecting such a conservation.

Kidney tubules show much variation in vertebrates, although all possess a proximal convoluted tubule. In the aglomerular kidneys of the marine teleosts, mentioned above, the proximal convoluted tubule seems to be the only portion that is functional, yet the urine of these fishes does not differ basically from urine of fishes possessing glomeruli. The proximal convoluted tubule in such cases is undoubtedly excretory. Its excretory activity in glomerular kidneys has been clearly demonstrated under experimental conditions. In all vertebrates except birds and mammals, a ciliated neck connects Bowman's capsule with the proximal convoluted tubule. The ciliary current set up may facilitate glomerular filtration.

Elasmobranch kidney tubules differ from those of other vertebrates in possessing two portions, or segments, of peculiar structure. They are often referred to as the *elasmobranch segments*. The first is situated between Bowman's capsule and the proximal convoluted tubule; the second follows the proximal convoluted tubule. It has been known for a long time that elasmobranch fishes differ from all other vertebrates in the large percentage of urea present in tissues, blood, and other body fluids. The urine is particularly low in its concentration of urea, showing that most of the urea filtered through the glomeruli is absorbed. The unique segments in the elasmobranch kidney tubule seem to be specially adapted for selective reabsorption of urea. Not all species of elasmobranchs have the typical elasmobranch segments. They are notably absent in the lesser electric ray and the bat ray. In these species, however, the proximal convoluted tubules show a differentiation into two regions which are distinguished by the height of their cells and the thickness of their walls. These possibly serve a function similar to that of the elasmobranch segments (R. T. Kemptom, 1957, *Anat. Record,* **128**:575).

The activity of the distal convoluted tubule remains somewhat obscure. It seems that among amphibians, at any rate, a degree of selective reabsorption also takes place here, particularly of chloride and bicarbonate.

SUMMARY

1. The excretory organs of vertebrates consist of paired kidneys together with their ducts. Several types of kidney are recognized within the group.

2. The various kinds of kidneys have probably all been derived from a primitive structure referred to as the archinephros, or holonephros. This consisted of a pair of archinephric ducts which extended the length of the coelom and which were joined by segmentally arranged tubules, one pair to each segment. The free end of each tubule opened into the coelom by means of a ciliated, funnel-shaped nephrostome. Associated with each tubule was a small knot of interarterial capillaries known as a glomerulus. The larval stages of the hagfish and caecilians have an archinephros. The various types of vertebrate kidneys may be regarded as successive stages which have evolved in a cranio-caudal direction from the original archinephros.

3. In adult anamniotes (cyclostomes, fishes, and amphibians) the anterior portion of the primitive archinephros usually becomes modified or degenerates. In the embryo, however, it appears as a transitory structure called the pronephros. In a few lower vertebrates the pronephros persists in the adult stage and is called the head kidney. A head kidney is found in the adult hagfish and in certain teleosts.

4. The remainder of the anamniote kidney is the opisthonephros. It retains the original archinephric duct but differs from the pronephros in that several kidney tubules may be present in each segment and the tubules usually lose their peritoneal connections.

5. There is a general tendency toward a concentration of kidney tubules toward the posterior end of the opisthonephros. The anterior end loses its significance as an excretory organ and, in the male, is appropriated by the reproductive system, the archinephric duct becoming the ductus deferens. Accessory urinary ducts may then form by a fusion of two or more kidney tubules and serve to drain the urinary wastes.

6. In amniotes (reptiles, birds, and mammals) three types of kidneys are recognized: pronephros, mesonephros, and metanephros. These appear in succession in a cranio-caudal direction during embryonic development, but only the metanephros persists to become the adult kidney. Mesonephros and metanephros represent different levels of the opisthonephros of the anamniota, the metanephros being the equivalent of the posterior portion. The archinephric duct is now called the Wolffian duct.

7. The pronephros of amniotes, with its metamerically arranged tubules with peritoneal connections, is merely a transitory structure of no functional importance. Its duct, however, persists as the Wolffian duct.

8. The mesonephros appears only in the embryos of amniotes. Several kidney tubules are present in each segment and most, or all, peritoneal connections are lost. This kidney may function for a time but then degenerates and disappears. Vestiges of the mesonephros and Wolffian duct persist. In males the latter gives rise to the epididymis, ductus deferens, and certain other parts of the reproductive system.

9. The metanephros has no peritoneal connections. Its origin is twofold. The collecting portion, including the ureter, is derived from a diverticulum of the Wolffian

duct; the secretory portion arises from nephrogenous tissue, the metanephric blastema, which lies posterior to the mesonephros.

10. The important factors in kidney function involve glomerular filtration at the renal corpuscle, selective reabsorption by the secretory tubules, and excretion of wastes by cells of the secretory tubules.

11. Ducts from the kidneys lead to the cloaca in most forms. In teleost fishes they open directly to the outside. In mammals they enter the urinary bladder (monotremes are exceptions). Urinary bladders, when present, represent ventral outpocketings of the wall of the cloaca. In mammals, the bladder opens to the outside through a urethra, which in the males of all forms except monotremes is also utilized by the reproductive system.

12. The important factors in the evolution of the basic structure and function of the vertebrate kidney appear to have been associated with body-fluid regulation, involving the maintenance of a constant water and salt content of the body, regardless of the type of environment.

| 8 |

REPRODUCTIVE
SYSTEM

The two cardinal concerns of living things are the maintenance of self and the perpetuation of the species. All activities of plants and animals are, in the last analysis, related to one or the other of these two fundamental matters. Reproduction is essential only so that the species may continue to live upon the earth. Two methods of reproduction, *asexual* and *sexual,* are recognized.

Asexual reproduction. Asexual, or agamic, reproduction does not involve sex. In the animal kingdom it occurs only in members of certain lower phyla. Some animals reproduce asexually by a process called *fission,* in which the body is divided into two approximately equal parts, each part being capable of independent existence. Fission is a method of reproduction in such invertebrate animals as *Hydra* and *Planaria.* Budding, which may also be observed in *Hydra* and other coelenterates, is another form of asexual reproduction. Asexual reproduction is not known to occur in the chordates.

Sexual reproduction. Sexual, or gamic, reproduction is the most common method of propagation and takes place in animals which have the power of reproducing asexually as well. In all cases it involves development from an egg which may or may not be fertilized. In the rather rare eventuality of an egg developing without fertilization, the term *parthenogenesis* is used. This is usually considered to be a type of sexual reproduction because a sexual element, the egg, is involved. The word *amphigony* is used to denote the common type of sexual reproduction in which an egg cell must be fertilized by a sperm cell before development can occur. The fertilized egg is known as a *zygote* which, under proper conditions and by a series of complicated processes, develops into an *embryo*.

Reproductive organs. Egg cells, or *ova,* and sperm cells, or *spermatozoa,* are formed in the *primary reproductive organs,* which are spoken of collectively as the *gonads.* The gonads in the male are known as the *testes* and in the female as the *ovaries.* Besides forming *gametes,* or reproductive cells, both ovaries and testes give off endocrine secretions, or *hormones,* which pass into the blood or lymphatic streams and are carried to all parts of the body, where they bring about profound effects. A discussion of the endocrine activity of the gonads is included in Chap. 9.

The gonads are paired structures, although in such forms as cyclostomes (Fig. 8.2), certain fishes, and female birds of most species, what seems to be an unpaired gonad is the result either of fusion of paired structures or else of unilateral degeneration. It was formerly believed that the gonads were derived from tissues which showed evidence of segmentation, or metamerism. This concept has been discarded, and except for such

a primitive form as amphioxus in which the gonads are distinctly segmented, it is clear that no metamerism exists in the gonads of chordates.

The gonads are mesodermal derivatives. In both sexes they arise as paired thickenings of the coelomic epithelium along the dorsal aspect of the body cavity on either side of the dorsal mesentery of the gut. The underlying mesenchyme condenses and proliferates, and soon a pair of *genital ridges* can be observed coursing longitudinally along the medial sides of the opisthonephros or mesonephros close to the base of the dorsal mesentery (Fig. 8.1). It is remarkable that development of the gonads in the earliest stages is similiar in both sexes and that sexual differences cannot at first be recognized even under the microscope. The ovaries and testes eventually come to be attached to the dorsal body wall by mesenterylike bands of tissue, the *mesorchium* in the male and the *mesovarium* in the female. These bear the same

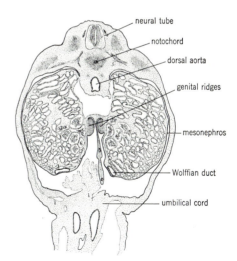

neural tube
notochord
dorsal aorta
genital ridges
mesonephros
Wolffian duct
umbilical cord

Fig. 8.1 Cross section through 12-mm pig embryo, showing relation of genital ridges to mesonephros.

relation to the gonads as does the dorsal mesentery to the gut.

The exact site of origin of *primordial germ cells,* or the cells from which eggs and sperm are basically derived, has long been a point of contention. Today, however, it is generally agreed that at least in amphibians, birds, and mammals the primordial germ cells first appear in close association with the yolk-sac endoderm or in the yolk-sac splanchnopleure. In a certain teleost fish some cells which can be identified in the fifth cleavage stage are known to be the forerunners of the future germ cells. It has also been recognized in the frog's egg, even before cleavage, that a particular area in the cytoplasm near the vegetal pole is destined to give rise to germ cells. In some species the posterior end of the primitive streak may possibly be involved. At any rate, the cells migrate from their point of origin, apparently by a sort of ameboid, or diapedesis-like, movement to the genital ridges. A histochemical test has been devised which distinguishes these cells from all others in the vicinity, thus making it possible to identify the primordial germ cells during their migration. Their presence is necessary for the proper development of the gonad in either sex. Those cells which fail to reach the genital ridges degenerate.

The reproductive elements formed in the gonads must be transported to the outside of the body either as ova or spermatozoa or as young which have developed from fertilized eggs within the body of the female. For this purpose, in most vertebrates, ducts are utilized, those of the male being known as the *deferent ducts* and those of the female as oviducts, or *Müllerian ducts.* In a few forms, such as cyclostomes, no ducts are present in either sex, and eggs and sperm escape from the body cavity through *genital* or *abdominal pores* (Fig. 7.7). The deferent ducts of the male are usually the archinephric, or the

Wolffian, ducts which in some cases also serve to carry urinary wastes from the opisthonephros or mesonephros in those animals in which these kidneys function either during embryonic or adult life. In reptiles, birds, and mammals in which the metanephros is the functional adult kidney and the mesonephros degenerates, the Wolffian duct on each side persists as the ductus deferens.

When the reproductive ducts first develop, in most cases they open posteriorly into the cloaca in both sexes. This relationship persists throughout life in many vertebrates, but in some, modifications in the cloacal region occur and the reproductive ducts either open separately to the outside, or else, in the male, join with the excretory ducts to emerge by a common orifice.

In many aquatic vertebrates, fertilization is external. However, in all terrestrial forms except anuran amphibians, and even in many aquatic species, internal fertilization is the rule. The passage of sperm from male to female is brought about in some by apposition of the cloacae of the two sexes. More commonly, however, copulatory organs are present in the male, which deposit spermatozoa in the reproductive tract of the female. Various types of copulatory organs are to be found among vertebrates.

In both sexes, all structures or organs which serve to bring the germ cells, or products of the primary sex organs, together are known as the *accessory sex organs.* They include the reproductive ducts, associated glands, and intromittent organs. *Secondary sex characters,* which are not directly concerned with sex, also play a part in the reproductive scheme. Sexual differences in such secondary sex characters as ornamental plumage, body size and strength, and vocal apparatus are indirectly related to reproduction. The development and maintenance of the secondary sex characters and accessory sex organs are controlled, at least in part, by the endocrine secretions (hormones) of the gonads.

The functioning of the gonads of numerous seasonal breeding birds and mammals is affected by the length of exposure to light. In some, as in the junco (page 366) and ferret, increase in the hours of daylight has the effect of stimulating the reproductive organs to functional activity; in others, as in sheep and goats, which normally breed in the autumn, it seems that the decrease in hours of daylight has a stimulating effect upon the gonads.

The sex of an individual is, in the last analysis, dependent upon the chromosomes received from both parents at the time that an egg is fertilized. The kinds of gonads and types of reproductive systems that form during embryonic development are basically determined by the chromosome makeup. That the balance between maleness and femaleness is a delicate one is reflected in the fact that environmental factors may assume an influential role in sexual development, and hormonal secretions may modify the extent to which various structures and even behavioral characteristics develop and are maintained. The distinctions between the sexes are usually not so firmly fixed in lower vertebrates as in higher groups. In lower vertebrates it is common to observe intersexes, hermaphroditic individuals, or even sex reversals.

The chromosomes present in the nuclei of the germ cells of an individual are of two general kinds: *autosomes* and *sex chromosomes*. In the human female, for example, each ovum or egg cell contains 22 autosomes and a single sex, or X, chromosome. The sperm cells of men contain 22 autosomes and a sex chromosome, but this may be either of the X or Y type since there are two types of sperm cells which differ from one another in

this respect. When an egg is fertilized, therefore, it will have 46 chromosomes: 44 autosomes and 2 X chromosomes, or 44 autosomes and an X and a Y chromosome. The former will develop into a female; the latter into a male. The situation is similar in other species although the number of autosomes differs in various forms.

One would expect that the cells of individuals which develop parthenogenetically would have only the haploid number, or half the number, of chromosomes characteristic of the species. This is frequently the case. Often, however, the double, or diploid, number is restored, this being accomplished by fusion of nuclei of a haploid cleavage or by absorption of a polar body with its haploid number of chromosomes.

The sex of an individual can be detected by studying the chromosome makeup of even highly specialized somatic cells. This was first discovered in a study of nerve cells of the cat (M. L. Barr and E. G. Bertram, 1949, *Nature*, **163**:676), and has since been described in many other kinds of body tissues of several other species of mammals. If the nuclei of somatic cells are examined under the microscope during the interphase, a small mass of chromatin material, lying adjacent or very close to the nuclear membrane, can be detected in female cells. This is not evident, or else is very inconspicuous, in cells of the male. The chromatin material of the female probably represents the two X chromosomes clumped together. Why the X chromosome of the male, paired with its Y mate, does not manifest itself distinctly in stained preparations is not known with certainty. The technique of identifying sex in cells from the skin, lining of the mouth, blood, etc., has proved to be of considerable importance in clarifying many medical, legal, as well as genetic questions.

Occasionally when the first cells of the embryo are dividing, something goes wrong, so that the resulting cells have an extra X or an extra Y chromosome, or both. Instead of the normal XY pattern there may be XXY, XYY, or XXYY configurations. Such incidents are rare, but there is evidence that in man an XXY individual may develop into a sterile, asthenic male, mentally retarded, and with some enlargement of the breasts; an XYY individual may turn out to be a supermale with an overaggressive and potentially criminal personality. The XXYY complex has been found in persons who, as inmates of institutions, are hard to manage or who are criminals. The abnormal sex-chromosome incidence probably is not transmitted to offspring.

THE FEMALE

Ovaries. It is difficult to generalize about ovarian structure among the chordates since so much variation exists. A typical mammalian ovary, for example, is essentially a solid structure made up of an inner medullary region and an outer cortex. The medulla is composed of connective tissue, blood vessels, lymphatic vessels, smooth muscle, and nerve fibers. Indistinctly separated from it is the cortex, consisting largely of connective-tissue stroma. It is here that cytogenic and endocrine elements are situated. The cortex in a mature ovary is usually irregular in shape because of *ovarian,* or *Graafian, follicles,* which are developing there. An ovum,* or

egg cell, lies within each follicle, and the follicular cells which surround it supply nourishment. After growing to a certain extent, and at the proper time, certain follicles push out to the surface of the ovary. Either the follicle ruptures, liberating the ovum into the coelom *(ovulation),* or else the follicle and its contained ovum degenerate. Degenerating ovarian follicles are said to be *atretic.*

In certain fishes and in amphibians, snakes, and lizards, the ovaries are of the saccular type and are hollow. In most teleost fishes the cavities are actually closed-off portions of the coelom into which the ripe ova are liberated. This is not the case in amphibians, snakes, and lizards, in which the ovarian cavity is *not* homologous with that of teleosts and in which ripe ova escape into the coelom through the *external* surface of the ovaries (page 285). It is a general rule in vertebrates that ova are shed into the coelom. Often this is not apparent since the opening into the oviduct is so close to the ovaries that there is little chance for them to escape into the main part of the body cavity.

It was long believed that formation of oögonia and primary follicles ceased shortly after birth. It is now generally held, however, that new ova are formed periodically by mitotic activity of oögonia which are present in the *germinal epithelium.* This is originally derived from the outer epithelial covering of the genital ridges, which is in continuity with the mesodermal layer lining the rest of the

---

Terminology. The process of development of a mature ovum from a primorial germ cell is known as *oögenesis.* The names applied to the various stages are often confusing. Primordial germ cells give rise by numerous mitotic divisions to *oögonia.* Each oögonium becomes surrounded by a single layer of neighboring cells, thus forming a *primary follicle.* In the two ovaries of the human female at birth there are several hundred thousand primary follicles. Most primary follicles become atretic and degenerate. When a primary follicle begins to develop further, the follicular cells multiply, usually forming numerous layers, and the oögonium enlarges, becoming a pri-

mary oöcyte. The follicle is then called a *secondary follicle.* As the follicle enlarges and develops further, the primary oöcyte undergoes a *meiotic* division, forming a *secondary oöcyte* and a small polar body. In each of these cells the chromosome number is but one-half that of the primary oöcyte. In many species it is actually the secondary oöcyte that is discharged at the time of ovulation. If fertilized it then undergoes a mitotic division, giving off another polar body, and becomes a mature *ovum.* A fertilized egg, regardless of the stage of its development at the time of fertilization, is referred to as a *zygote.* In this volume, *oögonia* as well as primary and secondary *oöcytes* are generally referred to as *ova.*

coelom. Each developing ovum is soon surrounded by one or more layers of small epithelial cells which make up the follicle.

The size of an ovarian follicle differs greatly in various vertebrates. It depends mostly upon the volume of the ovum, which is characteristic of each particular species, and upon the season of the year when the ovaries are examined. The mature ova of all forms below mammals are relatively large because of the yolk which they contain. They range in size from those of certain fishes, in which the ova are just visible to the unaided eye, to the enormous "yolk," or ovum, of the ostrich and the mackerel shark. In vertebrates below mammals in the evolutionary scale, the ovum completely fills the cavity of the follicle. The relationship of the microscopic mammalian ovum to the Graafian follicle is discussed elsewhere (page 285). The relatively large size of the ovum as compared with the spermatozoön (see page 313) reflects the nature of its contents. The enlarged nucleus is due to the presence of an increased quantity of *nuclear sap* which provides for a greater length of the chromosomes. This situation is favorable for the production by the genes, which are composed of deoxyribonucleic acid (DNA) and make up the chromosomes, of messenger ribonucleic acid (RNA). This passes into the cytoplasm where it brings about, or directs, the synthesis of ribosomal RNA and some proteins. The chemical elements of which DNA and RNA are composed are brought to the ovum from various sites in the body of the female by the circulatory system via the cells surrounding the ovum. The structure and size of ova of different species is distinctive, and this distinction is reflected in the type of embryonic development which will ensue after the egg is fertilized. These matters are emphasized here in order that it may be appreciated that an ovum, ready for fertilization, is not a homogeneous mass of cytoplasm but is highly structured. The yolk content of various types of eggs has already been referred to (page 57).

In most annual-breeding vertebrates there is a seasonal fluctuation in the size of the ovaries, which is maximum at the breeding season. After the eggs have been discharged, the ovaries become relatively small and remain in a "resting" condition until called into activity with the approach of the next breeding period, when a new series of ova and follicles develops. Most domestic animals are not limited to an annual breeding period but have definite times during the year when reproductive activity is manifested. Cyclic changes in the ovaries accompany these periods but are not of the same magnitude as is observed in annual breeders.

AMPHIOXUS. The ovaries of amphioxus are metameric structures consisting of approximately 26 pairs of gonads projecting from the inner surface of the body wall into the atrium, or peribranchial space. The most anterior gonads are located at about the middle of the pharyngeal region. Those in the center are larger than the ones at the anterior and posterior extremes. Each gonad has a layer of atrial epithelium reflected over its free surface. Under this is a double layer of coelomic epithelium which almost completely surrounds the gonad. Thus, each ovary lies in a closed pocket of coelom, the *gonocoele*. When the eggs are ripe, the coelomic pouches and atrial epithelium rupture. Eggs are forced into the peribranchial space from which they pass directly to the outside through the atriopore. Some observers report that spawning amphioxus often discharge their eggs through the mouth. If this is true, they must pass from the peribranchial space through the gill slits into the pharynx and then forward, through the mouth, to the vestibule.

CYCLOSTOMES. The adult female lamprey has a single gonad, representing a fusion of two, which runs the length of the body cavity

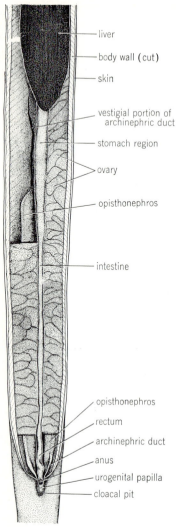

liver

body wall (cut)

skin

vestigial portion of
archinephric duct

stomach region

ovary

opisthonephros

intestine

opisthonephros

rectum

archinephric duct

anus

urogenital papilla

cloacal pit

Fig. 8.2 Ventral view of coelomic
viscera of female lamprey. Portions
of the ovary have been removed to
show details of kidney structure.

abdominal cavity (Fig. 8.2). In the ammocoetes larva two ovaries are present, separated by the dorsal mesentery of the gut. The
mesentery disappears, and the gonads of
the two sides fuse. In cyclostomes, ripe eggs
are shed directly into the coelom. Fertilization is external. When spawning is taking
place in the lamprey, the male wraps the posterior part of his body around the female and
sheds sperm over the eggs as they emerge.

The hagfish is *hermaphroditic,* the anterior
part of the single gonad being ovarian in nature and the posterior portion testicular.
Usually one region or the other matures and
becomes predominant. Only the left gonad
develops fully. In no case do both sexual
regions of the gonad of an individual function at the same time. A certain percentage
of individuals remain in a sterile, intersexual
condition throughout life.

FISHES. It is unusual to find unpaired
ovaries in fishes. Even when this condition
seems apparent, there is actually a fusion of
the two gonads. Sometimes the ovaries are
asymmetrical in position. In a number of
elasmobranchs only the right ovary becomes
fully developed, the left ovary becoming
atrophic during adult life.

The eggs of elasmobranchs are discharged
from the anteriorly located ovaries directly
into the body cavity. Few eggs are produced
at a time. In correlation with the small number, the ova contain relatively enormous
amounts of yolk. In both oviparous and ovoviviparous elasmobranchs certain bodies,
called *corpora lutea,* may form from nurse
cells of the ovarian follicles following ovulation. Presumably these structures have an
endocrine function (page 372) and may play
an important role in the longer-than-usual
retention of eggs in the oviducts of oviparous
species, and in providing for the development
and retention of the young in the uteri of ovoviviparous forms.

and is attached to the middorsal body wall
by a thin sheet of tissue continuous on either
side with the peritoneum. This is the *mesovarium.* At the height of the breeding season
the ovary occupies the greater part of the

Many diverse conditions in regard to ovarian structure are encountered among teleost fishes. In most, the ovaries are of the saccular type. During development they become folded or else form a pocket (Fig. 8.3) in such a way as to enclose part of the coelom within them. Thus all connection with the rest of the coelom is lost. The anterior end of the cavity is blind, but in most cases the posterior end connects directly with the cavity of the very short oviduct formed from peritoneal folds which are continuous with the ovaries. Teleost ovaries are very large during the breeding season, and ripe ova, which are sometimes numbered in the millions, are discharged directly into the central ovarian cavity from which they pass down the oviduct to the outside. In adult teleosts a cloaca does not exist, the oviducts having separate openings from those of the urinary and digestive systems. In some, the urinary and genital systems have a common opening but the anus opens independently. It would seem that the condition in these teleosts is an exception to the rule that ova are liberated into the coelom. However, since the hollow cavity within the ovary is really a portion of the coelom that has been cut off, the exception is only apparent.

Although most species of teleosts are oviparous, there are many ovoviviparous species. Corpora lutea have been identified in the ovaries of many of them following ovulation. In some, the eggs are fertilized and undergo development while still inside the ovarian follicles; in others, fertilization and development of the young take place within the cavities of the ovaries. In the former, no true ovulation actually takes place since it is the young themselves which leave the ovary at the time of birth. The garpike, *Lepisosteus,* is the only member of the so-called "ganoid" fishes that has a saccular type of ovary. In most other members of the group, the solid

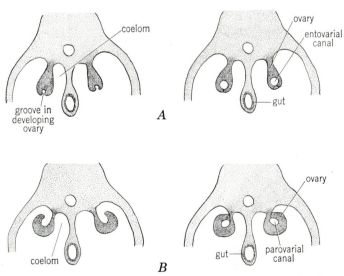

Fig. 8.3 Diagrams showing two methods of formation of hollow teleost ovaries. *A,* entovarial canal formation; *B,* parovarial canal formation.

ovaries are flattened, elongated structures from which mature ova break out into the coelom in the orthodox manner.

AMPHIBIANS. Although the paired amphibian ovaries are saccular structures, ova escape into the coelom through their *external* walls. The cavity within each ovary becomes lymphoid in character and does not correspond morphologically or functionally to the cavity of the teleost ovary. Amphibian ova are of the *mesolecithal* type, with a medium amount of yolk. Ovaries filled with ripe ova are, therefore, large irregular structures which at the breeding season occupy the greater part of the body cavity. The shape of the ovaries varies with the shape of the body. They are long and narrow in caecilians. In urodeles the ovaries are also elongated but to a lesser degree (Fig. 8.10). In anurans they are decidedly shortened and more compact. The cavity within the urodele ovary is single and continuous, but in anurans it becomes divided into a number of pockets. These correspond to the several lobes of which the ovary is composed. Each ovary is attached to the middorsal body wall by a mesovarium, a two-layered band of tissue continuous with the peritoneum. The

peritoneum is reflected over the entire ovarian surface where it is referred to as the *theca externa*. A second layer, the *theca interna*, lies directly under the theca externa and lines the ovarian cavity. It is actually in the form of great numbers of small sacs, each almost completely surrounding a developing oöcyte (Fig. 8.4) and provided with smooth-muscle fibers. The sacs bulge into the ovarian cavity. Between the oöcyte proper and the theca interna is a layer of follicular cells which provide nourishment to the developing ovum. The theca interna surrounding each oöcyte is lacking only from a small area, which is destined to be the point of follicular rupture at the time of ovulation. Here the layer of follicular cells is covered only by the single-cell-layered theca externa.

Hormones from the anterior lobe of the pituitary gland control ovarian development and function. It has been demonstrated in amphibians that, when the pituitary gland is removed during early life, the ovaries fail to develop. If an animal is hypophysectomized during adult life, the ovaries fail to function and gradually become atrophic. Corpora lutea have been reported to be

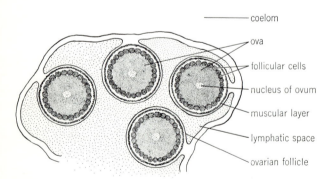

coelom

ova

follicular cells

nucleus of ovum

muscular layer

lymphatic space

ovarian follicle

Fig. 8.4 Diagrammatic representation of a small portion of frog ovary showing relationships of developing ova to ovarian follicles, ovary, and coelom.

present during pregnancy in the ovaries of the ovoviviparous African toad, *Nectophrynoides occidentalis.*

Closely associated with the ovaries in amphibians are the *fat bodies,* or *corpora adiposa.* In anurans they are yellowish structures situated anterior to the gonads and consisting of a number of fingerlike processes (Fig. 7.10). In urodeles they are long and slender and run parallel to the ovaries along their medial edges. Apparently the fat bodies serve as storage places for nutriment. They undergo profound changes during the year and are smallest at the approach of and during the breeding season, which is usually in the spring. After this they gradually increase in size until a maximum is reached.

A peculiar structure in the male toad, known as *Bidder's organ,* may under certain conditions develop into a true ovary (page 372).

REPTILES. Snakes and lizards possess saccular ovaries of the amphibian type, whereas turtles and crocodilians have solid ovaries. In the latter animals the gonads of the female are rather broad and symmetrically disposed, but in snakes and lizards there is a tendency toward elongation. This is more pronounced in snakes, where the right ovary lies in advance of the left.

Eggs of reptiles are of the *polylecithal* type. Only that part of the reptilian egg which is called the "yolk" is formed in the ovaries, and this represents the true ovum. The size of the large eggs is generally in proportion to the size of the animal which lays them.

Most reptiles are seasonal breeders, their ovaries being functional only during a restricted period. Exceptions include certain tropical forms which are capable of breeding throughout the year. No manifestations of seasonal, rhythmic reproductive activity are evident in such reptiles.

In certain ovoviviparous snakes and liz- ards, following ovulation, corpora lutea form from the ruptured follicles and persist throughout pregnancy. It is possible that these structures, as in mammals, secrete a hormone (progesterone or some other progestogen?) necessary for the maintenance of pregnancy.

BIRDS. Although both ovaries are present during embryonic development, in most birds (excepting birds of prey) the right ovary degenerates and the left becomes the functional adult gonad. This is in keeping with the elimination of unnecessary ballast observed in connection with other structures of birds. An important structural difference between the ovary of the bird and that of the mammal, to be described below, is that the ovarian follicles of birds are borne on stalks which extend from the surface of the ovary. Each stalk contains many follicles in various stages of development or atresia. Mature ova of the polylecithal type escape from the ovarian follicles into the coelom through a preformed, nonvascular band, the *stigma,* or *cicatrix,* located on the surface of the follicle opposite the stalk (Fig. 8.15). As in the case of the reptiles, only the yolk of the egg represents the true ovum formed in the ovary. Following ovulation, the wall of the discharged follicle, or *calyx,* is retained for a certain period, apparently exerting some influence upon the time that the ovum remains within the reproductive tract.

Among birds are to be found both seasonal and continuous breeders, but most are of the former type. Among domesticated species the goose and turkey have retained a seasonal, rhythmic reproductive pattern; others are of the continuous type. In many birds an increase in the number of hours of daylight has been shown to stimulate ovarian activity. Under experimental conditions it has been found that even brief exposure to intense and bright light during hours of sleep increases egg production in the domestic fowl. The effect is probably mediated through stimulation of the pituitary gland. Increased or prolonged activity of the ovaries may also be induced by removing eggs from the nest soon after they are laid and before the parent has had an opportunity to incubate them. Under such conditions the total number of eggs laid may be increased manyfold. Even such a psychological effect as limited exposure to a mate may have a profound effect upon the functioning of the reproductive system via stimulation of gonadotrophic hormone secretion by the anterior lobe of the pituitary gland. A female ring dove, for example, comes into full reproductive condition and ovulates as a result of 7 days exposure to a male, even if the two are separated by a glass partition (R. J. Barfield, 1966, *Amer. Zoologist*, **6**:518).

Experiments have shown that if the functional left ovary is removed from the domestic fowl, the rudimentary right gonad will develop into a testislike organ (page 373).

MAMMALS. Cortical and medullary portions of the mammalian ovary are more clearly marked off from each other than in most other vertebrates. The cortex consists chiefly of a compact *stroma* made up, for the most part, of connective tissue and some interstitial cells. The Graafian follicles lie in the cortex and form its most significant components. The relationship of the small mammalian ovum to the ovarian follicle differs somewhat from conditions in other vertebrates. Follicles vary in size according to their state of development and can be grouped into *primary, growing,* and *mature* follicles. Generally, the youngest lie near the surface of the ovary; those which are growing advance toward the medulla; and mature follicles appear as large vesicles which may extend throughout the thickness of the cortex and even produce rounded elevations of the outer ovarian surface.

A young, or primary, follicle consists of a centrally disposed ovum which is surrounded by a single layer of epithelial *follicular cells*. The stroma separates the follicles from each other. In a growing follicle the follicular cells have increased greatly in number, and form several layers about the ovum. A sort of capsular arrangement of cells known as the *theca* now surrounds the follicle. The theca is arranged in two layers: an outer *theca externa,* the cells of which are long and spindle-shaped, and an inner, vascular *theca interna,* made up of more-rounded cells which, however, are sharply delineated from the follicular cells. It is highly probable that the cells of the theca interna secrete the female sex hormone, *estradiol.* The ovum, which has been growing up to this point, now ceases to enlarge. It is surrounded by a clear, thick, transparent envelope, the *zona pellucida,* which apparently is of dual origin. The outer portion is composed of an acid mucopolysaccharide material formed by the follicular cells; the inner part, formed by the ovum, is composed of neutral polysaccharide material (H. Wartenberg and H. E. Stegner, 1960, *Z. Zellforsch.,* **52**:450). Recent studies, using the electron microscope, reveal that microvillous processes from the ovum extend outward through the zona pellucida and that similar but longer and branching processes from the surrounding follicular cells extend inward through this envelope, making contact with the surface of the ovum. It is highly probable that this interdigitation of microvilli is of importance in the transfer of metabolic substances between the ovum and the surrounding follicular cells. The follicular cells making up the first layer around the zona pellucida are more columnar in shape, stain somewhat differently, and are more conspicuous than the others. They form what has been termed the *corona radiata.* So far the follicle is solid but it is soon converted into a fluid-filled vesicle by intercellular vacuolation and a rearrangement of the follicular cells. The cavity thus formed, known as the *antrum,* or *follicular cavity,* is filled with a semiviscous fluid, the *liquor folliculi,* or *follicular fluid.* A small hillock of follicular cells, the *cumulus oöphorus,* or *germ hill,* within which lies the ovum surrounded by the zona pellucida and corona radiata, protrudes into the antrum (Fig. 8.5). The several layers of follicular cells surrounding the ovum make up the *discus proligerus;* those bordering the rest of the antrum constitute the *stratum granulosum.* The discus proligerus at its base is continuous with the stratum granulosum.

Many follicles in the mammalian ovary begin to grow, but most of them become atretic and degenerate before long. In a mature female, at periodic intervals which are typical of each species, one or more follicles grow to maturity. Usually many more ova are shed than become implanted. A notable example is the elephant shrew which may release over a hundred ova at a time, of which only one or two implant and undergo further development. When a Graafian follicle has become fully developed, the cumulus with its contained ovum separates from the remainder of the follicle; the follicle ruptures; and the ovum, with its surrounding layers, is extruded into the coelom. It then passes through the *ostium tubae* into the *Fallopian tube.* It is of interest that in the rabbit, ferret, cat, mink, shrew, and possibly others, ovulation will not take place unless the animal copulates. In these forms, then, ovulation is not spontaneous, and the nervous stimulus of copulation brings about changes which result in the rupture of the Graafian follicles (page 392).

In some mammals the antrum of the follicle fills with a clot of blood coming from the small, ruptured blood vessels in the theca. Such a body is called a *corpus hemorrhagicum.* In other mammals a corpus hemorrhagicum does not form. In either case the cells of the stratum granulosum and theca interna undergo a transformation. They enlarge and push into the antrum, which is thus gradually obliterated. If a corpus hemorrhagicum is present, the blood is soon resorbed. Finally, a relatively large, rounded mass of cells is formed which is known as a *corpus luteum,* or *yellow body.* This name is derived from the fact that in the human being, cow, and some other mammals the corpus luteum has

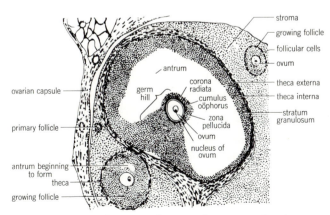

Fig. 8.5 Section of portion of cortex of rat ovary, showing one primary follicle, two growing follicles, and a single large mature follicle.

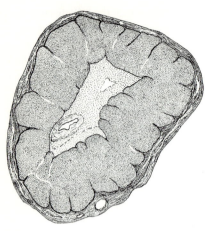

Fig. 8.6 Section through human corpus luteum of pregnancy.

a characteristic yellow-orange color. In most mammals, however, the corpora lutea are of a pinkish hue. The corpus luteum of the human being rarely becomes solid. The luteal cells form the thick wall of a hollow vesicle (Fig. 8.6) which fills with the secretion of former follicular cells. This later either acquires a gelatinous consistency or is replaced by an ingrowth of connective-tissue cells.

If pregnancy does not occur, the corpus luteum persists for only a short time and is known as a *corpus luteum spurium,* or *corpus luteum of ovulation.* If pregnancy does ensue, the corpus luteum grows larger and is called a *corpus luteum verum,* or *corpus luteum of pregnancy.* The corpora lutea vera in such species as the rabbit, mouse, goat, etc., persist and function throughout pregnancy but degenerate as the time of parturition approaches. In certain other mammals, as in the guinea pig, ewe, dog, and cat, they apparently function only during approximately the first half of pregnancy, and then degenerate. In such forms the placenta undoubtedly takes over the secretion of progesterone or other essential steroids. Much variation

exists in this respect among the many species of mammals. A corpus luteum in the ovary of a blue whale is enormous, measuring approximately 9 in. in diameter.

In the pregnant mare, accessory corpora lutea form in the ovaries *during* pregnancy. Some develop from follicles which ovulate; others form from follicles which undergo luteinization without ovulating. A uterine hormone (page 399) is probably responsible for the formation of the accessory corpora lutea, which persist only for about the first six months of the 11-month gestation period. Accessory corpora lutea also appear during pregnancy in the blue antelope of India and the Indian elephant. In any case, following degeneration of a corpus luteum, all that remains is a whitish, scarlike remnant of connective tissue, the *corpus albicans.*

The mammalian corpus luteum is of primary importance as an endocrine gland. Its hormone, *progesterone,* is essential for the proper preparation of the reproductive tract for pregnancy and, in many forms, for the maintenance of pregnancy. In several species, as indicated above, progesterone, or some other progestogen elaborated by the placenta, maintains pregnancy when the corpora lutea degenerate. Another hormone, *relaxin,* of uncertain origin, has been identified. It has been extracted from the corpora lutea of sows and is believed to be formed by the corpora lutea of the mouse and rat as well. This hormone, under certain conditions, brings about a relaxation of the pubic ligaments of the pregnant guinea pig (page 375). The corpora lutea should be thought of as structures of transitory appearance, the function of which is concerned with changes that take place during pregnancy or in preparation for pregnancy.

Ovaries in mammals are located in the lumbar or pelvic regions. Each is attached to the middorsal body wall by a double peritoneal fold, the *mesovarium.* A *round ligament,* or *ovarian ligament* (Fig. 8.25), connects the ovary with the Fallopian tube or uterus, as the case may be. During development the ovaries arise at a level anterior to the metanephros, but later descend to their permanent and more posterior location.

In monotremes the left ovary shows a much greater degree of development than the right and it is generally believed that only the left gonad may be functional. Eggs with shells are laid by the monotremes. The ovum of the platypus at the time of ovulation measures approximately 3 mm in diameter. Albumen and shell are added later when the ovum is passing down the oviduct.

Mammalian ovaries are usually small in relation to the size of the body. Blood vessels and nerves enter at a point called the *hilum.* The ovaries are usually so situated that the surface is exposed to the coelomic cavity, but the ostium tubae, or opening into the oviduct, is so close to the ovary that ova generally enter the oviduct directly rather than getting lost in the body cavity. Experiments on the sow indicate that external migration of ova is possible, since ova from one ovary have been shown to pass across the body cavity and enter the ostium on the other side. Whether this occurs under normal conditions or in other species has not been determined so far as the author is aware. In forms with closed ovarian capsules (rat, mouse), external migration of ova is impossible.

Ovulation, as has been mentioned previously, occurs at rather regular intervals. Most vertebrates are annual breeders, and ovulation in such forms takes place but once a year. In many mammals, particularly in domesticated forms, ovulatory cycles occur at shorter intervals. At such times the females are in a condition referred to as *estrus,* or "animal heat," and at this time only are they physiologically and psychologically willing to accept the male in copulation. The primates seem to be an exception to this rule. The activity of the female reproductive system in certain mammals has been shown to be affected by the length of exposure of the animals to light.

The following list gives the average length of the ovulatory cycle in some familiar mammals. If pregnancy supervenes, ovulation, with few exceptions, is suspended until after birth of the young and in many cases even during the period of lactation.

Opossum	28 days
Guinea pig	16 days
Rat	5 days
Mouse	5 days
Dog	6–7 months
Sheep	Twice a year (many variations)
Sow	2–4 weeks
Cow	21 days
Mare	21 days
Woman	28 days
Macaque monkey	24–26 days
Chimpanzee	37 days

In certain laboratory mammals, for example, rats and mice, having short estrous cycles, a condition known as *pseudopregnancy* can be induced in which estrous cycles are held in abeyance for as long as 13 to 14 days. Pseudopregnancy may be induced by permitting copulation with a vasectomized or sterile male, stimulating the cervix of the uterus when the animal is in "heat" with a glass rod or an electric current, or by certain other procedures. During pseudopregnancy, conditions of ovary and uterus resemble those of pregnancy but are less extensive and of shorter duration. Undoubtedly nerve impulses originating in the cervix affect the hypothalamus which, via the anterior lobe of the pituitary gland in turn, brings about the release of gonadotrophic hormones (see page 392). These are responsible for prolonging the life and hormone-secreting capacities of the corpora lutea.

Although in species having several young at a time both ovaries may form numerous follicles and corpora lutea, it has been reported that in certain others, in which only one young normally develops during a given gestation period, there is a sequence between the two ovaries in regard to follicular development, ovulation, and corpus luteum formation. This may or may not be uniformly true.

Oviducts. The oviducts of the female, except in teleosts and some other fishes, are modifications of the *Müllerian ducts.* The latter are formed in either of two ways during embryonic development. In elasmobranchs and urodele amphibians, the archinephric duct splits longitudinally, one part remaining as the kidney duct and the other forming the Müllerian duct. In elasmobranchs, after the splitting occurs, the Müllerian duct opens anteriorly into the coelom by means of one, or possibly more, of the very much enlarged pronephric tubules and nephrostomes. In other vertebrates the usual method of Müllerian duct formation is by the appearance of a groove or an invagination of the peritoneum covering the ventrolateral part of the opisthonephros or mesonephros, as the case may be, near its cephalic end and near the archinephric, or Wolffian, duct (Fig. 8.7). The edges of the groove come together and fuse, thus forming a tube, the Müllerian duct, which grows caudad to join the cloaca. The anterior end does not close, and the opening becomes the *ostium tubae abdominale.* In certain amphibians, pronephric tubules and nephrostomes play a part in forming the ostium tubae as well as the anterior end of the Müllerian duct. Whether the oviducts of elasmobranchs and urodele am-

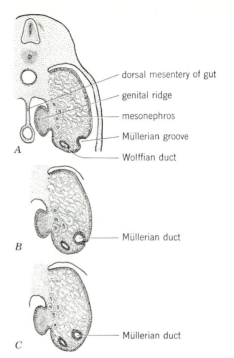

dorsal mesentery of gut
genital ridge
mesonephros
Müllerian groove
A
Wolffian duct

Müllerian duct
B

Müllerian duct
C

Fig. 8.7 Successive stages, *A*, *B*, and *C*, in the formation of the Müllerian duct.

phibians are actually homologous with those of higher forms is a matter for speculation. It is generally assumed, however, that the elasmobranch and urodele condition is the more primitive and that rather definite homologies exist. At any rate, in both male and female the close relationship of the reproductive ducts to the excretory organs and ducts is obvious.

In much the same manner as the Wolffian ducts of amniotes degenerate in the female except for certain vestiges, the Müllerian ducts in the male usually become atrophic and only a few remnants persist. However, in the males of some groups, amphibians and certain fishes in particular, the Müllerian ducts may persist in the adult as rather prominent structures. It is doubtful if any function can be assigned to these vestigial oviducts, but their presence is strongly indicative of the bisexual nature of certain amphibians.

Although the reproductive ducts of the male are invariably paired, and the paired condition is typical of the female, in birds only one oviduct, the left, generally persists. The oviducts in most vertebrates become differentiated into regions which are associated with particular functions. Commonly the lower portion is expanded to form a *uterus*. This may serve as a temporary storage place for eggs and in many vertebrates affords a site in which the young develop.

AMPHIOXUS. No reproductive ducts are present in amphioxus. Ripe ova which have been discharged into the atrium, or peribranchial space, aided by muscular contractions, are carried along by the respiratory current out the atriopore into the surrounding sea water. In some cases the ova may be discharged through the mouth (page 279). Fertilization is external.

CYCLOSTOMES. Reproductive ducts are lacking in cyclostomes. Ova make their way from the coelom through the genital pores at the posterior end of the abdominal cavity into the urogenital sinus and out the aperture at the tip of the urogenital papilla (Fig. 7.7). External fertilization also occurs in this group.

FISHES. Structures for transport of ova from coelom to the outside show a great diversity in fishes. In some teleosts of the family Salmonidae, as well as in occasional members of other groups, eggs escape through modified abdominal pores. The abdominal pores are frequently continued a slight distance within the body cavity by short, funnel-like projections. It is not certain whether these are homologous with the oviducts of other classes. In other fishes, as in certain elasmobranchs, abdominal pores may be present, but their function is unknown. It is possible that they may not be homologous with those of teleosts. In addition to abdominal pores, elaborately developed ducts for the transport of ova are usually present in fishes.

The two Müllerian ducts in elasmobranchs fuse at their anterior ends so that a single ostium tubae connects with the coelom (Fig. 8.8). A narrow, but distensible, oviduct leads from the ostium on each side. An enlarge-

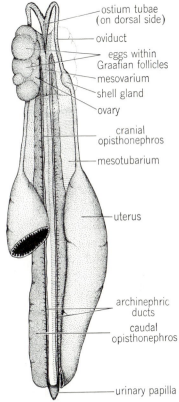

Fig. 8.8 Ventral view of urogenital organs of mature, pregnant female dogfish. One ovary is shown only in outline, and part of one uterus has been removed in order to expose structures lying dorsal to them.

Reproductive System

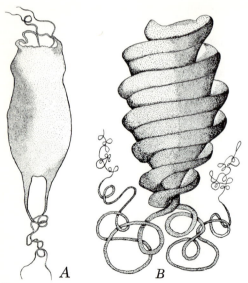

Fig. 8.9 *A*, egg case of dogfish, *Scyllium; B*, egg case of bullhead shark, *Heterodontus galeatus.* (*After Waite.*)

masses of seaweed in such a manner as to fix the egg firmly in position. In *Scyllium,* depending upon temperature and other factors, it takes from 6 to 9 months for the embryo to develop. The shell is ruptured by the fully developed embryo at the time of hatching.

The shell formed about the ova in ovoviviparous species is very much reduced. In accord with this, the shell gland in oviparous species is better developed than in ovoviviparous forms. On the contrary, the uteri of ovoviviparous elasmobranchs are larger and more highly developed than those of egg-laying species, for they serve as structures in which the young undergo a considerable part of their development.

In the dogfish *Squalus acanthias,* which is ovoviviparous, a beautifully clear, amber-colored, temporary shell at first surrounds several eggs in the uterus. The eggs undergo early development within the shell. At about the time that the transient external gills of the embryo are being absorbed, the shell breaks down. Numerous villuslike folds appear in the lining of the uterus during pregnancy. They are in close contact with blood vessels in the yolk sac of the embryo, and an exchange of metabolic substances is believed to take place between the circulatory systems of the mother and her developing young. In oviparous species the lining of the uterus is almost smooth.

It has been reported that the spiny dogfish, *Squalus acanthias,* has a 2-year gestation period. In the Woods Hole region of Massachusetts the animals migrate northward in late April and early May toward the coast of Newfoundland. Pregnant females caught at this time fall into two groups: those having "pups" with large yolk sacs (Fig. 5.1) and those in the so-called "candle" stage. The latter have large undeveloped ova in the uteri. In October and

ment called the *shell gland* is present on each oviduct. Beyond the shell gland the Müllerian duct enlarges on each side to form a uterus which opens into the cloaca.

In all elasmobranchs, fertilization takes place internally. However, the method of reproduction is varied, there being both oviparous and ovoviviparous species. Oviparous forms lay eggs encased in elaborate horny shells formed by the shell gland. The cases protect the embryos during development. An egg case usually contains one egg, but in some species three or four eggs may be present. In general, there are two types of egg cases (Fig. 8.9*A* and *B*). That of the dogfish *Scyllium,* for example, is rectangular with a tendril-like projection extending from each corner. The second type of egg case is screw-shaped and has long, filamentous projections coming off the apex. The Port Jackson shark has an egg case of this type. The tendrils are usually coiled about upright

November the fish again pass Woods Hole, migrating to the South. The females that bore candles in May have pups that are 3 or 4 in. long. Those which carried pups in May either have given birth or else the yolk sacs of the almost fully developed young have been practically resorbed, indicating that birth is about to occur. In the following spring the females which were in the candle stage the previous year are carrying pups and those which have given birth to young are once more in the candle stage.

In still other elasmobranchs, as in the ovoviviparous *Mustelus laevis*, there is an even closer relation between mother and young, for the yolk sac gives off branched processes which are closely attached to the uterine wall.

Holocephalian eggs are essentially similar to those of oviparous elasmobranchs, each being encased in a horny capsule. In the California chimaera, or spookfish, the spindle-shaped capsule is about 6 in. long, one end of which is narrow and rather pointed. This end sticks in the mud of the ocean floor when the egg is laid. The interior of the capsule is more or less divided into three compartments, each corresponding in size and shape to a certain portion of the fish that is to develop within. In this species (*Hydrolagus colliei*), the embryo has just begun to develop at the time the egg is laid and consists only of a few cells.

Teleosts with saccular ovaries have short oviducts which are directly continuous with the ovaries. They are derived from peritoneal folds continuous with the gonads (Fig. 8.3) and should probably be distinguished from true Müllerian ducts since there is some question about their homologies. The two oviducts often fuse, continuing posteriorly as a single structure which may open by a genital pore either between the rectum and urinary aperture or else at the tip of a papilla. Such papillae are occasionally elongated to form tubelike *ovipositors* through which ova are discharged. A cloaca is lacking in teleosts. In some forms the oviducts enter a urogenital sinus in common with the archinephric ducts. Most teleosts are oviparous, but numbers of ovoviviparous species are known. Internal fertilization in these species is accomplished by copulatory organs which are generally modifications of the anal fin of the male. Development of young may take place within the cavities of the ovaries in some cases, but more commonly it occurs within the oviducts.

In most of the "ganoid" fishes and in many teleosts, as previously indicated, the ovaries are solid and eggs are shed into the general coelomic cavity. The oviducts in these species are not continuous with the ovaries, although the ostia are in close proximity. Whether such oviducts are true Müllerian ducts has been questioned.

The oviducts in Dipnoi are long, paired, coiled structures which are true Müllerian ducts. They open anteriorly by funnel-shaped apertures. At the posterior end they unite and have a common opening into the cloaca. A gelatinous coating is secreted about the eggs as they pass down the oviducts.

AMPHIBIANS. Oviducts in amphibians are of the same general pattern throughout the class. They are paired, elongated tubes with ostia situated well forward in the body cavity (Fig. 8.10). Posteriorly, each Müllerian duct is enlarged somewhat to form a short uterus, which in most species opens independently into the cloaca. In certain toads, however, the oviducts unite before entering the cloaca. The uteri in most amphibians serve only as temporary storage places for ova that are soon to be laid. The lining of the oviducts is glandular. Prior to

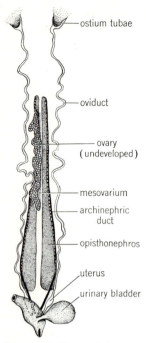

ostium tubae

oviduct

ovary
(undeveloped)

mesovarium

archinephric
duct

opisthonephros

uterus

urinary bladder

Fig. 8.10 Urogenital system of female urodele amphibian, *Necturus*, during the nonbreeding season, ventral view. The left ovary has been removed to show the opisthonephros above.

the breeding season the ducts become greatly enlarged (Fig. 8.11) and markedly coiled and the glandular lining epithelium begins to secrete a clear, gelatinous substance. Eggs escaping from the ovaries into the body cavity are directed toward the ostia by cilia located on the peritoneum lining the body wall, liver, and adjacent structures. After entering the ostia, the eggs pass down the oviducts in a twisting, spiral motion and are forced along by muscular peristaltic waves. The oviducal glands deposit several layers of jellylike material about each ovum. This swells when the egg enters the water (Fig. 8.12).

In most anurans fertilization is external. The male grasps the female in a process called *amplexus,* and as the eggs emerge from the cloaca, spermatozoa are shed over them. No copulatory organs are present. Nevertheless, one group of African frogs (*Nectophrynoides*) is ovoviviparous, the young developing within the uterus over a 9-month period.

The reproductive habits of many frogs and toads are of extraordinary interest. The Surinam toad of South America, *Pipa americana,* is probably one of the most re-

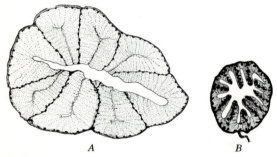

A   B

Fig. 8.11 Cross section of oviduct of salamander, *Eurycea bislineata:* A, during the breeding season (Apr. 11); B, after the breeding season is over (June 21).

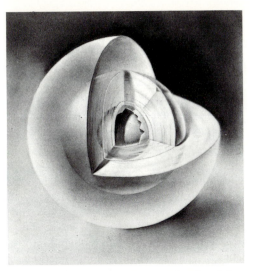

Fig. 8.12 Stereogram showing fertilized frog ovum surrounded by three layers of "jelly." (*Drawn by G. Schwenk.*)

markable of the group. When the eggs are laid, the male places them on the back of the female. The eggs adhere by means of a sticky secretion. Gradually they sink into the skin in small, pitlike depressions which are finally closed over by a membrane. Here they are kept moist. The young develop in the cell-like depressions and do not emerge until they have metamorphosed. After the young have departed, the depressions on the back of the female gradually flatten out.

In the "pouched," or marsupial, frog of South America, *Gastrotheca pygmaea,* the female has a large permanent pouch on her back, not possessed by the male. Here the eggs are placed and carried about while the young develop. The opening to the pouch is inconspicuous and hidden among the folds of the skin.

The obstetrical toad of western Europe, *Alytes obstetricans* is curious in that the male wraps strings of eggs about his body and hind legs, protecting them and keeping them moist until the tadpoles emerge, well advanced in development.

In a certain species of frog, *Rhinoderma,* found in Chile, the eggs are carried in the large vocal sacs of the male, where they are protected during development. The eggs in this case are large and fewer in number than usual.

A tree frog of Paraguay, *Phyllomedusa hypochondrialis,* is known to glue its eggs to the folded leaves of trees. The leaves are in a position overhanging the water, so that, when the tadpoles hatch, they fall into the water below.

Tree frogs of the genus *Hyla* in Jamaica place their eggs in little pools of rain water which have collected between the large leaves of certain tropical plants.

Internal fertilization occurs in most urodeles, but no copulatory organs are present. In these amphibians the males deposit *spermatophores,* which are actually small packets of spermatozoa held together by secretions of the cloacal glands. The spermatophores are placed here and there in pond or stream and even on the land in certain species. The structure of the spermatophores shows some variation. In most species the spermatozoa are gathered together at the tip of the spermatophore (Fig. 8.13). The females search for the spermatophores and take the spermatozoa, gathered together at the tip, into the cloaca. This is accomplished by muscular movements of the cloacal lips. A dorsal diverticulum of the cloaca, the *spermatheca* (Fig. 8.14), serves as a receptacle for the spermatozoa, which are thus available for fertilizing the ova as they pass down the oviducts to the cloaca. In several species of salamander the female protects the eggs by

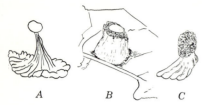

Fig. 8.13 Three types of salamander spermatophores: *A, Diemictylus (Triturus) viridescens; B, Desmognathus fuscus; C, Eurycea bislineata.* (*From Noble, "Biology of the Amphibia," McGraw-Hill Book Company. By permission.*)

coiling her body about them. Fertilization is external in *Cryptobranchus,* in the Asiatic land salamanders, and also, apparently, in the Sirenidae.

The breeding habits of *Salamandra atra,* the black salamander of Switzerland, are of special interest. The animal lives in the rushing waters of mountain streams. Two eggs are retained in the reproductive tract of the female, where they undergo development. When the young hatch, they are well enough developed to be safe in the turbulent waters. Another exception is *Salamandra maculosa,* the fire salamander. About 15 young develop within the oviducts, but at the time they are born gills are still present and metamorphosis occurs considerably later. Some plethodontid salamanders are ovoviviparous.

Internal fertilization is the rule in caecilians. The eversible cloaca of the male is considered by some authorities to serve as a copulatory organ. In a certain species of caecilian, *Ichthyophis,* native to Ceylon, breeding takes place in the spring. A burrow is prepared by the female in the moist ground in close proximity to running water. She coils her body about the relatively large-yolked eggs, which number 20 or more, and guards

them during development, protecting them from predaceous snakes and lizards. The eggs gradually swell until they are double their original size, and the embryo, when about to hatch, weighs about four times as much as the original egg. External gills are present at first but are lost soon after hatching. The larvae, which are fishlike in form, metamorphose into the burrowing, limbless adults which will drown if kept under water. Another genus, *Typhlonectes,* is aquatic and gives birth to living young.

REPTILES. The oviducts of reptiles open into the coelom by means of large, slitlike ostia. Each oviduct is differentiated into regions which mediate different functions in forming the envelopes deposited about the ova prior to laying. In *Sphenodon,* turtles, crocodiles, and alligators, oviducal glands lining the upper part of each oviduct secrete albumen about the ovum. Eggs of snakes and lizards lack albumen, and corresponding glands are missing in these forms. Ova are forced down the oviducts by ciliary action and muscular contraction until they reach the uterus, or shell gland, at the posterior end where the shell is deposited. The uteri enter the cloaca independently.

The eggshell in most oviparous reptiles is of a parchmentlike consistency, but in some lizards and in crocodilians it is hard and rigid

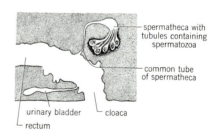

Fig. 8.14 Diagram showing relation of spermatheca to cloaca in the salamander, *Eurycea bislineata.*

like that of a bird's egg. Turtle eggs are usually round, but most reptilian eggs are elongated and elliptical.

The size of the oviducts varies with the season and is at its maximum at the breeding period. In a number of reptiles the right oviduct is longer than the left. This is particularly true of snakes, in which the right ovary is situated farther anteriorly. Vestiges of the degenerate Wolffian duct are frequently found in close association with the ovaries of snakes and turtles and to a lesser degree in other reptiles.

In all reptiles, fertilization is internal and occurs in the upper part of the oviduct. The lining of this portion of the Müllerian duct is ciliated, but the beat of the cilia is in an abovarian direction, i.e., away from the ovary and toward the uterus. It would seem that this would tend to prevent the ascent of spermatozoa. However, it has been discovered that in the reproductive tract of the turtle there is a narrow band of ciliated cells along the side of the oviduct where the mesotubarium is attached in which the ciliary beat is in a pro-ovarian direction. It is believed that the ascent of spermatozoa in the oviduct, at least in the turtle, is brought about by the action of this narrow band of ciliated epithelium. It is not certain whether a similar condition exists in other reptiles. The males of all reptilian species, with the exception of *Sphenodon*, possess copulatory organs by means of which spermatozoa are introduced into the cloaca of the female.

Most reptiles are oviparous. In such forms the eggs are usually incubated by the warmth of the sun. When the young hatch they must fend for themselves.

An alligator may have from 20 to 40 eggs in a nest. The female prepares a hollow in the sand in the bank of a river or stream, places some eggs in the hollow, and then covers them over with leaves, grass, and mud. Another batch of eggs is placed on top of this, and these in turn are covered as before. Several such layers may be prepared. It takes about 2 months for the eggs to hatch. The warmth of the sun and the heat generated by the decaying vegetation aid in the incubation of the eggs. The female guards the nest, and, when the young are hatched, leads them to water and watches over them until they are strong enough to care for themselves.

In some egg-laying snakes, such as the python, the mother coils her body about the eggs and guards them throughout the period of incubation. During this time her body temperature may be from 3° to 4° higher than that of the environment.

Some snakes and lizards furnish good examples of ovoviviparity, since their eggs are retained in the oviducts until the young are born.

Cloacal glands are present in many reptiles. They are, for the most part, hedonic in nature, but in some forms they give off a secretion with a nauseating odor and thus serve as organs of defense.

BIRDS. The right Müllerian duct as well as the right ovary in most birds is degenerate, and only the left oviduct is functional. Both oviducts are formed in the embryo at the same time, but that on the right side fails to develop further and only vestiges remain. Both ovaries and oviducts are functional in certain raptorial birds.

The left oviduct is long, coiled, and made up of several regions. The ostium tubae is bordered by the *fimbriated funnel*, or *infundibulum*. This is followed by a *glandular portion* (*magnum*) in which albumen is secreted. Then follows a short *isthmus* leading to a dilated *uterus*, or *shell gland*, which opens into the cloaca by a short terminal *vagina*.

The reproductive tract of the hen (Fig.

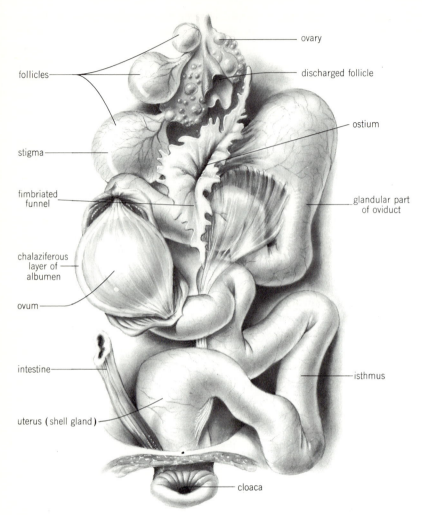

follicles

ovary

discharged follicle

ostium

stigma

fimbriated funnel

glandular part of oviduct

chalaziferous layer of albumen

ovum

intestine

isthmus

uterus (shell gland)

cloaca

Fig. 8.15 Reproductive organs of hen. Only the left ovary and oviduct are functional. A portion of the oviduct has been cut away to expose a descending ovum. Actually only one ovum is in the reproductive tract at any one time. (*Drawn by G. Schwenk, after Duvall.*)

8.15) has been studied thoroughly. An egg which has ruptured from the ovarian follicle through the stigma is literally swallowed by the infundibulum. Cases have been observed in which the infundibulum was seen to surround the follicle and exert a sort of tugging action, possibly facilitating the emergence of the ovum from the follicle. The egg passes down the oviduct by peristaltic contractions of the oviducal muscles and pursues a spiral course. In the upper part of the glandular region a thin but very dense layer of albumen is deposited about the *vitelline membrane* which encloses the yolk. This is known as the *chalaziferous layer*. At either end, in the longitudinal axis, the chalaziferous layer is twisted into a spiral strand, the *chalaza*. As the ovum travels farther down the oviduct, a thick, *dense layer* of albumen is wrapped about the chalaziferous layer. This forms about half of the ultimate albumen content of the egg of the hen. At this point the egg enters the isthmus. *Inner* and *outer shell membranes* (Fig. 8.16), made up of a fibrous meshwork, are deposited in the isthmus. The so-called *fluid albumen* enters the egg in both isthmus and uterus and actually passes through the shell membranes. The hard calcareous, porous shell is formed in the uterus. It has been estimated that under normal conditions, the time occupied by an egg in passing down the oviduct, from the time of ovulation until it is ready to be laid, is approximately 21 to 23 hours. Of this, 3 hours are spent in the glandular part of the oviduct, 2 to 3 hours in the isthmus, and 16 to 17 hours in the uterus. Passage through the vagina takes no more than a minute. Here a thin coat of mucus is applied which seals the pores of the shell. This is effective in preventing undue evaporation of water and, to some extent, in protecting against invasion by bacteria. Eggs which have been washed, with removal of the mucous coat, generally do not hold up so well in storage as unwashed eggs. Conditions in other birds are undoubtedly similar to those observed in the domestic fowl.

The formation of abnormal eggs, which are occasionally encountered, can be readily understood on the basis of the foregoing description. Double-yolked or extremely rare triple-yolked eggs are the result of almost simultaneous ovulation of more than one ovum. In passing down the oviduct in close proximity to one another, a single mass of albumen is wrapped about the yolks and a single membrane and shell are formed about the mass. Such eggs are much larger than ordinary hen's eggs. Sometimes a fully formed egg is found within another and larger egg. This unusual state of affairs can be explained by assuming that a reversal of peristalsis has occurred and that a fully formed egg from the uterus has been forced back into the glandular portion of the oviduct. When it

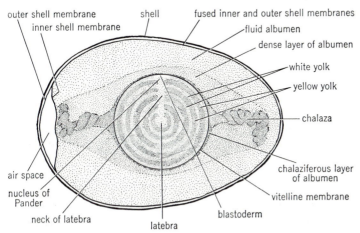

Fig. 8.16 Diagram of sagittal section of hen's egg showing the stratification of yolk frequently encountered.

again passes down the oviduct in the usual manner, the normal reactions of the various parts of the oviduct are evoked.

Experiments in which small cork balls were placed in the abdominal cavities of hens resulted in eggs being laid with cork balls instead of yolks. Blood clots and tissue fragments entering the oviducts may, in a similar manner, stimulate the various parts to activity. Very small or abnormally shaped eggs are frequently the result of such processes.

Internal fertilization, which is characteristic of birds, is accomplished in the majority of species by apposition of the eversible cloacae of the two sexes. Intromittent organs in the form of *penes* are present in male ostriches, swans, geese, ducks, and some other birds. Fertilization takes place in the upper end of the oviduct. In the oviduct of the pigeon a band of cilia beating in a pro-ovarian direction has been described. The condition is similar to that of the turtle. Movement of spermatozoa up the oviduct is apparently mediated by the action of these cilia.

MAMMALS. Paired Müllerian ducts are present in all mammals. In this group various degrees of fusion occur between the two sides and the ducts become differentiated into regions. Fertilization is internal and generally takes place in the upper part of the oviducts.

The most primitive condition found in mammals is that of monotremes, in which the Müllerian ducts remain separate and terminate independently in a urogenital sinus anterior to ureters and bladder (Fig. 8.17). Each duct consists of a narrow anterior *Fallopian tube* and a posterior expanded *uterus*. Although both oviducts are present in the platypus, apparently only the left side is functional. No eggs have been observed in the right oviduct so far as the author is aware. The Fallopian tube opens

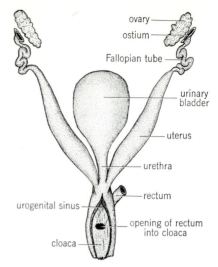

Fig. 8.17 Reproductive tract of female monotreme (semidiagrammatic).

into the coelom by an ostium bordered by a fimbriated, or fringed, infundibulum. A very thin layer of albumen is formed about the small ovum in the Fallopian tube. A horny shell is deposited when the ovum reaches the uterus. When the shell of the platypus egg is first formed, it is round and measures only 4 mm in diameter but while still in the uterus it increases in size until its diameter is approximately 10 mm. Calcium salts are then deposited in the shell, which thickens, continues to increase in size, and becomes elliptical in shape. The long axis is similar to that of the developing embryo. The size of the egg just prior to hatching is about 14.5 by 17 mm. The shell is thin and easily indented. The continued growth of the monotreme egg, after the shell has been deposited, is unique. It is due to the absorption of uterine fluids and to the plasticity of the horny shell. Usually two eggs are laid, although in some cases one or three have been observed. At the time of laying, the embryos are well along in their

Anatomy of the Chordates

development and at about the same stage as a 36-hour chick embryo. They probably have been nourished to some extent by the imbibed uterine fluid. The eggs are incubated outside the body for about 2 weeks before hatching takes place. During this time the mother does not leave the nest and the eggs are kept warm by being placed between the abdomen and upturned tail. Apparently, after hatching, the young are not fed by the mother for a week or more. Not until that time are the mammary glands of the parent developed sufficiently to secrete milk. The young are born naked and rather incompletely developed. At 6 weeks of age they are about a foot long, their eyes have opened, and they are able to crawl about the burrow. Before long they begin to swim.

The eggs of echidnas, like those of the platypus, grow while in the uterus. Nutriment from the uterus is absorbed through the shell. An average-sized egg, just after laying, measures 15 by 17 mm. The females have a depressed mammary area on the ventral side of the abdomen. A temporary brood pouch develops about this area, and here the eggs are placed and incubated until they have hatched. A group of mammary glands opens on each side of this temporary structure. After the young animal has grown to a certain size, the mother removes it from the brood pouch but returns it from time to time for feeding. At night when she is out searching for food, the baby is left in a special burrow which has been prepared for that purpose.

In other mammals the Müllerian ducts become differentiated into three distinct regions. The upper, narrow Fallopian tube, which is frequently spoken of as the oviduct, is sharply marked off from the expanded uterus. It opens into the coelom by a funnel-shaped opening which is sometimes fimbriated. It seems, at least in the rabbit, that the fimbriated funnel plays an important and active role in bringing about ovulation which, in this species, is not spontaneous. In experiments in which the fimbria has been attached surgically to some site away from the ovary, the germ hills of the Graafian follicles, each with its contained ovum, remain intact and attached to the follicles. In the meantime, mature follicles of the other ovary close to its own normal fimbriated funnel have been swept clean (T. H. Clewe, 1961, *Anat. Record,* **139:**217). In certain species, as in the rat and mouse, the free end of each Fallopian tube is modified so as to form a complete closed capsule about the ovary (Fig. 8.20). In such forms there is no communication between the general coelom and the outside. The uterus leads to a terminal vagina which serves for the reception of the penis of the male during copulation. The lower part, or neck, of the uterus is usually telescoped into the vagina to a slight degree. This portion is referred to as the *cervix.* A canal passing through the cervix communicates with the uterine cavity by a small opening, the *os uteri internum.* Its opening into the vagina is the *os uteri externum.*

Marsupials retain the primitive paired condition of the Müllerian ducts, and two vaginae open into a urogenital sinus. In some, the vaginae fuse at their upper ends so as to form a *vaginal sinus* which extends posteriorly as a blind pocket or tube (Fig. 8.18). This caecumlike structure may even open independently into the urogenital sinus and is sometimes referred to as a third vagina. Young, which become lodged in this structure at the time of birth, pass directly into the urogenital sinus. In the event that the pouchlike caecum has no opening, a rupture takes place at its blind end.

The method of reproduction in the opossum is fairly typical of the group. The eggs

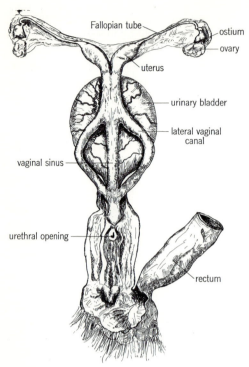

Fig. 8.18 Dorsal view of reproductive tract of female kangaroo which had recently given birth. A single sucking young, 9 cm long (crown–rump) was in the marsupial pouch at the time of the mother's death. The urogenital sinus region has been slit longitudinally and the edges spread apart.

ine milk, nevertheless, is highly nutritious. The number of young developing at any one time varies from 5 to 14. Two or three litters are generally reared in a year.

When the young are born they are far from being fully developed (Fig. 2.39) and are slightly less than ½ in. long. It has been estimated that it would take about 1,750 newborn young to weigh a pound. Nevertheless, the forelimbs at birth are large and provided with strong claws which are later shed, whereas the hind legs are scarcely further developed than the embryonic limb buds of other forms. The young opossums, probably because of a well-developed olfactory sense, are able to find their way to the fur-lined marsupial pouch, climbing or crawling along the fur on the underside of the mother. Inside the pouch is the mammary area where they attach themselves to nipples. It is possible that glands other than mammary glands give off an odor to which the young respond. It was formerly believed that the immature young exhibited a negative geotropism (a reaction causing them to crawl upward, or against the pull of gravity). This is highly unlikely since the semicircular ducts of the inner ear are not even developed at the time of birth in these extremely immature young opossums and, hence, their sense of equilibrium is not yet established.

The nipples soon enlarge somewhat and become firmly fastened in the mouths of the young, which hold on to them securely by means of a well-developed sphincter muscle. At first the young are too weak and undeveloped to suck actively, and the mother, by contracting her abdominal muscles, actually forces milk down their throats. The milk does not get into the trachea and lungs, because at this time the opening of the trachea extends upward into the back part of the nasal passages. As the young animals

of the opossum, when liberated from the ovary, are microscopic in size and contain but a small amount of yolk material. Development takes place in the uteri, but the period of gestation is short and extends only over a 12-day period. There is no true placental attachment, and the young derive their nourishment from secretions coming from glands which line the uterine walls. The uterine secretion which bathes the developing embryos is often referred to as uterine "milk." It bears no resemblance to true milk formed in the mammary glands. Uter-

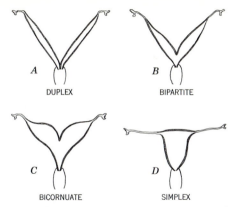

A — DUPLEX   B — BIPARTITE

C — BICORNUATE   D — SIMPLEX

Fig. 8.19 Diagram illustrating degrees of fusion of the uterine portions of the two Müllerian ducts in four types of mammalian uteri.

grow and develop further, they begin to suck of their own accord.

Reproduction in the kangaroo is similar to that of the opossum. The gestation period is about 3 weeks in the large kangaroos. When the young are born, they are approximately the size of the little finger in man.

Eutherian mammals are characterized by possessing a single vagina which represents a fusion of two. A cloaca is present only in the embryo except in the pika, belonging to the order Lagomorpha. The uterine portions of the Müllerian ducts may fuse to varying degrees, resulting in different types of uteri (Fig. 8.19). The *duplex* uterus (Fig. 8.20), found in many rodents, elephants, some bats, conies, and the aardvark, is the most primitive of these types, with two *ora uterorum* opening separately into the vagina. In the *bipartite* uterus (most carnivores, pigs, cattle, few bats, some rodents) the two sides fuse at their lower ends and open by a single os uteri (Fig. 8.21). The *bicornuate* uterus (sheep, whales, insectivores, most bats, some carnivores, many hoofed animals) is the re-

sult of a still greater degree of fusion (Fig. 8.22). In the *simplex* uterus of the armadillo, apes (Fig. 8.23), and man the fusion of the two sides is complete. Only the bilaterally disposed Fallopian tubes indicate the paired origin of the simplex type of uterus. Anomalous uteri of the duplex, bipartite, and bicornuate types are occasionally encountered in the human being.

The urethra coming from the bladder may join the vagina to form a urogenital sinus, or *vestibule*, which opens to the outside. In marsupials the urogenital canal is considerably elongated. In other forms it is short, and in still others, urethra and vagina have independent openings (rats, mice). Often a more or less complete fold of mucous membrane called the *hymen* marks the border of the vagina where it opens into the vestibule.

The external part of the female reproductive system is called the *vulva*. In primates, two folds of skin, the *labia minora*, are located about the margins of the opening of

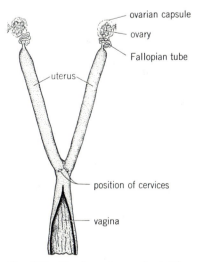

Fig. 8.20 Duplex uterus of rat. The position of the cervices is indicated.

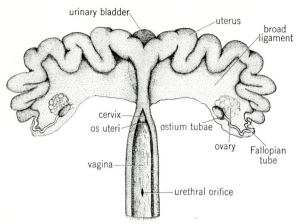

Fig. 8.21 Bipartite uterus of sow as viewed from the dorsal side. The two uterine cornua are fused at their lower ends.

the vestibule. In some apes and in the human female (Fig. 8.24) two additional outer folds, the *labia majora,* make up part of the vulva. They correspond to the scrotal swellings of the male (page 311).

On the ventral wall of the vestibule a small *clitoris* is situated. It bears a rudimentary *glans clitoridis* at its tip, which is homologous with the glans penis of the male but is much smaller. The clitoris, like the penis, is partly composed of erectile tissue and becomes erect and distended with blood during sexual excitement. It differs from the penis, however, in that it has no connection with the urethra except in a few cases, as in rats and mice. In some forms a *clitoris bone,* homologous with the *os penis* in males of certain species, is found within the clitoris.

In addition to the above structures, certain glands, as well as rudiments remaining from the degenerated mesonephros, are associated with the female reproductive system. The glands are generally homologues of similar structures present in the male reproductive system. *Glands of Bartholin* correspond to Cowper's glands (bulbo-urethral glands) of males. They open into the vestibule near

the hymen and secrete a clear, viscid fluid under sexual excitement. This serves as a lubricant during copulation. *Paraurethral glands (of Skene),* corresponding to the

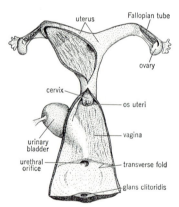

Fig. 8.22 Bicornuate uterus of mare, dorsal view. The uterine cornua are fused to a greater degree than in the bipartite type. (*After Ellenberger in Leisering's Atlas, from Sisson, "The Anatomy of the Domesticated Animals," W. B. Saunders Company. By permission.*)

Anatomy of the Chordates

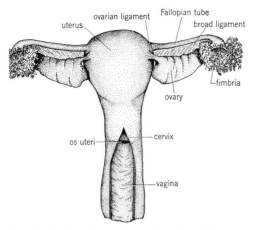

Fig. 8.23 Simplex uterus of gorilla. The only evidence of its bilateral origin is indicated by the paired Fallopian tubes.

prostate glands of males, are occasionally encountered in females but are inconstant in appearance and of doubtful function. Small *vestibular glands,* homologous with the glands of Littré, are located around the opening of the urethra and on the clitoris. They are mucus-secreting structures.

Among the remnants of the degenerated mesonephros which are found in association with the reproductive organs of the female is the *epoöphoron,* a complex of degenerate anterior mesonephric tubules connecting to a persistent portion of the Wolffian duct. The *paraoöphoron,* a similar group of more posterior tubules, is located farther caudad. Both epoöphoron and paraoöphoron are situated in the *broad ligament* and near the ovaries. A *canal of Gärtner,* located in the wall of uterus or vagina, represents a ves-

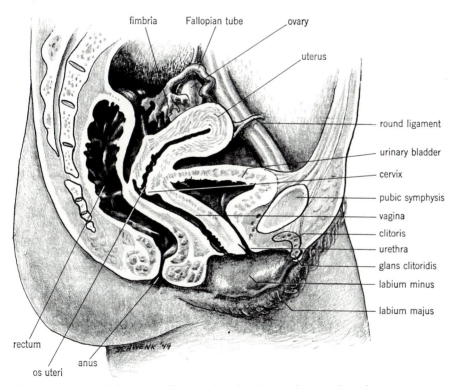

Fig. 8.24 Urogenital system of human female. (*Drawn by G. Schwenk*).

tige of the Wolffian duct proper. None of these structures is functional.

The wall of the Fallopian tube consists of an outer serous layer, an intermediate muscular layer, and an inner mucous membrane. The muscular coat is made up of inner circular and outer longitudinal fibers. The mucous membrane is rather thick and thrown into ridges or folds.

The epithelium lining the Fallopian tube consists of two kinds of cells: ciliated cells, which are most numerous on the fimbriated funnel and upper part of the Fallopian tube, and others of a glandular nature. No true glands are present in this part of the oviduct, however. The beat of the cilia is abovarian in direction, and this, together with peristaltic muscular contractions, serves to transport ova down the Fallopian tube. In most species of mammals it takes 3 or 4 days for fertile ova to pass down the Fallopian tube to the uterus. Experiments have shown that in the rabbit newly ovulated ova have migrated half the length of the Fallopian tube within two hours after ovulation, whereas it takes about 70 hours to pass down the remainder of the tube (G. S. Greenwald, 1959, *Anat. Record,* **133**:386). If fertilization does not occur, the ova die and probably disintegrate before they have even reached the uterus. How spermatozoa can ascend the Fallopian tube against the beat of the cilia has been the subject of much conjecture. Since the tubes of most mammals are coiled to some extent and undergo rhythmic peristaltic contractions, it is possible that a churning of the contents, together with movements of the spermatozoa themselves, and constantly changing ciliary eddies, currents, and whorls may result in the passage of spermatozoa up the tube against the movement of the cilia. No cilia beating in a proovarian direction, similar to those described for the pigeon and turtle, have been observed in mammals.

The wall of the uterus is also made up of three layers: a thin outer layer of *visceral peritoneum,* or *serous membrane;* a thick middle layer of smooth muscle, the *myometrium;* and an inner layer of mucous membrane, the *endometrium.* The myometrium consists of an inner layer of interwoven circular and oblique muscle fibers and an outer longitudinal layer next to the serous membrane. The endometrium, with its *lamina propria* of reticular fibers and blood vessels, is characterized by the presence of *uterine glands* which extend down toward the muscular layer. They are simple tubular glands which often show some branching toward their blind ends. The epithelium of the uterine glands is continuous with that lining the uterine cavity. Cilia occur only sporadically.

The state of development of the endometrium with its uterine glands varies during the reproductive cycle and shows a definite periodicity correlated with changes taking place in the ovaries. The ovarian hormones *estrogen* and *progesterone* have been shown to control these endometrial changes. In primates, at rather definite intervals, a phenomenon known as *menstruation* occurs. This involves a degeneration and sloughing of the superficial portion of the endometrium accompanied by hemorrhage from the blood vessels of the lamina propria. This process lasts from 3 to 5 days, on the average. The endometrial lining is then regenerated. Further details are discussed in Chap. 9.

The vagina consists of three layers: an outer thin layer of connective tissue which joins the vagina to surrounding tissues; a middle muscular layer made up of inner circular and outer longitudinal smooth-muscle fibers; and an inner mucosa. Between the mucosa and muscular layers is a zone of erectile tissue. The lining of the mucosa is thrown into folds which may be in contact under normal conditions. Longitudinal folds or columns may be present as well as transverse ridges known as *rugae*. The mucosa is devoid of glands. The mucus which lubricates the vagina comes from glands located on the cervix of the uterus. The vaginal epithelium is of a stratified, squamous type. It undergoes periodic changes in correlation with the stage of the sexual cycle. So clearcut is this correlation that in many forms an examination of scrapings of the vaginal wall can be used with precision in determining the exact stage of the reproductive cycle. In primates these changes are not so definite as in rodents and some others.

During pregnancy profound changes occur in the uterus. The fertilized ovum becomes embedded, or implanted, in the endometrium, where it undergoes development. The *placenta* (Fig. 8.25) also forms at the implantation site. This is an organ which establishes communication between the mother and developing young by means of the umbilical cord.

Among the armadillos there is one phase of reproduction which is unique. Four young are born at a time. They are always of the same sex and are identical quadruplets. There is only one placenta, but four umbilical cords are present. All four young come from the same fertilized egg, which during the early developmental stages has divided into four parts. Occasionally eight young are born. In such cases they represent two sets of quadruplets.

The connective tissue of the uterus increases in amount during pregnancy, and histological changes occur in both myometrium and endometrium. There is a marked augmentation and extension of the blood vessels which supply and drain the uterus. The entire organ grows until it is many times

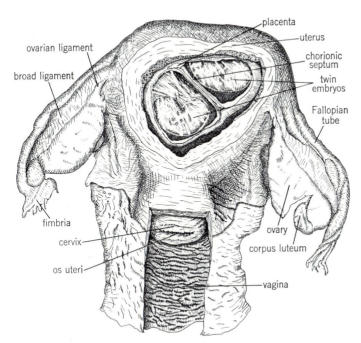

Fig. 8.25 Reproductive tract of pregnant woman, containing twin embryos. In this specimen only a single corpus luteum is present.

its original size. In animals that normally have several young at a time, there is a fairly even spacing of the embryos along the cornua of the uterus (Fig. 8.26). It has been reported that in the mouse the ovarian end of the uterus is less favorable to normal development than the lower portion. Embryos developing at the ovarian end are usually smaller, and such anomalies as cleft palate and cleft lip are more likely to appear in them (D. G. Trasler, 1960, *Science,* **132**:420).

The length of the *gestation period,* which extends from the time of fertilization to *parturition,* or time at which the young are born, shows great variation among mammals. The number of young born at a single delivery also is highly variable. The following table of averages lists these data for certain familiar mammals:

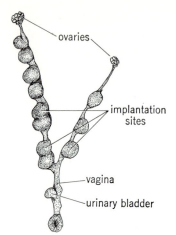

Fig. 8.26 Ventral view of rat uterus on fourteenth day of pregnancy; eight implantation sites in right horn, three in left.

Animal	Gestation period, days	Number of young
Dog (collie)	63	6–7
Black bear	208	2–3
Rat	22	6–10
Guinea pig	63–70	5–6
Rabbit	30	4–6
Elephant	641	1
Cow	270	1
Sow	112–115	4–10
Mare	335	1
Chimpanzee	231	1
Man	266	1
Monkey (rhesus)	213	1
Opossum	12½	12
Kangaroo	40	1
Cat	63	4–7
Lion	106	2–3

The phenomenon of delayed implantation or delayed pregnancy deserves some mention here. Although the gestation period of the rat, for example, is normally 22 days it may, under certain circumstances, be extended for as long as 36 or even more days.

Anatomy of the Chordates

Delay of this sort occurs when a rat becomes inseminated immediately following the birth of a litter and is, therefore, pregnant while she is suckling the litter from a former mating. The length of the prolonged gestation period is correlated with the number of young being suckled. It has been shown that a delay in the time when the blastocysts implant in the uterus occurs under such circumstances, the blastocysts lying free but inactive in the uterine lumen for extended periods. That some endocrine mechanism is involved in this phenomenon is indicated by the fact that minute amounts of the hormone *estrogen* (page 370) administered to pregnant lactating rats will readily evoke the implantation of such blastocysts (C. K. Weichert, 1942, *Anat. Record,* 83:1). Delayed implantation unrelated to lactation occurs in numerous mammals including certain marsupials, bears, badgers, seals, minks, armadillos, and others.

The state of development of mammalian young at the time of birth shows a great diversity. Many marsupials are developed only to a stage that is comparable to that of an embryo of higher forms. The young of rats, mice, rabbits, and even of the human being are born in a helpless condition and require a great deal of parental care before being capable of fending for themselves. On the other hand, a guinea pig at birth has its eyes open, is covered with fur, and does not even depend upon milk from its mother for food.

THE MALE

Testes. The typical testis is a compact organ, the shape of which shows much variation in members of different vertebrate classes. In all except a few low forms, each testis is composed of numbers of *seminiferous ampullae* or *tubules* which connect by means of ducts to the outside. The rounded ampullae or the elongated tubules, as the case may be, at first consist of solid masses of cells which later develop lumina, or cavities. Two types of cells make up the walls. The first of these are large, but not too abundant, *Sertoli cells,* which are supporting and nutrient elements. The others are *sex cells (spermatogonia)* which, by a complicated series of cell divisions, and finally a metamorphic change (*spermiogenesis*), give rise to spermatozoa. The several layers of cells that can be seen in a cross section of a fully developed seminiferous tubule (Fig. 9.12) represent different stages in the development of mature spermatozoa which finally are set free in the lumen of the tubule. At a certain stage of spermatogenesis the developing sperm cells become embedded in the Sertoli cells. It is believed that the latter serve in some manner as "nurse" cells, but the exact significance of this relationship is obscure. The term *spermiation* is applied to the release of spermatozoa from the Sertoli cells. By use of injected radioactive material, it has been determined that in the rat it takes between 48 and 51.6 days to complete all phases of spermatogenesis. In the mouse and man the corresponding times are 33.5 and 74 days, respectively (C. G. Heller and Y. Clermont, 1963, *Science,* 140:184).

In addition to their sperm-producing function, the testes of vertebrates are endocrine organs which elaborate the male hormone *testosterone.* The actual tissue in the testes which secretes testosterone has not been definitely determined, at least in the lower classes of vertebrates. In mammals, groups of cells which lie in the interstices between adjacent seminiferous tubules undoubtedly represent the endocrine elements of the testes. They are called *interstitial cells,* or *cells of Leydig.* In the lower forms this tissue has been reported to be obscure or lacking, or else to appear only at certain times. Conflicting reports are undoubtedly based on the fact that it is rather difficult to distinguish newly forming interstitial cells from connective-tissue cells unless special staining techniques are employed.

In most vertebrates the testes are located in the dorsal part of the body cavity, where they first appear during embryonic development. In the majority of mammals, however, they undergo a descent to a position outside the coelom proper and come to lie in a special, pouchlike structure called the *scrotum.*

In most annual breeders the size of the testes, like that of the ovaries, fluctuates. In such forms they are larger just before the breeding period, but after the spermatozoa have been discharged they shrink to only a fraction of their former size. Spermatogenesis begins once more, following a "resting period," and the testes return to their fully mature condition.

AMPHIOXUS. The gonads of the male amphioxus (Fig. 2.5) are similar in form to those of the female. They are metameric structures which consist of about 26 pairs of testes projecting from the body wall into the peribranchial, or atrial, cavity. Each is almost completely surrounded by a closed coelomic pouch, the *gonocoele.* With the rupture of the wall of the gonad, spermatozoa pass into the peribranchial space and thence to the outside via the atriopore.

CYCLOSTOMES. The gonad of the male lamprey differs but little in general appearance from that of the female. In the adult it is an unpaired structure which represents a fusion of two. The testis, even when fully developed, does not become so voluminous as does the ovary in the female. This is because of the small size of spermatozoa as compared with ova. Sperm cells break through the outer wall of the testis and are shed directly into the coelom (Fig. 8.27).

The hermaphroditic condition of the gonad

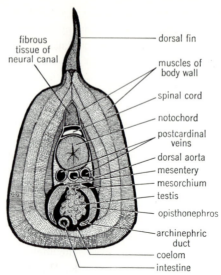

Fig. 8.27 Cross section of male lamprey at level of first dorsal fin.

of the hagfish has already been alluded to (page 280).

FISHES. In elasmobranchs, the testes are paired, symmetrical structures situated at the anterior end of the coelom (Fig. 8.32). There is a tendency for the right testis to be somewhat larger than the left and for the two gonads to fuse at their posterior ends. As in other vertebrates, each is suspended from the middorsal body wall by a mesorchium. In shape, the testes are roughly oval or elongated. In most other fishes they are elongated and often lobulated as well. The position of the testes corresponds in general to that of the ovaries in the female. In Dipnoi the testes are enveloped in a mass of lymphoid tissue. The male gonads vary greatly in size during the year and are extremely large at the breeding season.

Interstitial cells in teleost fishes apparently exist in two forms: (1) in some species they are typical, lying in spaces among the seminiferous ampullae, or lobules; (2) in others they form parts of the walls of the ampullae. In either condition there are seasonal variations in their appearance and activity.

AMPHIBIANS. The shape of amphibian testes shows a rough correlation with body shape. Thus, in caecilians each testis is an elongated structure which appears like a string of beads. The swellings consist of masses of seminiferous ampullae, and these are connected by a longitudinal collecting duct. In urodeles the testes are somewhat shorter and irregular in outline (Fig. 8.28); in anurans they are more compact and of an oval or rounded shape (Fig. 7.10). A pronounced difference in size is apparent during breeding and nonbreeding seasons. Fat bodies are also associated with the gonads of male amphibians. They resemble those of the females in position and appearance. The size of the fat bodies varies or fluctuates with the seasons. They become smaller as the breeding season approaches and enlarge again after this season has passed.

REPTILES. The testes of reptiles are compact structures of an oval, rounded, or pyriform shape. Seminiferous tubules within the testes are long and convoluted. In snakes and lizards there is a tendency for one testis to lie farther forward in the body cavity than the other. Periodic fluctuations in size of the gonads are clearly indicated in most reptilians.

BIRDS. The round or oval shape of the bird's testis is characteristic. There is a tendency for the left gonad to be larger than the right. In such birds as the domestic fowl the testes are functional throughout the year and no periodic variations in size are to be noted. In others, as in the sparrow and junco, which have a limited breeding period, the testes enlarge conspicuously at the approach of the mating season. They become many times larger than the inactive gonads of the nonbreeding period. Increase in the number of hours of daylight has a very definite effect in stimulating spermatogenesis in certain birds, and hence in bringing about testicular enlargement which, therefore, generally occurs during the spring of the year. This reaction is undoubtedly mediated through the pituitary gland.

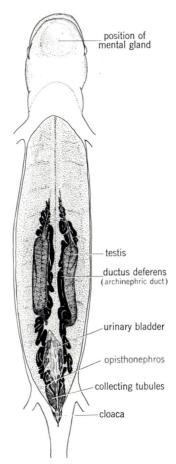

Fig. 8.28 Urogenital organs of male salamander, *Eurycea bislineata*, during breeding season. The other abdominal organs have been removed. In this species the opisthonephros is reduced and practically confined to the posterior part of the body cavity.

MAMMALS. The smooth, oval-shaped mammalian testes are enclosed in a tough, fibrous envelope, the *tunica albuginea*. Each testis is divided internally into a number of *lobules* which show occasional intercommunications. Several seminiferous tubules lie within each lobule. The tubules are very much convoluted and compactly arranged. In man a single tubule when straightened out measures 12 to 27 in. in length. It has been estimated that their combined length in the human being is 750 ft or more. The seminiferous tubules of the bull measure about 3 miles in combined length. It is therefore not surprising that enormous numbers of spermatoza are produced by such an extensive system of tubules. In all mammals except monotremes the testes move from their place of origin to the pelvic region of the body cavity where they may remain permanently or else descend farther into the scrotum. In some mammals the testes are located in the scrotum only during the breeding period and are withdrawn into the body cavity after this season has passed. Such testes shrink in size and become inactive. At the approach of the next breeding period they enlarge and again descend into the scrotum. In several species a relation between the number of hours of daylight and testicular activity has been demonstrated. In many other mammals the testes remain permanently in the scrotum and are capable of functioning at any time.

The general condition in each of the major groups of mammals is given in the list below:

I. INTRA-ABDOMINAL TESTES
   A. Monotremata
   B. Some Insectivora
   C. Most Edentata
   D. Many Pinnipedia (true earless or hair seals—*Family Phocidae;* walrus—*Family Odobenidae*)
   E. Cetacea
   F. Proboscidea
   G. Hyracoidea
   H. Sirenia
   I. Some Perissodactyla (rhinoceri)
II. PERIODIC WITHDRAWAL INTO BODY CAVITY
   A. Some Insectivora
   B. Chiroptera
   C. Most Rodentia
   D. Some Carnivora (otters)
   E. Tubulidentata
   F. Some Artiodactyla (llamas)
III. PERMANENTLY IN SCROTUM
   A. Metatheria
   B. Some Insectivora
   C. Primates
   D. Some Edentata
   E. Most Carnivora
   F. Some Pinnipedia (eared or fur seals —*Family Otariidae*)
   G. Perissodactyla
   H. Most Artiodactyla

The scrotum actually represents a fusion of two *scrotal pouches* which develop during embryonic life. The two halves are separated internally by a partition, the *septum scroti*. Externally the line of fusion is represented by a scarlike *raphe*. Each scrotal pouch contains a diverticulum of the peritoneum known as the *vaginal sac*, or *process*. The testis lies under the peritoneum and, as it descends into the scrotum, pushes against the peritoneum of the vaginal process in such a manner as to become partly covered by a reflected fold which comes in close contact with the tunica albuginea. This peritoneal fold is called the *tunica vaginalis*. It may become entirely separated from the peritoneal lining of the remainder of the coelom. Thus the testis lies entirely outside the coelomic cavity but within the scrotum (Fig. 8.29). The descent is brought about primarily by unequal rates of growth of the

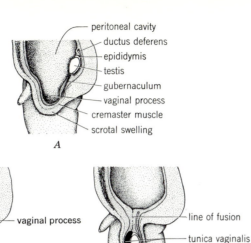

A

vaginal process

gubernaculum

B

line of fusion

tunica vaginalis

gubernaculum

C

Fig. 8.29   Diagram illustrating the descent of the testes in man.

body proper and certain structures closely related to the developing testis such as the ligamentum testis and the scrotal ligament, which together form the *gubernaculum*. This is actually a cord of fibrous tissue which extends from the fetal testis to the scrotal swelling on each side, and which guides the testis during its descent. No muscle fibers are present in the gubernaculum which, therefore, does not serve as a contractile unit (K. M. Backhouse, 1959, *Anat. Record,* 133:246). It appears first in an embryo as a mass of mesenchyme and is of significance only during developmental stages. After testicular descent has occurred, and by the time of birth, it ceases to exist as a definite structure, becoming blended into the connective tissue forming part of the scrotal wall.

The process vaginalis remains open in those species in which the testes periodically withdraw into the body cavity. It closes in most forms in which the testes lie permanently in the scrotum. In the latter, the tunica vaginalis becomes a closed sac with a *vis-*

ceral layer partially surrounding the testis and a *parietal layer* lining the remainder of the scrotal cavity.

A fairly common condition encountered in human males is an *inguinal* or *scrotal hernia.* This may be congenital, in which case it is technically referred to as the *indirect,* or *oblique, type;* or it may later be acquired and spoken of as the *direct type.* The latter may occur as the result of a rupture or tearing of the tissues in the region where the cavity of the vaginal sac became separated from the rest of the peritoneal cavity. A loop of the intestine may then descend into the pouch (Fig. 8.30), resulting in an inguinal hernia or scrotal hernia, depending upon the extent to which the intestine descends.

The wall of the scrotum consists of layers of fascia, muscle, and skin. The integument is thin, folded, and often rather deeply pigmented. Within the dermis of the scrotum is found a relatively continuous and oriented layer of smooth-muscle fibers referred to as the *dartos tunic.* It reacts to variations in temperature in such a manner that under the

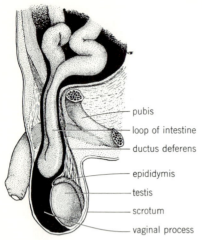

Fig. 8.30 Diagram showing a scrotal hernia in man. A loop of the intestine has pushed through the vaginal process into the scrotum.

- pubis
- loop of intestine
- ductus deferens
- epididymis
- testis
- scrotum
- vaginal process

influence of warmth the scrotum is relaxed and flaccid, but under the influence of cold it becomes contracted, thick, and corrugated, and pushes the testes up close to the body. A layer of muscle fibers and connective tissue called the *cremasteric fascia* lies between the dartos tunic and the parietal layer of the tunica vaginalis.

In various mammals the testes descend into the scrotum at different relative times. In some, descent takes place before birth; in others a considerable time may elapse before the testes arrive at their final destination. It is of interest that the testes of several kinds of mammals, including man, are fairly large at birth but shrink in size conspicuously within a few days. The larger size is due to the presence of quantities of interstitial cells. These have undoubtedly developed in response to maternal hormones originating in the placenta. When the influence of the maternal hormones is removed the interstitial cells shrink and normally do

not again become active until the time of puberty. Androgenic hormones (page 368), secreted by the interstitial cells of the fetal testis, are probably responsible for the changes in the inguinal canal which result in the descent of the testis.

In mammals which normally possess scrotal testes, if a gonad is abnormally retained in the body cavity, it is said to be *cryptorchid*. Cryptorchidism may be unilateral or bilateral. It may be induced experimentally. The single scrotal testis of the unilaterally cryptorchid individual is morphologically and functionally normal. A bilaterally cryptorchid man or animal is sterile. Cryptorchid testes are ordinarily considerably smaller than normal gonads, and spermatozoa are absent. Occasionally spermatogonia and spermatocytes may be present in the seminiferous tubules, but the process of spermatogenesis has obviously been interfered with. The small size of the cryptorchid testis is thus due to a reduction in the seminiferous tubules. The interstitial cells of Leydig are apparently unaffected except under certain conditions in which the cryptorchid condition is unduly prolonged. Bilaterally cryptorchid individuals are usually normal in every respect except that they lack the ability to procreate.

It has been definitely established that the scrotum serves as a temperature regulator and produces an environment for the testes which is several degrees lower than that of the body. A slightly lower temperature seems to be a requirement for spermatogenesis in most mammals. For normal spermatogenesis to occur in mammals, it is also essential that the testes be properly stimulated by gonadotrophic hormones (page 392) and that an adequate amount of vitamin E be available. If the testes are made cryptorchid experimentally by securing them in the body cavity, they quickly undergo the profound changes that are characteristic of the condition. Restoring them to their scrotal environment will generally bring about a return to normal within a few weeks. It has been found that if the undescended testes of cryptorchid boys are lowered into the scrotum by surgical procedure or, in some cases, by giving injections of a hormone extracted from the urine of pregnant women, the testes often begin to function in producing spermatozoa. Under normal conditions the thin wall of the scrotum, its exposed position, and the reaction of the dartos tunic in responding to warmth and cold, all serve to keep the temperature of the testes relatively constant.

Birds, the only other homoiothermous vertebrates, have relatively high body temperatures and have intra-abdominal testes. However, there is some indication that even in birds a lowered temperature may be a requirement for full spermatogenesis to occur. In sparrows, for example, greatest mitotic activity in the seminiferous tubules occurs at night when the body temperature usually drops several degrees. In certain passerine birds a nodule, formed by coils of the ductus deferens, protrudes markedly into the cloaca during the height of the breeding season (page 320). Furthermore, it has been demonstrated that in many birds there is a preseasonal migration of the testes which push against the abdominal air sacs. This may exert a cooling effect upon the gonads.

Those mammals which under normal conditions have intra-abdominal testes would appear to be exceptions to the rule that a lowered temperature is a requirement for spermatogenesis. It is probable that the thermoregulatory mechanism of such mammals is not so highly organized as in others and that their body temperatures are not normally so high. Furthermore, in such mammals the testes usually lie in shallow pockets where the covering skin is unusually thin. In mammals in which there is a periodic withdrawal of the testes into the abdominal cavity during the nonbreeding season, the smaller size of the testes and the cessation of spermatogenesis during this period are undoubtedly caused to some extent by the cryptorchid condition.

In marsupials the scrotum is situated anterior to the penis, but in other mammals it lies posterior to that organ. In a few (rhinoceros, tapir, armadillo, etc.) the scrotum is not typically pendulous and the descended testes lie in recesses in close proximity to the integument. The scrotum of the male kangaroo is homologous with the marsupial pouch of the female. This has been proved by castrating young males which then are injected with estrogenic hormones. A marsupial pouch develops instead of a scrotum.

Spermatozoa. The size of the spermatozoön is extremely small as compared with that of the ovum. In vertebrates having eggs with an abundance of yolk (polylecithal) the ovum may be several hundred thousand times larger than the spermatozoön, yet both are single cells. Spermatozoa show a great variety of form. In certain invertebrates, such as crustaceans and nematode worms, the sperm cells are ameboid in appearance and move along the substratum on which they are deposited by a creeping ameboid movement. The spermatozoa of many invertebrates and of all vertebrates, however, possess long, filamentous, flagellumlike tails. These are used in a whiplike manner in moving through the fluids in which they are deposited.

A spermatozoön, despite its small size, is an exceedingly complex cell. It is made up of a *head, middle piece,* and *tail.* The nucleus is located in the head surrounded by a small amount of cytoplasm. The head sometimes narrows down to a short *neck* to which the middle piece is attached. The tail is continuous with the middle piece. An *axial thread,* or *filament,* is continued throughout the length of the tail. Studies of spermatozoa under the electron microscope indicate that the axial thread is actually composed of numerous fine fibers. Their contractile properties are undoubtedly responsible for the whiplike movement of the tail. The axial thread is enveloped by a thin, cytoplasmic capsule which does not extend to the tip. Other details have been described, but are of greater interest to the cytologist than to the comparative anatomist. The spermatozoön should be considered primarily as a motile nucleus, the chief function of which is to furnish the zygote with half its complement of chromosomes.

Many curious shapes are to be seen in the spermatozoa of vertebrates (corkscrew, oval, spiral, rod, hook, cone, etc.), particularly in the region of the head (Fig. 8.31). The tail, in some cases, bears an undulating mem-

brane. The length also shows much variation and ranges from 0.018 mm in amphioxus to 2.25 mm in a certain toad. In the dwarf siren, an aquatic Floridian salamander, each spermatozoön has two axial filaments, each of which bears an undulating membrane bordered by a flagellum (C. R. Austin and C. L. Baker, 1964, *J. Reprod. Fertility,* **7**:123).

Spermatozoa are usually not motile within the reproductive organs of the male. Movement is dependent upon the relative acidity and alkalinity of the sperm environment. Contact with water, in certain lower forms, excites them to activity. In other cases chemicals liberated by the eggs seem to be essential in activating the spermatozoa. In most mammals spermatozoa do not become active until mixed with secretions of the accessory sex glands. Contrary to many early reports, sperm cells do not live for an extended period after they have become freely motile. It is not surprising that the energy of such a small cell should soon be dissipated. Motility is not the only criterion of fertilizability, however. A spermatozoön may be

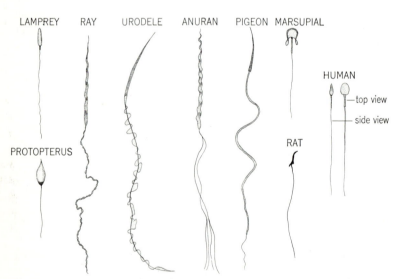

Fig. 8.31 Various types of vertebrate spermatozoa.

motile but still not be able to fertilize an egg. Fertilizing ability is lost more quickly than motility. Loss of energy or of essential chemical constituents may possibly be the explanation.

In vertebrates with internal fertilization, secretions in the female reproductive tract apparently have a deleterious action on spermatozoa, the life of which is measured by hours rather than by days as is generally believed. In artificial insemination, so widely practiced with domestic animals, it is better to collect semen fresh from the male rather than to recover it from the vagina of the female. The higher temperature of the vagina in mammals may also have a harmful effect on the spermatozoa. Storage of seminal fluid to be used in artificial insemination presents many problems. Human spermatozoa apparently differ from those of other mammalian species in that they will survive and regain motility after having been frozen at a temperature of $-195°C$ in liquid nitrogen for as long as 60 to 70 days. Generally, however, a temperature of $12°C$ is optimum for sperm storage. Lower temperatures are deleterious to the spermatozoa of most species unless precautions are taken to lower the temperature very gradually. High concentrations seem to survive better than low concentrations, probably because essential chemical substances are lost in dilute suspensions. Practical measures taken to ensure sperm survival during storage include: use of isotonic media, hydrogen-ion concentration approaching neutrality, addition of glucose or fructose to provide energy, and the presence of some antibiotic substance to prevent growth of bacteria.

Survival of spermatozoa in the reproductive tract of the female bat is exceptional among mammals. Since copulation occurs in the autumn but ovulation and fertilization are delayed until the following spring, spermatozoa live for several months in the female reproductive tract without losing their fertilizing ability. During the period of inactivity they are nonmotile.

In a number of other vertebrates a long period of survival within the female reproductive tract has been reported. In the salamander, *Eurycea bislineata,* for example, spermatophores are taken into the cloaca of the female in October although ovulation does not take place until near the end of the following March. The spermatozoa within the spermatheca are presumably nonmotile. Certain fishes are fertile several months after being isolated from males. In the domestic fowl, fertile eggs may be laid from 15 to 20 days after copulation, but in the duck the period is only about half as long.

The number of sperm cells produced by the male is enormous. In man a single ejaculate of semen measures about 4 cc in volume. An average number of approximately 300 million spermatozoa is present in this quantity of seminal fluid. It has been estimated that an average man in his lifetime discharges about 400 billion sperm cells. This would be equivalent to a billion spermatozoa for every ovum liberated by the ovaries of the female. It is difficult to understand why there should be such a discrepancy in number of germ cells produced by male and female. The necessity for the motile sperm cell to seek out the egg and the many hazards which it encounters on its way would explain the large numbers of spermatozoa, at least to some extent. An enzyme, *hyaluronidase,* present in extracts of mammalian testes or sperm, is capable of dispersing the follicular cells around freshly ovulated mammalian ova. It has been suggested that the large numbers of sperm cells in an ejaculate of semen may be necessary in order to obtain a high enough concentration of hyaluronidase to disperse the closely adhering follicular cells from the ovum so that the sperm may have opportunity to penetrate the egg. In the absence of hyaluronidase, according to this idea, no fertilization occurs. Some authorities doubt the importance of hyaluronidase in this connection. At any rate, it would seem that nature is more than prodigal in supplying an abundant quantity of spermatozoa.

Male ducts. The ducts which in most vertebrates serve to transport spermatozoa to the outside of the body are the archinephric ducts or the Wolffian ducts formed in connection with the development of the kidneys. It will be recalled that the term *archinephric duct* is applied to the kidney duct in the anamniotes. The name *Wolffian duct* is given to the duct in amniotes, which forms in connection with the pronephros and mesonephros. These are really different names for the same thing (page 262). The original function of these ducts is elimination of urinary wastes. In a number of fishes and amphibians certain modified kidney tubules are employed in carrying spermatozoa from the testis to the archinephric duct. They are known as *efferent ductules,* and the archinephric duct then becomes the *ductus deferens.* Even in the amniotes, in which the mesonephros degenerates, its duct persists to become the *ductus epididymidis* (*epididymis*) and the ductus deferens, the former establishing connections with the testis via efferent ductules which are modified and persistent mesonephric tubules.

AMPHIOXUS. Reproductive ducts are lacking in amphioxus. Spermatozoa, which have been discharged into the peribranchial chamber, are swept to the outside along with the respiratory current of water through the atriopore.

CYCLOSTOMES. The reproductive system of the male lamprey functions in a manner similar to that of the female. Again, no ducts are present, and sperm cells, which have been liberated from the testis into the body cavity, leave through genital pores which enter the sinus in the urogenital papilla and thence pass out the urogenital aperture. At the approach of the breeding season the urogenital papilla becomes considerably longer and narrower.

FISHES. A variety of conditions is encountered in the reproductive systems of male fishes. In elasmobranchs (Fig. 8.32), small efferent ductules, leading from the testis, course through the mesorchium and connect with certain anterior kidney tubules along the medial border of the opisthonephros. Spermatozoa are thus conveyed from the testis through the efferent ductules and kidney tubules into the archinephric duct which now serves almost entirely as a ductus deferens. It courses along the ventral side of the opisthonephros and in young specimens is a straight tube with a urinary function. In older specimens it becomes convoluted, its anterior end sometimes being referred to as the *ductus epididymidis.* The posterior portion is markedly dilated to form a *seminal vesicle.* The two seminal vesicles open into a common *urogenital sinus,* which in turn communicates with the cloaca through an aperture at the tip of a *urogenital papilla.* A pair of blind *sperm sacs* passes forward from the ventral wall of the urogenital sinus. The sperm sacs are possibly remnants of the Müllerian ducts which persist in the male. When the archinephric duct is utilized for sperm transport its function in draining urinary wastes from the kidney is practically lost. Accessory, ureterlike, urinary ducts are then formed, particularly in the posterior part of the opisthonephros, and these take over the urinary function.

Although in elasmobranchs the connection of testis and archinephric duct occurs at the

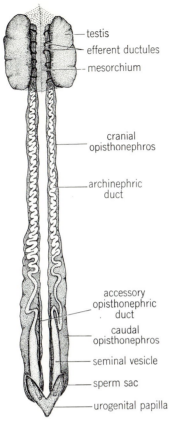

testis
efferent ductules
mesorchium

cranial
opisthonephros

archinephric
duct

accessory
opisthonephric
duct

caudal
opisthonephros

seminal vesicle

sperm sac

urogenital papilla

Fig. 8.32 Ventral view of urogenital organs of mature male
dogfish, *Squalus acanthias.*

anterior end of the opisthonephros, in other
fishes conditions may be quite different. For
example, in the chondrostean *Polypterus,*
only the posterior part of the opisthonephros
becomes involved. The glomeruli of the posterior kidney tubules frequently degenerate,
so that this portion of the kidney is concerned
mostly with reproduction and loses much of
its excretory character. A sperm duct, running the length of the testis, connects, by
means of posteriorly located efferent ductules, to a longitudinal canal within the
kidney. This is *not* the archinephric duct. The

two, however, join posteriorly to open into a
common urogenital sinus (Fig. 8.33*A*).

In *Protopterus* a longitudinal sperm duct
also extends the length of the testis. Each
sperm duct leaves its testis at the posterior
end, and those from the two sides enter the
cloaca through a median genital papilla (Fig.
8.33*B*). No relation to the archinephric duct
is apparent even though the sperm duct,
which is not considered to be a true ductus
deferens, may pass through the posterior end
of the opisthonephros. Only when the archinephric duct is utilized for sperm transport is
it appropriate to use the term ductus deferens.

The male ducts of teleosts (Fig. 8.33*C*) in
many cases are entirely different in origin
from those of most fishes. Their relation to
the gonads is similar to that of the oviducts in
females. Folds of peritoneum enclose a portion of the coelomic cavity, which is thus
continuous with the gonad. In the male the
tube connects with the gonad by a number of
anastomosing canals and is not quite so simple in arrangement as that of the female. It is
not a true ductus deferens. A connection
with the archinephric duct is established posterior to the opisthonephros, but the two
ducts in some teleosts have independent
openings.

The anterior connection of the testes and
archinephric ducts, as seen in elasmobranchs,
seems to be most primitive. The posterior
connections found in *Polypterus* and the
gradual separation of the testis duct and
archinephric duct, as observed in *Protopterus*
and many teleosts, probably represent deviations and specializations from the primitive
condition.

AMPHIBIANS.   The relationship of the
reproductive and excretory systems in male
amphibians is closer than is found in most
fishes. It most closely resembles the condition found in elasmobranchs. Efferent ductules are usually connected inside the testis

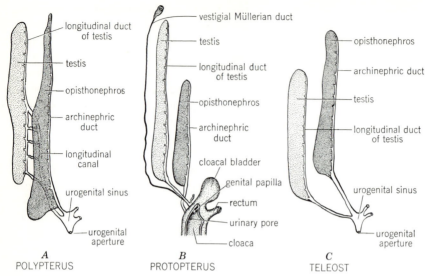

Fig. 8.33 Three types of male fish urogenital organs: *A, Polypterus; B, Protopterus; C,* teleost. *(Modified from Goodrich.)*

or along its medial border by a longitudinal canal. They course through the mesorchium, enter the anterior part of the opisthonephros on its medial side, and may either form direct connections with the archinephric duct or join certain kidney tubules which in turn connect with the archinephric duct.

In caecilians the longitudinal collecting duct, which connects the lobules of the testis, gives off small transverse canals between successive lobules. These pass toward the kidneys to join another longitudinal duct which runs along the lateral edge of the opisthonephros. Spermatozoa then pass through a second series of tranverse canals to join kidney tubules which transport them to the archinephric duct.

In urodeles, efferent ductules join a narrow longitudinal canal, called *Bidder's canal,* which runs along *outside* the medial edge of the kidney but within the mesorchium. Bidder's canal connects by a number of short ducts with kidney tubules in the narrow,

anterior part of the opisthonephros which is called the epididymis. Although Bidder's canal is present in females, it is rudimentary. Certain kidney tubules emerge from the lateral edge of the epididymis and join the archinephric duct, which courses posteriorly (Figs. 7.9 and 8.34). The anterior part of the archinephric duct, or ductus deferens, is concerned primarily with transport of spermatozoa, but the posterior portion may serve for elimination of urinary wastes as well. The archinephric ducts enter the cloaca independently.

In anurans (Fig. 7.10) conditions are quite similar to those of urodeles, but minor variations are present. Efferent ductules enter the anterior end of the opisthonephros along its medial edge. In some forms they connect directly with the archinephric duct but in others join Bidder's canal which, in these animals, lies *within* the opisthonephros close to its medial border. Spermatozoa are then conveyed from Bidder's canal through kidney

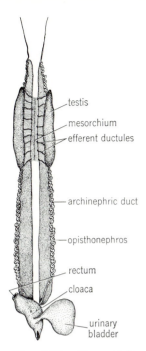

Fig. 8.34 Urogenital system of male *Necturus*, ventral view.

- testis
- mesorchium
- efferent ductules
- archinephric duct
- opisthonephros
- rectum
- cloaca
- urinary bladder

then undergo regression. Their rise and wane, in all probability, are controlled by seasonal changes in the secretion of testosterone, the hormone secreted by the testes.

The ductus deferens proper, particularly in urodeles, also varies greatly in size and is largest just prior to the breeding season. During the nonbreeding period it may be reduced to a mere thread (Fig. 8.35). The cloacae of male urodeles have glandular linings, and these, too, become markedly enlarged as the mating season approaches. The cloacal glands secrete a jellylike material which is used in forming spermatophores deposited by the males.

REPTILES. When the mesonephros of the embryo degenerates in male reptiles its duct persists as the male reproductive duct. The end of the Wolffian duct near the testis becomes greatly convoluted to become part of the epididymis. Certain persistent mesonephric tubules become modified to form efferent ductules which connect the seminiferous tubules of the testis with the epididymis. In some reptiles the epididymis is even larger than the testis. The Wolffian duct is continued posteriorly as the ductus deferens, which is sometimes straight but

tubules to the archinephric duct, which also courses within the opisthonephros but along its *lateral* border. A Bidder's canal is present in the opisthonephros of the female, but its function, if indeed it has any, is obscure. The archinephric duct emerges from the kidney near its posterior end and passes to the cloaca. In males of several species a dilatation of the archinephric duct, as it nears the cloaca, forms a *seminal vesicle* in which spermatozoa may be stored temporarily. Seminal vesicles are very poorly developed in *Rana pipiens* and *Rana catesbeiana*, species which are commonly studied in introductory courses in zoology.

In those species having seminal vesicles, these structures are at the height of their development during the breeding season and

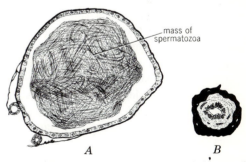

- mass of spermatozoa

A *B*

Fig. 8.35 Cross section of ductus deferens of male salamander, *Eurycea bislineata: A,* from specimen captured Mar. 21, just prior to the breeding season; *B,* from specimen obtained May 16, several weeks after the breeding season.

more often convoluted. In most reptiles the ductus deferens on each side joins the metanephric ureter, so that the two ducts enter the cloaca through a common aperture at the tip of a urogenital papilla.

Müllerian ducts commonly persist in male reptiles, but are generally very much reduced in size, and a lumen may be lacking. In the European lizard, *Lacerta viridis,* however, the Müllerian ducts of the male are as well developed as those of the female.

The epididymides and deferent ducts of such reptiles as have been studied show seasonal modifications and are apparently under endocrine control. In many lizards and other reptiles some of the posterior urinary tubules of the metanephros become enlarged periodically and produce an albuminous secretion which contributes to the seminal fluid in which spermatozoa are suspended.

Other accessory genital organs in male reptiles include glandular structures in the walls of the cloacae of snakes and lizards. In snakes these consist of a single pair of glands, but in lizards a lateral pair is present in addition. Their secretion passes into the groove formed by the hemipenes (page 326). No accessory glands are found in turtles. Scent glands in some snakes and crocodilians open into the cloaca, but these have more to do with defense or sexual allurement than with the sexual act itself. In chelonians and crocodilians a single, protrusible, penial copulatory organ is present (page 326). It bears a longitudinal groove on its upper surface. The deferent ducts from the testes open at the proximal end of the groove which conveys spermatozoa to the free end of the penis.

BIRDS. Efferent ductules connect with a small epididymis composed of a long coiled portion of the Wolffian duct which serves for the passage of spermatozoa from the testis to the highly convoluted ductus deferens.

The deferent ducts in birds open independently into the cloaca and have no relation to the metanephric ureters (Fig. 8.36). In some passerine birds, as previously indicated, a nodule, composed of an intensely coiled portion of the ductus deferens, protrudes into the cloaca. By use of thermocouples it has been demonstrated that the temperature of these nodules is somewhat lower than body temperature. It is quite possible that this may be of importance to the spermatozoa before they can become completely mature and functional. The reproductive tracts of male birds lack accessory glands. In the few birds possessing copulatory organs (page 327), a groove on the upper surface of the single penis carries spermatozoa to its apex.

MAMMALS. The male duct in mammals is again the Wolffian duct. Efferent ductules, which are persistent and modified kidney tubules homologous with the similarly named structures of lower forms, connect by means

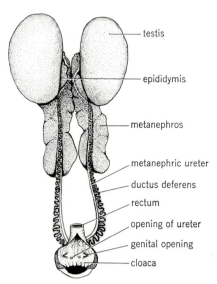

Fig. 8.36 Urogenital system of male fowl.

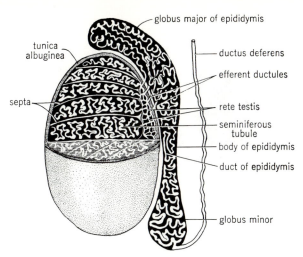

globus major of epididymis

tunica albuginea

ductus deferens

efferent ductules

septa

rete testis

seminiferous tubule

body of epididymis

duct of epididymis

globus minor

Fig. 8.37 Diagram showing relation of mammalian testis to epididymis and ductus deferens.

of a long, compactly coiled duct, the epididymis, to the ductus deferens. That portion of the epididymis to which the efferent ductules are attached is spoken of as the *globus major,* or the *caput epididymidis;* the central part is the *body,* or *corpus epididymidis;* the remainder forms the *tail, globus minor,* or *cauda epididymidis* (Fig. 8.37). It has been estimated that in the guinea pig the total length of the portion of the Wolffian duct included in the epididymis is 9 ft. In a man it measures approximately 20 ft. Experiments have shown that when spermatozoa pass through the epididymis they undergo a physiological maturation. The secretions of the glandular lining of the epididymis may contribute to this process. In experiments on artificial insemination in guinea pigs and rats, spermatozoa from the caput epididymidis have been found to be less effective in fertilizing ova than those from the cauda epididymidis (R. J. Blandau and R. E. Kumery, 1961, *Anat. Record,* 139:209). It is probable that acquisition of full motile power is one of the chief developments which result from

passage through the epididymis. Observations on the mature rat and hamster indicate that both peristaltic contractions and segmentation movements occur in the globus major and body of the epididymis at regular and rather frequent intervals. These undoubtedly facilitate the transport of spermatozoa away from the testis proper (P. L. Risley and C. Turbyfill, 1957, *Anat. Record,* 128:607) and are mediated through the action of gonadotrophic hormones from the anterior lobe of the pituitary gland controlling the secretion of male sex hormone.

The portion of the ductus deferens which leads from the epididymis is convoluted. It then straightens out to join the urethra a short distance below the bladder. In mammals in which the testes are located in a scrotum, the ductus deferens enters the pelvic region of the body cavity where it lies between the peritoneum and the lateral wall of the pelvis. It crosses in front of the ureter, loops over that structure (Fig. 8.38), and then courses posteriorly for a short distance before entering the urethra. This peculiar state of

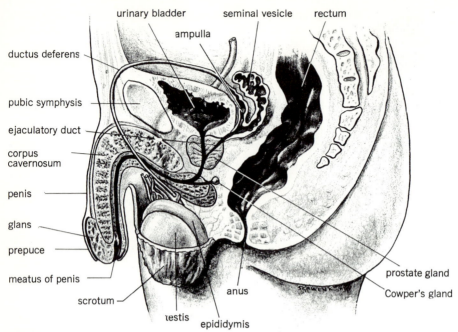

urinary bladder · seminal vesicle · rectum

ampulla

ductus deferens

pubic symphysis

ejaculatory duct

corpus
cavernosum

penis

glans

prepuce

meatus of penis

scrotum

testis · epididymis

anus

prostate gland

Cowper's gland

Fig. 8.38 Urogenital system of the human male. (*Drawn by G. Schwenk.*)

affairs is brought about by the descent of the testis from its original abdominal position. It is related to the location of the ductus deferens during development in reference to the ureter and urethra.

In a number of forms there is an enlargement of the ductus deferens near its posterior end, termed the *ampulla.* This may function as a temporary storage place for spermatozoa and is found in such mammals as ruminants, shrews, certain rodents, carnivores, and primates. An *ampullary gland* may also be connected to the ductus deferens at this point.

A glandular structure called the *seminal vesicle* may arise as a saccular diverticulum of each ductus deferens near its point of junction with the urethra. Its secretion forms an important constituent of the seminal fluid in which spermatozoa are suspended. In such mammals as monotremes, marsupials,

carnivores, and cetaceans, seminal vesicles are wanting. Spermatozoa are not generally stored in mammalian seminal vesicles but have frequently been observed in those of the ram and male deer.

The lower portion of the ductus deferens between seminal vesicle and urethra is sometimes called the *ejaculatory duct* because of its muscular walls which aid in ejecting semen.

The urethra, which extends throughout the length of the penis, opens at the tip of that structure by means of the small *external urethral orifice,* or *meatus.* It thus serves a dual function, since both seminal fluid and urine are expelled through it. Accessory sex glands are also found associated with the urethra. The most prominent of these is the *prostate gland.* It consists of a number of large lobules opening separately into the urethra by numerous small ducts. The lobules are generally bound together by a mass of connective tissue. The prostate gland surrounds the urethra near its junction with the bladder, and its ducts enter the urethra just above the openings of the deferent ducts. Monotremes, marsupials, edentates, and cetaceans lack a prostate gland. The prostate furnishes the remainder of the seminal fluid. When the secretions of seminal vesicles and prostate gland are intermingled, they become coagulated. In the rat a special lobe of the prostate on either side lies close to the seminal vesicle (Fig. 8.39). It is called the *coagulating gland,* or coagulating portion of the prostate. In some species, as in the rat, just after mating a so-called *copulation plug* is present in the vagina. This is nothing more than a mass of coagulated semen. The copulation plug prevents further copulations for a time, aids in the retention of spermatozoa, and thus increases the chances of successful fertilization.

The *bulbo-urethral glands of Cowper* join the urethra at the base of the penis. During sexual excitement they secrete a clear, viscid fluid with a slightly alkaline reaction. This secretion tends to neutralize any acid which may be present in the urethra after the previous passage of urine and also helps to neutralize any acid condition in the vagina. It also facilitates the sexual act by means of its lubricating properties. Small quantities of the secretion of Cowper's glands are added to the semen. In monotremes, Cowper's glands are the only accessory reproductive glands.

In addition, numerous small *urethral glands,* or *glands of littré,* open into the floor of the urethra. They are mucus-secreting glands.

Homologues of many of the above-mentioned accessory glands may be found in females.

The urethra is conveniently divided into three portions: (1) a *prostatic region,* surrounded by the prostate gland; (2) a short, narrow *membranous portion;* and (3) a *cavernous portion* traversing the penis from base to apex.

It has been definitely established that the normal functioning of the accessory sex organs in mammals is controlled by the hor-

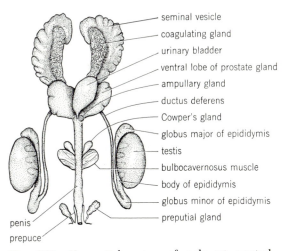

seminal vesicle
coagulating gland
urinary bladder
ventral lobe of prostate gland
ampullary gland
ductus deferens
Cowper's gland
globus major of epididymis
testis
bulbocavernosus muscle
body of epididymis
globus minor of epididymis
preputial gland
penis
prepuce

Fig. 8.39 Urogenital organs of male rat, ventral view.

mone secreted by the testes. All these structures undergo a profound regression after castration. Injection of androgenic substances (page 368) restores them to normal size, structure, and functional activity.

Other glands, which are not directly related to the sexual act but which are of secondary sexual importance, include scent glands located in the anal and inguinal regions and *preputial glands,* or *Tyson's glands,* which secrete a sebaceous substance called *smegma* about the base of the glans penis underneath the prepuce. In the mouse it has been demonstrated that preputial glands have a sex-attractant function.

In the male a few rudimentary vestiges of the degenerate mesonephros may be present. These include the *paradidymis,* a small mass of surviving mesonephric tubules frequently found above the globus major of the epididymis. One or two long, narrow *aberrant ductules* are occasionally present, connected to the upper and lower parts of the epididymis. A small vesicle, the *appendix of the epididymis,* lying against the surface of the epididymis, represents such a detached and persistent mesonephric tubule.

Persistent homologues of the Müllerian ducts are also present in males. These include the small *appendix of the testis (hydatid of Morgagni),* which lies just under the globus major of the epididymis. It is a remnant of the upper end of the Müllerian duct. The *prostatic utricle (vagina masculina, uterus masculinus)* is a short, saclike structure which opens into the posterior portion of the prostatic urethra. It represents the fused lower, or posterior, ends of the Müllerian ducts.

Copulatory organs. Fertilization always takes place in a fluid medium. Spermatozoa, which swim by a whiplike movement of the long, flagellumlike tail, must have some medium in which they can travel in order to gain access to the ova. In many aquatic animals external fertilization takes place and water furnishes the medium by means of which spermatozoa reach the eggs. In terrestrial forms a liquid environment must be provided. In such animals internal fertilization is the rule, and fluids supplied by both male and female furnish the means by which spermatozoa may come in contact with ova. In a number of terrestrial vertebrates, spermatozoa are transferred from male to female by cloacal apposition. However, in most terrestrial forms and even in many aquatic species, intromittent, or copulatory, organs are present in the male by means of which spermatozoa suspended in seminal fluid are deposited directly within the reproductive tract of the female.

Fertilization usually takes place in the upper part of the oviduct. In those forms in which the egg is surrounded by a shell, fertilization must take place before the shell has been formed about the ovum. Sperm cells do not generally reach the upper part of the oviduct entirely under their own motive power. They are aided by churning muscular movements of the female reproductive organs, action of cilia, and other factors.

A number of types of intromittent organs exists among vertebrates. Since they are by no means all homologous, it is best to consider them separately.

FISHES. Although most fishes are oviparous, many lay fertile eggs or else give birth to living young. In fishes, copulation with internal fertilization occurs only in elasmobranchs, holocephalians, and some teleosts. Among elasmobranchs are found oviparous, ovoviviparous, and viviparous species. Copulation is accomplished by means of *clasping organs* (Fig. 8.40), modifications of the medial portions of the pelvic fins of males,

Anatomy of the Chordates

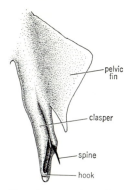

Fig. 8.40 Pelvic fin of male dogfish, *Squalus acanthias*, showing clasper in detail. (*After Petri.*)

used for transport of spermatozoa. Each clasper is provided with a medial groove, the middle portion of which actually serves as a closed tube, since the edges overlap in the manner of a scroll. The anterior opening into the clasper tube is termed the *apopyle;* the posterior exit is the *hypopyle*. Spermatozoa enter the apopyle which is situated close to the cloaca. In sharks and dogfish a sac, the *siphon,* with heavy muscular walls lies on each side just beneath the skin in the posterior, ventral abdominal region. In *Squalus acanthias* the lining of the sac is of an ochre-yellow color. Its anterior portion ends blindly, but posteriorly it communicates with the apopyle. It has been reported that gobletlike cells in the lining epithelium secrete a mucopolysaccharide material which may serve as a lubricant at the time of intromission (G. Heath, 1956, *Anat. Record,* **125**:562). Some observations indicate that the siphon may be used with some force in the ejection of spermatozoa. Under normal conditions the apopyle is open and sea water is drawn into the siphon. The clasper tube gradually fills with spermatozoa.

During the copulatory act the clasper is bent in such a manner as to close the apopyle. Contraction of the siphon forces water down the clasper tube, ejecting the spermatozoa already there into the uterus of the female.

Not all elasmobranchs have hollow siphons. In skates and rays the space normally occupied by the siphon is almost filled with glandular tissue. The function of this integumentary *pterygopodial gland* is still obscure.

It is probable that during copulation only one clasper is inserted into the cloaca of the female. This has been noted during observation of copulation in *Scyllium* in an aquarium. The coition lasted for 20 minutes. Motion pictures of rays, taken during the act of mating, have also revealed that but a single clasper is inserted at a time. Unqualified data are lacking, however, since the difficulty of observing such a process under natural conditions is obvious.

In holocephalians, clasping organs similar to those of elasmobranchs are present. In addition, however, a pair of *anterior claspers* extends from a pouchlike depression in front of the pelvic fins. Also a *frontal clasper* protrudes from the top of the head. It is not clear how these are used.

In those teleosts in which internal fertilization occurs, the anterior border of the anal fin of the male may be elongated posteriorly to form an intromittent organ, the *gonopodium* (Fig. 8.41). The common tropical aquarium fish called the "guppy" is an example of an ovoviviparous teleost possessing a modified fin of this type. Other copulatory organs in teleosts include modifications of the hemal spines of certain caudal vertebrae or of an outgrowth posterior to the anus.

AMPHIBIANS. Although fertilization in anurans is external, a peculiar process called *amplexus,* or false copulation, takes place. The male mounts upon the back of the fe-

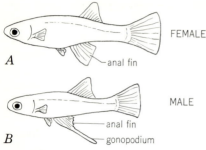

FEMALE

A ⎯ anal fin

MALE

B ⎯ anal fin

⎯ gonopodium

Fig. 8.41 *A*, showing anal fin of female poeciliid fish; *B*, illustrating how the anterior portion of anal fin of male is modified to form the gonopodium.

male and clasps her just behind the forelegs. Swollen glandular thumb pads aid in making his hold secure. So strong is the clasp reflex that it is very difficult to dislodge a male during amplexus. The body of the female is compressed by the strong grasp of the male, and the resulting increase in internal pressure aids in extrusion of the eggs. Spermatozoa are shed over the eggs as they pass from the cloaca.

Internal fertilization in urodeles has been discussed previously (page 293). No copulation or amplexus takes place in this group.

A peculiar method of internal fertilization is noted in the caecilians. The muscular cloaca of the male is protrusible to a marked degree and serves as a sort of copulatory organ when the cloacae of the two sexes are in apposition.

REPTILES. The only reptile entirely lacking in copulatory organs is *Sphenodon*. In other reptiles two types of structures are recognized, one being typical of snakes and lizards and the other of turtles and crocodilians.

In snakes and lizards, peculiar paired structures called *hemipenes* are employed in an intromittent manner. They consist of saclike structures normally lying under the

skin adjacent to the cloaca at the base of the tail and frequently containing rigid spines. Although hemipenes are also present in the female they are very small and of doubtful function. Each hemipenis bears a spiral furrow which may be so deep as to give the organ a notched appearance (Fig. 8.42). The hemipenes are everted during copulation, and spermatozoa pass down the grooves into the cloaca of the female. Retractor muscles are present by means of which the hemipenes may be withdrawn. These structures do not contain erectile tissue as do the penes of higher forms and are not homologous with them. In some male lizards small conelike projections from the integumentary *femoral pores,* or *glands,* which are situated on the undersurfaces of the thighs, are used in clasping the females during copulation.

A single penis is present in turtles and crocodilians. It is apparently derived from paired thickenings or ridges in the anterior and ventral walls of the cloaca and is made up of connective and erectile tissues. The paired masses of erectile tissue are called *corpora cavernosa.* The penis can be extruded and retracted. A homologous structure, the *clitoris,* is present in females. At its proximal end the penis divides, indicating a bilateral origin, but the distal end is single and free, terminating in a spongy, rounded *glans.* A groove along the dorsal surface provides for the passage of spermatozoa.

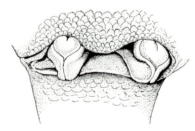

Fig. 8.42 Hemipenes of lizard, *Platydactylus.* (*After Unterhössel.*)

Anatomy of the Chordates

During the act of mating the corpora cavernosa are filled and distended with blood; the groove, in effect, becomes a tube, and the penis is firm and enlarged. It is then said to be *erect*. This property of erectile tissue makes it possible for the penis to serve as an intromittent organ during the act of copulation. Unless the penis becomes erect, copulation is impossible. In crocodilians the penis is longer and the groove is deeper than in turtles.

BIRDS. A penis is present in only a few birds such as ducks, geese, swans, and ostriches. It is a single structure built upon the same plan as that of crocodilians. A clitoris is present in the females of these species. In many other birds a rudimentary penis can be identified. Most birds copulate by cloacal apposition, the cloacae of both sexes being eversible and, during copulation, placed in juxtaposition.

MAMMALS. A single penis is typical of mammals, and its homologue, the clitoris, is present in all females. In monotremes, under normal conditions, the penis lies on the floor of the cloaca. It is similar to the organ in turtles, crocodilians, and birds except that the groove on the dorsal side has become a closed tube. Moreover, the tube is surrounded by erectile tissue known as the *corpus spongiosum*, which differs somewhat in structure from that of the paired corpora cavernosa. The canal in monotremes is believed to carry only spermatozoa, since the urethra has a separate opening into the cloaca (Fig. 8.43). In this respect the penis of the monotreme differs from that of all other mammals. The end of the penis is enlarged to form a sensitive, swollen *glans*. The glans penis in monotremes and other mammals differs from that of reptiles in being surrounded by a fold of skin termed the *prepuce*, or *foreskin*. Erectile tissue is present in the glans. This structure in

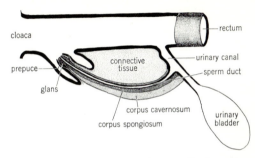

Fig. 8.43 Diagrammatic longitudinal section through cloacal region of male monotreme, showing penis in retracted position. (*After Boas, based on Wiedersheim, "Comparative Anatomy of Vertebrates," copyright 1907 by The Macmillan Company and used with their permission.*)

echidnas is peculiar in being divided into two double-lobed knobs (Fig. 8.44*A*). In the platypus the glans is bifurcated at its tip and covered with soft spines. The sperm canal in monotremes is reported to open onto the surface of the glans through a number of small apertures. Spermatozoa are scattered over a considerable surface at the time of copulation, which, in the platypus, occurs in the water.

Marsupials lack a cloaca, at least in the adult stage. The marsupial penis, therefore, does not lie within a cloaca as in monotremes. It is covered by a sheath and opens to the outside of the body just underneath the anus. It may be protruded and retracted. As previously noted, the scrotum in marsupials is anterior to the penis. The urethra in this and subsequent groups carries both urine and seminal fluid. The urinary duct, which in the monotremes leads to the cloaca, has become closed off in marsupials. The two corpora cavernosa are separated by a septum. The urethra is surrounded by the erectile corpus spongiosum. In several marsupials the tip of the penis is bifurcated (Fig. 8.44*B*). Usually three pairs of bulbourethral, or Cowper's, glands open into the urethra at the base of the penis. These and the preputial glands are the only accessory glands found in marsupials.

In the Eutheria, there is a tendency for the penis to be directed forward, and in all forms possessing a scrotum, the penis is located anterior to that structure. It is situated along the midline of the abdomen and usually in a horizontal position. In most forms, the penis lies within a sheath from which it can be protruded and retracted. In primates, however, it is permanently exerted, and a preputial sheath, or foreskin, covers and protects only the sensitive glans. A dartos tunic, composed of smooth-muscle fibers and similar to that in the scrotal wall, is present in the skin covering the penis. It reacts to temperature variations, contracting under the influence of cold and relaxing under the influence of warmth. In a number of mammals, a *penis bone (os penis, os priapi)* develops in the septum between the corpora cavernosa. This helps to increase the rigidity of the penis, which normally becomes erect by distension with blood. Penis bones are present in members of the orders Rodentia, Carnivora, Chiroptera, Cetacea, and lower Primates. In a walrus, the os penis is extremely large. In some species of whales it may attain a length of 6 ft. Only a single corpus cavernosum is present in cetaceans.

The glans penis is really the enlarged distal end of the corpus spongiosum covered with thin and delicate skin. It is supplied with numerous sensory nerve end bulbs and is extremely sensitive to certain stimuli. Many variations in shape and structure are to be observed (Fig. 8.44). In the cat the glans bears numerous horny papillae or spines which undoubtedly function as a sexual irritant during copulation. Lions and tigers possess similar spines, but they are not so prominent. In certain ruminating animals,

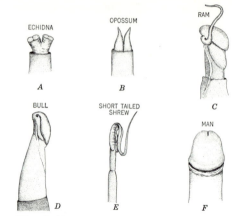

Fig. 8.44 Various types of mammalian glans penis. (*C and D after Böhm, "The Anatomy of the Domesticated Animals,"* W. B. Saunders Company. By permission.)

such as sheep, giraffes, and certain antelopes, a long filament, traversed by the urethra, extends from the tip of the glans. In sheep the filiform appendage is attached to the left side, but in other ruminants it is median in position. It contains erectile tissue which is continuous with the corpus spongiosum.

Anomalous development of the vertebrate penis is encountered now and then. Most frequently this takes the form of a failure of the urethra to continue to the tip of the penis, so that the urogenital opening may be located anywhere along its under (dorsal) surface. Such a condition is known as *hypospadias.* A hypospadic penis is comparable to the grooved penis of reptiles. More rarely the urethra opens along the upper (ventral) side of the penis (*epispadias*).

HERMAPHRODITISM

The term *hermaphroditism* is used to designate a condition in which the reproductive organs of both sexes are present in a single

    Anatomy of the Chordates

individual. In many invertebrates and lower vertebrates hermaphroditism is not uncommon, but in higher forms it appears only as a sporadic anomaly. Two types of hermaphroditism are recognized. In the first type, known as *true hermaphroditism,* an individual possesses gonads of both sexes either as separate ovaries and testes or as an *ovotestis,* in which elements of the two gonads are combined within a single organ. A true hermaphrodite is said to be *monoecious.* The opposite term, *dioecious,* is used to refer to the more usual condition in which the sexes are separate. *False, spurious,* or *pseudohermaphroditism* is a condition in which the individual possesses the gonads of only one sex but the external genitalia and secondary sex characteristics may closely resemble those of the opposite sex. Since the genital organs of both male and female have a common embryonic origin, and because close homologies exist between the various structures, it is not surprising that faulty development, resulting in the various forms of hermaphroditism, should occasionally be encountered.

Cyclostomes. The hagfish is a true hermaphrodite. Only the left gonad develops fully. The single gonad consists of anterior ovarian and posterior testicular tissue. A single individual, however, functions as either male or female, since either one or the other region becomes functional. It is reported that a single animal may at times function as a male and at another time as a female. If this is true, it is an example of what is called *protandrous hermaphroditism.*

Lampreys are normally dioecious, but specimens have been found which show evidence of hermaphroditism. Ammocoetes larvae, on the other hand, show a high incidence of hermaphroditic gonads.

Fishes. Although hermaphroditism occurs but rarely in fishes, it has been observed in such teleosts as the ling, herring, cod, and mackerel. The sea basses *Serranus scriba, hepatus,* and *cabrilla* are considered to be hermaphroditic under normal conditions and have been reported to fertilize their own eggs. In the daurade, *Chrysophrys auratus,* another teleost, eggs and sperm ripen alternately within the gonad. Specimens of other teleosts are sometimes found with an ovary on one side and a testis on the other.

Amphibians. Among anurans occasional hermaphrodites are encountered. The newt, *Triton taeniatus,* is an example of the condition in urodeles. Bidder's organ in the male toad (page 372) has been shown to be capable of developing into a functional ovary. Numerous cases of sex reversal in amphibians have been reported under experimental conditions.

Reptiles. Functional hermaphroditism is not known to occur in reptiles.

Birds. The bisexual potencies of birds have been subjected to much experimental analysis. After removal of the functional left ovary, the capacity of the rudimentary right gonad of the female domestic fowl to develop into a testislike structure has been demonstrated. Such modified gonads may elaborate male hormone but have not been shown to be capable of producing mature spermatozoa. A single case of true sex reversal in birds was reported in 1923. A fowl which at one time had laid eggs and reared several broods was found at a later period to have mated successfully with a hen and fathered two chicks. This is the only case of the sort that has been scientifically verified.

A peculiar condition referred to as *gynandromorphism* has been observed in the domestic fowl, pheasant, and other birds. Of four known cases among certain European finches, three had male plumage on the right and female plumage on the left, whereas the fourth was just the opposite. Only one of the first three cases was subjected to anatomical study, and in this specimen a testis was found on the right side and an ovary on the left. The explanation of the origin of such an unusual condition is obscure.

Mammals. True hermaphroditism among mammals is extremely rare. It has frequently been described in pigs in which it occurs about once in a hundred thousand animals. Only about 25 cases of true hermaphroditism in human beings are known in all medical literature. In no case is there any record of production of both eggs and spermatozoa.

Gynandromorphs among mammals are rarely observed. A Syrian hamster gynandromorph has been described which had male reproductive organs on the right side and female organs on the left (H. Kirkham, 1957, *Anat. Record,* **127**:317).

Pseudohermaphroditism in the male is the most common sexual abnormality among mammals. In human beings the presence of a hypospadic penis, accompanied by a failure of the testes to descend, brings about a condition in which the external genitalia of the male come to resemble those of the female.

Sometimes in a female the ovaries descend to the labia majora and the clitoris enlarges so that the masculine condition is simulated. Accompanying these anomalous developments, unusual endocrine factors may be at work which bring about various degrees of development of the accessory sex organs and secondary sex characters of the opposite sex. On the other hand, hormonal functioning may be perfectly normal.

All the above changes may be traced back to faulty development during embryonic life, but the actual cause of the abnormal development is unknown. Experiments involving the injection of certain hormones at various times during prenatal and early postnatal life have been successful in evoking permanent abnormalities in the reproductive organs of both sexes. Certain of these changes are comparable to conditions sometimes encountered in hermaphrodites.

Of special interest is the *freemartin.* When a cow has heterosexual twins, in a large percentage of the cases the female co-twin is sexually abnormal and is known as a freemartin. The reproductive organs of the freemartin have been modified in a masculine direction. In such cases, during early embryonic development, there is an anastomosis of the blood vessels of the two. Some humoral factor, undoubtedly carried to the female member, has a profound influence on her developing reproductive organs, bringing about an abnormal persistence and further development of Wolffian-duct derivatives and a suppression of those structures derived from the Müllerian duct. The exact factors responsible for this are obscure. On the other hand, the marmoset, a small platyrrhine monkey, also may have heterosexual twins with an anastomosing circulation, yet the female co-twin is sexually normal and a condition comparable to the freemartin does not occur.

SUMMARY

1. The reproductive system consists of primary and accessory sex organs. The primary organs are the paired gonads: testes in the male and ovaries in the female.

Spermatozoa are formed in the testes, and ova in the ovaries. Gonads are derived from mesoderm and arise from the medial sides of the opisthonephros or mesonephros as thickened genital ridges which project into the coelom. The accessory sex organs are the ducts and glands which provide for the transport of eggs or spermatozoa to the outside of the body. The archinephric duct (in anamniotes) or the Wolffian duct (in amniotes) becomes the reproductive duct, or ductus deferens, in the male. The Müllerian duct, or oviduct, in the female of elasmobranchs and urodele amphibians arises by a splitting of the archinephric duct. In other forms an invagination of the peritoneum covering the opisthonephros or mesonephros, as the case may be, closes over to form the Müllerian duct, which, nevertheless, is generally considered to be a phylogenetic derivative of the kidney duct. Ducts are lacking in amphioxus and cyclostomes.

2. Eggs develop in ovarian, or Graafian, follicles within the ovaries. When fully developed, they break out of the ovaries into the coelom. Each oviduct generally opens into the coelom by a funnel-shaped ostium, which has been homologized with one or more peritoneal funnels of the pronephric kidney. The oviducts of teleost fishes are derived in a different manner and may not be homologous with those of other vertebrates.

3. In forms below mammals the paired oviducts are separate and usually open independently into a cloaca. In the higher mammals a cloaca no longer is present, except in the pika. Each oviduct differentiates into three regions: Fallopian tube, uterus, and vagina. Various degrees of fusion occur from distal to proximal ends, and different types of uteri result. The Fallopian tubes always remain paired. In higher mammals, vagina and urethra usually have a common opening to the outside. In a few groups, however, separate openings are present. The young in viviparous and ovoviviparous species develop within the uterus.

4. The Wolffian duct in amniotes disappears or is nonfunctional in the female, but its remnants, together with those of the degenerated mesonephros, are frequently found associated with the female reproductive organs.

5. Certain accessory glands associated with the female reproductive tract secrete mucus, which serves to lubricate the vagina during copulation. Fluids secreted by the Fallopian tubes and uterus may aid in the passage of ova down the reproductive tract and, in those forms with internal fertilization, in the passage of spermatozoa to the upper part of the oviduct where fertilization takes place.

6. In males, spermatozoa are formed within seminiferous ampullae or tubules in the testis. Each ductus deferens establishes connections with the testis generally by means of an epididymis and persistent kidney tubules called efferent ductules. In the anamniota the epididymis and ductus deferens consist of portions of the archinephric duct which may serve in some cases as an excretory duct as well as for the passage of spermatozoa. In the amniota, however, the epididymis and ductus deferens are formed from the persistent Wolffian duct, which is entirely dissociated from the ureter of the metanephros.

7. The testes are located in the abdominal cavity except in a number of mammals. In most mammals they are either temporarily or permanently situated in a

a pouchlike scrotum outside the body proper. The scrotum serves as a temperature regulator.

8. Remnants of the Müllerian duct persist in the males of many species and are found associated with the reproductive organs.

9. Accessory glands, particularly in mammals, secrete seminal fluid in which spermatozoa are suspended and which is necessary not only for the viability of the spermatozoa but for their transport through the reproductive tracts of both male and female.

10. In most vertebrates with internal fertilization, copulatory organs are used in transferring spermatozoa from male to female. In fishes these usually consist of modifications of fins. In snakes and lizards, paired hemipenes are present; but in turtles, crocodilians, certain birds, and all mammals, a single penis serves the purpose. The penis consists largely of erectile tissue which, when distended with blood, causes the organ to become firm and erect, and only then can it serve as an intromittent organ. In some forms a penis bone contributes to the rigidity of the organ. The urethra coming from the urinary bladder passes through the penis in all mammals except the monotremes. It thus serves for the passage of both urinary and seminal fluids.

11. The parallel development of male and female reproductive organs is striking. The various structures in each sex have definite homologies with those of the opposite sex.

12. The close structural relationship between excretory and reproductive organs is very apparent in vertebrates, but the evolutionary trend is toward a complete separation of the two systems.

19

ENDOCRINE SYSTEM

All the glands in the body may be classified as *exocrine* or *endocrine*. Exocrine glands have ducts which convey their secretions to epithelial surfaces of the body where they are discharged. Endocrine glands, on the other hand, have no ducts to carry off the secretory product or products. Both types of glands, with the exception of the interstitial cells of Leydig of the testis (see page 307), develop from epithelial surfaces as groups of cells which grow into the connective tissue beneath the epithelial surface and there proliferate and differentiate into glandular structures of one type or another. In the case of an exocrine gland, the original connection between the gland and the surface is retained and differentiates into the lining of a duct. When endocrine glands develop, however, all connection with the epithelial surface is lost. The secretions, which are known as *hormones,* instead of passing through a duct, go directly into the blood or lymph and are carried to all parts of the body by the circulating fluid. The secretions are of a chemical nature, bringing about certain changes in other parts of the body. These may exert physiological effects of various kinds as well as act rather specifically on a certain organ or part of an organ, such structures being referred to as "target" organs or "target" tissues. Hormones affect growth and development of specific embryonic structures and are most important in influencing these processes in postembryonic life. They are involved in the very first steps leading to development, from the time gametes are formed and shed, to the implantation of the zygote, through morphological transformations including growth, movement of germ layers, and differentiation of cells and tissues during development, metamorphosis in those forms in which it occurs, and on to the death of cells.

A vast amount of research has been carried out in attempting to discover the functions of the endocrine glands, or glands of internal secretion. Many extremely important discoveries have been made concerning the activities of these glands in health and disease. Of special interest to the comparative anatomist is the fact that the endocrine glands occur with some uniformity in all vertebrates. Even though the organs making up the endocrine system vary widely as to embryonic origin and are derived, for the most part, from other organ systems, we shall consider them collectively as composing a definite organ system of the body. This is because of their functional relationships by means of the hormones they produce, even though each gland has its particular and specific functions. Some of the glands, however, show closer interrelationships than do others. The science of *endocrinology* is concerned with the structure, function, and interrelationships of these glands. It appears that no single hormone acts entirely by itself but, rather, interacts with other hormones in many ways to bring about total effects. The balance among hormones, their dual or multi-ple actions, their sequence of action, etc., are all rather complicated phenomena. When one considers, for example, that for full expression of mammary-gland function five or more hormones are involved, estrogen and progesterone from the ovaries or the placenta, prolactin and somatotrophin from the anterior lobe of the pituitary gland, and one or more hormones from the adrenal cortex, it is evident that proper coordination of endocrine factors affecting a body, an organ, or a part, is most complex. From a structural viewpoint the component parts of the endocrine system have undergone little change during the evolution of the several vertebrate classes. There have been rather striking phylogenetic changes, however, in the chemical nature and activity of hormones produced by the glands as evolution has progressed. Recent findings (S. Liao, 1968, *Amer. Zoologist*, 8:233) indicate that the biological actions of hormones may actually be secondary to the primary control of the synthesis of ribonucleic acid (RNA) by these hormones.

In addition to the gonads of both sexes and the placenta of pregnancy (at least in mammals), the endocrine system includes the thyroid, parathyroid, adrenal, and pituitary glands, and the islet tissue of the pancreas.

Among the endocrine glands are examples which have differentiated from all three germ layers. Those which are mesodermal derivatives (gonads, placenta, adrenal cortex) secrete hormones which are steroids; those derived from ectoderm or endoderm give off hormones which are modified amino acids, peptides, or proteins.

The pineal body and thymus gland were formerly considered to be endocrine organs. Then, for a time, it was believed that the pineal body was nothing more than a vestigial structure. Recent experiments (page 623) indicate that it probably belongs in the endocrine category after all. The thymus gland, apparently, is a lymphoid organ which is primarily of use to the body in establishing mechanisms which are of great significance in combating infection and protecting it from invasion by bacteria or foreign tissues. Evidence that thymus tissue may be the source of a blood-borne factor is discussed on page 592.

Certain glands play a dual role as regards their method of secreting. For example, the pancreas is an exocrine gland, with a duct or ducts opening directly or indirectly into the intestine. Its exocrine secretion is rich in enzymes which are important in the process of digestion. Scattered among the alveoli, or groups of cells which form the exocrine secretion, are irregular masses composed of an entirely different type of cell. Such groups of cells are known as the *islets (islands) of Langerhans.* They are believed to elaborate the internal, or endocrine, secretions of the pancreas. The hormones, of which there are probably two, are known as *insulin* and *glucagon,* respectively. The ovaries and testes are cytogenic organs since they produce the reproductive cells. They have, however, an endocrine function or functions in addition, the endocrine secretions being formed in tissues which are quite different from those that give rise to ova and spermatozoa.

Numerous chemical substances formed by cells in various organs in the body are carried by the circulatory system to other parts of the body where they bring about reactions of one sort or another. Since the cells which produce these secretions do not form discrete or circumscribed glandular structures and show no functional interrelationships, they are more properly discussed in connection with the various organs of which they form a part. Among the secretions included in this category are gastrin (page 184); serotonin, secretin, pancreozymin, enterogastrone, and cholecystokinin (page 191); renin, angiotensin, and erythropietin (page 268), as well as others.

It is not improbable that unicellular endocrine elements exist which are capable of migrating through tissues, giving off regulatory chemical substances into the body fluids.

The nervous and endocrine systems, now usually spoken of together as the *neuroendocrine system,* are the dominating and coordinating systems of the body. The nervous system exerts its control through nerves which are distributed directly to the various body structures. The endocrine system has a simpler method in that its means of control is humoral, the secretions, or hormones, being carried passively to all parts of the body via the blood and lymphatic vessels. Whereas the nervous system brings about or controls rapid coordinations of the body, the actions of hormones are characterized, in general, by duration rather than speed. Examples of hormone action include such phenomena as growth, reproduction, alterations in blood chemistry, regeneration, molting, etc. It is generally agreed that, in the last analysis, hormones, at the molecular level, modify the reactions of enzymes within the cells which they affect. They may alter cell permeability or the permeability of some intracellular organelle; they may activate preexisting proteins so as to increase their enzymic activity; they may affect the rate of synthesis of a particular enzyme so as to increase its amount; they may serve as co-substrates with specific enzymes in causing certain activities; they may even suppresss or activate certain genes and alter the synthesis of a specific kind of ribonucleic acid (RNA) (C. A. Villee, 1967, *Amer. Zoologist,* 7:109).

It is of interest that discrete endocrine glands are even to be found among certain invertebrates. The corpora allata and pro-thoracic glands of insects, and the androgenic glands as well as the Y organs of crustaceans, are examples. Neurosecretory cells and neuroendocrine regulation appear in turbellarians of the phylum *Platyhelminthes*.

THYROID GLAND

Perhaps the most familiar of all the structures making up the endocrine system is the thyroid gland. In man it is located in the lower part of the neck region, lying ventral and lateral to the trachea (Fig. 9.4) just caudad the larynx. Ordinarily the thyroid gland is inconspicuous, but when enlarged, as in certain abnormal conditions, it may protrude prominently. An enlarged thyroid gland is commonly known as a *goiter,* of which there are several types.

The thyroid may be considered to be almost as typical a vertebrate organ as the notochord. Its histological structure is similar in all vertebrates. The gland is composed of varying numbers of irregularly rounded

thyroid follicles of different sizes (Fig. 9.1). In man they range in magnitude from 0.05 to 0.5 mm in diameter, the larger ones being nearer the periphery. The follicles are closely packed together and each is surrounded by a basement membrane. A fine network of reticular connective-tissue fibers, supporting a very rich capillary supply, lies between the follicles. With the exception of the adrenal gland, it is probable that, in relation to its size, more blood passes through the thyroid gland than any other structure in the body. Small clumps of cells are occasionally observed lying between the cells of a follicle and its basement membrane. These *parafollicular cells* have a clear cytoplasm and are larger than the follicular cells proper. Their function, if any, is unknown. Each follicle is normally lined with a single layer of low cuboidal cells enclosing a mass of viscid material known as *colloid.* Use of the electron microscope has revealed that short and blunt microvilli extend from the follicular cells into the colloid and that certain follicular cells may even be ciliated. The

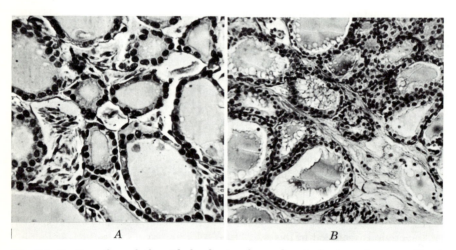

Fig. 9.1 Sections through thyroid glands: *A*, salamander, *Ambystoma texanum; B*, rat (same magnification).

Anatomy of the Chordates

epithelial cells are low and flattened when the thyroid gland is inactive, and tall and columnar when it is secreting to capacity. There is much variation in its microscopic appearance. Sufficient amounts of iodine must be available if the gland is to function normally. The colloid represents the stored-up product of the secretory epithelial cells, the amount present fluctuating with changes in the physiological activity of the gland.

The colloid is, for the most part, made up of an iodized glycoprotein called *thyroglobulin*. This is a large protein molecule with a molecular weight of approximately 680,000. Like all proteins it is composed of numerous amino acids linked together. Thyroglobulin is actually the form in which the thyroid hormones, which are amino acids, are stored. It is not ordinarily present in the bloodstream. Under normal physiological conditions thyroglobulin within the follicles is hydrolyzed by certain enzymes, yielding a number of amino acids having the element iodine as part of their chemical structure. Iodine is removed from all except two of these. The amino acids pass through the epithelial follicular cells and into the bloodstream via capillaries surrounding the follicles. It was formerly believed that one of the iodinated amino acids, *thyroxine*, was the only hormone produced by the thyroid gland. It is now known that in addition to thyroxine, another iodinated amino acid, *triiodothyronine*, representing a second thyroid hormone, is formed during the hydrolysis of thyroglobulin and passes into the bloodstream.

Thyroxine contains approximately 65 per cent iodine. Triiodothyronine contains less iodine than thyroxine but is about seven times more potent than the latter. Furthermore, its action on the body takes place more rapidly. These hormones become linked with blood proteins and circulate through the body in that form.

Some investigators are of the opinion that triiodothyronine represents merely thyroxine which has been altered chemically, possibly by the kidney, and is the actual form of the hormone affecting the tissues. More, however, share the belief that both hormones may be liberated directly into the bloodstream by the thyroid gland. When iodine is plentiful, thyroxine seems to be the major hormone liberated. When iodine supplies are deficient a greater amount of triiodothyronine is formed. Since the latter is far more potent than thyroxine, it would seem that its production portrays an effort by the body to use the available iodine in the most efficient manner possible.

The thyroid hormones are important in controlling the rate of metabolism of the entire body. Exactly how this is accomplished is not known. It is possible that deaminated analogues of these hormones may be the actual chemical compounds which affect the peripheral tissues in regulating their basal metabolic rate by stimulating the activation or synthesis of oxidative enzymes.

Reproductive processes are profoundly affected by abnormal thyroid function, suggesting that thyroid hormones may in part control neurosecretory cells in the hypothalamus, which in turn affect the release of gonadotrophic hormones by the anterior lobe of the pituitary gland (page 390). Brain development is generally retarded in vertebrates rendered hypothyroid early in life.

The nerves supplying the thyroid gland are, for the most part, vasomotor fibers. Since thyroid tissue may be successfully transplanted to other parts of the body, it is apparent that its secretory function is not dependent upon a nervous mechanism.

Administration of certain so-called antithyroid or goitrogenic drugs, such as thiourea, thiouracil, and sulfonamides, prevents the cells of the thyroid gland from synthesizing the hormones, even when an abundant supply of iodine is available.

Invertebrates. Nothing corresponding to a thyroid gland has been identified among invertebrates. The hormone thyroxine and some of its precursors have been found to exist in the exoskeletal structures of many invertebrates. Triiodothyronine has been detected in some of the gorgons, or sea fans. The latter are marine forms, belonging to the phylum Coelenterata. The presence of these hormones in invertebrates does not indicate that they have any special function and is of no particular significance. Even in mammals, thyroxine may be found in body tissues in the absence of the thyroid gland. It seems clear, however, that a special organ for their storage first appears in vertebrates.

Protochordates. No thyroid gland is present in the protochordates. In amphioxus and urochordates a glandular, ciliated groove in the ventral wall of the pharynx, known as the *endostyle,* is clearly a structure which aids in trapping food particles which are then passed on to the remainder of the digestive tract. Some comparative anatomists have homologized the endostyle of these lower forms with the thyroid gland of the vertebrates. Until recently there has been little justification for this. However, in experiments in which amphioxus has been immersed in sea water containing radioactive iodine, it has been found that iodine, bound in organic form, tends to concentrate in the region of the endostyle. There is also evidence that the endostyles of urochordates possess some iodine-binding capacity and can probably synthesize monoiodotyrosine, diiodotyrosine, and thyroxine.

Cyclostomes. A *subpharyngeal gland,* some times referred to as the endostyle, is present in the ammocoetes larva of the lamprey. It lies beneath the floor of the pharynx, roughly between the first and fifth gill pouches. The subpharyngeal gland is a complicated structure which possibly functions as a mucous gland. Its orifice is associated with a system of ciliated grooves in the pharyngeal wall. Five types of cells have been described in the gland. At the time of metamorphosis certain of these cells lose their cilia, change their shape, and become arranged in the form of typical thyroid follicles. It is for this reason that the endostyle of the protochordates has in the past been regarded as the forerunner and homologue of the thyroid gland of higher forms. Experiments in which radioactive iodine has been used as a tracer substance show that this element accumulates rapidly in certain cells of the ammocoetes subpharyngeal gland even before they become organized as follicles. Other cells apparently are unaffected. These findings indicate that only a portion of the ammocoetes subpharyngeal gland, and not the entire organ, may be the homologue of the thyroid gland, and that the homologies of the endostyles of the protochordates with the thyroid glands of higher forms are only partially indicated. The fact that the thyroid first makes its appearance in the embryos of higher forms as a median, ventral diverticulum of the pharynx is also of special interest in this connection. Administration of thyroid hormone is without effect in bringing about metamorphic changes in the ammocoetes larva of the lamprey.

In adult lampreys the thyroid follicles do not form a compact, encapsulated structure. They show a tendency to be distributed along the ventral aorta and the arteries leading to the gills.

In the hagfish the thyroid substance is composed of rounded or ovoid follicles embedded in fat. They are median in position, are arranged separately or in groups between the pharynx and ventral aorta, and also lie between the gill pouches on either side.

Fishes.   In all fishes a thyroid gland appears early in embryonic development. Its location in the adult shows considerable variation. The thyroid of elasmobranchs is a single, compact organ located posterior to the mandibular symphysis and just anterior to the point where the ventral aorta bifurcates into the first pair of afferent branchial arteries.

In teleosts it is usually paired and lies near the first branchial arch on either side. In some, as in the perch, it is more diffuse and consists of small masses which lie under the ventral aorta and scattered along the paths taken by the afferent branchial arteries. Thyroid tissue in certain teleosts is found in such unusual localities as the spleen, kidney, brain, and eye. It seems that in such cases tissue from the pharyngeal region is probably somehow distributed by the bloodstream to these remote areas.

Although the thyroid gland in the Dipnoi is arranged in a single mass, it shows some indication of being a paired structure, since it is made up of two prominent lateral lobes connected by a constricted central portion. It lies under the epithelium covering the tongue, anterior to the muscles of the visceral skeleton and just above the symphysis of the hyoid apparatus.

The thyroid glands of fishes reared on inadequate diets may become goitrous. Fish hatcheries must take precautions to see that enough iodine is present in the water to prevent abnormalities in growth due to improper function of the thyroid gland.

Amphibians.   In urodeles the development of the thyroid gland has been studied in several species. It appears very early in a developing embryo in the ventral wall of the pharynx as an unpaired structure which

soon divides into two. In the adult salamander the glands lie on either side of the throat region on the ventral side and slightly anterior to the aortic arches. The thyroid vein, a tributary of the external jugular vein, passes ventral to the gland. By following the course of this vessel the thyroid gland may easily be found (Fig. 9.2). The number of follicles in the thyroid varies considerably. Within a given species the number seems to be correlated with the size of the animal. For example, in a small species, such as *Eurycea bislineata,* as few as five follicles have been observed in the gland of a medium-sized individual, but as many as 14 may be present in a larger specimen. The average number for this species is 8.8.

Among anurans the thyroid of the frog has been most carefully studied. Here, too, the gland is paired and on each side appears as an oval-shaped structure located lateral to the hyoid apparatus in the notch formed by the junction of the lateral and thyrohyoid processes. It is rather deeply concealed, and this, together with the small size

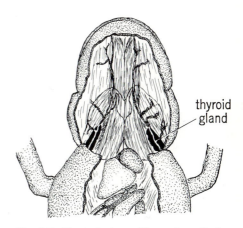

Fig. 9.2   Ventral view of lower jaw of salamander, *Ambystoma texanum.* The skin has been removed to show the position of the thyroid glands.

of the structure, accounts for the difficulty in finding it by ordinary dissection.

It has been demonstrated that if the thyroid rudiment is removed from amphibian tadpoles complete metamorphosis fails to occur. The animals continue to grow; in fact they may become larger than normal tadpoles and their lungs and reproductive organs will ultimately develop. If such animals are fed thyroid substance they soon undergo metamorphosis, the tail being resorbed, the limbs developing, and the mouth undergoing a pronounced transformation. It has long been known that when small tadpoles are fed with thyroid substance they stop growing and undergo a spectacular early metamorphosis, quickly being transformed into tiny frogs or toads (Fig. 9.3). Even today this reaction is considered to be one of the most sensitive tests for the presence of thyroid hormone. It appears, furthermore, that administration of elemental iodine to tadpoles will hasten metamorphosis in both normal animals and those from which the thyroid gland has been removed. It would seem, therefore, that when an adequate amount of iodine is present the animals can utilize it even in the absence of the gland. However, such tadpoles must have attained a certain degree of development before they are able to respond to iodine. Administration of iodine alone probably results in formation of thyroxine by combination with the amino acid, tyrosine, present in body tissues. Recent findings (K-H Kim and P. P. Cohen, 1968, *Amer. Zoologist,* 8:243) indicate that the natural and thyroxine-induced metamorphosis of the tadpole may basically be caused by an increased synthesis of ribonucleic acid (RNA) by the nuclei of liver cells. This, in turn, induces the formation of certain enzymes, the level of which increases dramatically at the time of metamorphosis. These enzymes are involved primarily in the biosynthesis of urea from ammonia. It is still too early to speculate as to the precise effects which the enzymes and the increase in urea have upon the striking biochemical and morphological changes in those body structures which are altered at the time of metamorphosis.

There is quite a marked variation in the response of different amphibian species to thyroid hormones. In some frogs, the usual aquatic tadpole stage of development is passed over entirely and the young hatch from the eggs as completely formed miniatures of the adults. Administration of thyroxine has no effect upon development in such species. The American axolotl, a salamander inhabiting high altitudes in the

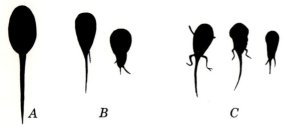

Fig. **9.3** Toad tadpoles in which acceleration of metamorphosis has been brought about by thyroid feeding: A, control; B, effect of 5 days of thyroid feeding; C, effect of 15 days of thyroid feeding. (*After Gudernatsch.*)

Anatomy of the Chordates

western United States, normally retains its gills throughout life. Treatment with thyroxine, or iodine alone, brings about prompt resorption of the gills, and the animal then becomes identical with adults of the species *Ambystoma tigrinum.* Failure of the axolotl to metamorphose in its normal habitat may be the result of (1) a lack of iodine in the environment, (2) inability of the thyroid gland to manufacture its hormone, (3) failure of the general body tissues to respond normally to the hormone, or (4) a defect in the pituitary gland insofar as its ability to secrete the thyroid-stimulating hormone (TSH) is concerned. The urodele, *Necturus,* retains its gills throughout life and seems incapable of metamorphosing under any conditions. This may possibly be explained as a lack of responsiveness of the tissues of the animal to thyroid hormone. It seems more likely, however, that there is a morphological explanation for the failure of *Necturus* to metamorphose (page 565).

The process of ecdysis, or periodic shedding of the corneal layer of the skin, is also at least partly under control of the thyroid gland. Amphibians in which the thyroids have been extirpated fail to undergo ecdysis. The dead layers of skin pile up, and the animal appears to assume a much darker color than normal (page 387). Thyroid administration enables such animals once more to cast off the dead layers of epidermis.

Reptiles. In snakes, turtles, and crocodilians the thyroid gland is unpaired, whereas in lizards it is bilobed in the young but paired in the adult. In lizards it lies ventral to the trachea, approximately halfway along its course. In other reptiles it lies farther posteriorly and immediately in front of the pericardium.

Experiments have shown that in reptiles, as in amphibians, the process of ecdysis is partly under control of the thyroid hormone. However, it seems that in snakes, removal of the thyroid enhances ecdysis, whereas thyroid administration interferes with the normal shedding of the corneal layer (W. H. Schaefer, 1933, *Proc. Soc. Exp. Biol. Med.,* **30**:1363). Why the action of the thyroid hormone in respect to corneal shedding in snakes is just the reverse of its effect on amphibians is not clear. Lizards apparently are similar to amphibians in this regard.

Birds. The paired thyroid glands of birds possess no noteworthy features not previously considered. They lie on either side of the trachea near the region where the trachea divides into bronchi and just anterior to the large vessels leaving the heart. Experiments in which thyroxine has been injected into brown leghorn chickens show that the color pattern and shape of certain feathers can be decidedly altered in the presence of unusual amounts of the hormone.

Mammals. The typical mammalian thyroid consists of right and left lobes connected across the ventral side of the trachea by a narrow *isthmus* (Fig. 9.4). The lateral lobes are firmly attached to the larynx, and the isthmus is fastened to the trachea. Thus if the larynx and trachea move, the gland moves also. The isthmus is the only part of the normal thyroid that is ordinarily palpable in man. If the finger is gently pressed against the trachea at the level of the suprasternal notch and the movement of swallowing is executed, the isthmus of the thyroid will be felt to move up and back again.

The thyroid gland arises in the mammalian embryo as a median ventral diverticulum of the pharynx between the first and second pharyngeal pouches. Thus, for a time, it is connected to the pharynx by a duct known as the *thyroglossal duct.* Under normal condi-

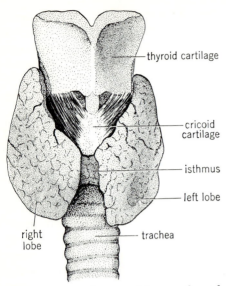

thyroid cartilage

cricoid cartilage

isthmus

left lobe

right lobe

trachea

Fig. 9.4 Ventral view of human thyroid gland shown in relation to trachea and larynx.

tions this connection is soon lost and the only indication of its former point of origin is a small depression at the base of the tongue called the *foramen caecum* (Fig. 5.8). Occasionally small remnants of thyroid tissue are embedded in the substance of the tongue, and if these should undergo hypertrophy, a *lingual goiter* is the result.

In human beings, as well as in certain other mammals, several abnormal conditions are known to result from improper function of the thyroid gland. As was mentioned previously, the thyroid gland plays an important role in regulating the metabolism of the body. Excessive production of thyroxine increases heat production, sensitizes the nervous system, and accelerates the rate of heartbeat. All activities of the body are speeded up.

The thyroid gland may, on the other hand, be lacking entirely or may not give off a sufficient amount of the hormone, and the resulting effects are quite the opposite of those dis-

cussed. Thyroid deficiency in the young is marked by a retardation of growth and development. In severe cases intelligence is subnormal, the teeth are slow to form and erupt, the sex organs fail to develop, the abdomen protrudes, and body temperature is low. The tongue is thick and may extend out of the mouth, the skin is wrinkled, and the hair is scant and coarse. A child having such extreme lack of thyroid is known as a *cretin.* Some cretins are born without thyroids.

If the thyroid gland degenerates during adult life, the symptoms are, in general, similar to those of cretinism except that, since growth has been completed, only degenerative changes are observed. This condition, called adult cretinism, or *myxedema,* occurs much more frequently in women than in men. The body temperature is lowered, and the person may continually complain of cold. Mental processes become dulled; the skin is thick and puffy in appearance; the tongue thickens; and speech becomes labored. The hair is dry and sparse. Persons suffering from myxedema usually gain weight and may even become quite obese.

Mild degrees of over- or undersecretion of the thyroid gland show less severe symptoms than those described above.

Fortunately, the administration of thyroid substance either by mouth or by injection can bring persons who suffer from thyroid deficiency back to normal. Usually the treatment must be continued indefinitely; otherwise the patient may lapse into his former state. Persons who take thyroid without the supervision of a physician run a great risk, because the drug is so powerful that harmful effects may result from its use.

There are, in general, two types of thyroid enlargement, or goiter, one marked by an undersecretion of hormones and the other by an oversecretion. In both, the thyrotrophic, or thyroid-stimulating, hormone (TSH), believed to be secreted by basophilic cells of the anterior lobe of the pituitary gland, is involved. There is a reciprocal, or "push-pull," relationship between the thyroid hormones and TSH. If sufficient thyroid hormone is present, the secretion of TSH by the pituitary is inhibited. If there is a thyroid deficiency, the amount of TSH increases, and the thyroid gland is stimulated to greater activity. The relationship may be direct. It has been demonstrated that in rats the output of TSH is stimulated by low environmental temperatures. This in turn increases the output of thyroxine and triiodothyronine by the thyroid gland, enabling the animal better to utilize more food and keep its body temperature constant. The hypothalamus is undoubtedly involved in initiating this sequence of events (page 390).

The most common form of thyroid enlargement, *endemic goiter,* is actually the result of thyroid deficiency. Because of a paucity or lack of iodine, the gland cannot manufacture its hormones. The pituitary gland then secretes greater quantities of TSH; the thyroid gland is stimulated; the colloid store becomes reduced; the epithelial cells of the follicle become more columnar in shape and much more numerous because of mitotic divisions. The actual tissue of the gland, therefore, increases in volume, despite the lack of colloid. An enlarged thyroid gland of this type is called a simple *parenchymatous goiter.* The requirements of the body for thyroid hormones vary under different conditions. At the time of puberty or during pregnancy they are usually greater than at other times. It is at such periods that a parenchymatous type of goiter is most apt to develop, especially if iodine supplies are low. When the demands of the body are decreased, or if the iodine supply again becomes adequate, the concentration of thyroid hormones in the bloodsteam increases. The production of TSH is then suppressed, and the follicles of the thyroid gland once more begin to store colloid. The follicular epithelium changes to the low cuboidal type as the follicles become distended with colloid. As a result, the gland becomes somewhat larger than it was in the parenchymatous state and is referred to as a *simple colloid goiter.* Both parenchymatous and colloid goiters are actually the result of a deficiency of iodine. A cycle such as that just described may be repeated many times in the life of a given individual and the thyroid gland may ultimately become relatively enormous in size. This type of goiter occurs much more frequently in women than in men.

Endemic goiter occurs more often in certain regions of the world than in others. These areas, known as goiter belts, are alike in that the soil and drinking water are deficient in iodine. Some of the best-known goiter belts are in the Himalaya Mountains in Asia, the Swiss Alps, the Andes Mountains in South America, and, in North America, the Great Lakes basin as well as the Cascade Range in Oregon, Washington, and British Columbia. Less important regions are found in the Appalachians and in the Rocky Mountains. In general, these are either well-drained areas or places where the soil, deposited by glaciers in ancient times, has been leached thoroughly by melting ice. In either case the iodine content of soil and water is unusually low. Goiter rarely occurs along the seacoast. Before the days of extensive travel and before the treatment of goiter was fully established, those who lived in goiter belts were so accustomed to the presence of goiter that in some cases it was considered a natural and normal development.

Administration of small amounts of iodine to persons inhabiting these areas is of great value in preventing the appearance of goiter.

A number of years ago an important and interesting experiment was carried on among the schoolgirls of Akron, Ohio, which lies in the Great Lakes goiter belt. The girls were divided into two groups. One group was untreated but observed before and after the experiment which extended over a period of 2½ years. To the other group, for 2 weeks twice a year, small amounts of a chemical compound containing iodine were given by mouth. At the end of the experiment, out of 2,305 untreated children, 445 had developed goiter, while of the 2,190 receiving treatment, only 5 developed a thyroid enlargement. This significant work indicates in a striking manner how iodine may be used in preventing goiter. Even if a goiter has developed, the administration of iodine under the supervision of a physician may often result in its disappearance. Iodized salt is frequently used to supply small amounts of iodine which otherwise may be lacking in the diet.

Another type of thyroid enlargement, accompanied by protrusion of the eyeballs, is known as *exophthalmic goiter*. This is a more complex condition than endemic goiter and all the factors involved are not completely understood. Despite the availability of an adequate supply of iodine, the thyroid gland presents a histological picture similar to that of parenchymatous goiter. The gland is secreting actively, or even overactively, producing more thyroid hormones than the body requires. Yet, for some reason or other, the secretion of TSH by the anterior lobe of the pituitary gland is not suppressed. It seems that the primary seat of this condition lies in the abnormal functioning of the anterior lobe of the pituitary rather than in the thyroid gland. Excessive secretion of TSH is responsible both for hyperactivity of the thyroid and protrusion of the eyeballs. Abnormal action of the sympathetic nervous system in contracting the orbital muscle behind the eyeball may possibly be involved. Surgical removal of all or part of the thyroid is often resorted to in order to reduce the amount of thyroid hormone being secreted and thus relieve the hyperthyroid symptoms even though the exophthalmos continues to be maintained. It has been demonstrated experimentally that the exophthalmic condition can be brought about by injecting large doses of TSH into animals whose thyroid glands have been extirpated. There is reason for believing that a long-acting thyroid stimulator (LATS) may be involved in the development of exophthalmic goiter. Little is known about this factor. Administration of such goitrogenic drugs as thiourea or thiouracil to patients with exophthalmic goiter may relieve the hyperthyroid symptoms by preventing synthesis of thyroid hormones. In some cases, however, protrusion of the eyeballs may become even more accentuated, since the thyroid hormones are reduced and even greater amounts of TSH are then elaborated.

The relationship of the thyroid gland to the other glands of the endocrine system is not entirely clear. The reciprocal action between the production of thyroid hormones and the thyrotrophic hormone of the anterior lobe of the pituitary gland has already been stressed. Another hormone secreted by the anterior lobe of the pituitary gland is the growth, or somatotrophic, hormone (STH). Apparently, in cases of thyroid deficiency, the elaboration of this hormone by the pituitary gland is also interfered with. When young animals from which the thyroid gland has been removed, and in which growth has ceased, are given injections of STH they will continue to grow but at a very much reduced rate. On the contrary, if the pituitary gland is removed, administration of thyroxine will not restore normal body growth even though it increases the metabolic rate. It would seem that normal thyroid secretion is related to normal production of somatotrophic hormone by the pituitary gland and that the two have a synergistic action in promoting growth of the skeleton.

Some impairment of the functional activity of the cortex of the adrenal gland is indicated in cases of thyroid insufficiency. This may be due to suppression of secretion of the adrenocorticotrophic hormone (ACTH) by the anterior lobe of the pituitary gland.

The proper functioning of the reproductive organs also depends upon normal activity of the thyroid gland. There is, however, considerable variation among species in this respect. The age of the animal is an important factor, however, the gonads of the young being more affected by thyroid deficiency than those of the old. Production of ovarian and testicular hormones, as well as oögenesis and spermatogenesis, are affected.

Some interesting facts have been brought to light by experiments in which small amounts of radioactive iodine have been injected into laboratory animals. Within an hour after radioiodine is administered, most of it is concentrated in the thyroid follicles between the epithelial cells and the colloid. By the end of 24 hours it is evenly distributed throughout the colloid. This illustrates the rapidity with which new thyroglobulin is formed, replacing that which has undergone hydrolysis and liberated hormones. Administration of radioiodine is utilized in certain diseased conditions of the thyroid gland since it becomes concentrated in the gland itself rather than being generally distributed throughout the body.

Chemists have prepared numerous analogues of thyroxine which are not found in nature and which differ considerably in potency from either of the natural hormones. For example, a preparation known as triiodothyropropionic acid has been shown to be 300 times as effective as thyroxine in bringing about metamorphosis of tadpoles of the common leopard frog, *Rana pipiens*.

In recent years a hormone has been discovered which has the effect of rapidly lowering the level of calcium in the blood. Whether this hormone, which is a polypeptide named *thyrocalcitonin*, or *calcitonin*, originates in the thyroid or parathyroid glands (page 349) is at present a matter of controversy.

PARATHYROID GLANDS

The parathyroid glands get their name from the fact that in mammals they lie alongside of or dorsal to the thyroid gland. Frequently they are wholly or partially embedded in the thyroid or thymus glands. In early stages of development in all vertebrates small bud-like masses of epithelial cells proliferate from both dorsal and ventral portions of the pharyngeal pouches. The number varies in different vertebrates and vertebrate groups. It is clear that in the higher tetrapods certain of these epithelial buds give rise to parathyroid glands which are thus of endodermal origin. In cyclostomes and fishes, however, because of the small size, diffuse nature, and difficulty in extirpating these glands completely, it has not been conclusively determined whether any of the buds actually represent parathyroid tissue. No homologues of parathyroid glands have been identified in invertebrates.

Cyclostomes. The lampreys possess the most complete set of these small epithelial structures, since all pharyngeal pouches give rise to both dorsal and ventral buds. Their significance is unknown.

Fishes. Glandular pharyngeal bodies of varying number have been identified in fishes, but their functional identity remains obscure. Irregular masses of soft tissue in the region above the gill slits, however, are believed to represent thymus tissue.

Amphibians. The parathyroid glands are first identified as such in amphibians. These, as well as the thymus glands, are usually derived from the third and fourth pairs of pharyngeal pouches. In anurans, however, the thymus arises almost entirely from the second pouches. Whereas the thymus glands arise as dorsal buds, the parathyroids come from the ventral portions of the pouches. In urodeles they lie in a position lateral to the aortic arches and ventral to the thymus glands. Usually two or three parathyroids appear on each side, but in some forms only a single pair may be present. They are rather widely separated from the thyroids in urodele amphibians. Perennibranchiates seem to lack discrete parathyroid glands. In anurans the parathyroid glands are small, rounded, reddish bodies, two of which usually lie on either side of the posterior portion of the hyoid cartilage next to the inner ventral surface of the external jugular vein. Some authors in the past have referred to them as *pseudothyroid glands.*

Apparently the parathyroids in amphibians function in a manner similar to those of mammals, controlling the level of calcium and phosphorous salts in the blood.

Reptiles. The parathyroids of reptiles arise from the ventral ends of pharyngeal pouches II, III, and IV. In turtles only those from pouches III and IV persist, being embedded in thymus tissue; in crocodilians those from pouch III are the only ones to survive. In most other reptiles all three pairs of parathyroid elements are present in the adult. They are frequently confused with other bodies of pharyngeal origin and are located on the sides of the neck, somewhat lateral and posterior to the thyroids. In snakes the parathyroids are situated nearer the skull than in other members of the class.

Birds. The parathyroid glands of birds are derived from ventral diverticula of pharyngeal pouches III and IV. In the adult animal they may be observed as small, paired bodies, slightly posterior to the thyroid gland on each side. There are one or two pairs. There is a tendency for the two elements on each side to fuse. Occasionally the parathyroids lie on the dorsal surface of the thyroid. They are frequently situated close to the carotid body. Accessory elements have even been reported to be located inside the small ultimobranchial bodies.

The eggs of birds are enclosed by a hard calcareous shell. Much calcium is needed for shell formation at egg-laying periods. Some experiments on pigeons (L. G. Raisz and D. F. Hammock, 1959, *Proc. Soc. Exp. Biol. Med.,* **100**:411) are of special interest. Injections of the female sex hormone, estrogen, simulating ovarian activity, cause a great increase in blood calcium, whereas administration of PTH (page 348) brings about only a moderate increase. Nevertheless, in the absence of the parathyroid glands, estrogen administration has no effect on the blood calcium.

Mammals. Much more is known about the parathyroids in mammals than in any other group. In this class, contrary to the condition in lower forms, the parathyroids arise from the *dorsal* ends of pharyngeal pouches

Anatomy of the Chordates

III and IV (Fig. 9.5). In man (Fig. 9.6) there are usually two glands on each side supplied by branches of the inferior thyroid artery. Occasionally accessory glands may be found, as many as 12 having been reported. The typical glands are roughly oval in shape, about the size of a grain of corn, and of a yellowish-brown color. The names *superior* and *inferior parathyroids* are generally applied to them, the superior pair being more cranial in position. During embryonic development their position becomes shifted so that those coming from the third pair of pharyngeal pouches are dragged along by the developing thymus rudiments and assume the more caudal position. Those coming from the fourth pair of pouches move slowly in a cranial direction and become the superior pair. The superior parathyroids are much more constant in position than the inferior ones, which are sometimes found

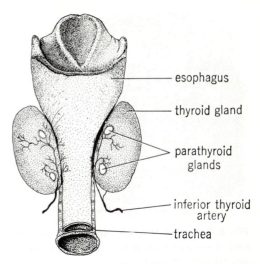

Fig. 9.6 Dorsal aspect of human esophagus and trachea, showing position of thyroid and parathyroid glands. (*After Halstead and Evans.*)

some distance caudad of the thyroid gland, associated with the thymus gland, and lying fairly deep in the anterior portion of the mediastinum. In dogs they are occasionally found as far back as the point of bifurcation of the trachea.

The nerve supply to the parathyroid glands is scanty and probably confined to vasomotor fibers of the blood vessels supplying them. Glands transplanted to vascular sites apparently function normally even though lacking a direct nerve supply.

In mammals the parathyroid glands, particularly the superior pair, may be embedded in the thyroid gland (Fig. 9.7). In the rat, mouse, and several other small mammals only one pair of glands is usually present. Accessory glands occur more frequently in such herbivorous forms as cattle, rabbits, sheep, and goats. The parathyroids of the horse most closely resemble those of man in histological appearance.

The parathyroid glands are composed of rather densely packed, polygonal cells which

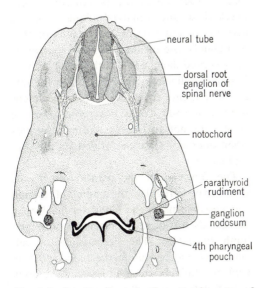

Fig. 9.5 Section through pharyngeal region of 10-mm pig embryo, showing the parathyroid rudiments arising as dorsal diverticula from the ends of visceral pouch IV.

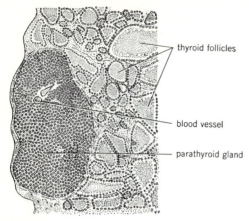

thyroid follicles

blood vessel

parathyroid gland

Fig. 9.7 Section through parathyroid and thyroid glands of rat, showing parathyroid partially embedded in thyroid tissue.

are separated into cords and clumps by delicate connective-tissue septa. In adults two types of cells have been described: *principal,* or *chief, cells* and *oxyphile cells.* In man the latter do not make themselves evident until the fourth to seventh years of life. They are reported to increase in number after the age of puberty has been attained. The chief cells are small, the nucleus occupying the greater portion of the cell. Granules are lacking in the cytoplasm which, therefore, stains rather lightly. Nevertheless, some cells stain more deeply than others so that both light and dark chief cells may be identified. Certain of the light chief cells are referred to as "clear cells." Oxyphile cells, which are found only in certain mammals, are larger than the chief cells and have a more abundant cytoplasm containing acidophilic granules. The significance of this dual arrangement of cells is not clear. No function has been ascribed to the oxyphile cells and most investigators are of the opinion that the chief cells alone are concerned with the elaboration of the parathyroid hormone. The latter is commonly known as *parathormone,* or

PTH. In older individuals the chief cells frequently become arranged in the form of follicles enclosing small amounts of colloidal material. The colloid is lacking in iodine; otherwise little is known about it.

The parathyroid glands are essential to life unless, in their absence, remedial measures are instigated. If they are removed or are badly diseased the subject suffers from severe spasms of certain muscles, a condition known as *tetany.* Death usually follows in a few days after removal. Certain species of animals are more susceptible to the effects of parathyroidectomy than others, the variations being due, in all probability, to the presence or absence of accessory parathyroid tissue. It should be mentioned in passing that not all forms of tetany are related to parathyroid deficiency.

It was formerly believed that the thyroid gland was necessary for life because when it was removed death resulted. We now know that in such cases the parathyroids were overlooked and removed along with the thyroid. In old museum specimens of thyroid glands, the parathyroids can sometimes be observed dangling from the other tissues. It is not difficult to speculate as to the fate of the patients from whom the specimens were obtained. Today, surgeons exercise great care not to remove the parathyroids when operations are performed on the thyroid gland.

The function of the parathyroid glands seems to be to regulate the amount of calcium and phosphorus in the blood serum. Hence the glands are most important in bone and tooth formation. The percentages of calcium and phosphorus in the blood must be kept at relatively constant levels if the body is to function normally. There is a peculiar reciprocal relationship between the amount of blood calcium and blood phosphorus: if the calcium level rises, the phosphorus level falls; if the calcium level decreases, the phosphorus level rises. Actually the effect of the parathyroid glands seems to be primarily directed at the calcium level alone.

The calcium level in the blood falls when the parathyroid glands are removed, and the symptoms of tetany supervene when it reaches a certain level. It seems that calcium ions act in some way to prevent excessive irritability of the nerves and neuromuscular mechanisms of the body. If the calcium level of the blood is raised to the proper level the tetanic symptoms disappear. This may be accomplished in three ways: (1) by giving calcium salts by intravenous injection; (2) by administering vitamin D, which facilitates absorption of calcium from the digestive tract; and (3) by injection of extracts of the parathyroid glands.

Collip, in 1924, was the first to prepare an active extract of the parathyroid glands which was effective in elevating the level of blood calcium. The active constituent, or hormone (parathormone, or PTH), has not yet been isolated in pure form. In recent years, with the introduction of new chemical and biological techniques, much has been learned about PTH and its functions. The hormone seems to be a straight-chain polypeptide with a molecular weight of approximately 9,000. The sequence of amino acids in the chain is not completely known. The chief cells are believed to manufacture this hormone. Until recently it was believed that PTH was the only hormone secreted by the gland. Another hormone which lowers the level of serum calcium in the blood has already been mentioned (page 345). Whether *calcitonin*, or *thyrocalcitonin*, originates in the parathyroid or thyroid glands or both needs clarification. It has been claimed that the parathyroid glands may produce a factor (calcitonin) which brings about the release of thyrocalcitonin by the thyroid gland. Thyro-calcitonin seems to be a polypeptide unrelated to thyroxine and triiodothyronine. Extracts of a number of other organs have been shown to have a hypocalcemic effect, thus confusing the issue. At any rate it is of interest that the parathyroid-thyroid complex may possibly produce a hormone which counteracts the hypercalcemic effect of PTH. This also provides a possible reason for the close anatomical relationship of the thyroid and parathyroid glands.

If there is too much PTH in the body, as in certain cases involving tumors of chief cells or when an excess of the hormone is administered, the level of blood calcium is raised above normal. This brings about nausea, loss of appetite, and, if prolonged, will cause serious skeletal defects, such as softening of the bones, because of withdrawal from them of these mineral elements. Deposits of calcium in the soft tissues of the body are often observed in such cases. The bones become very fragile and are easily broken. In severe cases bone resorption takes place and the bony tissue is replaced by fibrous connective tissue. Such a condition is referred to as *osteitis fibrosa*, or *Von Recklinghausen's disease.*

The symptoms of parathyroid insufficiency are more pronounced in the growing young and in pregnant or lactating females. In all these the calcium requirements of the body are in excess of normal. In growing individuals the effect of calcium withdrawal is most pronounced in the bony trabeculae on the diaphysial sides of the epiphysial plates of the long bones. In man, dogs, and rats, cataracts of the lenses of the eyes almost invariably occur in long-standing cases of parathyroid insufficiency.

It is worthy of note that the parathyroid glands diminish in size and activity when excessive amounts of PTH are administered. Furthermore they become enlarged and overactive in cases of calcium deficiency, as in rickets. It would seem that there is a direct relationship between the amount of calcium in the blood and the activity of the parathyroid glands. Some evidence has been presented to indicate that PTH also exerts its effect by stimulating the excretion of phosphorus by the kidneys. There is no clear-cut evidence that parathyroids are controlled by the pituitary gland. That there may possibly be some relationship is indicated by the appearance of numerous mitotic figures in the parathyroid glands of animals given extracts of the anterior lobe of the pituitary gland.

ADRENAL GLANDS

The adrenal glands are endocrine organs which derive their name from the fact that they are situated in close proximity to the kidneys. In man, with his upright posture, they are often referred to as *suprarenal glands* because they lie *above* the kidneys and are in contact with them. In mammals they consist of a pair of compact bodies, each made up of two distinct portions: an inner *medulla* of ectodermal origin, derived, along with certain ganglia of the sympathetic nervous system, from neural-crest cells; and an outer *cortex,* derived from mesoderm. The cortex arises on each side from mesenchymatous tissue located between the dorsal mesentery of the gut and the medial surface of the mesonephric kidney. The genital ridges, or gonad primordia, also make their appearance in this same general region. Although in mammals medulla and cortex are so closely related anatomically, it has not been established that a functional relationship exists between them unless it be that both are concerned in "stress reactions," to be described later (page 391).

Most of the cells of the medulla contain fine cytoplasmic granules that stain a brownish-yellow color with the salts of chromic acid. This is called the *chromaffin reaction.* The staining reaction is correlated with the presence of the hormone *epinephrine,* or *adrenalin,* since the depth of the stain and the content of hormone rise and fall together. A similar reaction may be obtained in a test tube when epinephrine and chromium salts are mixed. It seems clear, therefore, that the granules either represent epinephrine itself, or its precursor. The reaction depends upon the oxidation of epinephrine by the salts of chromic acid or some other oxidizing factor or agent. Chromaffin tissue may be found in such other parts of the body as the prostate gland, seminal vesicles, cervix of the uterus, and carotid body. In these regions there is also a close association of the chromaffin tissue with that of the sympathetic nervous system (see page 654).

A substance called *serotonin* (5-hydroxytryptamine) has been identified in the central nervous system, in thrombocytes, and mast cells of certain species. It seems to serve as a chemical transmitter and gives a positive chromaffin reaction. This test, therefore, is not specific for epinephrine.

In the lower classes of vertebrates the homologues of medulla and cortex are usually completely divorced from one another. The structures which are homologous with the medullary regions are referred to as *suprarenal,* or *chromaffin, bodies.* They also give the typical chromaffin reaction. Those structures corresponding to the cortex are frequently referred to as *interrenal bodies.* Only in fishes of the class Chondrichthyes do they actually lie *between* the kidneys. The term "interrenal" is, therefore, misleading, and the name *steroidogenic tissue* is considered to be more appropriate. Certain steroid hormones are secreted by this tissue.

Nothing corresponding to adrenal glands has been described in amphioxus. Chromaffin tissue occurs in certain invertebrates but it is very doubtful whether this bears any relation to the structures being considered here.

Cyclostomes. Adrenal elements are found for the first time in cyclostomes. In *Petromyzon*, for example, there are two clearly distinguished series of bodies which represent adrenal homologues. The first consists of small, irregular, lobelike structures situated along the postcardinal veins, renal arteries, and the arteries running dorsal to the opisthonephros. They may even extend into the lumina of these vessels. This series of small bodies is the steroidogenic, or cortical, series. The other, or chromaffin, series extends from near the anterior end of the gill region (opposite the second gill cleft) to the tail. Here the chromaffin bodies consist of small strips of tissue which extend along the course of the dorsal aorta and its branches. It has been demonstrated experimentally that extracts of tissues from these regions, when injected into cats, will cause a rise in blood pressure comparable to that induced by epinephrine.

Although chromaffin tissue has been found in the hagfish, no steroidogenic tissue has been described.

Fishes. In elasmobranchs the steroidogenic bodies lie at the posterior end of the opisthonephros. They are usually paired in the Batoidea and unpaired in the Selachii. The ochre-yellow color of these structures in the living animal resembles that of the adrenals in anuran amphibians. The chromaffin bodies are paired and located on the segmental branches of the aorta on either side of the spinal column. Some of the anterior bodies may be elongated and united. A close relationship to sympathetic ganglia is clearly shown in this group.

Among actinopterygian fishes there is much variation in arrangement of steroidogenic and chromaffin bodies. In many they are separate entities arranged in a manner similar to that of elasmobranchs; in others the two types of tissue are intermingled. Certain paired bodies of a pinkish color may be present on the surface of the opisthonephros in teleosts or partially embedded in that structure. The name *corpuscles of Stannius* has been used to designate these bodies which were formerly considered to be composed of steroidogenic tissue. This has recently been questioned, and the corpuscles of Stannius may not be homologous with the adrenal cortex after all. Several workers have demonstrated the presence of various steroids in the corpuscles of Stannius. Whether these structures actively metabolize certain steroid hormones or merely serve as steroid storage organs is not clear (J. Nandi, 1967, *Amer. Zoologist,* 7:123). Chromaffin bodies in actinopterygians lie in the walls of the postcardinal veins and their branches, particularly on the right side and toward the anterior end.

In the Dipnoi no tissues corresponding to steroidogenic bodies have been described with certainty. Chromaffin bodies are arranged segmentally around the intersegmental arteries and in the walls of the anterior part of the postcardinal and right azygos veins.

Amphibians. Beginning with this group of vertebrates, the two components of the adrenal glands are much more closely associated. The familiar orange or yellow bands on the ventral surfaces of the opisthonephros of the frog are composed of two types of tissue closely intermixed. The steroidogenic elements are arranged in cell columns of different shapes and sizes. The chromaffin cells are irregularly distributed throughout the gland but are often found along the ends or borders of the cell columns of steroidogenic tissue. They show the typical chromaffin reaction.

In urodeles the adrenal glands are by no means so compactly arranged as in anurans. Steroidogenic and chromaffin components are intermingled and located in strips of tissue which extend the entire length of the opisthonephros and even anterior to it, sometimes being located as far forward as the subclavian arteries.

Reptiles. In most members of this class the steroidogenic and chromaffin constituents are intricately intermingled. In some reptiles, however, the dorsal side of the gland is composed largely of chromaffin cells which penetrate into the mass of steroidogenic tissue only to a slight degree. In Crocodilia and Chelonia the arrangement is practically the same as in birds. In turtles the adrenal glands are yellowish in color and irregular in outline, situated on the anterior ventral

sides of the kidneys; in snakes they are located anterior to the kidneys and at some distance from them.

Birds. In birds the adrenal glands are ochreyellow bodies lying on either side of the vena cava immediately anterior to the kidneys and close to the reproductive organs. The two components are closely interwoven, the chromaffin tissue lying in crevices between masses of steroidogenic tissue.

Mammals. In mammals the adrenals lie at the cranial ends of the kidneys (Fig. 9.8). In some forms they are in close apposition to the kidneys; in others they are separated by a space. The right gland lies closer to the inferior vena cava than does the left. These glands in mammals differ greatly from those of other vertebrates in having a true cortex and medulla. The medulla occupies the central portion of the gland, and its cells show the typical chromaffin-staining reaction. The outer cortex is homologous with the steroidogenic tissue of members of the other vertebrate classes. Accessory "corti-

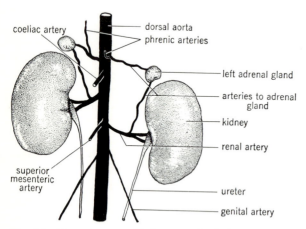

Fig. 9.8 Ventral view of kidneys and adrenal glands of rat, showing arrangement of arteries which supply them with blood.

cal" and "medullary" bodies are frequently found in the vicinity of the adrenal glands but may sometimes lie at a considerable distance from them. There is a great deal of variation in size and number of accessory bodies among individuals and species.

In some species, as in the rat, the adrenals of females are considerably larger than those of males. In the nutria, a rodent, the left gland is 50 per cent larger than the right (E. D. Wilson, M. X. Zarrow, and H. Lipscomb, 1964, *Endocrinology,* **74:**515). Even though medulla and cortex are easily distinguishable in all mammals, the microscopic appearance of these regions varies and an experienced histologist can even distinguish species by these variations.

In the Prototheria, cortex and medulla are not so distinctly separated as in other mammals and cords of medullary tissue penetrate into the cortex. In some, as in the mouse and ant bear (*Tamandua*), a layer of connective tissue separates the two types of cells. Even the proportion of cortex to medulla varies. In the porcupine and guinea pig the cortex is preponderant, whereas in the porpoise, chimpanzee, and others the medulla forms the greater part of the gland. The adrenals may appear lobulated because of strands of connective tissue from the enclosing *capsule* which may penetrate deep into the cortex.

When viewed under the microscope, the typical adrenal cortex is found to consist of three or more layers, or zones (Fig. 9.9). Under the capsule of the gland lies the *zona glomerulosa*. The cells of this relatively narrow layer are arranged in alveolarlike groups and stain rather deeply. The middle and widest zone in the *zona fasciculata*. Here the cells are arranged in long columns, one to three or four cells in width, which extend from the zona glomerulosa toward the center of the gland in a radial sort of arrangement.

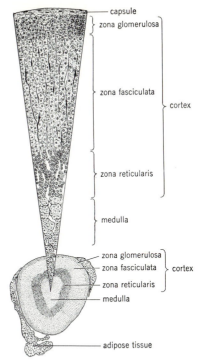

Fig. 9.9 Segment of adrenal gland of rat, showing zonation. The gland is shown below in cross section. The wedge above indicates the microscopic appearance of the medulla and cortical layers.

The innermost zone of the cortex is the *zona reticularis*, the cells of which are arranged irregularly. It is this zone which lies next to the medulla. Often cords of chromaffin tissue penetrate into it.

The typical zonation of the adrenal cortex is not obvious in all mammals. In the opossum, for example, such a division is not discernible. In man and the higher apes complete differentiation of the layers is not obvious until some time after birth. According to one theory, about which there is some contention, cells which are formed in the inner part of the zona glomerulosa migrate toward the zona reticularis where the older cells die and are removed.

The cells of the medulla are more homogeneous in appearance than those of the cortex. It is usually stated that they are arranged irregularly about blood sinuses, which are abundant in this portion of the adrenal gland since it is among the most vascular organs of the body. By the use of special techniques it has been determined that the cells show a very special arrangement in relation to blood vessels. They resemble columnar epithelial cells in shape and are so arranged that the basal end of each cell, where the nucleus is located, borders upon a capillary; the other end abuts upon a vein. Secretion granules, or droplets, are more abundant at the latter end (H. S. Bennett, 1941, *Amer. J. Anat.,* **69:**333), suggesting that epinephrine passes from the cells into the venous system directly rather than into capillaries. The medullary portion of the gland is innervated by preganglionic sympathetic fibers. It is generally held that the cells of the medulla are modified postganglionic cells of the sympathetic nervous system which have become secretory in nature.

In human fetuses the adrenal glands are relatively very large. During the third month of development they even exceed the kidneys in size. At birth they are about one-third the size of the kidneys. After birth, however, they fail to grow at the same rate as neighboring structures. The large size of the adrenal glands of the fetus is due to the presence of an additional zone, or layer, which lies between the medulla and the zona reticularis. It is variously referred to as the *fetal zone, X zone, boundary zone,* and *androgenic zone.* During early development this is the first cortical structure to appear and is known as the *fetal cortex.* It later becomes surrounded by a second mass of cells which originates from the same general region as the first, between the dorsal mesentery and the medial surface of the mesonephros. The second mass of cells becomes the *definitive, true* or *permanent cortex* which later undergoes the zonation described above. The fetal cortex differs histologically from the true cortex and its cells exhibit a distinctive staining reaction. After birth there is a rather rapid involution of this zone, which usually disappears during the first year of life.

A comparable zone is present in the mouse, hamster, and a few other mammals. In the male mouse the X zone persists only until about the sixth week after birth, whereas in the female it may persist for 3 months or more. Castration in the male mouse causes this zone to persist, but in the female removal of the ovaries seems to have no effect.

The full significance of the X zone is not clear. Some investigators are of the opinion that its persistence, often associated with tumors or enlargement of the adrenal cortex, may result in certain abnormalities of the reproductive system. In young girls and also in older women a pronounced *virilism* may occur in which various masculine characters may develop. The body hair increases in amount, being distributed like that of the male, the voice deepens, muscular development becomes more pronounced, the clitoris enlarges to penislike proportions, and menstruation either fails to occur or ceases, depending upon the age when these abnormal manifestations first appear.

A similar condition in young boys results in a precocious puberty in which the child may resemble an infant Hercules, with a high degree of muscular development and having fully developed accessory sex organs and secondary sex characters. In such boys, however, the testes remain infantile, indicating that the adrenal cortex is the source of an androgenic hormone similar to that normally secreted by the testes. Such a hormone, known as *adrenosterone,* has been extracted from the adrenal cortex.

It has been suggested that the chorionic gonadotrophin secreted by the placenta during pregnancy may be an important factor in bringing about the development and maintenance of the fetal zone in man. This may possibly be effected by bringing about the release of andrenocorticotrophic hormone (ACTH) by the fetal pituitary gland, which in turn would have a profound effect upon the developing adrenal cortex.

More experimental work must be done before the role of the adrenal cortex in causing such abnormalities is clearly defined and understood.

A British physician, Thomas Addison, in 1855 published a paper in which he described a disease which ever since has borne his name. Addison's disease is characterized by anemia, loss of appetite, great languor with muscular weakness, and feeble heart action. Most peculiarly the skin becomes dark and of an unusual bronze color. The bronzing of the skin is more pronounced in parts exposed to the air, such as the face and hands, and also in parts normally subjected to pressure of the clothing (see page 387). The severity of the symptoms gradually increases, and before long death ensues. Addison discovered that almost all the 11 patients he had observed possessed lesions of the adrenal glands due either to tumors or to tuberculosis.

If the adrenal glands are removed from a cat, within a few days the animal weakens, becomes prostrate, and dies. If the cortex alone is removed, the same thing occurs. If, however, only the medulla is removed and the cortex left intact, harmful effects are not observed. In Addison's disease it is the destruction of the adrenal cortex that is important, since the proper functioning of this portion of the gland is necessary for life.

The original description of the symptoms of Addison's disease in 1855 was the first time that deficient function of an endocrine gland was related to clinical disorders. Since that time, but particularly in recent years, great advances have been made in our knowledge of the functioning of the adrenal cortex in health and disease.

About 50 different organic crystalline compounds, known as steroids and closely allied chemically to the ovarian, testicular, and certain placental hormones, have been isolated from the adrenal cortex. They are referred to as *adrenocorticoids* (Fig. 9.10). All have a similar basic chemical structure consisting of a 4-benzene-ring nucleus called the cyclopentanoperhydrophenanthrene nucleus. The sex hormones, to be studied later, also possess this nucleus. Chemical groupings of one kind or another are attached at various positions on the basic structure. It is remarkable that variations in those groupings attached at certain positions greatly affect the type of biological action that the steroid substance or hormone has upon the body.

As our knowledge of the chemistry of steroids advances, it becomes increasingly evident that interconversions from one kind to another are constantly going on in the body under the influence of numerous enzymes, the activity of which varies under different conditions. No single organ is solely responsible for the production of any particular steroid. The adrenal cortex, testis, ovary, and placenta have many enzymes in common which affect the synthesis and interconversions of steroid hormones. Manifestations of the activities of these hormones are thus based upon quantitative rather than qualitative factors.

Only a relatively few (approximately eight) of the adrenocorticoids that have been identified are physiologically active. Others may merely represent precursors of those possessing definite physiological activity. It is not improbable that additional adrenocorticoids of importance may yet be discovered as new

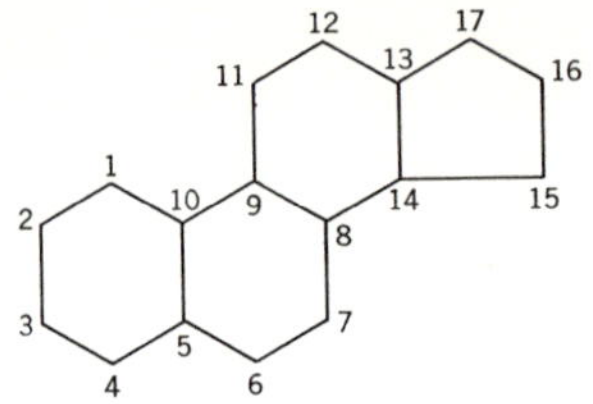

12 17
11 13
16
1
10 9 14 15
2 8
3 7
5 6
4

A. CYCLOPENTANOPERHYDROPHENANTHRENE NUCLEUS

CH_2OH
$C=O$
HO

CH_2OH
$C=O$

O
O

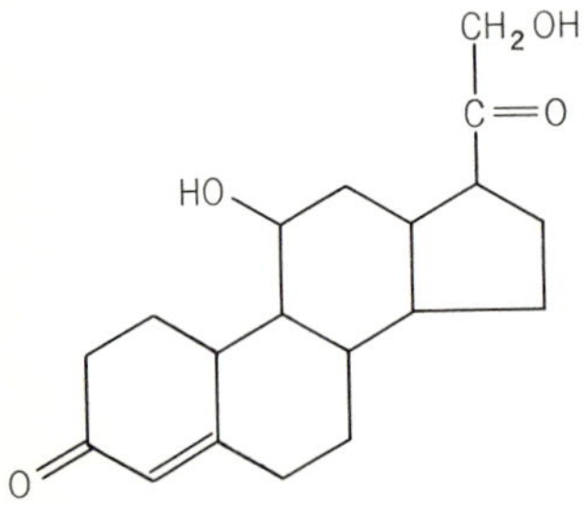

B. CORTICOSTERONE 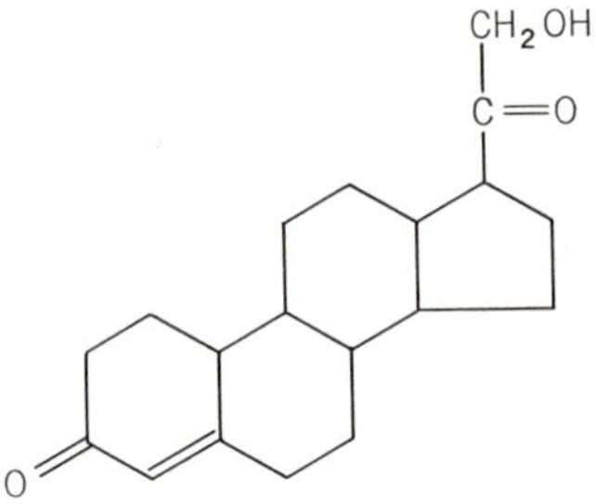   *C.* 11-DEOXYCORTICOSTERONE

Fig. 9.10 A, showing the positions of the various carbon atoms in the cyclopentanoperhydrophenanthrene nucleus; B, chemical structure of corticosterone, an 11-oxygenated adrenocorticoid; C, chemical structure of 11-deoxycorticosterone, which lacks an oxygen atom at position 11.

chemical techniques are developed. It is of interest that chemists have produced artificial steroids in the laboratory which differ from any of the naturally occurring adrenocorticoids. Some of them bring about more profound and exaggerated effects than do the natural adrenocorticoids and may prove to be of considerable value in medical practice. Certain adrenocorticoids have been used on animals that were at the point of death after removal of their adrenal glands. Within a short time after injection, rapid and astonishing recovery ensued. The hormones must be administered at intervals in proper combination and amounts, or the animals will soon revert to their former state of collapse. Animals from which the adrenal glands have been extirpated have been kept alive and in good health for extended periods by injecting

the proper adrenocorticoids. Good results have also been obtained by the use of these hormones on patients suffering from Addison's disease and numerous other conditions. Although extensive experimental work has been carried on in this field, the problem of the complete function of the adrenal cortex has not as yet been completely solved. Each new discovery, however, further emphasizes the fact that the adrenal cortex, excepting only the pituitary body which controls it, is the most important of the endocrine glands.

A drug called *amphenone* is known to inhibit the activity of the adrenal cortex and its production of hormones. Amphenone has toxic effects and is, therefore, of no clinical value, but it is useful in the laboratory as an experimental tool.

According to their chemical structure and

Anatomy of the Chordates

the type of biological effect which they induce, the various adrenocorticoids fall into four main groups. Group I includes such hormones as *cortisone, cortisol (compound F), corticosterone* (Fig. 9.10*B*), and others. They are known as the 11-oxygenated adrenocorticoids, or *glucocorticoids,* and are important in carbohydrate and protein metabolism. Their action on water and electrolyte metabolism is relatively insignificant. Group II includes those which lack oxygen at position 11 in the cyclopentanoperhydrophenanthrene nucleus. The chief hormones in this group are 11-*deoxycorticosterone* (DOC) (Fig. 9.10*C*) and 17α-*hydroxy-11-deoxycorticosterone (compound S).* Their activity centers in the manner in which they affect the sodium and potassium content of the blood and control electrolyte and fluid shifts in the body. Group III includes *aldosterone,* another adrenocorticoid controlling electrolyte metabolism but which is much more potent than those in group II insofar as sodium retention is concerned. Like the glucocorticoids, it has an oxygen atom at position 11 and has an important influence on carbohydrate, fat, and protein metabolism in addition to its electrolytic effects. Aldosterone has an aldehyde group at position 18 (not shown in Fig. 9.10*A*). A substance from the kidney, called *renin* (page 268), apparently causes a release of aldosterone by the cortex of the adrenal gland. The hormones in groups II and III are often referred to as *mineralocorticoids* because of their action in affecting the sodium and potassium content of the blood. Group IV is composed of those adrenocorticoids having androgenic, estrogenic, or progesteronelike properties similar to those of hormones elaborated by the gonads. Among them is *adrenosterone,* mentioned above. It would seem that the adrenal cortex may supplement the gonads in producing sex hormones.

When the adrenal cortex is removed there is an increased elimination of salt in the urine and a loss of sodium from the blood. As the sodium in the body is depleted, there is an increased elimination of water. This causes a lowering in the volume of circulating blood, and the tissues of the body become dehydrated. The blood is consequently more concentrated and blood pressure is lowered. These symptoms are also found in patients with Addison's disease. Other symptoms are, without doubt, also related to the loss of sodium and water. Administration of quantities of salt (sodium chloride) in the diet is of benefit in these cases. Experimental adrenalectomized animals are usually maintained by adding salt to the drinking water in a concentration of about 1 per cent. Under such conditions only minimal amounts of the proper adrenocorticoids need be administered to the animals.

Certain investigators, using cytochemical techniques, believe that they have demonstrated that two secreting zones exist in the adrenal cortex, each serving a different function. They believe that the zona glomerulosa may secrete the corticoid(s) regulating sodium balance, whereas the zona fasciculata secretes those affecting carbohydrate metabolism. This is not conclusive. Although far from completely understanding its functioning, most authorities believe that the adrenal cortex does not secrete many different substances, each a finished chemical entity in itself. It would seem more logical to assume that it acts as a sort of chemical factory, synthesizing certain substances in a series of complicated reactions. At intermediate stages certain products may be liberated to meet bodily requirements, thus accounting for the multiplicity of steroids which have been extracted from the gland. It is possible that each cortical zone has a different enzyme system. These may thus be responsible for the formation of the various kinds of steroid hormones.

Dramatic results have been obtained by physicians who have used cortisone and cortisol on patients suffering from such diseases as rheumatoid arthritis, Addison's disease, certain leukemias, and other conditions. Administration of ACTH, the adrenocorticotrophic hormone of the anterior lobe of the pituitary body (page 390), which controls the output of the adrenal cortex, is also often used successfully in such conditions.

Since treatment for adrenocortical insufficiency means supplying the body with specific hormones which its intact glands are unable to furnish, administration of adrenocorticoids must be continued indefinitely. In some cases unfortunate side reactions may take place which preclude their further use.

The adrenal cortex also plays a very important role in enabling the body to adjust itself to many kinds of stress. This is discussed briefly as the general-adaptation syndrome (GAS) on page 391.

Epinephrine, or adrenalin, from the adrenal medulla, was the first of all the hormones to be discovered. Chemists have been able to synthesize it, and today epinephrine is one of the most widely used drugs in the practice of medicine.

The injection of even minute quantities of epinephrine into the body produces striking results. It causes a rise in blood pressure, due primarily to its constricting effects on peripheral blood vessels. This forces the blood to other parts of the body. The skin is blanched; the blood supply to the muscles is increased and less goes to the visceral organs. The heart beats more forcefully, and there is an increase in the amount of sugar in the blood, providing more fuel for increased muscular activity. The air passages to the lungs dilate, and breathing is made easier. The processes

of digestion stop temporarily, the hair stands on end (gooseflesh in man), and the mouth becomes dry.

These reactions usually occur during periods of fear or anger and are referred to in their aggregate as the "emergency mechanism" of the body. During emotional stresses of this sort the body is better adapted for self-protection.

The increased amount of epinephrine, furthermore, may augment the production of the adrenocorticotrophic hormone (ACTH) of the anterior lobe of the pituitary gland, which, in turn, causes certain hormones to be released from the adrenal cortex. These then aid in supplying additional fuel to the body via their influence on carbohydrate, fat, and protein metabolism. ACTH apparently has little or no effect in bringing about the release of aldosterone, its influence being primarily on the glucocorticoids.

It is probable that little, if any, epinephrine is secreted by the adrenal medulla under ordinary circumstances. This is apparently not secreted rapidly enough, under normal conditions, to increase its concentration in the bloodstream to the level required to produce important physiological effects. During emotional stresses, in all likelihood, its secretion is increased. All the reactions brought about by epinephrine, however, can occur when the adrenal medulla has been removed. Other chromaffin tissues of the body, when properly stimulated, secrete epinephrine, or a similar substance. Stimulation of the sympathetic nervous system (page 653) will also bring about the reactions mentioned above. The postganglionic fibers of the sympathetic nerves are known to secrete *norepinephrine* (page 359), which elicits reactions normally attributed to epinephrine even in the absence of the adrenal medulla. It should be obvious that the presence of the medulla is not essential for life, whereas the cortex is essential unless accessory cortical bodies are present which may function in its absence. It was many years before it became clearly understood that the severe effects resulting from extirpation of the adrenal glands were due to the removal of the cortical portions alone.

What, in the past, have seemed to be pure extracts of the adrenal medulla are now known to contain two different substances, epinephrine and norepinephrine (noradrenalin). They belong to a chemical category known as *catecholamines*. Some authorities have held the opinion that norepinephrine may be a precursor of epinephrine. It is more probable that these substances actually represent two separate medullary secretions. The proportions of the two, which are slightly different chemically, are relatively constant for each species of animal. Preparations made from the adrenal glands of oxen contain approximately 20 per cent of norepinephrine. Adult human adrenal glands secrete approximately ten times as much epinephrine as norepinephrine. Although the physiological properties of the two substances are very similar, it seems that the norepinephrine content of the adrenal medulla is considerably higher in those species possessing a naturally aggressive temperament (cats, lions, tigers) as opposed to those which possess a more placid nature (guinea pigs, hamsters, rabbits). There is some indication that in man the proportions of epinephrine and norepinephrine that are secreted from time to time vary according to whether an individual is provoked to intense anger or is subject to fear of a passive sort. It seems likely, furthermore, that the two hormones may not even be secreted by the same cells but that separate cells exist for the production of each. This is borne out by fairly recent histochemical studies. Production of norepinephrine does not seem to be confined to the adrenal medulla and other chromaffin tissues since it is found in different concentrations in many mammalian structures. Its presence seems to be associated with the nerve supply to a tissue or organ. There is some evidence that it is synthesized by nerve cells and stored in their axons. It has been reported that conversion of norepinephrine to epinephrine is facilitated by administration of glucocorticoids or ACTH. This indicates that the close anatomical relationship of medullary and cortical components of the adrenal glands of mammals may not be merely fortuitous since there is a functional reason for close association of steroidogenic and catecholamine-producing tissues. Despite the chemical and functional similarities of epinephrine and norepinephrine, there are significant differences between them insofar as the degree to which they produce their effects. Norepinephrine has little effect upon metabolic activities but has a stimulating influence on the heart and causes a general vasoconstriction except for the coronary arteries.

Practical applications for the use of epinephrine include its action as a styptic to stop bleeding from superficial blood vessels and its use in asthma to dilate the respiratory passages so that breathing can be more easily accomplished.

Several other substances have been developed which have properties similar to those of epinephrine. The most common of these are *ephedrine*, prepared from a Chinese herb, and two synthetic compounds named *synephrine* and *neosynephrine*.

In those fishes, amphibians, and reptiles capable of changing color so as to blend in with the environment (metachrosis), epinephrine is believed to be responsible for the concentration of pigment granules around the nucleus of the integumentary chromatophores, or pigment-bearing cells of the skin, thus causing a generalized pallor.

PANCREAS

The dual role of the pancreas as both an exocrine and endocrine gland has already been mentioned. This structure is made up of two distinct kinds of tissues which, although so intimately associated anatomically, are functionally entirely separate. The comparative anatomy of the pancreas as an exocrine gland has been discussed in the chapter on the digestive system. In the exocrine portion the secretory cells making up the alveoli pour digestive juices into ducts which empty directly or indirectly into the duodenum. In most vertebrates, scattered apparently at random among the alveoli are masses of epithelial cells of different appearance and different staining properties. These isolated masses, which are of variable size, are known as the *islets* (*islands*) of *Langerhans* (Fig. 9.11). The cells which

compose the islands have an embryonic origin in common with the cells making up the alveoli. Small clumps of cells which become detached from the duct system during development give rise to the endocrine units. It is improbable, once differentiation has occurred, that cells of the duct system, or exocrine portion, can become transformed into islet tissue.

Amphioxus. Although amphioxus has no true pancreas, certain cells in the wall of the anterior portion of the intestine are reported to resemble pancreatic cells.

Cyclostomes. No discrete pancreas is present in cyclostomes. A few small masses of cells have been described, buried in the substance of the liver and in the wall of the intestine. It is believed that they represent pancreatic tissue and are comparable to the islets of Langerhans since no duct is present. Whether, because of this situation in primitive forms, the endocrine portion of the pancreas should be considered to be phylogenetically older than the exocrine portion is a matter for conjecture.

Fishes. It has been reported that the islet cells of certain elasmobranchs are arranged along the ducts of the exocrine portion of the pancreas. This undoubtedly represents a primitive condition. In teleosts the pancreas is usually very diffuse. The islets are fewer in number and proportionally larger than in mammals and, even when minute, may be identified by the unaided eye. They are pale in color and, since they are rather thick, can easily be distinguished against the more translucent sheet of the surrounding pancreatic alveoli. In many species of teleosts there is an encapsulated island of Langerhans called the *principal island*, which is of relatively large size and constant occurrence. It

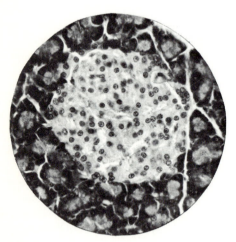

Fig. 9.11 Photograph of a section through the pancreas of a rat, showing an islet of Langerhans surrounded by alveolar tissue.

is sometimes referred to as the *endocrine pancreas*. It lies in the region of the gall-bladder, pyloric caeca, or spleen, its relation to the rest of the pancreas being slight. In the angler fish, *Lophius piscatorius*, the principal island may reach the size of a pea. In some fishes the principal island may be the only islet tissue present. Because of this fact, in the early days of *insulin* research, the principal islands of certain teleosts were considered as a possible commercial source of the hormone. They were of particular value since no other tissues or secretions were present to contaminate the extracts of insulin during their preparation. The supply of these principal islands was of necessity extremely limited. Removal of the principal island in teleosts causes a marked increase in the amount of sugar in the bloodstream.

Other classes. In amphibians, reptiles, birds, and mammals, the islets of Langerhans are scattered among the alveoli in the typical manner, and no unusual features are to be noted. In mammals they are most numerous in that portion of the pancreas which lies nearest the spleen. It has been estimated that in the human pancreas there are from 200,000 to 1.8 million islets. In some mammals the number is considerably larger.

Vertebrates from which the pancreas has been completely removed soon develop characteristic symptoms. The glycogen stores of the liver quickly become depleted, and excess carbohydrates can no longer be stored in the liver cells in the form of glycogen. Oxidation of blood sugar in producing heat and energy seems to be impaired to some extent. Glycogen stored in the muscles is also soon reduced below normal levels. As a result of these factors, and with the conversion of stored glycogen to glucose, the amount of sugar in the blood increases to abnormally high levels and the kidneys excrete large amounts of it in the urine. A loss of weight follows since body proteins and fats are drawn upon as sources of energy. The body is less able to combat infection. Ketone bodies, which are intermediate products formed during the oxidation of fats, are produced in such large quantities that the body cannot cope with them in breaking them down as rapidly as is desirable. As a result, they accumulate in the blood. Two of these are acids (*acetoacetic acid* and *β-hydroxybutyric acid*). They gradually deplete the *alkaline reserve* of the body. *Acetone,* a third ketone body, can be detected in the breath and also appears in the urine. The blood becomes relatively more acid, and a condition called *acidosis* occurs. This is accompanied by diminution of perception, a reduction in the carbon dioxide-combining power of the blood leading to "air hunger," and finally, unless remedial measures are taken, there is a loss of consciousness known as *diabetic coma,* and death occurs. These reactions actually are the result of removal of the islets of Langerhans for, although the entire pancreas has been removed, the portion which secretes the digestive enzymes has been shown not to be involved in this condition. It is of interest that herbivorous mammals are less severely affected by removal of the pancreas than are carnivorous forms.

The symptoms described above are similar to those of human beings suffering from a disease known as *diabetes mellitus*. In this condition the islets of Langerhans become diseased or are affected so that insulin secretion either ceases or else normal quantities of insulin are no longer secreted into the bloodstream.

When an experimental pancreatectomized animal, or an animal or person suffering from diabetes mellitus, is given injections of insulin, the ability of the body to utilize carbohydrates and fats in an approximately normal manner is restored. The excessive breakdown of protein ceases and ketone bodies disappear. The liver once more begins to store glycogen, and the blood sugar and the glycogen in the muscles return to their normal levels. The body again becomes resistant to infection. It is generally agreed that the exact mode of insulin action is to promote the transfer of glucose from extracellular to intracellular localities.

Three types of cells, which differ in staining and histochemical properties, have been identified in the islets of Langerhans of man and other mammals. They are referred to as *alpha* or A *cells, beta* or B *cells,* and *delta* or D *cells.* A fourth type, the C *cell,* has been described in the islets of guinea pigs. Estimates of the relative numbers of A, B, and D cells in the dog indicate that they make up approximately 20, 75, and 5 per cent of the cells, respectively. It is well established that the B cells are the source of insulin secretion. When a chemical called *alloxan* (the ureide of mesoxalic acid) is administered to an animal for a short period, a permanent diabetes results. It has been found that in alloxan diabetes and diabetes mellitus it is the B cells of the islets that have been damaged or destroyed.

The existence of a second pancreatic hormone, *glucagon,* is now fairly well established. It is also referred to as the hyperglycemic-glycogenolytic factor, or as HGF. The A cells of the islets are believed to be responsible for its secretion and elaboration. Histochemical studies reveal that the granules of A and B cells differ markedly. Those of the B cells are smaller and are soluble in alcohol, whereas the larger granules of the A cells are resistant to alcohol but soluble in water. The granules of the two types of cells also exhibit different reactions to certain stains.

Glucagon, a polypeptide with a molecular weight of approximately 3,500, is a blood-sugar-raising factor which encourages the conversion of liver glycogen into sugar. Epinephrine, from the adrenal medulla, also brings about this effect but the two hormones, even though sharing this ability, differ widely in other respects. Whether glucagon is liberated directly from the A cells in response to lowered levels of blood sugar or whether its secretion is controlled by some hormone from the anterior lobe of the pituitary gland is not entirely clear. Extracts of the anterior lobe, administered over a sufficiently long period to experimental dogs, render them permanently diabetic. There is a degranulation and finally a degeneration of the B cells, which may become permanently impaired. The A cells are not visibly affected by such extracts. Evidence has been presented which indicates that the activity of the islets of Langerhans may be subject to some degree of nervous control via parasympathetic fibers located in the right vagus nerve. Furthermore, the existence of *neuroinsular complexes* has been demonstrated (L. C. Simard, 1942, *Rev. Can. Biol.,* 1:2). These are composed of ganglion cells in close association with islet cells.

Since certain extracts of the anterior lobe of the pituitary gland have blood-sugar-raising properties, the existence of a separate diabetogenic pituitary hormone has been claimed in the past. It has not, however, been established that such a separate factor actually exists. It is more probable that the somatotrophic, or growth, hormone of the anterior lobe is responsible. Rather than being directly affected by a pituitary-gland stimulus, it would seem more likely that the B cells degenerate after long-continued overwork in producing insulin to counteract the high blood-sugar level induced by the pituitary factor. Furthermore, as previously indicated, the symptoms of diabetes caused by removal of the pancreas do not supervene if the pituitary gland is removed.

It has long been known that if the pituitary gland is removed from a young animal there is a cessation of growth and a retention of juvenile features. This is usually attributed to the fact that the somatotrophic, or growth, hormone secreted by the anterior lobe is no longer being formed. If insulin is administered to such animals, however, there is a sudden stimulus to growth as reflected by a thickening of the epiphysial plates of bones and by an increase of body fat and protein. Insulin and the somatotrophic hormone undoubtedly differ in the manner in which they exert their effects.

It is of interest that the islets of Langerhans of certain urodele amphibians lack A cells. Urodeles also fail to respond to injections of glucagon by a rise in the blood-sugar level. Nevertheless, they are sensitive to insulin injections.

The discovery of insulin in 1921 was of paramount importance to mankind, for the use of this hormone has prolonged the lives of innumerable diabetic persons who otherwise would have succumbed. Insulin is now prepared in pure crystalline form but its use is by no means a cure for diabetes. It does, however, enable those suffering from the disease to lead normal lives. Even when using insulin, the diabetic patient must watch his diet carefully so that the intake of carbohydrate does not exceed his impaired ability to utilize it. He must, nevertheless, receive an adequate amount of carbohydrate to maintain normal, healthy conditions.

The exact structure of the insulin molecule was finally determined in 1954. It is a small protein with a molecular weight of approximately 6,000. The chemical synthesis of insulin was first announced in 1963 by Katsoyannis and his colleagues. This represented the first synthesis of a naturally occurring protein. Human insulin was first produced artificially in 1966. Most work has been done on fishes and mammals, and little is known of the chemistry of the hormone in amphibians, reptiles, and birds. It has been well established that the molecule consists of two polypeptide chains, one of 21 amino acids, the other of 30 amino acids. The chains are connected by two interchain disulfide bridges. The insulins of different species vary somewhat in the sequence of amino acids within the molecule. Variations in mammalian insulins, the structure of which is known with certainty, are relatively few, occurring mainly at positions 8, 9, and 10 of the A chain, and at position 30 of the B chain. The sequence found in pig insulin is probably basic. Variations found in other forms have probably been derived from it by mutation during the course of evolution. It is of interest that the rat has two insulins, differing only in the amino acid located at position 29 of the B chain. One has lysine at this position; the other has methionine. A fish, the bonito, also has two insulins. Administration of insulin by mouth has little effect, as it is destroyed by the action of digestive juices. An overdose of insulin given by injection must be guarded against because it reduces the blood sugar below the normal level and may result in a condition known as *insulin shock*. This first makes its appearance as an uncomfortably weak or faint feeling; the patient breaks out in a profuse perspiration, the eyeballs are unusually firm, and the skin becomes pale. This may lead to complete unconsciousness followed by convulsions. The typical signs of insulin shock are due to a loss of activity of nerve cells in the higher centers of the brain. In insulin shock, injection of sugar, or feeding sugar in some form, is necessary so as to raise the sugar concentration in the bloodstream rapidly. A prompt recovery usually follows if sugar is given. Insulin shock is deliberately induced as a therapeutic measure in patients having a certain type of mental disease, and remarkable recoveries have been effected.

The cause of diabetes mellitus is not definitely known, but it is becoming apparent that other endocrine organs besides the pancreas may be involved.

TESTES

The male gonads play a dual role in the body, being both *cytogenic* (cell-producing) and endocrine in nature. The comparative anatomy of the testes as spermatozoa-producing organs has already been discussed (page 308). The importance of the endocrine secretion of the testes lies primarily in the fact that not only does the development and maintenance of functional seminiferous tubules depend upon it, but the development of the *secondary sex characters* of the male, which in many cases distinguish him markedly from the female, are dependent upon this secretion, or hormone. In some animals the secondary sex differences are very striking as, for example, the presence of antlers in the male deer and their absence in the female in many species. In other animals such as the rat and the mouse, the differences are less apparent. The normal development, growth, and functioning of the accessory sex organs (page 276) are almost entirely under control of the male sex hormone (androgen). Since the endocrine secretion of the testes plays so important a part in the development and maintenance of many structures of the male body, it is important that the student of anatomy have an understanding of the tissues responsible for its elaboration.

The nature of the cells which actually pro-

duce the endocrine secretion of the testes is rather obscure in members of the lower vertebrate classes. The cytogenic portion of the testes is made up of seminiferous ampullae, or tubules, in which spermatogenesis takes place. It is still a question whether in such vertebrates the endocrine secretion is formed by germinal cell constituents, mature spermatozoa, connective tissue, or interstitial cells lying in spaces, or interstices, between the seminiferous tubules. It is almost without doubt that it is the latter cells which are responsible even though they may be rather inconstant in appearance.

Amphioxus and Cyclostomes. No work has been done which demonstrates an endocrine activity on the part of the testes in these animals. Only the cytogenic portion has been described. The fact that in the lamprey the urogenital sinus assumes somewhat different forms in male and female during the breeding season would lead one to infer that sex hormones were at work. In the male, with the approach of sexual maturity the urogenital sinus becomes rather narrow and opens by means of a small pore at the end of a long tube.

Fishes. Most fishes are seasonal breeders, having a definite reproductive cycle and breeding but once a year. In many forms sexual dimorphism is not very clear cut, whereas in others marked secondary sex characters are apparent. These take the form of color patterns, modifications of fins, appearance of excrescences on the skin, etc. Such features are most prominent at the breeding season and usually regress at the termination of the nuptial period. Changes in such internal structures as kidney tubules in certain places, ductus deferens or archinephric duct, and cloacal glands occur concomitantly with the external changes mentioned above. These structures also undergo a regression at the close of the breeding season.

Observations on the testes of several species show that as the mating period approaches, interstitial cells, which are normally absent, appear among the seminiferous tubules. Following the breeding season, these cells gradually disappear, and there is a regression of the secondary sex characters which were so prominent at the height of the reproductive period. The correlation of the appearance and disappearance of the interstitial cells with the waxing and waning of the nuptial dress is circumstantial evidence that the internal secretion of the testes may be formed by the interstitial cells.

Castration, or removal of the testes, in certain species of fishes results in a complete failure of the nuptial changes to appear. If the operation is performed while such characters are at the height of their development, there is a much more rapid regression than would normally take place.

The fact that, in other species of fishes, secondary sex characters appear with sexual maturity and are retained as such throughout life, apparently in the absence of interstitial cells, has led some investigators to believe that germinal elements, especially mature spermatozoa, may be the source of the male hormone in these forms. In fishes, therefore, it has not been determined definitely whether germinal elements or interstitial cells are the seat of hormone production.

Amphibians. Among amphibians there is usually an annual breeding period which is narrowly restricted in some forms and spread out over a rather long period in others. The season also varies from early spring, as in *Rana pipiens, Eurycea bislineata, Ambystoma texanum,* and most others, to early fall in case of the salamander *Ambystoma*

opacum. As in the case of seasonal breeders among the fishes, with the approach of the breeding season certain secondary sex characters appear in male amphibians. These consist of various modifications, such as swelling of the thumb pads (*Rana pipiens*), enlargement of the mental gland (*Eurycea bislineata*) (Fig. 4.9), changes in color pattern and development of a dorsal crest (*Triton cristatus*), and enlargement of the cloacal glands (*Ambystoma texanum* and others). Changes in the reproductive ducts accompany these modifications. After the breeding season is over, there is a regression of the various organs and structures mentioned above.

The testes of amphibians, in addition to their sperm-producing function, have been shown to be responsible for the development of the male accessory sex organs and secondary sex characters. Castration prevents the appearance of these nuptial features. The seat of hormone production in the testes of amphibians is still a matter for conjecture. Whether germinal elements, mature spermatozoa, or interstitial cells, which are very inconstant in appearance, are responsible for male-hormone elaboration remains undetermined. Interstitial cells appear in most forms only after the ampullae of the testes have been emptied of spermatozoa. They consist largely of connective-tissue cells, and it would seem unreasonable to believe that these should be responsible for the development of characters which make their appearance sometime previous to the actual breeding period.

Reptiles. Seasonal reproductive activity is characteristic of most reptiles, and periodicity in the appearance and disappearance of secondary sex characters is also evident in this class of vertebrates. These characteristics are not so strikingly evident as in members of other classes. In lizards, on which most work has been done, changes in skin pigmentation and development of femoral glands and hemipenes, as well as such internal organs as epididymides, have been reported to be seasonal. Interstitial cells are present in the testes throughout the year but show a cyclic modification, increasing in number following the discharge of germ cells toward the end of the breeding period. Certain cytological studies of the interstitial cells of lizards indicate an activity which is closely correlated with the appearance of the secondary sex characters. Very little, relatively, is known concerning reproductive physiology and detailed morphology of reptilian testes.

Birds. Just as in the other classes of vertebrates, the testes of birds have an endocrine function in addition to their cytogenic activity. In the domestic fowl interstitial cells seem to be relatively few in number in the adult cock, whereas they are more numerous in the testes of young, growing males. Nevertheless there is no indication that the output of androgen by the adult testis is diminished. Experiments in which certain pituitary extracts have been injected show that such treatment brings about an increase in the number of interstitial cells.

Castration of the male domestic fowl has long been practiced. The prepuberally castrated *capon* fails to develop comb and wattles, does not crow, and has a more quiet demeanor. If the operation is performed after these characteristics have developed, they show a marked regression. The flesh of the capon is richer, of better flavor, and of higher quality than that of the uncastrated rooster, and for that reason capons command a higher market price. In many birds the brilliant plumage of the male as contrasted to his rather drab mate is evidence of the activity of the male hormone, although in a number of cases certain genetic factors may also be involved. Most birds are seasonal breeders, and fluctuations in the appearance of certain characteristics (beak color, plumage changes, etc.) may be correlated with the breeding and nonbreeding seasons. In other birds, particularly the domesticated species, fully developed secondary sex characters are in evidence throughout the year. The accessory sex organs are likewise under control of the male hormone.

Experiments with birds have revealed that in a number of species an increase in the total number of daylight hours brings about a marked stimulation of testicular development. Conversely, reduction of the normal daylight period delays the regular seasonal increase. In the junco, for example, the testes are at a minimal size in November. At the height of the breeding season they weigh several hundred times as much as the inactive organs. It has been demonstrated that if juncos in Canada, kept out of doors at subzero temperatures in winter, are subjected to progressive increases in electric lighting after sunset, there is a marked increase in size and activity of the testes. Not only are the spermatogenic elements of the testes affected, but the interstitial cells respond as well. The premature changes in accessory sex organs and secondary sex characters of the experimental animals are indicative of an awakened activity of the endocrine portion of the testes, which is undoubtedly brought about by stimulation from hormones elaborated by the anterior lobe of the pituitary gland.

Mammals. In the testes of mammals the interstitial cells, or *cells of Leydig,* are a constant feature. Masses of these cells, with their abundant vascular supply, are present in the interstices formed by adjacent seminiferous tubules (Fig. 9.12). The interstitial

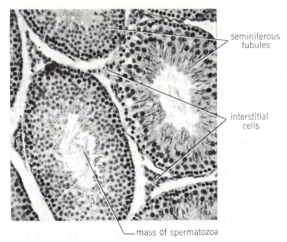

seminiferous tubules

interstitial cells

mass of spermatozoa

Fig. 9.12 Photograph of a section through the testis of a rat, showing cells of Leydig lying in the interstices between adjacent seminiferous tubules.

cells originate during fetal development from mesenchymal cells of the stroma which lies among the seminiferous tubules. It represents an unusual type of endocrine tissue since it is diffuse and does not arise from an epithelial surface as do the others. In the human fetus the interstitial tissue is more conspicuous from the fourth month to birth than it is in the interval between birth and puberty.

Electron microscopic studies have shown that the cytoplasm of the interstitial cells contains a network of very fine interconnecting tubular structures as well as droplets of lipoid substance and some crystalline elements. The network is probably the site of testicular hormone secretion; the crystalline elements, which are known as the *crystalloids of Reinke*, exhibit a highly organized microstructure.

Much experimental work has been done on the mammalian testis, and the assignment of male-hormone production to the interstitial cells is fairly well established. Even in cryptorchid individuals in which the semi-niferous tubules are degenerate and spermatozoa have not formed (page 312), the secondary sex characters and the accessory sex organs develop normally. In such individuals the interstitial cells appear to be normal and are apparently not influenced by the higher body temperature as are the seminiferous tubules. Only after cryptorchidism has been in existence for a long time is there any evidence that the secretion of the interstitial cells may be diminished. It seems obvious, then, that masculine characteristics are not dependent upon the presence of sperm cells but must be controlled by some other factor in the testes.

Castration of various domestic mammals has long been practiced to improve the quality of flesh or to make them more docile. A castrated bull is known as a *steer;* a castrated stallion is called a *gelding.* A castrated boar is a *barrow,* and a castrated man is a *eunuch.* If such animals or human beings are castrated *before* the age of puberty, typical masculine structures and characteristics fail to develop. Castration *after* the age of puberty invariably leads to a regression of such accessory sex organs as seminal vesicles, the prostate, and Cowper's glands. Certain secondary sex characters of the post-puberal castrate, such as voice and beard in human beings, are altered little, if at all.

The male hormone, *testosterone,* was first extracted from bull testes. It can now be prepared synthetically. Substances having the physiological, masculinizing properties of testosterone are referred to as *androgens.* It has been mentioned (page 334) that the actual biological action of hormones may be secondary to the primary control of ribonucleic acid (RNA) synthesis by hormones. The action of androgen on the nuclei of the prostate gland, for example, has been found to enhance the synthesis of RNA at nucleolar or perinucleolar regions (or both) of prostatic chromatin material. Furthermore, certain biological effects of androgens can be suppressed by substances known to interfere with synthesis of RNA or proteins in general. Low concentrations of the drug, actinomycin-D, have such an effect (S. Liao, 1968, *Amer. Zoologist,* 8:233). Injections of testosterone are fully capable of maintaining the activity of the accessory sex organs of the castrated male and will prevent any change in the secondary sex characteristics. Treatment, however, must be continued indefinitely in such cases.

It is worthy of note that certain skeletal muscles seem to be selectively stimulated by androgens. The levator ani muscle of the rat and the masticating muscles of the male guinea pig are examples of muscles which show pronounced atrophy following castration and which may be restored by injection of androgens. Such muscles seem, however, to be limited in their response since they cannot be induced to exceed their normal size by giving excessive amounts of androgen or by extending the period of injection.

It is interesting that the internal secretion of the testes has a rather profound effect on the development of certain parts of the skeleton. Modifications of the pelvis of the castrated male, when compared with a normal individual, are often obvious. Comparison of the pelves of males and females in different mammalian species often reveals differences of the nature of secondary sex characters. The resorption of the pubic bones of the female pocket gopher with the approach of sexual maturity and the failure of this process to occur in normal or castrated males are good examples of hormonal control of development of some skeletal parts (Fig. 10.85). The large larynx and deep voice of the human male contrasted with the small larynx and higher-pitched voice of the female, the presence of antlers in male deer of many species, their absence in the female, and their failure to develop in the prepuberally castrated male, all bear witness to the fact that the male hormone may determine the presence or absence of bony or cartilaginous structures. The larger size of the male mammal as compared with the female is undoubtedly a reflection of the influence of male hormone on skeletal growth. In prepuberally castrated chimpanzees it has been demonstrated that the epiphyses remain open much longer than in the normal male. It has been reported that the activity of the sebaceous glands of the skin in man is regulated by androgenic hormones.

Experiments indicate that in certain mammals, as in birds, increased exposure to light has a stimulating effect upon the activity of the reproductive system. Many species, however, including guinea pigs, squirrels, and rabbits, do not seem to respond.

Androgens, like other steroid compounds, possess the 4-benzene-ring cyclopentanoperhydrophenanthrene nucleus. The chief androgen, testosterone, is normally secreted by the interstitial cells of the testes. Another androgen, *androstenedione*, is also produced by the testes. The former bears an OH group at position 17 in the nucleus; the latter has oxygen at this position. Androstenedione normally undergoes reduction to testosterone. Production of androgens, however, is not confined to the testes. They are formed also in such other steroid-producing organs as the cortex of the adrenal glands (adrenosterone — page 355), ovaries, and possibly in the placenta. The differences among the ovaries, adrenal cortex, and testes, so far as androgen production is concerned, are quantitative rather than qualitative. The fact that androgens may be formed in organs other than the testes accounts for the abnormal masculinizing effects of certain adrenal and ovarian tumors.

When testosterone is liberated in the bloodsteam it becomes chemically bound with blood proteins and, therefore, cannot be filtered through the glomeruli of the kidney. Excess testosterone is altered chemically by the liver and is excreted in degraded form largely as *androsterone* in the urine, feces, and bile. Androsterone has an OH group at position 3 and hydrogen at carbon 5. As compared with testosterone, androsterone is a relatively inactive androgen.

Androgens are usually ineffective when administered orally, either being destroyed by digestive juices or rendered inactive by the liver, to which they are conveyed by the hepatic portal vein. They are, therefore, usually given by subcutaneous or intramuscular injection. Some orally effective androgens, such as methyl testosterone, apparently are absorbed directly into the lymphatic system.

Many tests may be used to determine the hormonal potency of various extracts and preparations. Most involve bioassay methods, measuring the degree to which various

structures in castrated animals are restored to normal, morphologically or physiologically. It is of interest that the comb of the capon may be stimulated to grow by direct application of an ointment containing testosterone. If testosterone is injected into the capon, the amount needed to bring about a certain degree of response is approximately 100 times more than that required when it is administered by inunction.

The secretion of testosterone by the interstitial cells of the testes is regulated by a gonadotrophic hormone secreted by the anterior lobe of the pituitary gland (page 392). In both male and female there are three gonadotrophic hormones, known respectively as FSH (*follicle-stimulating hormone*), LH (*luteinizing hormone*), and *prolactin*. The latter is sometimes referred to as the *luteotrophic hormone* (LtH). It is obvious that these names were originally applied to the gonadotrophic hormones in connection with their relationship to the female reproductive system. One of these, LH, is responsible for the activity of the interstitial cells of the testes in secreting testosterone. For this reason, in the male, the name *interstitial-cell-stimulating hormone* (ICSH) is more properly applied to it, despite the confusion brought about by the dual nomenclature.

There is a reciprocal relationship between the interstitial cells of the testes and the anterior lobe of the pituitary gland insofar as the production of testosterone and ICSH are concerned. ICSH stimulates the interstitial cells to secrete testosterone which, when it reaches a certain level in the bloodstream, in turn inhibits the secretion of ICSH. This is, for the most part, probably effected via a neuroendocrine mechanism involving the hypothalamus (pages 391 and 624).

The activity of the gonadotrophic hormone FSH in the male is focused chiefly on the seminiferous tubules of the testis and is believed to be of special importance in connection with certain phases of spermatogenesis. On the other hand, androgen also plays an important role in normal spermatogenesis. It would seem that FSH, ICSH, and androgen are all concerned in promoting the normal development and function of the seminiferous tubules. The degree to which each of these hormones acts in this capacity seems to vary considerably in different species.

Dietary deficiencies, particularly an inadequate supply of vitamin B, may result in a regression of the accessory sex organs in the males of certain species. That such an effect is actually due to the inability of the pituitary gland and the testes to manufacture their hormones is indicated by the fact that administration of androgen or ICSH will restore the accessory sex organs to their normal condition in the face of continued dietary deficiency. The term *pseudohypophysectomy* is sometimes applied to the effects of nutritional disturbances of this sort.

Estrogens are commonly found in testicular extracts and in male urine. The amount, however, is far less than in the female. This gives further emphasis to the concept that the distinctions between maleness and femaleness are based upon quantitative rather than qualitative factors. The source of testicular estrogens remains obscure. It is possible that the cells of Sertoli may be involved.

OVARIES

As is true of the testes of the male, the ovaries in the female are both cytogenic and endocrine in nature. These two functions are mediated through tissues in the ovaries which have a much closer relationship than do the corresponding parts of the testes.

The endocrine secretions of the mammalian ovary include *estrogens, progestogens,* *androgens,* and *relaxin.* The first three of these are steroids; the last is not.

It is still a matter for conjecture as to which cellular components of the mammalian ovary are responsible for the secretion of estrogen. The cells of the theca interna, and possibly those making up the stratum granulosum and discus proligerus of the Graafian follicle, seem to be the source of estrogen secretion. The principal estrogenic compounds elaborated by these cells are the steroids, *estradiol* and *estrone.* All substances having the physiological properties of estradiol are referred to as estrogens. Estradiol is the most potent of the naturally occurring estrogens. Many other estrogens have been isolated from such tissues as the placenta, adrenal cortex, ovaries, testes, and from various body fluids and have been prepared synthetically. All natural estrogens are steroids having the 4-benzene cyclopenthanoperhydrophenanthrene nucleus (Fig. 9.10A) as a base, with three double bonds in the first benzene ring, the absence of a methyl group at position 10, and the presence of a hydroxyl group at position 3. Natural estrogens become bound to blood proteins in the liver and are thus transported to "target" organs. They are also inactivated by the liver, or, at least, their chemical state is altered in preparation for their excretion through the urine or bile. Several other chemicals, which are not steroids, have estrogenic properties. The most important of these are *diethylstilbestrol, hexestrol, dienestrol,* and *benzestrol.* It is of interest that diethylstilbestrol administered to turkeys induces dissecting aneurysms and rupture of blood vessels. Concomitant treatment with iodinated casein, which alters the histological structure of the thyroid glands, alleviates such effects (C. F. Simpson and R. H. Harms, 1968, *Proc. Soc. Exp. Biol. Med.,* **128**:863).

Many estrogens are sold commercially under a variety of trade names. Ovarian estrogens are responsible for the development of the secondary sex characters of the female. The development and maintenance of the organs making up the female reproductive tract also are primarily under estrogenic control. Estradiol bears an OH group at carbon 17 of the steroid nucleus; estrone has oxygen at this position. Estrone normally undergoes reduction to estradiol. It has been demonstrated that estrogen enhances the synthesis of RNA by uterine nuclei in a manner similar to that of androgen acting upon prostatic nuclei. This is a further indication that the ultimate biological action of a hormone upon an organ may be secondary to the primary control of RNA synthesis by the nuclei of the organ.

The term *progestogen* is used collectively to include steroid substances which bring about progestational changes in the mammalian uterus. In general, estrogens should be thought of as being primarily effective in inducing *growth* of tissues, whereas progestogens are influential in bringing about tissue *differentiation. Progesterone* is, in the main, secreted by the cells of the corpus luteum and is the chief naturally occurring progestogen. It may also be present in the testis, adrenal cortex, and placenta. In addition to having its own functional attributes, this steroid seems to be an intermediate in the biosynthesis of such other steroids as 11-deoxycorticosterone and cortisone (page 357). The progesterone molecule has a methyl group at position 10 and a two-carbon side chain on carbon 17. The corpus luteum also secretes some estrogenic steroids. Progesterone is usually not effective when acting alone. Normally it works synergistically with estrogen. Several synthetic progestogens have been prepared, some of which are effective when taken orally. One of these, 19-*norprogesterone,* when injected, is five times as effective as progesterone itself.

The source of ovarian androgens remains obscure. It is possible that interstitial cells in the stroma of the ovary have an endocrine function and under certain conditions may secrete androgens as well as estrogens. Whether normal ovaries secrete appreciable amounts of androgens is questionable. Some ovarian androgens are actually steroid substances which have estrogenic or progestogenic effects on the female but which, in massive doses, may stimulate portions of the male genital tract. It should again be emphasized, in connection with administration of steroid hormones, that there may be conversion in the body from one to another by the action of various enzymes. The differences among them, so far as their secretion is concerned, are essentially quantitative rather than qualitative. The actual chemical structure of a particular steroid does not necessarily signify that its biological effect is on one sex only or that it is restricted to a single activity.

It is of interest that one injection of 1.25 mg of testosterone propionate in female rats 21 days of age will bring about premature vaginal opening on day 27 instead of day 42. Precocious estrus also may be induced by testosterone injections under varying conditions (R. H. Naqvi and M. X. Zarrow, 1966, *Amer. Zoologist,* 6:569).

The exact site of secretion of the nonsteroid hormone, *relaxin* (page 286), is not known. Its physiological potencies have been demonstrated in a number of mammals.

Amphioxus and Cyclostomes. Although the cytogenic activity of the ovaries of these animals is well understood, little is known of their function as endocrine glands. In the female lamprey, with the approach of sexual maturity, the urogenital sinus forms a large vestibule with a vulva and a rather wide opening. This would appear to be a response of the urogenital organs to the presence of a female sex hormone elaborated by the ovary.

Fishes. The development of the reproductive ducts in female fishes with the approach of sexual maturity or with the advent of the breeding season indicates that endocrine factors are at work. A few experiments have been reported on fishes from which the ovaries have been removed. A small minnow, *Phoxinus laevis,* in which both male and female exhibit a red nuptial coloration along the dorsal side during the breeding season, has been the subject of some investigation. Removal of the ovaries either brings about a rapid regression of the nuptial coloring, if it is already present, or else prevents it from appearing. In this particular species an endocrine secretion of the ovary undoubtedly regulates the nuptial coloring of the female.

There is much variation among the ovaries of fishes. In most species, following ovulation, the emptied ovarian follicles shrink and are resorbed. In certain oviparous and ovoviviparous elasmobranchs and teleosts, however, corpora lutea form from nurse cells of the ova at the sites of follicular rupture. Their presence seems to be in some way related to extended retention of eggs in the oviducts in oviparous forms, and with the retention of developing embryos in the uteri of ovoviviparous species. Corpora lutea are rather conspicuous in certain elasmobranchs. More experimental work must be done before the man-

ner in which these corpora lutea function can be fully established and before it can be determined whether their presence is actually essential. Control of the growth of the ovipositor in some teleosts is under the influence of progestogens or similar substances derived from their corpora lutea.

Extracts of the ovaries of pregnant sharks have been found to contain relaxin, indicating the early appearance of this hormone in vertebrate phylogeny.

Amphibians. The actual seat of hormone production in the amphibian ovary is not definitely known, but evidence is clear cut that this organ functions as an endocrine gland. If the ovaries are removed from a female frog, the oviducts atrophy markedly. When estrogen is injected into such an ovariectomized female, the oviducts may be caused to enlarge to even a greater extent than is found in the normal animal during the breeding season. This indicates that estrogen controls the anatomical and functional development of the female reproductive ducts. Amphibians as a group are notable for their oviparity, and generally no structures resembling corpora lutea are to be found. A few species of salamander and a certain African toad, *Nectophrynoides occidentalis,* give birth to living young. In the latter, corpora lutea develop in the ovaries and the young are retained in the uterus of the mother for as long as 9 months.

In the *male* toad, a peculiar lobelike structure called *Bidder's organ* is present at the anterior end of each testis (Fig. 7.10). It has been shown that if the testes are removed, Bidder's organs will, in the course of approximately 2 years, develop into functional ovaries, thus bringing about a complete sex reversal in the male. Under such conditions the otherwise rudimentary oviducts and uteri enlarge, seemingly in response to female sex hormone elaborated by the transformed Bidder's organs.

In normal female amphibians, with the approach of the breeding season the oviducts and uteri enlarge greatly and the glandular cells of the lining epithelium begin to secrete actively (Fig. 8.11). Development of ova within the follicles precedes the growth changes in the oviducts by several weeks.

Reptiles. Those reptiles which hibernate have a breeding period that commences shortly after the period of hibernation has terminated. Reptiles living under tropical conditions, on the other hand, have recurrent breeding periods which may extend over many months. Very little is known about the endocrine secretions of reptilian ovaries and their action on the secondary sex characters and accessory sex organs. It is safe to assume, however, that such activity is similar to that of other vertebrates.

In certain snakes of the genera *Storeria, Potamophis, Thamnophis,* and *Natrix,* which give birth to living young, well-developed corpora lutea form in the ovaries after ovulation. It has been demonstrated that the presence of these corpora lutea is essential for the successful completion of pregnancy, for when the ovaries with their corpora lutea are removed the pregnancy is invariably terminated. The corpora presumably secrete the hormone progesterone (page 371). Some ovoviviparous lizards are also known to have corpora lutea in their ovaries during pregnancy.

Birds. In most birds only the left ovary develops into a functional adult organ. In birds of prey both ovaries develop, but this is the exception rather than the rule.

If both the functional left ovary and the rudimentary right gonad are removed, the animal, after a time, takes on characteristics which are practically indistinguishable from those of a capon. If only the left ovary is removed, the single left oviduct becomes greatly reduced and the convolutions, which are present in the oviduct of the normal animal, are lacking. In addition, a remarkable transformation takes place. The rudimentary right gonad becomes modified into a testislike organ. Although spermatozoa seldom appear in such gonads, they apparently secrete androgen, which causes the appearance of such masculine characteristics as comb and wattle growth, development of spurs, assumption of male plumage, and even such phenomena as crowing and exhibition of a pugnacious attitude. At a later period reversion back to the female type of plumage may occur. It has been demonstrated that in such animals certain fluid-filled nodules appear on the testislike gonads, and it is probable that estrogen given off by them is responsible for the plumage change.

It is assumed that the follicular cells surrounding the large, yolk-filled ova are the seat of estrogen secretion in the bird. Corpora lutea do not develop in birds' ovaries although degenerating or atretic follicles are sometimes mistaken for them.

Administration of estrogen has a stimulating effect upon the oviduct of the domestic fowl. In order for the oviduct to become fully and functionally developed, however, a complex of steroid hormones is apparently required. In experimental animals the glandular part of the oviduct (magnum), or the region in which albumen is secreted about the ovum, can be brought to full functional development only by administration of estrogen followed by either progesterone or androgen. It has been found that progesterone is present in the blood of laying hens, but its exact source is unknown. After ovulation in the domestic fowl, the collapsed follicle (calyx) is retained for some time and has some effect upon the timing of the subsequent ovulation. If the calyx is removed surgically, the egg which came from it is retained in the uterus, or shell gland, for an abnormally long time. Progesterone, or a kindred hormone, may possibly be secreted for a time by the ovarian follicle, even after ovulation when only the calyx remains.

It is of interest that in the phalarope the nuptial plumage of females is more colorful than that of males. Females are also more aggressive than males. Phalarope ovaries, in comparison with those of other birds, have been shown to have a relatively high androgen content (E. O. Höhn and S. C. Cheng, 1965, *Amer. Zoologist*, **5**:658). The source of androgen or an androgenlike hormone in the female bird is far from clear. Some peculiar cells near the theca of the follicles have been suspected as a possible source, as have cells in the rudimentary right gonad.

The hormone relaxin has also been extracted from the ovaries of birds. If its presence in the bird has any significance, it is unknown.

As in the case of the testis of the male, it has been demonstrated that increased exposure to light has a marked stimulating effect upon both the gamete-producing and endocrine functions of the bird ovary. This is in all probability mediated through the agency of the anterior lobe of the pituitary gland.

Mammals. Estrogens are known with certainty to be secreted by the ovaries of most vertebrates below mammals in the evolutionary scale. In mammals and possibly in ovoviviparous elasmobranchs and other fishes, as well as in ovoviviparous amphibians, lizards, and snakes, progesterone or some other progestogen is secreted by the corpora lutea. The hormone relaxin, previously mentioned, is primarily a hormone of pregnancy, occurring in various species of mammals. It is found in the ovaries of nonpregnant sows and rats, but not in appreciable amounts. It has, however, been extracted in quantity from ovaries, corpora lutea, and placentas of pregnant females. This nonsteroid hormone, the chemical structure of which is not known, is water soluble. Further information concerning relaxin is given on page 375.

Estrogen has been shown to be responsible for the development of the secondary sex characters and accessory sex organs of the female. Progesterone and, in certain cases, relaxin are concerned only with changes occurring in preparation for and during pregnancy.

Removal of the ovaries in the prepuberal female mammal prevents the further development of uterus, vagina, and Fallopian tubes, all of which remain in an infantile condition. Postpuberal castration brings about atrophy of these organs. The secondary sex characters react in much the same manner as do the accessory sex organs. Administration of estrogen prevents regressive changes and restores structures to their full size even after they have atrophied.

One very interesting response to estrogen is the resorption of the pubic bones of the female pocket gopher with the approach of sexual maturity (Fig. 10.85). In this destructive rodent of the western United States, males and young females have well-developed pubic bones which unite ventrally at the pubic symphysis. In adult females the pubic bones have been resorbed, resulting in an enlargement of the restricted pelvic passageway, thus making it possible for the young to pass through the birth canal at the time of parturition. Experiments show that the hormone estrogen is responsible for the resorption of the pubic bones as the time of puberty approaches. Even the pubic bones of males, which normally never undergo resorption, can be made to disappear if injections of estrogen are given.

In females of certain species of monkeys, one of the secondary sex characters is the brilliantly colored "sexual skin" in the region of the buttocks. If the ovaries are removed, this color quickly fades but is soon restored if injections of estrogen are given. Moreover, in a young female monkey the skin is pale before the age of puberty, but as sexual maturity is reached the bright color of the sexual skin appears for the first time. The color rises and wanes in accordance with the cyclic changes in amounts of estrogen secreted by the ovaries.

Much research has been carried on to determine the function of the corpus luteum (Fig. 9.13). It is considered to be a true

Fig. 9.13 Highly magnified photograph of the ovary of a pregnant rat, showing numerous corpora lutea.

endocrine gland which secretes some estrogenic steroids as well as progesterone, the action of which is necessary in preparing the uterus for pregnancy and in the maintenance of pregnancy. It has also been demonstrated that the placentas of certain mammals secrete progesterone. Functional corpora lutea must be present during the early stages of pregnancy in all mammals and, in many species, throughout the entire gestation period. Removal of the ovaries, or of corpora lutea alone, results in resorption or abortion of the developing young. In some mammals, however, as in the human being, mare, and guinea pig, extirpation of the ovaries or corpora lutea during the latter part of pregnancy has no such effect. In these animals, and probably certain others, progesterone secreted by the placenta may take over, supplementing that from the ovaries and thus enabling the pregnancy to continue without interruption. Since the corpus luteum hormone is necessary for bringing about certain essential changes in the uterine mucosa, for the early development of the placenta, and for complete development of the mammary glands, its importance can readily be appreciated. A previous sensitization of the uterus and mammary glands by estrogen is necessary before progesterone can become fully effective.

Whether the hormone relaxin is a true ovarian hormone is questionable since the actual site of its formation is obscure. The general consensus is that in such mammals as the sow, mouse, and rat, in which the presence of ovaries is required for the continuance of gestation, these organs are the primary source of the hormone. In others, in which the ovaries can be dispensed with during the latter part of the gestation period, the placenta may be important in relaxin production. The hormone is effective in relaxing the pubic ligaments of female guinea pigs which have been previously sensitized with estrogen. It is also effective to some extent in relaxation of the human pelvis during the latter part of pregnancy. In the normal pregnant guinea pig the enlargement of the birth canal, resulting from relaxation of the pubic ligaments, makes it possible for the guinea pig to deliver her relatively large young, which could not otherwise pass through the narrow pelvic opening. It is interesting that in the female pocket gopher the resorption of the pubic bones seems to be controlled by estrogen alone, whereas in the guinea pig relaxation of the pubic ligaments is controlled by the cooperative action of estrogen and relaxin. Other functions attributed to relaxin include: (1) some stimulating effect on lobule-alveolar growth of the mammary glands, (2) an inhibitory effect upon spontaneous uterine contractions, (3) a softening of tissues in the cervix of the uterus, and (4) bringing about certain biochemical changes in the uterus involving water, carbohydrate, and protein composition.

The activity of the reproductive organs in the female is cyclic in all species of mammals. In seasonal breeders such activity may occur but once a year. In other mammals, as in rats and mice, periods of activity take place every 5 days. In some species the endocrine activity of the ovaries is influenced by the length of time the animal is exposed to light. Not only do the hormones estrogen and progesterone bring about periodic changes in the reproductive organs, but the very periodicity itself depends upon a reciprocal action of these hormones and the gonadotrophic hormones of the anterior lobe of the pituitary gland.

The cyclic uterine changes which result in menstruation in the human being and other primates are brought about by the rise and wane of estrogen and progesterone

with their subsequent effects on the uterine endometrium. Under the influence of estrogen, which increases in amount as the Graafian follicle grows to maturity, the uterine endometrium undergoes changes. It increases in thickness and receives a richer blood supply. The uterine glands also increase in size, but here the response ceases. This is the condition at the time of ovulation when the endometrium is said to be of the *midcycle type*. After ovulation, which takes place approximately halfway between two menstrual periods, and with the development of the corpus luteum, further changes occur which are brought about under the influence of progesterone. The endometrium becomes thicker, its blood vessels become extremely dilated, and the uterine glands not only become twisted and coiled but begin to secrete actively. The endometrium is now ready to receive the fertilized egg and is said to be a *progestational type* of endometrium. If pregnancy follows, the fertilized egg becomes implanted in the endometrium and develops into an embryo. If fertilization does not occur, part of the endometrium sloughs, accompanied by a loss of blood. Menstruation takes place approximately 2 weeks after ovulation. Following menstruation, a regeneration of the surface epithelium occurs and the endometrium increases in thickness. With the growth of a new ovarian follicle, followed by ovulation and corpus luteum formation, the whole process is repeated. Menstruation is held in abeyance during pregnancy.

The actual explanation of the mechanism responsible for the sloughing of the inner part of the endometrium with its accompanying bleeding has been clarified by studying pieces of monkey endometrium transplanted into the anterior chamber of the eye, where direct observations can be made. The observed changes have been substantiated by histological studies. The mechanism involves (1) vascular fluctuations in the coiled arteries supplying the inner third of the endometrium, (2) cyclic variations in the levels and activity of the ovarian hormones acting upon the endometrial tissues, and (3) the cyclic activity of the anterior lobe of the pituitary gland in producing and releasing the gonadotrophic hormones which in turn affect the ovaries. Menstruation appears to be precipitated by a decrease or withdrawal of the hormonal stimulus which is necessary to maintain the active condition of the endometrium, at the end of the cycle. Cases of menstruation without previous ovulation and corpus luteum formation are known to occur occasionally. In such instances it is the midcycle type of endometrium that is shed, but in these cases also, withdrawal of the maintaining hormonal stimulus (estrogen) is believed to bring about sloughing.

No satisfactory suggestion has been advanced which would indicate that there is any evolutionary significance to menstruation or that it has any adaptive value. Why it should occur only in primates and in no other group of mammals remains an obscure point. The uterine bleeding observed in dogs when in estrus, or occasionally in human beings at the time when an embryo is implanting in the endometrium, should not be confused with menstrual phenomena. It is the result of an entirely different process. Menstruation might be considered to be the result of a biological failure. The uterine endometrium has been built up as though in preparation for implantation of an embryo. When fertilization does not occur and no embryo is present, the endometrium is no longer maintained and shedding and bleeding occur. It then repairs itself and starts to build up once more.

In human beings the menstrual cycle averages 28 days in length. The rhesus monkey's cycle is of similar duration; that of the chimpanzee is 35 days.

In all female mammals at about middle age, the ovaries cease to function in the normal manner. The time at which this occurs is variously known as the *menopause, change of life,* or *climacteric*. In human beings it generally occurs between the ages of forty-five and fifty. In other mammals the time of the menopause is proportionately correlated with the life span. Changes in certain secondary sex characters occur at the menopause, and there is an accompanying atrophy of the genital organs and a cessation of periodic manifestations. In monkeys, when estrogen secretion by the ovaries ceases, there is a marked fading of the brilliantly colored sexual skin. The most important use of estrogens in medical practice is to alleviate the symptoms of the menopause, which in some cases are severe.

Extirpation of the ovaries at any time before the menopause will bring about an artificial or premature climacteric in which all the symptoms of the natural menopause are manifested.

The reciprocal action of the ovarian hormones, estrogen and progesterone, and the gonadotrophic hormones of the anterior lobe of the pituitary gland, in bringing about periodic manifestations in the reproductive organs of the female mammal in polyestrous species and in primates, has already been alluded to (page 376). When the pituitary gland, or the anterior lobe alone, is removed, all cyclic activities cease and the ovaries and other reproductive organs become atrophic. There is much variation among the many species of mammals so far as frequency of periods of reproductive activity is concerned.

At least three gonadotrophic pituitary hormones affecting the ovaries are usually recognized. These include FSH (*follicle-*

stimulating hormone), LH (luteinizing hormone), and prolactin, or the lactogenic hormone, since its main effect is upon the secretion of milk by properly developed mammary tissue. The only reason for including prolactin among the gonadotrophic hormones is that in rats and mice it seems to be essential for maintaining corpora lutea and for their secretion of progesterone. The name luteotrophic hormone (LtH) has been applied to this hormone in the past, but since its luteotrophic function is confined to so few species it is scarcely appropriate to use the term generally.

FSH, as the name implies, affects the cytogenic activity of the ovary, the growth and development of Graafian follicles, and, acting synergistically with LH, induces secretion of the hormone estrogen by the follicular cells. When the estrogen level in the bloodstream reaches a certain height, it suppresses further release of FSH by the anterior lobe of the pituitary gland, but at the same time augments the secretion of LH. During the time that LH secretion is in the ascendancy, a preovulatory swelling of the follicles occurs and the secretion of estrogen is augmented. In animals with restricted mating periods, it is at this time that the female is receptive to the male. The ripened Graafian follicles then rupture, ovulation takes place, and corpora lutea begin to develop. The exact method by which LH brings about ovulation is not known. It is not due to increased pressure within the follicle, but rather to microscopic changes in the follicular wall. Liberation of the ovum, along with follicular fluid, is not a cataclysmic process but rather a slow oozing of the contents of the follicle through a break in the follicular wall. It is not certain at present whether LH in itself brings about ovulation. For this reason the possible, but not probable, existence of another hormonal

factor, an ovulation-inducing hormone, has been postulated. In any case, it seems that both FSH and LH are necessary for ovulation and subsequent formation of corpora lutea. There seems to be an acute release of FSH at approximately the same time that the ovulatory surge of LH occurs. At least in rats and mice, the third gonadotrophic hormone, luteotrophin (LtH or prolactin), released at this time, stimulates the corpora lutea to secrete progesterone. Progesterone, in turn, has an inhibitory effect upon LH production by the pituitary gland. Unless pregnancy occurs, secretion of progesterone by the corpora lutea soon diminishes and its inhibitory influence upon the pituitary gland ceases. FSH production and release begin once more and the whole cycle is repeated.

A neuroendocrine mechanism which is initiated in the hypothalamus (page 624) plays an important role in the release of LH. This seems to be particularly obvious in such mammals as the rabbit, ferret, and cat, in which the nervous stimulus of copulation is normally necessary for ovulation to occur. Even in polyestrous forms and in primates in which there are no restricted periods of sexual receptivity, it seems that the actual effect of estrogen and progesterone on the anterior lobe is mediated via the hypothalamus and a neuroendocrine mechanism.

It is of interest that after the ovaries cease to function at the time of the menopause they no longer are capable of being restored to functional activity by administration of gonadotrophic hormones. They appear to be exhausted.

It is well known that during pregnancy all cyclic manifestations of the reproductive system are held in abeyance. As is indicated later (page 398), the placenta in certain mammalian species is the source of estrogenic and progestogenic hormones responsible in one way or another for the inhibition of ovarian activity via the pituitary gland.

The problem of contraception has become of world-wide importance in connection with the so-called "population explosion," particularly in underdeveloped countries. An inexpensive and reliable method of preventing conception and one not involving routine injections or other unpleasant procedures has become readily available. The principles involved are based upon hormonal interactions as outlined above and include the suppression of development of Graafian follicles and hence failure of the ovaries to release ova. This is accomplished through the oral administration of synthetic steroid hormones in the form of a small pill. It is a method of birth control referred to as *oral contraception*. This has proved to be the most popular and effective method of contraception known to man. It is estimated that in the United States today over 7 million women are taking oral contraceptives. The reduction in birth rate in this country in recent years (H. M. Rosenberg, 1968, *Obstet. and Gynec. Survey,* **23**:74) may in large measure be attributed to the use of oral contraceptives.

Numerous preparations are now on the market under various trade names. They are composed of a combination of synthetic steroid hormones which have an effect similar to estrogen and progesterone in inhibiting the release of gonadotrophic hormones by the pituitary gland. Furthermore, administration of progestogens affects the mucus secreted by the glands of the uterine cervix, increasing its viscosity and rendering it unfavorable to the life of spermatozoa. One pill is taken orally each day for 20 or 21 consecutive days, beginning with the fifth day of the menstrual cycle. After the twentieth or twenty-first daily dose, administration is discontinued. Menstruation generally occurs on the third day after the last pill is taken. Whereas at first a 20-day regimen for taking "the Pill" was prescribed, the 21-day sequence is now becoming more common. This means that there can be a regular routine in which a woman takes the tablets for three weeks and then goes without for one week. Thus, a new course of medication always starts on the same day of the week. This serves as an aid to the memory and is particularly helpful to the uneducated.

There are actually two methods of oral contraception in use. One, known as the *combined method,* consists of a combination of estrogen and progestogen in the same tablet, administered for 20 or 21 days; the other, called the *sequential method,* involves the administration of an estrogenic pill for the first 15 days and another pill containing estrogen combine with progestogen, for the last five or six days. The estrogen commonly used is mestranol (17α-ethynyl-estradiol-3-methyl-ether). Progestogen is sometimes administered in the form of norethindrone (17-α-ethynyl-19-nortestosterone), derived from testosterone. This may exhibit some androgenic effects. Others, including 17α-hydroxyprogesterone acetate, 17α-hydroxyprogesterone caproate, and 6α-methyl-17α-hydroxyprogesterone acetate, are progestogens derived from basic progesterone which do not exhibit androgenic effects. One kind of pill formerly used very widely is composed of norethynodrel (17α-ethynyl-17-hydroxy-5(10) estren-3-one) combined with mestranol.

Women vary somewhat in their response to "the Pill," a few exhibiting unfortunate side effects. A slight enlargement of the breasts may occur but this is not generally considered to be detrimental. Reports in the literature indicate that some women who take oral contraceptives exhibit a deposition of pigment in the skin, causing a cutaneous discoloration; others are troubled with tenderness of the breasts, but this rarely continues for more than a month or two; some exhibit nausea and have to discontinue the medication; gain in weight may sometimes be pronounced; pills containing testosterone-derived progestogens occasionally cause some androgenic effects such as increase in acne; certain women may experience thromboembolic phenomena in which a blood vessel may be obstructed by a blood clot formed elsewhere (it has not been proved that this is caused by taking "the Pill"). Patients with marked varicose veins should avoid taking oral contraceptives as should those with a tendency toward carcinoma of the breast, since these conditions may be aggravated by steroid hormones. It has been reported (M. L. Voorhess, 1968, *Obstet. and Gynec. Survey,* **23**:70) that there may possibly be some danger to an individual who, forgetting to take "the Pill" for a few days, becomes pregnant and then continues to take pills containing a testosterone-derived progestogen. Such a person may give birth to a female child exhibiting masculinization of the external genitalia. Some authorities indicate a possible relation between steroid therapy and hypertension.

On the other hand, relief of such phenomena as dysmenorrhea (painful menstruation), premenstrual tension, and excessive bleeding at the menstrual period are beneficial effects which are experienced by those who take oral contraceptives routinely. The same material may be of possible value in treating certain cases of infertility, since when administration is stopped and the pituitary gland is no longer inhibited, the ovaries become more active than they were formerly. There are some indications that women taking these steroids routinely show a lower incidence of cancer of the breast and reproductive organs than do those of the general population.

When first introduced, "the Pill" was administered at a dosage level of 5, 10, or even more mg. Today smaller doses of the magnitude of 1 to 2.5 mg are given. They seem to be just as effective as contraceptives and are much less likely to cause unfortunate side effects.

The *combined method* seems to be the one most favored and is practically 100% effective in preventing conception. An occasional individual may require a higher level of estrogen to inhibit ovulation than most. In such a case a small dosage may prove to be insufficient to prevent conception.

The U. S. Food and Drug Administration and the Planned Parenthood Federation have approved the use of combinations of steroids, as described, for use as a contraceptive. For the vast majority of women "the Pill" is safe.

Experiments involving long-term administration of a norethynodrel combination to rats showed no effect upon reproductive performance (S. M. Husain and G. Pincus, 1965, *Amer. Zoologist,* **5**:660).

The possibility of producing long-term effects by implantation of a pill in some vascular site in the body (e.g., subcutaneously), has been explored but at the time of this writing has not progressed to the point of practicality. The length of time such treatment remains effective depends largely upon the vehicle in which the steroids are administered. In such cases periodic manifestations, such as menstrual phenomena, would be ruled out or would be very erratic because of the *constant* presence of the hormones in the blood serum.

In recent years it has been determined that the insertion in the uterus of metallic or plastic intrauterine devices (IUD) for long periods of time is an effective method of contraception. Such devices prevent implantation via mechanical means. It is beyond the scope of this book to describe these in detail.

The end products of metabolism of the ovarian hormones are excreted in the urine in water-soluble form. In some species estrogens are also excreted by the liver via the bile duct into the intestine. The estrogenic metabolites in the urine vary among different species. *Estriol*, a degradation product of former steroids, is the chief estrogen present in the urine of pregnant and nonpregnant women. *Estrone* is the corresponding hormone in the pregnant mare. *Pregnanediol* and *pregnanetriol*, in the form of glucuronides, are inert substances found in the urine and have none of the potency of the parent hormone, progesterone.

Several methods are used in detecting the presence and strength of ovarian hormones in extracts and other preparations. No good chemical tests are available at the present time and bioassays must, therefore, be relied upon. The material to be tested is generally administered to ovariectomized rodents. A simple method for determining the presence of estrogen is based upon a study of cornification of the vaginal epithelium. This can easily be detected by the appearance of cornified epithelial cells in vaginal smears a certain number of hours after the initial injection. A more sensitive indication of estrogenic activity is based upon a significant increase in weight of the uterus 6 hours after a single subcutaneous injection of the substance being tested. The increase in uterine weight is due to a rapid uptake of water by the tissues. A second increase, involving tissue hypertrophy and hyperplasia, occurs about 24 hours after the single injection.

Progestogenic activity is usually tested by administering substances to female rabbits previously primed with estrogen. A histological study of progestational changes is made and the uterine response graded accordingly. Tests such as this are time-consuming and tedious. It is hoped that eventually a fairly simple chemical test may be devised.

PITUITARY GLAND

The great importance of the pituitary gland is generally recognized. In the past it was frequently referred to as the "master" gland. It has become increasingly evident, however, as more discoveries have been made, that practically all the endocrine glands are dependent upon the action of others and that such a designation for the pituitary gland is misleading. Furthermore, the manner in which neurosecretory products from the central nervous system control the activity of the pituitary gland, via the hypothalamus, gives increasing emphasis to the concept of a neuroendocrine designation to the nervous and endocrine systems in their joint coordination of various body functions. An outstanding feature of the pituitary gland is that, in addition to secreting agents which exert their effects on tissues other than those of an endocrine nature, it influences many other endocrine glands and is in turn influenced by them.

The pituitary gland, or *hypophysis cerebri*, lies at the base of the brain in the region of the diencephalon, connected to the brain by a *hypophysial* or *infundibular stalk*. It is a compound organ which in higher forms is situated in a depression in the upper face of the basal portion of the sphenoid bone (basisphenoid) called the *sella turcica* (Fig. 9.14. Because of its well-protected and inaccessible location, early anatomists thought it might be the "seat of the soul."

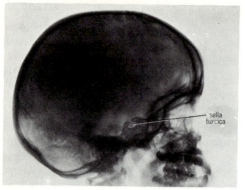

Fig. 9.14 X-ray photograph of human skull, showing well-defined sella turcica in basal portion of the sphenoid bone in which the pituitary gland lies.

There has been some confusion in the literature in regard to the terminology of the various parts of the pituitary gland. Some authorities use the term "hypophysis" to denote only a portion of the pituitary; others use the two terms interchangeably. We shall follow the latter scheme here.

The pituitary gland is composed of three main parts, or lobes: anterior, intermediate, and posterior. A fourth component referred to as the *pars tuberalis* is really a modified part of the anterior lobe. The pituitary body has a dual embryonic origin and is entirely ectodermal. The *neurohypophysis* originates from the *infundibulum*, a ventral evagination of the embryonic diencephalon. From this is derived the *posterior lobe, neural lobe,* or *pars nervosa.* It is this part which remains attached to the brain by the infundibular stalk. The neurohypophysis actually includes the median eminence of the *tuber cinereum* of the hypothalamus, a number of nuclei in the hypothalamus, the infundibular stalk containing axons of nerve cells in these nuclei, and the posterior lobe itself. From the ectodermal epithelium of the primitive mouth

cavity, or stomodaeum, arises a dorsal evagination called the *adenohypophysis*, or *Rathke's pocket.* This usually constricts off the stomodaeum and becomes a closed vesicle which comes in contact with the neurohypophysis. The anterior portion of Rathke's pocket enlarges and becomes modified to form the *anterior lobe*, or *pars distalis*, from which the pars tuberalis is also derived. The posterior portion, which makes contact with the neurohypophysis, enlarges to a lesser extent to become the *intermediate lobe*, or *pars intermedia.* The original cavity of Rathke's pocket usually persists as a small space, the *hypophysial cleft*, or *residual lumen.*

The pars tuberalis is a vascular, collarlike mass of tissue partially surrounding the infundibular stalk and situated close to the tuber cinereum at the base of the hypothalamic portion of the diencephalon. It develops as a pair of lateral, lobelike extensions of the anterior lobe. Since the pars tuberalis is inconstant in appearance, being notably absent in snakes and lizards, and because no specific function has been assigned to it, no further mention of this part of the pituitary complex will be made. Whether it possesses a definite function or is merely rudimentary in character cannot be stated at the present time. Occasionally a remnant of Rathke's pocket becomes separated from the main mass and gives rise to a *pharyngeal hypophysis.* This lies above the nasopharynx. Its cellular components are similar to those of the anterior lobe.

The pituitary gland receives its blood supply from branches of the internal carotid artery and from the *circle of Willis*, which encircles the pituitary gland and is formed from branches of the internal carotid and basilar arteries. Some vessels from the circle of Willis break up in capillaries in the tuber cinereum. The *hypophysio-portal vein* originates from this capillary network and passes to sinusoids of the anterior lobe (page 391).

Each of the anterior, intermediate, and posterior lobes of the pituitary gland gives off hormones which produce different effects in the body. Although the three lobes are so closely related anatomically, they are functionally quite unrelated.

The terminology used for the various hormones in the body is confusing, even to endocrinologists. Different authors sometimes have given diverse names to what has later proved to be the same thing. Abbreviations, or symbols, are frequently used, particularly for certain pituitary and placental hormones. These perhaps add to the confusion. The student who has difficulty in identifying the pituitary hormones may find the following list of symbols to be of some help. Only certain hormones to which abbreviations have been assigned are included.

ACTH   adrenocorticotrophic hormone; corticotrophin
ADH   antidiuretic hormone; vasopressin; pitressin
FSH   follicle-stimulating hormone
ICSH   interstitial-cell-stimulating hormone; luteinizing hormone; LH; ovulation-inducing hormone
LH   luteinizing hormone; interstitial-cell-stimulating hormone; ICSH; ovulation-inducing hormone
LtH   luteotrophic hormone; luteotrophin; prolactin; lactogenic hormone; galactin
MSH   melanophore-stimulating hormone; chromatophorotrophic hormone; intermedin
OIH   ovulation-inducing hormone
STH   somatotrophic hormone; somatotrophin; growth hormone
TSH   thyroid-stimulating hormone; thyrotrophic hormone; thyrotrophin

Posterior lobe. The posterior, or neural, lobe does not have the histological appearance of an endocrine gland. It is composed largely of the terminal portions of gray, non-medullated, or sparsely medullated nerve fibers having their origin in the supraoptic and paraventricular nuclei of the hypothalamus. The axons of these nerve cells course in tightly packed, parallel bundles down the hypophysial stalk. A considerable amount of intercellular substance is present in the posterior lobe. Here are also found branched, cellular elements called *pituicytes*. They appear singly or in groups and seem to be modified neuroglia cells which are more of the nature of supporting tissue than glandular secretory elements. In some mammals, including man, the neuroglia cells contain pigment granules. It is now generally conceded that the hormones attributed to the posterior lobe are not actually produced in this portion of the pituitary gland but are formed within the bodies of neurosecretory cells of certain nuclei in the hypothalamic portion of the brain. Secretions containing the hormones pass down the axons of the nerve cells to the posterior lobe where they enter the hypophysio-portal vascular plexus there. Thus, the posterior lobe of the pituitary gland should be looked upon as a structure in which endocrine products of neurosecretory cells are stored, rather than as a truly secreting endocrine gland. The term *neurohemal organ* is applicable to such a structure.

At least two general kinds of hormones from the posterior lobe are recognized. Both have been synthesized and are rather closely related chemically. They are cyclopeptides or, more specifically, octapeptides, each containing eight different amino acids. Being so similar in structure, it is not surprising that there is some degree of overlapping, so far as their functions are concerned (page 384). Each, however, differs greatly from the other in the extent to which it brings about certain physiological effects.

The first hormone is variously known as

vasopressin, pitressin, the *antidiuretic hormone*, or *ADH*. It has the effect of elevating blood pressure by acting upon the smooth-muscle cells of blood vessels in a manner somewhat similar to epinephrine, although its action is more prolonged. Vasopressin also has an antidiuretic effect since, when administered, it seems to act upon the epithelial cells of distal convoluted kidney tubules, increasing the resorption of water and probably decreasing the resorption of salt. An excessive loss of water, or dehydration, in mammals results in an increase in the osmotic pressure of the blood. When blood in this condition circulates through the brain, certain receptors in the hypothalamus are stimulated. These, in turn, induce an increase in output of vasopressin. This antidiuretic principle, acting upon the kidney tubules, induces the return of water to the circulatory system, and the urine that is excreted is more concentrated and its volume reduced. Extracts containing vasopressin are very effective in reducing to normal levels the enormous output of urine in a disease known as *diabetes insipidus* in which as many as 20 liters of urine are voided during a 24-hour period, and a corresponding amount of water taken in. The disease can be produced under experimental conditions in several species of mammals by severing the hypophysial tracts near the hypothalamic nuclei or by inducing injuries in the hypothalamic nuclei themselves.

It is of interest that the peculiar adaptability of desert rodents to environmental conditions marked by a paucity, or even lack, of water is probably related to the vasopressin-producing capacity of the neurohypophysis.

The other hormone from the posterior lobe is *oxytocin* (*pitocin*). This acts rather specifically in contracting the smooth muscle of the uterus. It also is important in the ejection of milk from secreting mammary tissue by inducing contraction of the branched, ectodermal myoepithelial cells which surround the alveoli of the mammary glands (Fig. 9.15). In numerous species it is commonly known that during the first half minute or so of sucking or milking but little milk is obtained. Suddenly there is a "letdown" and the milk begins to flow freely. The reaction is caused by the release of oxytocin which, in turn, brings about contraction of the myoepithelial cells surrounding the alveoli. Oxytocin will induce a *decrease* in blood pressure of birds when this hormone is administered in the proper amounts. Vasopressin also induces these effects but to a much slighter degree. On the other hand, oxytocin can bring about antidiuretic and pressor effects normally induced by vasopressin but only to the extent of 1 per cent or less of the latter.

The response of the uterus to oxytocin differs under varying conditions. It reacts best when sensitized with estrogen. In the normal mammal at the time of mating, a neurohumoral reflex probably occurs, releasing oxytocin. This increases the contractility of the uterus, a factor of importance in bringing about a rapid movement of spermatozoa from the lower to the upper portions of the uterus. When corpora lutea in the ovary are functioning and secreting progesterone, or in the presence of progesterone of placental origin, there is an inhibiting effect upon uterine contractions and the uterus does not respond to injections of oxytocin. The hormore aids in expelling the fetus during childbirth, or parturition, by inducing uterine contractions. It is effective at this time because of the reduction of progesterone at the end of pregnancy. Oxytocin has long been used as an aid to childbirth because of its powerful effect in inducing uterine contractions. Spawning in *Fundulus heteroclitus,* a small marine fish, has even been brought about by oxytocin injections.

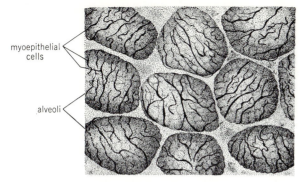

myoepithelial
cells

alveoli

Fig. 9.15 A group of mammary-gland alveoli, showing their relationship to myoepithelial cells surrounding them.

In view of the powerful effect of oxytocin on the uterus, it would be logical to assume that it could not be dispensed with at the time of parturition. It has been demonstrated, however, that parturition can take place normally even when the pituitary gland has been extirpated. Possibly under such conditions enough oxytocin is furnished by the hypothalamus to induce uterine contractions. The role of oxytocin as a necessary factor in parturition awaits further clarification. An extract of the posterior lobe of the pituitary gland, sold under the name *pituitrin,* has been used by physicians for many years. Pituitrin is really a mixture of the two hormones, vasopressin and oxytocin.

Recent studies indicate that the pituitary content of both vasopressin and oxytocin fluctuates during the normal estrous cycle. In rats, the content rises during the follicular phase of the cycle and then falls rapidly following estrus. During pregnancy in the rat there is little or no fluctuation in the pituitary content of these hormones, but during the subsequent lactation period, when the female suckles its young, there is a significant decrease.

Variations in chemical structure of the neurohypophysial hormones, vasopressin and oxytocin, are to be found among members of the several vertebrate classes and are based upon the kinds of amino acids located at certain positions in the molecule. The biologic potencies of the hormones are profoundly influenced by these differences. Authorities have elucidated evolutionary trends in the arrangement of amino acids in these molecules. In mammals, three neurohypophysial hormones have been identified: oxytocin, which has the amino acid isoleucine at position 3 of the molecule and leucine at position 8, and two kinds of vasopressin, *argenine vasopressin* and *lysine vasopressin,* both of which have phenylalanine at position 3 but differing in that the amino acid argenine is located at position 8 in one, whereas lysine occupies this position in the other. Argenine vasopressin is the more widely distributed among mammals. Its presence in the lowly echidnas and in marsupials indicates that it is probably the more primitive of the two. Lysine vasopressin is found in pigs and hippopotami; both hormones are present in peccaries and warthogs. Five kinds of oxytocin have been identified among lower vertebrates. All have isoleucine at position 3. The most widely distributed is *argenine vasotocin,* with argenine at position 8. It is present in cyclostomes, elasmobranchs, and holocephalians and, therefore, is considered to be the most primitive of these hormones. It is not found in mammals. *Mesotocin,* also referred to as *8-isoleucine oxytocin,* with isoleucine at position 8, is found in *Polypterus,* lungfishes, amphibians, and rattlesnakes. *Isotocin* is present in the posterior pituitary glands of fishes of the superorders Holostei and Teleostei. Serine and isoleucine occupy positions 4 and 8, respectively, in isotocin. Oxytocin proper is present in holocephalians, lungfishes, and tetrapods in general. A fifth oxytocic principle, *glumitocin,* with glutamine at position 8, has recently been identified in certain rays.

It is not surprising that in the actions of these numerous and rather similar octapeptide neurohormones there is a considerable degree of overlapping. For example, the vasopressins of mammals are clearly antidiuretic; argenine vasotocin is antidiuretic in amphibians, reptiles, and birds. Neither of these principles affects water conservation in fishes in which water conservation is unnecessary. Our ever-increasing knowledge of the presence of these hormones of varying chemical structure, and the evolutionary trends indicated by them, is a most fascinating new phase of endocrinological investigation. Much more work must be done before the significance of structural differences in the neurohypophysial hormones of vertebrates is fully understood.

Intermediate lobe. As was mentioned in Chap. 4, Integumentary System, chromatophores, which are special pigment-bearing cells, are largely responsible for the color of the skin in many of the lower vertebrates. These cells, which originate from neural crests, are located in the outer part of the dermis. They differ from melanocytes, also of neural-crest origin, which in birds and mammals are located in the lower layers of the epidermis and bear pigment granules. In most fishes, amphibians, and reptiles, the ability to change color within limits so as to blend in with the environment is in large part due to changes in the integumentary chromatophores (Fig. 4.5). The cytoplasm of these cells is usually spread out into a number of complex branches. The pigment granules may be dispersed evenly throughout the cytoplasm, giving the animal a darkened appearance, or they may be concentrated about the nucleus, thus making the animal appear lighter in color (Fig. 9.16). Variations in color and of patterns are often made possible by the action of chromatophores bearing pigments of different colors. The mechanism controlling the dispersion of pigment granules in these changeable chromatophores varies considerably. In some, the autonomic nervous system and neurosecretory cells are responsible. However, in elasmobranchs, amphibians, and many reptiles, the chromatophores are under purely endocrine control. A hormone from the intermediate lobe of the pituitary gland has the effect of causing dispersion of the pigment granules and thus brings about the darkening effect. The hormone is variously known as *intermedin, melanophore-stimulating hormone, MSH,* and *chromatophorotrophic hormone.* Epinephrine and norepinephrine from the medulla of the adrenal gland, or norepinephrine secreted by postganglionic fibers of the sympathetic nervous system, bring about an aggregation, or condensation, of pigment granules. Thus, the color of the animals at different times appears to depend upon the relative activity of intermedin and epinephrine or norepinephrine.

One of the hormones from the anterior lobe of the pituitary gland, already mentioned,

Fig. 9.16 Photograph of salamanders, *Ambystoma texanum:* (*left*) normal control animal; (*center*) a few hours after removal of pituitary gland, with pigment granules concentrated around nuclei of chromatophores; (*right*) several weeks after pituitary removal, the dark color being due to the failure of the animal to shed the corneal layer of epidermis which has become very thick. The last effect is undoubtedly caused by the failure of the thyroid gland to function in the absence of the pituitary gland.

is the adrenocorticotrophic hormone, ACTH, also known as *corticotrophin*. This is a polypeptide made up of a long chain of 39 amino acids. It seems that there are actually two intermedins, both being polypeptides consisting of shorter chains of amino acids than that of ACTH. It is of extraordinary interest that in the two intermedins, as well as in ACTH, there is a common sequence in the arrangement of certain amino acids forming part of their polypeptide chains. This point

is emphasized here because ACTH has been found to possess some melanophore-stimulating activity, whereas the intermedins do not have any effect in stimulating the adrenal cortex. The fact that both the intermediate and anterior lobes of the pituitary gland have a common embryonic origin, being derived from Rathke's pocket, makes it less difficult to understand why the hormones elaborated by them may have some similarities. Birds and certain mammals which lack an intermediate lobe per se nevertheless secrete intermedin, which seems to be derived, for the most part, from that portion of the anterior lobe adjacent to the posterior lobe. This further emphasizes the close relationship that exists between ACTH and intermedin.

It has been mentioned (page 106) that the pigment in the skin of birds and mammals is formed in certain cells called melanocytes, of neural-crest origin, present in the lower layers of the epidermis. These cells manufacture their own pigment but, unlike the chromatophores of the lower vertebrates, do not respond to stimulation by changing the distribution of pigment granules within the cells. They are, however, capable of synthesizing pigments and transferring them to other epidermal cells. If intermedin has any function in birds and mammals, it has not as yet been discovered. Long-continued injections of ACTH are known to bring about a gradual darkening of the human integument. Also a tendency toward an increase in skin pigmentation during human pregnancy is frequently observed. Whether these phenomena are related to MSH activity on integumentary melanocytes remains to be determined. It is possible that the bronzing of exposed areas of the human skin in persons suffering from Addison's disease may in some way be related to abnormal pituitary activity involving excess production of ACTH or MSH. Apparently the intermediate-lobe cells which secrete MSH are basophils.

Anterior lobe. The anterior lobe of the pituitary gland is particularly important and in the past has been reported to be the source of some 11 or 12 different hormones. The actual existence of several of these has not been convincingly demonstrated. Today it is generally conceded that the anterior lobe probably produces six hormones. In some species, as mentioned above, a seventh hormone, MSH, is also produced by the anterior lobe in the absence of an intermediate lobe. The degree to which these hormones are related still requires some clarification. Only three kinds of secretory cells are present in the anterior lobe, the distinctions among them being based upon staining properties. They include *acidophils, basophils,* and *chromophobes.* The chromophobes are believed to be the precursors of both acidophilic and basophilic cells. Variations in the appearance and staining properties of these cells may be observed during different physiological states. In some species they are seasonal. The origin of six or seven hormones from these three kinds of cells evokes some interesting speculations. It is most probable that a given type of cell may alter the character of its secretion under different conditions. Control of the activity of these cells is evidently humoral rather than nervous since secretory nerve endings seem to be lacking in the anterior lobe.

The six hormones from the anterior lobe are all large proteins or polypeptides. Three of them, the *thyroid-stimulating hormone* (*TSH*), the *follicle-stimulating hormone* (*FSH*), and the *luteinizing hormone* (*LH*), also referred to as the *interstitial cell stimulating hormone* (*ICSH*) in the male, are glycoproteins, or proteins containing some carbohydrate. It is believed that these hormones are elaborated by the basophilic cells. A fourth hormone secreted by the basophils of the anterior lobe is the *adrenocorticotrophic hormone* (*ACTH*), a polypeptide. The *chromatophorotrophic hormone* (*MSH*), if formed in the anterior lobe, is also of basophilic cell origin.

Two hormones are believed to be secreted by the acidophilic cells of the anterior lobe: (1) *somatotrophic,* or *growth, hormone* (*STH*) and (2) *lactogenic hormone,* or *prolactin,* also sometimes known as the *luteotrophic hormone* (*LtH*).

The somatotrophic hormone, STH, is frequently referred to as the growth hormone. As more and more has been learned about this endocrine secretion, it has become evident that the term "growth hormone" is inappropriate. In addition to stimulating skeletal growth, it seems to affect protein and fat metabolism, the level of blood sugar, and other metabolic phenomena. Moreover, it is very difficult to define exactly what is meant by growth, since many diverse phenomena are involved. Other endocrine glands also play important roles in growth processes. Growth is actually the result of many quantitative changes occurring in the body. Removal of the pituitary gland from a young animal causes an immediate cessation of skeletal growth. Extracts containing the growth hormone cause a resumption of growth when injected into such animals. The small size of the so-called *ateliotic,* or *pituitary, dwarf,* which resembles an adult in miniature and has normal body proportions, is believed to be due to a lack of sufficient somatotrophic hormone. On the other hand, oversecretion during early life or before normal growth has ceased, may result in *gigantism.* Again, the body proportions are normal but the bones are excessively large. One such boy at the age of eighteen was 8 ft 3¾ in. tall and still growing. If oversecretion of growth hormone begins after the long bones of the body have stopped growing, a different condition known as *acromegaly* occurs. The long bones do not increase in length, since the epiphyses have already closed (page 408). Rather, a change takes place in the bones of the head and face. Other bones increase in diameter, and the spine may become bowed.

It has been clearly shown that prolonged administration of STH results in a permanent diabetes in certain species, as in the dog. Furthermore, in dogs rendered diabetic by removal of the pancreas, extirpation of the pituitary gland relieves the diabetic symptoms. The existence of a separate diabetogenic pituitary hormone was formerly considered a possibility but has never been indisputably demonstrated. ACTH and prolactin, as well as STH, play a role in carbohydrate metabolism and may be used as diabetogenic agents. At any rate, the somatotrophic hormone, in addition to its effects upon skeletal growth, possesses blood-sugar-raising, or diabetogenic, properties. It may possibly stimulate the alpha cells of the islets of Langerhans to secrete glucagon, but it is more likely that the beta cells, under prolonged administration of STH, are eventually destroyed, probably being overworked by the unusually high blood-sugar levels evoked by the hormone.

Much has been learned about the growth hormone itself. Researches conducted during the last four or five years have markedly changed concepts previously held. It seems that the growth hormones of various species differ considerably. They are all proteins consisting of long chains of amino acids. The complete sequence of amino acids is known only for the growth hormone of man. Human STH contains 188 amino acids

(C. H. Li, W.-K. Liu, and J. S. Dixon, 1966, *J. Amer. Chem. Soc.,* **88**:2050). The approximate molecular weights of highly purified growth hormones of nine different mammals have been determined (I. I. Geschwind, 1967, *Amer. Zoologist,* **7**:89). All are of the order of 21,000 to 22,000. The variations in sequence of amino acids in the several mammalian growth hormones probably account for the fact that the hormone of one species may be ineffective when administered to another species. It would seem that here, as in the case of the lactogenic hormone (page 393), there has been some progressive evolution in structure of STH. Most studies have been done on rats. Extracts of growth hormone from the pituitary glands of *Squalus acanthias* and five species of teleosts have been found to be ineffective in the rat although some of them stimulate growth in the fish *Fundulus heteroclitus.* Pituitaries of *Protopterus,* on the other hand, give positive results in the rat. Another point of interest is that although the growth hormone of cattle has no effect upon the chicken or upon tadpoles of *Rana pipiens* and *Rana catesbeiana,* it is effective in the pigeon. It is important in this connection to realize that *lactogenic hormone* from sheep pituitaries promotes growth in hypophysectomized tadpoles of these species (see page 394). This, together with a number of other reactions, indicates that the growth and lactogenic hormones may not be too divergent chemically. No growth hormones tested, even that of the guinea pig, are effective on the hypophysectomized guinea pig. Guinea pig extract, however, is potent when applied to the rat. None of the nonprimate hormones has any effect upon primates.

The synergistic action of STH and such other anterior-lobe hormones as ACTH, FSH, LH, and TSH, which act primarily on "target" organs, enhances the effects of the latter, whereas acting by itself it would have little or no effect upon these organs. Particularly noteworthy is the fact that in order to obtain full growth response from STH, the presence of thyroid hormone(s) is essential. STH apparently promotes growth per se, whereas thyroid hormone induces maturation of tissues. It is quite possible that STH is effective in bringing about the proper tissue environment necessary to enable other hormones to express fully their capabilities. It is evident that we are just beginning to appreciate the overall significance of the growth hormone and its evolution. Its basic function would seem to be to retard the breakdown of amino acids in the body and to enhance the passage of extracellular amino acids across cell membranes, especially those of muscle cells.

Most of the other hormones secreted by the anterior lobe of the pituitary gland affect specific structures in the body. The thyrotrophic, or thyroid-stimulating, hormone (TSH), for example, stimulates the cells of the thyroid gland to secrete thyroglobulin. As the amounts of the thyroid hormones, thyroxine and triiodothyronine, in the blood rise, they in turn inhibit the secretion of TSH (page 343). Reciprocal action of this sort has also been referred to in connection with the ovarian and testicular hormones in their relation to pituitary gland activity. The level of these hormones in the bloodstream, therefore, is controlled by a self-regulating mechanism. The effect of lowering environmental temperature on the output of TSH in rats has already been referred to (page 343). In exerting reciprocal effects upon the anterior lobe of the pituitary gland, it is clear that most hormones can reach hypothalamic centers via the general systemic circulatory system. It has recently been determined that the anterior pituitary thyrotrophic hormone may follow a shorter and more direct route via an internal, ascending, vascular, feedback loop from the anterior lobe to the hypothalamus (K. M. Knigge, 1967, *Amer. Zoologist,* **7**:141). When the thyroid gland is removed or destroyed it is not uncommon for tumors of the anterior lobe of the pituitary to develop. These produce unusually large amounts of TSH since the inhibiting effect, normally brought about by the thyroid hormones, does not exist. When drugs such as thiouracil or thiourea are administered (pages 337 and 344), they prevent the synthesis of the thyroid hormones and the normal effect of the latter in inhibiting secretion of TSH is diminished or lost. TSH is then elaborated in greater quantity, stimulating the thyroid gland to enlarge. The enlarged gland is, of course, unable to synthesize its hormones. Because of this property, which causes an enlargement of the thyroid gland, thiouracil and other closely related compounds are often referred to as *goitrogenic agents.* Tumors of the anterior lobe are sometimes produced when goitrogenic drugs are routinely administered. The detailed chemical structure of the thyroid-stimulating hormone is not known with exactitude, nor has it been prepared in pure form. It is a glycoprotein containing sulfur in the molecule. Its molecular weight has been variously estimated as ranging from 10,000 to 30,000. TSH from cattle is effective in representatives of all classes of vertebrates. There may be species variation in structure of the hormone, but this has not been confirmed so far as the author is aware.

The adrenocorticotrophic hormone (ACTH) stimulates the cells of the adrenal cortex to secrete adrenocorticoids but, in addition, may have other effects upon the body not related to the adrenal glands. When the pituitary gland is removed, the cortex of the adrenal gland shrinks to only a frac-

Anatomy of the Chordates

tion of its normal size. When adrenocorticoids reach a certain level in the bloodstream, they inhibit or suppress the production of ACTH, thus furnishing another example of reciprocal hormonal action.

Since the six anterior-lobe hormones seem to bear some relation to one another in respect to their sites of origin and also in the general similarity of their chemical structure, it is quite probable that the action of such hormones as those from the thyroid and adrenal glands, as well as those from the gonads, may not act specifically or directly in inhibiting the production of one or another of the anterior-lobe hormones. They may, instead, affect neurosecretory cells in the hypothalamus. These, by liberating neurosecretions into the hypophysio-portal system of veins, may influence the cellular elements of the anterior lobe in one way or another. The neurosecretory cells often terminate within storage-and-release centers referred to as neurohemal organs, adjacent to a blood vessel such as the hypophysio-portal vein. The posterior lobe of the pituitary gland thus serves, in part, as a neurohemal organ.

Certain so-called *release factors* (RF) have been extracted from the hypothalamus. They are peptides. In addition to stimulating secretion of ACTH by the anterior lobe, they are effective in bringing about release of four other anterior-lobe hormones, FSH, LH, TSH, and STH. The effect upon prolactin (LtH), however, seems to be inhibitory.

A potent substance, which has been called *hypothalamic D,* or the *ACTH-hypophysiotrophic hormone,* has been extracted from the hypothalamus. This substance primarily stimulates the production of ACTH by the pituitary gland and is possibly an important link in the general-adaptation syndrome, described below.

ACTH and the adrenocorticoids play a most important part in helping the body to react to prolonged, nonspecific stresses of various kinds. When subjected to such stresses as cold, burns, hemorrhage, starvation, social and economic pressures, psychological and emotional strains, etc., the body responds in a manner which serves to counteract the harmful effects induced by the stressing cause or agent. As was noted earlier, the sum of responses made by the body is referred to as the general-adaptation syndrome (GAS). The endocrine system is of paramount importance here since one of the first reactions is an increase in output of ACTH. This causes an enlargement of the adrenal cortex and an increase in output of adrenocorticoids. These in turn, because of their various properties, increase the resistance of the body to the stressing stimulus. Overproduction or imbalance of ACTH and of cortical steroids may lead to various diseases collectively referred to as *diseases of adaptation.* Gastrointestinal ulcers, colitis, hypertension, rheumatic fever, heart attacks of certain types, and other similar conditions are examples of such diseases.

Social pressures of one kind or another, operating via the brain-hypothalamus-pituitary-adrenal cortex pathway are believed to account for sporadic increases and decreases among mammalian populations.

Other actions brought about in the body by ACTH, apparently unrelated to cortical activity of the adrenal glands, include metabolic changes of various types involving carbohydrates, fats, etc. In some respects the action of ACTH is similar to that of the somatotrophic hormone. The chemical structure of ACTH has been elucidated. Reference has already been made to the fact that it is a polypeptide consisting of 39 amino acids linked together. It is the smallest of the anterior-pituitary hormones. ACTH has been isolated only from mammalian pituitary glands (sheep, beef, pig, and man). It has been synthesized in the laboratory, and much is known about its structure. Although in all four animals mentioned the hormone consists of 39 amino acids, species differences exist. Of particular interest is the fact that in all, only the first 20 amino acids in the molecular chain are necessary to evoke the full adrenal-stimulating activity. Species differences are reflected in the portion of the molecule that is inactive. It has been mentioned (page 387) that ACTH and MSH, the melanophore-stimulating hormones, have a certain sequence of amino acids in common in their molecules. This sequence is undoubtedly responsible for the activity of ACTH on melanin dispersion in melanophores of test animals and, in forms lacking melanophores, upon its activity in causing a darkening of the skin. It had formerly been believed that this response to ACTH injections was caused by contamination with MSH.

It has already been mentioned that three gonadotrophic hormones are known to be secreted by the anterior lobe, FSH, LH, and prolactin (LtH). FSH and LH are glycoproteins; prolactin is not. The exact chemical structure of the gonadotrophic hormones is not as yet known, and there are probably species differences among them. Prolactin and LH have been obtained in pure form; FSH has not, and hence little is known of its structure. Gonadotrophic hormones present in the pituitary glands of fishes, and the lower classes of vertebrates in general, differ markedly from those of mammals about which most is known. It has become increasingly evident that there is a sex difference in the hypothalamus, at least in mammals. In the male it affects the anterior lobe of the pituitary gland so as to cause the release of

gonadotrophic hormones in a tonic fashion, whereas in the female the pattern of secretion is cyclic as determined by the hypothalamus.

The follicle-stimulating hormone (FSH) in the female affects the cytogenic activity of the ovary as well as the growth and development of the ovarian follicle. It acts synergistically with LH in promoting the secretion of estrogen in maturing follicles. In the male it is partially responsible for development of mature sperm cells in the seminiferous tubules. Androgen from the interstitial cells is also essential for complete spermatogenesis. Until pure FSH is available, it will be difficult to determine its exact role in the physiology of the male. Apparently it has no effect upon the interstitial cells.

The luteinizing hormone (LH) not only acts synergistically with FSH in promoting the secretion of estrogen by follicular cells but is responsible for the preovulatory swelling and rupture of the Graafian follicles, ovulation, and the formation of corpora lutea. With prolactin (page 378), in rats and mice it stimulates the corpora lutea to secrete progesterone and some estrogen. In the male, LH directly stimulates the interstitial cells of Leydig to secrete androgen. For this reason it is frequently referred to as ICSH. It may indirectly or even directly have some effect upon the seminiferous tubules. When administered to the male frog, it effects a release of mature spermatozoa from the Sertoli cells of the testis. Homoimplants of pituitary glands will induce ovulation in female amphibians of numerous species. LH seems to be the ovulation-inducing hormone here. LH bears some chemical similarity to the thyroid-stimulating hormone of the anterior lobe of the pituitary gland and is a glycoprotein. The molecular weight of human LH approximates 26,000; that of sheep LH is between 28,000 and 30,000. It appears that the luteinizing hormones of different species differ rather widely in their chemical structure. It has recently been suggested that the action of LH on corpora lutea is indirect, operating initially on ovarian interstitial cells which, in turn, maintain the corpora lutea (G. R. Davenport, P. Rennie, and S. Longley, 1968, *Proc. Soc. Exp. Biol. Med.*, **128**:728).

The third gonadotrophic hormone is the lactogenic hormone, prolactin, which, in a small number of mammals only, affects the ovaries by stimulating the corpora lutea to secrete progesterone and maintains them in functional condition for an extended period. It is for this reason that the term luteotrophic hormone (LtH) is sometimes applied to this endocrine secretion. In numerous birds and mammals, however, prolactin administration actually causes repression of the ovaries. The term lactogenic hormone, or prolactin, was first applied to the hormone because of its action in mammals in causing properly prepared mammary tissue to secrete milk. Since then it has been found that the hormone has widely differing activities in representatives of the various classes of vertebrates. Furthermore, the pituitary glands of different vertebrates vary widely as to secretion of prolactin and the types of activity it will evoke. It seems obvious that during the course of evolution, the use to which the hormone was put has been greatly altered (I. I. Geschwind, 1967, *Amer. Zoologist*, 7:89).

Among amphibians, the newt, *Diemictylus* (*Triturus*) *viridescens*, assumes a terrestrial existence for a few years and then returns to the water for breeding. Newts in the terrestrial, or red eft, stage may be induced to migrate to water by administering prolactin. This suggests that the pituitary gland, through its manufacture and release of this hormone, may be the controlling factor. With the assumption of an aquatic environment there is some development of secondary sex characters as well as of courtship patterns. These seem to be independent of gonadal activity. The pituitary glands of representatives of *all* classes of vertebrates have been tested for their water-drive effect on efts and *all* have given positive results. This suggests that promoting eft water-drive activity is one of the earlier effects of prolactin which appeared in the course of evolution. It has recently been found that high doses of prolactin administered to bullfrogs about to metamorphose arrests the metamorphic changes as long as the hormone is administered (W. Etkin and A. Gona, 1967, *Amer. Zoologist*, 7:5).

In pigeons of both sexes prolactin is responsible for the development and normal functioning of the crop glands (Fig. 5.22). There is some dispute as to whether the pituitary glands of chondrichthyes and teleosts contain a factor capable of stimulating the crop glands. Those of *Protopterus* and tetrapods unquestionably contain such a factor.

Prolactin causes properly prepared mammary tissue to secrete milk. Pituitary glands of urodeles, turtles, and pigeons are as effective as those of sheep in eliciting this reaction, but it is doubtful whether those of teleosts, *Protopterus*, and anurans can bring about more than a minimal response. It would seem that the crop gland action on pigeons, the milk secretion effect on mammals, and the luteotrophic effect on the ovaries of rats and mice all indicate that in the course of evolution prolactin itself has evolved to the extent of taking on additional functions. This does not necessarily signify any change or increase in size of the lactogenic hormone molecule.

Apparently the hormone is related to broodiness in the domestic fowl, a behavioral pattern exhibited prior to and during incubation of the eggs. Prolactin also seems to have an effect upon the development of new feathers in birds. Furthermore, it seems to suppress gonadal activity in birds of both sexes. The hormone has been shown to promote growth in hypophysectomized tadpoles and to aid in survival of certain hypophysectomized euryhaline fish transferred to fresh water. It also stimulates melanogenesis in a few teleosts. Numerous investigators have reported that prolactin brings about a variety of other metabolic effects in the body. Its diabetogenic properties have already been mentioned (page 362).

Some investigators have suggested that a *parathyrotrophic hormone* from the anterior lobe of the pituitary gland may control the secretion of the parathyroid glands. There is little evidence for this.

Comparative anatomy of the pituitary gland. Amphioxus. Practically every structure in the head region of amphioxus has been examined with the view of homologizing it with the pituitary gland of higher forms. The most promising suggestion indicates that a depression in the ectoderm, called the *preoral pit,* which lies under the head and in front of the mouth of a young individual may be the hypophysial homologue. The left head cavity in an adult specimen opens into the preoral pit, which by that time has formed a connection with the buccal cavity. The walls of the pit become ciliated, and the structure is then known as the *wheel organ,* or *organ of Müller.* If this is indeed the forerunner of the adenohypophysis, it is of interest that an organ which functions in creating a current of water should give rise to so specialized and important an organ in higher forms.

Urochordata. In an adult tunicate, such as *Molgula manhattensis,* a single nerve ganglion lies embedded in the mantle between the incurrent and excurrent siphons. On the ventral side of the ganglion lies a body, the *adneural,* or *subneural, gland,* which some investigators have homologized with the pituitary gland of vertebrates. A duct leads from the gland into the cavity of the pharynx (Fig. 2.6*A* and *B*). The products of secretion of the adneural gland, which are said to be cellular in nature, pass through the duct, indicating that the gland is not one of internal secretion. Furthermore, it does not rise from a stomodaeal outgrowth but rather represents a degenerated cerebral vesicle which is present in the larval stage. The homologies of the adneural gland and the pituitary of higher forms are, therefore, far from clear. If there is any homology it would seem that this structure of tunicates possibly represents the pars nervosa, or posterior lobe, of the pituitary gland (page 383).

The origin of the pituitary gland from the lowest vertebrates to the highest is very constant in its fundamental constituents and method of development. The consistency shown in the differentiation of the parts of the gland may be traced from cyclostomes to the highest mammals.

Cyclostomes. In the lamprey the single nostril on top of the head leads by a short passageway into the olfactory sac which lies just in front of the brain. A large *nasopharyngeal pouch* extends in a ventroposterior direction from the olfactory sac and terminates blindly beneath the anterior end of the notochord. Between the ventral part of the diencephalon and the nasopharyngeal pouch lies the pituitary gland (Fig. 5.18), consisting of the usual component parts. The homologies of the structures in this region, about which there has been much confusion in the past, have been fairly well clarified. During early development (Fig. 14.23) a *nasohypophysial stalk* appears in close association with the developing olfactory sac at some distance from the dorsal lip and stomodaeum. The solid stalk extends posteriorly beneath the forebrain. From the caudal tip of this stalk certain cells are budded off which are to become the intermediate lobe of the pituitary gland. These intermingle with the nervous tissue forming the floor of the third ventricle. The anterior lobe tissue is budded off somewhat later from a dorsal thickening of a caudal portion of the remaining nasohypophysial stalk. Soon the anterior lobe becomes completely detached. The nasopharyngeal pouch is an *adult* structure formed during metamorphosis from the persistent remnant of the larval nasohypophysial stalk at some time after the pituitary gland has become a definite entity. It hollows out and extends caudally beneath the notochord. Although no relation of nasohypophysial stalk or nasopharyngeal pouch to the stomodaeum is evident, it is believed that at some time a separation of the stomodaeal cavity into two parts may have occurred. The dorsal and anterior portion may then have been drawn inward with the adjacent nasal sac and have given rise to the nasopharyngeal pouch, which has no counterpart in other vertebrates.

In the hagfish, *Myxine,* the nasopharyngeal pouch opens into the pharynx by an aperture which appears late during larval life. The pituitary gland is represented by clusters of cells lying between the infundibulum and the nasopharyngeal pouch. Instead of neural and stomodaeal components coming in close contact as in other forms, they are separated by a layer of connective tissue.

FISHES. In elasmobranch fishes, posterior to the optic chiasma and ventral to the diencephalon is a conspicuous projection, the *hypothalamus.* Its midventral portion consists of a narrow *infundibulum* with its two rounded, laterally placed *inferior lobes* (Fig. 13.3). The infundibulum, which appears as a thin, median stalk, extends posteriorly between the inferior lobes and then expands

into a thin-walled saclike structure, the *saccus vasculosus,* of unknown function. It is lined with a neurosensory epithelium. Since the saccus vasculosus is best developed in actively moving marine fishes, it may possibly have a sensory function associated with awareness of change in fluid pressures. The ventral part of the saccus vasculosus is attached to the small hypophysis, or pituitary gland, which often remains in the sella turcica when the brain is removed in dissection. In the Selachii the posterior lobe of the pituitary gland is a diffuse structure and not consolidated as a distinct entity as in other vertebrates.

Not a great deal has been published concerning the structure of the pituitary gland in the higher fishes. In *Polypterus* and *Latimeria,* the gland exhibits its most primitive condition. In *Polypterus* it connects to the mouth cavity by a persistent opening of the hypophysial cleft, or the original cavity of Rathke's pocket. The duct passes through a foramen in the parasphenoid bone. In *Latimeria* a glandular cord about 10 cm long unites the pituitary gland with its point of origin in the roof of the mouth. In the haddock, the gland is flattened and closely pressed to the brain and does not lie in a sella turcica. In another fish, the angler, *Lophius piscatorius,* the pituitary is at the end of a very long infundibular stalk which extends forward so that the gland is situated about an inch in front of the expected position. In many teleosts, a large infundibulum, inferior lobes, and a saccus vasculosus are present. In some species, the adenohypophysis completely surrounds the neurohypophysis. Teleosts lack a hypophysial cleft. Many minor variations may be observed in the pituitaries of fishes, but the fundamental structure is similar in all.

AMPHIBIANS. In the frog, the anterior lobe is prominent but paradoxically lies pos-

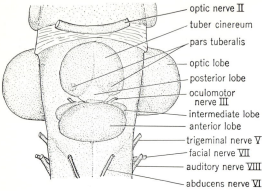

optic nerve II
tuber cinereum
pars tuberalis
optic lobe
posterior lobe
oculomotor nerve III
intermediate lobe
anterior lobe
trigeminal nerve V
facial nerve VII
auditory nerve VIII
abducens nerve VI

Fig. 9.17 Ventral view of hypophysis and adjacent parts of the brain of an adult frog, *Rana pipiens.* *(After Atwell, "The Pituitary Gland," Chap. 20, Assoc. Research Nervous Mental Disease. By permission of the Association.)*

terior to the other parts of the pituitary gland (Fig. 9.17). The intermediate lobe is shaped somewhat like a dumbbell and lies transversely above the anterior lobe. The posterior part of the infundibulum is thickened to form the posterior lobe. As in teleosts, no hypophysial cleft is present in the pituitary gland of amphibians. The stomodaeal evagination which corresponds to Rathke's pocket is, from the beginning, a solid outgrowth. The urodele amphibians show no marked deviations from anurans in respect to the pituitary gland.

It has been shown that if the rudiment of Rathke's pocket is removed early in larval life, the neurohypophysis fails to develop normally.

REPTILES. In this class the infundibular stalk is unusually long and slender and extends posteriorly. The pituitary gland, therefore, lies in a position ventroposterior to the diencephalon. The anterior lobe is loosely connected to the rest of the pituitary gland. Otherwise the pituitary of reptiles is fairly typical of that of other vertebrates. In snakes

and lizards the pars tuberalis is conspicuously absent.

BIRDS. The sella turcica of birds is comparatively deep, and the pituitary gland, which lies within this depression of the sphenoid bone, is farther removed from the brain than in members of the other vertebrate classes. The most unusual feature of the avian pituitary body is the lack of an intermediate lobe and hypophysial cleft. Search for an intermediate lobe in the fowl, duck, and pigeon has failed to reveal any trace of this structure. Although a hypophysial cleft is present during embryonic development, it becomes smaller and smaller and finally disappears altogether. The tissue which would ordinarily form the pars intermedia, lying between the hypophysial cleft and posterior lobe, fails to develop when the hypophysial cleft disappears.

Despite the fact that an intermediate lobe is lacking in birds, the hormone intermedin is present in the *anterior* lobe, the greatest concentration being in its cephalic portion.

MAMMALS. The mammalian pituitary

body is typical of vertebrates in general (Fig. 9.18). Minor variations are to be observed in various species. The central cavity of the infundibular stalk is a continuation of the third ventricle of the brain. This cavity may dip deep into the posterior lobe, as in the cat, or may be short and extend only to a slight degree into the stalk of the infundibulum, as in the ox. In the ox and pig, a peculiar structure projects from the intermediate lobe into the hypophysial cleft. This is called the *cone of Wulzen*. Its cellular structure is more like that of the anterior lobe than of the pars intermedia. The pituitary glands of whales, Indian elephants, and armadillos have no discrete intermediate lobe, and the anterior and posterior lobes are completely separated by connective tissue.

UROHYPOPHYSIS

A peculiar structure associated with the posterior end of the spinal cord, and to which the name *urohypophysis* (*urophysis*) has been applied, is present in numerous fishes but is particularly well developed in teleosts. Unusually large neurosecretory cells are present in the spinal cord in this area. Their axons course posteriorly and, in teleosts, terminate in small bulbs on the basement membranes of blood vessels (capillaries) in the urohypophysis. Secretion droplets have been observed at the ends of the terminal

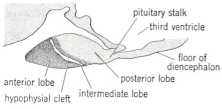

Fig. 9.18 Sagittal section of pituitary gland of rat, showing its relation to the brain.

vesicles. The chemical composition of the neurosecretion has not been specifically identified but seems to be of a protein nature and hormonal in its method of functioning. Apparently, the neurosecretion affects the salt concentration of the blood and is important in osmotic regulation. It has also been suggested that it may bear some relation to the swim bladder in regulating the buoyancy of the animal. The urohypophysis, like the posterior lobe of the pituitary gland, may serve to store and release the endocrine products of neurosecretory cells. Elasmobranchs have large neurosecretory cells in the posterior end of the spinal cord even though a urohypophysis is lacking. Whether this is a primitive arrangement is a matter for speculation. No small terminal bulbs have been observed in the neurosecretory cells of elasmobranchs.

PLACENTA

The placenta might be spoken of as a structure which provides for the exchange of metabolic substances between the mother and her developing young. It consists of two essential portions: a *maternal placenta*, derived from the endometrium of the uterus, and a *fetal placenta*, composed of extraembryonic membranes of fetal origin. It is essentially a mammalian structure, transitory in nature, developing within the uterus of the female early in pregnancy and being expelled at the time of parturition. In those lower nonmammalian forms which retain their developing young in the uterus, there is no true placental attachment. In such ovoviviparous species, secretion of uterine glands may provide nourishment for the developing young to supplement that furnished by the yolk of the egg. In a few species, close apposition of blood vessels in the yolk sac of the embryo or fetus, with vascular, villuslike folds of the uterine lining, provide for a transfer of metabolic substances between mother and young. Study of fetal membranes in mammals gives as clear a concept of mammalian phylogeny as that derived from comparative anatomical or paleontological studies. It should be understood that there are no *direct* connections of maternal and fetal blood vessels. Gases, wastes, nutritive substances, etc., pass from one circulatory system to the other via tissues interposed between the two.

The placenta, in addition to its other functions, is, in most species, an important endocrine organ. Investigations have shown that hormones are elaborated by the placentas of various mammals, but much more systematic study must be made before all the components responsible for their formation and secretion are definitely known. There is much variation among mammalian species as to types of placentas, kinds of placental hormones, and times when they are elaborated. Their physiological properties also vary. Placental hormones are, in general, similar to ovarian hormones and pituitary gonadotrophins. It is often difficult, however, for the student to appreciate the detailed differences in endocrine functions among the placentas of various mammalian species.

Because the placenta is a transitory structure, its role in the endocrine system is that of a highly specialized organ rather than one which plays an important role in the general body economy.

Several types of placentas are to be found among mammals. Whether the more primitive types, such as the yolk-sac and allantoic placentas of metatherians, produce endocrine secretions has not been established so far as the author is aware. Even in some of the higher mammals, the functions of the placenta as an endocrine organ remain rather obscure.

The placenta in women, mares, cows, and many other mammals is the source of the extraovarian estrogen produced during pregnancy. In these species, large quantities of estrogenic steroids are excreted in the urine during the period of gestation. The urinary estrogens are, for the most part, in conjugated form and have little physiological potency. In certain species, estrogen continues to be secreted in large amounts, even after the ovaries have been extirpated from pregnant females, thus indicating that the placenta does not merely serve as a storage place for ovarian estrogens but is very active in producing its own hormones. Soon after parturition there is a rapid drop to normal values, indicating that with the expulsion of the placenta the source of secretion has been removed from the body. Extracts of fetuses show rather low estrogen values, whereas extracts of the placentas of women, mares, cows, sheep, macaques, chimpanzees, and pigs have been shown to possess a much higher estrogen content.

Progesterone is also secreted by the placenta in a number of species. It has been extracted from human and some other placentas. The inert steroids, pregnanediol and pregnanetriol, appear in large quantities in the urine of pregnant women. They reach their highest level shortly before parturition, after which there is an abrupt drop. These substances are believed to be the end products of progesterone metabolism. If the ovaries of pregnant women are removed after a certain stage of pregnancy has been reached, pregnanediol and pregnanetriol continue to appear in the urine, indicating that some extraovarian source, unquestionably the placenta, is producing progesterone.

In the rabbit, on the other hand, removal of the ovaries toward the end of pregnancy results in an abrupt drop in the level of progesterone in the blood with subsequent abortion of the young. This would indicate that progesterone is not formed in the rabbit placenta but is strictly the product of ovarian luteal activity.

The placenta is also the source of gonadotrophic hormones found in the blood and urine of certain mammals during pregnancy. In these species it would appear that the placenta serves as an accessory to the anterior lobe of the pituitary gland, furnishing sufficient gonadotrophic hormones to maintain pregnancy. Placental gonadotrophins, however, are chemically and physiologically different from those produced by the pituitary gland. In the cat, dog, and rabbit, removal of the anterior lobe of the pituitary gland at any time during pregnancy results in resorption or abortion of the young. In several other mammals, including the monkey, guinea pig, rat, and mouse, pituitary ablation during the second half of pregnancy does *not* interfere with the continuance of gestation. Presumably, in the latter group, the placenta is capable of maintaining pregnancy, either by furnishing estrogenic and progestogenic steroids or by maintaining the animals' ovaries in a functional state through the action of its own gonadotrophic hormones. In the cat, dog, and rabbit, this ability is apparently lacking. Again, there are numerous species variations so far as the chemical structure and physiological activities of placental gonadotrophins are concerned.

In man and other primates the term *chorionic gonadotrophin* is used to denote the gonadotrophic hormone elaborated by the placenta. It is believed to be secreted by certain cells of the chorionic villi known as *Langhans' cells.* Human chorionic gonadotrophin is excreted in large quantities in the urine early in pregnancy. It reaches its height at about the sixth week and then drops to a relatively low level. It is a glycoprotein with properties similar to both LH and, in a few species, prolactin (LtH) in maintaining the corpus luteum in a functional state, secreting progesterone until the placenta itself reaches a point at which it can provide sufficient quantities of ovarian steroids. The presence of excreted chorionic gonadotrophin in human pregnancy urine (PU) provides material which is used as a basis for several pregnancy tests (page 400).

Another gonadotrophic hormone is present in the blood serum of pregnant mares. It is known either as *equine gonadotrophin* or PMS (pregnant mare serum) and has properties of a mixture of FSH with a small amount of LH. It is a glycoprotein. This hormone does not appear in the urine but is confined to the circulatory system. It may be recalled that the gestation period of the mare averages 335 days. Equine gonadotrophin appears in the bloodsteam of the mare around the 40th day of pregnancy and disappears around the 180th day, being at its height at about the 120th day. The hormone actually is uterine, rather than placental, in origin, being formed in the endometrial cups of the pregnant uterus. Under the influence of equine gonadotrophin, additional ovarian follicles form during pregnancy. These rupture and develop into accessory corpora lutea. The accessory corpora lutea degenerate around the 180th day of pregnancy when the hormone disappears and the placenta can provide the steroid hormones needed to maintain pregnancy. PMS has been shown to have a profound effect upon protein and nucleic acid synthesis of ovarian tissue.

Equine gonadotrophin seems to have a profound effect upon the ovaries and testes of developing fetuses of the mare. The fetal gonads are much larger during the time that the hormone is present than they are later on.

It is noteworthy that the large blue antelope of India, *Boselophus tragocamelus,* as well as the Indian elephant, possess accessory corpora lutea during pregnancy. Whether in these species a hormone comparable to equine gonadotrophin is secreted remains to be determined.

The pituitary gonadotrophic hormones acting upon the ovaries, and the ovarian hormones, estrogen and progesterone, acting upon the uterus, are essential in bringing about the uterine changes necessary for implantation of the embryo and the early development of the placenta. In all probability the placental hormones either exert an additive effect or replace the pituitary and ovarian hormones.

The placenta also seems to be the source of the hormone relaxin, at least in certain species in which the ovaries may be dispensed with during the latter part of pregnancy. This hormone can be extracted in quantity from the placenta of the rabbit. Certain steroids similar to adrenocorticoids have been identified in placental extracts.

It has been mentioned that androgens, normally produced by the testis, may also be elaborated by the ovaries and adrenal cortex under certain conditions. They are probably present in the placenta as well, but their significance in this connection is obscure.

The presence of chorionic gonadotrophin in the urine of pregnant women is the basis for the Friedman pregnancy test, commonly known as the "rabbit test." A morning sample of urine from the patient to be tested is neutralized and filtered, and 5 cc is injected into the marginal ear vein of an immature female rabbit, or one which has been isolated for at least two weeks. After 24 hours the ovaries of the rabbit, which do not ovulate spontaneously, are examined for the presence or absence of corpora lutea. If developing corpora lutea (or corpora hemorrhagica) are present, the woman from whom the urine was obtained is pregnant. If no signs of corpus luteum formation are evident, the test is negative.

This test is the most widely used pregnancy test in the practice of medicine. It is 99 per cent accurate. Inaccuracies may appear when certain abnormal uterine growths are present, known, respectively, as *chorioepithelioma* and *hydatidiform mole*. These abnormal chorionic tissues also give off the chorionic gonadotrophic hormone which is excreted in the urine.

Another widely used pregnancy test is the Aschheim-Zondek test in which immature female mice are used as test animals. Here the action on the ovaries takes 4 or 5 days, but the principles involved are the same as those of the Friedman test. In both cases the injected gonadotrophic hormone causes the rapid development of corpora lutea from follicles already present in the ovaries. The urine of men suffering from a cancerous condition, known as chorioepithelioma of the testis, will also give a positive "pregnancy" test.

Other tests, using frogs and toads, give much faster results than the Friedman and Aschheim-Zondek tests, and are equally reliable. In the Galli-Mainini test, pregnancy urine is injected into the dorsal lymph sac of a male frog. Spermatozoa are released from the Sertoli cells of the testes and appear in the cloacal fluid about 3 hours after injection. Ovulation in female frogs and toads will occur around 6 or 8 hours after pregnancy urine is administered via the dorsal lymph sac.

Commercial preparations of the gonadotrophin found in the urine of pregnancy are often used successfully in lowering the undescended testes of cryptorchid boys into their position in the scrotum. Hypertrophy of the interstitial cells is the primary result, and the descent into the scrotum is secondary.

The role played by the placenta as an endocrine organ thus appears to be that of an accessory ovary and pituitary gland, at least insofar as the gonadotrophic properties of the latter are concerned. The degree to which it takes over such functions varies from species to species, and at the present time, no accurate generalization as to placental function can be made.

SUMMARY

1. The endocrine system consists of a number of ductless glands which secrete substances called hormones into the blood or lymph. They are carried by the circulating fluid to all parts of the body. The secretions are of a chemical nature and bring about certain changes which are either of a specific or general character.

2. Each glandular component of the system has its particular and specific functions, yet there is a close interrelationship among them.

3. The endocrine glands are all derived embryonically from epithelial surfaces, with the exception of the interstitial cells of the testis.

4. Certain endocrine organs have a dual function and serve either as exocrine and endocrine glands (pancreas) or as cytogenic and endocrine glands (ovaries and testes).

5. It is not certain whether the pineal body and thymus gland should be included among the endocrine glands.

6. The thyroid gland, consisting of a number of rounded thyroid follicles, secretes two hormones, thyroxine and triiodothyronine, which control the rate of metabolism of the entire body.

7. The parathyroids, which first make their appearance with certainty in the class Amphibia, usually lie in close proximity to the thyroids. The hormone of the parathyroids is known as parathormone. It is concerned with maintaining normal levels of calcium and phosphorus in the bloodstream.

8. The adrenal glands of mammals consist of two portions: an inner medulla of ectodermal origin and an outer, mesodermal cortex. In other forms the homologues of these regions do not show similar anatomical relationships. In the lower classes they are separate entities called the chromaffin bodies and interrenal bodies, respectively. In anurans, reptiles, and birds the two types of tissues are inextricably intermingled.

The hormones of the medulla (chromaffin tissue) are called epinephrine, or adrenalin, and norepinephrine, or noradrenalin. Among other things they bring about an increase in blood pressure and amount of sugar in the blood. They dilate the air passages leading to the lungs. Hormones of the cortex or interrenal tissue are known as adrenocorticoids. They have to do with carbohydrate metabolism, maintenance of the volume of circulating blood, and the control of sodium balance and fluid shifts in the body. The normal functioning of this portion of the adrenal complex is necessary for life.

9. The islands of Langerhans in the pancreas are believed to secrete two hormones, both affecting carbohydrate metabolism. One of these, insulin, has the effect of lowering blood sugar; the other, glucagon, has the opposite effect.

10. The interstitial cells of the testes secrete a hormone called testosterone which controls the normal development and functioning of the accessory sex organs of the male and the masculine secondary sex characters.

11. The ovaries of all vertebrates secrete a hormone referred to as estrogen. It is responsible for the development and maintenance of the accessory sex organs and secondary sex characters of the female. In mammals and in several ovoviviparous species among the lower vertebrates, one and often two additional hormones have been recognized: progesterone and relaxin. Both hormones, at least in some species, are probably formed by the cells of the corpus luteum, although the exact origin of relaxin is still in doubt. Progesterone is concerned only with changes occurring prior to and during pregnancy. Relaxin, in general, brings about changes which facilitate the delivery of young. The cyclic activity of the female reproductive system is controlled by the ovarian hormones and their interaction with the anterior lobe of the pituitary gland.

12. The pituitary gland plays a dominant role in the endocrine system. It is usually made up of three parts, or lobes, which have close anatomical relationships. Each lobe elaborates one or more hormones which either affect the body generally or stimulate other endocrine glands to activity. A reciprocal relationship exists between the pituitary and several of the other endocrine glands.

13. The placenta, in addition to its other functions, secretes hormones. Its role is that of an extremely specialized organ concerned only with certain phenomena associated with pregnancy. The hormones which it elaborates are similar to certain pituitary and ovarian hormones.

14. The endocrine and nervous systems (now frequently referred to as the neuro-endocrine system) are the dominating and coordinating systems of the body. The more simple method of humoral distribution of secreted chemical substances is used by the endocrine system in bringing about its effects.

15. With minor variations the several endocrine organs are similar in all classes of vertebrates. Increasing knowledge of the chemical structure of the hormones they produce has made it evident that progressive evolution of hormonal structure has occurred. The appearance of functional corpora lutea in mammals, ovoviviparous elasmobranchs, lizards, and snakes, and their relation to placental nourishment of intrauterine young, has been a significant evolutionary development.

| 10 |

SKELETAL
SYSTEM

The term *skeleton* refers to the framework of the animal body. In vertebrates it is composed of cartilage, bone, or a combination of the two and serves for support, attachment of muscles, and the protection of certain delicate vital organs which it more or less completely surrounds. Furthermore, it maintains the definite form of the animal and is provided with joints so that movement is made possible.

The word *endoskeleton* is used to denote internal skeletal structures. The presence of an endoskeleton is one of the distinguishing characteristics of chordates. Although in most chordates the integument is soft and contains no hard skeletal parts, there are many members of the phylum in which bony elements are present in the skin, being derived from the dermis and providing the animals with a protective armor. Dermal scales are present in most fishes living today, in a few amphibians, in crocodilians, turtles, and even armadillos among mammals. The term *dermal skeleton* is used in referring to such structures and their derivatives. Sometimes the dermal skeleton is spoken of as the exoskeleton, but the latter term is more properly used in connection with the skeleton of invertebrates. It is difficult to decide whether the endoskeleton or the dermal skeleton is phylogenetically older. The presence of bony dermal plates in the fossil remains of extinct ostracoderms and placoderms indicates that they appeared very early in the evolutionary history of vertebrates and should be considered to be primitive structures. However, during the embryonic development of animals possessing both an endoskeleton and dermal skeletal structures, the endoskeleton appears much earlier than the dermal skeleton. This would lead one to infer that the endoskeleton is phylogenetically older. Although many modern vertebrates at first sight appear to lack a dermal skeleton, the membrane bones of their skulls, as well as their teeth, are believed to represent dermal skeletal elements.

In certain animals cartilaginous or bony tissues may be present in various organs of the body, quite dissociated from the rest of the skeleton. They are spoken of as *heterotopic skeletal elements*. Examples include *sesamoid bones*, which develop within tendons at points where tendons move over bony surfaces; bony plates found in the diaphragmatic muscle of the camel and in the hearts of ruminants; snout bones of hogs; and the *baculum*, or *penis bone* (*os penis, os priapi*), found in the copulatory organs of many male mammals. Such heterotopic bones are special structures, and they are generally not considered to belong to either the endoskeleton or dermal skeleton.

Cartilage. Cartilage, or "gristle," is a type of connective tissue which forms an important part of the endoskeleton in all vertebrates, although in higher forms it is much reduced in amount, most of it having been replaced by bone. In an embryo, much of the endoskeleton first appears in cartilaginous form. In such vertebrates as elasmobranchs the endoskeleton does not go beyond the cartilage stage. In higher vertebrates the cartilage which is later replaced by bone is spoken of as *temporary cartilage*, whereas that which is retained throughout life is *permanent cartilage*. The histological structure of cartilage has been described in a former chapter (page 86).

During development the first indication of *hyaline cartilage* formation is to be seen as a condensation of mesenchymal cells in the place where the cartilage is going to develop. Gradually the mesenchymal cells undergo a change, losing their branching processes and becoming rounded cartilage cells, or *chondrocytes*. These cells begin to secrete a clear matrix about themselves which forces the cells apart so that they are more or less isolated from each other by the matrix which they have secreted. During young, or developmental, stages the chondrocytes retain their capacity to divide. Since the newly formed cells then produce additional intercellular matrix, which soon separates them, the cartilage grows. This form of growth, called *interstitial growth*, is necessarily restricted to early stages of cartilage development when there is only a moderate amount of intercellular matrix produced. The surface of the developing cartilage becomes covered by a dense layer of connective tissue, the *perichondrium*, to which muscles, tendons, and ligaments may later become attached. The outer part of the perichondrium is composed of dense fibroelastic tissue and is known as the *fibrous layer*. The inner portion is at first made up of many connective-tissue cells arranged in rows parallel to the surface of the cartilage. These cells gradually differentiate into typical chondrocytes to form the *chondrogenic layer* of the perichondrium. The very actively proliferating and secreting chondrocytes of

Anatomy of the Chordates

the chondrogenic layer are responsible chiefly for the increase in diameter of the developing structure. This is termed *growth by apposition.* In some cases no further change takes place and the cartilage exists as such throughout life. Examples of permanent hyaline cartilage in the body of man include the lower parts of the ribs (costal cartilages), the articular surfaces of bones, and the cartilages of the larynx, trachea, and bronchi. Other types of cartilage are also present in certain parts of the body. *Elastic cartilage* gives support to the external ear, epiglottis, and other structures. *Fibrous cartilage* is found in the intervertebral discs and in places where tendons and ligaments are attached to cartilage of the hyaline or elastic types.

Cartilage is essentially an avascular tissue since blood vessels do not penetrate it. Chondrocytes receive nourishment and dispose of waste products by diffusion through the intercelluar matrix. Deposition of insoluble calcium salts in the matrix of hyaline cartilage is common, enabling the tissue to bear more weight because of its increased strength. Although calcification, for this reason, might seem to be a desirable feature, it actually has the adverse effect of interfering with the diffusion of nutrient and waste materials throughout the matrix, with the result that the chondrocytes ultimately die. Although this might seem to be of little or no importance since a rigid matrix is already present to provide proper support, nevertheless the actual result is that in the absence of living chondrocytes the calcified matrix gradually breaks down and is resorbed. The intercellular substance in permanent cartilage normally is not calcified to any degree, whereas that of temporary cartilage usually becomes calcified, and upon its dissolution is replaced by bone. The size of the chondrocytes is in some way related to the calcification process, since the matrix surrounding large cells is much more likely to become calcified than that produced by cells of smaller size.

Bone. Bone, which forms most of the skeleton in higher vertebrates, is another type of connective tissue, the structure of which has been previously described. It will be recalled that bone exists in either spongy (*cancellous*) or *compact* form (page 88). Bone is formed in two ways: (1) by direct ossification in connective tissue without an intervening cartilage stage, such bone being referred to as *membrane,* or *dermal, bone,* and (2) by replacement of preexisting cartilage, in which case it is known as *cartilage bone* or *endochondral bone.*

MEMBRANE BONE. Certain flat bones of the face and of the skullcap, or vault of the cranium (*calvarium*), are typical membrane bones. The frontal and parietal bones are examples. They actually represent dermal plates, or scales, which have moved from their original peripheral location and have sunken inward to a deeper position, where they have become attached to the true endoskeleton. Although they are often considered to be parts of the endoskeleton, the manner in which they develop differs. Such bones are formed by a method known as *intramembranous ossification.*

The first indication of the formation of a membrane bone appears as an aggregation or condensation of mesenchymal cells in the region where the bone is to develop. Gradually these cells form small groups which tend to string out in various directions so as to form a sort of interlacing network commonly referred to as *membrane.* Bundles of fine collagenous fibers course in all directions among the mesenchymal cells. A semifluid, *amorphous ground substance* is present between the cells and fibers. The developing bone may first be observed as thin bars of a dense *inter-cellular substance* which appears between the mesenchymal cells and which is apparently secreted by them. These bars thicken and join each other in a rather haphazard manner, forming a sort of meshwork in which the original mesenchymal cells with their branching processes become trapped. These bone-forming cells are called *osteoblasts.* They gradually change their shape, but the branching processes of adjacent cells for a time retain their connections. The intercellular substance, now known as the *matrix,* continues to increase, gradually taking the place of the amorphous ground substance. Soon the matrix is impregnated with calcium salts and becomes hard and rigid, at which time it is properly referred to as *bone.* Each individual cell, now known as an *osteocyte,* lies in a space, the *lacuna,* surrounded by matrix. Later on, as the bony tissue matures, the connecting cytoplasmic processes are withdrawn, leaving in their places tiny canals, called *canaliculi* (Fig. 3.22). These ramify throughout the bony matrix, or calcified intercellular substance, and provide a means by which tissue fluid may gain access to the osteocytes trapped in the lacunae. New osteoblasts, arising by a transformation of mesenchymal cells, arrange themselves in a continuous layer over the surface of the small

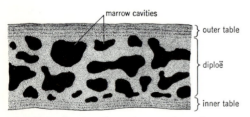

Fig. 10.1 Section through a portion of a membrane bone (parietal bone of man) showing cancellous bone bounded by compact periosteal bone on both inner and outer surfaces.

bony bar, and as a result of their activity the bar increases in thickness. Each osteoblast soon becomes surrounded by matrix and is transformed into an osteocyte lying in its own lacuna. In this manner numerous, irregular bony plates and bars are formed. As they enlarge and thicken they gradually coalesce, forming what is known as *primary cancellous bone.* This has a spongy appearance. The small spaces between the plates and bars are filled with very vascular connective tissue which gradually becomes transformed into bone marrow. The spaces are then referred to as marrow cavities.

While the primary cancellous bone is developing and taking on the general configuration of the bone which is to form, the mesenchyme surrounding the developing membrane bone, above and below, differentiates into a dense layer of collagenous connective tissue, the *periosteum.* Osteoblasts appear on the inner surface of the periosteum and arrange themselves in a single layer. These osteoblasts begin to lay down strata of *compact,* or *periosteal, bone,* thus forming *inner* and *outer* tables, fusing with and enclosing the cancellous bone, or *diploë,* between them (Fig. 10.1). The periosteum becomes tightly applied to the surface. Further growth of the bone in thickness depends upon the activity of osteoblasts which lie beneath the periosteum and are derived from its innermost layer. Because of the rapid calcification of the matrix and its subsequent lack of malleability, interstitial growth of bone is precluded. Growth of bone, therefore, unlike that of cartilage, is by apposition alone. There is no clear-cut distinction between periosteal and cancellous bone. They actually represent different arrangements of the same type of tissue.

CARTILAGE BONE. Cartilage bones, in contrast to membrane bones, go through a cartilaginous stage in their development.

The cartilage is usually a miniature of the bone which is to replace it. The term *endochondral bone formation* is used to describe this type of bone development, since it takes place within preexisting cartilage.

The main part, or shaft, of a long cartilage bone such as the tibia is called the *diaphysis*. The first indication of cartilage-bone formation is given by the appearance of a ring of bony tissue around the center of the cartilaginous diaphysis (Fig. 10.3*A*). The original perichondrium becomes the *periosteum*. Osteoblasts, derived originally from the inner, or chondrogenic, layer of the perichondrium, give rise to compact, *periosteal bone*, similar in every respect to the periosteal bone which forms the inner and outer tables of membrane bones as described above. At about the same time that this bony ring is forming, degenerative changes occur within the cartilage in the center of the diaphysis. The cartilage cells undergo certain peculiar changes and then start to degenerate. The cells die, and the intercellular matrix, which has become calcified in the interim, begins to break down, with the result that small, irregular cavities appear. Capillaries from the periosteum together with osteoblasts, in the form of *periosteal buds,* push into the spaces in the eroding central portion of the cartilage, forming a center of ossification. Soon the area is composed of a number of small intercommunicating channels which become filled with embryonic bone marrow. The irregular, calcified, cartilaginous bars, or *trabeculae*, which lie among the eroded spaces, serve as centers about which the invading osteoblasts aggregate. These cells deposit bone around the cartilaginous bars (Fig. 10.2). This forms a bony meshwork of trabeculae, with the result that the center of the diaphysis is composed of a mass of spongy bone. In the meantime osteoblasts beneath the periosteum are depositing compact bone

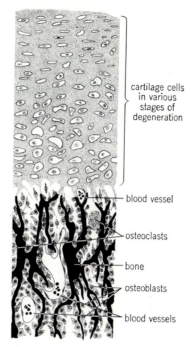

cartilage cells in various stages of degeneration

blood vessel

osteoclasts

bone

osteoblasts

blood vessels

Fig. 10.2 Section through a portion of the end of a long cartilage bone, illustrating endochondral bone formation.

which surrounds the cancellous bone. While these processes have been going on, the cartilage at the ends of the developing structure continues to grow in length. Increase in diameter is brought about by continued deposition of bone by osteoblasts located beneath the periosteum. The center of ossification, which begins in the middle of the diaphysis, gradually extends toward the ends of the bone which, however, remain cartilaginous and continue to grow until much later. In mammals, and to a lesser degree in reptiles, secondary centers of ossification later appear at the ends of the bone. These are known as *epiphyses* (Fig. 10.3*D* and *E*). The time of appearance of the epiphysial centers of ossification varies in different bones of the body. In some they

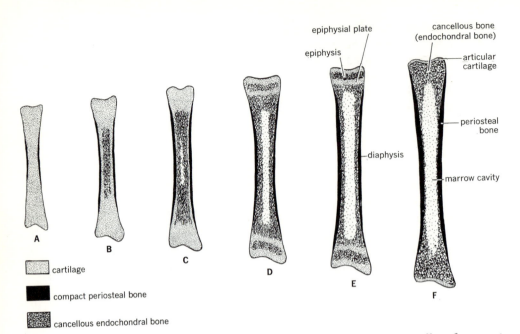

epiphysial plate

epiphysis

cancellous bone
(endochondral bone)

articular
cartilage

periosteal
bone

diaphysis

marrow cavity

A

B

C

D

E

F

cartilage

compact periosteal bone

cancellous endochondral bone

Fig. 10.3   Diagrams indicating progressive ossification of a long bone: *A*, collar of compact periosteal bone surrounding cartilage in middle region of diaphysis; *B*, endochondral bone appearing as primary center of ossification; *C*, marrow cavity appearing as a result of dissolution of cancellous bone in center; *D*, appearance of secondary centers of ossification, the epiphyses, at either end of structure, separated from primary center by epiphysial plate; *E*, later stage, similar to *D*; *F*, closure of epiphyses with disappearance of epiphysial plate; articular cartilage at either end of bone—no further increase in length can take place.

appear toward the end of fetal life, but in others they do not make their appearance until some time after birth. Some bones, as in the case of certain phalanges, have only one epiphysis, although most have one at either end. Still others may have two or three separate ossification centers. The proximal end of the femur has three such centers of ossification.

Between each epiphysis and the diaphysis there remains a cartilaginous plate, or pad, known as the *epiphysial plate*. Until adult stature is attained, increase in length of the developing bone takes place in the region of the epiphysial plate. Its terminal portion grows by the formation of new cartilage, whereas the inner part is continually being

replaced by developing bone from the diaphysis. Eventually the epiphysial plates are entirely replaced by bone and the epiphyses are united firmly to the diaphysis (Fig. 10.3*F*). Further growth in length of the bone can then no longer occur. The ends of the bones are covered with articular cartilage throughout life. The rate of growth at the two epiphysial plates of a long bone may differ markedly. In the femur, for example, the main increase in length takes place at the distal end.

Growth in diameter of cartilage bones takes place by the addition of new periosteal bone on the outer part of the developing diaphysis. It gradually extends toward the epiphyses. While growth is taking place, constant internal alterations and reconstruction are going on, leading to the formation of a large marrow cavity in the center of the bone. Not only does this cavity form a space to house the bone marrow, but the resulting cylindrical shape of the bone has the mechanical advantage of providing greater strength. As new bone is forming, old bone is being destroyed. This has been demonstrated under experimental conditions in which madder leaves have been fed to growing animals. Bone which is developing during the period of madder feeding is red in color. If times of madder feeding are alternated with intervals in which no madder is given, colored and noncolored rings of bone deposits can easily be discerned. It has thus been observed that old bone is being removed in the center of the diaphysis as new bone is being added to the circumference (Fig. 10.4). Destruction or resorption of old bone appears to be the function of certain multinucleated giant cells called osteoclasts which are sometimes confused with megakaryocytes (page 85). Megakaryocytes possess a single, but lobed, nucleus. They are present in bone marrow and are concerned with the production of blood platelets. Osteoclasts, on the other hand, with their numerous nuclei, abut upon naked bony surfaces. They are possibly identical with so-called foreign-body giant cells, which may be found in many other parts of the body in association with foreign bodies that somehow gain entrance to the tissues. If this view is accepted, it means that these cells react to a naked bony surface as though it were a foreign body. Osteoclasts are mesenchymal derivatives. Their multinucleated condition has been interpreted as being due to a fusion of numerous cells. Their ability to bring about bone resorption may be attributed to enzymic action which affects the retention of mineral elements previously deposited in the intercellular matrix. Osteoclasts are responsible for the dissolution of the central part of the diaphysis, removing the original cancellous bone and forming the marrow cavity. As a result, the shaft of the bone finally becomes a cylinder of compact bone containing a marrow cavity in its center filled with bone marrow.

We have thus seen how the center of the diaphysis of a long bone is at first composed of hyaline cartilage. This becomes eroded, but small, irregular, and calcified cartilaginous trabeculae remain, serving as centers about which bone is deposited. Spongy bone replaces the cartilage and later,

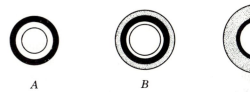

A   *B*   *C*

Fig. 10.4 Diagrammatic representation of the manner in which a long bone increases in diameter: *A*, from an animal recently fed with madder, colored layer of bone (black) deposited on outside; *B*, madder feeding discontinued, new colorless bone (stippled) deposited over colored layer; *C*, still later, colorless outer layer is thicker, inner layer and part of colored layer removed by osteoclastic activity.

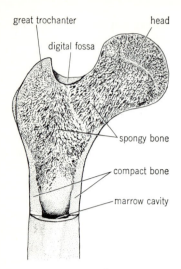

great trochanter   head

digital fossa

spongy bone

compact bone

marrow cavity

Fig. 10.5 Section through proximal end of fully developed human femur, showing relation of
spongy bone, compact bone, and
marrow cavity.

through osteoclastic activity, is itself destroyed, so that finally only a marrow cavity
remains, surrounded by compact bone of
periosteal origin (Fig. 10.5).

It is of interest that once the epiphyses
have united with the diaphysis, no further
increase in length of a bone can occur.
However, increase in diameter is still possible. This is well illustrated in the disease
known as *acromegaly*. Under excessive stimulation by somatotrophic hormone from the
anterior lobe of the pituitary gland, after
epiphysial closure has occurred, there may
be a marked increase in the diameter of the
long bones as well as a thickening of other
types, such as the bones of the face and
of the vertebral column.

Bone Marrow. *Hemopoietic tissue,* in
which the various kinds of blood cells are
formed, occurs in lower vertebrates in a
variety of structures. Likewise, during developmental stages in higher vertebrates,
several structures serve as blood-cell-form-
ing organs (page 595). It is in the anuran
amphibians that the bone marrow, which occupies the cavities within certain bones, is
first concerned with blood-cell formation. In
these amphibians, as well as in reptiles and
birds, all types of blood cells are formed in
the bone marrow. In mammals, however,
the hemopoietic function of the bone marrow is restricted to the production of *red
blood corpuscles* (*erythrocytes*), *blood platelets* (which are derived from megakaryocytes), and three kinds of *white corpuscles,*
the *granular leukocytes* (*heterophils, eosinophils,* and *basophils*). Other kinds of white
corpuscles are formed, for the most part, in
lymphoid tissue located in various parts of
the body. In mammalian embryos and in
newborn young the only type of bone marrow that is present is *red bone marrow,* or
myeloid tissue. It is the red bone marrow
that has hemopoietic properties. As an individual becomes older, however, the red bone
marrow is replaced in many regions by
yellow, or *fatty, bone marrow.* This type of
bone marrow contains a quantity of fat-
storing cells, or adipose tissue, and is not
actively engaged in hemopoiesis. Nevertheless, it retains its ability to produce red
blood cells should the requirements of the
body again demand it. Under such conditions the yellow bone marrow changes back
to red bone marrow. Estrogenic substances
exert an inhibiting action on the production
of erythrocytes, whereas another substance,
erythropoietin, probably from the kidneys,
stimulates red-cell production (page 268).
In adult human beings and other mammals,
red bone marrow is confined to the diploë
in the membrane bones of the skull, the
proximal epiphyses of the humerus and
femur, the ribs, sternum, and vertebrae.

Other marrow cavities serve primarily as storage places for fat.

It should be obvious that in order for bone to grow and for bone marrow to develop and maintain its hemopoietic properties, an abundant blood supply must be received. Despite its rigid consistency, bone is a very vascular tissue. It has numerous small canals, called *nutrient foramina,* throughout its substance which furnish passageways for the blood vessels which supply and drain the internal portions of the bone. Bone is one of the most active tissues in the body, even in an adult in which it has become completely differentiated. Its internal structure is continually undergoing alteration as old bone is renewed and replaced by new bony tissue. Changes in calcium, phosphate, and other inorganic salts are constantly going on in an effort to maintain the proper level of these salts in the blood stream so necessary for the proper functioning of many vital processes. These processes are controlled by various enzymes, vitamins, and hormones. The importance of bone marrow in generating blood corpuscles has already been emphasized.

Joints. The term *joint* refers to an articulation between cartilages or bones. Two general types are recognized: (1) *synarthroses,* or immovable joints, and (2) *diarthroses,* or freely movable joints. There are several types in each category. Among synarthroses are included such immovable joints as *sutures,* which are the lines of junction of facial and cranial bones; *gomphoses,* in which a bony projection fits into a socket, as in the case of teeth in the mandible, premaxillary, and maxillary bones; *schindyleses,* in which one bone fits into a slit in another, as in the articulation of the *lamina perpendicularis* of the ethmoid bone with the vomer. Among diarthroses are *ball-and-socket* joints

(*enarthroses*), *hinge joints* (*ginglymi*), *pivotal,* or *rotary, joints* (*rotatoria*), and *gliding joints* (*arthrodia*).

In the formation of synarthroses three conditions may be encountered: (1) the mesenchyme between the bones may differentiate into connective tissue proper which, as in the case of sutures, binds them firmly together (*sutural ligament*); (2) the mesenchyme may differentiate into cartilage, as in the pubic symphysis; and (3) the mesenchyme may differentiate into bone, as in the union of the two halves of the mandible in the cat.

In the development of diarthroses, spaces appear in the mesenchyme between adjacent bones ultimately to form the *synovial,* or *joint, cavity* (Fig. 10.6). This is lined with a *synovial membrane* composed of dense, irregular connective tissue. The capsule which surrounds the joint is continuous with the periosteum of the two articulating bones. It is composed of an outer, dense, *fibrous layer* and a less dense, more cellular inner *synovial layer* which encloses the synovial cavity. The synovial cavity contains a sticky, glairy, *synovial fluid* believed to be secreted by the synovial membrane. It serves to lubricate the articulating surfaces of the bones in question. The ends of the bones, it should be recalled, are covered with hyaline articular cartilage. Although the synovial membrane is said to be reflected over these cartilages during embryonic development, it is lacking here in the adult.

Ligaments and tendons, which sometimes appear to pass through the synovial cavity, actually lie outside the cavity, since the synovial membrane is reflected over them. The *ligamentum teres femoris,* which joins the head of the femur to the acetabulum, is an example of such a structure.

In some cases the joint may be divided, completely or incompletely, by an *articular*

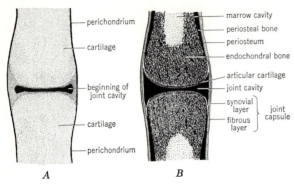

Fig. 10.6 Diagrams showing method of formation of a diarthrodial joint: A, beginning of joint cavity appearing in the mesenchyme between two adjacent cartilages; B, section through fully formed joint.

disc or meniscus. Under such conditions (Fig. 10.7) the edges of the disc are continuous with the fibrous portion of the capsule and there may actually be two synovial cavities, one at the end of each of the articulating surfaces.

DERMAL SKELETON

The dermal skeleton of vertebrates originates in the corium, or dermis, of the skin and is therefore of mesenchymal origin. It should not be confused with *horny* scales or other structures derived from the epidermis, which is, of course, of ectodermal origin.

The various types of fish scales, already referred to (pages 125 to 128), represent remnants of the dermal skeleton. They are composed of membrane bone or of substances very closely related to bone. In the past it was believed that the placoid scales, or dermal denticles, of elasmobranchs represented a very primitive form of dermal derivative. With the realization that dermal bony plates were already present in the ancient ostracoderms and placoderms, it became necessary to adopt a different inter-

pretation. It is now assumed that the elasmobranchs, with their cartilaginous skeletons, have lost most of the dermal skeleton and that the placoid scales simply represent persistent remnants of such structures.

That teeth are undoubtedly dermal derivatives is strongly indicated in the elasmobranch fishes in which the transition between

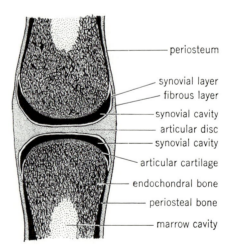

Fig. 10.7 Diagram of diarthrosis with an articular disc and two separate synovial cavities.

placoid scales and teeth is to be seen clearly at the borders of the mouth. This relationship is not surprising when one considers that the stomodaeal lining is actual invaginated integument which has retained its capacity for scale (tooth) formation. The fact that both placoid scales and teeth in elasmobranchs develop under the influence of enamel organs even though enamel is not formed over the exposed surface of the placoid scale or elasmobranch tooth further suggests a homology between teeth and placoid scales.

Fin rays, which support the peripheral portions of the fins of fishes, are also included in the dermal skeleton. However, the cartilaginous or bony elements, which support them at the base, belong to the endoskeleton.

In the evolution of the vertebrates it would seem that there has been a tendency to eliminate the dermal skeleton, which is fundamentally a protective device. Epidermal derivatives such as epidermal scales, hair, and feathers have in many cases become elaborately developed, serving to protect the body in lieu of dermal structures. The dermal skeleton has been referred to in some detail in the chapter on the integument and requires no further discussion here. The dermal plates of extinct labyrinthodont amphibians and the integumentary scales of caecilians hark back to a fishlike ancestry. Likewise, the bony plates in the integument of crocodilians, turtles, and a few snakes and lizards, together with the ventral dermal "ribs," or *gastralia*, encountered in *Sphenodon*, crocodilians, certain lizards, and in *Archaeopteryx*, also indicate a piscine origin.

The homology of the membrane bones of the skull of modern teleosts, amphibians, reptiles, birds, and mammals to the bony dermal plates, or armor, of the earliest-known vertebrates is well established.

Except for the membrane bones of the skull, dermal skeletal elements are usually lacking in birds.

The same thing is true of most mammals. The antlers of deer are outgrowths of the frontal bones, formed under influence of the integument. They may be considered to belong to the dermal skeleton. Bony plates are present in some of the whales, in which they appear along the back and in the dorsal fins. The edentate armadillos are the only living mammals in which the dermal skeleton is well developed. The large, extinct glyptodons, which were also edentates, possessed a rigid armor somewhat like that of the armadillo but lacked movable rings. Whether the dermal skeleton of the edentates has arisen *de novo* or has been derived from reptiles has not been determined.

ENDOSKELETON

It seems somewhat artificial to make a clearcut distinction between dermal skeleton and endoskeleton because of the fact that membrane bones, which are clearly dermal elements, contribute to what we generally consider to be the endoskeleton. In the following account we shall not stress these differences.

The endoskeleton of vertebrates is made up of *axial* and *appendicular portions*. As its name implies, the axial skeleton is that which forms the main axis of the body. It is made up of (1) the skull, (2) the vertebral column, (3) the ribs, and (4) the sternum. The latter is considered to be part of the axial skeleton mainly because of its topographical relationships. Its embryonic origin suggests that it is really a portion of the appendicular skeleton (page 474). The *visceral*, or *branchial, skeleton*, consisting of elements which in fishes support the gills and jaws, contributes to the skull in higher forms. For this reason it is here included in the axial skeleton although some authors, for various reasons, make a distinction between the visceral skeleton and the somatic skeleton, which includes all the remaining endoskeletal structures. The appendicular skeleton is composed of the skeletal elements of the paired anterior *pectoral* and posterior *pelvic appendages* together with their respective *girdles* by means of which the appendages are connected directly or indirectly with the axial skeleton.

Axial Skeleton

Notochord

The primitive axial skeleton consists of the notochord, present during early development in all chordates but replaced by the vertebral column and base of the skull in all but a few of the lower vertebrates. The notochord is derived, during early development, from the region of the dorsal lip of the blastopore. Embryologists refer to this tissue as *chorda-mesoderm*. The notochord first makes its appearance as a long, rodlike structure extending longitudinally from the infundibulum, on the ventral side of the diencephalon, to the posterior end of the body. It lies directly beneath the neural tube. It is composed of vesicular connective tissue and is unsegmented. The cells composing the notochord are usually encased within an *outer sheath* of dense fibrous connective tissue. The large, turgid cells originally inside the sheath undergo a change as development progresses. The peripheral cells secrete an *inner elastic sheath* beneath the outer sheath. The centrally located cells become vacuolated, and only their walls persist, so that this portion of the notochord has a pithy appearance (Fig. 1.5).

In amphioxus, cyclostomes, and a few fishes, the notochord persists unchanged throughout life as the main axial support. Even in the lamprey, however, the protection which the notochord gives to the spinal cord lying above it is supplemented by the presence of small, paired, segmentally arranged cartilages which abut against the dorsal side of the notochord and form a sort of incomplete arch over the spinal cord. They foreshadow the appearance of neural arches in the vertebrae of higher vertebrates.

The notochord of *Latimeria* furnishes its chief axial skeletal support. It is large and unconstricted. It extends anteriorly to below the front part of the brain (also true of some Rhipidistia), and posteriorly into the small "supplementary" caudal fin. It seems that in this large fish, which in the adult stage attains a length of 4 to 5 ft, the notochord is between 4 and 5 cm in diameter. The sheathing layers are unusually thick and strong, and the notochord is actually a fluid-filled tube, its cellular character having been lost, at least in the adult stage, except at the very posterior end. Some cartilage cells invade the notochordal sheath but without any regularity. They come from adjacent neural and hemal arches.

Spinal column

The portion of the axial skeleton which protects the spinal cord is, in most chordates, composed of a series of segmentally arranged skeletal structures, the *vertebrae*. In their aggregate they make up the *spinal*, or *vertebral*, *column*. This extends posteriorly from the base of the skull and continues to the tip of the tail. The vertebral column gives rigidity to the body and furnishes a place for the direct or indirect attachment of the appendage girdles and numerous muscles. A *neural arch*, on the dorsal side of each vertebra, surrounds and protects the spinal cord. Various projections extend outward from the vertebral column, thus furnishing surfaces for attachment of muscles and ribs. The separate elements, or vertebrae, are bound together in such a manner as to provide

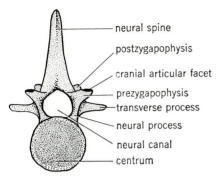

neural spine
postzygapophysis
cranial articular facet
prezygapophysis
transverse process
neural process
neural canal
centrum

Fig. 10.8 Anterior view of typical mammalian vertebra.

rigidity together with a certain degree of flexibility. In no case is the column so flexible as to permit damage to the delicate spinal cord. In some fishes the vertebral column is entirely cartilaginous, but in most forms it is bony, having developed, for the most part, by the endochondral method of ossification.

In the lower vertebrate classes there is much variation in regard to the elements of which each vertebra is composed, and considerable confusion exists concerning the homologies of the various components. In higher forms, such as mammals, the structure is less complex.

A typical mammalian vertebra (Figs. 10.8

and 10.9) is composed of a solid, ventral, cylindrical mass, the *centrum*, or *body*, above which is the *neural arch* surrounding the *neural (vertebral) canal*. The ends of the centrum are smooth and joined to the centra of adjacent vertebrae by means of intervening fibrocartilaginous *intervertebral discs*. The lateral portions of the neural arch consist of two upright *pedicles*, or *neural processes*, fastened to the dorsal part of the centrum on either side. In most forms a flattened plate of bone, the *lamina*, extends medially from the upper part of each pedicle to form the roof of the neural arch. A dorsal median projection, the *neural spine*, or *spinous process*, is formed as the result of fusion of the two laminae. Anterior and posterior articular processes, the *prezygapophyses* and *postzygapophyses*, are projections which serve to join the neural arches of adjacent vertebrae. The prezygapophyses of one vertebra articulate with the postzygapophyses of the vertebra immediately anterior to it. Smooth *articular facets* are present on both prezygapophyses and postzygapophyses; those on the prezygapophyses usually are on the dorsomedial surface, whereas those of the postzygapophyses are generally on the ventrolateral surface. When examining a

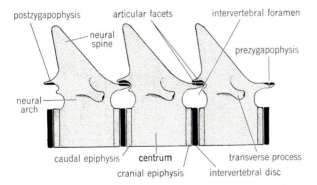

postzygapophysis
articular facets
intervertebral foramen
neural spine
prezygapophysis
neural arch
caudal epiphysis
centrum
cranial epiphysis
transverse process
intervertebral disc

Fig. 10.9 Lateral view of three typical vertebrae from young mammal, showing epiphyses (semidiagrammatic).

single vertebra, it is thus possible to distinguish anterior and posterior ends merely by noting the position of the articular surfaces of the zygapophyses. Various other projections, often referred to as *transverse processes,* extend outward from the lateral sides of the vertebrae. There are several types: (1) *diapophyses,* arising from the base of the neural arch, serve for attachment of the tuberculum, or upper head of a two-headed rib; (2) *parapophyses,* from the centrum, serve for attachment of the lower, or capitular, head of a rib (Figs. 10.61 and 10.63); (3) *basapophyses,* extending ventrolaterally from the centrum, are believed to be remnants of the hemal arch; (4) *pleurapophyses,* projecting laterally, are extensions with which ribs have fused. Another type of projection, the *hypapophysis,* is found in certain forms, extending ventrally from the median part of the centrum. The anterior and posterior ends of the neural arches are notched so as to form anterior and posterior *interver-*

tebral notches. When two vertebrae are in apposition, the anterior notch of one together with the posterior notch of the next form an *intervertebral foramen.* This serves for the passage of a spinal nerve as it emerges from the neural canal.

The vertebrae, as well as the ribs, are derived from the sclerotome regions of the mesoblastic somites during embryonic development. The mesenchymal cells of the segmentally arranged sclerotomes early come to lie in paired masses on either side of the notochord. The metameric arrangement of the sclerotomes is indicated by the presence of small intersegmental arteries which come from the dorsal aorta and course between adjacent somites. Each sclerotome soon differentiates into a compact caudal region and a less dense cranial region. Next, the cranial and caudal portions split apart vertically. The less dense cranial part of one sclerotome unites with the dense caudal portion of the sclerotome in front of it to form a single

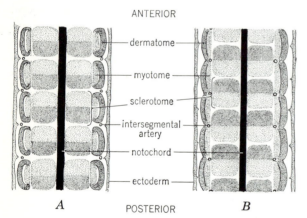

Fig. 10.10 Diagrammatic frontal section through somite region of a typical vertebrate embryo, illustrating the differentiation of sclerotomes and vertebrae: *A,* sclerotomes differentiated into caudal compact and cranial less compact regions; *B,* union of portions of adjacent sclerotomes to form the beginnings of vertebrae. Note the relations of myotome to sclerotome in each case.

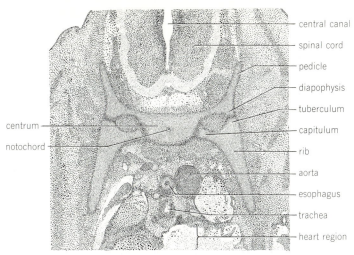

central canal
spinal cord
pedicle
diapophysis
tuberculum
capitulum
centrum
notochord
rib
aorta
esophagus
trachea
heart region

Fig. 10.11 Cross section through the thoracic region of 16-day rat embryo, showing the appearance of a developing vertebra. (*Drawn by R. Speigle.*)

vertebral mass (Figs. 10.10 and 10.13). Thus each vertebra is actually derived from two sclerotomes. The myotome regions of the somites, by reason of the above, come to alternate with the vertebrae, and the intersegmental arteries now pass under the middle of the centrum. The muscle fibers derived from a single myotome thus come to pass from one vertebra to the next. Such an arrangement must exist if movement of the vertebral column is to take place. Furthermore, in lower forms in which locomotion depends for the most part on powerful axial muscles, attachment of these muscles to two adjacent vertebrae and corresponding ribs enables them to exert force and to provide a degree of strength not otherwise obtainable.

Further development consists of a medial extension of mesenchymal cells so that they surround the notochord, forming the primordium of the centrum; some of the cells extend dorsally on either side of the neural tube, which they gradually surround, forming the primordium of the neural arch; still others

migrate in a ventrolateral direction, pushing between the myotomes, to become the *costal processes* concerned with the development of ribs (Fig. 10.11). It is the more dense cranial portion of each vertebra primordium which is primarily concerned with this differentiation.

Mesenchyme between the centra of adjacent vertebrae, and derived from them, differentiates into the intervertebral discs. In the meantime, the notochord gradually disappears, but remnants persist between the vertebrae as the *pulpy nuclei* of the intervertebral discs.

Centers of chondrification soon appear in each developing vertebra, and before long a solid cartilaginous structure is formed. The transverse processes, neural spines, and zygapophyses are outgrowths of the cartilaginous mass. Resorption of tissue occurs at the junction of the costal processes and the remainder of the vertebral mass, so that the rib cartilages become separate from the vertebra, at least in the thoracic region. Later on, several centers of ossification appear, and the cartilage is replaced by bone by the usual method of endochondral ossification. In many vertebrates certain portions may form by direct ossification in the mesenchyme of the sclerotomes. In mammals, cartilage persists at either end of the centrum. Secondary centers of ossification later appear in these cartilaginous areas to form the disclike *epiphyses* of the vertebra. An *epiphysial plate* of cartilage persists between each epiphysis and the bony centrum until the epiphyses finally fuse with the centrum when full growth is attained.

From the foregoing account it would appear that the formation of the vertebral column is relatively simple. However, comparative studies of the spinal column in different vertebrates point to a rather complex evolutionary history. In certain lower forms the vertebrae originate from a number of separate cartilaginous or bony elements which may or may not retain their integrity. The presence of these separate components has been more or less obscured in the development of vertebrae of higher forms in which the only evidence of their existence may be indicated by the appearance of separate centers of chondrification and ossification during embryonic development. Paleontological studies indicate that primitively each sclerotome on both sides, right and left, gave rise to

four separate cartilaginous elements, the *arcualia*, two dorsal and two ventral. The dorsal ones are referred to as *basidorsal* and *interdorsal cartilages*, and the ventral ones as *basiventral* and *interventral cartilages* (Figs. 10.12 and 10.13). The basidorsals and basiventrals are derived originally from the compact caudal regions of the sclerotomes which become the anterior parts of the separate vertebrae. Interdorsals and interventrals, which are the less conspicuous elements, are derived from the less dense cranial regions of the sclerotomes, thus contributing to the posterior parts of the vertebrae. The basidorsals of a given sclerotome fuse dorsally, giving rise to the *neural arches*. The basiventrals, at least in the tail region, similarly give rise to the *hemal arches*, which serve to enclose and protect the caudal artery and vein which lie ventral to the notochord in the tail region. Anterior to the tail region the basiventrals usually contribute only to the formation of the centrum. Likewise the interdorsal and interventral cartilages may contribute to centrum formation, although in some cases the interdorsals may extend dorsally and fuse to form an arch, the *interneural*, or *intercalary, arch,* over the spinal cord (Fig. 10.14). The cartilaginous arcualia may eventually be replaced by bone.

Although cells from the arcualia may contribute to centrum formation, the centrum is

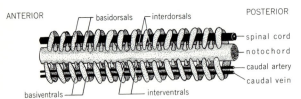

Fig. 10.12 Diagram showing primitive arrangement of cartilaginous arcualia derived from the sclerotomes in relation to spinal cord, notochord, caudal artery, and caudal vein.

Anatomy of the Chordates

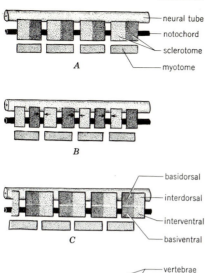

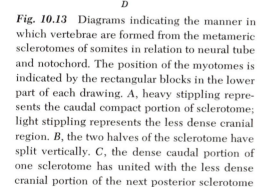

Fig. 10.13 Diagrams indicating the manner in which vertebrae are formed from the metameric sclerotomes of somites in relation to neural tube and notochord. The position of the myotomes is indicated by the rectangular blocks in the lower part of each drawing. *A*, heavy stippling represents the caudal compact portion of sclerotome; light stippling represents the less dense cranial region. *B*, the two halves of the sclerotome have split vertically. *C*, the dense caudal portion of one sclerotome has united with the less dense cranial portion of the next posterior sclerotome to form the primordium of a vertebra. Four centers of ossification represent arcualia as indicated. *D*, fully formed vertebra.

also derived in part from sclerotomal mesenchyme which appears between the arcualia and the notochord which it surrounds. This *perichordal mesenchyme* may undergo direct ossification without passing through an

intervening cartilage stage. In teleosts and amphibians the perichordal component gives rise to the greater part of the centrum and the arcualia contribute but little.

In a few lower fishes a so-called *chordal centrum* is present. This is formed by an invasion of cells from the arcualia into the notochordal sheath, which thickens greatly as the cells increase in number within the sheath. The sheath then undergoes chondrification and becomes a cylindrical structure surrounding the compressed or constricted notochord. A purely chordal centrum occurs but rarely. Usually other components derived from perichordal mesenchyme and arcualia are present in addition.

The vertebrae of numerous fossil adult tetrapods possess two bony elements, the exact homologies of which are somewhat questionable. A pair of dorsolateral *pleurocentra* and a pair of ventral *hypocentra* are present in many forms. These seem to have originated in the perichordal mesenchyme but to have fused with the interdorsal and basiventral arcualia, respectively. Many comparative anatomists are of the opinion that the pleurocentra, by further elaboration, and with a corresponding reduction of the hypocentra, have given rise to the centrum as found in reptiles and mammals. On the other hand, the hypocentra have played the more important role in centrum formation in amphibians. Here the pleurocentra have been reduced or have disappeared (Fig. 10.15). Because of insufficient embryological and paleontological evidence, the exact evolutionary history of these vertebral components continues to be a matter for conjecture.

Cyclostomes. The notochord persists in cyclostomes as the main part of the axial skeleton. It is enclosed by a thick *chordal sheath*. In the Myxinoidea small pieces of cartilage, which appear to be neural-arch

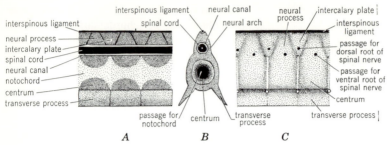

Fig. 10.14 Trunk vertebrae of dogfish, *Squalus acanthias: A,* sagittal section through three vertebrae; *B,* end view of vertebra; *C,* lateral view of three intact vertebrae.

elements, are present in the tail region. In the lampreys, on the other hand, small cartilaginous plates, two pairs per segment, are present throughout the length of trunk and tail. They abut upon the notochordal sheath but do not meet above the spinal cord. These cartilages, which probably represent the basidorsals and interdorsals, form rudimentary neural arches. Since they first appear in the primitive cyclostomes, it is possible that neural arches were the earliest skeletal elements to make their appearance in vertebrates. Similar cartilages in the tail region extend ventrally in lampreys to form hemal arches.

Fishes. Among fishes the spinal column is composed of two types of vertebrae. These consist of *trunk* vertebrae, all of which are practically alike, and *caudal* vertebrae, bearing hemal arches and confined to the tail. The position of the anus marks the point of transition between the two regions. The form of the caudal part of the vertebral column varies in the heterocercal, diphycercal, and homocercal types of tail (Fig. 10.67).

Considerable variation is to be found in the vertebral columns of fishes. In some forms, which are primitive in this respect,

the condition is not much in advance of the lamprey. In the sturgeon, for example, the notochord persists unchanged and centra are lacking. Pieces of cartilage are present dorsal and ventral to the notochord. The basidorsal cartilages, forming the neural arch, fuse together above the spinal cord to form a neural spine; the interdorsal cartilages lie between the basidorsals. The basiventral cartilages form an incomplete hemal arch in the tail region, the interventral cartilages lying between them. In the Holocephali and Dipnoi a similar condition exists, but cartilage has begun to invade the notochordal sheath in which ringlike calcifications may appear, four or five to each segment. Perichordal cartilage covers the notochordal sheath to a minor degree. Interdorsals and interventrals may be lacking. In *Latimeria,* protecting the spinal cord, which lies dorsal to the notochord, is a series of neural arches, each consisting of a thin layer of bone with a core of cartilage and topped by a neural spine. Since cartilage is not preserved in fossil remains, the neural spines originally appeared to be hollow, hence the name *Coelacanthus* (hollow spine).

The cartilaginous spinal column is much better developed in elasmobranchs than in

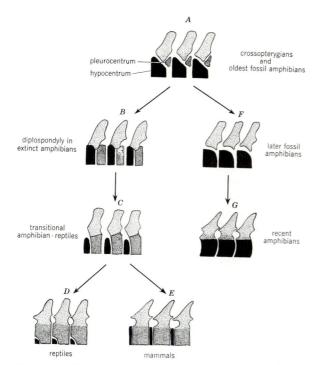

Fig. 10.15 Diagrammatic hypothetical scheme indicating two possible lines of evolution of vertebrae from ancestral crossopterygian fishes and the oldest known fossil amphibians; that on the left leading to the condition found in reptiles and mammals; that on the right leading to the condition found in recent amphibians. Black areas represent the hypocentrum; heavily stippled areas represent the pleurocentrum.

those fishes mentioned above. Each vertebra bears a biconcave, or *amphicoelous, centrum* (Fig. 10.14), through the center of which the constricted notochord runs. Between the centra, however, the notochord is not constricted. If it alone were removed, it would resemble a string of beads because of the alternating arrangement of constricted and unconstricted portions. The centrum is formed by the invasion of the notochordal sheath with cartilage cells derived from the primitive cartilaginous plates as well as from mesenchyme surrounding the noto-

chord. Neural arches are present, formed by the basidorsal cartilages, or *neural plates,* between which are interposed *intercalary plates,* representing the interdorsals. The neural plate is perforated by a foramen for the passage of the ventral root of a spinal nerve; the intercalary plate bears a foramen for the passage of the dorsal root. In elasmobranchs dorsal and ventral roots of spinal nerves unite *outside* the vertebral column. In some species the intercalary plates and even the neural plates are further subdivided. In the tail region a

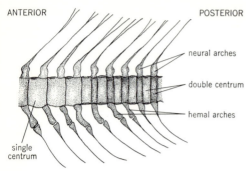

ANTERIOR — POSTERIOR

neural arches

double centrum

hemal arches

single centrum

Fig. 10.16 Diplospondyly in *Amia.* Lateral view of portion of vertebral column, showing three full body vertebrae with single centra to the left and the first five caudal vertebrae with double centra to the right. Only the posterior part of a double centrum bears neural and hemal arches.

complete hemal arch is formed and a *hemal spine* is present, formed by the fusion of the two hemal processes. The hemal arch, of course, is not present in the trunk region. Here a pair of ventrolateral projections, the transverse processes (basapophyses), representing the remnants of the hemal arch, extends toward the lateral skeletogenous septum. Small cartilaginous ribs in the skeletogenous septum extend from the basapophyses.

In the caudal vertebrae of some selachians

and certain other fishes, such as *Amia,* there are two centra per vertebra. This condition is referred to as *diplospondyly.* In *Amia* only the posterior centrum bears neural arches and hemal arches (Fig. 10.16). Paleontological evidence is incomplete in regard to the specific elements represented in the diplospondylous condition, and there is little agreement among comparative anatomists as to their exact homologies. Diplospondyly undoubtedly permits greater flexibility in movement of the tail during locomotion.

In the garpike, *Lepisosteus,* the centra of adjacent vertebrae form well-defined articulations of a type not found in other fishes. The posterior end of each centrum is concave, and the anterior end convex. Thus the convex portion of one centrum fits into the concavity of the vertebra anterior to it in the manner of a ball-and-socket joint. Vertebrae of this type are said to be *opisthocoelous* (Fig. 10.17D). No remnants of the notochord are to be found in the adult.

Teleost fishes, for the most part, have vertebrae with biconcave, or *amphicoelous,* centra (Figs. 10.17C and 10.18). A small canal in the center represents the location of the remnant of the notochord which is very much reduced and compressed. The concavities are filled with a pulpy material

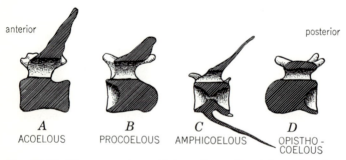

anterior   posterior

A	*B*	*C*	*D*
ACOELOUS	PROCOELOUS	AMPHICOELOUS	OPISTHO-COELOUS

Fig. 10.17 Diagrammatic sagittal sections of vertebrae showing four types of centra.

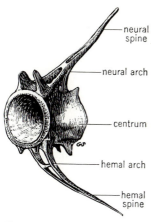

Fig. 10.18 Caudal vertebra of teleost fish showing biconcave, amphicoelous centrum.

probably derived from the degenerate notochord. In the eel group the centra are more or less flattened or even slightly convex at the anterior end. Rather poorly defined prezygapophyses and postzygapophyses, arising from the neural arch, and otherwise appearing only in amphibians and amniotes, first make their appearance in certain bony fishes. Similar processes, but less well developed, may be associated with the hemal arches. In the trunk vertebrae of teleosts a pair of ventrolateral projections (basapophyses) represents the basal remnants of the hemal arch. Slender ventral ribs are attached to these basapophyses. The transition between ventral ribs and the hemal arches is very clearly indicated in many teleosts although it is not so apparent in others.

In some fishes, as in skates and chimaeras, there is a well-defined joint between the skull and the first vertebra. In dogfishes, however, there is usually an actual fusion.

Amphibians. It has already been mentioned that it is generally believed that from a primitive ancestral condition in which two bony elements on each side, pleurocentrum and hypocentrum, made up the centrum of a vertebra, the evolutionary trend has taken place in two directions. The posterior pleurocentrum has disappeared in amphibians, and the hypocentrum has become the true centrum. Complete paleontological evidence to support this theory is lacking.

With the appearance of limbs in the evolutionary scale, a further differentiation of the spinal column into regions has occurred. This is clearly shown in amphibians with the exception of the caecilians and certain extinct forms. A single *cervical* vertebra is present which articulates with the skull at its anterior face. This is followed by a series of *trunk* vertebrae. A single *sacral* vertebra, which is connected to the pelvic girdle by means of its transverse processes (ribs), lies between the trunk vertebrae and the caudal vertebrae, which are, of course, found only in tailed forms. All the caudal vertebrae, except the first, bear hemal arches with hemal spines (Fig. 10.19).

Amphibian vertebrae bear transverse processes and well-developed zygapophyses. There is some variation within the class in regard to the method of articulation of the centra of adjacent vertebrae. The vertebrae

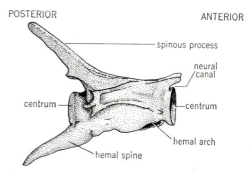

Fig. 10.19 Lateral view of caudal vertebra of *Necturus.*

of caecilians are of the amphicoelous type. The same is true of several urodeles, although most have opisthocoelous vertebrae. Anurans, for the most part, have *procoelous vertebrae* (Fig. 10.17B). In this type the anterior face of the centrum is concave and the posterior face convex. Thus, adjacent vertebrae articulate in a manner just opposite to that of the opisthocoelous type.

The most striking change to be noted within the amphibian class is to be observed in anurans, since they lack a tail in the adult stage. The number of vertebrae, therefore, is less than in urodeles. In most higher modern anurans there are 10 bones in the vertebral column. The number of vertebrae may, however, be reduced by fusion. Some primitive species have an additional presacral vertebra. In such common frogs as *Rana pipiens* and *Rana catesbeiana,* with 10 elements in the vertebral column, a single cervical vertebra, the *atlas,* articulates with the skull. It has no transverse processes or prezygapophyses. Instead it bears on its anterior face two concave depressions which serve for the reception of the occipital condyles of the skull. The next six vertebrae are procoelous trunk vertebrae with no unusual features. The eighth vertebra differs from the other trunk vertebrae in being amphicoelous. The ninth is the sacral vertebra, in which the cranial end is convex and fits into the concavity of the caudal end of the eighth vertebra. The caudal end of the sacral vertebra bears two convex projections which fit into corresponding concavities of the *urostyle,* or tenth vertebra. The transverse processes (ribs) of the sacral vertebra are large and strong and slope posteriorly in an oblique direction (Fig. 10.20A). The pelvic girdle articulates with the ends of these transverse processes. The urostyle is very much modified. It is a long bone which extends from the sacral vertebra

to the posterior end of the pelvic girdle, where it rests in a depression formed by the union of the two halves of the girdle. The urostyle was formerly believed to be formed by a fusion of several caudal vertebrae and to be equivalent to the tenth, eleventh, and twelfth. There is little or no embryological evidence for this, since the urostyle actually forms by ossification of a cartilaginous rod which, for a time, lies beneath the notochord. A pair of small openings near the anterior end is sometimes considered to be the equivalent of intervertebral foramina. These connect with the neural canal. In a few forms, small, rudimentary, transverse processes are present at the anterior end of the urostyle.

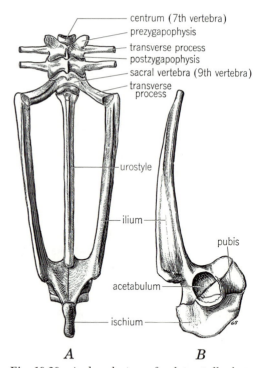

Fig. 10.20 A, dorsal view of pelvic girdle, last three vertebrae, and urostyle of bullfrog, *Rana catesbeiana;* B, lateral view of the right innominate bone.

Anatomy of the Chordates

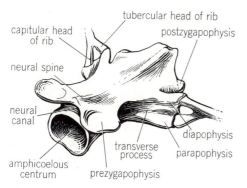

capitular head of rib
tubercular head of rib
postzygapophysis
neural spine
neural canal
amphicoelous centrum
transverse process
prezygapophysis
diapophysis
parapophysis

Fig. 10.21 Trunk vertebra of *Necturus*.

The transverse processes of the trunk vertebrae are tipped with small, cartilaginous projections, the *ribs* (Fig. 10.21). In some species the ribs attain considerable size, but in most cases they are very small.

Reptiles. In reptiles, contrary to amphibians, the pleurocentrum is the primitive element which has given origin to the true centrum. The hypocentrum has been reduced or has disappeared in most forms, but in *Sphenodon* it persists as a small ventral remnant.

Because of the diversity in form and habit of various reptiles, differences of considerable magnitude are to be observed in their vertebral columns. The vertebrae are ossified, but remains of the notochord may occasionally be found in the centra.

In turtles the number of vertebrae is less than in other groups. Snakes have the greatest number, over 400 being present in certain large species.

Perhaps the most noteworthy advance over fishes and amphibians to be observed in the vertebral column of reptiles is its division in most lizards and in crocodilians into different regions: *cervical, thoracic, lumbar, sacral,* and *caudal.* Cervical, or neck, vertebrae are those lying between the skull and the thoracic region. They are usually more freely movable than those in the remainder of the spinal column. Small ribs are frequently attached to the cervical vertebrae, but they do not extend ventrally to meet the sternum. Thoracic vertebrae bear ribs which articulate ventrally with the sternum. These are succeeded by the larger and more freely movable lumbar vertebrae which do not bear ribs. Next come the sacral vertebrae, to the transverse processes and ribs of which the pelvic girdle is attached. Strength and rigidity are attained in the sacral region by a fusion of *two* separate vertebrae to form a *sacrum.* Caudal, or tail, vertebrae follow the sacrum. Small V-shaped *chevron bones,* on the ventral side of the caudal vertebrae of reptiles and other amniotes, probably represent the remnants of the more primitive hemal arches. They are of hypocentral origin.

Turtles, snakes, and limbless lizards do not exhibit the five divisions of the vertebral column seen in crocodilians and lizards with legs. Turtles, for example, lack a true lumbar region. Snakes and limbless lizards have only *precaudal* and *caudal* regions, and no cervical vertebrae or sacrum are present. The separate vertebrae show great similarity in structure.

All types of centra are encountered among reptiles. Amphicoelous vertebrae are present in some primitive forms such as turtles and *Sphenodon.* Opisthocoelous vertebrae occur in certain regions in a few reptiles. Most, however, have procoelous vertebrae. In some extinct forms the centra of certain vertebrae had flat ends, such vertebrae being referred to as *acoelous* or *amphiplatyan* (Fig. 10.17*A*). The sacral vertebrae of living crocodilians are of this type. Two or more types of vertebrae may thus be found in the same individual.

Among reptiles is found for the first time a modification of the first two cervical vertebrae, a condition which is encountered in all birds and mammals. The second vertebra, the *axis,* or *epistropheus,* in crocodilians, some snakes, lizards, and chelonians, bears a projection, the *odontoid process,* at the anterior end of the centrum. This is actually the centrum (pleurocentrum) of the *atlas,* or first vertebra, which has become united secondarily to the centrum of the axis. It serves as a pivot which permits considerable freedom of movement of the head. The atlas bears an anterior concavity for the reception of the single occipital condyle of the reptilian skull. The ventral part of the atlas represents the hypocentrum. A separate bone, the *proatlas,* in the form of a neural arch, is found in *Sphenodon,* alligators, and a few other reptiles. It is located between the atlas and the occipital bone of the skull.

Although two sacral vertebrae are present in living crocodilians and lizards, in many prehistoric forms as many as five or six were present. These were firmly ankylosed.

In snakes, certain lizards, and *Sphenodon,* in addition to pre- and postzygapophyses, other articular facets are present in the neural arches. These are called *zygosphenes* and *zygantra.* Zygosphenes are located at the anterior ends of the neural arches; zygantra are at the posterior ends. The zygosphenes of one vertebra articulate with the zygantra of the next anterior vertebra.

In *Sphenodon* and many lizards the centrum of each caudal vertebra is cut across by an unossified partition. The vertebra is rather easily broken along this septum when the tail is cast off by an animal when attempting to escape from an enemy. The condition is similar in a way to diplospondyly observed in the tails of some fishes, but probably represents a specialized rather than a primitive condition.

The 10 thoracic vertebrae of turtles differ from others in that they lack transverse processes. Their neural arches are rigidly united with the carapace. The two sacral and first caudal vertebrae are also fused to the carapace (Fig. 10.62).

Some of the extinct reptiles possessed unusual variations of the vertebral column. In *Dimetrodon,* a primitive pelycosaur in the line of evolution toward mammals, the spinous processes of the thoracic vertebrae were 2 ft long, although the centra of these vertebrae are no more than an inch in diameter.

Birds. Developmental stages of bird vertebrae indicate that they are formed in the typical manner from basidorsal, basiventral, interdorsal, and interventral cartilages. The most characteristic feature of the spinal column of birds is its rigidity. The cervical region, however, is extremely mobile, but the remainder is, in most cases, capable of little, if any, movement. Thoracic, lumbar, and sacral regions are firmly united (Fig. 10.22), to such an extent that it is difficult to delimit one region from another. Rigidity is of advantage in flight.

The cervical vertebrae are variable in number, there being anywhere from 8 or 9 to 25 separate vertebrae in the neck. Swans have the largest number of cervical vertebrae. The atlas differs from the others in its small size and its narrow, ringlike shape. Its centrum has become secondarily fused to the axis, forming the odontoid process of the latter. The anterior face of the atlas bears a deep depression for articulation with the single occipital condyle of the skull. The transverse processes of the cervical vertebrae are pierced by *foramina transversaria,* the presence of which is characteristic of neck vertebrae. They serve for the passage of the vertebral arteries and veins. The centra of birds' vertebrae do not usually bear epiphyses at their articular ends. In parrots,

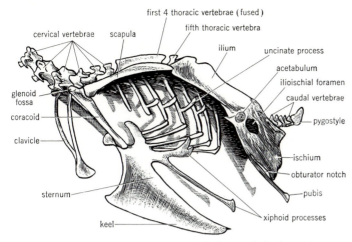

Fig. 10.22 Lateral aspect of part of skeleton of chicken, showing union of vertebrae in thoracic, lumbar, sacral, and caudal regions. The wing and leg bones are not shown. (*Drawn by G. Schwenk.*)

however, epiphyses are present, a condition which is otherwise confined to mammals and a few reptiles. *Archaeopteryx* had amphicoelous vertebrae, a characteristic which is not encountered in modern forms except in the caudal region. The ends of the centra in most modern birds are saddle-shaped, or *heterocoelous* (Fig. 10.23). In such vertebrae the anterior face is convex in a dorsoventral direction but concave from side to side. The posterior face, of course, has its curves in just the opposite direction, so that proper articulation is provided. Penguins, parrots, and a few others possess opisthocoelous vertebrae. Cervical ribs are frequently rather well developed in birds, and hypapophyses are of common occurrence.

Thoracic vertebrae bear ribs which unite ventrally with the sternum. The rather prominent transverse processes of the thoracic vertebrae have articular surfaces at their tips. These, together with small projections on the sides of the centra, serve for the attachment of ribs. The vertebral end of the rib, there-

fore, is attached at two places. Not all the thoracic vertebrae are always fused together. In the fowl, for example, there are seven thoracic vertebrae. The first four are fused, but the fifth is free. The sixth and seventh are ankylosed to the first lumbar vertebra. Posterior thoracic, lumbar, sacral, and the first few caudal vertebrae are fused into a single bony mass, the *synsacrum*. The num-

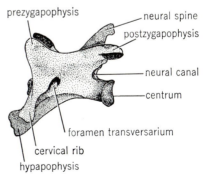

Fig. 10.23 Heterocoelous cervical vertebra of bird.

ber of vertebrae making up the synsacrum varies in different birds. Twenty or more elements may be included in it. Examination of the ventral side of the synsacrum will reveal the number of separate vertebrae of which it is composed, since the transverse processes of these vertebrae retain their integrity to some degree despite the fusion. It has been determined from developmental studies of bird embryos that only two of the vertebrae are originally sacral, the ostrich being an exception since it has three.

Only a few free caudal vertebrae are present in birds, although *Archaeopteryx* had 20 separate vertebrae in its tail (Fig. 2.31). In most birds the posterior caudal vertebrae are fused together to form the "plowshare" bone, or *pygostyle,* which supports the large tail feathers. The pygostyle may be composed of from 6 to 10 fused vertebrae. It is poorly developed or absent in the Paleognathae. The free caudal vertebrae are usually amphicoelous.

Mammals. Fully developed vertebrae of mammals show little evidence of their ancestral origin from arcualia and such perichordal components as pleurocentra and hypocentra. During development in mammals, many ancestral stages are glossed over, and this seems to be true of vertebra development. The various separate centers of chondrification in the mesenchyme in which a vertebra is first blocked out are believed to represent the arcualia of lower forms with which they have been homologized. It will be recalled that the mammalian centrum, like that of reptiles, has probably been derived from the pleurocentrum of ancestral forms and that the hypocentrum has been lost or reduced.

Before adult stature is attained, the centra of mammalian vertebrae, with the exception of monotremes and sirenians, bear an epiphysis at either end. In most mammals these finally become fused to the centrum. Intervertebral discs of fibrous cartilage, probably representing remains of hypocentra, are also typically present in mammals.

The mammalian vertebral column, like that of many reptiles and of birds, is divided into cervical, thoracic, lumbar, sacral, and caudal regions.

Perhaps the outstanding feature of the cervical vertebrae in mammals is their constancy in number, 7 vertebrae being present with few exceptions. Length of neck, then, is determined by the length of individual vertebrae rather than by an increase in number. Three of the four mammals which are exceptions to this rule belong to the order Edentata. They include the two-toed sloth, *Choloepus,* with 6 cervical vertebrae; the three-toed sloth, *Bradypus,* with 9; and *Tamandua,* the ant bear, with 8. The other exception is the manatee, *Trichechus manatus,* which is provided with 6 cervical vertebrae. It belongs to the order Sirenia. Fusion of some or all of

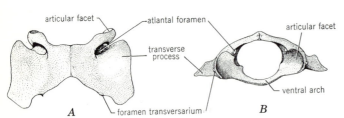

Fig. 10.24 Atlas vertebra of cat: *A,* dorsal view; *B,* anterior view.

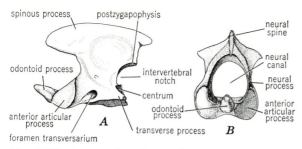

Fig. 10.25 Axis vertebra of cat: *A*, lateral view; *B*, anterior view.

the cervical vertebrae occurs in some mammals. This is true of many of the whales, armadillos, manatees, jerboas, and the marsupial mole, *Notoryctes*. The atlas is provided with a pair of concavities for articulation with the two occipital condyles of the skull (Fig. 10.24). The axis bears an odontoid process (Fig. 10.25) which is peg-shaped except in some of the ungulates, in which it assumes the form of a trough. The cervical vertebrae of mammals (Fig. 10.26) may bear small ribs, but these never reach to the sternum. The transverse processes characteristically arise from two roots, one from the base of the neural arch and the other from the centrum. These transverse processes (pleurapophyses) consist for the most part of ribs which are fused to the vertebra. The foramen transversarium (vertebrarterial canal)

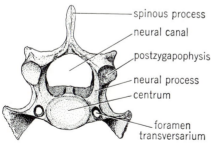

Fig. 10.26 Typical cervical vertebra of cat, posterior view.

lies between the two points of attachment of the rib. The foramen serves for the passage of the vertebral artery and vein and a sympathetic nerve plexus. The presence of foramina transversaria is diagnostic of cervical vertebrae. The seventh vertebra, however, and sometimes the sixth, may lack these foramina. In most mammals the centra of the cervical vertebrae are acoelous (amphiplatyan), but in the Perissodactyla they are distinctly opisthocoelous. This is also true of elephants but to a lesser degree. Considerable variation is to be observed in different species in regard to length of the separate vertebrae, height of neural spines, and other similar features. The seven cervical vertebrae of the giraffe are notable for their extreme length.

Thoracic vertebrae (Fig. 10.27) articulate with ventrally projecting ribs which connect directly or indirectly with the sternum. Some of the posterior ribs, however, may not reach the sternum, in which case they are known as *floating ribs*. Special articular facets are present on thoracic vertebrae for articulation with the ribs. These consist of (1) *tubercular facets* on the ventral side of the free ends of the transverse processes (diapophyses), for articulation with the *tubercle* of the rib, and (2) costal demifacets (parapophyses) at the dorsolateral angles of the centrum, for reception of the *head*, or *capitulum*, of the rib. The

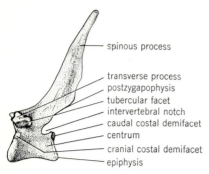

spinous process
transverse process
postzygapophysis
tubercular facet
intervertebral notch
caudal costal demifacet
centrum
cranial costal demifacet
epiphysis

Fig. 10.27 Thoracic vertebra of cat, lateral view.

caudal costal demifacet of one vertebra together with the cranial costal demifacet of the next may form a depression which serves as the point of articulation. Thus the capitulum may actually articulate with the centra of two thoracic vertebrae, although this is not always true. In the cat, for example, each of the last three of the 13 thoracic vertebrae bears an entire facet on each side of the centrum. Tubercular facets are also lacking in these three vertebrae. The number of thoracic vertebrae varies considerably in different mammalian species. The smallest number is 9, as found in one of the whales, *Hyperoödon;* the greatest number is 25, as in some sloths. Fusion of thoracic vertebrae is not known to occur in mammals except in the extinct glyptodons. Slightly opisthocoelous vertebrae are found in the thoracic region of some of the ungulates, but in most mammals they are amphiplatyan.

Lumbar vertebrae also show great variation in number among various mammals, 4 to 7 being the usual number encountered. The extremes in number are to be found among the whales, in which one species, *Neobalaena,* has but 2, and another, the dolphin, from 21 to 24. Usually the number of lumbar vertebrae varies inversely with the number in the thoracic region. Lumbar vertebrae are

usually large and strong. They tend to be longer than other vertebrae. Their transverse processes, or pleurapophyses, are prominent and directed forward. The anterior lumbar vertebrae may bear *mammillary processes* (*metapophyses*) dorsolateral to the prezygapophyses. The mammillary processes of the armadillo are particularly large and blunt, aiding in support of the carapace. Accessory processes may also be present, situated between the postzygapophyses and transverse processes (Fig. 10.28).

The sacrum consists of several vertebrae firmly ankylosed together and serving for articulation with the pelvic girdle. Again, the number of separate elements which fuse to form the sacrum is variable. There are 5 in the horse, 4 in the pig, 3 in the dog and cat (Fig. 10.29), and 5 in man. Whales lack a sacrum. Anomalous conditions are occasionally found in which more or less than the usual number form the sacrum. Extra elements are taken over by the sacrum from either the lumbar or caudal regions. The presence of *sacral foramina* on both dorsal and ventral sides provides for the passage of dorsal and ventral rami of sacral spinal nerves. These foramina communicate with the neural canal within the sacrum but

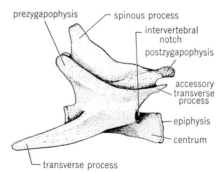

prezygapophysis
spinous process
intervertebral notch
postzygapophysis
accessory process
transverse process
epiphysis
centrum
transverse process

Fig. 10.28 Lumbar vertebra of cat, lateral view.

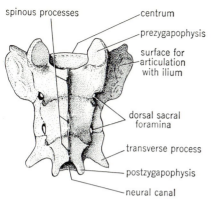

spinous processes
centrum
prezygapophysis
surface for articulation with ilium
dorsal sacral foramina
transverse process
postzygapophysis
neural canal

Fig. 10.29 Sacrum of cat, dorsal view.

do not correspond to the usual intervertebral foramina. Their location indicates the boundaries of the separate sacral vertebrae. The first sacral vertebra is most important in supporting the pelvic girdle and hind limbs. Large *lateral masses* bear cartilage-covered *auricular surfaces* for firm articulation with the ilium. They are composed of transverse processes and sacral ribs which have become indistinguishably fused together. In man, with his upright posture, a loosening or slipping of the sacroiliac articulation sometimes occurs, accompanied by considerable pain.

Caudal vertebrae in mammals may vary from 3 or 4, as in man, to around 50 in the scaly anteater, *Manis,* and the insectivore, *Microgale.* The anterior caudal vertebrae possess neural arches, spines, zygapophyses, etc. These gradually diminish in size toward the end of the tail, so that the terminal caudal vertebrae consist of little more than centra. Chevron bones, homologous with those of reptiles (page 425), are frequently present in the form of small V-shaped elements articulating with the ventral side of the centrum. They are particularly prominent in whales and edentates. The three or four caudal vertebrae in man usually unite to form a single *coccyx.* Sometimes, how-

ever, the first one or two of the coccygeal vertebrae remain free.

The number of vertebrae in any one species of mammal is quite constant except in regard to the caudal region. For convenience it is customary to use a *vertebral formula* to express succinctly the number of vertebrae found in each region. The kind of vertebrae is represented by the letters C, T, L, S, Cy, indicating cervical, thoracic, lumbar, sacral, and caudal (or coccygeal), respectively. A few vertebral formulas are given below to illustrate the kinds of variations that are met with:

Horse	$C_7T_{18}L_6S_5Cy_{15-21}$
Cow	$C_7T_{13}L_6S_5Cy_{18-20}$
Sheep	$C_7T_{13}L_{6-7}S_4Cy_{16-18}$
Pig	$C_7T_{14-15}L_{6-7}S_4Cy_{20-23}$
Dog	$C_7T_{13}L_7S_3Cy_{20-23}$
Man	$C_7T_{12}L_5S_5Cy_{3-4}$

Skull

The skeletal framework of the vertebrate head is referred to as the skull. The comparative anatomy of the skull is extremely complex, and homologies are often ascertained with difficulty. Confusion arises from two facts: (1) that such large numbers of separate elements enter into the formation of the skulls of various vertebrates and (2) that they are derived from such different sources. Furthermore, certain gaps in the paleontological record, together with the fact that groups living today usually represent specialized or degenerate conditions, make it difficult to construct the true phylogenetic history of skull development.

The part of the skull which surrounds and protects the brain is called the *cranium (neurocranium).* Sense capsules associated in development with the olfactory and auditory sense organs become attached to the cranium, as do numerous components of the visceral skeleton which in fishes support the gills and jaws. All these structures in their aggregate make up the skull.

The skull of most vertebrates is derived essentially from three different embryonic components: (1) the *chondrocranium*, composed of cartilage, which contributes chiefly to the base of the skull and which in most vertebrates is replaced by bone; (2) the *dermatocranium*, made up of membrane bones which roof over the chondrocranium,* or are closely applied to its lower surface; and (3) the *splanchnocranium*, derived from the visceral, or branchial, skeleton and which is primitively cartilaginous although some of it becomes invested with or replaced by membrane bones.

Early Development. CHONDROCRANIUM. Soon after the appearance of the central nervous system during embryonic development, the mesenchymal cells which surround it begin to differentiate, forming a membranous investing layer. In the region of the brain this is referred to as the *membranous cranium*. The notochord extends forward beneath the brain, terminating near the infundibulum. The membranous layer also surrounds the anterior portion of the notochord. It furnishes the material in which the cartilaginous chondrocranium develops by histological differentiation of cartilage within the *ventral* region of its substance. The chondrocranium is made up of several components. First there appears a pair of flat, curved cartilages, the *parachordal*

---

* The use of these terms by different authors varies somewhat. Some use "skull" and "cranium" interchangeably. They include all parts of the skull derived from cartilage under the heading "chondrocranium" and all membrane bones under the category of "dermatocranium." The author is of the opinion that the terminology which he has employed is less confusing to the student who is attempting to understand a rather complex subject. The skeletal elements of the lower jaw are here considered to be parts of the skull.

plates, which flank the notochord on either side (Fig. 10.30*A*). They extend laterally as far as the otic capsules (page 699) and posteriorly to the point where the tenth cranial nerve emerges. In back of this point are two to four *occipital vertebrae* with which the parachordal plates later fuse and which are incorporated in the skull. The parachordal cartilages grow larger, their posterior portions fusing together in the midline to form the *basilar plate*, which encloses the tip of the notochord and becomes the floor beneath the midbrain and hindbrain. The anterior ends form a separate connection, being united by a transverse *acrochordal bar*. A small opening, the *basicranial fenestra*, often remains in the center (Fig. 10.30*B*).

Next there appears in front of the parachordals another pair of cartilages, the *prechordal cartilages*, or *trabeculae cranii*. The anterior ends of these fuse with each other, forming a transverse bar, the *ethmoid plate*. This grows anteriorly to become the *rostrum*, which later contributes to the formation of the *internasal septum* between the nasal capsules. Posteriorly the prechordal cartilages unite with the parachordals in such a manner as to leave a prominent opening in the center, the *hypophysial fenestra*, or *pituitary space*, in which the pituitary gland comes to lie and through which the internal carotid arteries pass.

At the same time as the parachordal and prechordal cartilages are developing, three pairs of capsules make their appearance about the developing olfactory, auditory, and optic sense organs. They are known, respectively, as the *nasal, otic,* and *optic capsules*. With the growth of the parachordal and prechordal cartilages, a union is effected with the cartilaginous nasal and otic capsules, the entire mass forming the chondrocranium. The optic capsules remain as independent elements, thus permitting unre-

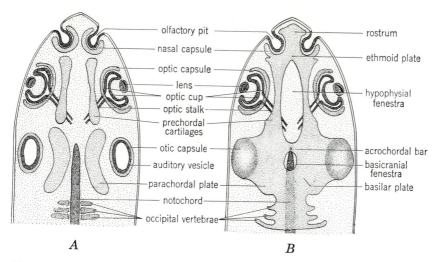

Left diagram labels (A):
olfactory pit
nasal capsule
optic capsule
lens
optic cup
optic stalk
prechordal cartilages
otic capsule
auditory vesicle
parachordal plate
notochord
occipital vertebrae

Right diagram labels (B):
rostrum
ethmoid plate
hypophysial fenestra
acrochordal bar
basicranial fenestra
basilar plate

A   *B*

Fig. 10.30   Diagrams illustrating development of the chondrocranium: *A*, separate
chondrocranial cartilages surrounding the special sense organs and flanking the
notochord and brain; *B*, all the cartilages except the optic capsules have fused to
form the early chondrocranium.

stricted movement of the eyes. The optic capsules are seldom cartilaginous. In mammals they are of a fibrous nature.

The basal cartilaginous cranium, composed of the parts just mentioned, does not exist unchanged for very long. The side portions soon begin to grow dorsally for a short distance on either side of the brain. In such vertebrates as elasmobranchs these dorsally extending parts tend to meet above the brain and fuse, thus enclosing that structure rather completely with a true chondrocranium. In most vertebrates, however, only the *otic capsules* and *occipital region* become roofed over as described. The floor and sides of the more anterior region are cartilaginous, but the roof is composed of membrane. This portion of the cranium will later be covered, or roofed over, by membrane bones of the dermatocranium. The large opening at the posterior end of the chondrocranium is the *foramen magnum.* It is

through the foramen magnum that the spinal cord emerges from the skull.

Openings for the passage of the various cranial nerves and for certain important blood vessels represent spaces which persist following fusion of the major cartilaginous components. Cranial nerves V and VII emerge in front of the otic capsule; nerves IX and X have their exits posterior to the otic capsule, between it and the occipital complex.

DERMATOCRANIUM.   In addition to the chondrocranium, which appears uniformly throughout the vertebrate series, several dermal plates or membrane bones contribute to the formation of the skull. These appear first in the head region of bony fishes in the form of large scales. They are similar to bony scales on other parts of the body. The dermal scales gradually sink down into the head, where they roof over the more anterior region of the chondrocranium with which they fuse, thus completing the protective envelope surrounding the brain. In the skull of a young individual, before the growth of the dermal parietal bones is complete, there are gaps between the incompletely developed angles of these and neighboring bones. Such temporary gaps are referred to as *fontanelles* (Figs. 10.31 and 10.32). In man the fontanelles usually close during the first half of the second year of life. They are situated at the four angles of the parietal bones, so that two are median and two are lateral. The median ones are by far the most prominent. The frontal fontanelle is the last to close during the second half of the second year. This is an important landmark in the practice of midwifery and obstetrics in determining the position of the head of the fetus just prior to delivery. The bone which fills the anterior lateral fontanelle occasionally persists as a separate *epipteric* bone. It usually, however, unites with the parietal, although it may join the sphenoid, temporal, or frontal bones, all of which are contiguous. Numerous dermal bones, other than those mentioned, become closely applied to the ventral portion of the chondrocranium; others are associated with portions of the visceral skeleton. These are discussed more fully below.

SPLANCHNOCRANIUM. In such primitive aquatic forms as fishes, a series of cartilag-

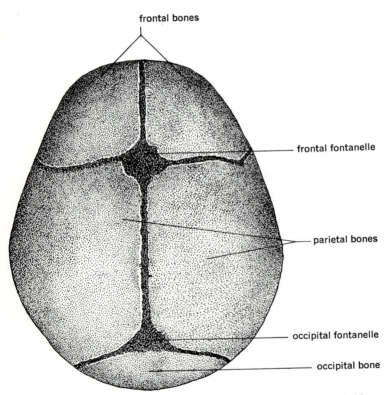

frontal bones

frontal fontanelle

parietal bones

occipital fontanelle

occipital bone

Fig. 10.31 Dorsal view of developing cranial bones of human child at birth, showing fontanelles. (*After Piersol.*)

Anatomy of the Chordates

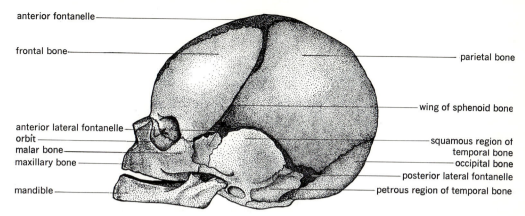

Fig. 10.32 Lateral view of developing cranial bones of human child at birth, showing fontanelles and other essential features. (*After Piersol.*)

inous or bony visceral, or pharyngeal, arches encircles the pharyngeal portion of the digestive tract (Fig. 10.33). They serve primarily to support the gills and are arranged between the gill slits, one behind the other. They arise from mesenchyme surrounding the gut. It appears that neural crests, at least in amphibians and birds, contribute to the mesenchyme from which the visceral arches are derived, which is sometimes referred to as mesectoderm. The first arch is the *mandibular arch.* It becomes divided into dorsal and ventral portions, the

palatopterygoquadrate bar and *Meckel's cartilage,* which contribute to the upper and lower jaws, respectively. The second is the *hyoid arch.* It usually furnishes support to the anterior hemibranch, if present, but in many fishes its upper division, the *hyomandibular cartilage,* by means of a ligamentous connection, also serves to attach the jaws to the cranium in the otic region. This is referred to as the *hyostylic* method of jaw attachment (Fig. 10.35*B* and *F*). The remaining visceral arches vary in number in different species, furnishing support to the

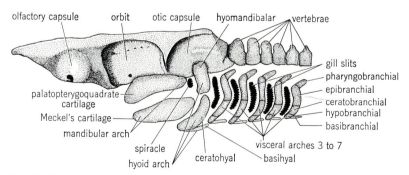

Fig. 10.33 Diagram showing relation of visceral arches of an elasmobranch to the chondrocranium, vertebral column, and gill slits.

remaining gills. The last visceral arch, however, usually does not bear gills. The typical number of visceral arches in fishes is seven. Each arch is usually divided into several separate cartilages.

In higher forms there has been a reduction in number of visceral arches and the entire visceral skeleton has become greatly modified. The parts of which it is composed are primitively made up of cartilage, but the cartilage, for the most part, tends to disappear, being replaced by cartilage bone or surrounded by membrane bones.

Further development. In cyclostomes, elasmobranchs, and a few of the higher fishes, no dermal bones (except those associated with scales) are present and the entire skull remains cartilaginous. Even the cranium proper may be partially roofed over with cartilage in these forms. In most vertebrates, however, the primitive chondrocranium is replaced, at least to some extent, by cartilage bones. So many variations are met with that it is difficult to generalize in discussing the origin of the various bones of the skull. Nevertheless, a more or less common pattern of development is found throughout the vertebrate series.

Separate centers of ossification appear in the basal region of the chondrocranium in what are called, beginning from behind, the *basioccipital, basisphenoid (otic), presphenoid (orbital),* and *ethmoid* regions (Fig. 10.34). The ethmoid region lies anterior to the cranial cavity. Other centers appear in the cartilage which extends a short distance dorsally on either side of the brain. Each center of ossification typically gives rise to a separate bony element which may or may not fuse with others, as the case may be.

Three bony segments, *occipital, parietal,* and *frontal,* form below and along the sides of the brain; a fourth, or *nasal, segment* develops anterior to the brain. The most posterior, or the *occipital segment,* is made up of four cartilage bones: a *basioccipital*

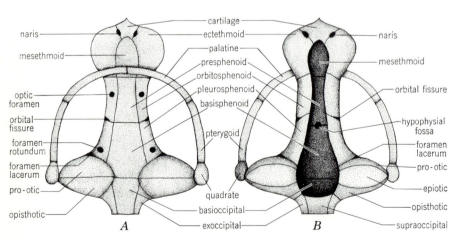

Fig. 10.34 Diagrams of *A,* ventral, and *B,* dorsal, views of the typical tetrapod chondrocranium, showing the main regions which will later ossify to form cartilage bones. The palatopterygoquadrate bar is also included. (*After Kingsley, "Comparative Anatomy of Vertebrates," The Blakiston Division, McGraw-Hill Book Company. By permission.*)

below, two lateral *exoccipitals,* and a dorsal *supraoccipital.* It will be recalled that in addition to the otic capsules, only the occipital region of the chondrocranium becomes roofed over by cartilage. Anterior to the occipital segment the basisphenoid region, at least in reptiles and birds, is composed of three cartilage bones: a *basisphenoid* below, and two lateral *pleurosphenoids.* Two dorsal membrane bones, the *parietals,* form a roof over this region which then is referred to as the *parietal segment.* In mammals the pleurosphenoids are lacking. Instead, a pair of bones, the *alisphenoids,* representing the *epipterygoids* of certain reptiles (page 464) and originating as centers of ossification in the palatopterygoquadrate bar, comes to lie on either side of the basisphenoid. The alisphenoids are not homologous with the pleurosphenoids. The presphenoid region, lying anterior to this complex, also is composed of three cartilage bones, a *presphenoid* below, and an *orbitosphenoid* on either side. In addition, two membrane bones, the *frontals,* roof over this region, which is then known as the *frontal segment.* Several cartilage bones originate in the region of the olfactory, or nasal, capsules. It will be recalled that these capsules fuse with the ethmoid plate formed by a union of the anterior portions of the prechordal cartilages. This makes up the ethmoid region. An ossification center in the middle of the ethmoid plate gives rise to the *mesethmoid* bone which contributes to the nasal septum. In some forms this area remains unossified. In fishes the nasal capsules give rise to *lateral ethmoids,* or *ectethmoids. Turbinate* bones (page 207) are the chief derivatives of the nasal capsules. Also from this region in mammals arises the *cribriform plate* with its numerous perforations, the *olfactory foramina.* Several of these bones may unite to form a single *ethmoid bone.* Two sets of

membrane bones usually develop in this area: a pair of *nasal* bones, roofing over the mesethmoid, and a pair of *vomers* lying below it. Often there is a single vomer. The nasal bones lie medial or posteromedial to the external nares. The entire region of the skull anterior to the cranial cavity is commonly spoken of as the *nasal segment.* Many variations are encountered in this portion of the skull.

It should be noted that the nasals, frontals, and parietals, being membrane bones, are not comparable to the supraoccipital, which is a cartilage bone. Sometimes some membrane bones, representing ancestral elements, are added to the dorsal portion of the occipital. For example, an *interparietal,* or *postparietal,* is usually fused to the supraoccipital in man. In the Inca Indians it was a separate bone. In such vertebrates as modern amphibians, the basioccipital and supraoccipital regions remain unossified. In mammals all four regions finally fuse to form a single *occipital bone.* The cartilage bones of the occipital segment trace their origin to the occipital vertebrae and possibly to the posterior portion of the parachordal plates; those of the parietal segment are derived from the anterior portions of the primitive parachordal cartilages. The cartilage bones of the frontal segment arise from the prechordal cartilages; those of the nasal segment are derivatives of the prechordal cartilages as well as of the nasal capsules.

Certain other bones of the skull trace their origin to the otic sense capsules which were parts of the original chondrocranium. Three separate centers of ossification appear in the cartilaginous otic capsules, forming three bones: (1) an anterior *pro-otic,* (2) a posterior *opisthotic,* and (3) a dorsal *epiotic.* These bones may unite with each other or with neighboring bones. The complex comes to lie on either side between the pleurosphenoid and exoccipital. In bony fishes two additional bones, the *sphenotic* and *pterotic,* develop in the otic capsule.

Several membrane bones, in addition to those already mentioned, are frequently present in the vertebrate skull. These include a *parasphenoid,* lying ventral to the basicranial axis; *prefrontals* and *postfrontals,* situated anterolateral or posterolateral, respectively, to the frontal bone proper; and *lacrimals,* which help to complete the inner wall of the orbit, or eye socket.

From the primitive visceral skeleton is also derived a number of separate skeletal elements. The mandibular, or first, visceral arch is very much modified. It is divided on each side into a dorsal *palatopterygoquadrate bar* and a ventral *Meckel's cartilage.* The mouth cavity is bounded by these components. The angle of the mouth is located posteriorly where the palatopterygoquadrate bar and Meckel's cartilage join at a sharp angle.

In bony fishes, centers of ossification in the more posterior portion of the palatopterygoquadrate bar give rise to such cartilage bones as the *metapterygoid* * and *quadrate.* The anterior part of the bar is reduced and replaced by such membrane bones as the *palatine* and *pterygoid.** The lower jaw articulates with the quadrate. In these fishes the quadrate forms a connection with a *symplectic* bone, derived from the hyomandibular and attached to it. The hyomandibular in turn unites with the otic region of the chon-

drocranium. This is a variation of the hyostylic method of jaw suspension (Fig. 10.35F).

In tetrapods other than mammals a quadrate bone, with which the lower jaw articulates, is also formed by direct ossification of the posterior end of the palatopterygoquadrate bar. An *epipterygoid,** corresponding to the metapterygoid of fishes, develops anterior to the quadrate. The quadrate forms a firm union with the otic region of the chondrocranium. The hyomandibular cartilage is reduced in tetrapods and is not concerned with jaw suspension (Fig. 10.35C). It becomes the *stapes,* one of the auditory ossicles. This method of suspending the jaws via the quadrate is termed the *autostylic* method, in contrast to the hyostylic method in which suspension is by means of the hyomandibular. Another type of jaw suspension found in a few primitive sharks is called the *amphistylic method* (Fig. 10.35A). Here both the palatopterygoquadrate bar and hyomandibular cartilage of the hyoid arch are united to the otic region of the chondrocranium.

In some tetrapods, as in *Sphenodon,* lizards, and turtles, the epipterygoid becomes a separate bone. In mammals, however, it comes to lie on either side of the basisphenoid and is commonly referred to as the *alisphenoid.* It will be recalled that pleurosphenoids, which in reptiles and birds develop on either side of the basisphenoid, are lacking in mammals.

In bony vertebrates another arch, the *maxillary arch,* composed entirely of membrane bones, forms outside of and roughly parallel to the palatopterygoquadrate bar (Figs. 10.36 and 10.37). This gives rise to the greater part of the functional upper jaw which is fused to the rest of the skull in these forms. The most anterior bones to develop in this region are the paired *premaxillaries.* They are followed by the *maxillaries,* and these in turn by the *jugals (malars, zygo-*

---

* Much confusion exists in regard to whether the pterygoid is a cartilage bone or a membrane bone. Misunderstandings are largely a matter of terminology. In this account we shall consider the *pterygoid* proper to be a membrane bone. *Ectopterygoids* (transpalatines, mesopterygoids) and *endopterygoids,* when present, and which are subdivisions of the pterygoid, are also membrane bones. Such bones as the *metapterygoids* of fishes and the *epipterygoids* of certain tetrapods, which develop from ossification centers in the palatopterygoquadrate bar, are clearly cartilage bones.

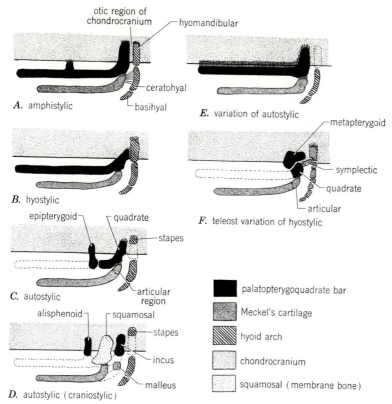

Fig. 10.35 Diagrams indicating various methods of jaw suspension: *A*, amphistylic, found in a few elasmobranchs; *B*, hyostylic, found in sharks and sturgeons; *C*, autostylic, found in tetrapods other than mammals; *D*, craniostylic, typical of mammals; *E*, variation of autostylic, found in *Polypterus*, holocephalians, and lungfishes; *F*, teleost variation of hyostylic, found in *Amia*, *Lepisosteus*, and teleosts. The membrane bones investing Meckel's cartilage are not indicated.

matics). Next in succession come the *quadratojugals*, *squamosals*, and *supratemporals*. Only the premaxillaries and maxillaries in this arch bear teeth.

The term *palate* is used to indicate the roof of the mouth and pharynx. In fishes and amphibians this is a flattened area directly beneath the floor of the cranium. In most reptiles and in birds a pair of longitudinal *palatal folds* grows medially for a short distance on either side. These folds do *not* meet in the median line. In crocodilians and mammals, horizontal shelflike projections of the premaxillary, maxillary, palatine, and, in some cases, of the pterygoid bones extend medially to meet their partners of the opposite side. In this manner a complete *secondary palate* is formed separating the nasal passages above from the mouth cavity below. As a result, the nasal passages communicate with the mouth cavity much farther posteriorly than they would otherwise. The portion

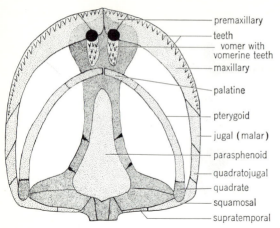

premaxillary
teeth
vomer with
vomerine teeth
maxillary

palatine

pterygoid

jugal (malar)

parasphenoid

quadratojugal

quadrate

squamosal

supratemporal

Fig. 10.36 Diagrammatic ventral view of skull of typical tetrapod, showing relationship of maxillary arch to chondrocranium and palatopterygoquadrate bar. Membrane bones are lightly stippled; cartilage bones are heavily stippled. The palatines and pterygoids, as represented here, are membrane bones which have invested the anterior portion of the palatopterygoquadrate bar. (*After Kingsley, "Comparative Anatomy of Vertebrates," The Blakiston Division, McGraw-Hill Book Company. By permission.*)

of the secondary palate which is bony is called the *hard palate.* In mammals the secondary palate is continued for some distance posteriorly by a *soft palate,* composed mostly of connective tissue without any bony foundation.

The original Meckel's cartilage forming the lower jaw becomes considerably modified in higher vertebrates with bony skeletons. The posterior region is usually replaced by a cartilage bone, the *articular.* This forms an articulation with the quadrate. In anuran amphibians at the point where the two halves of the lower jaw join in front, a small cartilage bone, the *mentomeckelian,* develops on either side. The remainder of Meckel's cartilage serves as a core about which a number of membrane bones form a sheath (Fig. 10.38). The separate elements which, for the

most part, enter into this complex are (1) an anterior *dentary* (*dental*), surrounding Meckel's cartilage, (2) a medial *splenial,* (3) a ventral *angular,* (4) a posterolateral *surangular,* (5) a dorsolateral *coronoid,* and (6) a *gonial,* lying below and to the medial side of the articular. Many modifications of the above are to be found in vertebrates. Only rarely are all these bones present. They exist as separate elements in amphibians and reptiles, but in birds they begin to unite. In mammals the lower jaw consists of a single bone, the *mandible.* It represents the dentary of lower forms, the other elements having been lost except for the angular, which is said to give rise to the *tympanic* bone, and the articular, which has become the *malleus,* one of the auditory ossicles, or ear bones. Study of the fossil remains of

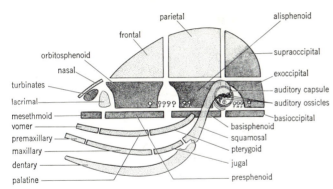

Fig. 10.37 Diagrammatic lateral view of mammalian skull, showing relationships of the various bones. Membrane bones are lightly stippled; cartilage bones are heavily stippled. The palatines and pterygoids, as represented here, are membrane bones which have invested the greater portion of the palato-pterygoquadrate bar. Numbers refer to foramina through which cranial nerves emerge. (*After Flower, from Kingsley, "Comparative Anatomy of Vertebrates," The Blakiston Division, McGraw-Hill Book Company. By permission.*)

some of the mammal-like reptiles indicates clearly how, in all probability, the mammalian mandible evolved.

In tetrapods below mammals in the evolutionary scale, the articular bone of the lower jaw articulates with the quadrate of the upper jaw. Both of these are cartilage bones. In mammals the articular and quadrate are modified to form the *malleus* and *incus* bones, respectively. A new articulation of the lower jaw with the squamosal,

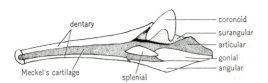

Fig. 10.38 Showing the position of membrane bones forming about Meckel's cartilage in the development of the lower jaw. (*After Kingsley, "Comparative Anatomy of Vertebrates," The Blakiston Division, McGraw-Hill Book Company. By permission.*)

a membrane bone of the maxillary arch, is then established (Figs. 10.35 and 10.37). This occurs only in mammals. The method of jaw suspension in mammals is considered by many authorities to be autostylic even though the articulation of the lower jaw is with the squamosal bone rather than with the quadrate. Other anatomists believe that a distinction should be made and have applied the terms *craniostylic* and *amphicraniostylic* to the mammalian method.

The hyoid, or second visceral, arch does not become so highly modified as the mandibular arch during its evolutionary history. In elasmobranch fishes it is usually composed of three cartilages, a dorsal *hyomandibular*, a lower *ceratohyal*, and a small, ventral, median *basihyal* which serves to unite the ceratohyals of the two sides. In elasmobranchs the hyomandibular forms a ligamentous union with the otic region of the skull and serves as a suspensorium of the jaws (hyostylic method). In higher fishes several cartilage bones may be derived from the hyoid arch although their appearance and number are highly variable. In general they are named as follows in a dorsal-ventral sequence: *hyomandibular, interhyal, epihyal, ceratohyal,* and *hypohyal*. A small median *basihyal* unites the hypohyals of the two sides. A *symplectic* bone, derived from the hyomandibular and found only in teleost fishes (Fig. 10.35*F*), extends forward from the ventral end of the hyomandibular near its junction with the interhyal and articulates with the quadrate. In tetrapods the hyomandibular bone is reduced and gives rise to the *columella,* or to the *stapes* bone of the middle ear. The ceratohyal is important in elasmobranch fishes as the chief support of the gill filaments of the most anterior, or hyoid, hemibranch. In higher forms it gives rise to a portion of the hyoid apparatus which furnishes support to the tongue and larynx.

The operculum, a posteriorly directed flap of tissue which covers the gill chamber in most fishes, begins in the region of the hyoid arch. In bony fishes it is strengthened by a number of thin membrane bones, the *opercular bones,* of variable size and number. The lower borders of the two opercula are joined by a membranous structure which, in certain primitive fishes, contains one or two *gular* bones. In teleost fishes these become *branchiostegal rays* which give support to the branchiostegal membrane. No homologues of opercular and gular bones are to be found in tetrapods.

The visceral arches posterior to the hyoid serve an important function in fishes in which they support the remaining gills. Their ventral ends are united by separate median cartilages or bones, the *basibranchials,* or *copulae*. In higher forms, the posterior visceral arches are much reduced, being of importance mainly in contributing

elements to the hyoid apparatus and in furnishing cartilages to the larynx.

Speculations regarding phylogenetic relationships of ancestral vertebrates which became extinct millions of years ago are based upon study and comparisons of hard or bony structures preserved as fossils. Our knowledge of skull evolution in the *bony* fishes is rather incomplete and very complex. Much of what we know concerning primitive skull structure has come from study of the comparatively recent discovery of the fossil remains of so-called labyrinthodont amphibians, which first appeared upon the earth during the latter part of the Devonian Period, over 300 million years ago. The skull structure of fossil ancestral crossopterygian fishes, also from the Devonian Period, has been found to be basically similar to that of the labyrinthodont amphibians, clearly indicating the probability that it was through the crossopterygian fishes that evolutionary progress was made. A significant feature which relates these two groups was the presence in both of a movable articulation, or hinge, between what are essentially two main regions of the skull, anterior and posterior. The anterior portion includes those structures derived from the prechordal cartilages or associated with them. It includes the presphenoid, ethmoid, and nasal regions. The palatal elements and the upper jaw were firmly united to this anterior portion. The posterior region is composed of derivatives of the parachordal cartilages, or basilar plate, as well as those from the occipital and otic regions. Furthermore, there is evidence that the notochord was located in a canal incorporated in the floor of the sphenoid (or basilar plate) region and that it extended as far forward as the posterior portion of the fused prechordal cartilages, or presphenoid region. The ancient crossopterygians and the labyrinthodont amphibians, therefore, furnish a clue to the basic skull pattern of later tetrapods. The skull of the living *Latimeria* is essentially similar to that just described, with the notochord extending under the hinge joint in the anterior region where it terminates in a socket. Behind the hinge level it is contained within a large cylindrical groove in the floor of the skull which, toward the posterior end, is actually in the form of a short tube completely surrounded by bone. The cephalic extension of the notochord and its attachment to the skull at two points makes it possible to serve as a sort of shock absorber. On either side of the notochord a prominent pair of muscles unites the two parts of the base of the skull, enabling the animal to raise its upper jaw.

The contribution of the visceral skeleton to skull formation in ancestral forms remains somewhat obscure. The mesenchyme from which it is derived seems to be partly of ectodermal origin since neural-crest cells which have migrated to the gill region apparently contribute to it. Our knowledge of the arrangement of the various elements comprising the visceral skeleton of ancestral forms is incomplete. Paleontologists have, nevertheless, advanced what appears to be a logical story concerning the role of the visceral skeleton in the evolution of the skull. The fact that in cyclostomes the skeletal support of the gill region is in the form of a branchial basket (page 214) composed of continuous cartilage and not broken up into separate elements as in higher forms, not only raises doubts concerning its homologies, but makes it difficult to speculate as to the true ancestral origin of the visceral skeleton as it appears in primitive cartilaginous and bony fishes.

Skull in different classes of vertebrates.

CYCLOSTOMES. The skull of hagfishes is very primitive, consisting only of a floor of cartilage with side walls and roof of connective tissue. Parachordal cartilages are united with the otic capsules. The prechordal cartilages terminate abruptly at the anterior end.

In lampreys the skull is better developed. Dorsolateral extensions of the prechordal cartilages are present, forming side walls and a roof over the brain in this region. The remainder of the brain is enclosed in fibrous connective tissue. Prechordal and parachordal cartilages as well as otic capsules are homologous with similar structures in higher forms. Other cartilages of doubtful homology are present, supporting the round, suctorial mouth, the olfactory organ, and the tongue. *Posterior* and *anterior dorsal cartilages* are located anterior to the chondrocranium. They too appear to be without homologues in higher vertebrates.

The gill region in lampreys is supported by a peculiar cartilaginous *branchial basket*. The cartilage of which it is composed is continuous and not divided into separate parts. Lateral openings for the gill slits are present in addition to other openings above and below. The branchial basket (Fig. 10.39) lies just beneath the skin. Its posterior end furnishes support for the pericardium. It does not appear to be homologous with the visceral skeletons of other vertebrates. In hagfishes the visceral skeleton is much reduced, but a hyoid arch and third and fourth visceral arches are recognizable. They are located far within the head and are interpreted as true visceral arches.

FISHES. *Elasmobranchs*. The skull in elasmobranch fishes is entirely cartilaginous. The cartilage is frequently calcified, however. Absence of dermal bones is presumably a secondary development and does not represent a primitive condition. The chondrocranium consists of a single piece of cartilage, the separate elements of which it is formed having fused together. Nasal and

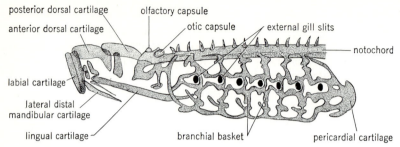

posterior dorsal cartilage

anterior dorsal cartilage

olfactory capsule

otic capsule

external gill slits

notochord

labial cartilage

lateral distal
mandibular cartilage

lingual cartilage

branchial basket

pericardial cartilage

Fig. 10.39 Lateral view of chondrocranium and branchial basket of lamprey, *Petromyzon marinus. (After Parker, from Schimkewitsch.)*

otic capsules are indistinguishably united with the braincase. A large, depressed *orbit* on either side serves to receive the eyeball. Openings are present for the passage of blood vessels and cranial nerves (Figs. 10.40 and 10.41). The floor of the chondro-

cranium is complete. A cuplike depression, the *sella turcica,* serves to receive the pituitary body. The occipital region of the chondrocranium on either side bears a posterior projection, the *occipital condyle,* which usually forms an immovable articulation with

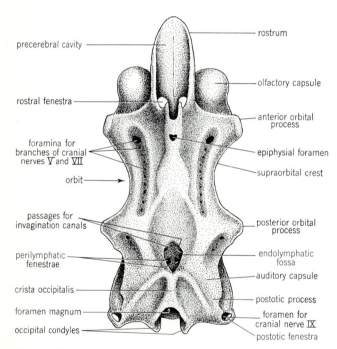

precerebral cavity

rostrum

olfactory capsule

rostral fenestra

anterior orbital
process

foramina for
branches of cranial
nerves V and VII

epiphysial foramen

orbit

supraorbital crest

passages for
invagination canals

posterior orbital
process

perilymphatic
fenestrae

endolymphatic
fossa

auditory capsule

crista occipitalis

postotic process

foramen magnum

foramen for
cranial nerve IX

occipital condyles

postotic fenestra

Fig. 10.40 Dorsal view of skull of elasmobranch, *Squalus acanthias. (After Senning, outline drawings for "Laboratory Studies in Comparative Anatomy," McGraw-Hill Book Company. By permission.)*

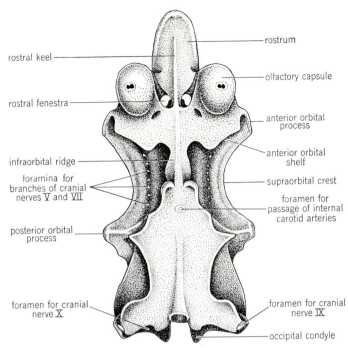

rostral keel

rostral fenestra

infraorbital ridge

foramina for
branches of cranial
nerves V and VII

posterior orbital
process

foramen for cranial
nerve X

rostrum

olfactory capsule

anterior orbital
process

anterior orbital
shelf

supraorbital crest

foramen for
passage of internal
carotid arteries

foramen for cranial
nerve IX

occipital condyle

Fig. 10.41 Ventral view of skull of elasmobranch, *Squalus acanthias. (After Senning, outline drawings for "Laboratory Studies in Comparative Anatomy," McGraw-Hill Book Company. By permission.)*

the first vertebra. Perhaps the most peculiar feature of the elasmobranch chondrocranium is the *rostrum,* an anterior, trough-like extension, formed originally from the prechordal cartilages. The presence of the rostrum causes the mouth to be subterminal and ventral in position. In the sawfish the rostrum is a very extensive projection, supplied along the sides with numerous sharp teeth and forming the formidable, sawlike weapon with which this fish is supplied (Fig. 4.16). In some of the larger sawfishes the rostrum may reach 6 ft in length and 1 ft in width.

The visceral skeleton of elasmobranchs (Figs. 10.33 and 10.42) is typically composed of seven cartilaginous visceral arches

surrounding the anterior part of the digestive tract and furnishing a firm support to the gills.

The first, or mandibular, arch is larger and more conspicuous than the rest. From it are derived both upper and lower jaws. On each side the mandibular arch divides into a dorsal palatopterygoquadrate bar and a ventral Meckel's cartilage. These articulate with each other posteriorly. Both cartilages are suspended, by means of a ligamentous attachment, from the hyomandibular cartilage of the hyoid arch which in turn is joined to the otic region of the chondrocranium. This is the hyostylic method of jaw attachment (Fig. 10.35*B*). The palatopterygoquadrate bars of the two

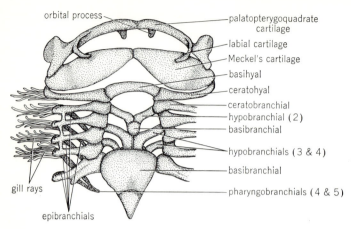

orbital process
palatopterygoquadrate cartilage
labial cartilage
Meckel's cartilage
basihyal
ceratohyal
ceratobranchial
hypobranchial (2)
basibranchial
hypobranchials (3 & 4)
basibranchial
pharyngobranchials (4 & 5)
gill rays
epibranchials

Fig. 10.42 Ventral view of visceral skeleton of elasmobranch, *Squalus acanthias*. The hyomandibular cartilage is not shown.

sides make up the upper jaw. They are heavy cartilages which unite with each other anteriorly by means of a ligamentous connection. The anterior, or palatine, portion of each palatopterygoquadrate bar usually bears a *palatine*, or *orbital*, *process* which extends upward along the inner wall of the orbit. The two Meckel's cartilages, forming the lower jaw, are united anteriorly by a ligament. Both palatopterygoquadrate bar and Meckel's cartilages bear teeth. Small *labial cartilages* of doubtful homologies are embedded in the tissue just outside the upper and lower jaws near the anterior end.

The second, or *hyoid, arch* is composed of three parts. An upper *hyomandibular cartilage* articulates with the chondrocranium behind the orbit in the otic region. Below this is the *ceratohyal cartilage*. The ceratohyals of the two sides are joined together ventrally by means of a median plate, the *basihyal*. The posterior edges of both hyomandibular and ceratohyal cartilages give off numerous slender cartilaginous *gill rays* which support the tissue bearing the gill lamellae composing the first, or hyoid, hemibranch.

The number of visceral arches posterior to the hyoid varies in elasmobranchs, but usually there are five. Each consists of several cartilages on each side. Named in order from dorsal to ventral borders they are (1) *pharyngobranchial*, (2) *epibranchial*, (3) *ceratobranchial*, and (4) *hypobranchial*. The pharyngobranchials slope posteriorly and terminate just below the vertebral column. They are attached to the vertebral column by fibrous bands of tissue. The hypobranchials may meet their partners in the midventral line or are connected by unpaired *basibranchial cartilages*. The epibranchials and ceratobranchials bear the gill rays which support the remaining gills. Gill rays are wanting on the last visceral arch which, as previously mentioned, does not bear a gill. Short but strong, pointed cartilaginous projections called *gill rakers* extend into the pharynx from the inner edges of the visceral arches. They are covered with mucous membrane and are used to strain food particles from the water.

Other fishes. It may be recalled from Chap. 2 that the bony fishes belonging to the class Osteichthyes first appeared on earth during Devonian times, around 350 million years ago. In evolutionary terms they are much older than members of the class Chondrichthyes. The presence of a bony skeleton, present even in ancestral placoderms, therefore, must be considered to be a primitive feature which has been retained. Paleontological evidence indicates that very early in their history the bony fishes had already divided into two branches, the ray-fins, or Actinopterygii, and the lobe-fins, or Sarcopterygii. The subclass Sarcopterygii contains members of the order Crossopterygii, primitive relatives of which were undoubtedly the ancestors of tetrapods.

Since crossopterygians were near the bottom of the ladder, so to speak, a study of their skull structure is of particular interest in drawing parallels between the elements making up the skulls of modern fishes and those of tetrapods.

The skulls of the ray-finned fishes, with the exception of *Polyodon*, the spoonbill or paddlefish, are invested with numerous membrane bones, several of which are homologous with similarly placed and similarly named bones of crossopterygians and tetrapods. There has been much confusion in the past in interpreting such homologies, and some bones have been incorrectly named. A chief difference between the skulls of the ray-fins and crossopterygians is that in the former group the front part of the skull is relatively long, and the posterior, or cheek, region is short. The opposite condition was evident in the crossopterygians, in which the anterior region was short and the posterior region tended to be elongated. Another, and significant, difference is that in the ray-fins there is normally a lack of development of a homologue of the squamosal bone present in crossopterygians and tetrapods.

Among the skulls of fishes are to be found all the transitional stages from those which are completely cartilaginous to those which are almost completely bony.

In holocephalians the skull is composed entirely of cartilage. Noteworthy is the fact that the entire palatopterygoquadrate bar is immovably fused to the cranium. The lower jaw is suspended from the quadrate region, and the hyoid arch plays no part in suspending the jaws (Fig. 10.35E). This is a variation of the autostylic method of jaw suspension. The skull articulates with the vertebral column by means of a movable joint. The hyoid arch is developed only slightly more than the remaining visceral arches. The hyomandibular portion is degenerate. The ceratohyal component, with its gill rays, supports the *operculum,* which first makes its appearance in this group of fishes.

The sturgeon, *Acipenser,* has a skull almost completely formed of cartilage. No cartilage bones are present although ossifications appear in the otic and orbitosphenoid regions. Several well-developed membrane bones have made their appearance. On the dorsal side they include frontals, postfrontals, parietals, supraorbitals, and others, which form a protective shield of dermal scales over the cartilaginous cranium beneath. On the ventral side are two other membrane bones, an anterior *vomer* and a more posterior *parasphenoid.* The rostrum, covered with scales, is prominent. The method of jaw attachment in the sturgeon is hyostylic. Maxillary and dentary bones are associated with the jaws, and dermal opercular bones lie over the hyomandibular cartilage. Jaws and visceral skeleton are poorly developed. In *Polyodon,* a close relative of the sturgeon, membrane bones are less prominent. Nevertheless, the snout or paddle, which is actually the enormously elongated rostrum, is covered with dermal scales.

The ancient ray-fin, *Polypterus,* exemplifies an advance over the primitive condition toward the more highly developed teleostean skull. The cartilaginous skull is retained for the most part, but a veritable armor of numerous membrane bones covers it. Among these are frontals, parietals, nasals, supratemporals, posttemporals, and *dermal* supraoccipitals. Since the palatopterygoquadrate bar is fused to the cranium, the method of jaw attachment is autostylic (Fig. 10.35*E*). Several important membrane bones are developed in connection with the palatopterygoquadrate bar. They include palatines and a number of pterygoids. Vomer bones are also present. The outer arch of membrane bones includes premaxillaries, maxillaries, and jugals. In connection with Meckel's cartilage are developed dentary, angular, and splenial bones. Other membrane bones appear on the operculum.

In the garpike, *Lepisosteus,* and in the bowfin, *Amia,* both cartilage bones and numerous membrane bones are present. Exoccipitals, pro-otics, and lateral ethmoids are among the cartilage bones which appear. In the gar, teeth are borne by the maxillary bones in the upper jaw. Teeth are present in *Amia* on the palatines, pterygoids, and vomers as well as on premaxillaries and maxillaries. The lower jaw of the gar consists mainly of the tooth-bearing dentary. Small angular and articular bones are also present. In *Amia* there are several splenials, an angular, surangular, coronoid, and articular. The dentary, bearing teeth, however, is the chief bone of the lower jaw. The hyostylic method of jaw attachment is found in these fishes (Fig. 10.35*F*).

A rather wide variation is to be observed among the teleosts. Their skulls have more bones than those of any other vertebrates. Numerous membrane bones are present, but most characteristic is the presence of several cartilage bones which have formed from ossification centers in the primitive chondrocranium, which still persists to some degree. These bones include basioccipitals, exoccipitals, supraoccipitals, basisphenoids, pleurosphenoids, or alisphenoids (not homologous with the alisphenoids of mammals), orbitosphenoids, pro-otics, epiotics, and opisthotics. Sphenotics and pterotics, present in addition, are derived from the auditory capsules. Mesethmoid and lateral ethmoids develop anteriorly. Among the more prominent membrane bones on the roof of the skull are parietals, frontals, and nasals. A series of *circumorbital* bones affords protection to the eyes. Several other membrane bones are present in addition to those mentioned.

Forming the roof of the mouth are a large, single parasphenoid and a pair of vomers. The palatopterygoquadrate bar in these fishes ossifies to form such cartilage bones as the metapterygoid and quadrate. The posteriorly located quadrate articulates with the *symplectic.* The latter joins the hyomandibular, which in turn is fastened to the cranium. This is a variation of the hyostylic method of jaw suspension and is characteristic of teleosts (Fig. 10.35*F*). Other bones (palatine, ectopterygoid, metapterygoid) usually aid in holding the quadrate in place. Palatines and pterygoids (ectopterygoids and endopterygoids) are membrane bones formed in relation to the palatopterygoquadrate bar. The outer arch of membrane bones includes premaxillaries and maxillaries. Both usually bear teeth, but in most teleosts the maxillaries are reduced in size, do not form part of the margin of the jaw, and do not bear teeth. Teeth are frequently present on vomers, palatines, and pterygoids. The number of bones in the lower jaw of teleosts shows a reduction. The dentary, which usually bears teeth, is the chief component. An angular bone may be present. The articular bone is small and reportedly derived from both membrane and cartilage elements. It articulates with the quadrate.

The hyoid arch is ossified in several places. A hyomandibular bone articulates with the sphenotic and pterotic bones of the auditory capsule and with the metapterygoid of the palatopterygoquadrate bar. The symplectic bone, found only in bony fishes, and segmented off the hyomandibular cartilage during development, is interposed between hyomandibular and quadrate bones. The remainder of the hyoid arch is composed of a chain of bones. First there is a small *interhyal* which articulates at the junction of hyomandibular and symplectic. This is followed ventrally and in order by an *epihyal,* a large *ceratohyal,* and a small *hypohyal.* Several sturdy, curved, bony *branchiostegal rays* project posteriorly from the ceratohyal. They are probably homologous with the gular bones which appear ventral to the operculum in some primitive fishes. The branchiostegal rays support the branchiostegal membrane (page 215). In some fishes, as in the salmon, a large, forward-projecting *glossohyal* (*entoglossal*), bearing teeth, serves to unite the hyoid arches of the two sides and to support the tongue. A posteriorly directed *urohyal* may connect with the basibranchial of the next visceral arch. Both these bones are probably hyoid-arch derivatives. In bony fishes the hyoid hemibranch is missing. It may possibly be represented by opercular gill lamellae on the inner wall of the operculum. These, however, may actually be remnants of the pseudobranch located on the anterior wall of the spiracle of lower fishes. The homologies of the opercular gill are uncertain. Of importance in the bony fishes are the opercular bones. They usually consist of four large membrane bones which give support to the operculum, helping to protect the gill chamber. They lie directly posterior to the hyomandibular region.

The five remaining visceral arches are often considerably reduced. They are usually composed of several separate elements named in the same manner as those of elasmobranchs. The last one consists only of a single dorsal element, the pharyngobranchial, which bears teeth but does not support a gill. Gill rakers, or strainers, project from the inner surfaces of the visceral arches. No membrane bones develop in connection with the hyoid or the remaining visceral arches.

The skull in Dipnoi is not so far advanced as that of teleosts. It is cartilaginous, for the most part, but both cartilage bones (exoccipitals) and a few large membrane bones are found. The palatopterygoquadrate bar, like that of holocephalians, is indistinguishably fused to the cranium. No cartilage bones are derived from it. The method of jaw suspension is autostylic. Marginal teeth are lacking, but *tooth plates* are present on the vomers and pterygoids, both of which are membrane bones.

In the evolution of the skull from primitive fishes to mammals, there has been a gradual reduction in number of the separate bony elements by elimination and fusion. As many as 180 skull bones are present in certain primitive fishes. The human skull consists only of 28, including the auditory ossicles. In some cases cartilage bones and membrane bones fuse together to form a single structure. The autostylic method of jaw attachment is the rule among the tetrapods.

AMPHIBIANS. It is generally considered by paleontologists that early tetrapods evolved from rhipidistian crossopterygian fishes or some of their very close relatives. It is quite probable that the early ancestors of amphibians were little more than fish possessing primitive "legs" rather than fins. Modern amphibians are far removed from the primitive condition. Nevertheless, the dependency of most modern amphibians upon water, particularly as regards their reproductive habits, is a further indication of their ancestral relationships to fishes.

The ancient labyrinthodonts, considered to be the earliest amphibians, were descendants of ancestral crossopterygians. They gave rise not only to modern amphibians but to primitive reptiles as well. As previously pointed out (page 30), their fossil remains show striking similarities to those of ancestral crossopterygians in many ways, but particularly in details of skull structure.

Although the skulls of early tetrapods retained certain characteristics of their piscine progenitors, those of modern amphibians show considerable deviation. Among the most important changes are a reduction in the number of bones and a general flattening of the skull. In addition, the length of the skull has apparently been reduced, particularly in the occipital region. The otic capsule bears a ventral opening, the *fenestra ovalis,* into which a cartilaginous or bony plug fits. This is the *stapedial plate* of the columella which, it will be recalled, is derived from the hyomandibular cartilage. It is possible that the columella proper is derived from the symplectic. The columella (stapes) has developed in connection with the evolution of the sense of hearing and with the change from the hyostylic to the autostylic method of jaw suspension in which the hyomandibular bone loses its significance as a suspensorium.

The embryonic chondrocranium persists to a considerable extent in amphibians, but some of it has been replaced by cartilage bones. The basioccipital and supraoccipital regions are not ossified. Articulation of the skull with the atlas vertebra is accomplished by means of a pair of occipital condyles, projections of the exoccipital bones. Basisphe-noid and presphenoid regions also are not ossified. Pro-otics, and in some cases opisthotics, are ossified and fused to the exoccipitals. Much variation in the ossification of this region is encountered.

Membrane bones form the greater part of the roof of the skull. They are no longer closely related to the integument and occupy a deeper position in the head than they do in fishes. In some of the extinct labyrinthodonts, grooves for the accommodation of branches of the lateral-line canal were present on some of the membrane bones of the skull, giving evidence of their dermal origin. A rather large membrane bone, the parasphenoid, covers the ventral part of the chondrocranium.

In the labyrinthodonts the dorsal surface of the skull was completely covered by bone, leaving only openings for the nares and eyes. In some extinct forms, however, another opening, the *interparietal foramen,* was evident. This accommodated the stalk of the median parietal eye. The interparietal foramen has disappeared in living species. Modern apodans, or caecilians, also show great solidity of the skull. There has, however, been a reduction in the number of skull bones. In other amphibians large spaces are present on both dorsal and ventral sides of the skull.

It has been mentioned that in forms above fishes in the evolutionary scale the autostylic method of jaw attachment is the rule. The quadrate, in amphibians, is fused to the otic region of the cranium. Palatine and pterygoid membrane bones, which form about the anterior portion of the palatopterygoquadrate bar, are well developed. The outer arch of membrane bones is represented by premaxillaries and maxillaries. Anurans have a quadratojugal in addition. Premaxillaries and maxillaries, together with the vomer and sometimes the palatines, bear teeth in most species.

The lower jaw consists of a core of Meckel's cartilage surrounded by membrane bones. In early amphibians as many as 10 bones formed about Meckel's cartilage on each half of the jaw. In modern forms the number has been considerably reduced.

Urodeles. In the tailed amphibians the skull is less well developed than in anurans. The chondrocranium proper becomes ossified in only two regions. Anteriorly there is a pair of elongated orbitosphenoids, each bearing a foramen for the passage of the optic nerve. Posteriorly are found the paired exoccipitals which are fused with the otic bones. Two foramina in each exoccipital near the condyle provide for the passage of the glossopharyngeal and vagus nerves. Placed farther anteriorly are two additional foramina for the facial and trigeminal nerves. The membrane bones on the dorsal side of the cranium consist of parietals, frontals, and *prefrontolacrimals.* A pair of nasal bones develops in the olfactory region. An unpaired parasphenoid bone, often bearing teeth, lies on the ventral surface. Anterior to this is a pair of *vomeropalatines,* usually bearing a row of teeth. The choanae open just posterior to the vomerine portion. The upper jaw consists of premaxillary and maxillary membrane bones. The premaxillaries in some cases are united and articulate with the vomeropalatines on the ventral side and with the nasals, and frequently with the frontals, on the dorsal side. Laterally the premaxillaries articulate with the maxillaries. The latter terminate freely at their posterior ends, failing to meet the quadrates. Jugals and quadratojugals are lacking. The quadrate bone, representing the remainder of the palatopterygoquadrate bar, is fused to the otic capsule and articulates with the stapes. It serves for articulation with the lower jaw. The squamosal, a membrane bone, invests the outer surface of the quadrate. A cartilage extending anteriorly from the quadrate toward the maxillary is invested ventrally and laterally by the pterygoid. In some cases the pterygoid may form a cartilaginous connection with the maxillary.

The lower jaw of urodeles usually consists of two tooth-bearing membrane bones, an anterior dentary and a posterior splenial, which encase Meckel's cartilage, and a single cartilage bone, the articular, which articulates with the quadrate of the upper jaw. Another membrane bone, the angular, is also frequently present.

Although the above description applies in general to most urodeles, many variations are encountered. Nasals, maxillaries, and sometimes palatines may be lacking. In *Necturus* (Figs. 10.43 and 10.44) the skull retains many primitive features. The nasal capsules are not united with other parts of the skeleton. Maxillary bones are lacking. A *palatopterygoid* membrane bone with teeth on its anterior portion unites the vomer with the quadrate on each side.

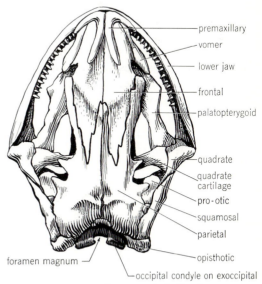

premaxillary
vomer
lower jaw
frontal
palatopterygoid
quadrate
quadrate cartilage
pro-otic
squamosal
parietal
opisthotic
foramen magnum
occipital condyle on exoccipital

Fig. 10.43 Dorsal view of skull of *Necturus.*

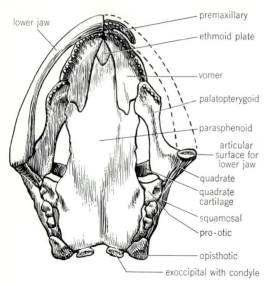

Fig. 10.44 Ventral view of skull of *Necturus*.

(labels, clockwise from top)
premaxillary
ethmoid plate
vomer
palatopterygoid
parasphenoid
articular surface for lower jaw
quadrate
quadrate cartilage
squamosal
pro-otic
opisthotic
exoccipital with condyle
lower jaw

The remaining portion of the visceral skeleton in urodele amphibians is reduced in comparison with fishes (Fig. 10.45). Larval forms generally possess portions of the hyoid and four other visceral arches, but these are reduced at the time of metamorphosis. In adults the hyoid arch usually consists of two pieces on each side, a lower hypohyal and an upper ceratohyal, which is frequently ossified. The hypohyals are in most cases united ventrally to a median basibranchial or *copula.* The next two arches consist of proxi-

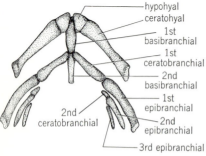

Fig. 10.45 Hyoid apparatus of *Necturus*.

(labels)
hypohyal
ceratohyal
1st basibranchial
1st ceratobranchial
2nd basibranchial
1st epibranchial
2nd epibranchial
3rd epibranchial
2nd ceratobranchial

mal ceratobranchials and distal epibranchials. The ceratobranchials unite medially with the basibranchial, which is usually composed of a single piece. After metamorphosis the last two arches are much diminished or disappear altogether. The hyoid is not always the best developed of the postmandibular visceral arches. In the salamander *Eurycea,* for example, one of the branchials is better developed than the hyoid. The entire apparatus serves to support the tongue and regions posterior to the tongue.

Anura. The broad, flat skull of anuran amphibians (Fig. 10.46) is noteworthy for the fact that the elements composing the jaws are so widely separated from the cranium. The cranium retains much of its original cartilaginous character. Part is replaced by cartilage bone; part is covered with membrane bone. The cartilage bones consist of (1) an unpaired *sphenethmoid,* a ringlike bone which encircles the anterior portion of the cranial cavity; (2) two *exoccipitals,* forming the margins of the foramen magnum, each bearing a condyle for articulation with the atlas; and (3) *pro-otics,* of irregular shape, situated anterolaterad the exoccipitals. The fenestra ovalis, a well-defined opening in the cartilage lying ventral to the pro-otic, is plugged by the minute stapedial plate of the columella. The epiotic and opisthotic regions of the auditory capsule are united with the exoccipitals, of which they are often considered to be a part. The membrane bones of the cranium include paired *frontoparietals* on the dorsal side and an unpaired *parasphenoid,* located ventrally. There is no real evidence that the frontoparietal bones of anurans represent a fusion of frontal and parietal bones. Each develops as a single entity. Three openings, or *fontanelles,* are present in the roof of the chondrocranium, but these are covered over by the frontoparietals. The nasal capsules, although united with

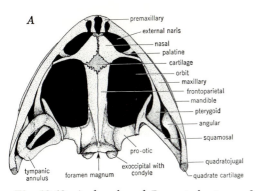

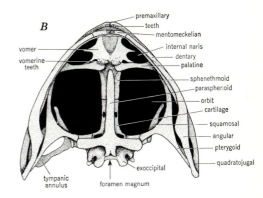

Fig. 10.46 A, dorsal, and B, ventral, views of skull of bullfrog, *Rana catesbeiana*. In A, the right tympanic annulus has been omitted for the sake of clarity; in B, the left one has been omitted.

the cranium, usually remain unossified. Two pairs of membrane bones, however, form in connection with the nasal capsules. They are the *nasals* above and the *vomers* below, the latter bearing *vomerine teeth*. The posterior lateral borders of the vomers bound the medial borders of the internal nares.

The upper jaw consists largely of an outer maxillary arch articulating anteriorly and posteriorly with the rest of the skull but being rather widely separated from it in the middle. The large space thus formed on each side is the *orbit* which accommodates the eyeball. The upper jaw is composed of a series of membrane bones. From anterior to posterior ends these include the premaxillaries, maxillaries, and quadratojugals. The latter are lacking in the Aglossae. Only the first two bones of this series bear teeth. At the posterior angle of the upper jaw is a small, unossified area, sometimes called the *quadrate,* with which the quadratojugal articulates. The quadrate is connected to the cranium by a short bar of cartilage. To this, in turn, is attached a cartilaginous ring, the *tympanic annulus* (Fig. 10.46), which encircles and supports the tympanic membrane. The tympanic annulus appears to be derived from the palatopterygoquadrate bar and is, therefore,

not homologous with the dermal tympanic bone of higher forms, which has a different origin (page 440). The columella, derived from the hyomandibular cartilage of fishes, meets the tympanic membrane distally. Its base articulates with the small stapes (stapedial plate) and also with a slight depression in the pro-otic bone. Three other bones complete the upper jaw. The *pterygoid* is a triradiate bone extending forward from the quadrate and pro-otic to a point near the junction of the maxillary and palatine bones. The *palatine* is a small, transverse membrane bone located ventrally and crossing from sphenethmoid to maxillary. The *squamosal* is a T-shaped bone. Its base articulates at the angle of the jaw with the quadratojugal. It is separated from the pterygoid below by a strand of cartilage. The posterior arm of the T joins the pro-otic, whereas the anterior arm extends toward the middle of the pterygoid.

The lower jaw consists of a very small pair of anteriorly located cartilage bones, the *mentomeckelians,* together with two membrane bones on each side, the *dentary* and *angular,* which form a sheath about Meckel's cartilage. Teeth are lacking on the lower jaw. There is no separate *articular* ossification which in other vertebrates forms the point of articulation with the quadrate. Articulation is by means of an *articular cartilage.*

In anurans the remainder of the visceral skeleton, or hyoid apparatus, is greatly modified. In larval forms it is much like that of urodeles. In adults, however, the hyoid apparatus consists of a broad, cartilaginous *basilingual plate* with anterior and posterior projections, the *cornua.* The basilingual plate represents the united ventral portions of the visceral arches. The anterior cornua are homologous with the ceratohyals of lower forms. They curve anteriorly, then backward and upward to articulate with the auditory capsules near the fenestra ovalis. The posterior cornua, or *thyrohyals,* corresponding to the fourth visceral arches, are ossified.

The chief differences between the skulls of urodele and anuran amphibians are as follows: (1) the latter possess an unpaired sphenethmoid, homologous with the paired orbitosphenoids of urodeles; (2) the frontals and parietals, which in urodeles are separate bones, have fused to form the frontoparietals in anurans; (3) the maxillary bone in anurans is connected posteriorly with the cranium by means of a quadratojugal; (4) small cartilage bones, the mentomeckelians, are present at the tip of the lower jaw in anurans; (5) the skeletal elements supporting the middle ear are present in anurans, in which a middle ear appears for the first time.

REPTILES. The "stem" reptiles, or cotylosaurs, from which all reptiles, extinct and modern, are believed to have evolved, bore such close resemblance to labyrinthodont amphibians that it has not been easy to distinguish the two groups. The general pattern of the roof of the skull is essentially the same in both. Descendants of the cotylosaurs, in general, show a loss or reduction of bones of the skull roof, and numerous changes in structure are apparent.

The chief differences to be noted in the reptilian skull as compared with that of amphibians is the greater degree of ossification and the increased density of bones. Little remains of the embryonic chondrocranium, which has become well ossified. Only the ethmoid region has retained its primitive cartilaginous character. All four parts of the occipital complex are ossified, but all do not necessarily bound the foramen magnum. A single occipital condyle is present, formed from the basioccipital and usually by a contribution from each exoccipital. A cartilaginous or bony interorbital septum, forming a partition between the orbits, is present in many reptiles. In others, in which the brain extends farther anteriorly, the anterior part of the cranial cavity lies in this region and the septum is reduced. The pro-otic region of the auditory capsule is ossified and usually remains separate from epiotics and opisthotics. The parietal bones are paired in some species but frequently are fused. The same is true of the frontals. An *interparietal foramen* is prominent in *Sphenodon* and many lizards but is lacking in other modern reptiles. Membrane bones are more numerous in reptiles than in amphibians. Prefrontal, postfrontal, postorbital, and lacrimal bones are often well developed.

Chelonia. The skulls of turtles and tortoises (Fig. 10.47) are characterized by their solidity. The roof of membrane bones is unusually complete. Teeth are lacking within the group, although vestiges have been described in embryos of the soft-shelled turtle. Otherwise the jaws are covered with a sharp, horny beak. In chelonians the single occipital condyle is formed from contributions of the basioccipital and the two exoccipitals. The latter bound the side and greater part of the floor of the foramen magnum. All four parts of the occipital segment form cartilage bones. Only the basisphenoid region of the re-

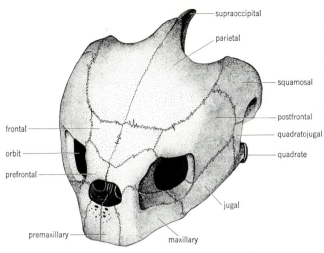

Fig. 10.47 Dorsolateral view of skull of sea turtle.

mainder of the chondrocranium is ossified, however, the presphenoid and orbitosphenoid regions remaining cartilaginous. Membrane bones of the frontal segment consist of frontals, prefrontals, and postfrontals. Prefrontals join ventrally with vomers and palatines. They help to bound the external nares. The postfrontals, which articulate dorsally with the frontals, prefrontals, and parietals, extend posteriorly and ventrally to join squamosals, quadratojugals, and jugals. All three frontal bones form part of the wall of the orbit which in chelonians is completely encircled with bone. In some primitive chelonians a false roof is located above the true cranial roof, the two being separated by a wide gap. The false roof is composed of portions of the postfrontals, parietals, and squamosals.

The three elements of the otic capsule are ossified. Epiotics and opisthotics are united with the supraoccipitals, and the latter with the exoccipitals in addition. The pro-otic, in which the inner ear lies, is situated anterior to the opisthotic and supraoccipital. A large opening between the pro-otic and opisthotic on the side toward the brain is called the *internal auditory meatus*. It serves for the passage of the auditory nerve from inner ear to brain. Between these bones on the lateral surface of the head is the fenestra ovalis, which opens into the cavity of the middle ear. A large opening, the *external auditory meatus*, leads from this to the outside. One end of the bony columella fits into the fenestra ovalis. The other end articulates with a small, cartilaginous *extracolumella*, which in turn meets the tympanic membrane. A ring of bony scales, the *sclerotic bones*, encircles the eye. They are not fused with the skull. The unpaired vomer is a membrane bone in the nasal region. The mesethmoid remains cartilaginous and lies above the vomer. Nasal bones are lacking.

Only one cartilage bone is present in the upper jaw. This is the quadrate, united with exoccipital, opisthotic, and pterygoid. It joins the squamosal dorsally and the quadratojugal anteriorly. Premaxillaries are small. They articulate laterally with the maxillaries. Jugal, or malar, bones are interposed between maxillaries and quadratojugals. A dorsal extension of each jugal meets the postfrontal to complete the posterior boundary of the orbit. A parasphenoid bone is lacking. The small palatine bones are united posteriorly with strong pterygoids and in some cases with the median basisphenoid. The internal nares are bordered posteriorly by the anterior edges of the palatines. The pterygoids may join each other medially, or else the basisphenoid lies between them. Posteriorly they join the quadrate and the occipito-otic complex. A dorsal projection meets the parietal.

The bones of the lower jaw are generally fused into one piece, and the two sides are firmly ankylosed. As many as six pairs of bones enter into the formation of the lower jaw.

The remainder of the visceral skeleton is represented by components of the hyoid and next two visceral arches. The latter may be considerably larger than the part contributed by the hyoid (Fig. 10.48).

Rhynchocephalia. The skull of *Sphenodon,* the only living representative of this ancient group of reptiles, is peculiar in that it possesses two *temporal fossae,* or *fenestrae,* on each side (Fig. 10.49). The *supratemporal fossae* are separated above by the parietals.

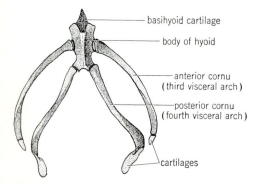

basihyoid cartilage

body of hyoid

anterior cornu
(third visceral arch)

posterior cornu
(fourth visceral arch)

cartilages

Fig. 10.48 Hyoid apparatus of turtle, showing reduction of hyoid arch. The two pairs of posteriorly projecting cornua are remnants of visceral arches 3 and 4.

Each is bounded laterally by an arch of bone referred to as the *supratemporal arcade* and formed by the postfrontal, postorbital, and squamosal bones. The postorbital, interposed between postfrontal and jugal, makes up the greater part of the posterior wall of the orbit. Another large opening, the *lateral temporal fossa,* is situated below the supratemporal arcade. It is bounded by the ascending branch of the jugal, postorbital, squamosal, and quadratojugal bones. The jugal, quadratojugal, and part of the squamosal form the *infratemporal arcade.* A pair of *posttemporal fossae* is located posteriorly, bounded by the parietal, squamosal, exoccipital, and opisthotic bones. The quadrate bone is fused immovably to the skull. Except for certain extinct forms, only *Sphenodon* and the crocodilians, among reptiles, possess supratemporal and infratemporal arcades.

There are four conditions regarding temporal fossae which help in understanding the direction in which evolution has taken place: (1) the primitive *anapsid* condition in which no temporal vacuities exist, as in the ancient "stem" reptiles and turtles; (2) the *diapsid* condition, such as exists in *Sphenodon* and crocodilians in which supratemporal and lateral (inferior) temporal fossae are present with a supratemporal arcade passing laterally between the two; (3) the *synapsid* condition as found in mammals and mammal-like reptiles in which only the lower fossa persists; and (4) the *parapsid* condition, occurring in a few extinct forms, in which only the upper fossa remains. It is believed that the appearance of the temporal fossae, whether diapsid, synapsid, or parapsid, is a deviation from the primitive condition permitting greater freedom of action of the musculature of the jaws. The posttemporal fossae of *Sphenodon* are not of special importance in this connection since they are of fairly common occurrence in reptiles and even in monotremes, among mammals.

Anatomy of the Chordates

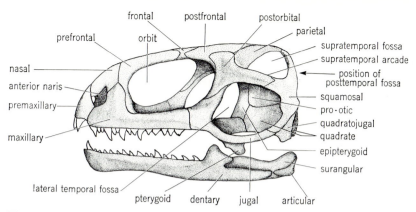

Fig. 10.49 Lateral view of skull of *Sphenodon punctatum.*

Other noteworthy features of the skull of *Sphenodon* include the presence of a rather large interparietal foramen, and teeth on premaxillary, maxillary, palatine, vomer, and dentary bones. The premaxillary teeth of the adult are large and appear much like the chisel-shaped incisors of mammals.

Squamata. An infratemporal arcade is lacking in the Squamata. The quadrate, except in chameleons, forms a movable union with the squamosal, and a quadratojugal is lacking. The nares are separate. In addition to the internal nares, which are anterior in position, large spaces, the *palatal vacuities,* are present in the roof of the mouth. All four parts of the occipital complex surround the foramen magnum.

SAURIA. The presence of an imperfect interorbital septum characterizes the lizard skull. In only one small family of limbless lizards (Amphisbaenidae) does the cranial cavity extend forward between the orbits. Most lizards possess an interparietal foramen. In some forms the parietals and supraoccipitals are modified, in part, to form a pronounced *sagittal crest.* This serves for the attachment of the temporal muscles that are used to close the jaws. The orbit and temporal fossa are not confluent, nor is a complete

infratemporal arcade present. Thus a modified diapsid condition exists. The quadrate bone forms an articulation with the pterygoid. A lacrimal bone, perforated by a lacrimal canal, is situated anteriorly just within the orbit. In many lizards a parasphenoid bone, homologous with that of amphibians, is present in much reduced form. It lies between the pterygoids anterior to the basisphenoid to which it is ankylosed. The secondary palate in lizards is not at all complete. As in all other reptiles a single occipital condyle is present. On the upper jaw, teeth are borne by premaxillaries and maxillaries. Occasionally they are also present on pterygoids and palatines.

The lower jaw is formed of six pairs of membrane bones surrounding Meckel's cartilage. The anterior *dentary* bears all the teeth of the lower jaw. The *coronoid* forms an upward-projecting *coronoid process* immediately behind the last tooth. The coronoid process extends into the temporal fossa. The two halves of the lower jaw are firmly united by a mandibular symphysis.

SERPENTES. The skull of snakes differs somewhat from that of other reptiles in correlation with their habit of swallowing large prey. Many of the facial bones are rather

Skeletal System   457

loosely connected, thus permitting a considerable degree of adjustment during the act of swallowing. The quadrate connects indirectly with the skull by means of the squamosal, which is interposed between quadrate and parietal (Fig. 10.50). The pterygoid articulates posteriorly with the quadrate and anteriorly with the palatine. Jugals and quadratojugals are absent. The maxillary bone is connected indirectly with the pterygoid by the *transpalatine* (*ectopterygoid*). It is in the arrangement of the premaxillary and maxillary bones that the skulls of poisonous and nonpoisonous snakes differ mostly. In venomous snakes the premaxillary is very small and does not bear teeth. The maxillary is also small in venomous forms. Anteriorly it unites by a movable articulation with the lacrimal, which is in turn loosely connected with the frontal.

In such poisonous snakes as vipers and pit vipers, these bones are so arranged that when the mouth is closed the fangs on the maxillary bone lie back against the roof of the mouth. This position is assumed because of traction exerted by the pterygoid on the transpalatine and hence on the maxillary. When the snake strikes, its mouth is opened and the quadrate is thrust forward. This in turn causes the pterygoid and transpalatine to be forced forward. As a result, the position of the maxillary changes so that its toothed surface turns downward and is in the most effective position for striking. In cobras, coral snakes, and sea serpents the fangs are permanently erect.

In snakes the two halves of the lower jaw are united anteriorly by an elastic ligament which is capable of stretching to a marked degree when large prey is swallowed whole.

In the snake skull the cranial cavity extends between the orbits so that the interorbital septum is reduced. Snakes lack both supratemporal and infratemporal arcades as well as interparietal foramina. Nevertheless, this is considered to be a modified diapsid condition. A reduced parasphenoid bone is usually present as an anterior projection of the basisphenoid. It lies at the base of the interorbital septum. The lower jaw consists of several component bones surrounding Meckel's cartilage. In most species the parts of the original hyoid apparatus are lost except for an occasional remnant.

Crocodilia. The crocodilian skull is dense and massive, with a firm union between the

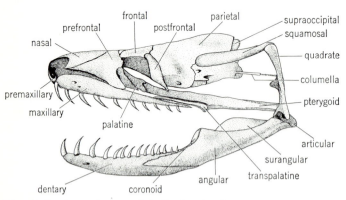

Fig. 10.50 Lateral view of skull of snake (boa constrictor). (*After Schimkewitsch.*)

separate bones. The outer surface of the skull bones is unusual in being rough and pitted. Of the four occipital bones only the basioccipital and exoccipitals bound the foramen magnum. The supraoccipital is small and united with the epiotic. The basioccipital bears a single condyle to which the exoccipitals have contributed portions. The exoccipitals bear foramina for the passage of the last four cranial nerves and for the internal carotid artery. The passageway for the hypoglossal nerve lies just outside the foramen magnum.

The parietal segment is well developed with well-ossified pleurosphenoids uniting on either side with the basisphenoid. The parietal, in the adult, is a single bone representing the fusion of two embryonic components. No interparietal foramen is present.

Basal and lateral portions of the frontal segment are absent or are very imperfectly developed. An interorbital septum of cartilage is prominent. The single frontal bone is large and conspicuous. It also represents a fusion of two parts. The frontal is overlapped anteriorly by the nasals which roof over most of the nasal passages. The prefrontals form part of the inner walls of the orbits. Postfrontals are also represented by small bones lateral and posterior to the frontals. Each postfrontal sends out a postorbital process which unites with a dorsal projection of the jugal to form a *postorbital bar,* separating the orbit from the lateral temporal fossa. Crocodilians, like *Sphenodon,* have two temporal fossae on each side. This, it will be recalled, is the diapsid condition. The supratemporal fossa is located above the lateral temporal fossa, the two being separated by the supratemporal arcade. An infratemporal arcade is continuous anteriorly with the postorbital bar previously mentioned. There is no posttemporal fossa.

No unusual features are to be observed in connection with the sensory capsules. The lacrimals and supraorbitals are large bones associated with the orbit. The lacrimal forms a conspicuous part of the anterior border of the orbit. Thin vomer bones form a vertical plate which separates the two nasal passages.

The upper jaw is very well developed in crocodilians. Premaxillaries, bearing thecodont teeth, form almost the entire boundary of the external nares. These bones are partially separated on the midventral line by small *anterior palatine vacuities.* The maxillary bones are relatively large and bear the remainder of the teeth of the upper jaw. Premaxillaries and maxillaries, meeting their partners in the midline, form the anterior portion of the broad and extensive secondary palate. Palatines are long and narrow, articulating anteriorly with the maxillaries and posteriorly with the pterygoids and transpalatines. Maxillaries, palatines, pterygoids, and transpalatines bound the large *posterior palatine vacuities.* The internal nares are completely surrounded by the pterygoids. Their extremely posterior location is noteworthy. The jugal, or malar, on each side is a large bone articulating anteriorly with lacrimal and maxillary and posteriorly with the quadratojugal to form the infratemporal arcade or zygomatic arch. The jugal sends a postorbital process dorsally to meet a downward projection of the postfrontal bone, the two processes together forming the postorbital bar, previously referred to. The jugal is also united medially with the transpalatine. The quadratojugal is small and joins the quadrate posteriorly. The quadrate, in turn, articulates with exoccipital and squamosal. The latter is the chief component of the supratemporal arcade.

The two halves of the lower jaw are firmly

united at the symphysis. Each is composed of six separate bones. The dentary, which unites with its mate to form the mandibular symphysis, is the largest and bears the teeth of the lower jaw. A large opening, the *external mandibular foramen,* perforates the lateral side of the mandible near its posterior end. It is bounded by the dentary, surangular, and angular. On the medial side is a smaller *internal mandibular foramen,* bounded by splenial and angular bones. These two foramina serve for the passage of blood vessels and branches of the inferior dental nerve. A small coronoid bone connects angular and splenial above the internal mandibular foramen.

The hyoid apparatus of crocodilians is represented by a cartilaginous *basilingual plate* and a pair of *cornua.* The latter articulate with the basilingual plate at about the center of its outer border. The other end passes backward and upward.

BIRDS. It is generally agreed that birds arose in evolution from the branch of archosaurs, or "ruling reptiles," which included dinosaurs with bipedal locomotion, rather than from the flying pterosaurs as one might at first suspect. The scales covering the body have been modified into feathers, used to resist the air in flight. The skull of birds (Figs. 10.51, 10.52, and 10.53) has not deviated far from the reptilian type. It resembles that of lizards in a number of ways. Some differences in structure of the palate serve to distinguish members of the superorders Paleognathae and Neognathae. Otherwise, there is actually, within the entire class, little variation from a common pattern. A single occipital condyle is present which, instead of being located at the posterior end of the skull, lies somewhat forward along the base. The skull articulates with the vertebral column at almost a right angle, thus accounting for the more anterior

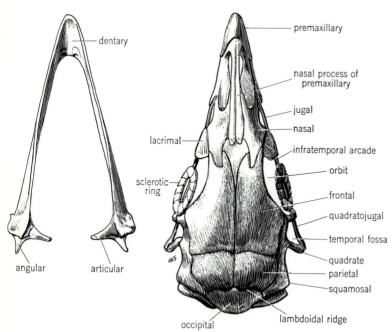

Fig. 10.51 Dorsal view of skull of chicken. (*Drawn by G. Schwenk.*)

Anatomy of the Chordates

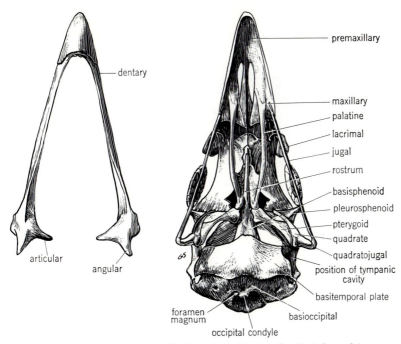

Labels on figure:
- dentary
- premaxillary
- maxillary
- palatine
- lacrimal
- jugal
- rostrum
- basisphenoid
- pleurosphenoid
- pterygoid
- quadrate
- quadratojugal
- position of tympanic cavity
- basitemporal plate
- basioccipital
- articular
- angular
- GS
- foramen magnum
- occipital condyle

Fig. 10.52 Ventral view of skull of chicken. (*Drawn by G. Schwenk.*)

position of the condyle. With the greater development of the brain in birds, the cranial cavity has increased in size, mostly by lateral expansion. The orbit is large to accommodate the relatively massive eye. It is continuous with the single temporal fossa. A complete supratemporal arcade is lacking, but the infratemporal arcade, though slender, is completely formed. The condition in birds is essentially diapsid, although modified. The orbits are situated somewhat anterior to the cranium. Each is bounded dorsally by the frontal bone, to which the lacrimal is attached, and posteriorly by the pleurosphenoid. A thin interorbital septum is present. It is better developed, however, than in most reptiles. The chief components of the interorbital septum are the anterior mesethmoid and the posterior orbitosphenoid. A large optic foramen lies on each side of the interorbital septum in addition to open-

ings for the passage of oculomotor and trochlear nerves. An olfactory foramen is located dorsal and anterior to each optic foramen. There is no interparietal foramen. A ring of thin membrane bones, the *scleral ossicles,* or *sclerotic bones,* embedded in the sclerotic coat of the eye, provides protection to the eye. They are sometimes included in the category of heterotopic bones.

In correlation with their adaptation for flight, birds have skulls unusually light in weight. The cranial bones are fused together to a high degree (Fig. 10.53). All four components of the occipital complex surround the foramen magnum. Where the supraoccipital joins the parietal, a prominent crest, the *lambdoidal ridge,* is present. Epiotics and opisthotics are fused to the occipital bones. Most conspicuous is the great development of the premaxillary bones, which together with the maxillaries form

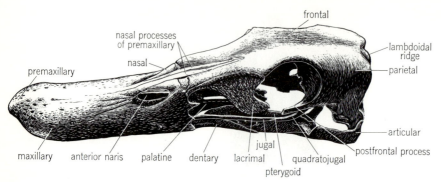

Fig. 10.53 Dorsolateral view of skull of duck. (*Drawn by G. Schwenk.*)

the upper portion of the beak, or bill. Although *Archaeopteryx, Hesperornis,* and *Ichthyornis* possessed true teeth, teeth are, of course, lacking in modern birds, the beak being covered with a horny sheath derived from the epidermis. Premaxillaries, maxillaries, and nasals are completely fused. All three bones bound the external nares which, except in the kiwi, are located near the proximal end of the beak.

Contrary to the condition in crocodilians, there is no complete bony secondary palate. The inner arch, or palatopterygoquadrate bar, is slender and not connected, in most cases, with its partner across the midline. The quadrate bone is well developed and moves freely. The outer arch is also slender. It consists of a process of the maxillary together with the jugal and quadratojugal bones. The latter joins the quadrate posteriorly. Thus inner and outer arches are united at the quadrate.

An anterior prolongation of the basisphenoid, the *rostrum*, forms the base of the interorbital septum. The rostrum is believed to be homologous with the anterior part of the parasphenoid of lower forms. The posterior portion of the parasphenoid is represented by the *basitemporal plate* which lies beneath the basioccipital and basisphenoid. The pterygoids in some birds

articulate with the rostral portion of the basisphenoid.

Since the quadrate is movable and the palatopterygoid can slide along the rostrum of the basisphenoid, the upper jaw can be raised or lowered within limits. The infratemporal arcade may aid in bringing about such movements. Movement of the upper jaw is best seen in parrots. In these birds the upper beak is not firmly united with the cranium but forms a flexible attachment, so that a movable joint is present. The nasals, and the loose connections of the premaxillary bones, attach the beak to the cranium.

The lower jaw consists of two rather flattened halves, united anteriorly and articulating with the quadrate posteriorly. Actually each half is composed of five bones fused together. Only the articular is a cartilage bone. The anterior dentary makes up the greater part of the lower jaw. A foramen is frequently present between the dentary and the more posterior bones. The lower jaw is covered with a hard, horny scale, which makes up the lower portion of the beak, or bill, used in lieu of teeth.

In birds the hyoid apparatus consists of a median portion composed of three bones placed end to end, and one or two pairs of cornua, or horns. The three median bones are called, respectively, the *entoglossal, basihyal,* and *urohyal.* The anterior entoglossal, which extends into the tongue, represents the fused lower ends of the ceratohyals. The median basihyal corresponds to the basihyal of lower forms. The posterior urohyal is probably the basibranchial of the third visceral arch. The anterior cornua, which project from the entoglossal, are the free ends of the ceratohyals. The long and posteriorly directed posterior cornua, which articulate at the junction of basihyal and urohyal, represent the ceratobranchials and epibranchials of the third visceral arch. The posterior cornua are extremely long in woodpeckers in correlation with the feeding habits of these birds in which the tongue, supported by the hyoid apparatus, can be thrust out for some distance.

MAMMALS. Mammals, like birds, have sprung from ancestral reptilian stock but their history indicates a very early divergence from the main reptilian line. They have little in common with reptiles living today, nor do they show any close relationships to the "ruling reptiles" which at one time dominated the earth. The cynodont, or mammal-like, reptiles are generally considered to be in the direct line of ascent of mammals but even they were rather far removed in time from the oldest-known reptiles, the primitive pelycosaurs. The skulls of the latter differed little from those of the most ancient amphibians but already showed a trend toward mammalian evolution. Unlike most reptiles, which have a single occipital condyle, the cynodonts possessed two, as do amphibians and mammals.

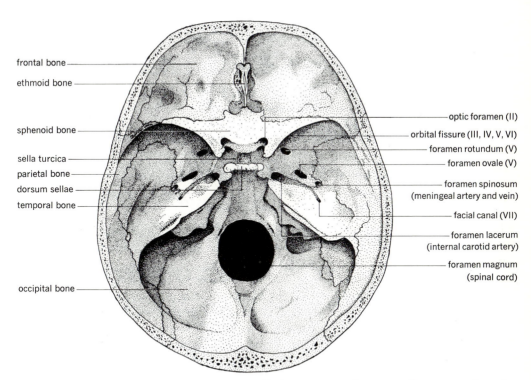

Fig. 10.54 Dorsal view of interior of human skull, showing ventral region with numerous foramina.

Although much variation exists in the skulls of different groups of mammals, there are certain features which they share in common and which serve to distinguish them from lower forms.

Most noteworthy is the increased capacity of the cranium, which has expanded in dorsal and lateral directions in correlation with the much greater size of the mammalian brain. A firm union occurs between all the bones of the skull with the exception of mandible, hyoid, and auditory ossicles. All four components of the occipital complex are present, surrounding the foramen magnum. Two occipital condyles, formed mostly by the exoccipitals, articulate with the atlas vertebra. In higher apes and man the occipital condyles are decidedly ventral in position. Articulation of mandible and upper jaw is at the *glenoid,* or *mandibular, fossa* of the squamous region of the temporal rather than at the quadrate. This is the craniostylic or amphicraniostylic method of jaw suspension. The squamosal usually forms the squamous region of the temporal bone. The quadrate is reduced to the incus. A bony tympanic ring, apparently derived from the angular bone of the reptilian lower jaw, may form part of the temporal bone. The developing skull of the opossum embryo provides evidence for this. The tympanic bone is, therefore, not homologous with the tympanic annulus of anuran amphibians. Otic bones are fused to form the *petrosal,* which encloses the inner ear. The petrosal may fuse with squamosal and tympanic and thus becomes the *petrous region* of the temporal bone. The opisthotic is represented by the spongy *mastoid region* of the temporal.

Teeth of the upper jaw are borne only by premaxillary and maxillary bones. The premaxillaries frequently fuse together. The thecodont method of tooth attachment is the rule. Except in the toothed whales, which have homodont dentition, the teeth of mammals are heterodont.

The number of bones in the mammalian skull (Figs. 10.55, 10.56, and 10.57) is less than in most lower forms. There are usually about 35 of them. This is the result of loss of certain bones and fusion of others. Among the elements that have disappeared in mammals are the prefrontals, postfrontals, and postorbitals. The temporal bone, just referred to, is an example of the result of fusion. Another common area of fusion is in the cartilage bones of the sphenoid complex (Fig. 10.54). The basal portion of the frontal segment frequently unites with the orbitosphenoids to form a presphenoid bone. Similarly the basisphenoid and alisphenoids may fuse. In mammals there are no pleurosphenoids. Their place is taken by the epipterygoids, cartilage bones derived from the palatopterygoquadrate bar. These are then known as the alisphenoids. In some cases, as in man, all six elements of the sphenoid complex unite to form a single *sphenoid* bone. An *orbital fissure (anterior lacerate foramen)* in the sphenoid bone provides for the passage of the eye muscle nerves (III, IV, and VI) (see pages 436 and 516) and the anterior part of the trigeminal nerve (V). The maxillary and mandibular branches of the trigeminal pass through the *foramen rotundum* and *foramen ovale.* A small *foramen lacerum (middle lacerate foramen),* which may fuse with the foramen ovale at the posterior edge of the alisphenoid or the petrous region of the temporal bone, provides for the passage of the internal carotid artery into the cranial cavity. In most mammals a single occipital bone containing the foramen magnum is present. This also represents a fusion, although the four separate elements of which it is composed can easily be identified during development. Fusion of the membrane bones of the calvarium is common, particularly in

Anatomy of the Chordates

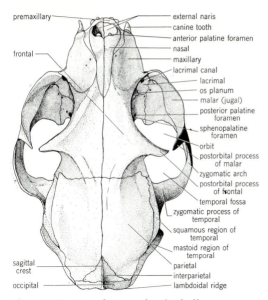

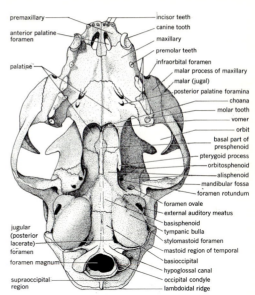

premaxillary	external naris
	canine tooth
	anterior palatine foramen
frontal	nasal
	maxillary
	lacrimal canal
	lacrimal
	os planum
	malar (jugal)
	posterior palatine foramen
	sphenopalatine foramen
	orbit
	postorbital process of malar
	zygomatic arch
	postorbital process of frontal
	temporal fossa
	zygomatic process of temporal
	squamous region of temporal
	mastoid region of temporal
sagittal crest	parietal
occipital	interparietal
	lambdoidal ridge

Fig. 10.55 Dorsal view of cat's skull.

premaxillary	incisor teeth
anterior palatine foramen	canine tooth
	maxillary
	premolar teeth
palatine	infraorbital foramen
	malar process of maxillary
	malar (jugal)
	posterior palatine foramina
	choana
	molar tooth
	vomer
	orbit
	basal part of presphenoid
	pterygoid process
	orbitosphenoid
	alisphenoid
	mandibular fossa
	foramen rotundum
	foramen ovale
	external auditory meatus
jugular (posterior lacerate) foramen	basisphenoid
	tympanic bulla
	stylomastoid foramen
foramen magnum	mastoid region of temporal
	basioccipital
supraoccipital region	hypoglossal canal
	occipital condyle
	lambdoidal ridge

Fig. 10.56 Ventral view of cat's skull.

the case of the frontals and parietals. A membrane bone, the *interparietal,* may develop between the parietals and supraoccipital. It may later fuse with the supraoccipital or with the parietals. The facial bones in mammals are more firmly united with the cranium than in other vertebrates. This is true both of the original maxillary arch and palatopterygoquadrate bar components. A supratemporal arcade is lacking, but the infratemporal arcade, or zygomatic arch, is invariably present. As a result, mammals possess but a single temporal fossa, an example of the synapsid condition. The zygomatic arch is composed of a posterior projection of the jugal (malar) which joins an anteriorly projecting process of the squamosal. Temporal fossa and orbit are often confluent but in many cases are separated by a bar formed by a downward-projecting postorbital process of the frontal which joins a dorsal projection of the jugal. A lacrimal bone is always present in the anterior wall of the orbit. Other bones helping to form the orbit are the maxillary, orbito-

sphenoid, palatine, and sometimes the ethmoid.

A hard or bony secondary palate is typically found in mammals. It is composed of premaxillary, maxillary, and palatine bones. In whales and certain edentates the pterygoids form a portion of the hard palate, but these bones are much reduced in most mammals, so that the choanae are usually bordered by the palatines alone.

The nasal cavities of mammals are large in correlation with the highly developed olfactory sense. Premaxillaries, maxillaries, and nasals surround the greater part of the nasal cavities. The ethmoid complex is well differentiated in most species. The mesethmoid forms a vertical cartilaginous plate which separates right and left nasal passages. The posterior portion becomes the bony *lamina perpendicularis*. The ethmoid also contributes the *cribriform plate*, separating nasal and cranial cavities. The cribriform plate bears numerous small *olfactory foramina* for the passage of fibers of the olfactory nerve. Ethmoturbinates are scroll-like outgrowths of the ectethmoids. Other turbinates are contributed by nasal and maxillary bones. The vomer of mammals is an unpaired median element which lies ventral to the median nasal septum. In mammalian embryos it is formed by a fusion of two ossification centers. It passes downward from the basal portion of the cranium to join the hard palate. According to some authorities, the mammalian vomer is homologous with the parasphenoid bone of lower forms rather than with their *paired* vomers, the latter often being referred to as *prevomer* bones.

Many interesting variations from the above general description of the mammalian skull are to be observed. Among these is the presence of hollow, bony projections of the frontals in many members of the order Artiodactyla. They are covered with horn derived from the epidermis. The antlers of deer, which develop under the organizing influence of the integument in response to stimulation by certain hormones, when fully developed are naked projections of the frontal bones.

In some mammals a prominent sagittal crest extends forward from the posterior part of the skull in the middorsal line. A lambdoidal ridge may pass transversely across the skull along the line of junction of parietals and supraoccipital. Outgrowths of this sort

provide surface areas for muscle attachment.

Considerable variation occurs in the relative sizes of frontals and parietals, the latter being unusually large in primates (Fig. 10.57). In some of the whales a projection of supraoccipital and interparietal extends forward to the frontals, thus separating the parietals in the midline. Of particular interest is the fact that in some of the large whalebone whales the skull may comprise as much as one-third the length of the body. Premaxillaries and maxillaries are excessively elongated. This accounts for location of the external nares, far back on the dorsal surface.

Large air spaces, or sinuses, in communication with the respiratory passages, are present to varying degress in the skulls of mammals. The most common are the frontal, ethmoid, sphenoidal, and maxillary sinuses, the last being often referred to as the *antra of Highmore*.

An ossified derivative of the dura mater, the *tentorium* (page 632), in such forms as the dog and cat becomes secondarily attached to the skull on the inner surface of the parietals. It serves to separate partially the cerebral and cerebellar fossae.

The lower jaw, or mandible, of mammals (Figs. 10.57 and 10.58) differs from that of other vertebrates in its articulation with the squamosal rather than with the quadrate

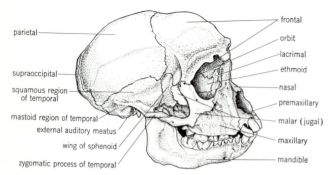

Fig. 10.57 Lateral view of skull of monkey, *Macacus rhesus.*

parietal

supraoccipital

squamous region of temporal

mastoid region of temporal

external auditory meatus

wing of sphenoid

zygomatic process of temporal

frontal

orbit

lacrimal

ethmoid

nasal

premaxillary

malar (jugal)

maxillary

mandible

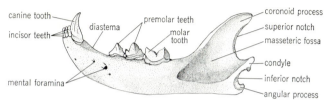

Fig. 10.58 Lateral view of mandible of cat.

(page 441). This variation is linked with the developmental history of the base of Meckel's cartilage and of the quadrate. It will be recalled that these structures become independently ossified and become the *malleus* and *incus* bones of the middle ear, respectively. The third auditory ossicle, the *stapes*, is a derivative of the hyomandibular portion of the hyoid arch. The point of junction of malleus and incus thus actually represents the point of articulation of lower and upper jaws in forms below mammals in the vertebrate scale. The remainder of Meckel's cartilage becomes invested by the dentary, a membrane bone. Other membrane bones such as are present in lower forms have disappeared in mammals. The articulating surface of the mandible is the *condyle*. A *coronoid process* from the posterior portion, of various degrees of development in different species, extends into the temporal fossa. A shallow depression on the lateral surface of the mandible serves as an area for insertion of the masticating muscles. An *angular process* at the posterior ventral angle of the mandible in many species forms a prominent projection, but in others it is reduced. The two halves of the mandible are united in front at the symphysis. In primates, bats, and members of the order Perissodactyla an inseparable fusion occurs between the two halves of the mandible.

The mammalian hyoid apparatus shows much variation. Typically it consists of a body, or *basihyal,* and two pairs of *cornua,* or horns. The anterior cornua, or *styloid processes,* are attached to the otic region of the skull. In some forms (Fig. 10.59) the hyoid consists of a chain of bones extending from the basihyal to the otic region and known, respectively, as the *ceratohyal, epihyal, stylohyal,* and *tympanohyal.* The posterior cornua are sometimes smaller, each consisting of a single *thyrohyal* which forms a connection between the basihyal and the thyroid cartilage of the larynx. In other mammals the separate bones may lose their identity and are represented only by fibrous bands. The hyoid apparatus serves to support the tongue and furnishes an area for muscle attachment.

The basal portion of the hyoid apparatus

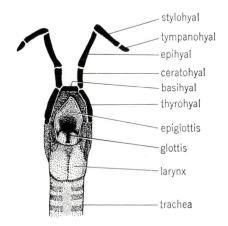

Fig. 10.59 Hyoid bones of cat, shown in relation to larynx and trachea, dorsal view.

together with the anterior cornua are derivatives of the original hyoid arch, which has also contributed the stapes of the middle ear. The posterior cornua represent remnants of the third visceral arch. The remaining visceral arches contribute to the laryngeal cartilages.

Despite the fact that mammalian skulls are, in general, built upon the same pattern, there are some very evident differences which distinguish monotremes, marsupials, and the placental mammals. Monotremes more closely resemble the ancient reptilian stock, being the only mammals which have retained the posttemporal fossa and paired vomer (prevomer) bones. Pterygoids and palatines, unlike those of higher mammals, contribute to the cranium proper. The cranial cavity of marsupials is relatively very small in comparison with that of the placental mammals; openings, or fenestrae, are present in the palate; the lower jaw, or mandible, has a peculiar angle.

Ribs

Ribs consist of a series of cartilaginous or bony elongated structures attached at their proximal ends to vertebrae. In the primitive condition there is a pair of ribs for each vertebra. In some vertebrates the ribs consist of little more than small cartilaginous tips on the transverse processes of the vertebrae. They may be fused permanently with the transverse processes (pleurapophyses), or else a joint may appear between the two. In higher forms they may be stout, strongly arched structures surrounding the thoracic cavity and uniting ventrally with a median bony structure, the *sternum,* or *breastbone.* The ribs in these cases protect such vital organs as the heart and lungs by enclosing them with a basketlike framework.

Fish ribs. Ribs first appear in fishes and are of two kinds, dorsal and ventral. Many fishes possess either one type or the other, but in *Polypterus* and many teleosts, both types are present.

DORSAL RIBS. The upper, *intermuscular,* or *dorsal,* ribs (Fig. 10.60) extend out as costal processes from the transverse processes of vertebrae into the lateral skeletogenous septum which separates epaxial and hypaxial muscles (page 508). They occur metamerically at points where the myocommata, or septa between adjacent myotomes, cut across the lateral septum. Some authorities are of the opinion that true dorsal ribs are present only in fishes. Most, however, believe that the ribs of tetrapods are homologous with the dorsal ribs of fishes.

VENTRAL RIBS. The ventral, or *pleural, ribs* of fishes are attached to the ventrolateral angles of the centrum. They lie beneath the muscles of the body wall just outside the peritoneum and are situated along the lines where the myocommata pass toward the coelom. There is general agreement that the ventral ribs represent the two halves of hemal arches of caudal vertebrae which have split ventrally and spread apart. However, this conception may be erroneous, for the caudal vertebrae of some fishes are reported to possess both ventral ribs and hemal arches. Additional intermediate bones having a rib-like appearance are frequently found embedded in the myocommata of fishes.

Tetrapod ribs. There is some dispute among comparative anatomists as to whether the ribs of tetrapods are homologous with the dorsal or with the ventral ribs of fishes. Most believe that they are the homologues of dorsal ribs. There is, however, some evidence that the ventral ribs, by a shifting of position during development in relation to muscles, have come to occupy a different position in the adult from that in the embryo. Some investigators are of the opinion that in tetrapods there has been a fusion of dorsal and

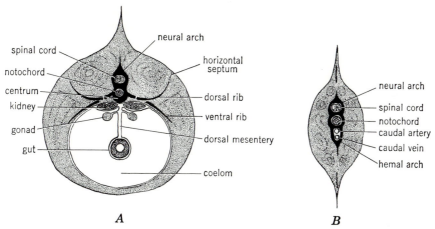

Fig. 10.60 A, diagrammatic cross section of typical vertebrate, showing position of dorsal and ventral ribs; B, section through tail of fish, showing hemal arch. (*After Parker and Haswell, "Textbook of Zoology," Vol.* II, *used with permission of The Macmillan Company.*)

ventral ribs, thus accounting for the two-headed, or bicipital, condition typical of tetrapod ribs. This view is not generally accepted. Regardless of homologies, each tetrapod vertebra typically bears a single pair of ribs. Except in the thoracic region, the ribs are usually short and are immovably fused to the vertebrae. In most amniotes the ribs in the thoracic region are considerably modified, separating from the vertebrae with which they then form movable articulations. These ribs are elongated and unite ventrally with the sternum. The ends attached to the sternum are usually cartilaginous and are spoken of as *costal cartilages.*

Tetrapod ribs usually form articulations both with the centra and neural arches of vertebrae. In these *bicipital ribs* (Fig. 10.61) the upper head of the rib, or *tuberculum,* articulates with a dorsal process, the *diapophysis,* coming from the neural arch. The lower head, or *capitulum,* joins a projection of the centrum, the *parapophysis.* The bicipital condition is considered to be primitive in tetrapods, since it exists in some of the oldest

fossil amphibians and reptiles. It undoubtedly aids in terrestrial locomotion by strengthening the trunk.

In higher forms the diapophysis may be represented by a small *tubercular facet* on the ventral side of the transverse process; the parapophysis in such cases may consist of nothing more than a small facet on the centrum for the reception of the capitulum. The capitulum may articulate at the junction of two centra, in which case each centrum may bear a half facet, or *demifacet,* at either end. Sometimes tetrapod ribs have but a

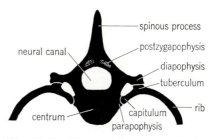

Fig. 10.61 Diagram showing method of articulation of tetrapod rib with vertebra.

single head. This may represent either the tuberculum or capitulum or a fusion of the two.

In the cervical region, where ribs are fused to the vertebrae, the space formed between the two articulations of the rib with the vertebrae is known as the *foramen transversarium*, or *vertebrarterial canal* (page 429). It serves as a passageway for the vertebral artery and vein and a sympathetic nerve plexus. In the lumbar, sacral, and caudal regions the ribs are fused with the vertebrae (pleurapophyses), and there is little or no evidence of the primitive bicipital condition.

Ribs in different classes of vertebrates.
FISHES. In elasmobranchs, small ribs, attached to the ventrolateral angles of the centrum (basapophyses), extend laterally into the skeletogenous septum. It would at first sight appear as though these are dorsal ribs, but it is generally agreed that they more probably represent ventral ribs which have moved secondarily into this position. The caudal vertebrae bear hemal arches with which the ventral ribs are homologous.

In most of the higher fishes only ventral ribs are present. However, the chondrostean *Polypterus*, previously mentioned, as well as the salmon and a number of other teleosts, possess both dorsal and ventral ribs.

AMPHIBIANS. In some of the fossil labyrinthodont amphibians the ribs were strong structures which extended ventrally around the body for some distance. In modern amphibians they have been reduced and are small and poorly developed.

In most urodele amphibians all the vertebrae from the second to the caudal bear ribs which are typically bicipital (Figs. 10.21 and 10.61). The dorsal tubercular head forms a union with the diapophysis, or base of the neural arch; the lower, or capitular, head articulates with the parapophysis, a projec-

tion of the centrum. These points of attachment actually represent connections with the primitive basidorsal cartilage and the hypocentrum, respectively. In a few primitive salamanders the pointed ribs may protrude through the skin as small, naked projections of a protective nature. The strong sacral ribs are attached to the pelvic girdle. The caudal vertebrae lack ribs.

The ribs in anurans are lacking entirely except on the sacral vertebrae, or else they take the form of minute cartilages attached to the transverse processes.

Ribs in caecilians are somewhat better developed than in other amphibians. They occur on all except the first and a few of the more posterior vertebrae and are attached by means of a double or single head, as the case may be. Ribs of amphibians never form connections with the sternum.

Some of the extinct fossil amphibians possessed ventral, dermal "ribs," or *gastralia*. These consisted of V-shaped, riblike, dermal bones located on the ventral abdominal region and extending in a posterodorsal direction on either side. Gastralia should not be confused with true ribs, which are endoskeletal structures.

REPTILES. Rib development in reptiles has deviated far from the condition observed in amphibians. Ribs may be borne by almost all the vertebrae in trunk and tail regions. Attachment of ribs to a sternum is first observed in the class Reptilia.

Turtles lack ribs in the cervical region. The 10 trunk vertebrae bear ribs which are flat and broad and, like the vertebrae in this region, are immovably fused to the underside of the carapace (Fig. 10.62). The carapace is thus formed partly of dermal skeletal bony plates and partly of endoskeletal structures derived from vertebrae and ribs. It will be recalled that epidermal scales overlie these skeletal elements (page 131). Each rib has

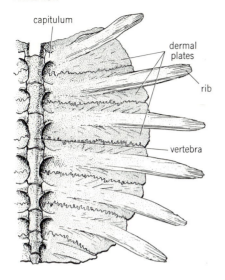

ANTERIOR

capitulum

dermal plates

rib

vertebra

POSTERIOR

Fig. 10.62 Underside of portion of carapace of young turtle, showing fusion of ribs and vertebrae to dermal bones. The marginal dermal plates in this specimen had not as yet become ossified.

but a single head, the capitulum, for articulation with the vertebral column. The point of articulation is usually at the boundary of adjacent vertebral centra. The two sacral and first caudal vertebrae are also united to the carapace. The ribs of the sacral vertebrae form a union with the pelvic girdle. Those of the anterior caudal vertebrae consist of small projections, their union with the vertebrae being indicated by the presence of distinct sutures.

The ribs of *Sphenodon* are extensively developed, being present even in the caudal region. A few of the anterior thoracic ribs join a typical sternum. Most of the remaining ribs, however, join a median ventral *parasternum* which extends from sternum to pubis. The parasternum is an endoskeletal structure derived from cartilages which

develop in the ventral portions of the myocommata. They should not be confused with gastralia, or dermal "ribs," previously mentioned, which are dermal derivatives and which in *Sphenodon* are present in the same region. Each rib is typically composed of three sections, an upper, ossified, *vertebral section;* a ventral cartilaginous *sternal,* or *costal, section;* and an intermediate section also composed of cartilage. In *Sphenodon* each rib bears a flattened, curved *uncinate process* which projects posteriorly from the vertebral section, overlapping the next rib and thus providing additional strength to the thoracic body wall.

Small cervical ribs occur in lizards except on the atlas and axis. In the geckos, however, even the first two vertebrae bear ribs. In most lizards a few of the anterior thoracic ribs curve around the body to meet the sternum. Each of these ribs is divided into two or three sections, but only the vertebral section is bony in most cases. Articulation with the vertebra is by capitulum alone, the tuberculum being much reduced. Uncinate processes are lacking. The ribs of the flying dragon, *Draco volans,* are of particular interest, since the posterior ribs are greatly extended, supporting an extensive fold of skin on each side of the body. This provides a wide surface which, when extended, enables the animal to soar. When at rest, the "wings" are folded against the sides of the body.

In snakes all trunk vertebrae except the atlas and axis bear ribs. They articulate loosely by means of a single head, the capitulum. Since no sternum is present in these reptiles, the ribs terminate freely. The lower ends, however, have muscular connections with ventral scales which are used extensively in locomotion (Fig. 11.21).

In crocodilians the ribs are typically bicipital. All five regions of the vertebral column are rib-bearing, although the ribs are much reduced in the cervical, lumbar, and caudal regions. Even the atlas and axis bear ribs in these reptiles. The cervical ribs increase in length as they progress in an anteroposterior direction. The two heads of the cervical ribs unite with the vertebrae so as to surround an opening, the foramen transversarium. Eight or nine thoracic ribs connect with the sternum. These are composed of vertebral, intermediate, and sternal sections, of which only the vertebral section is completely ossified. Uncinate processes are present on the vertebral sections. The last two or three thoracic ribs are *floating ribs* with no sternal connection. Only the vertebral section is present in these. The two pairs of sacral ribs are strong projections which articulate with the ilium of the pelvic girdle. Some of the anterior caudal vertebrae bear ribs which are fused to the transverse processes, of which they appear to be parts. Dermal gastralia, similar to those of *Sphenodon,* are well developed in the ventral abdominal region of crocodilians.

BIRDS. In *Archaeopteryx* the ribs resembled those of lizards more closely than those of modern birds. They were slender, articulated by means of a single head, and lacked uncinate processes. *Archaeopteryx* also possessed gastralia, not encountered in birds of today. In modern birds some of the posterior cervical vertebrae bear movable ribs. Strong, flattened, bicipital ribs connect most of the thoracic vertebrae with the sternum. Each is typically composed of vertebral and sternal sections, both ossified. A prominent uncinate process from each vertebral section (Fig. 10.22) overlaps the rib behind, furnishing a place for muscle attachment and increasing the rigidity of the skeleton, an important factor in flight. In the posterior thoracic, lumbar, sacral, and anterior caudal regions the ribs are fused to the large synsacrum.

MAMMALS. Mammalian ribs are usually bicipital. In the cervical region the transverse processes are composed in part of the remains of ribs which have fused with the vertebrae. The foramina transversaria, present only in cervical vertebrae, represent the original space between the points of attachment. Thoracic ribs are well developed in mammals. The tubercular head articulates with the tubercular facet (diapophysis) on the ventral side of the transverse process; the capitulum typically articulates at the point of junction of two adjacent vertebrae, the centrum of each bearing a costal demifacet. In some posterior ribs, only the capitular head remains. Each rib consists of two sections: an upper, bony vertebral section and a lower, usually cartilaginous sternal section. The latter joins the sternum directly or indirectly and is commonly referred to as the *costal cartilage.* The ribs of monotremes are composed of three sections as in primitive reptiles. Mammalian ribs do not bear uncinate processes.

The part of the rib between the two heads is called the *neck;* the remainder is the *shaft* (Fig. 10.63). The area where the shaft curves most markedly is the *angle* of the rib. Those ribs which make a direct connection with the sternum are called *true ribs.* *False ribs* are located posteriorly, and their costal cartilages either unite with the costal cartilages of the last true rib, or they terminate freely. The latter are termed *floating ribs.* Mammals lack ribs in the lumbar and caudal regions, although the transverse processes of the lumbar vertebrae are, in all probability, pleurapophyses.

The number of ribs in mammals shows considerable range, from 9 pairs in certain whales to 24 in the sloth. True ribs, however, range in number only from 3 to 10 pairs.

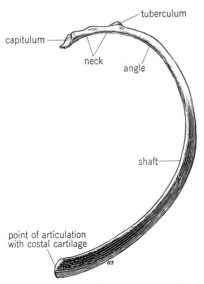

Labels on figure:
capitulum
tuberculum
neck
angle
shaft
point of articulation with costal cartilage

Fig. 10.63 Vertebral section of human rib.

Sternum

The sternum, or breastbone, found only in tetrapods, is composed of ventral skeletal elements closely related to the pectoral girdle of the appendicular skeleton and, except in amphibians, to the thoracic ribs. Not all tetrapods have a sternum. It is lacking in some of the elongated salamanders such as *Proteus* and the "Congo snake," *Amphiuma.* Caecilians, snakes, and limbless lizards lack a sternum, as do turtles. In the latter the plastron serves as a strengthening support in the absence of a sternum. The sternum seems to be a skeletal structure which provides greater support and a place for muscle attachment in terrestrial vertebrates, thus aiding in locomotion on land.

The origin of the sternum in evolution is obscure. Several theories have been advanced to explain its origin, none of which is wholly acceptable at the present time. According to one conception the earliest in-dication of a sternum is to be found in the perennibranchiate salamander, *Necturus,* in which it is referred to as the *archisternum.* In this amphibian, cartilages occur in a few of the myocommata in the ventral thoracic region. The largest and most prominent of these irregular skeletal elements is found at precisely the same region in which, in higher salamanders, a sternal plate overlaps the coracoids of the pectoral girdle. Such a theory of the origin of the sternum is plausible, relating it to the *parasternum* of such forms as *Sphenodon,* lizards, and crocodilians (page 471). The parasternum is a posterior continuation of the sternum. Fusion of some of the anterior archisternal or parasternal elements may thus possibly account for the origin of the true sternum. If this theory is correct, the connection of ribs to the sternum must be secondary.

According to another theory the sternum arises as a fusion of the ventral ends of some of the anterior ribs. This type of sternum, found in amniotes, has been spoken of as the *neosternum.* Objection to this theory is based upon the fact that although a sternum is present in most amphibians, in no case are ribs associated with it. It is possible, however, that the ribs have become shortened secondarily in amphibians in the course of evolution and have lost their sternal connections. There is no paleontological evidence for this.

Another theory seeks to derive the sternum from the median portion of the pectoral girdle of fishes. This concept has least support of all from a paleontological or an embryological point of view.

Some authorities are of the opinion that the archisternum of amphibians and the neosternum of amniotes, with its rib connections, are in no way related and are of independent origin. Since, however, there is considerable evidence from embryology that the ribs of amniotes connect with the sternum secondarily, there is no particular reason for assuming a lack of homology between the two types of sterna.

Recent evidence from experiments in which marking techniques were used (T. Seno, 1961, *Anat. Anz.,* **110**:97) shows that the sternum of birds and mammals is a derivative of the lateral plate mesoderm, whereas ribs, along with the vertebrae, are clearly somite derivatives. This indicates that the sternum should properly be classified as a portion of the appendicular skeleton rather than of the axial skeleton.

Amphibians. In urodele amphibians the sternum appears for the first time and in its most primitive form. That of *Necturus* has already been referred to. In no case, however, is the sternum very well developed. In most salamanders it consists of little more than a small, median, triangular plate, lying behind the posterior, medial portions of the

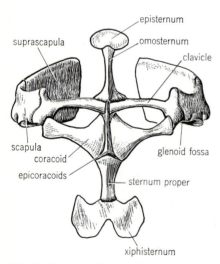

Fig. 10.64 Ventral view of sternum and pectoral girdle of bullfrog, *Rana catesbeiana.*

coracoids of the pectoral girdle (Fig. 10.77).

In anurans it is better developed. In the common frog, for example, the anterior clavicles and posterior coracoids of the pectoral girdle are separated from their partners of the opposite side only by narrow cartilaginous strips, the *epicoracoids* (Fig. 10.64). Anterior to the junction of the clavicles with the epicoracoid cartilages lies the bony *omosternum,* with an expanded, cartilaginous *episternum* joined to it anteriorly. Posterior to the junction of coracoids and epicoracoids lies the *sternum proper.* An expanded cartilage, *the xiphisternum,* is attached to it posteriorly. The sternum is thus considered to be composed of four median elements. It is not clear which part corresponds to the sternum in urodeles, but most anatomists consider the section referred to as the *sternum proper* to be the actual homologue.

Reptiles. Snakes, most limbless lizards, and turtles lack a sternum, but in many other reptiles this portion of the skeleton is more fully developed than it is in amphibians. In turtles, the plastron, composed of dermal plates covered with epidermal scales, is closely associated with the pectoral girdle. The membrane bones of the plastron are usually considered to be homologous with the gastralia of such forms as *Sphenodon,* crocodilians, certain other reptiles, and *Archaeopteryx.*

In *Sphenodon* a few of the anterior thoracic ribs join a typical midventral sternum. Most of the remaining ribs, however, join the median ventral parasternum which extends from the true sternum to the pubis. It will be recalled that this endoskeletal structure is derived from cartilages which appear in the midventral portions of the myocommata in this region.

In lizards the sternum usually consists of

a large, cartilaginous, flattened plate to which the sternal sections of several anterior thoracic ribs are united. A membrane bone, referred to as the *episternum,* or *interclavicle,* in some cases lies ventral to the sternum. It is *not* considered to be part of the sternum.

The sternum of crocodilians is a simple cartilaginous plate which splits posteriorly into two *xiphisternal cornua.* The sternal sections of the thoracic ribs are attached to the sternum and to the xiphisternal cornua. Numerous gastralia of dermal origin are present posterior to the cornua but form no connections with them.

Birds. The sternum of birds is a well-developed bony structure to the sides of which the ribs are firmly attached. In flying birds and penguins it projects rather far posteriorly under a considerable portion of the abdominal region. The ventral portion is drawn out into a prominent *keel,* or *carina* (Fig. 10.22), which provides a large surface for attachment of the strong muscles used in flight. It is of interest to note in this connection that the extinct flying reptiles, the pterosaurs, had carinate sterna as do bats among mammals. A sternum in *Archaeopteryx* is unknown. From the posterolateral borders of the sternum extends a pair of elongated *xiphoid processes.* Running birds of the superorder Paleognathae have rounded sterna which are not carinate. In swans and cranes a peculiar cavity of the anterior end of the sternum houses a loop of the trachea.

Mammals. In mammals the sternum is typically composed of a series of separate bones arranged one behind the other. Three regions may be recognized: an anterior *presternum,* or *manubrium;* a middle *mesosternum,* consisting of serveral *sternebrae;* and a posterior *metasternum,* or *xiphisternum,* to the end of

which a *xiphoid cartilage* is attached (Fig. 10.65). The costal cartilages of the ribs articulate with the sternum at the points of union of the separate bones. In bats the presternum is conspicuously keeled, as is the mesosternum in some cases. The number of sternebrae is variable in mammals, as is the number of true ribs which articulate directly with the sternum. In some cases, as in certain cetaceans, sirenians, and primates, the separate elements may fuse. Thus, in the human being, the entire structure, which is considerably flattened, consists of only three parts: an anterior *manubrium,* a *body* (*gladiolus*), and a small posterior *xiphoid process* (Fig. 10.66). The last is long and thin and essentially cartilaginous. In older individuals the proximal portion may become ossified. The clavicles, and the costal cartilages of the first pair of ribs, join the manubrium, the second pair of costal cartilages articulating at the junction of the manubrium and body. Five more pairs

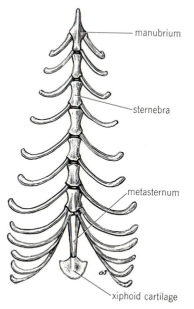

Fig. 10.65 Sternum of cat, showing relation to costal cartilages.

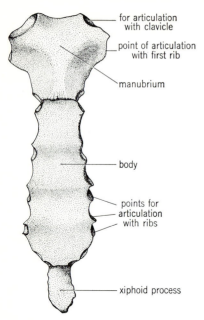

for articulation
with clavicle

point of articulation
with first rib

manubrium

body

points for
articulation
with ribs

xiphoid process

Fig. 10.66 Sternum of man.

of ribs join the sides of the body of the sternum at rather evenly spaced intervals. No ribs articulate with the xiphoid process. Study of development of the sternum in man shows it to be composed of separate structures, each derived from separate centers of ossification. These fuse together at various times from puberty to old age. The mammalian sternum arises quite independently of the ribs which unite with it secondarily. In such aquatic mammals as whales and sirenians the sternum has lost much of its importance. It is composed of a single bone which forms articulations with only a few of the most anterior ribs.

Median appendages

In discussing the axial skeleton, certain unpaired, median appendages must be taken into consideration. Only aquatic vertebrates possess median appendages, which are always finlike in character. They are found in cyclostomes, fishes, larval amphibians, some adult urodele amphibians, and cetaceans. Typically they are present in dorsal, anal, and caudal regions. In some forms the median fins are continuous, but most frequently the continuity is interrupted and the various fins are separated by gaps. In some of the newts the dorsal fin of the male becomes developed only during the breeding season. At other times it is rather inconspicuous. Fin development and regression in these forms are controlled by the endocrine activity of the testes. In cyclostomes and fishes the median fins are supported by skeletal structures (fin rays) and special muscles, but in higher forms they are merely elaborations of the integument. Dorsal and anal fins are used chiefly in directing the body during locomotion. The caudal fin also helps in this respect in addition to being the main organ of propulsion.

The skeleton supporting the unpaired dorsal and anal fins in fishes consists of radial cartilaginous or bony *pterygiophores* supporting slender *fin rays* at their distal ends. Union of several pterygiophores may take place at the base of the fin to form one or more *basipterygia,* or *basalia.* These are ossified in bony fishes. Pterygiophores and basalia may form secondary connections with the neural spines of the vertebral column. There may or may not be a segmental correspondence between the skeletal elements of the fin and the vertebrae. Distal to the basalia the pterygiophores continue as cartilaginous or bony *radialia.* From the radialia extend numerous fin rays of dermal origin, supporting the greater part of the fin. Frequently the fin rays connect directly with the basalia, the radialia being absent. Various types of joints permit the fin to be moved in a complex, undulatory manner or merely to be raised or lowered, as the case may be. In the males of many fishes the anal fin and its skeletal elements are modified to form a *gonopodium* (Fig. 8.41), which serves to aid in the transport of sperm from male to female in internal fertilization. Development of the gonopodium has been shown to be under control of the male hormone, testosterone.

Caudal fins are of several shapes (Fig. 10.67). Presumably the most primitive type is the *protocercal* tail of adult cyclostomes. The notochord is straight and extends to the tip of the tail, the dorsal and ventral portions of which are of practically equal dimensions. A second type, the *diphycercal* tail, found in Dipnoi, *Latimeria* (Fig. 2.18), and *Polypterus,* appears at first sight to be similar to the protocercal type. However, paleontological and embryological evidence indicates that the diphycercal tail may really be a secondary modification of a third type, the *heterocercal* tail. In this type the skeletal axis bends upward to enter the dorsal flange of the tail, which thus becomes more prominent than the ventral flange. A heterocercal tail is typical of elasmobranchs and lower fishes in general. The *homocercal* type of tail is the most common. Superficially it appears

to be composed of symmetrical dorsal and ventral flanges. The posterior end of the vertebral column, the *urostyle,* is, however, deflected into the dorsal flange. This type of tail, therefore, is not far removed from the heterocercal type.

In caudal fins the fin rays on the dorsal side may connect with basal pterygiophores or with neural spines directly. On the ventral side, however, they form connections with modified hemal spines which are spoken of as *hypurals.*

Appendicular Skeleton

Speculation as to the phylogenetic origin of the paired appendages of vertebrates has long occupied the attention of comparative anatomists. Several theories have been advanced, some of which have fallen into discard with new paleontological discoveries. Others have been modified from time to time as new lines of evidence have been brought to light. It is universally agreed that the limbs of tetrapods have arisen in evolution from the fins of fishes. The *fin-fold theory* of the origin of both unpaired and paired vertebrate ap-

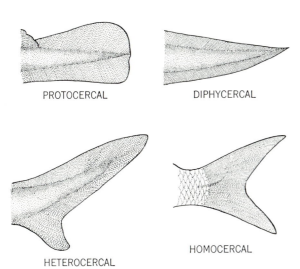

PROTOCERCAL   DIPHYCERCAL

HETEROCERCAL   HOMOCERCAL

Fig. 10.67 Four types of fish tails.

pendages seems to be one of the most plausible and in the past has been quite generally accepted.

According to some authorities, the fin-fold theory (Fig. 10.68) goes back to amphioxus as a starting point. In this animal the single dorsal fin continues around the tail to the ventral side as far forward as the atriopore. At this point it divides in such a manner that a *metapleural fold* extends anteriorly on either side almost to the mouth region. Gaps appearing in the dorsal fin and in the metapleural folds, for one reason or another, may have resulted in the appearance of median and paired fins. Certain rather valid objections have been advanced which would make it seem improbable that the metapleural folds of amphioxus were in any way concerned with the origin of the paired appendages of vertebrates. Among these are the fact that the folds terminate at the atriopore, a structure without homologues in higher forms. Furthermore, amphioxus has no skeletal structures other than a notochord. The so-called "fin rays" of amphioxus, which support the fins, are small rods of gelatinous connective tissue not homologous with the cartilaginous or bony fin supports of fishes. Usually the cyclostomes are dismissed in this connection as a specialized, limbless, divergent side shoot of the ancestral vertebral stock.

According to another idea, fin folds similar to the metapleural folds of amphioxus were present in some hypothetical ancestral fish but terminated at the anus rather than at the atriopore. In some of the very primitive acanthodian sharks, a row of six or seven spiny fins extended on each side between pectoral and pelvic fins (Fig. 10.69). They have been interpreted as remnants of fin folds. Traces of lateral fin folds can be observed in embryos of certain elasmobranchs as mesodermal proliferations which develop extensively only in the pectoral and pelvic regions as metameric myotomic buds (Fig. 11.14). The pectoral and pelvic fins may thus be regarded as persistent remnants of primitive fin folds. Furthermore, the basic skeletal structure of the paired fins is essentially like

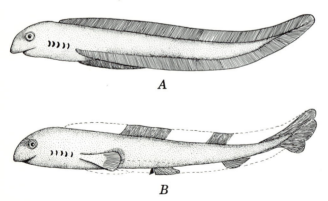

A

B

Fig. 10.68 Diagrams illustrating the fin-fold theory of the origin of paired appendages: A, the undifferentiated condition; B, the manner in which permanent fins might be formed from the continuous fin folds. (*After Wiedersheim, "Comparative Anatomy of Vertebrates," copyright 1907 by The Macmillan Company and used with their permission.*)

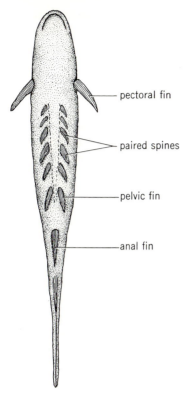

Fig. 10.69 Ventral view of acanthodian fish, *Euthacanthus macnicoli,* showing paired rows of ventral spines (fins). (*After Watson.*)

pectoral fin

paired spines

pelvic fin

anal fin

Those at the base, the *basalia,* show a tendency to fuse. The *radialia* may form two or three rows of short cartilages distal to the basalia. Dermal fin rays in turn are distal to the radialia. According to one theory, fusion of the anterior basalia with their partners in the midline resulted in the formation of a transverse bar. This is the rudiment of the pectoral or pelvic girdle, as the case may be.

Another more recent theory explaining the origin of the paired appendages and girdles does not assume a primitive fin-fold origin. According to this idea the appearance of paired appendages goes back to the ostracoderms. Ostracoderms may have been the remote ancestors of existing cyclostomes. Some of the early ostracoderms had a cephalothorax covered with a shield of plates with posteriorly directed cornua. The plates in some cases are known to have contained bone cells. Certain forms possessed lateral fleshy lobes, projecting from each side, medial to the cornua of the thoracic shield. A bony, skeletogenous septum passed behind the gill chamber and behind the pericardium. The fleshy lobes may have become pectoral fins, and the endoskeletal part of the pectoral

that of the unpaired fins, indicating that the two have a common origin. The structure of the fins of the extinct shark, *Cladoselache* (Fig. 10.70), has been cited as providing additional support to the fin-fold theory. The fins are broad at their bases and contain numerous parallel pterygiophores which by their very arrangement suggest a primitive fin-foldlike origin.

The pterygiophores are primitively metameric, appearing in the fin along with metameric myotomic buds. Each pterygiophore is subdivided into several small pieces.

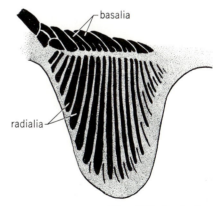

basalia

radialia

Fig. 10.70 Pectoral fin of *Cladoselache.* (*After Dean.*)

girdle may have developed as an ingrowth of basal portions of the pectoral fins. The dermal bones of the pectoral girdle may have been derived from the bony plates of the thoracic shield.

Another group of ostracoderms is known to have possessed a paired row of dermal spines on the ventral side of the body not unlike the extra fins or spines of the acanthodian sharks referred to above. The spines may have been used merely as an aid in clinging to the bottom of rapidly flowing fresh-water streams. Loss of armor and reduction of spines, but their persistence in pectoral and pelvic regions, might well account for the origin of the paired appendages from ostracoderm ancestors. With the growth of muscles from the metameric myotomes into the bases of the spiny fins, their movement may first have been accomplished. Appearance of radial skeletal elements, the pterygiophores, could have accompanied the segmental muscles as they spread outward in a fanlike manner, in opposing groups, to the peripheral parts of the fin. According to this concept the basal pieces of the endoskeleton of the fin may have pushed farther into the body to give rise to the girdles. This idea holds that the paired fins were not origi-nally parts of fin folds but were, from the beginning, separately spaced ridges supported by spines. Much paleontological evidence has been cited in support of the ostracoderm theory.

Paired fins and girdles of fishes. PEC-TORAL FINS AND GIRDLES. Certain extinct sharks, such as *Cladoselache,* possessed the simplest fins known to occur in vertebrates (Fig. 10.70). The skeletal structures of which they are composed consist of several basalia from which radialia extend in a fanlike manner. The basalia of the two sides did not unite as in a true pectoral girdle.

In existing elasmobranchs the pectoral girdle (Fig. 10.71) is a U-shaped cartilage in the form of an arch, located just posterior to the branchial region and open dorsally. It is not connected with the axial skeleton except in skates in which the upper portion, or *suprascapula,* joins the vertebral column on each side. The girdle consists of one piece, the *scapulocoracoid cartilage.* The ventral portion is the *coracoid bar* from which a long *scapular process* extends dorsally on each side beyond the *glenoid region* where the pectoral fin articulates. Each scapular process has a *suprascapular cartilage* attached at

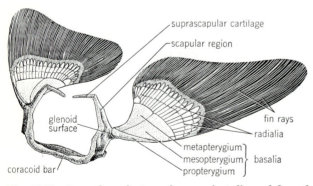

Fig. 10.71 Anterolateral view of pectoral girdle and fins of *Squalus acanthias.*

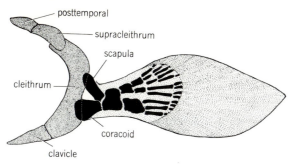

Fig. 10.72 Diagram showing the membrane bones of the pectoral girdle of a primitive fish. Membrane bones are stippled; cartilages or cartilage bones are black.

its free end. The fin itself consists of three basal cartilages (basalia) which articulate with the girdle at the glenoid region. They are an anterior, lateral *propterygium;* an intermediate *mesopterygium;* and a posterior, medial *metapterygium.* Numerous segmented radial cartilages extend distally from the three basal cartilages. They are arranged in rows. From the radialia many dermal fin rays extend peripherally.

In primitive bony fishes certain membrane bones have become associated with the pectoral girdle. They have been derived, in a manner similar to the membrane bones of the skull, from dermal scales which have assumed a deeper position in the body. There are usually at least two such membrane bones on each side. A pair of *clavicles* meets on the midline close to the coracoid region. A *cleithrum* overlies the scapula. In many fishes the cleithra as well as the clavicles join each other medially, thus providing additional support. In *Amia, Lepisosteus,* and teleosts, the clavicle has disappeared. Other membrane bones, *supracleithrum* and *posttemporal,* are present in many forms, the latter forming a connection with the skull (Fig. 10.72). The original part of the pectoral girdle may remain cartilaginous but usually

forms two separate bones on each side, the scapula and coracoid, articulating at the glenoid region. The two halves of the original cartilaginous girdle, as seen in elasmobranchs, no longer connect ventrally. The membrane bones now make up the greater part of the pectoral girdle, the original girdle being much reduced.

In bony fishes the pectoral fins themselves exhibit much variation in their detailed structure. In the chondrostean, *Polypterus,* the fin is much like that of elasmobranchs except that ossification of the separate elements has begun to take place. In most bony fishes the number of separate skeletal pieces is reduced, the dermal fin rays taking over much of the supporting function. Many separate basalia, however, may articulate with the pectoral girdle. In the dipnoan *Epiceratodus,* the pectoral fin consists of a main, segmented, metapterygial, radial axis from the sides of which numerous segmented secondary rays project (Fig. 10.73). Dermal fin rays extend from the secondary rays to the periphery. In *Protopterus* and *Lepidosiren* only the main radial axis persists. The type of fin seen in *Epiceratodus* is sometimes referred to as the *archipterygium,* since it was formerly thought to be the

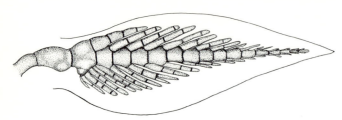

Fig. 10.73 Pectoral fin of *Epiceratodus*.

archetype from which the tetrapod limb evolved. This view is no longer held. Instead, the type of fin found in living Dipnoi is interpreted as being a specialized derivative of the primitive, tribasic, multiradial fin of elasmobranchs. The link between the fins of fishes and the limbs of tetrapods is to be sought among the fossil crossopterygians, only a few specimens of which are known. Study of the fin structure of *Eusthenopteron* (Fig. 10.76) strongly suggests that this group will provide the clue to the true explanation of the origin of tetrapod girdles and limbs (pages 485 to 486).

PELVIC FINS AND GIRDLES. The pelvic fins and girdles of fishes are usually more primitive and simple in structure than those of the pectoral region. In *Cladoselache* there is little difference between the two, although the pelvic structures are somewhat reduced. The sturgeon shows little change over the condition in *Cladoselache*. The pelvic girdle of fishes is always free of the axial skeleton. Fusion of the anterior basalia of the two sides occurs in elasmobranchs to form the *ischiopubic bar* (Fig. 10.74), comparable to the coracoid bar of the pectoral girdle. Attachment of the fin is at the *acetabular* rather than at the glenoid region. A small *iliac process*, projecting dorsally to a slight degree at either end of the ischiopubic bar, corresponds to the scapular region of the pectoral girdle. In the Dipnoi the pelvic girdle gives off anterior and posterior medial and lateral processes. The

anterior medial projection, spoken of as the *epipubis*, has homologies with similar structures of higher forms. The posterior lateral processes serve for fin articulation. The other processes offer a surface for muscle attachment. It is rare to find any membrane bones in the pelvic region of fishes comparable to those associated with the pectoral girdle. The pelvic girdle of teleost fishes is reduced to a small element on each side with no connections between the two. It may even be lacking, as in the eels which have no pelvic fins. Often the pelvic fins have shifted so far forward that they come to lie just behind the pectoral fins and are attached at their bases to the cleithra. In elasmobranchs each pelvic fin typically consists of two basalia: a medial, posterior *metapterygium* and a small anterior *propterygium*. The latter may be absent.

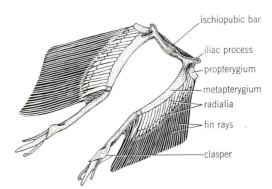

Fig. 10.74 Pelvic fins and girdle of male *Squalus acanthias*.

Anatomy of the Chordates

Numerous radialia extend outward from these two basal pieces. Dermal fin rays complete the skeletal structure of the fins. In male elasmobranchs the skeleton supporting the clasper is a continuation of the metapterygium which is broken up into several segments (Fig. 10. 74). In the Dipnoi the pelvic fins are of the same archipterygial type as the pectoral fins. The central axis in some cases is split into two. Teleosts possess rather degenerate pelvic fins. Usually a single basal bone gives off a few poorly developed radials from which the fin rays project. Often the fin rays arise directly from the basal. A curious situation exists in certain fishes, such as the cod, in which the pelvic fins have pushed forward so as to lie *anterior* to the pectoral fins and have become attached to the throat region. The fossil crossopterygian *Eusthenopteron* shows a better-developed pelvic-fin structure than is to be found in forms living today (Fig. 10.76*B*). Although somewhat smaller in size than the pectoral fins, the basic skeletal structure is similar. A strong pelvic girdle is also present.

Paired girdles and limbs of tetrapods. The appendicular skeleton in all tetrapods shows a fundamental similarity in structure. There is a shoulder, or pectoral, girdle to which the forelimbs are attached and a hip, or pelvic, girdle supporting the hind limbs. The pectoral girdle does not form a connection with the vertebral column except through muscles and ligaments. The pelvic girdle, however, attaches directly to the vertebral column in the sacral region. The limbs are typically pentadactyl (having five fingers or toes) or have been modified from the primitive pentadactyl type. Each girdle is composed of two halves, and each half in turn consists of three bones which show a fairly comparable arrangement in the two girdles.

In the pectoral girdle of tetrapods there has been a reduction of membrane bones, and of these only the *clavicle* sometimes persists. Connection of the pectoral girdle with the skull, as occurs in teleost fishes, is lost. The cartilage bones of the girdle in tetrapods have become the dominant elements. The bones of the pectoral girdle typically consist of a ventral *coracoid*, which meets the sternum medially; a *scapula*, extending dorsally; and a *clavicle*, which lies on the ventral side between scapula and sternum and anterior to the coracoid. In some of the lower tetrapods an additional cartilage bone, the *precoracoid*, lies anterior to the coracoid and close to the clavicle (Fig. 10.75*A*). The precoracoid probably is derived from an ossification center in a ventral projection of the scapula. At the junction of scapula and coracoid is a depression, the *glenoid fossa*, which serves as the point of articulation of the forelimb with the pectoral girdle.

The pelvic girdle (Fig. 10.75*B*) is composed of a ventral *ischium* comparable to the coracoid, a dorsal *ilium* similar to the scapula, and an anterior, ventral *pubis* corresponding in position to the clavicle and precoracoid. Clavicle and pubis are not considered to be homologous since the former is a membrane bone * and the latter a cartilage bone. Generally, pubis and precoracoid are considered to be homologous. The ilium is the portion which forms a union with the sacrum. The two pubic bones usually unite ventrally at the *pubic symphysis*. The ischia also often come together to form an *ischial symphysis*. The two may be separated by a

* In some higher forms, as in the rat and other placental mammals, developmental studies of the clavicle show that it is not entirely a membrane bone but some parts develop by the endochondral method of ossification and others by the intramembranous method. Therefore it is not clear just what the mammalian clavicle actually represents, and homologies are somewhat uncertain. It is possible that both clavicles and precoracoids are represented in a single element.

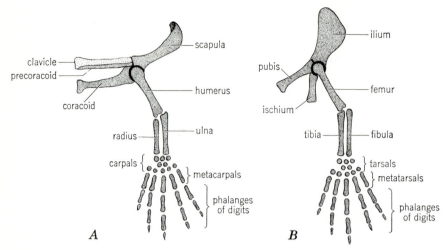

Fig. 10.75 Diagram showing skeletal structures of tetrapod limbs and girdles: *A*, pectoral girdle and limb; *B*, pelvic girdle and limb. The clavicle is lightly stippled to indicate that it is a membrane bone.

puboischial foramen or may join to form a *puboischial symphysis.* A depression, the *acetabulum,* located at the junction of the three bones serves as the point of articulation of the hind limb with the pelvic girdle. It is comparable to the glenoid fossa of the pectoral girdle. A fourth and very small *acetabular,* or *cotyloid,* bone often enters in formation of the acetabulum in mammals. Its homologies are uncertain, but it may possibly be comparable to the precoracoid of the pectoral girdle.

The skeletal elements of the forelimbs and hind limbs are also arranged on the same general plan. In the forelimb there is a single bone, the *humerus,* in the upper arm, or *brachium.* The head of the humerus joins the pectoral girdle at the glenoid fossa. Two bones arranged parallel to each other are present in the forearm, or *antebrachium.* They are a lateral *radius* and a medial *ulna.* Distal to radius and ulna, in the primitive condition, are 9 or 10 *carpal,** or *wrist,* bones; 5 longer *metacarpals* in the hand;

and then a few rows of small *phalanges* in the fingers. The carpal bones are primitively arranged in three rows. There are 3 bones in the proximal row, 5 in the distal row, and 1 or 2 in the middle row. Corresponding bones in the hind limb are the *femur,* which joins the acetabulum; *tibia* and *fibula,* comparable to radius and ulna; 9 or 10 *tarsal,** or *ankle,* bones; 5 *metatarsals;* and rows of *phalanges* in the toes.

Some remarkable specializations have taken place in the evolution of the limbs of tetrapods. These consist, for the most part, of reductions or fusions of various parts of the primitive, basic, pentadactyl limb skeleton, some notable examples of which are discussed below.

PHYLOGENETIC ORIGIN OF TETRAPOD GIRDLES AND LIMBS. Some reference to the origin of tetrapod girdles and limbs from

---

* Apparently the carpal and tarsal bones of early tetrapods were originally 12 in number: 3 proximal bones, 4 central bones, and 5 distal bones. The latter articulated with the 5 primitive metacarpals or metatarsals.

piscine ancestors has been made previously. Several theoretical possibilities have been advanced in the past to explain such evolutionary changes. There is general agreement among authorities today that connecting links are to be sought among fossil crossopterygians of the extinct suborder Rhipidistia. The theory of Gegenbaur, deriving the tetrapod limb from the archipterygium of the Dipnoi, has been generally discarded. A more recent theory, based upon an exhaustive study of the fossil remains of the rhipidistian crossopterygian *Eusthenopteron*, rather convincingly points out that the various skeletal parts of the fins of this primitive fish can be closely homologized with those of the pentadactyl limbs of tetrapods.

Each fin of *Eusthenopteron* consists of a chain of bones along the postaxial side, from which a series of radials comes off as indicated in Fig. 10.76. It is believed that the first skeletal element at the base of the fin is equivalent to the humerus (femur) and the second element to the ulna (fibula). The first radial is comparable to the radius (tibia). The rest of the radials are believed finally to have become carpal (tarsal) bones, whereas metacarpals (metatarsals) have arisen as new distal outgrowths from the margin of the fleshy, muscular portion of the paddlelike fin. *Eusthenopteron* itself is not considered to be the direct ancestor of tetrapods. Some close rhipidistian relative, rather, was probably the tetrapod progenitor.

Joints are believed to have developed during the evolution of such limbs at points in the appendages where bending was required to produce effective locomotion.

It has already been mentioned that in the evolution of the pectoral girdle of tetrapods there has been a loss or reduction of those

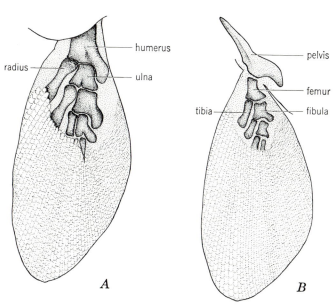

Fig. 10.76 Skeletal structure of *A*, left pectoral, and *B*, left pelvic fins of the crossopterygian fish, *Eusthenopteron*. (*After Gregory and Raven, Ann. N.Y. Acad. Sci. By permission.*)

portions derived from the dermal skeleton. The clavicle, however, persists in most forms, although in mammals it is frequently much reduced or absent. A new element, the *interclavicle*, a membrane bone, has been added in many tetrapods. This is an unpaired, median portion of the skeleton, lying between the coracoids just posterior to the clavicles and joining the anterior end of the sternum. It is best developed in lizards but appears also in birds as the median part of the *furcula*, or wishbone. An interclavicle is also present in monotremes among mammals. The elements derived from the original, cartilaginous pectoral arch are of most importance in the evolution of the pectoral girdle. Scapula and usually the coracoid persist. In most mammals, however, the coracoids are reduced. Precoracoids, except in mammals, are usually present but frequently remain unossified, the clavicles taking their place. The glenoid fossa is usually at the point of junction of scapula, precoracoid, and coracoid. In monotremes both precoracoids and coracoids persist, but in higher mammals the precoracoid is lost and the coracoid is represented only by a small *coracoid process* attached to the scapula.

The pelvic girdle, which in fishes is poorly developed and not attached to the vertebral column, is a much more important structure in tetrapods in which, in most groups, it forms a firm union with the sacrum. It has been suggested that the pelvis of tetrapods could have logically been derived from the primitive pelvic girdle of such a fish as *Eusthenopteron*. Changes may have involved a shifting of the acetabulum to the lateral surface, an elongation of the ischial region, and a dorsal extension of the iliac process on each side. A gradual widening of the ischial processes with the resultant formation of the ischial symphysis, posterior to the pubic symphysis, was a probable de-

velopment. The union of the iliac processes with sacral ribs was the next important change to occur but probably did not take place until a later geological era. Ossifications in pubic, ischial, and iliac regions apparently resulted in the appearance of the three bony elements making up each half of the pelvic girdle.

AMPHIBIAN GIRDLES AND LIMBS. In some of the primitive amphibians the pectoral girdle was very similar to that of piscine ancestors except for the addition of an interclavicle. In modern amphibians the dermal elements have been lost except for the clavicle which persists in anurans. Limbs and limb girdles are much better developed in anurans than in urodeles, but in both groups the basic structural plan is similar. Caecilians lack limbs and limb girdles.

The limbs and girdles of urodeles are small and weak. In *Siren* both pelvic limb and girdle are absent altogether. The urodele pectoral girdle is of simple structure (Fig. 10.77) and is apt to remain cartilaginous except in the region of the glenoid fossa. The coracoids, which are united on each side with the scapula to form a single piece, are broad and overlap each other medially in a loose manner just anterior to the sternal cartilage. A precoracoid process may project anteriorly from each coracoid. Clavicles are absent. Each scapula is connected dorsally with a broad suprascapular cartilage.

The urodele pelvis is firm, the irregular platelike puboischia being united in a median symphysis. An iliac process on each side unites with the transverse process of the sacral vertebra (Fig. 10.79). Ossification centers appear in ilium and ischium but not in the pubis. A Y-shaped *ypsiloid cartilage* lies anterior to the puboischium but is not generally regarded as part of the pelvic girdle. This cartilage and the muscles attached to it

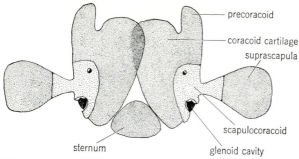

precoracoid

coracoid cartilage

suprascapula

scapulocoracoid

sternum

glenoid cavity

Fig. 10.77 Pectoral girdle of the salamander *Ambystoma jeffersonianum*, shown as though it were flattened out. Actually the suprascapula arches dorsally. (*After Noble, "Biology of the Amphibia," McGraw-Hill Book Company. By permission.*)

have been shown to be concerned with the hydrostatic function of the lungs.

In the urodele forelimb (Fig. 10.78), humerus, radius, and ulna are distinct. Carpal bones are reduced in number by fusion. No more than four digits are present, the first, or *pollex* (thumb), probably being the one that has been lost along with its carpal and metacarpal. In some species, only two or three digits remain. In the hind limb (Fig.

10.79), femur, tibia, and fibula remain separate, and in most cases the primitive pentadactyl condition is retained. In a few species, however, only two or three digits persist.

The anuran pectoral girdle shows several modifications within the group. In frogs (Firmisternia) the two halves are firmly united in the midline and are closely related to the sternum (Fig. 10.64). In toads (Arcifera) the two halves overlap in the middle.

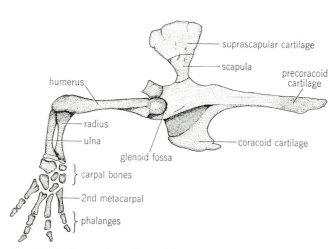

suprascapular cartilage

scapula

precoracoid cartilage

humerus

radius

ulna

glenoid fossa

coracoid cartilage

carpal bones

2nd metacarpal

phalanges

Fig. 10.78 Pectoral girdle and limb of *Necturus*, lateral view.

Skeletal System

487

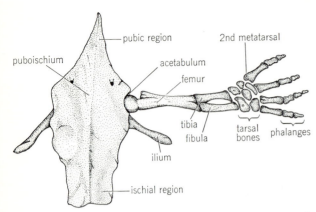

Fig. 10.79 Pelvic girdle and limb of *Necturus*, ventral view.

Coracoids, precoracoids, clavicles, scapulae, and suprascapulae are present. The clavicles are fused to or cover the cartilaginous precoracoids in the firmisternal type, whereas in toads the cartilaginous precoracoids and coracoids join their partners medially. All stages between the two types of girdles have been identified within the group. The pelvic girdle of anurans is V-shaped and consists on each side of a long ilium and a small ischium and pubis. The ilium is attached anteriorly to the transverse process of the sacral vertebra. All three bones join at the acetabulum. The limbs of anurans are more specialized than those of urodeles (Fig. 10.80). Humerus and femur are typical, but radius and ulna in the forelimb, and tibia and fibula in the hind limb, tend to fuse. The hind limb is pentadactyl, but the forelimb usually has only four digits. The tarsal bones are modified and consist of two rows. The proximal row contains two long bones, an inner *astragalus* (*talus*) and an outer *calcaneum.* The distal row of tarsals is reduced to two or three small cartilaginous or bony pieces. Metatarsals and phalanges of the hind limbs are long and webbed. A small additional

bone, the *prehallux,* or *calcar,* occurs on the tibial side of the tarsus in most anurans. It is sometimes interpreted as being the rudiment of an extra digit. It is more probable, however, that the prehallux represents an additional tarsal bone.

REPTILIAN GIRDLES AND LIMBS. The girdles and limbs of reptiles are well developed except in snakes and limbless lizards. In snakes there is no trace of pectoral girdle or forelimb, but vestiges of a pelvis and hind limb are found in members of the families Glauconiidae and Boidae. In the former group the three components of the pelvis, and even the femur, are represented. In the Boidae the pelvic vestiges appear externally in the form of small spurs on either side of the cloacal opening. Remnants of the girdles persist to some extent in most of the limbless lizards, but in a few no traces remain.

Reptilian pectoral girdles usually possess the typical three cartilage bones: coracoid, precoracoid, and scapula. Two membrane bones, clavicle and interclavicle, are frequently present. In turtles the location of the pectoral girdle is peculiar in that it lies within the arch formed by the ribs or is ventral to

Anatomy of the Chordates

them. A sternum is lacking in turtles. Coracoids and precoracoids are not united medially except by fibrous bands. Clavicles and interclavicles are lacking unless certain ossifications in the plastron represent these structures. Lizards have a well-developed pectoral girdle united medially by the sternum. Clavicles and interclavicles are present except in chameleons. Crocodilians, on the other hand, have an incomplete pectoral girdle. Clavicles and precoracoids are lacking. A small interclavicle lies between the coracoids ventral to the sternum but projects beyond it anteriorly. The suprascapula is represented by the small cartilaginous dorsal border of the scapula.

Pelvic girdles of reptiles are typically composed of the usual three bony elements which retain their integrity throughout life. The ilium is firmly united with the two sacral ribs. Both pubic and ischial bones form median symphyses. A small foramen in the

pubic bone, serving for the passage of the obturator nerve, was the only opening or foramen in the pelvic girdle in primitive amphibians and reptiles. Next, as in *Sphenodon,* a puboischial foramen appeared between pubis and ischium. In most reptiles these two foramina have become confluent and a new name, *obturator foramen,* is then applied. In crocodilians only an ischial symphysis is present. The two pubic bones are reduced, and a pair of *epipubic bones* projects ventrally. These are separated by a membranous area and do not form a symphysis. The epipubic and pubic bones are frequently confused. According to some authorities the epipubic bones are actually pubic bones which do not enter into formation of the acetabulum and which form a movable articulation with the rest of the pelvis. In most lizards and turtles a median fibrocartilaginous bar connects the rather widely separated pubic and ischial symphyses and separates the obturator foramina of the two sides. In these groups of reptiles a median, cartilaginous *epipubis* lies anterior to the pubic symphysis. A posterior prolongation from the ischial symphysis in *Sphenodon* and in many lizards and turtles is called the *hypoischiac process,* or *cloacal bone.*

The limbs in reptiles show no very unusual features. They are typically pentadactyl, even in sea turtles, the limbs of which are in the form of paddlelike flippers. The chief differences in the limbs of the various groups are to be observed in the carpal and tarsal bones, in which the number may be reduced because of fusion. In crocodilians three tarsal bones unite just distad the tibia to form an *astragalus* (*talus*). A single bone, the *calcaneum,* lies distad the fibula. The calcaneum bears a heel-like projection which appears, for the first time in phylogenetic history, in this group of

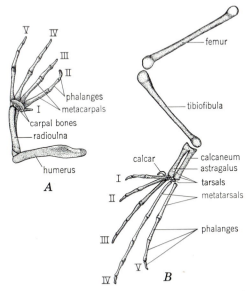

Fig. 10.80 Arrangement of bones in *A*, pectoral, and *B*, pelvic, appendages of the bullfrog, *Rana catesbeiana.*

reptiles. A feature of the reptilian hind limb is the presence of an intratarsal joint between the two rows of tarsal bones. Movement takes place here rather than at the junction of the tarsals with tibia and fibula. A *patella,* or kneecap, which is a sesamoid bone, appears in certain lizards for the first time.

In some of the prehistoric reptiles interesting modifications are present. The paddle-like limbs of plesiosaurs and ichthyosaurs have a very large number of phalanges (hyperphalangy). As many as a hundred phalanges to a digit have been observed. In some instances additional rows of phalanges are present in excess of the usual pentadactyl number. In the flying pterosaurs humerus, radius, and ulna are of normal proportions. One heavy metacarpal and three slender bones lie distal to radius and ulna. The slender metacarpals support the first three digits which are small and free and bear claws at their tips. The phalanges of what is presumably the fourth digit are enormously elongaged. They supported the large integumentary fold which extended outward from the body from shoulder to ankle and was used as a wing. A spurlike sesamoid bone, the *pteroid,* which is *not* a modified digit, projects toward the shoulder from the base of the metacarpals. It is believed that the pteroid provided additional support to the wing. Flexure of the pterosaur wing occurred at the junction of the fourth metacarpal and the first phalanx of that digit.

GIRDLES AND LIMBS OF BIRDS. The appendicular skeleton of birds shows a remarkable uniformity within the group. The pectoral girdle consists on each side of a large coracoid, a thin, narrow scapula, and a slender clavicle. The two clavicles which are fused medially to a small interclavicle form the *furcula,* or "wishbone" (Fig. 10.22). The precoracoid has practically disappeared. The coracoids form a firm union with coracoid grooves on the sternum. The glenoid fossa is formed by an imperfect union of scapula and coracoid. In the Paleognathae the pectoral girdle is relatively small, the clavicles being much reduced or absent. In the extinct moas a pectoral girdle seems to have been missing altogether, and in the kiwi it is extremely small.

Except in *Archaeopteryx* the three bones of the very large pelvis are fused together to form the *innominate* bone. The ilium is large and projects forward and backward from the acetabulum for some distance. It is completely fused to the synsacrum along the entire length of the latter. A suture, however, indicates the line of fusion. The synsacrum, it will be recalled, is the single bony mass composed of the posterior thoracic, lumbar, sacral, and first few caudal vertebrae, all fused together. Embryological studies make it clear that only two of the vertebrae are truly sacral. The ischium extends posteriorly, paralleling the ilium and fusing with it throughout the greater part of its length. A gap, the *ilioischial foramen,* varying in size in different birds, separates the two bones for some distance (Fig. 10.22). Ischia and pubes do not form symphyses. The two pubic bones, projecting in a postero-ventral direction, usually terminate freely, thus permitting passage of the relatively large eggs of birds. In a number of birds the distal end of the pubis unites with the distal end of the ischium. Only in the ostrich is there a true pubic symphysis. An ischial symphysis occurs in the rhea. *Archaeopteryx* may have had an ischial symphysis. The large space between pubis and ischium, whether or not these bones unite distally, represents the obturator foramen. The pelvic girdle of birds is much like that of some extinct dinosaurs.

The forelimb (wing) of birds is much modified from the primitive condition. In flying birds, humerus, radius, and ulna are

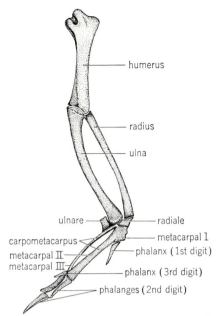

humerus

radius

ulna

ulnare

radiale

carpometacarpus

metacarpal 1

metacarpal II

phalanx (1st digit)

metacarpal III

phalanx (3rd digit)

phalanges (2nd digit)

Fig. 10.81 Wing bones of right wing of pigeon.

exceptionally well developed, since in strong fliers the wings are much longer than the hind limbs. Modification of carpals, metacarpals, and phalanges is primarily responsible for the specialized condition of the endoskeleton of the wing (Fig. 10.81). Carpal bones, due to fusion, are at first reduced in number to four. These are arranged in two rows of two each. The two distal carpals become fused to the corresponding metacarpals, forming the *carpometacarpus*, thus leaving the proximal carpal bones free. They are referred to as the *radiale* and *ulnare*. Three metacarpals persist in birds, but there is a difference of opinion as to which three are actually represented. In order to avoid confusion, we shall refer to them merely as metacarpals I, II, and III. Metacarpal I is much reduced. All three are united proximally, and numbers II and III unite distally in addition. The first meta-

carpal bears one phalanx, the second bears two, and the third only one, except in the ostrich in which two phalanges are present on the third digit. In *Archaeopteryx*, claws were present at the ends of all three digits, in which the metacarpals remained separate. In other birds, claws are sometimes encountered on digit I. The young hoatzin, *Opisthocomus*, has claws on the first two digits, but they disappear in the adult animals.

The various bones of the forelimb form a base supporting the large wing feathers which are called *remiges*. There are three sets of remiges: (1) *primaries*, or *metacarpodigitals*, attached to the various bones of the wrist and hand, (2) *secondaries*, or *cubitals*, attached to the ulna, and (3) *humerals*, supported by the humerus. The primaries are further subdivided into groups, depending upon the particular bone to which they are attached.

The hind limbs of birds are modified for *bipedal* locomotion in contrast to the wings, which have become adapted for flight. In addition to their use in locomotion, the feet of various birds are adapted for swimming, perching, wading, running, scratching, nest building, clinging, fighting, and sundry other purposes (Fig. 2.35). Although there is so much variation, nevertheless a striking uniformity exists in the basic structure of the hind limbs of birds. A strong femur articulates with the pelvis at the acetabulum. The fibula is usually much reduced and often represented only by a small bony splint. The tibia is strong and fused to the proximal tarsal bone to form a *tibiotarsus*. A sesamoid bone, the *patella*, is found in most birds anterior to the junction of femur and tibia. The distal tarsal bones unite with the second, third, and fourth metatarsals, forming a single *tarsometatarsus*, which is thus a compound bone (Fig. 10.82). Grooves

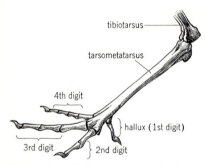

tibiotarsus

tarsometatarsus

4th digit

hallux (1st digit)

3rd digit

2nd digit

Fig. 10.82 Skeleton of hind limb of chicken.

at its distal end indicate its composite origin. Because of the fusion of the tarsals with other bones, the ankle joint is said to be *intratarsal* in position. An oblique bar of bone crosses the anterior surface of the distal end of the tarsometatarsus. It is lacking in the ostrich. The first metatarsal is represented by a free projection from the distal end of the tarsometatarsus. The spur of the domestic fowl is a bony projection of the tarsometatarsus and is directed posteriorly. It is covered with an epidermal cap of horny material. No more than four digits are present in birds, the fifth metatarsal and the accompanying phalanges being lacking. The first digit (hallux) is usually directed backward, and the other three are directed forward. Some variation is met with in the direction in which the toes point. The typical number of phalanges, going from the first to the fourth digit, is 2, 3, 4, 5. The terminal phalanges bear claws.

A few variations from the above description are encountered in birds. Tibia and fibula in *Archaeopteryx* were distinct and of approximately equal length. In penguins, too, the fibula is complete. Members of the superorder Paleognathae, with the exception of the kiwi, have only three toes, the first, or hallux, being absent. In ostriches the

second toe is also lacking. Of the two remaining digits in the ostrich, the third is by far the larger. It bears a claw, but the fourth digit is clawless.

MAMMALIAN GIRDLES AND LIMBS. The appendicular skeleton of mammals shows considerable range from a primitive, reptile-like condition to one of a very high degree of specialization. As would be expected, the monotremes are the most primitive in this respect.

In monotremes the cartilage bones of the pectoral girdle consist on each side of scapula, coracoid, and precoracoid (epicoracoid). The coracoids form a ventral connection with the manubrium (presternum). They furnish the greater portion of the glenoid fossa. Precoracoids join a median episternum which lies anterior to the sternum. Clavicles and interclavicle are present, representing the dermal portion of the reptilian pectoral girdle. In most mammals, however, the interclavicles and precoracoids are lost and the coracoid is reduced to a small *coracoid process* on the scapula adjacent to the glenoid fossa (Fig. 10.83). The latter lies entirely on the scapula. In some mammals the clavicle persists as a strong bony arch from scapula to manubrium. In others it is lost or else remains as a small, unimportant bony vestige embedded in muscle. Persistence of the clavicle is correlated with freedom of movement of the pectoral limb. In the absence of the clavicle all direct connection between axial skeleton and pectoral girdle is lost. The scapula is thus the most important part of the mammalian pectoral girdle.

The pelvic girdle of mammals is made up of the usual three bony elements which, in most cases, are united at the acetabulum. Although the three parts are distinguishable in the young, they are fused in adults to form a single *innominate* bone on each side (Fig. 10.84). A small *acetabular* (*cotyloid*) bone is

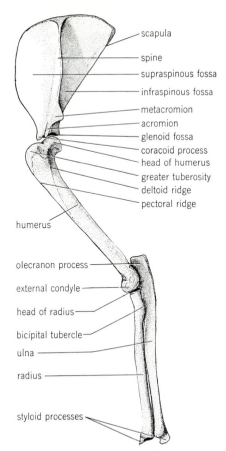

scapula
spine
supraspinous fossa
infraspinous fossa
metacromion
acromion
glenoid fossa
coracoid process
head of humerus
greater tuberosity
deltoid ridge
pectoral ridge

humerus

olecranon process
external condyle
head of radius
bicipital tubercle
ulna
radius

styloid processes

Fig. 10.83 Lateral view of left pectoral girdle and forelimb bones of cat.

the estrogenic hormone of the ovaries, thus facilitating passage of the young from the reproductive tract at the time of parturition. The presence of a large obturator foramen on each side, bounded by pubis and ischium, is characteristic of mammals. The ilium in primates, particularly in man, has become broad and flat in connection with the assumption of an upright posture. The weight of the body is then supported by ilia, sacrum, and the two femurs where they join the innominate bones. Pelvic girdle and hind limbs are lacking in whales and sirenians. However, paired remnants of pubis and ischium remain in much reduced form.

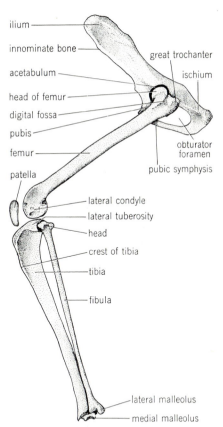

ilium
innominate bone
acetabulum
head of femur
digital fossa
pubis
femur
patella

great trochanter
ischium

obturator foramen
pubic symphysis

lateral condyle
lateral tuberosity
head
crest of tibia
tibia
fibula

lateral malleolus
medial malleolus

Fig. 10.84 Lateral view of left pelvic girdle and hind limb bones of cat.

frequently present in the acetabulum, substituting for the pubis and sometimes the ilium in forming the acetabular depression. The ilium and sacrum are firmly ankylosed. Both pubes and ischia, in many mammals, form symphyses, but in others the ischia do not meet ventrally, so that only a pubic symphysis is present. Strong posterior projections of the ischia may develop. These support the body when in a sitting position. Even a pubic symphysis is lacking in certain mammals. In the female pocket gopher (Fig. 10.85) it undergoes resorption under the influence of

ADULT MALE   ADULT FEMALE

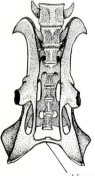 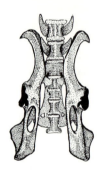

pubic symphysis

Fig. 10.85 Ventral view of pelves of adult male and female pocket gophers, showing presence and absence, respectively, of the pubic symphysis. (*Original by F. L. Hisaw, J. Exp. Zool., 42:437. By permission of the Wistar Institute of Anatomy and Biology.*)

In monotremes and marsupials an additional pair of bones, preformed in cartilage, extends forward from the pubes in the ventral wall of the abdomen. They are the *marsupial,* or *epipubic*, bones. A movable articulation occurs between them and the pubes. The homologies of these bones are uncertain. Some authorities state that they are sesamoid bones which develop in the tendons of the external oblique muscles of the abdomen. Apparently they have no homologues in other vertebrates although it is possible that they represent ossifications of the epipubic cartilages of such reptiles as lizards and turtles and even are comparable to the epipubic bones of crocodilians, if indeed these reptiles actually have epipubic bones (page 489).

The forelimbs of mammals, which in many cases are highly specialized, usually deviate less than the hind limbs from the primitive pentadactyl form. A mammalian humerus may bear a *supracondyloid foramen* near its distal end. This serves for the passage of the brachial artery and median nerve of the arm. The brachial vein does not pass through the supracondyloid foramen. In man this foramen is usually lacking but may be encountered in an occasional individual. Radius and ulna connect with the distal end of the humerus by a hinge joint which permits movement in only one plane. In primates and a few other mammals with prehensile forelimbs, radius and ulna are not arranged in a fixed position but, rather, articulate in such a manner that the distal end of the radius can rotate about the ulna so as to turn the hand in either a prone or supine position. Articulation with the humerus is chiefly through the ulna, which bears a notch and a process for junction with the *olecranon fossa* of the humerus. A projection from the proximal end of the ulna is called the *olecranon process,* commonly spoken of as the elbow. In many cases radius and ulna are fused to some extent. The ulna is most important in forming the elbow joint. The radius is chiefly concerned in forming a support for the hand.

In primates the pollex is usually quite independent of the other digits and more freely movable. The fact that it can be brought into opposition to the remaining digits and to the palm of the hand makes it possible for animals with opposable thumbs to pick up minute objects and to use their hands for a variety of purposes. This is one factor which has led to the superior position held by man among his contemporaries.

Carpal bones in mammals consist of several separate elements, although some fusion may occur. Metacarpals are elongated. Usually only two phalanges are present on the first digit, with three in each of the remainder. Reduction in number of digits is common in mammals, the tendency being toward reduction in the following order: 1, 5, 2, 4. In the horse and its close relatives, only the third digit remains. It is this digit

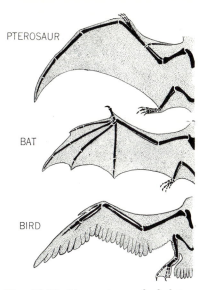

PTEROSAUR

BAT

BIRD

Fig. 10.86 Comparison of skeleton supporting wing in pterosaur, bat, and bird. (After Romanes.)

Its distal end is fused to the radius, which is thus the main skeletal element of the lower arm. The first digit is short and free and bears a long claw. The metacarpals and phalanges of the remaining digits are greatly elongated and support the web, or wing membrane. The third digit is the longest (Fig. 10.86). Comparison of the wings of bats and of pterosaurs is of interest. The same principle is involved in the development of wings in these two groups, but the manner in which it is accomplished differs greatly (page 490).

The pentadactyl limbs of sirenians and cetaceans are webbed to form flippers. The pentadactyl nature of the flipper is very evident in the developing embryo although not superficially apparent in the adult. The skeletal structure differs essentially from other mammals only in the excessive number of phalanges in the central digits.

Mammalian pelvic appendages in general show more variation than is found in the pectoral region. A patella, or kneecap, is present in most mammals. The tibia is the chief bone of the lower leg, the fibula usually being much reduced in size or else fused to the tibia as a small splint. Tarsal bones are distinct entities. A projection of the calcaneum forms the heel. In some of the primates the large toe, or hallux, is opposable in the same manner as the pollex.

Unlike the condition in most mammals, the head of the femur of the orangutan is not attached by a *ligamentum teres femoris* to the acetabulum. This permits greater freedom of movement of the hind limbs but makes them

which bears the hoof. Remnants of the metacarpals of digits 2 and 4 remain as the *small metacarpals*, or *splint bones*. The evolution of the limb of the horse from a primitive pentadactyl ancestor has been traced with a degree of completeness almost without parallel in paleontological investigations. In cattle there is one large and one small metacarpal. The larger of the two is actually a fusion of the third and fourth metacarpals, the line of fusion being clearly indicated by a groove. The small metacarpal is a vestige of the fifth digit. Only digits 3 and 4 are developed. Each consists of three phalanges, the terminal phalanx bearing the hoof in these split-hoofed animals. Several sesamoid bones may be present in addition to the bones just described.

Among the many interesting modifications of the mammalian forelimb is the condition found in bats. Most of the elements are elongated. The ulna is very much reduced, only the proximal portion being present in most cases.

Fig. 10.87 Skeleton of hand of kangaroo.

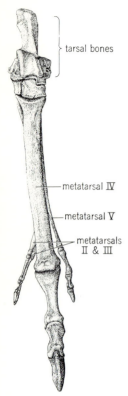

tarsal bones

metatarsal IV

metatarsal V

metatarsals II & III

Fig. 10.88 Skeleton of hind foot of kangaroo.

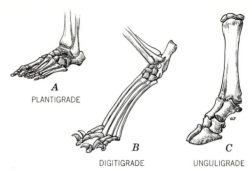

A
PLANTIGRADE

B
DIGITIGRADE

C
UNGULIGRADE

Fig. 10.89 Three types of foot posture in mammals: A, man (plantigrade); B, cat (digitigrade); C, cow (unguligrade).

less strong. These apes walk with difficulty.

In mammals which show a reduction in number of digits, the hind limbs, in most cases, show a greater reduction than do the forelimbs. This is well shown in the kangaroo (Figs. 10.87 and 10.88) in which the hand is pentadactyl but the foot has lost the hallux. In addition, metacarpals 2 and 3 are very slender. The fourth toe is greatly developed and bears a formidable claw at its tip, which is of great value in defense. The fifth toe is less pronounced. The metacarpals of digits 2 and 3 are closely applied to each other, and their basal portions are enclosed in a single integumentary sheath. These two modified toes are used by the animal as a sort of comb in grooming its fur.

Among mammals, three types of foot posture are recognized (Fig. 10.89). The most primitive is the *plantigrade* posture, observed in the hind limbs of man, in bears, and certain insectivores. In the plantigrade animal the entire foot is in contact with the ground during locomotion. *Digitigrade* animals such as cats, dogs, etc., place only their digits on the ground, the wrist and ankle being elevated. In *unguligrade* forms (horse, cow, deer, etc.), only the hoof is in contact with the ground. This type of foot posture is the most specialized of the three, since it deviates most from the primitive condition and is to be found only in the most swiftly running mammals.

Detailed consideration of the various features of the appendicular skeleton in mammals is beyond the scope of this book. A comparative study of mammalian appendages is a most interesting field of biology. Only a few of the salient features have been presented above.

SUMMARY

1. All chordates have a supporting endoskeleton which, in the lowest forms, may consist only of a notochord. In a few groups some bony structures are present in addition, which are referred to as the dermal skeleton. Isolated elements, called heterotopic bones, are present in a few groups of vertebrates. Bones in the hearts of ruminants and penis bones of numerous mammals are examples.

2. The skeleton is composed of cartilage, bone, or a combination of the two. Bones are of two types: (*a*) membrane bones which do not go through a cartilaginous stage in development, and (*b*) cartilage bones which do. The terms intramembranous ossification and endochondral ossification are applied, respectively, to the two types of development.

3. Joints are present at points where bones join, or articulate. There are two main kinds of joints: (*a*) synarthroses, or immovable joints; (*b*) diarthroses, or movable joints. Several types are included under each category.

4. The *dermal skeleton* includes skeletal structures derived from the dermis of the skin. There has been a tendency in evolution to eliminate the dermal skeleton, which is fundamentally a protective device. Among structures included in the dermal skeleton are the scales and fin rays of fishes; the bony plates underlying the epidermal scales of crocodilians, turtles, and a few snakes and lizards; riblike gastralia found in certain reptiles; the bony armor of the armadillo; and the many membrane bones forming parts of the skull and pectoral girdle.

5. The *endoskeleton* consists of axial and appendicular portions. The axial skeleton includes the spinal column, skull, and ribs. The sternum is also usually included. The appendicular skeleton is composed of the skeletal elements of the limbs and limb girdles.

Axial skeleton. *The notochord* is the primitive axial skeleton. In amphioxus, cyclostomes, and a few fishes it persists throughout life, but in most chordates it is replaced by the centra of vertebrae.

The vertebrae are metameric structures extending from the base of the skull to the tip of the tail. They enclose and protect the spinal cord, give rigidity to the body, furnish a place for attachment of the limb girdles, and provide surfaces to which ribs and muscles are attached. A typical vertebra consists of a solid ventral body, or centrum, above which is a neural arch enclosing a neural canal in which the spinal cord lies. Various projections from the vertebrae provide for articulation with adjacent vertebrae and ribs as well as for muscle attachment. In lower forms the vertebrae tend to be similar throughout, but higher in the scale they are grouped in several regions in each of which they show certain peculiar characteristics. Cervical, thoracic, lumbar, sacral, and caudal regions are recognized. Fusion of several vertebrae is encountered in many forms but is most pronounced in birds in which rigidity of the body is important in flight. Otherwise, fusion is most common in the sacral region to which the pelvic girdle is attached in tetrapods.

The skull is derived from three different sources: (1) a chondrocranium, composed of cartilage, (2) a dermatocranium, made up of dermal bones which become attached to the chondrocranium secondarily, and (3) the visceral skeleton, or splanchnocranium, derived from the visceral arches which support the gills in lower forms and which also become closely associated with the chondrocranium.

The chondrocranium is derived by histological differentiation of cartilage in a mesenchymatous condensation which early surrounds the brain. It first appears as a pair of parachordal cartilages on either side of the notochord ventral to the base of the brain. This is followed by the appearance of a pair of anterior prechordal cartilages. Three pairs of sense capsules next make their appearance in the nasal, optic, and otic regions, respectively. A union of prechordal, parachordal, nasal, and otic cartilages results in the formation of the primitive chondrocranium. The optic capsule does not enter into union with the others. The side portions of the chondrocranium grow dorsally for a short distance on either side of the brain. The posterior, or occipital, portion becomes roofed over by cartilage, but the rest remains open. Ossification centers in various parts of the chondrocranium bring about the formation of the several cartilage bones of the cranium.

The dermatocranium consists of membrane bones derived originally from dermal plates of lower forms which have assumed a deeper position in the body. Dermal bones roof over the more anterior part of the chondrocranium to form such bones as the parietals and frontals. Other dermal bones may be closely applied to the lower surface of the chondrocranium, the parasphenoid of bony fishes and lower tetrapods being a notable example. Still other membrane bones become associated with the visceral skeleton.

The visceral skeleton is derived orginally from the cartilaginous visceral arches which support the gills. The first, or mandibular, arch divides into dorsal and ventral portions which become the upper and lower jaws, respectively. Membrane bones come to invest the original cartilage which may ultimately disappear, although endochondral ossification in certain regions may result in the formation of certain cartilage bones. The second, or hyoid, arch in lower forms acts as a suspensorium of the jaws. In higher forms it is reduced and serves merely to support tongue and larynx. The remaining visceral arches are usually lost or reduced. They may contribute certain cartilages to the hyoid apparatus and larynx.

The skulls of cyclostomes and elasmobranchs do not go beyond the cartilage stage. In other fishes all stages can be observed, from those which are entirely cartilaginous to those which are almost completely bony. Bony fishes have the largest number of skull bones of any vertebrates. In amphibians there is a notable reduction in the number of bones. Articulation of the skull with the first vertebra is by means of a pair of occipital condyles. Anuran amphibians have better-developed skulls than urodeles. Reptilian skulls show an increased degree of ossification, and several new bones make their appearance. Fusion of various bones is most pronounced in turtles. Articulation with the first vertebra is by means of a single, median occipital condyle. This is also true of birds, in which the skull is light in weight, but the various bones are fused to a high degree. In all forms below mammals, articulation of the lower jaw with the rest of the skull is at the quadrate bone. In mammals, however, the lower jaw articulates with another bone, the squamosal. Formation of the malleus and incus bones of the middle ear is the result of this shifting in position. Mammalian skulls have an increased cranial capacity, thus accommodating the larger brain. Two occipital condyles are utilized in articulation with the first vertebra. The visceral skeleton shows a gradual reduction as the evolutionary scale is ascended. The various elements of the vertebrate skull show interesting homologies, there being a surprising uniformity in the basic plan of development and arrangement.

The ribs are cartilaginous or bony rods attached at their proximal ends to vertebrae. Primitively there is one pair for each vertebra. There are two kinds of ribs, upper, or dorsal, and ventral, or pleural. The latter represent the spreadout halves of the hemal arch which, anterior to the anus, fail to unite ventrally. They lie just outside the peritoneum. Dorsal ribs appear within the myocommata and lie between hypaxial and epaxial muscle regions. They may be absent or reduced to small tips on the transverse processes of vertebrae. There is disagreement concerning homologies of tetrapod ribs. Many authorities believe that they represent the dorsal ribs of fishes. In most amniotes the ribs in the thoracic region form strong arches which unite ventrally with the sternum. Sacral ribs form a union with the pelvic girdle.

A sternum exists only in tetrapods. It is a midventral structure closely related to the pectoral girdle and, except in amphibians, to the ribs. A parasternum, which is a posterior continuation of the sternum in some reptiles, is derived from ossifications in the ventral portions of the myocommata. The sternum typically consists of a row of separate bones, but these may fuse, as in birds and man. It probably should be considered as a portion of the appendicular skeleton rather than of the axial skeleton.

Appendicular skeleton. The appendicular skeleton consists of the anterior, or pectoral, appendages and girdle and the posterior, or pelvic, appendages and girdle. In fishes the appendages are in the form of fins; in tetrapods they are limbs. The limbs of tetrapods are believed to have evolved from the fins of fishes. The connecting links are to be sought among the fossil rhipidistian crossopterygian fishes. Several theories have been advanced to account for the origin of the paired limbs of vertebrates. The "fin-fold" and "ostracoderm" theories offer plausible suggestions.

FISHES. Neither pectoral nor pelvic girdle in fishes is connected with the vertebral column. Cartilages and cartilage bones, homologous with portions of the girdles of tetrapods, can be identified in the girdles of fishes. Several membrane bones become associated with the pectoral girdle, but not with the pelvic girdle. Of these only the ventral clavicles are represented in higher forms. Large cartilaginous or bony plates, the basalia, are located at the base of the typical fin. From these extend numerous radialia which connect distally with dermal fin rays.

TETRAPODS. The girdles and limbs of all tetrapods show clear-cut homologies in fundamental structure. Pectoral girdles consist of two or three cartilages or cartilage bones. There is a single scapula and usually a coracoid and precoracoid on each side. A glenoid fossa serves as the point of articulation of girdle and forelimb. Membrane bones consisting of clavicle and interclavicle may be added. Coracoids and clavicles unite with the sternum. In higher forms the clavicles may be lost or reduced. The coracoids of mammals, except monotremes, are reduced to a small process on each scapula. The scapula, therefore, becomes the chief component of the pectoral girdle in mammals. Pectoral girdle and vertebral column have no direct connection.

The tetrapod pelvic girdle is united firmly to the sacral region of the vertebral column. It is composed on each side of three cartilage bones: ilium, ischium, and pubic. A depression, the acetabulum, usually located at the point of junction of the three bones, furnishes the articular surface for the femur. Pubes and usually the ischia meet their fellows ventrally to form symphyses.

The limbs of tetrapods are fundamentally similar. Forelimbs and hind limbs are also alike. The upper portion of each limb contains a single bone which articulates with the girdle. The lower portion of each limb contains two parallel bones. Next comes the ankle or wrist, as the case may be, with several small carpal or tarsal bones. This is followed by the hand or foot consisting primitively of five long, narrow bones and of five digits, each of which contains two or three small bones, the phalanges. The tetrapod limb is primitively pentadactyl. Reductions and fusions account for the many variations from the primitive condition that are encountered. Sesamoid bones, which represent ossifications in tendons, are frequently present in various parts of the limb.

| 11 |

MUSCULAR
SYSTEM

All movements of the body or parts of the body are brought about by muscular activity, at least in the higher forms of animal life. Muscular tissue, more than any other tissue in the body, has developed the power of contractility to a very high degree. With few exceptions (page 89) it is of mesodermal origin. Three general types of muscle tissue are recognized: (1) *smooth,* or *involuntary;* (2) *striated, voluntary,* or *skeletal;* and (3) *cardiac,* which is *striated* though *involuntary.* The detailed structure of these three types of muscle tissue has been described in Chap. 3 and need not be repeated here.

In addition to possessing the property of contractility, muscle tissue also has the property of conductivity developed to a greater extent than other forms of protoplasm with the exception of nervous tissue. Muscle cells are commonly referred to as *muscle fibers.* The fact that they are elongated, and that contraction results in a decrease in their length, makes it possible for effective action to take place resulting in movement of one kind or another. The contraction of muscle fibers, furthermore, is of great importance in the production of body heat.

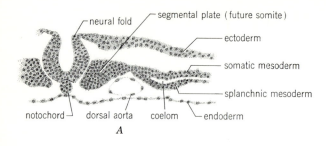

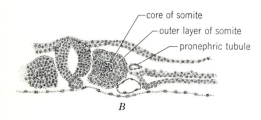

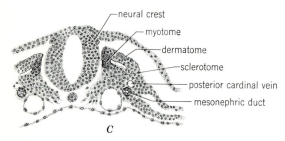

Fig. 11.1 Transverse sections through 33-hour chick embryo at three different levels, illustrating the manner of differentiation of somites: *A*, posterior section through segmental plate; *B*, section through middle of somite region; *C*, one of the more anterior somites, showing differentiation of dermatome, myotome, and sclerotome. The formation of neural tube and neural crests is also shown.

Development. It is important at this point to recall that during early development when proliferating mesodermal cells grow outward between ectoderm and endoderm, they gradually differentiate in such a manner that three different levels or regions may be recognized: (1) a dorsal *epimere*, (2) an intermediate *mesomere*, and (3) a lower *hypomere* (Fig. 3.11). It is with the epimeric and hypomeric portions that we are particularly concerned in our consideration of the muscular system. The mesomere is of importance in the development of the excretory and reproductive organs.

The epimere soon becomes marked off by a series of dorsoventral clefts which form in succession from anterior to posterior ends of the body. Thus a series of blocklike masses of mesoderm, the *mesoblastic somites*, is formed in the epimeric region. The mesomere and hypomere do not usually undergo a similar segmentation. When a somite is first formed it is made up of a practically solid mass of cells, although sometimes a small coelomic cavity is present in the center. With further differentiation of the somite the cells show a sort of radial arrangement around a central core of cells (Fig. 11.1). Next, the ventromedial cells of the somite lose their radial character and together with the cells in the central core form a mesenchymatous mass referred to as the *sclerotome*. These cells continue to proliferate and extend toward the notochord, ultimately to give rise to the vertebral column and proximal portions of the ribs. The dorsolateral part of the somite, which lies directly beneath the ectoderm, is then known as the *dermatome*. The cells of the dermatome later become mesenchymatous and contribute to the dermis of the skin. Some authorities report that the dermatome also may give rise to some muscle cells. The dorsomedial portion of the somite gives rise to the *myotome*. The cells of the myotome proliferate and grow laterally, away from the neural tube but ventral to the dermatome, finally occupying a position parallel to and ventromediad the dermatome (Fig. 11.2). It is the myotomes which undergo the most extensive development of any parts of the somites since they, possibly together with mesoderm of the somatic layer of the hypomere (see below), give rise to the greater part of the skeletal musculature of the body. It is with the myotomes, therefore, that we shall be primarily concerned in our discussion of the muscular system.

The *hypomere* is divided into two layers of mesoderm, somatic and splanchnic, sep-

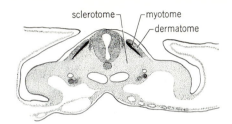

Fig. 11.2 Transverse section of 48-hour chick embryo, showing further differentiation of a somite.

arated by a space, the coelom, or body cavity. The somatic layer is closely applied to the ectoderm, the two together giving rise to the definitive body wall. The splanchnic layer surrounds the endoderm of the gut. This layer is of importance in forming the smooth muscle of the gut and also the cardiac muscle. Furthermore, certain *branchial, or branchiomeric,* muscles in the gill region of aquatic forms, which are voluntary and striated, originate from the splanchnic layer of the hypomere rather than from myotomes. They are thus homologous with the smooth muscle of the gut (visceral muscles) despite their striated appearance and the fact that they move under voluntary control. They are used in moving the visceral arches. Derivatives of branchial muscles are also present in higher vertebrates.

The cells of the dorsally located myotomes grow down in the lateral body wall between the outer ectoderm and the somatic layer of hypomeric mesoderm (Fig. 11.3). The sheets of cells thus formed on the two sides almost meet at the midventral line, being separated from each other only by a longitudinal band of connective tissue, the *linea alba*. Connective tissue also comes to occupy the spaces between adjacent myotomes. These partitions are called *myocommata*, or *myosepta*. They extend from the

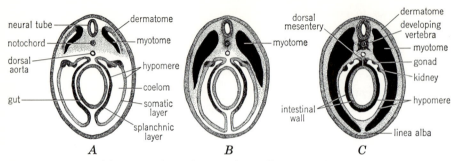

Fig. 11.3 Series of diagrams of vertebrate embryo, illustrating the epimeric origin of the myotomes and the manner in which they grow ventrally to form the parietal musculature of the body wall. The splanchnic layer of the hypomere surrounds the gut and contributes to the muscle and other mesodermal structures of the intestinal wall.

spinal column and wall of the coelom (peritoneum) outward to the dermis of the skin. As a result of the foregoing processes, the basic musculature of the trunk region of the body is clearly defined. The cells of the original myotome which give rise to muscle tissue are called *myoblasts*. They begin to differentiate, become spindle-shaped and arranged in bundles. Mitotic divisions then occur, resulting in an increase in mass. Further differentiation takes place until the cells give the typical syncitial histological picture of striated muscle. The muscles derived from the myotomes, and possibly from contributions from the somatic layer of the hypomere, are referred to as *somatic,* or *parietal,* muscles. These may be further subdivided into *axial* and *appendicular* muscles, depending upon whether they are confined to the region of the body wall or whether they are associated with the appendages. In fishes the axial muscles are of greater importance than the appendicular muscles since their contraction is responsible for locomotion in these animals. In tetrapods, however, in which locomotion depends largely upon movements of the limbs, the appendicular muscles have assumed greater importance and the axial muscles,

although still functional, have taken a minor role.

Terminology. Smooth, or involuntary, muscles are arranged in continuous sheets as in the walls of the digestive tract, blood vessels, and the like. They are not readily dissected. Voluntary muscles, however, are arranged as separate masses which are fairly easily separated from one another. In some cases identity is difficult to establish because of fusion of two or more muscles to form a single mass or because of separation of a single mass into two or more components. Each end of a voluntary muscle is attached to some structure of the body, but the middle, or *belly,* is usually free. Although muscles are generally attached to bony or cartilaginous skeletal parts, this is not always the case. Usually one end is attached to a less movable part than the other (Fig. 11.4). This is the *origin* of the muscle. The end attached to the more movable part is the *insertion.* A single muscle may have more than one origin or insertion. In many cases muscles are attached to skeletal structures by means of *tendons.* These are white, fibrous cords or bands of regularly arranged connective tissue, strong and inelastic. The

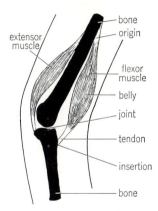

extensor muscle
bone
origin
flexor muscle
belly
joint
tendon
insertion
bone

Fig. 11.4 Diagram showing method of origin and insertion of flexor and extensor muscles.

term *aponeurosis* is applied to a broad, flat, ribbon-shaped tendon. In certain regions a small bone may develop within a tendon at a point where the latter moves over a bony surface. Such bones are called *sesamoid* bones. The kneecap, or *patella*, is an example of a sesamoid bone.

Sheets or bands of connective tissue called *fasciae* surround muscles, groups of muscles, and the body musculature as a whole. They tend to bind the parts of the body together and in some cases serve as points of muscle origin or insertion as well.

Muscles are usually present in groups of two, each group working in exactly the opposite manner of the other (Fig. 11.4). Such groups are named according to their action, thus:

Flexors tend to bend a limb or to bend one part of a limb against another.

Extensors tend to straighten a limb or one of its component parts.

Abductors draw a part away from a median line or from a neighboring part or limb. Abductors of a limb swing the limb in a direction away from the median longi-

tudinal axis of the body; abductors of the digits move the digits in a direction away from the median longitudinal axis of the limb.

Adductors draw a part toward a median line or toward a neighboring part or limb. Adductors of a limb swing the limb in a direction toward the median longitudinal axis of the body; adductors of the digits move the digits in a direction toward the median longitudinal axis of the limb.

Rotators are muscles which revolve a part on its axis. Some rotators are called *pronators* when, as in the case of the arm, they turn the palm of the hand downward. Others, referred to as *supinators,* serve to turn the palm forward and upward.

Elevators, or *levators*, raise, or lift, a part, as in the case of closing the mouth by raising the lower jaw.

Depressors are those which lower, or depress, a part, as when the lower jaw is depressed to open the mouth.

Constrictors draw parts together or contract a part. When constrictors surround an opening such as the mouth, anus, or pylorus, they are termed *sphincters*.

PARIETAL MUSCULATURE

Smooth-muscle fibers are not very highly differentiated cells. Their slow and rhythmic contraction is under control of the autonomic nervous system. In addition to bringing about active contractions, smooth muscle is also responsible for maintaining a condition of sustained, or prolonged, contraction known as *tonus*, by means of which the walls of tubular organs are kept at a relatively constant diameter. The manner in which smooth-muscle fibers encircle the cavities, or lumina, of such tubular organs as the intestine, arteries, etc., is of great physiological importance in the passage of

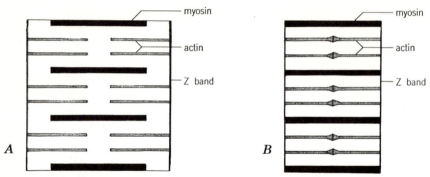

Fig. 11.5 Diagrammatic representation of portion of a sarcomere: *A*, in the relaxed position, two fine myofilaments of actin, lying between the coarser ones composed of myosin, are spread apart; *B*, in the contracted position, the fine myofilaments have slid toward each other and joined. T filaments are not indicated. (*Based on H. E. Huxley's concept.*)

food through the digestive tract, in the regulation of blood pressure, and the like.

The action of striated, voluntary, parietal muscle is of an entirely different character. It must be capable of extremely rapid contraction in response to the will, as well as maintaining a partial state of contraction, or *tonus*, which gives form to the body even when at rest and which is responsible for such phenomena as being able to stand up for prolonged periods, for keeping the head erect, etc. The minute structure of the myofibrils (page 91) of striated muscle fibers is very complex and will not be considered here in any detail. Suffice it to say that the alternating light and dark bands, or discs, are a reflection of the arrangement of two kinds of myofilaments, *coarse* and *fine*, which are arranged longitudinally within the confines of the *sarcomeres*, or the units of which the myofibrils are composed. Each sarcomere is marked at either end by a distinct, dark Z band which is probably not of the nature of a membranous septum. It has been found that striated muscle contains three proteins, *myosin, actin,* and *tropomyosin.* The coarse myofilaments are believed to be composed of myosin; the fine filaments seem to be made up of actin and probably also contain some tropomyosin. The coarse and fine myofilaments are so arranged that two fine myofilaments lie between a pair of coarse ones (Fig. 11.5). The number of myofilaments within a sarcomere is variable in different myofibrils. Upon proper nervous stimulation, the two kinds of filaments slide along each other in such a way as to shorten, or narrow, the sarcomere which, as a result, becomes somewhat thicker. Recent researches indicate the additional presence in striated muscles of very thin elastic filaments, or T filaments, 20 to 25 Å in diameter,* which course the entire length of the sarcomere and possibly of the myofibril. They run parallel to the actin and myosin myofilaments (G. Hoyle, 1967, *Amer. Zoologist,* 7:435). The existence of T filaments is an answer to certain questions which long have puzzled investigators in this field.

Food substances, of course, are carried to muscle cells by the bloodstream. Carbohydrate, in the form of glycogen, is stored

* An angstrom unit (Å) is 10^{-8} centimeters.

in muscle cells. It is this glycogen which basically provides the energy released by muscular activity. It is obvious, however, that the oxidation of glycogen cannot occur with the rapidity required in the contraction of striated muscle. A chemical substance called *adenosine triphosphate*, also present in striated muscle fibers, is of great importance in the rapid release of energy. It belongs in the category of nucleotides. These are the chemical units of which nucleic acids are composed; they contain nitrogen combined with phosphoric acid and a simple sugar. Adenosine triphosphate when properly stimulated breaks down very rapidly to *adenosine diphosphate*, accompanied by release of a large amount of energy. The glycogen store is used in re-synthesizing adenosine triphosphate from the diphosphate. This process is a much slower type of reaction.

Fundamentally the voluntary, striated, parietal musculature of chordates is made up of a linear series of myotomes extending from anterior to posterior ends. Each myotome lies opposite the point of articulation of two adjacent vertebrae and is supplied by the somatic motor fibers of a separate spinal nerve which emerges from the neural canal through an intervertebral foramen (Figs. 10.13 and 11.6). Whether or not the chordate head is primarily metameric has been the subject of much speculation on the part of comparative anatomists for many years. It is generally agreed that certain muscles of the head, like those of the trunk, are fundamentally metameric and are supplied by somatic motor fibers of cranial nerves rather than by spinal nerves (page 515). In higher vertebrates the metameric condition has been obscured by evolutionary changes, so that it is rather difficult to recognize homologies. Considerable confusion exists regarding homologies among the muscles of various vertebrates. In numerous cases certain muscles have been given names which are identical with those in the human being because they are located in similar positions. Such muscles may or may not be homologous with the similarly named muscles of man. Embryological studies, as a basis for ascertaining homologies, have received but scant attention. One of the most reliable criteria in determining muscle homologies is the nerve supply to a muscle. This, however, is not always infallible. Despite pronounced evolutionary changes, the nerve supply has remained relatively constant. The identification and tracing of small branches of nerves is a very difficult task, and for that reason the subject of muscle homology is most complex and is far from having been completely worked out. Only some of the broader aspects of the subject are treated in this volume.

Trunk musculature. AMPHIOXUS. In amphioxus the myotomes, when viewed from the side (Fig. 2.3), are observed to be V-shaped structures with the apex of the V pointing forward. The muscle fibers within the myotome course in a longitudinal direction. The fibers are interrupted from myotome to myotome by the presence of myo-

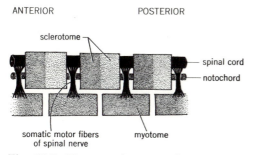

ANTERIOR   POSTERIOR

sclerotome

spinal cord

notochord

somatic motor fibers   myotome
of spinal nerve

Fig. 11.6 Diagram showing relation of myotome to sclerotomes of vertebrae and to the somatic motor portion of a spinal nerve. (*See also Fig. 10.13.*)

commata, to which they are attached anteriorly and posteriorly. The actions of the myotomes of the two sides alternate with each other in amphioxus. When the muscles of one side of the body are contracted, those on the other side are relaxed. By contracting fibers on alternate sides of the body in different regions, the animal is able to swim rather rapidly with a wriggling, undulatory motion. Because of the oblique arrangement of the myotomes, a cross section through the animal would cut through several of these structures (Fig. 2.5).

CYCLOSTOMES. In cyclostomes, in the absence of paired appendages, the regular segmental arrangement of the myotomes has undergone little change from that of amphioxus. The myotomes, however, instead of being V-shaped, are more nearly vertical. Each is bent forward slightly at its dorsal and ventral ends (Fig. 11.7). As a result the muscle fibers have a slightly radial ar-

rangement. Both myotomes and myocommata take a strongly oblique and posteriorly directed course in passing from the skeletal axis and peritoneum to the dermis of the skin. A cross section through the body wall, therefore, cuts across several myotomes (Fig. 8.27).

A series of *hypobranchial* muscles, not found in amphioxus, is present ventral to the gill region. These arise from the first few myotomes posterior to the gill region. During early development, outgrowths extend from these postbranchial myotomes in an anteroventral direction to give rise to the hypobranchial musculature.

FISHES. Beginning with the elasmobranch fishes, the myotomes of gnathostomes are separated into dorsal, or *epaxial,* and ventral, or *hypaxial,* portions. This division is brought about by the presence of a longitudinal *lateral septum* of connective tissue which extends laterally from the vertebral column to the

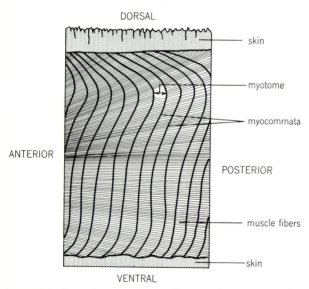

Fig. 11.7 Lateral view of several parietal myotomes of lamprey in region just anterior to the first dorsal fin. The skin has been removed to show the muscle fibers beneath.

Anatomy of the Chordates

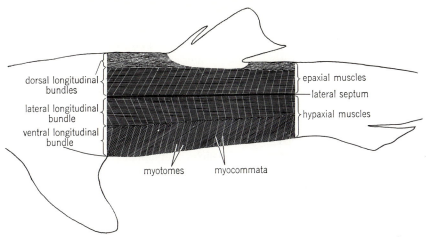

dorsal longitudinal bundles
lateral longitudinal bundle
ventral longitudinal bundle

epaxial muscles
lateral septum
hypaxial muscles

myotomes   myocommata

Fig. 11.8 Lateral view of portion of body wall of dogfish, *Squalus acanthias,* showing zigzag arrangement of the myotomes. The skin has been removed to expose the muscle fibers beneath.

skin at the point where the lateral line is located.

The myocommata now take a zigzag course (Fig. 11.8), but the muscle fibers in the myotomes, particularly those in the epaxial region, continue to run in a longitudinal direction. The epaxial muscles on each side are arranged in the form of two or three *dorsal longitudinal bundles* extending from the base of the skull to the end of the tail. They form rather large muscle masses used primarily in bending the body from side to side in swimming. The hypaxial muscles are divided into *lateral* and *ventral longitudinal bundles.* The lateral longitudinal bundle lies below the lateral septum and is attached anteriorly to the scapular process. In *Squalus acanthias* it is darker than the other muscles. Anterior to the pelvic fins, its fibers take a slightly oblique course anteriorly and upward, but posteriorly they course in a horizontal direction. The ventral longitudinal bundle, which in turn is divided into two parts, overlaps the lateral bundle to

a slight degree. The ventral longitudinal bundle is attached anteriorly to the ventral region of the pectoral girdle. Its fibers are arranged obliquely and course in an anteroventral direction. This is indicative of the direction in which the "pull" caused by contraction of the muscle fibers is being exerted. In some elasmobranchs the portion of the ventral longitudinal bundle on either side of the linea alba is further differentiated into a long but narrow *rectus abdominis* muscle in which the fibers are longitudinally arranged.

Each trunk myotome is supplied by somatic motor branches of a separate spinal nerve. With the appearance of the lateral septum, the epaxial muscles are innervated by the dorsal rami and the hypaxial muscles by the ventral rami of the spinal nerves.

As in cyclostomes, the ventral portions of the first myotomes posterior to the gill region grow forward in an anteroventral direction to form the hypobranchial musculature. Since the hypobranchial muscles are derived from myotomes, they are supplied by somatic motor branches of spinal nerves. They should not be confused with the *branchial* muscles, to be described later, which are derivatives of the anterior visceral musculature and which are innervated by visceral motor branches of certain cranial nerves (V, VII, IX, and X).

In fishes certain of the epaxial and hypaxial muscles become specialized, sending portions to the median fins, which thus can be moved in various directions.

The inner epaxial muscles of fishes are attached to the vertebral column. The deeper portions of the hypaxial muscles are attached to myocommata or to ventral ribs, if present. These muscles presage the appearance of intercostal, oblique, and rectus abdominis muscles of higher forms. As yet they do not form distinct muscle masses but have the usual metameric appearance, with myocommata being interposed between adjacent myotomes.

AMPHIBIANS. The axial trunk musculature of tetrapods has undergone considerable modification from that observed in lower aquatic forms. The original metameric condition becomes more and more obscure as the evolutionary scale is ascended. In most tetrapods the appendicular muscles have developed extensively and the axial musculature of the trunk, which has been reduced in volume, is relatively of minor significance in relation to body movement.

Urodele amphibians do not have the epaxial muscles so well developed as do fishes, nor are they used to the same extent in locomotion. The lateral septum is more dorsal in position as are the transverse processes of the vertebrae. Despite the reduction in epaxial musculature, the arrangement of the myotomes in this region is primitive. The number of myotomes corresponds to the number of vertebrae, and the muscle fibers run from one myocomma to the next, the entire dorsal, or epaxial, muscle mass forming the *dorsalis trunci* (Fig. 11.9). The myotomes are vertical in position and do not exhibit the zigzag arrangement observed in faster-swimming fishes. The muscle fibers

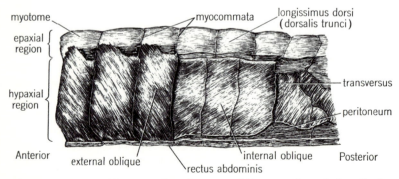

Fig. 11.9 Muscles of portion of trunk region of *Necturus,* lateral view. In the fourth, fifth, and sixth segments from the left, the external oblique layer has been removed to show the underlying internal oblique layer. In the seventh and eighth segments, both external and internal oblique layers have been removed to expose the transversus. The ventral portions of the internal oblique and transversus have been removed to show the underlying peritoneum.

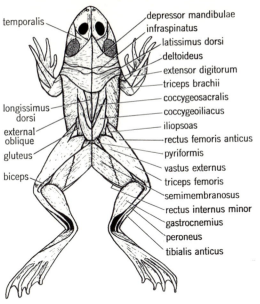

temporalis
depressor mandibulae
infraspinatus
latissimus dorsi
deltoideus
extensor digitorum
triceps brachii
coccygeosacralis
longissimus dorsi
coccygeoiliacus
external oblique
iliopsoas
gluteus
rectus femoris anticus
biceps
pyriformis
vastus externus
triceps femoris
semimembranosus
rectus internus minor
gastrocnemius
peroneus
tibialis anticus

Fig. 11.10 Dorsal view of bullfrog, *Rana catesbeiana*, with skin removed, showing the dorsal musculature.

adjacent to the vertebrae are attached to them and are referred to as *intersegmental bundles*.

In anurans the epaxial muscles are still further reduced. They are utilized in bending the vertebral column dorsally rather than laterally. The dorsalis trunci has become further differentiated into *intertransversarial* muscles, between the transverse processes, and *interneural* muscles, between the neural arches. In most frogs the outer fibers of the dorsalis trunci on each side form a long muscle, the *longissimus dorsi*, which passes all the way from the head to near the end of the urostyle (Fig. 11.10). These muscles take the form of a V, the apex pointing posteriorly. The outer fibers are not attached to the vertebrae to any extent. In both urodeles and anurans the anterior region of the dorsalis trunci has split up into several muscle masses,

of various names, which are attached to the skull and are effective in turning the head.

The hypaxial muscles of amphibians have undergone greater modification than those in the epaxial region. In larval urodeles the condition is much like that observed in fishes, but in adults these ventral trunk muscles are arranged in four distinct layers, or sheets. Beginning from the outside there are *superficial* and *deep external obliques*, in which the muscle fibers course in a posteroventral direction, an *internal oblique* layer, in which they extend in a posterodorsal direction, and a *transversus*, with fibers almost in a vertical position (Fig. 11.9). The transversus lies next to the peritoneum. On either side of the linea alba a rather well-developed *rectus abdominis*, derived from the obliques and with primitively arranged longitudinal fibers, extends from the head to the pelvic girdle. In some forms this muscle may split into superficial and deep layers.

In certain species of urodeles variations from the above occur, consisting for the most part of a reduction in the number of layers. This is brought about by fusion. Within the group there is also a tendency toward reduction or loss of the myocommata, so that the muscles begin to appear as distinct entities with independent actions.

The more specialized anurans go even further in reducing the number of layers in the hypaxial trunk musculature, there being only an outer superficial external oblique layer and an underlying transversus. The rectus abdominis is now a large muscle extending from sternum to pubis, its two halves being separated in the midventral line by the linea alba. Myocommata have disappeared from the external oblique and transversus but apparently are retained to some extent in the rectus abdominis as *tendinous inscriptions*, which are transverse in position and divide the muscle into segments.

The muscles which move the tongue in amphibians are derived from the hypobranchial musculature.

REPTILES. The trunk musculature of reptiles shows a still further deviation from the primitive condition leading to the more complex arrangement found in mammals. A distinct lateral septum, separating epaxial and hypaxial muscles, is lacking in amniotes. The most conspicuous change occurs in connection with the appearance of ribs.

The epaxial muscles of reptiles become differentiated into several groups. Immediately next to the vertebral column lie the *spinalis* and the more deep and laterally situated *semispinalis* muscles. These are rather long muscles originating on the dorsal portions of the vertebrae and inserting on more anterior vertebrae or on the skull. They are sometimes referred to collectively as the *transverse-spinalis system* of muscles. The deeper muscles in this region split into numerous short muscles, of various names, located between adjacent vertebrae or between vertebrae and ribs. The longissimus dorsi of amphibians now originates at the ilium and becomes separated into (1) a *longissimus dorsi proper*, confined to the lumbar region; (2) a *longissimus capitis*, located on either side of the neck and extending to the skull in the region of the temporal bone; and (3) a laterally situated *iliocostal dorsi*, passing to the proximal ends of ribs. In the neck region the iliocostal dorsi passes to the atlas vertebra and occipital portion of the skull. In a number of forms the presence of tendinous inscriptions indicates that these muscles have originated from segmental myotomes. As would be expected, the epaxial muscles of snakes are very well developed, whereas those of turtles are greatly reduced.

The hypaxial musculature of reptiles exhibits an even greater change from the primitive condition. In such forms as lizards and crocodilians, these muscles, in the abdominal region at any rate, are much like those of urodele amphibians, consisting of external and internal oblique and transversus layers. In the thoracic region, however, additional layers are present. Between external and internal obliques an *intercostal layer*, derived from the obliques, makes its appearance, with fibers connecting adjacent ribs. This is subdivided into *external* and *internal intercostals*. The fibers of these two sets of muscles run in opposite directions and are used in respiratory movements. The oblique muscles also give rise to the *scalene muscles*, which pass from the anterior ribs to the cervical vertebrae, and to the *serratus muscles* passing from the more posterior ribs to the inner surface of the scapula or to the thoracic and anterior lumbar vertebrae. The rectus abdominis is well developed in lizards and crocodilians, and tendinous inscriptions are prominent.

In snakes the ventral scales are used in locomotion. Small muscles derived from the obliques and rectus pass from the ribs to the skin underlying the scales. These, together with certain dermal muscles, are responsible for the movement of the ventral scales (Fig. 11.21).

The hypaxial muscles of turtles are poorly developed in association with the rigid shell which narrowly restricts movements of the body.

The fact that the number of hypaxial muscle layers, as found in urodele amphibians, is reduced in anurans but increases in reptiles arouses interesting speculations as to evolutionary trends. In both cases animals have taken to terrestrial existence, but this factor alone would seem to be of minor importance. The increased number of layers in reptiles has developed in association with the appearance of ribs, which, of course, are

Anatomy of the Chordates

lacking in anurans. This fact rather than emergence to life on land is of importance in considering evolutionary trends.

BIRDS. The trunk musculature of birds is quite different from that of other forms. This is not surprising when one considers their specialized mode of life. Epaxial muscles are poorly developed. Among the changes to be observed are the absence of the transversus in the abdominal region and reduction of the oblique muscles. A remnant of the transversus is recognized in the *triangularis sterni* on the inner portion of the ribs near where they are attached to the sternum.

MAMMALS. In mammals the trunk muscles do not deviate markedly from those of reptiles but the hypaxial musculature is somewhat reduced. With the greater importance of the limbs in locomotion there has been an increase in the muscles of the limbs and limb girdles. Some of these muscles overlie the greater part of the trunk musculature. The original metameric arrangement of the trunk myotomes is now only slightly indicated. The *intercostal, serratus, scalene, intervertebral,* and *rectus abdominis* are trunk muscles in which the original segmental arrangement is retained, at least in part.

The epaxial muscles of mammals show great similarity to the reptilian condition. A bewildering variety of names is applied to the numerous muscles of this region which, if described in detail, would only add to the confusion of the student. Only a few of the more obvious muscles will be mentioned. A *sacrospinalis* muscle originates from the sacrum and spinous processes of the posterior vertebrae. Anteriorly it splits into three columns on each side: a lateral *iliocostal,* an intermediate *longissimus dorsi,* and a medial *spinalis dorsi* muscle. Each of these continues forward to the cervical region, but only the longissimus dorsi and spinalis dorsi go as far as the skull.

A *multifidus* muscle represents the transverse spinalis system of reptiles. It fills in the groove on either side of the spinous processes all the way from the sacrum to the axis. The deeper epaxial muscles again form *interspinal* and *intertransversarial* muscles between adjacent vertebrae.

The hypaxial muscles of mammals are but little modified from the reptilian condition, at least in the abdominal region. The muscles of the abdominal wall consist of an inner *transversus abdominis,* a middle *internal oblique,* and an outer *external oblique* which lies beneath the integument. The large aponeurosis of the external oblique interlaces with that of its fellow of the opposite side at the linea alba. A portion of the aponeurosis is continued posteriorly as the *inguinal ligament.* The *quadratus lumborum* is a dorsoposterior derivative of the external oblique. The internal oblique is smaller than the external and best developed in its dorsal portion. Its large aponeurosis also passes to the linea alba. The transversus is a thin muscle layer quite similar to the internal oblique in arrangement. The *cremaster* muscle of the scrotum is continuous with the internal oblique and in some cases with the transversus. The rectus abdominis in mammals lies on either side of the linea alba just as in lower forms. It is enclosed in a sheath formed from the aponeuroses of the oblique and transversus muscles. A few remnants of the rectus abdominis, which originally extended the length of the body, are to be found in mammals in the neck region. They include the *sternohyoid, sternothyroid, geniohyoid, omohyoid,* and *thyrohyoid.* These are discussed further under the heading "Hypobranchial musculature." Tendinous inscriptions are clearly indicated on the rectus abdominis but only to a minor extent on its anterior derivatives.

In the thoracic region the mammalian

hypaxial muscles show a reduction from the reptilian condition. They are almost entirely covered by the large appendicular muscles associated with the pectoral girdles and limbs. *External intercostals* pass from one rib to the next, their fibers running posteroventrally in an oblique direction. *Internal intercostals* lie beneath the external intercostals. They also pass between adjacent ribs, but their fibers run anteroventrally in an oblique direction. The intercostals are used in respiration, the external intercostals serving to pull the ribs forward, whereas the internal intercostals may pull them back to their original position. Under the internal intercostals lies the transversus which, in the thoracic regions, is referred to as the *transversus thoracis.* It is confined to the inner surface of the anterior thoracic wall. *Scalene* muscles pass from the anterior ribs to the transverse processes of the cervical vertebrae. The *serratus* muscles pass from the angles of more posterior ribs to the inner surface of the scapula or to thoracic and anterior lumbar vertebrae. These are trunk muscles originally derived from the obliques, their hypaxial origin being indicated by the fact that they are supplied by the ventral rami of spinal nerves.

GENERAL. The trunk musculature of all vertebrates is built upon the same fundamental plan. The metameric arrangement of myotomes observed in adults of lower forms becomes less apparent with advance in the evolutionary scale. This variation from the primitive condition occurs as myotomes increase in size and thickness and as the connective-tissue septa, or myocommata, between them disappear. Many of the long trunk muscles have developed in this manner. In other cases there has been a degeneration of portions of certain myotomes leading to the formation of aponeuroses and fasciae. Other myotomes may be drawn out of position or may migrate to new regions, thus obscuring their original arrangement and position. In some cases portions of adjacent myotomes may fuse to form a single muscle.

Hypobranchial musculature. In cyclostomes, elasmobranchs, and some other fishes, as has been mentioned previously, the hypaxial portions of the first few myotomes posterior to the gill region send buds or branches anteriorly and ventrally to form the hypobranchial musculature. The hypobranchial muscles are actually continuations of the ventral longitudinal bundles. In *Squalus acanthias* (Fig. 11.20) these muscles consist of (1) the *common coracoarcuales,* arising from the coracoid cartilage and inserting on the strong connective tissue comprising the floor of the pericardial cavity, (2) a *coracomandibular,* originating from the fascia at the anterior end of the common coracoarcuales and inserting on the symphysis of the mandible, (3) paired *coracohyoids,* coming from the coracoarcuales and passing to the basihyal and ceratohyal cartilages, (4) several *coracobranchials,* which originate from fascia above the coracoarcuales and insert on the ceratohyal and other hypobranchial cartilages. These hypobranchial muscles are considered by some to be derived from the hypomeric branchial musculature, but they are undoubtedly of myotomic origin.

In tetrapods the hypobranchial muscles have been greatly modified. Several muscles in the neck and throat regions are hypobranchial muscles which, like all trunk muscles, are innervated by spinal nerves or their derivatives. A forward continuation of the rectus abdominis in the neck region is referred to as the *rectus cervicis.* From it are derived such muscles as the *sternohyoid, thyrohyoid, sternothyroid, genio-*

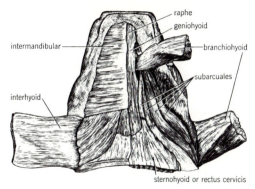

intermandibular

interhyoid

raphe
geniohyoid

branchiohyoid

subarcuales

sternohyoid or rectus cervicis

Fig. 11.11 Muscles of head of *Necturus*, ventral view.

hyoid, and *omohyoid* (Fig. 11.11). These muscles as well as the intrinsic muscles of the tongue, which are also of hypobranchial origin, are, in amniotes, supplied by the hypoglossal nerve which is a purely somatic motor cranial nerve (XII). It is generally conceded, however, that the hypoglossal nerve, which is found only in amniotes, represents a union of the first two or three spinal (spino-occipital) nerves such as are found in lower forms and which are somatic motor nerves. The fact that the above-mentioned muscles are supplied by the hypoglossal nerve furnishes additional evidence of their hypobranchial origin.

Study of the development of the intrinsic tongue muscles of mammals shows that they do not arise as anterior extensions of hypobranchial muscles, as do the tongue muscles of lower forms. Instead, they gradually differentiate directly (*in situ*) from mesenchyme within the developing tongue. Nevertheless, because the tongue muscles are innervated by the hypoglossal nerve, they are considered to be phylogenetic derivatives of hypobranchial muscles. Apparently this is an example in which a certain stage of development is omitted or glossed over so rapidly that it cannot be discerned.

Eye muscles. The hypoglossal nerve is not the only cranial nerve which bears somatic motor fibers. Certain others (III, IV, VI) are somatic motor nerves which supply the muscles moving the eyeball and which are derived from myotomes in the head region.

Primitively a series of somites is believed to have existed in the head region from the anterior part of the head to the trunk. Their presence is still indicated in cyclostome and elasmobranch embryos. In the ear region, however, these somites have been crowded out and have disappeared, for the most part. Three of the most anterior ones persist and appear uniformly throughout the vertebrate series. They are known as the *pro-otic somites,* referred to respectively as the *premandibular, mandibular,* and *hyoid somites* in reference to their location (Fig. 11.12). The myotomes of these somites contribute to the parietal musculature. As in the case of other myotomes of the body, each is supplied by a single nerve. The pro-otic somites are innervated by cranial nerves

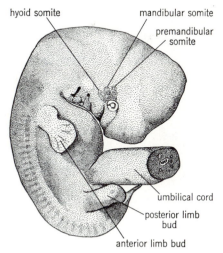

hyoid somite

mandibular somite

premandibular
somite

umbilical cord

posterior limb
bud

anterior limb bud

Fig. 11.12 Human embryo at an early age, showing position of pro-otic somites.

III, IV, and VI, respectively, rather than by spinal nerves.

The vertebrate eye first makes its appearance in cyclostomes. Its structure is essentially similar in all vertebrate groups. With the advent of the eye, six muscles, which move the eyeball, make their appearance. All six are derived from the three pro-otic myotomes. The first of these differentiates into four eye muscles: *superior rectus, inferior rectus, internal rectus*, and *inferior oblique* (Fig. 11.13), all of which are supplied by the oculomotor nerve (III). The second pro-otic myotome develops into the *superior oblique* muscle supplied by the trochlear nerve (IV). The third myotome, which becomes the *external rectus*, is innervated by the abducens nerve (VI).

In the lamprey the abducens nerve appears to supply the inferior rectus muscle as well as the external rectus. The oculomotor nerve, on the other hand, seems to innervate only the superior rectus, internal rectus, and inferior oblique. This peculiar and seemingly paradoxical situation raises questions of homology. It has been demonstrated, however, that in the lamprey the sixth cranial nerve, which leaves the brain rather far forward and close to the oculomotor nerve, actually contains fibers of the third nerve and the latter are the ones which pass to the inferior rectus.

Several other small muscles associated with the eye in numerous tetrapods are also derived from the pro-otic myotomes. One of these, the *levator palpebrae superioris*, used in elevating the upper eyelid, is innervated by the oculomotor nerve, thus indicating its probable origin from the premandibular pro-otic somite. Another muscle, the *retractor bulbi*, which in such species as the frog is used in retracting, or pulling, the entire eyeball deeper within the orbit, is a derivative of the third, or hyoid, pro-otic somite. The muscle fibers moving the nictitating membrane, if present, are also derived from the third pro-otic somite since, like the external rectus muscle, they are innervated by the abducens nerve.

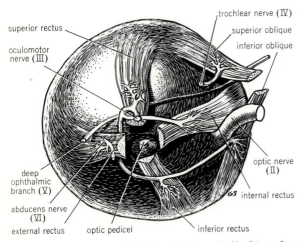

Fig. 11.13 Posteromedial view of left eyeball of *Squalus acanthias*, showing nerve supply to the six eye muscles. (*Drawn by G. Schwenk.*)

Other muscles of the head. The remaining muscles of the head, as well as many in the neck region, are derivatives of the dermal or of the branchial musculature. The latter are derived from the splanchnic mesoderm of the gut in the gill region and not from myotomes. It will be recalled that unlike the involuntary smooth muscle of the gut, with which they are homologous, the branchial muscles are of the voluntary, striated type. Since they are not of myotomic origin, they do not come under the category of parietal muscles and are discussed elsewhere (page 523).

Diaphragm. The muscular diaphragm, present only in mammals, is derived embryonically from the *septum transversum* and certain other membranes. The portion adjacent to the sternum, ribs, and vertebral column becomes invaded with muscle tissue to varying degrees in different species. This is contributed by the hypaxial portion of myotomes in the cervical region and seems to be derived mainly from the anterior portion of the rectus abdominis. The fact that the diaphragm is innervated by branches of certain cervical spinal nerves (phrenic nerves) indicates the embryonic origin of these muscular constituents (Fig. 13.19). The lower thoracic nerves also aid in innervating the diaphragm.

Appendicular musculature. One of the most important advances in the muscular system of vertebrates takes place in connection with the appearance of paired appendages first found in the pectoral and pelvic fins of elasmobranch fishes. During early development when the myotomes are growing ventrally in the body wall, hollow buds grow out from them into the folds from which pectoral and pelvic fins arise (Fig. 11.14). They develop into the muscles which move

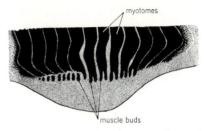

Fig. 11.14 Developing appendage of elasmobranch fish, *Pristiurus*, showing muscle buds arising from the myotomes. (*After Rabl.*)

the fins. These muscles are, therefore, derivatives of the myotomes of the trunk. Since the muscles are modified to course in different directions, their original metameric arrangement may not be apparent.

Two types of appendicular muscles are generally recognized, i.e., *extrinsic* and *intrinsic*. Extrinsic muscles attach the girdle or limb to the axial skeleton directly or indirectly. They originate in the axial musculature and are inserted on some skeletal element within the limb. They serve to move the entire appendage. Intrinsic muscles have both origin and insertion on parts of the limb skeleton itself, serving to move parts of the limb rather than the appendage as a whole. Actually the distinction between extrinsic and intrinsic muscles is not of too great importance since the appendicular muscles, at least in fishes, are all derived from myotomes.

In various species of elasmobranchs, different numbers of buds from the myotomes are concerned in the development of the fin musculature. In the electric ray, *Torpedo*, 26 myotomes are involved in forming the muscles of the pectoral fins. Each bud divides into anterior and posterior primary buds. These then give off dorsal and ventral secondary buds. The two dorsal buds of each myotome concerned give rise to the dorsal musculature of the fin; the ventral buds form the ventral fin muscles. The dorsal muscles become the extensors or levators of the fins; the ventral muscles become flexors or depressors.

In elasmobranchs, as illustrated by *Squalus acanthias* (page 525), the dorsal, or extensor, muscle of the pectoral fin consists of a single sheet, some fibers of which have their origin on the scapula while others originate from the fascia covering the hypaxial portions of the myotomes. These fibers insert, respectively, on the propterygium and on the distal radialia and connective tissue covering the dermal fin rays. Other fibers originate from fascia covering the medial surface of this sheet and all three of the basal cartilages. They also insert on the radialia and connective tissue covering the fin rays. The ventral, or flexor, musculature of the pectoral fin is more elaborately developed in *Squalus acanthias*. It makes up the anterior and ventral parts of the fin and consists of three divisions, all of which originate from various regions of the pectoral girdle. The *flexor protractor* inserts on the anterior border of the propterygium. The *flexor depressor* passes to the anterior half of the mesopterygium and to the radialia and dermal fin rays distal to it. The *flexor retractor* inserts on the posterior half of the mesopterygium and all of the metapterygium.

The various intrinsic muscles are grouped so as to move the pectoral fin in different directions.

Muscles moving the pelvic fins of elasmobranchs are also divided into dorsal, or extensor, and ventral, or flexor, groups. The dorsal muscle consists of superficial and deep portions. The superficial portion orginates from the connective-tissue fascia of the hypaxial regions of the myotomes and from the iliac region of the pelvic girdle; the deeper portion originates from the metapterygium. Both insert on the radialia and connective

tissue covering the dermal fin rays. The ventral muscles likewise form two masses, one originating from the ventral side of the ischiopubic bar and linea alba and passing to the metapterygium, the other originating on the metapterygium and inserting on the radialia and dermal fin rays. Complications are introduced in the pelvic musculature of male elasmobranchs because of the presence of the clasping organs. In other fishes the appendicular musculature is generally similar to that of elasmobranchs.

Attempts to homologize the fin musculature of fishes with conditions in tetrapods have led only to confusion.

The limbs of tetrapods originate embryonically from limb buds which develop in pectoral and pelvic regions. These at first consist of swellings of the lateral body wall, covered with superficial ectoderm and containing an undifferentiated mass of mesenchyme. It is from this mesenchyme that the appendicular muscles are differentiated (Fig. 11.15). Only in elasmobranchs can the development of appendicular muscles be traced to myotomic buds. In higher vertebrates it is practically impossible to trace the origin of mesenchyme in limb buds so far as any relation to definite myotomes is concerned. It would seem that the somatic layer of the hypomere is chiefly involved. However, the ultimate nerve supply to the appendicular muscles is precisely what one would expect were the muscles formed from outgrowths of specific myotomes as is true in elasmobranchs. It seems as though in tetrapods a certain stage in development has dropped out completely. Every other indication suggests that the appendicular muscles are of myotomic origin.

With emergence to life on land the limb musculature has undergone considerable change from the relatively simple condition observed in fishes. Separate muscles are now

Anatomy of the Chordates

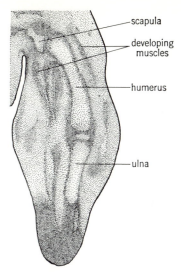

Fig. 11.15 Anterior limb bud of 16-mm human embryo, showing developing bones and muscle primordia. (*Redrawn from Arey, "Developmental Anatomy," W. B. Saunders Company. By permission.*)

differentiated from the dorsal and ventral muscle masses. The muscles are larger and sturdier and have broad attachments. A clear division into extrinsic and intrinsic groups is apparent. This provides for the greater freedom of movement in many directions required in terrestrial locomotion.

At a certain point in development the muscles begin to differentiate from the mesenchyme of the limb bud. This accompanies the development of the appendicular skeleton, which is also differentiated from limb-bud mesenchyme. Those muscles which arise in the dorsal part of the appendage in most cases become extensors, whereas those in the ventral aspect become flexors. Outgrowths of these muscles toward the trunk form abductors and adductors, respectively.

So many modifications are to be found in the appendicular muscles of various tetrapods that only a brief treatment of the subject can be given here. The tailed amphibians are the first vertebrates to possess limbs terminating in digits. Accordingly they give a clue to the origin of the appendicular musculature of higher forms.

The perennibranchiate salamander *Necturus* is considered to be the most primitive tetrapod living today. It is generally available for dissection and study in the laboratory. Hence, a description of the appendicular muscles as found in *Necturus* is apt to have more meaning to the student of comparative anatomy than if a general description of the muscles of other and less familiar forms were attempted. Whether the condition in *Necturus* may possibly represent a degenerate, rather than a primitive, state has not been determined with any degree of certainty.

EXTRINSIC MUSCLES OF THE PECTORAL APPENDAGES. In *Necturus* (Fig. 11.16) a rather prominent muscle, the *latissimus dorsi*, originates from myocommata and fasciae covering certain anterior myotomes of the dorsalis trunci. It passes to the humerus and vicinity of the shoulder joint. From the ventral side a *pectoralis* muscle extends from the linea alba to the ventral side of the proximal end of the humerus. Its fibers, near their origin, are continuous with the rectus abdominis and are attached to the myocommata. Deeper muscles, the *levator scapulae* and *serratus anterior,* inserting on the scapula, are clearly derived from the trunk, or axial, musculature.

The primitive condition encountered in *Necturus* becomes further modified in reptiles, which form an intermediate group as regards the evolution of the musculature of mammals. In a few reptiles and in mammals the anterior portion of the original latissimus dorsi has become a separate muscle, the

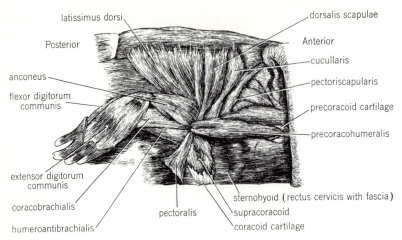

latissimus dorsi

Posterior

anconeus

flexor digitorum
communis

extensor digitorum
communis

coracobrachialis

humeroantibrachialis

pectoralis

dorsalis scapulae

Anterior

cucullaris

pectoriscapularis

precoracoid cartilage

precoracohumeralis

sternohyoid (rectus cervicis with fascia)

supracoracoid

coracoid cartilage

Fig. 11.16 Superficial muscles of right pectoral girdle and forelimb of *Necturus*, ventrolateral view.

teres major, which has been shifted so as to attach to the scapula. Posteriorly the latissimus dorsi comes to cover the greater part of the back (Fig. 11.18). The pectoralis is more highly developed and in some mammals forms superficial and deep layers. The superficial layer forms the *pectoralis major* of apes and man, the deeper layer giving rise to the *pectoralis minor*. A *subclavius* muscle may also be split off the anterior portion of the deeper layer.

The deeper extrinsic muscles, i.e., the *levator scapulae* and *serratus anterior,* have become more extensive in higher vertebrates. The latter now passes between the vertebral border of the scapula and the more anterior ribs. Furthermore, the levator scapulae and serratus anterior have given rise to a series of separate muscles known as the *rhomboideus group*. Only the crocodilia among reptiles possess a rhomboideus muscle. A *rhomboideus capitis,* passing from the occipital bone to the scapula, is present in most mammals but in man is encountered only occasionally and is referred to as the *rhomboideus occipitalis*. The rhomboideus of mammals is in man separated into two distinct elements, the *rhomboideus major* and *rhomboideus minor.*

The extrinsic muscles of the pectoral limbs in birds are extremely well developed in correlation with their habit of flight. The keeled sternum offers a broad base for their attachment.

Other muscles attached to the pectoral appendage of *Necturus* include the *cucullaris (trapezius)* which in higher forms gives rise to the *trapezius* and *sternocleidomastoid*. These, however, are derived from the branchial musculature and are not therefore included among the muscles of myotomic origin. They are discussed in further detail under the heading "Branchial musculature."

INTRINSIC MUSCLES OF THE PECTORAL APPENDAGES. Turning again to the primitive urodele *Necturus* (Fig. 11.16), we find a muscle, the *dorsalis scapulae,* lying immediately anterior to the latissimus dorsi. It originates on the suprascapula and inserts on the humerus. It is thus included among the intrinsic muscles of the forelimb. In higher forms the dorsalis scapulae is represented by the *deltoid* muscles, which lie between the trapezius and latissimus dorsi. The *teres minor* is a small muscle derived from the dorsalis scapulae. A *supracoracoid* muscle extends from the coracoid cartilage to the humerus in *Necturus*. This muscle has probably given rise to the *supraspinatus* and *infraspinatus* muscles of higher vertebrates. Another muscle, the *precoracohumeralis*, in *Necturus* passes from the precoracoid cartilage to the humerus. It is believed to contribute to the deltoid group.

On the dorsal side of the upper arm of *Necturus* is the large *anconeus* muscle made up of several parts, or *heads*, which originate on the pectoral girdle and humerus and insert by means of fascia on the ulna. The number of heads varies, there being four in *Necturus* and most reptiles but usually three in mammals. In the latter group the *triceps* represents the anconeus. It serves to extend the upper arm.

On the flexor side of the upper arm lies a small *humeroantibrachialis* muscle passing from humerus to radius. A large *coracobrachialis*, arising from the coracoid cartilage, inserts on the distal end of the humerus. The humeroantibrachialis probably has given rise to both the *biceps brachii* and *brachialis* muscles of higher forms. It is frequently referred to as the biceps brachii, but this name may not be entirely appropriate. The similarly named muscle in mammals, as in *Necturus*, inserts on the radius, but its origin, from the region of the glenoid cavity of the scapula, rather than from the humerus, raises some doubts as to its homology. On the other hand, the brachialis muscle of mammals which, like the humeroantibrachialis, originates from the humerus, inserts on the ulna rather than the radius. It is possible that both biceps brachii and brachialis have been derived from the humeroantibrachialis of lower tetrapods, but this has not been established

with certainty. The coracobrachialis muscles of *Necturus* and mammals are undoubtedly homologous.

The muscles of the forearm are more numerous than those of the upper arm. They provide for the more complex and diverse movements made by the forearm, hands, and digits. We shall not attempt to describe these muscles in detail since their main features are generally similar in all tetrapods. (For a detailed description of the muscles of the forearm and hand in *Necturus*, the interested student may refer to H. H. Wilder, "The Human Body," 1926, Holt, Rinehart and Winston, Inc.) In general the muscles may be grouped in two main divisions, *extensors* on the dorsal surface and *flexors* on the ventral side. *Supinators,* associated with the extensor group, and *pronators*, with the flexor group, are also present in many forms.

EXTRINSIC MUSCLES OF THE PELVIC APPENDAGES. The bones comprising the pelvic girdle are directly attached to the axial skeleton via the sacrum. This firm union is quite in contrast to the attachment of the pectoral girdle which is accomplished by means of muscles, tendons, and ligaments. Accordingly the extrinsic muscles of the pelvic appendage are much less highly developed than those of the pectoral appendages.

The only extrinsic pelvic muscles present in *Necturus* (Fig. 11.17) are three caudal muscles. They course along together for some distance in a common sheath. Two of these muscles clearly develop from the hypaxial portions of caudal myotomes a short distance posterior to the pelvic girdle. They include the *ischiocaudalis*, with its origin from caudal vertebrae and insertion on the posterior border of the ischium, and the *caudopuboischiotibialis,* also originating from caudal vertebrae but inserting on the fascia of the *puboischiotibialis* (described below) by means of a flat tendon. These muscles are

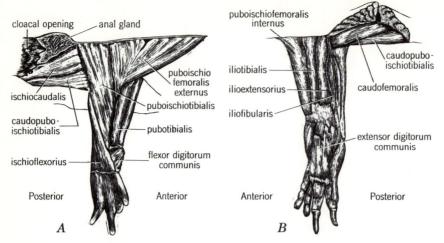

Fig. 11.17 Superficial muscles of left side of pelvic girdle and left hind limb of *Necturus: A,* ventral view; *B,* dorsal view.

used primarily in certain tail movements. In males, the ischiocaudalis is used also in compressing the glandular tissue in the cloacal region. The third caudal muscle, the *caudofemoralis,* originates from caudal vertebrae and inserts on the shaft of the femur. The homologies of these muscles are not clear but it is possible that the *pyriformis* in man and the higher anthropoids is a remnant of this caudal muscle complex.

INTRINSIC MUSCLES OF THE PELVIC APPENDAGES. The proximal intrinsic muscles of the pelvic appendages as observed in *Necturus* (Fig. 11.17*A* and *B*) are not so easily homologized with those of higher forms nor with the corresponding muscles of the pectoral appendages. Much confusion exists, and there is little agreement among comparative anatomists.

On the ventral side and originating from the midventral line of the pelvic girdle, or puboischial plate, are two superficial muscles, an anterior *puboischiofemoralis externus,* passing to the femur, and a posterior *puboischiotibialis,* inserting on the proximal

end of the tibia. Another muscle, the *ilioflexorius,* arising from the puboischial plate along with the puboischiotibialis, forms the posterior edge of the thigh. It inserts on fascia of the shank, A *puboischiofemoralis internus,* arising from the dorsal surface of the puboischium, passes to the shaft of the femur. Just posterior to this is the *pubotibialis,* coming from the pubic region in front of the obturator foramen and passing to the proximal end of the tibia. Three additional muscles, originating from the ilium, cover the pubotibialis. The most anterior of these is the slender *iliotibialis.* In back of it lies the *ilioextensorius.* Both these muscles insert on the tibia. Posterior to the ilioextensorius lies the *iliofibularis,* which passes to the proximal end of the fibula. Another small, deep, triangular muscle, the *iliofemoralis,* with its origin on the ilium, inserts on the posterior border of the femur. A *femorofibularis,* from the ventral face of the femur, also inserts on the fibula near the insertion of the iliofibularis.

Many investigators have attempted to

homologize the proximal muscles of the urodele thigh with those of higher vertebrates, but there is little general agreement. Wilder, in his "History of the Human Body," discusses some of the more probable homologies, a few of which are listed below:

Necturus	Mammals
Puboischiofemoralis externus	Obturator externus
	Quadratus femoris
Puboischiofemoralis internus	Obturator internus
	Gemellus superior
	Gemellus inferior
Puboischiotibialis	Adductor femoris
	Adductor longus
	Gracilis
	Sartorius
	Semimembranosus
	Semitendinosus
Ilioextensorius	Quadriceps femoris
	Vastus lateralis
	Vastus intermedius
	Vastus medialis
	Rectus femoris
Iliofemoralis	Gluteus maximus
	Gluteus medius
	Gluteus minimus
	Short head of biceps femoris
	Tenuissimus
Iliofibularis	Long head of biceps femoris

The mammalian muscles listed above exhibit many modifications associated with differences in posture and the uses to which limbs are put. Figure 11.18 illustrates some of the differences which may be observed in various muscles of the cat and man.

The distal intrinsic muscles of the pelvic limbs of *Necturus* are almost identical with those of the pectoral limbs. They too are grouped as extensors and flexors. In higher forms, however, the similarities between the two pairs of appendages, as observed in *Necturus*, are not very apparent. Among

mammals less variation is found in the distal muscles of the pelvic limb than in the proximal muscles.

BRANCHIAL MUSCULATURE

Turning back again to the elasmobranchs, we find an important feature in the *branchial*, or *branchiomeric*, muscles which move the visceral arches and jaws. They are voluntary striated muscles derived from the splanchnic mesoderm of the hypomere and *not* from myotomes. They are thus homologous with the smooth muscles of the gut. The branchial muscles begin anteriorly at the mandibular arch and run in series to each of the successive visceral arches. The segmental character of these muscles is related to the segmental character of the visceral skeleton. This is in turn related to the segmental character of the visceral pouches (see page 210). There is no indication, however, that this segmentation is in any way related to the fundamental metamerism of the body as revealed in the formation of somites and their derivatives, the spinal nerves, etc. For the above reasons, the branchial muscles are classified apart from the general axial musculature. As will be pointed out in more detail in Chap. 13, Nervous System (page 641), the branchial muscles are innervated by *visceral motor* branches of certain cranial nerves and not by somatic motor nerves or their branches which supply somatic, voluntary, striated muscles. The mandibular branch of the trigeminal nerve (V) is the nerve innervating the branchial muscles associated with the mandibular arch; the facial nerve (VII) is the nerve of the hyoid arch; the glossopharyngeal nerve (IX) innervates the branchial musculature of the third visceral arch located between the first and second *typical* gill slits; the vagus nerve (X) supplies the remainder of the branchial musculature.

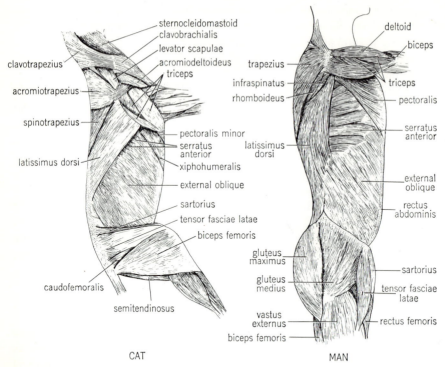

Fig. 11.18 Lateral view of superficial body muscles of cat and man, shown together for the sake of comparison.

The outer, or superficial, branchial muscles function as constrictors of the pharynx. They tend to close the gills and mouth during respiratory movements. They are divided into *dorsal* and *ventral* constrictors, lying above and below the gills, respectively. Each constrictor is separated from the next by a band of connective tissue, arranged vertically and known as a *raphe*.

In *Squalus acanthias* the dorsal constrictor of the first, or mandibular, arch is a small muscle which lies anterior to the spiracle. It has its origin on the otic capsule and inserts on the palatopterygoquadrate bar. The *adductor mandibulae*, which in elasmobranchs lies outside the angle of the mouth and acts in closing the jaws, is a derivative of the first

dorsal constrictor. The first ventral constrictor forms the rather broad *intermandibularis* muscle which originates from a midventral raphe. Its fibers pass obliquely forward on either side to insert on Meckel's cartilage. The constrictors of the remaining visceral arches, as illustrated by *Squalus acanthias* (Fig. 11.19), are bounded anteriorly and posteriorly by the gill slits. The second constrictor, lying between the first typical gill slit and the spiracle, is a large muscle with rather distinct dorsal and ventral portions which pass anteriorly above and below the jaws. The dorsal portion is known as the *epihyoideus* muscle. It has its origin on the otic capsule and its insertion on the hyomandibular cartilage. The anterior ventral

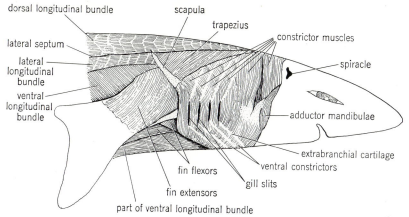

dorsal longitudinal bundle scapula trapezius constrictor muscles spiracle lateral septum lateral longitudinal bundle ventral longitudinal bundle adductor mandibulae extrabranchial cartilage fin flexors ventral constrictors fin extensors gill slits part of ventral longitudinal bundle

Fig. 11.19 Superficial muscles of shoulder and head of dogfish, *Squalus acanthias.* (*Original by Howell, J. Morphol.,* **54**:401. *By permission of the Wistar Institute of Anatomy and Biology.*)

portion of the second ventral constrictor gives rise to the *interhyoideus* muscle which is attached to the hyoid arch and which lies just above the intermandibularis. The last four constrictors are simply arranged, their dorsal and ventral portions being rather similar in appearance.

Other branchial muscles, covered for the most part by the constrictors, serve as levators of the visceral arches. The first levator is the *levator maxillae,* which, in selachians, acts in raising the upper jaw. The second is the levator of the hyoid arch. It lies directly under the epihyoideus. The *trapezius* (*cucullaris*) belongs to the levator group and probably represents the posterior portion of the remaining levators. It originates from dorsal fascia and overlying skin and extends in a ventrocaudal direction to insert on the epibranchial cartilage of the last visceral arch as well as on the ventral portion of the scapula. Four small *lateral interarcual* muscles originate from pharyngobranchial cartilages of visceral arches 3, 4, 5, and 6 and are inserted on the corresponding epibranchial cartilages (Fig. 11.20). Since the trape-

zius and lateral interarcualia insert on epibranchial cartilages in series, it is believed that they represent specialized portions of a single levator group. The remaining representatives of the branchial musculature include more deeply situated *dorsal,* or *medial, arcualia* which connect adjacent pharyngobranchial cartilages.

Some authors consider that certain muscles ventral to the gill region should be included in the category of branchial muscles. Among these are *common arcualia,* a single *coracomandibular,* paired *coracohyoids* and *coracobranchials.* These muscles, however, are supplied by spinal nerves (hypobranchial nerve) rather than cranial nerves, indicating that they are actually hypobranchial muscles derived from anterior myotomes of the trunk musculature.

Homologues of the branchial muscles as found in higher forms have been determined for the most part by tracing the visceral motor branches of the four cranial nerves which in fishes supply the branchial musculature.

The *adductor mandibulae,* derived from

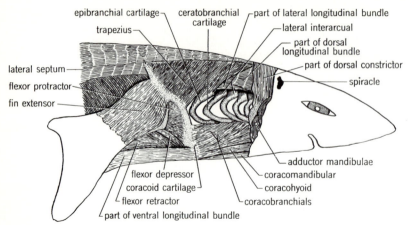

epibranchial cartilage
ceratobranchial cartilage
part of lateral longitudinal bundle
trapezius
lateral interarcual
part of dorsal longitudinal bundle
lateral septum
part of dorsal constrictor
flexor protractor
spiracle
fin extensor
adductor mandibulae
coracomandibular
flexor depressor
coracohyoid
coracoid cartilage
flexor retractor
coracobranchials
part of ventral longitudinal bundle

Fig. 11.20 Deeper muscles of shoulder and head of dogfish, *Squalus acanthias*. (*Original by Howell, J. Morphol.,* **54**:401. *By permission of the Wistar Institute of Anatomy and Biology.*)

the first dorsal constrictor of elasmobranchs, gives rise in higher forms to the *masseter, temporal, internal pterygoid, external pterygoid,* and *tensor tympani* muscles. The *intermandibularis,* derived from the first ventral constrictor, gives rise to the *mylohyoid* and part of the *digastric* muscle of mammals. The ventral constrictor of the hyoid arch also contributes to the digastric muscle, the double innervation of which, coming from the trigeminal and facial nerves, bears witness to its homologies. Some of the muscles of the face, as well as the *platysma* of mammals, are thought to be derivatives of the hyoid constrictors. The constrictors of the remaining visceral arches are apparently not represented in higher forms.

The levator muscles of the visceral arches are also imperfectly represented in tetrapods. The *trapezius* (*cucullaris* of lower forms) and the *sternocleidomastoid,* or any of its variations such as the *sternomastoid* or *cleidomastoid,* are probably such derivatives. The voluntary muscles of the pharynx and larynx are undoubtedly derived from the branchial

musculature, but their homologies are uncertain.

The anterior fibers of the eleventh cranial nerve (spinal accessory) of amniotes are so closely associated with the vagus that the muscles which it supplies are considered to be part of the branchial musculature. The sternocleidomastoid and the trapezius are such muscles. The trapezius in mammals is so obviously a muscle of the pectoral appendage that its inclusion among the branchial muscles might at first sight seem peculiar. This muscle may, however, have a twofold origin, since it is supplied by certain spinal nerves as well as by the spinal accessory (XI).

VISCERAL MUSCULATURE

The muscular tissue derived from the splanchnic layer of hypomeric mesoderm, is for the most part, of the involuntary, smooth type. An exception is the branchial musculature, already discussed, which is voluntary and composed of striated fibers. Smooth muscle is found in the walls of the digestive tract, in the ducts of digestive glands, in the walls of the respiratory passages, in the ducts of the urogenital system, in blood vessels, in lymphatic vessels in certain places, and in the spleen. The internal sphincter muscle of the anus is composed of smooth muscle as are such sphincters as the pylorus and ileocolic valve.

The peculiar cardiac muscle tissue, found only in the walls of the heart and the great vessels leaving the heart, is derived from visceral musculature. Its histological structure reveals it to be a somewhat intermediate type of muscle tissue having characteristics of both striated and smooth muscle.

The visceral muscles are more properly discussed in connection with the organs of which they form a part and will not be discussed further at this point.

DERMAL MUSCULATURE

Integumentary or dermal muscles appear, with few exceptions, only in amniotes. In some cases these muscles are attached firmly to some part of the skeleton and insert on the skin. Usually, however, both origin and insertion are in the integument. Most dermal muscles are derivatives of the parietal musculature from which they have become separated to varying degrees. Some, however, seem to have been split off the branchial musculature.

Fishes. Integumentary muscles, as such, are not present in fishes. The myocommata between myotomes are attached to the dermis, and the outermost muscle fibers of the myotomes are closely applied to the dermis. This, however, is not of the nature of a true insertion.

Amphibians. A few integumentary muscles may occur in anurans. The *gracilis minor* of

the hind leg inserts in part on the neighboring *gracilis major*. It also has an insertion on the skin of the posterior region of the thigh. Another muscle, the *cutaneous pectoris*, which is located on the ventral side of the anterior part of the body, comes from the body wall and is inserted anteriorly on the skin between the forelegs. Otherwise, integumentary muscles are usually present in amphibians only in the region of the external nares. In urodeles these are of the smooth involuntary type, serving to control respiratory currents. Rudimentary smooth muscles are also to be found about the external nares of anurans but, with few exceptions, are unimportant in opening and closing the nares. These amphibians utilize a different mechanism in occluding the nostrils. Movements of the premaxillary bones, which in turn move the nasal cartilages, are effective in closing the nares in most anurans.

Reptiles. Intrinsic dermal muscles in reptiles are of importance in moving certain scales. Locomotion in snakes, aside from the wriggling, undulatory movements of the body proper, is accomplished by alternately elevating and lowering the ventral and ventrolateral scales. Erecting the scales provides a greater friction with the ground, furnishing a temporary anchorage from which the body can be pulled or pushed forward. *Costocutaneous* parietal muscles (Fig. 11.21), extending from ribs to scales, are also of great importance in progressive locomotion.

Birds. Movement of individual feathers in birds is accomplished by contraction of cutaneous muscles. The feathers can be made to lie smoothly or else erected to different degrees, thus enabling the bird to regulate heat radiation from the skin to some extent. The patagium, or web of skin which

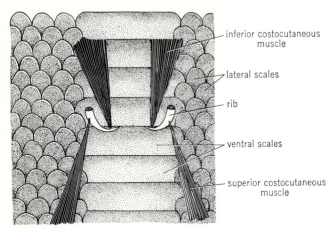

inferior costocutaneous muscle

lateral scales

rib

ventral scales

superior costocutaneous muscle

Fig. 11.21 Semidiagrammatic sketch of inner side of ventral integument of snake, showing costocutaneous muscles extending from ribs to scales. The dermal layer of skin is not shown. (*Modified from Buffa.*)

passes from the body to the wing, is provided with dermal *patagial* muscles. Stretching of the patagium increases the resistance of the wing to air during flight.

Mammals. Integumentary muscles are far more elaborately developed in mammals than in any other vertebrates. In such lower forms as monotremes and marsupials, a broad sheet of muscular tissue, the *panniculus carnosus*, almost envelopes the entire trunk and appendages. Twitching movements of the skin, used to dislodge insects and other foreign objects, are brought about by contraction of this muscular sheet. Defensive movements of the European hedgehog and armadillos, i.e., rolling into a ball when in danger, are brought about by contraction of an unusually well-developed panniculus carnosus muscle. Originally derived from the *latissimus* and *pectoralis* muscles, the panniculus carnosus becomes reduced in higher forms. Remnants are most often encountered in the shoulder, sternal, and

inguinal regions. In marsupials the inguinal portion forms the sphincter muscle of the marsupial pouch. The cloacal sphincter of monotremes is also such a derivative. In horses and cattle the skin of the forelegs and shoulders can be twitched to ward off annoying flies and other insects. The posterior part of the body, which can be reached by the swishing tail, is not, however, provided with integumentary muscles to the same extent.

In lower primates both axillary and inguinal remnants of the panniculus carnosus are encountered, but in the higher apes and man only a few axillary slips may remain. A *sternalis* muscle appears occasionally even in man, overlying the pectoralis major.

In the head and neck region is another group of integumentary muscles derived from a muscular sheet, the *sphincter colli*, present in turtles and birds. The fact that these muscles are supplied by visceral motor fibers of the facial nerve (VII) indicates that they have been derived from the branchial musculature in the region of the hyoid arch. The *mimetic* muscles, or facial muscles of expression, are derivatives of the sphincter colli. In most lower mammals these muscles are poorly developed, and as a result, the facial expression of such forms does not change. In carnivores and primates, however, they are better developed, but only in man do they reflect emotion to a high degree. The original sphincter colli apparently forms two layers, an outer, or superficial, *platysma* layer and an underlying sphincter colli. The muscle fibers of these two layers course in opposite directions. Both layers form slips which serve to move special parts of the face.

From the platysma layer are derived such muscles as the *auricularis frontalis, orbicularis oculi, mentalis, quadratus labii inferiores,* and *zygomatic.*

The deeper sphincter colli gives rise to the intrinsic muscles of the nose and to the *orbicularis oris, canine, buccinator,* and other muscles.

MISCELLANEOUS

Electric organs. Electric organs, consisting of modified muscular tissue, are to be found in numerous fishes. Noteworthy are those of the electric ray, *Torpedo*, and the South American electric eel, *Electrophorus (Gymnotus) electricus.* In *Torpedo* the electric organs lie on either side of the head between the gill region and anterior part of the pectoral fin (Fig. 11.22). In most fishes in which electric organs occur, however, they are confined to the tail. In *Electrophorus*, which grows to a length of 8 ft, the tail is very much elongated, forming approximately four-fifths of the total length of the animal. Here the electric organ is very well developed and occupies the position of the ventral longitudinal bundles. In both these fishes electric shocks of considerable inten-

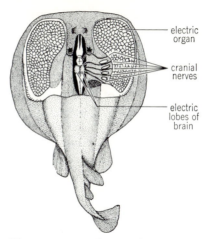

Fig. 11.22 Dorsal view of the electric ray, *Torpedo marmorata*, the upper surface being partially dissected away to show the electric organs and the brain with its electric lobes. Cranial nerves connect the electric lobes with the electric organs. (*After Dahlgren, Papers from the Department of Marine Biology of the Carnegie Institution of Washington,* 8:215.)

sity can be emitted at will, but those produced by *Electrophorus* are the stronger of the two. Charges of several hundred volts have been recorded. They are used by the animals in stunning or killing prey and also in self-defense. Several other fishes, including *Raja* and *Mormyrus,* possess this same ability, but the electric shocks which they produce are feeble in comparison with those mentioned above. It seems that in the latter fishes electric waves are sent forth by the electric organs and used in orientation in the water. Such waves are apparently reflected from objects in the water, the lateral-line organs (page 725) serving as sensory receptors.

One of the catfishes of Africa, *Malapterurus electricus,* possesses an electric organ situated between the skin and muscles. In this fish the electric organ has been reported to be a derivative of the dermis rather than of muscular tissue. It thus differs from the electric organs of other forms.

The electric organs of fishes, with the possible exception of *Malapterurus,* are built upon the same general plan, regardless of position. Each organ is composed of many parallel prismatic columns separated from one another by partitions of connective tissue. The columns in turn consist of large numbers of minute *electric plates,* or discs (electroplaxes), piled one upon the other. Each electric plate is made up of three parts: (1) an outer *electric layer,* (2) a middle *striated layer,* and (3) an inner *alveolar layer.* The electric layer is actually composed of an outer nervous layer and an inner layer of cells with very large nuclei. Between adjacent plates lies a thick layer of gelatinous material which is thought to be metamorphosed muscle substance. The nerve fibers leading to the electric plates ramify through the connective tissue to terminate in fine networks on one side of each plate. The nervous layers of all the plates face in the same direction. These appear to be similar to motor end plates, and their relationships are like those of the endings of motor nerves on muscle fibers.

In *Electrophorus* the electric plates form columns which run in a longitudinal direction, but in *Torpedo* the columns are vertically disposed. In the former the electric current passes from the tail toward the head; in the latter, from the ventral to the dorsal surface.

The exact manner in which electricity is formed and stored in the electric organs of these fishes is not known with any degree of certainty. The fact that muscular activity produces electricity is well known and of great practical importance in electrocardiography. In any case the modified muscular tissue comprising the electric organs has developed the ability to store electricity to a remarkable degree.

The phylogenetic relationships of the electric organs are obscure. Since they occur in various parts of the body in different species of fishes, they are not even homologous within the group. This is further borne out by a study of their nerve supply. In *Torpedo* the electric organ is innervated by branches of *cranial* nerves VII, IX, and X, together with a single branch of the trigeminal nerve (V). The nerves arise from an enlarged portion of the medulla oblongata spoken of as the *electric lobe.* It is believed that the adductor muscle of the mandible and the constrictor muscles of the visceral arches have given rise to the electric organ of *Torpedo.* In *Electrophorus* and other fishes *spinal* nerves are distributed to the electric organs which are derived from the caudal musculature. More than 200 nerves pass to the electric organ in *Electrophorus.*

It has been reported (page 20) that in the lamprey an electric field, produced in the water surrounding the head, enables the animal finally to localize its prey. No specific structures involved in producing such an electric field have been described, so far as the author is aware.

SUMMARY

The muscles of the vertebrate body are derived from the epimeric and hypomeric mesoderm of the embryo. The epimere, which becomes marked off into mesoblastic somites, forms segmentally arranged myotomes which give rise to the voluntary, striated, parietal musculature of the body. The hollow, unsegmented hypomere forms somatic and splanchnic layers enclosing the coelom between them. The splanchnic layer, which surrounds the gut, gives rise to the smooth, involuntary muscle of the gut and also to cardiac muscle. An exception is to be found in the voluntary, striated, branchial muscles of the gill region which are derived from the hypomeric mesoderm.

Parietal musculature. The parietal muscles consist primitively of a linear series of muscle segments, or myotomes, extending from anterior to posterior ends of the body. Each myotome lies opposite the point of articulation of two vertebrae and is supplied by a separate spinal nerve. Connective-tissue septa, the myocommata, occupy the spaces between adjacent myotomes. The myotomes of the two sides are separated at the midventral line by the linea alba composed of connective tissue.

TRUNK MUSCULATURE. In amphioxus, and to a lesser degree in cyclostomes, the myotomes tend to be V-shaped. In fishes they take more of a zigzag course. Beginning with elasmobranchs and in higher forms to a less conspicuous degree, the myotomes are separated into dorsal (epaxial) and ventral (hypaxial) portions. Each of these regions becomes subdivided into two or more longitudinal muscular bundles. In tetrapods the primitive metameric condition becomes more and more obscure as the evolutionary scale is ascended. This is brought about by a splitting into various layers, disappearance of myocommata, degeneration of portions of some myotomes, and a drawing out of position, a migration, or even fusion of others.

The epaxial musculature becomes reduced in higher forms but shows a tendency to split up into short and long systems of longitudinally directed muscles. The shorter ones connect adjacent vertebrae and are termed, according to position, interneural, intertransversarial, interspinal, multifidus, etc., muscles. The longer ones form such muscles as the longissimus dorsi, longissimus capitis, iliocostal dorsi, and spinalis dorsi.

The hypaxial muscles of tetrapods show greater modification than do those of the epaxial region. In urodele amphibians they split into four layers or sheets, the fibers in each sheet extending in a direction different from that of the others. These layers include the superficial and deep external obliques, an internal oblique, and a transversus. The last lies next to the peritoneum. On either side of the linea alba lies a rectus abdominis, derived from the obliques and with primitively arranged longitudinal fibers. In anuran amphibians the number of layers is reduced, but in reptiles additional layers are to be found in the thoracic region. An intercostal layer, derived from the obliques, makes its appearance. It in turn divides into external and internal intercostals which are attached to the ribs and are used in respiratory movements. Scalene and serratus muscles are also derived from the obliques. The hypaxial muscles in the thoracic region of mammals are somewhat reduced, being almost entirely covered by the large appendicular muscles of the pectoral girdle and limb.

HYPOBRANCHIAL MUSCULATURE. In cyclostomes and fishes, hypobranchial muscles, lying ventral to the gills, are derived from the hypaxial portion of the first few myotomes as buds which grow anteriorly. They are innervated by spinal nerves. In tetrapods these muscles have been greatly modified, but their innervation by spinal nerves bears witness to their origin. Among the muscles of hypobranchial origin are the sternohyoid, geniohyoid, thyrohyoid, sternothyroid, and omohyoid. The muscles of the tongue, supplied by cranial nerve XII, are considered to be hypobranchial in origin since this nerve, which is found only in amniotes, represents a union of the most anterior two or three spinal nerves of lower forms.

EYE MUSCLES. The six extrinsic eye muscles of vertebrates develop from three pro-otic somites supplied by cranial nerves III, IV, and VI, respectively. From the first come the superior, inferior, and internal rectus muscles and the inferior oblique; from the second comes the superior oblique; and from the third the external rectus is differentiated.

DIAPHRAGM. The diaphragm of mammals is composed partially of muscle fibers innervated by branches of certain cervical spinal nerves. The muscles, therefore, are believed to originate from cervical myotomes.

APPENDICULAR MUSCULATURE. In elasmobranchs hollow buds from certain myotomes grow out into the ventrolateral fin folds, accompanied by branches from corresponding spinal nerves. The original metameric arrangement of these buds is soon lost, but they give rise to the dorsal and ventral muscles moving the fins. No similar development of appendicular muscles from myotomic buds has been reported for higher forms, but in all other ways these muscles appear to be derivatives of myotomes. In tetrapods the dorsal and ventral masses differentiate into separate muscles. These are divided into extrinsic and intrinsic groups, which provide for the greater freedom of movement required for terrestrial locomotion. When the complex musculature of a typical mammalian limb is compared with that of a primitive urodele amphibian, such as *Necturus*, many fairly clear-cut homologies are indicated.

Branchial musculature. The striated, voluntary branchial muscles which move the visceral arches of fishes are derived from the splanchnic hypomeric mesoderm rather than from myotomes. They begin anteriorly at the mandibular arch and run in series to each of the successive visceral arches. The outer muscles serve as constrictors of the pharynx and are divided into dorsal and ventral portions. Deeper muscles serve as levators of the visceral arches. Even in fishes many of these muscles have deviated from the primitive condition, but their cranial nerve supply indicates their branchial origin. The mandibular branch of cranial nerve V supplies the muscles of the mandibular arch; the facial nerve (VII) goes to those of the hyoid arch; the third arch is innervated by the glossopharyngeal (IX), and the remaining arches by the vagus (X). In higher forms, in which the visceral arches have become greatly modified, the nerve supply to certain muscles of the neck and shoulder regions indicates that they have been derived from branchial musculature.

Visceral musculature. The smooth muscles of the hollow organs, those of the ducts of digestive glands, etc., as well as the cardiac muscle of the heart, all are developed from the unsegmented splanchnic layer of hypomeric mesoderm.

Dermal musculature. Integumentary, or dermal, muscles appear, with few exceptions, only in amniotes. They are, for the most part, derivatives of the parietal and branchial musculature from which they have become separated to varying degrees. In snakes the intrinsic muscles which erect the ventral scales, which are used so largely in locomotion, are dermal derivatives. Movement of the feathers of birds is accomplished by integumentary muscles. Greatest development of all is to be found in mammals. In lower mammals a broad sheet of muscular tissue, the panniculus carnosus, almost envelops the entire body. Its contraction causes twitching movements of the skin. In higher mammals this layer becomes reduced, so that remnants are found only in the axillary, sternal, or inguinal regions.

Another muscular sheet, the sphincter colli, present in turtles and birds, is most highly modified in primates, in which it gives rise to numerous mimetic muscles of facial expression.

Electric organs. Certain fishes such as the electric ray, *Torpedo,* and the electric eel, *Electrophorus,* possess electric organs which give off true electric shocks used to stun or kill prey and in self-defense. These organs are formed from highly modified muscular tissue. Their phylogenetic relationships and homologies are obscure.

| 12 |

CIRCULATORY
SYSTEM

Every living cell in the metazoan body is constantly undergoing metabolic change. Food materials in the form of glucose, amino acids, and fats must be supplied so that the anabolic phases of metabolism can take place. The cells must be furnished with oxygen so that they can utilize these food substances in the production of heat and other forms of energy and in carrying out their normal functions. Waste products of metabolism, e.g., carbon dioxide, urea, uric acid, etc., given off by the cells, must be removed before they exert any deleterious effects. Hormones, the secretion products of endocrine glands, must be carried from one part of the body to other regions in order to be effective. Moreover, living cells must be kept moist in order to remain alive and to carry on their activities. This is necessary because every substance which enters or leaves the cells is in solution and can pass only through moist membranes or media. Cells that are exposed to air are usually dry and dead and are converted into nonliving keratin, as in the corneal layer of the epidermis. It is constantly being worn off or shed and is replaced by keratinization of new cells which have proliferated from beneath.

In members of some of the lower phyla many of the above functions are carried out by individual cells which are in direct contact with the environment. However, in the evolution of higher metazoan animals, most of the cells making up their bodies have become farther and farther removed from the external environment, yet their metabolic requirements have undergone little or no change. Circulatory systems of various types have developed in the evolution of higher forms in response to demands created by the increasing complexity of their bodies. Fluids carried by the circulatory system transport substances to and from cells even in the most remote parts of the body. The fluids must circulate, for only by this means can supplies of food, oxygen, and water be replenished and harmful waste products removed.

The various cells of the body are not actually in *direct* contact with parts of the circulatory system and hence are not nourished directly, nor are they directly supplied with oxygen. Likewise, the waste products of cellular metabolism do not enter the circulatory system directly from the cells. The fluid surrounding, or bathing, the cells is referred to as *tissue fluid*, or *interstitial fluid*. It is this which serves as an intermediary in transporting nutritive materials, oxygen, and waste substances between the cells and the minute subdivisions of the circulatory system. Tissue fluid is intimately associated with various intercellular substances (page 79) but its relationship to them varies in different parts of the body and also depends upon the type of intercellular substance in question. Tissue fluid is actually composed of water in which crystalloids of various kinds are in solution. It does not usually contain colloidal material. The large molecules of colloids are generally confined to vessels of the circulatory system. Nevertheless, there normally seems to be considerable leakage of colloid into the tissue fluid probably because of pressure differences. Tissue fluid is thus much like sea water in its general composition (page 83).

The circulatory system in vertebrates is highly complex and actually made up of two systems of elaborately branched tubes which ramify throughout the entire body and thus are able to convey fluids indirectly to all living cells. They are the *blood-vascular* and *lymphatic systems*. The former is the better defined and the more obvious of the two.

Blood-vascular system. The blood-vascular system is composed of continuous tubes of various sizes, known as *blood vessels*, through which *blood*, a peculiar type of fluid connective tissue, is pumped to all parts of the body. This system consists of a muscular contractile organ, the *heart*, situated near the anterior end of the body and toward the ventral side. *Arteries* carry blood away from the heart. They subdivide into vessels of smaller caliber, the *arterioles*. These in turn branch into *capillaries*, which are extremely small vessels averaging from 0.004 to 0.012 mm in diameter and 1 mm or less in length. *Venules* collect blood from the capillaries. They combine to form *veins* which return the blood to the heart. Irrespective of the type of blood carried by the vessels, it should be clearly understood that *arteries carry blood away from the heart* and *veins carry it toward the heart*. The arteries and their branches make up the *arterial system*, and the veins with their tributaries form the *venous system*. The blood-vascular system is a closed system of blood vessels, having no direct connection with any other system of the body except the lymphatic system, with which it is closely associated.

The arterial system in higher vertebrates is

separated into *pulmonary* and *systemic divisions*, the former carrying blood to the lungs, the latter carrying it to all other parts of the body. The venous system likewise has pulmonary and systemic divisions, but, in addition, two or three *portal systems* may be present. A portal system is a complex of veins which begin in capillaries and end in capillaries before the blood which courses through them is returned to the heart. All vertebrates have a *hepatic portal system* in which blood collected from the digestive tract and spleen passes through capillaries (sinusoids) in the liver before reaching the heart. Adults of the lower vertebrate classes and embryos of higher forms have a *renal portal system* in addition, in which blood from the posterior part of the body passes through capillaries in the opisthonephros or mesonephros on its way to the heart. Another small but important portal system is found in association with the blood vessels draining the hypophysis, or pituitary gland.

Lymphatic system. The lymphatic system is composed in part of delicate tubes, the *lymphatic vessels.* They have their origin in very fine *lymph capillaries* which are blind at their free ends and drain the tissue spaces. Lymphatic vessels may form connections in various parts of the body with relatively large spaces called *lymph sinuses.* Some also join certain veins in the region of the heart, so that the connection between the blood-vascular and lymphatic systems is direct. The lymphatic system is a closed system, since the walls of even the lymphatic capillaries are composed of a delicate membrane made up of endothelial cells. The lymphatic system carries a fluid, *lymph,* the composition of which is, in general, rather similar to blood except that red corpuscles are lacking and the protein content is very much lower than that of blood plasma. In some animals, pulsating *lymph hearts* are present at points where lymphatic vessels enter certain veins. These aid in the propulsion of lymph into the blood-vascular system. In most forms, however, pulsations of adjacent blood vessels and movements of the viscera as well as of other parts of the body propel the lymph which, therefore, courses slowly through the vessels. Movement is in one direction since numerous valves are present which prevent backflow. In mammals, and to a lesser degree in birds, the lymphatic vessels are interrupted at certain places by the presence of *lymph nodes,* or *glands.* These are composed of a meshwork of reticular and collagenous connective-tissue fibers which support large masses of *lymphocytes.* Also present in lymph nodes are certain phagocytic *reticuloendothelial cells* as well as others known as *fixed macrophages.* These are derived from reticular cells. The lymphatic vessel breaks up into fine branches within a lymph node. The branches recombine on the other side of the node. Lymph filters slowly through the lymph nodes but in considerable quantity. Bacteria and other foreign substances are engulfed by the action of the fixed macrophages and reticuloendothelial cells. It is here, too, that lymphocytes develop. These pass into the lymphatic stream and ultimately enter the blood-vascular system. Some of the macrophages also become free, wandering cells. The lymphatic system is of importance also in connection with the regulation of both the quantity and quality of tissue fluid. It furnishes a means by which colloidal substances can get into the circulatory system since the endothelial walls of lymphatic vessels are somehow permeable to colloids, whereas those of blood capillaries are not. The lymph nodes are of further importance in the formation of specialized plasma cells, the function of which seems to be the production of antibodies. The actual function of lymphocytes is little understood. It is possible that they may serve to transport deoxyribonucleic acid (DNA) to cells in various parts of the body which are unable to synthesize this compound entirely by themselves. Deoxyribonucleic acid is a substance of great importance in determining the genetic constitution of cells.

BLOOD VESSELS

Origin. The vessels of the blood-vascular system arise in embryonic development from mesenchyme *in situ* and are included among mesodermal derivatives. The first indication of blood-vessel formation is seen when small clumps or cords of cells, called *blood islands*, appear in certain parts of the embryo (yolk sac, body stalk, etc.). In a chick embryo, for example, at first the blood islands are solid but they soon hollow out in such a manner as to form a thin, flat endothelium enclosing a fluid-filled space (Fig. 12.1). Some loose cells float about in the fluid and become blood cells, or corpuscles. The fluid, which is ap-

parently secreted by the cells of the blood island, is blood plasma. The scattered, hollow vesicles derived from the blood islands grow and coalesce with one another to form an irregular, anastomosing network of small blood vessels. The vessels of the main part of the embryo appear as clefts in the mesenchyme. The endothelium of the original vessels proliferates, and the vascular area expands. After a primitive blood-vascular system is once established, the formation of new vessels depends upon growth and branching of previously established channels. Every blood vessel, including the heart, has an endothelial lining.

Structure. At first there is no structural difference between arteries and veins, and the early blood vessels are histologically similar to the capillaries of later stages. From the mesenchyme surrounding the endothelium of the primitive vessels are derived the coating layers found in arteries and veins. The structures of the accessory coats differ in nature in these two types of vessels.

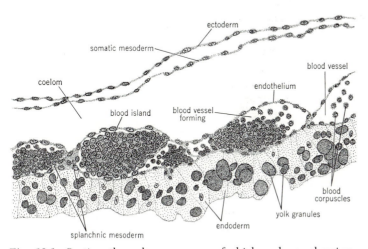

Fig. 12.1 Section through area opaca of chick embryo, showing formation of blood from blood islands.

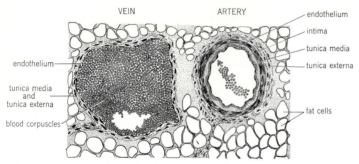

Fig. 12.2 Section through adipose tissue adjacent to ovary of rat, showing structure of a vein and an artery.

ARTERIES. Every artery has a wall (Fig. 12.2) composed of three coats: (1) an inner layer, the *tunica interna*, or *intima*, which is composed of the endothelium and an *internal elastic membrane*, the structure of which differs somewhat in arteries of different caliber; (2) a thick intermediate coat, the *tunica media*, consisting of smooth-muscle cells, arranged in a circular manner, together with an *external elastic membrane*, or elastic-fiber network, which also courses circularly; and (3) an outer layer, the *tunica externa*, or *adventitia*, of varying thickness in vessels of different caliber, made up of loosely arranged connective tissue as well as longitudinally arranged collagenous and elastic fibers. Large, medium, and small arteries differ to some degree in the detailed structure of their walls. In the tunica media of large arteries there is a preponderance of yellow elastic fibers in addition to smooth-muscle cells, the latter sometimes being referred to as the *muscular media*. In general, the smaller the artery, the greater is the relative amount of smooth muscle and the less is the relative quantity of elastic connective tissue. The main distinction between large and small arteries, aside from actual size discrepancies, lies in the thickness of the tunica media. The thick muscular and elastic walls of

arteries are of great physiological significance. Blood, forced into the arteries by contraction of the heart, distends these vessels. Since backflow is prevented by certain valves in the heart, tension of the elastic walls of the larger arteries tends to force the blood farther along the vessel. This accounts for the greater part of the *movement* of blood as well as for the maintenance of diastolic blood pressure * within the arterial system. The walls of the smaller, or *distributing, arteries*, which are primarily muscular, are of most importance in regulating the *amount* of blood supplied to a given area. This is accomplished by varying the size of the lumen under different conditions, the smooth-muscle cells being capable of responding to nervous stimuli, whereas elastic tissue can only react passively. Arterioles, with their relatively thick muscular walls and small lumina, are largely responsible for maintaining the high degree of blood pressure within the arterial system. They are responsible, also, for the fact that blood which enters a capillary bed is under greatly reduced pressure. If this were not

* The term *systolic blood pressure* refers to arterial blood pressure brought about by contraction of the heart. It is somewhat more than half again as high as the *diastolic blood pressure* which exists in the arterial system *between* contractions of the heart.

the case, the very thin and weak capillary walls would be damaged and incapable of functioning properly so far as diffusion of substances to and from tissue fluid is concerned.

VEINS. The walls of veins (Fig. 12.2) are much thinner than those of arteries but consist, nevertheless, of three similar coats. The tunica media is, in some cases, difficult to distinguish, particularly in the larger veins. Elastic and muscular elements are poorly developed in veins, but connective tissue is more clearly defined than in arteries. The tunica externa, or adventitia, makes up the greater part of the wall of the vein. Superficial, or subcutaneous, veins of the legs of mammals, particularly those with long hind legs, have a much thicker muscular portion of the tunica media than is usually the case. The walls of such veins, which lack the support of surrounding tissues, must be able to withstand the hydrostatic pressure produced by the relatively long column of blood being returned to the heart from the extremities. In long-legged birds and mammals, many of the medium-sized veins of the forelimbs and hind limbs are provided with paired semilunar valves on their inner walls (Fig. 12.3). They prevent backflow of blood in the veins and generally are located distal to the point where a tributary enters a larger vessel. The valves are formed by foldings of the tunica interna reinforced by connective tissue. In addition, they contain elastic fibers on the side toward the lumen of the vein. If the walls of superficial veins are inadequate to support the long column of blood, or if the semilunar valves break down, the veins become swollen and knotted and are referred to as *varicose veins*. Under certain conditions they are removed surgically. Valves have been demonstrated in the segmental veins of certain elasmobranchs and teleosts. It is believed that they may aid in increasing the

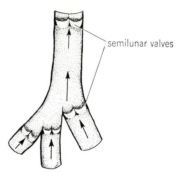

semilunar valves

Fig. 12.3 Diagram of portion of a vein and its tributaries, showing arrangement of semilunar valves.

return of venous blood to the heart. In general, the veins take the same course in the body as their corresponding arteries and usually are give similar names. A vein is always of greater diameter than its corresponding artery since its walls are thinner, are less rigid, and are more capable of being distended.

In both arteries and veins 1 mm or more in diameter, the cells making up their walls are supplied with nutrient blood vessels of their own. Only the cells in the walls of smaller vessels are able to utilize directly the nutrient materials and oxygen that are present in the blood coursing through the vessel proper. The nutrient blood vessels with which their walls are supplied are referred to as the *vasa vasorum* (vessels of the vessels). These arise from the same vessel or from a neighboring vessel some distance from the point where they are distributed. They are found in the tunica externa in both types of vessels. In general, the walls of veins have a more abundant supply of vasa vasorum than those of arteries. Furthermore, the vasa vasorum of veins may even penetrate to the tunica interna. The blood coursing through the veins is low in oxygen content and the cells in the walls can obtain very little oxygen by diffusion from the lumen.

Nerve fibers are abundant in blood vessels, particularly in arteries. They are of two types: medullated sensory fibers and nonmedullated, or sparsely medullated, autonomic motor fibers.

CAPILLARIES. The endothelium of the capillary is its only component. Since it is at the capillary bed (Fig. 12.4) that exchange of gases, food materials, and wastes takes place between the blood and the extravascular fluids, the thinness of the capillary wall is of obvious advantage. Diameter of the capillaries is relatively constant for each species and is related to the size of the red corpuscles which pass through them. In man the average diameter is about 0.008 mm. They are seldom more than 1 mm long. Lacking muscular and elastic coats, it has been a point of conjecture as to whether they are able to expand or contract of their own accord. In certain cases direct mechanical stimulation has been demonstrated to cause contraction of the endothelial cells, which have thus been thought to be contractile. It is possible, however, that changes in diameter of capillaries may be passive rather than active and a reflection of alterations in internal and external pressures. In amphibians and some other poikilothermous vertebrates, peculiar cells having a number of long processes, called *Rouget cells*, surround walls of capillaries in certain regions. Contraction of the Rouget cells has in the past been thought to bring about contraction of capillaries, but this is now considered doubtful. Rouget cells may be nothing more than connective-tissue elements which have become located on the capillary wall. Some cells similar to Rouget cells have been described for mammals. They are too few in number to have any significance in capillary contraction.

Capillaries are so numerous that it is practically impossible to prick the skin in any part of the body without drawing blood. There are approximately 2,000 in a cubic millimeter of human muscle. It has been

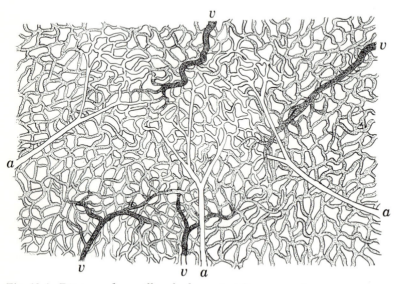

Fig. 12.4 Diagram of a capillary bed: *a*, arterioles; *v*, venules.

estimated that if all the capillaries in a man were stretched out in a single line they would measure nearly 150,000 miles.

The velocity of blood in arteries is high but when the blood reaches the arterioles, with their relatively thick muscular walls and small lumina, it encounters some resistance. This, coupled with the fact that the arterioles break up into great numbers of capillaries with a corresponding increase in the vascular area, brings about a considerable reduction in the velocity of the bloodstream. Blood, therefore, courses more slowly through capillaries than through the larger vessels and its hydrostatic pressure is diminished. Naturally this is of advantage in the exchange of nutrient materials, gases, and wastes between blood and tissue fluid at the capillary bed.

Electron microscopic studies have brought some interesting facts to light concerning the detailed structure of the cells of which capillaries are composed. The capillary wall is only one cell thick. It usually takes no more than two cells to surround the capillary lumen at a given point. Adjacent cells are joined together by a cement substance. Extremely small vesicles, from 400 to 650 Å in diameter, have been observed within the cytoplasm of these cells; some are clustered along the inner cell membrane and others against the outer cell membrane (Fig. 12.5). The vesicles actually seem to be tiny invaginations of the cell membranes. Apparently they are capable of separating themselves from the cell membrane on one side of the cell and of moving to the opposite side where they release their contents. It is possible that tissue fluid is absorbed or replenished by means of these vesicles. The process is termed *pinocytosis*. Pinocytotic vesicles have also been observed in the cells making up the proximal convoluted tubules of kidneys. Variations in the hydrostatic

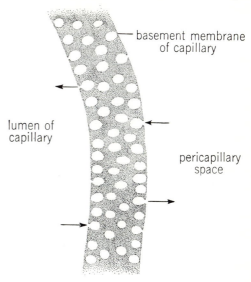

Fig. 12.5 Diagrammatic representation of portion of cytoplasm of an endothelial cell of capillary wall, illustrating pinocytosis, as suggested by electron microscope studies. Cytoplasmic vesicles may transport fluid from the capillary lumen to the outside of the cell or vice versa.

pressure of blood inside the capillaries, and changes in the osmotic pressures of blood and tissue fluid under different conditions, determine whether tissue fluid passes in or out of the capillaries.

Since the molecules of colloids are apparently too large to pass through the capillary endothelium, it is possible that the vesicles concerned in pinocytosis are too small to handle colloidal materials and confine their activity to the transport of water and crystalloids in solution. Nevertheless, it seems that some colloid is able to leak from the capillaries into the tissue fluid. How this is accomplished is not clear.

OTHER VESSELS. It is generally held that contractile smooth-muscle cells are not associated with capillaries and, therefore, cannot be responsible for changes in their diameter. It has been claimed, on the other hand, that there may be certain "preferred channels" through capillary beds. These are of capillary dimensions but have smooth-muscle cells here and there along their walls. These *arteriovenous bridges* are said to be direct continuations of arterioles and to give rise to regular capillaries as side branches. According to this idea, blood normally passes through the preferred channels but at times, when requirements demand that more blood be distributed to a given area, the other side-channel capillaries are brought into use.

In some regions there may be direct connections between arteries and veins without the intervention of a capillary network. Such vessels, which are called *arteriovenous anastomoses,* are to be found in the toes of birds, ears of rabbits, and such areas in man as the terminal phalanges of the fingers and toes, nail bed, lips, tip of tongue, nose, and eyelids. At ordinary temperatures they are closed, but when critical temperatures are reached, dilatation occurs. These vessels, according to one theory, serve in two capacities: (1) at high temperatures an increased volume of blood circulates at the surface, thereby permitting excess heat to be removed from the body by radiation; (2) at low temperatures increased circulation through these vessels prevents harmful effects which might be caused by cold. An arteriovenous anastomosis may be a straight or a tortuous branch of an arteriole connecting with a corresponding venule. The wall of the end connecting with the arteriole is similar to that of the arteriole, whereas that connecting with the venule is constructed like that of the venule. The midportion has a thicker and more muscular wall with an abundant supply of sympathetic nerve fibers which control the di-

ameter of the lumen of the vessel, thus regulating the amount of blood that passes through the anastomosis. Arteriovenous anastomoses should not be confused with arteriovenous bridges. Small arteriovenous anastomoses, known as *glomi,* are present in the dermis of the skin of the fingers and toes, as well as some other areas. Some authorities are of the opinion that arteriovenous anastomoses serve as a mechanism which regulates blood pressure and circulation in various parts of the body. Their actual function is not completely understood.

Sinusoids make up another kind of connection between arteries and veins. They differ from capillaries in structure and arrangement and are to be found in such organs as the liver, adrenal glands, bone marrow, parathyroids, pancreas, and spleen. They consist of relatively large, irregular, anastomosing vessels which, unlike capillaries with their rather constant diameter, are not lined with a continuous endothelial layer. Phagocytes and nonphagocytic cells, irregularly disposed, are present in their walls. Sinusoids are considered by some to represent a primitive form of capillary.

In certain parts of the body, arteries or veins break up into capillary networks but the small vessels, which may or may not be tortuous, recombine to form larger vessels of the same type, i.e., arteries or veins. Such connections of capillary dimensions are referred to as *retia mirabilia.* Examples of these structures within the course of arteries are to be found in the red bodies and red glands in the swim bladders of certain fishes, and in the glomeruli of kidneys. The portal vein of the hypophysis, and the hepatic portal and renal portal veins which break up into sinusoids or capillaries in liver or kidney, as the case may be, are examples of venous retia mirabilia, since the capillaries in these structures recombine to form veins.

In the penis and clitoris a peculiar type of tissue, called *cavernous,* or *erectile, tissue* is to be found. It consists of large, irregular, vascular spaces lined with endothelium and interposed between arteries and veins. They are supplied directly by arteries. *Erection,* or distention, of the structure containing this type of tissue is due to a filling of the spaces with blood under high pressure and a partial closure or compression of the venous outflow. Erectile tissue also occurs in the lamina propria of the mucous membrane covering the middle and ventral nasal conchae (page 207). Some authorities doubt whether the lamina propria of the nasal conchae should be classified as erectile tissue, which otherwise is confined to the penis and clitoris. This is because septa containing smooth-muscle fibers are lacking. At any rate, it consists of numerous thin-walled veins associated with smooth-muscle fibers which are arranged in both circular and longitudinal directions.

HEART

The hearts of chordates differ from those in members of the lower phyla in their ventral, rather than dorsal, location. Blood is pumped anteriorly through arteries and is then forced to the dorsal side. The greater part then courses posteriorly where the arteries terminate in capillaries in various parts of the body. The blood finally returns to the heart through veins.

In the lowest chordates the heart consists of little more than a pulsating vessel which is sometimes referred to as a one-chambered heart. In cyclostomes and fishes the heart is still a tubular structure but is divided into a series of compartments which, in a pos-

terior-anterior sequence, are called *sinus venosus, atrium (auricle), ventricle,* and *conus arteriosus,* respectively (Figs. 12.6, 12.10, and 12.12). The latter leads into a *ventral aorta.* Some authors refer to the hearts of cyclostomes and fishes as four-chambered hearts. This is confusing since the hearts of crocodilians, birds, and mammals, consisting of two atria and two ventricles, are also referred to as four-chambered hearts. The four chambers in the latter groups, therefore, do not correspond to the four compartments of more primitive forms. In the four-chambered hearts of higher forms there has been a division of both atrial and ventricular regions into two parts so that a double pumping mechanism is present. We shall, in the following pages, refer to atria and ventricles as *true* (in the sense of persistent) *chambers* and to the

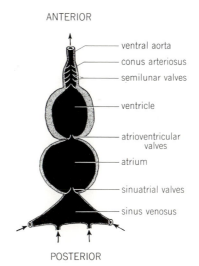

ANTERIOR

ventral aorta
conus arteriosus
semilunar valves

ventricle

atrioventricular valves
atrium

sinuatrial valves
sinus venosus

POSTERIOR

Fig. 12.6 Diagram of two-chambered heart as found in cyclostomes and most fishes, showing atrium and ventricle with accessory chambers, the sinus venosus and conus arteriosus.

sinus venosus and conus arteriosus as *accessory chambers*. In higher forms the sinus venosus, as such, disappears since it is incorporated in the wall of the right atrium. Furthermore, the proximal part of the conus arteriosus in higher forms has become incorporated in the ventricular walls. Its distal portion and the ventral aorta have become separated, or split, into two or three arterial trunks so that the conus no longer serves as a separate compartment. Only the atrial and ventricular regions, therefore, serve as chambers throughout the entire vertebrate series.

The heart in all vertebrates, during embryonic development, consists of a single tube which becomes twisted and compartmented in an increasingly complex manner as the evolutionary scale is ascended. Embryological studies make it clear that definite homologies exist and that all vertebrate hearts are deviations from a relatively simple fundamental plan.

We shall, in the light of the above, consider the hearts of cyclostomes and fishes (except Dipnoi) to be two-chambered. The conus arteriosus, coming from the ventricle, joins the ventral aorta anteriorly. In some forms, as in cyclostomes and teleosts, an enlargement, the *bulbus arteriosus*, is present at the proximal end of the ventral aorta. It does not contain cardiac muscle tissue and is actually a part of the ventral aorta rather than of the conus arteriosus. Sometimes the term *truncus arteriosus* is used as a synonym for ventral aorta. Dipnoans, amphibians, and most reptiles have three-chambered hearts with two atria and a single ventricle. Here the sinus venosus usually is reduced and joins the right atrium. In crocodilians, birds, and mammals the heart finally becomes four-chambered with two atria and two ventricles.

In animals having two-chambered hearts, the venous blood from all parts of the body is collected by the sinus venosus, which in turn enters the atrium. In those having three- or four-chambered hearts, unoxygenated blood (see footnote page 202) is returned to the heart through the sinus venosus or directly to the right atrium, as the case may be. Oxygenated blood, coming from the lungs, enters the heart through the left atrium.

In lower forms the heart is located far forward in the body, but there is a gradual backward shifting as the vertebrate scale is ascended. The heart lies in a *pericardial cavity*, which is surrounded by an investing membrane, the *pericardium*. Fluid, present in the pericardial cavity, serves as a lubricant, facilitating the movement of the heart when contracting and relaxing. The pericardial cavity is a portion of the coelom which has been cut off from the remainder of the body cavity. In such lower forms as elasmobranchs, the separation is incomplete and the two portions of the coelom are connected by a *pericardioperitoneal canal*. A thin, serous membrane, the *epicardium*, covers the surface of the heart. It is similar to, and continuous with, the outer lining of the pericardial cavity. The pericardium and epicardium, respectively, correspond to the parietal and visceral pleurae of the pleural cavities and to the parietal and visceral peritoneal membranes of the peritoneal cavity.

The heart is really a modified blood vessel. Like other blood vessels it is lined with endothelium, the *endocardium*, which covers all structures that extend into the cavity of the heart. Between the epicardium and endocardium is a thick muscular layer, the *myocardium*, composed of cardiac muscle. The endocardium is thickest where the myocardium is thinnest, and vice versa.

Just as the walls of blood vessels of greater caliber than 1 mm are supplied with blood via the vasa vasorum, the tissues of the heart itself are furnished with blood vessels which bring oxygen and nutrient materials to them and remove wastes. These vessels make up the *coronary circulation* which consists of coronary arteries and veins. The blood within the chambers of the heart is too far removed from most of the tissues to be available to them. Furthermore, cardiac tissue, in order to function properly, must receive an adequate supply of oxygen. The blood in two-chambered hearts, and on the right side of three- and four-chambered hearts, has given off its oxygen supply and is inadequate insofar as furnishing oxygen to cardiac tissues is concerned. The constantly beating heart muscle perhaps uses more oxygen, per gram of tissue, than any other tissue in the body.

Origin of the heart. Although the adult heart is an unpaired structure, its bilateral origin is, in most forms, indicated by its mode of development. The ventral mesentery of the gut is concerned in the development of the heart. In lower fishes and in amphibians a cavity appears in the ventral mesentery in the region of the foregut. This assumes a tubular form. From the surrounding mesoderm are differentiated the various layers of the heart. Since the ventral mesentery is formed by a coming together of the splanchnic mesoderm of the two hypomeres, the heart is basically a paired structure.

In the frog, for example, a pair of spaces appears in the mesoderm beneath the floor of the pharynx. These at first are separated from each other by a narrow strip of tissue. The lower layer of mesoderm (Fig. 12.7) is to become the parietal wall of the pericardial cavity. By a process of folding, the upper, or splanchnic, layer gives rise to the epicardium and myocardium of the heart. From the scattered cells beneath the floor of the gut comes the endothelium which forms the endocardium.

In the developing eggs of bony fishes, reptiles, and birds with their meroblastic cleavage, heart formation is somewhat different. At first the gut opens onto the yolk on the ventral side. Folds appear on either side of the midline and meet ventrally so as to close the gut (Fig. 12.8). In the region of the foregut, as this process is going on, the coelom widens to form a pair of *amnio-cardiac vesicles*, destined to give rise to the pericardial cavity. In the meantime a *vitelline vein* has developed from the mesoderm on either side of the foregut. With the closure of the foregut the two vitelline veins are brought together in such a manner as to form the endothelial lining of the heart. Myocardium and epicardium are derived from thickenings in the splanchnic mesoderm on the medial sides of the amino-cardiac vesicles. The fusion of the two sides leaves the primitive

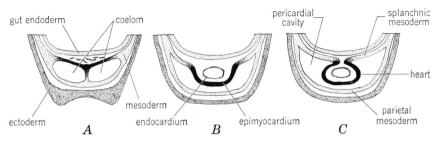

Fig. 12.7 Diagram showing method of development of heart of frog, *Rana temporaria.* (*After Schimkewitsch.*)

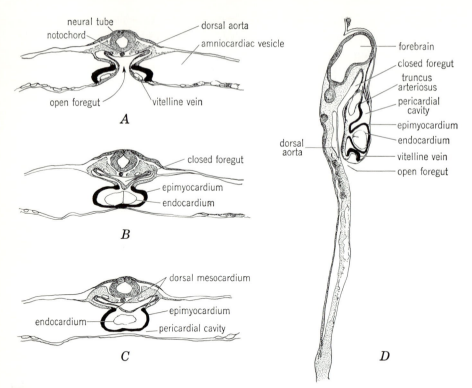

Fig. 12.8 A–C, transverse sections of 33-hour chick embryo, showing manner in which heart is formed; D, sagittal section of similar embryo.

heart suspended from the mesoderm surrounding the foregut by a mesenterylike band of tissue, the *dorsal mesocardium*, which shortly disappears. A *ventral mesocardium* also forms, but its appearance is even more transitory than that of the dorsal mesocardium.

Heart formation in mammals differs somewhat from the foregoing account, and a bilateral origin is not so strongly indicated. The heart develops from the *cardiogenic plate*, a crescent-shaped plate of mesoderm lying anterior to the embryo. The head of the embryo gradually grows over this area so that the latter assumes a more posterior position. A space, the pericardial cavity, lies within the mesoderm just above the cardiogenic plate. As development progresses a process of folding occurs (Fig. 12.9) so that the cardiogenic plate comes to lie *above* the pericardial cavity. Two strands of cells which soon acquire cavities appear in the upper part of the cardiogenic plate. They fuse to become the endocardium of the heart. The remaining mesoderm of the cardiogenic plate forms myocardium and epicardium.

In all cases cited, the heart, when first formed, is a simple tube. The appearance of constrictions, folds, and partitions of one kind or another results in the formation of two-, three-, or four-chambered hearts, as the case may be.

Comparative anatomy of the heart. Amphioxus. A single, median, contractile vessel lies ventral to the pharynx in am-

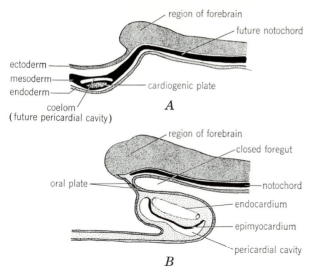

region of forebrain
future notochord
ectoderm
mesoderm
endoderm
cardiogenic plate
coelom
(future pericardial cavity)
A

region of forebrain
closed foregut
oral plate
notochord
endocardium
epimyocardium
pericardial cavity
B

Fig. 12.9 Diagrams showing two stages in heart formation of human embryo.

phioxus. By some authorities it is considered to be a one-chambered heart. Usually, however, it is spoken of as the *branchial artery, endostylar artery,* or *truncus arteriosus.* Its walls are provided with muscles which contract more or less rhythmically from posterior to anterior ends. On either side of the vessel lateral branches are given off which course through the primary gill bars. At the base of each lateral artery is a small, contractile enlargement, the *bulbillus* (Fig. 12.17). A single contraction of the "heart" followed by contraction of the bulbilli is sufficient to force blood through the entire body. It takes about 1 minute for the blood to make a complete circuit. Contractions of the heart also occur at approximately 1-minute intervals. Since the primitive embryonic condition of the heart in vertebrates is that of a single tube, the heart of amphioxus is considered to correspond to an early stage of development of the heart in higher forms.

CYCLOSTOMES. A two-chambered heart is present in cyclostomes. In the lamprey (Fig. 12.10) it consists of a large, thin-walled atrium which communicates with the ventricle by means of a small aperture guarded by *atrioventricular valves.* The ventricle lies on the right side of the atrium. It is smaller than the atrium and possesses thick muscular walls. The lining of the ventricle is irregular and provided with tough connective tissue cords, the *chordae tendineae,* which connect to atrioventricular valves. The cords prevent the valves from being pushed into the atrium when the muscular ventricle contracts. Blood

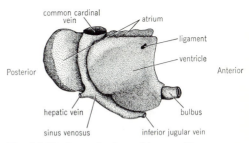

common cardinal vein
atrium
ligament
ventricle
Posterior
Anterior
hepatic vein
bulbus
sinus venosus
inferior jugular vein

Fig. 12.10 Lateral view of heart of lamprey as seen from the right side. The posterior part of the atrium has been deflected to expose the sinus venosus to view.

Circulatory System   547

is forced from the ventricle through a poorly developed conus into the ventral aorta which distributes it to the gills. A single set of two semilunar valves in the conus region prevents any backflow of blood. A small, thin-walled sinus venosus, which lies in the crevice between atrium and ventricle, opens into the atrium through a slitlike aperture guarded by a pair of *sinuatrial valves*. The sinus venosus receives three vessels: (1) a large, single, *common cardinal vein* on the dorsal side, (2) a small *inferior jugular vein* on its anteroventral side, and (3) a small *hepatic vein* from the posterior side. The heart, which lies posterior to the last pair of gill pouches, is located in a pericardial cavity surrounded by a thick, tough pericardium. The pericardium fits into a concavity at the anterior end of the liver. In the ammocoetes stage the pericardial cavity communicates with the rest of the coelom, but the connection disappears in the adult.

Only unoxygenated blood passes through the cyclostome heart. The heart sends blood to the gills where exchange of gases takes place. This is known as the single type of circulatory system (Fig. 12.11) since but one stream of blood passes through the heart. The heart of the hagfish is probably *aneural* since, unlike that of other vertebrates, it

receives no branches of the vagus nerve. As in other fishes, the hearts of cyclostomes lack innervation from the sympathetic nervous system (page 648).

FISHES. The hearts of most fishes are essentially similar in structure to the two-chambered heart described for the lamprey. In fishes the volume of the atrium is about the same as that of the ventricle. The single type of circulation is the rule. Variations in the relative positions of atrium and ventricle occur, but these are of minor significance. The arrangement of veins entering the sinus venosus also varies. The feature which perhaps deserves the greatest attention is the number and arrangement of the valves in the conus arteriosus (Fig. 12.12). These semilunar valves, which prevent backflow of blood into the heart, are most numerous in elasmobranchs and in the so-called ganoid fishes. In these forms the valves are arranged in three longitudinal rows, one dorsal and two ventrolateral in position. Several valves are present in each row. In *Epiceratodus* eight sets of valves are present. Many elasmobranchs have six sets. The valves of the anterior set are usually largest. The remainder decrease in size the more posterior their location. There is a tendency among fishes toward a reduction in the number of

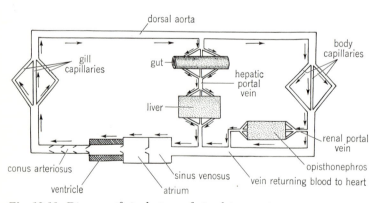

Fig. 12.11 Diagram of single type of circulatory system.

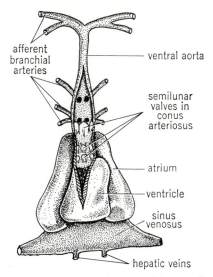

afferent branchial arteries

ventral aorta

semilunar valves in conus arteriosus

atrium

ventricle

sinus venosus

hepatic veins

Fig. 12.12 Ventral view of heart of dogfish, *Squalus acanthias.* The ventricle and conus arteriosus have been cut open to show the arrangement of the semilunar valves in the conus.

valves in the conus. Teleosts are characterized by having a single set, although in a few species (the ladyfish, *Albula vulpes*, and the tarpon, *Tarpon atlanticus*) two sets are retained. In teleosts and in cyclostomes, it will be recalled, the term *bulbus arteriosus,* rather than conus arteriosus, is used to denote the slightly enlarged region where the semilunar valves are located.

The first advance from a two- to a three-chambered condition is seen in the Dipnoi, or true lungfishes. Although the heart of *Epiceratodus* shows practically no partitioning, in *Protopterus* and *Lepidosiren* the atrium has become partially separated into right and left halves, or chambers, by an incomplete partition, or septum, confined to the posterior part of the atrium. In *Protopterus* unoxygenated blood from the body enters the right chamber via the sinus venosus, but oxygenated blood coming

from the swim bladder (lung) enters the left atrium. The ventricle possesses cavities in its walls. These, together with an incomplete interventricular septum made up of fibrous and muscular tissues, prevent mixing to a large extent. In *Protopterus* the conus arteriosus has become divided by spiral folds, much like the condition in amphibians (page 550), so that two streams of blood leave the heart. The oxygenated blood from the left side of the ventricle passes through the ventral channel of the conus on its way to the anterior gill region and thence directly to the dorsal aorta. It will be recalled that gill lamellae are lacking in the anterior gill region of *Protopterus*. The unoxygenated blood from the right side of the ventricle courses through the dorsal channel of the conus to the posterior gill region and swim bladder.

A coronary circulation, supplying and draining the walls of the heart itself, is particularly well developed in elasmobranchs. *Hypobranchial arteries* carrying oxygenated blood usually arise from efferent branchial arteries draining some of the anterior gills. The hypobranchials branch into various vessels, among which are the coronary arteries going to the heart muscle (Fig. 12.19). Coronary veins, draining the wall of the heart, enter the sinus venosus close to the sinuatrial aperture. Similar vessels have been described in certain teleost and ganoid fishes, but in others a special coronary circulation does not seem to be evident.

It is of interest that the heart of *Latimeria* is an S-shaped tube and not so compact an organ as even that of elasmobranchs. This, among other features, indicates an early evolutionary divergence of coelacanths from other groups of fishes.

The heartbeat of fishes is controlled by the parasympathetic portion of the autonomic nervous system via the vagus nerve.

A sympathetic innervation is lacking (page 656).

AMPHIBIANS. A double type of circulation, similar to that of *Protopterus,* is characteristic of adult amphibians. Two streams of blood, one oxygenated and the other partially oxygenated, enter the heart. In amphibians the sinus venosus has shifted in position so that it opens into the right atrium (Fig. 12.13).

The oxygenated stream coming from the lungs (also from gills via the lungs, in certain urodeles) enters the left atrium and thence passes to the left side of the ventricle. The partially oxygenated stream composed of unoxygenated blood, from most of

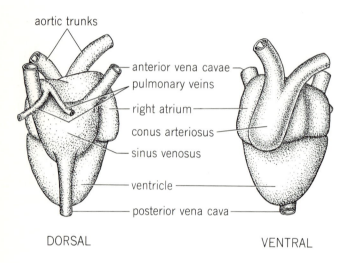

DORSAL   VENTRAL

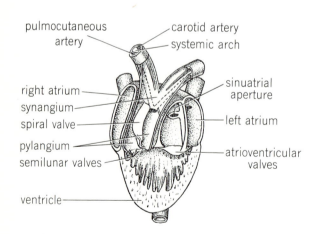

FRONTAL SECTION

Fig. 12.13 Heart of bullfrog, enlarged. (*After Storer, "General Zoology," McGraw-Hill Book Company. Drawn by R. Speigle.*)

the body, mixed with oxygenated blood, from the lining of the mouth and from the skin, enters the sinus venosus from which it passes to the right atrium and thence to the right side of the ventricle.

The system of vessels coming from the lungs to the heart and those leading from the heart to the lungs is referred to as the *pulmonary circulation.* That which is distributed to the body in general and then returned to the heart is called the *systemic circulation.*

The *interatrial septum* is a thin membrane of connective tissue covered with endothelium. Perforations are frequently present in the interatrial septum of urodeles but are of no special significance since little mixing of the two bloodstreams occurs here. The ventricle is not partitioned, but its lining is thrown into many pockets by muscular bands which, to a large extent, keep the blood coming from the right atrium separate from that coming from the left. Chordae tendineae fastened to the atrioventricular valves prevent blood from being regurgitated into the atria when the ventricle contracts.

Despite the presence of a single ventricle, the two bloodstreams, as previously mentioned, probably mix only to a slight degree. This, however, is not too important since the systemic blood entering the right side of the ventricle has, to a considerable extent, been oxygenated in the skin and lining of the mouth. Both these areas serve as important respiratory structures in amphibians, which utilize buccopharyngeal and cutaneous respiration in addition to pulmonary respiration.

Many amphibians have developed a rather complicated system of valves and partitions which ostensibly serve to keep the two streams fairly separate upon leaving the ventricle. There is some disagreement among biologists regarding the extent to which such

a separation actually occurs. Variable results have been obtained by different investigators who have injected India ink, radiopaque substances, and similar materials into different parts of the bloodstream and followed their course to and from the heart.

The conus arteriosus bears striated muscle fibers in its walls and is, therefore, capable of contraction. It is made up of two regions. That part next to the ventricle is the *pylangium.* It is more muscular than the distal portion, which is referred to as the *synangium.* When viewed externally the anterior end of the synangium appears to divide into two trunks, each of which in turn separates into three arteries. The most anterior is the *carotid artery,* going to the head region; the second is the *systemic artery,* or *arch,* which gives off a few branches before the two systemic arches join each other posteriorly to form the *aorta,* which in turn distributes blood to the rest of the body; the third, or *pulmocutaneous artery,* leads to the lungs and skin.

The internal structure of the conus is rather complicated. Two sets of semilunar valves are present in the conus, one set at the base of the pylangium, the other at the junction of pylangium and synangium. In most amphibians one of the latter has become modified to form the *spiral valve* (Fig. 12.13), which is generally believed to play a part in separating the two bloodstreams as they leave the heart.

The term *systole* is used to designate the contraction of the ventricle(s) of the heart. *Diastole*, on the other hand, refers to the state of dilatation of the heart, especially the ventricle(s), when it is being filled with blood coming from the atria.

The right atrium contracts slightly in advance of the left, and both streams of blood enter the ventricle during diastole. Little mixing of blood occurs in the ventricle.

During systole the blood in the ventricle is forced into the pylangium, which leads from the right side. The atrioventricular valves prevent backflow into the atria. The mixed blood from the right side of the ventricle is the first to enter the pylangium, which then contracts in turn. The semilunar valves at the base of the pylangium prevent backflow into the ventricle.

In frogs the blood flows over the free edge of the spiral valve to the left side of the pylangium to enter a single aperture located just posterior to the synangial valves. This leads to the two pulmocutaneous arteries. The mixed blood, apparently following the path of least resistance, passes through the relatively short pulmocutaneous arteries to lungs and skin, being forced forward by contraction of the muscular wall of the pylangium. At the same time the wall of the pylangium is pressed against the free edge of the spiral valve. Therefore the remainder of the mixed blood, and the oxygenated blood which follows from the left side of the ventricle, passes through the pylangium on the *right* side of the spiral valve, finally entering the systemic and carotid arteries. A spongy mass, the *carotid body* or *carotid "gland,"* is located at the base of the carotid arch (Fig. 12.22). Its histological appearance is much like that of an endocrine gland, consisting of clumps of cells resembling those of epithelia, as well as numerous small blood sinuses. The cells have an abundant nerve supply. They seem to respond to changes in the oxygen and carbon dioxide content of the blood. Messages are apparently relayed via the nerve endings to those centers in the brain which regulate the contractions of the heart and arteries. Thus the carotid body may be of great importance in keeping the blood in the carotid vessels leading to the head at a rather constant pressure and gaseous content. In salamanders all three pairs of arteries

lead from the synangium. One would expect that under such conditions mixing of the two bloodstreams would occur in this region. Again the results of experiments attempting to determine the extent to which separation occurs are controversial.

In perennibranchiate salamanders, such as *Necturus,* blood is sent to the gills for aeration. Blood going to the lungs through pulmonary arteries has already been oxygenated in the gills (Fig. 12.25). This would indicate that the lungs function but little as respiratory organs in such salamanders and may be used only in times of emergency.

Variations from the condition described above are to be found in the hearts of certain amphibians. The interatrial septum may be incomplete or furnished with numerous openings or a single, large aperture. This is true of caecilians, which use the skin as a major respiratory organ, the lungless salamanders, and those aquatic forms in which the lungs have been reduced in size and function. The sinus venosus may open into both right and left atria, and the left atrium may be reduced in size. The spiral valve may be rudimentary or wanting in those forms in which no precise separation of the two bloodstreams occurs. It seems that only if functional lungs exist are structures present in the heart which limit circulation of oxygenated and unoxygenated types of blood. Biologists are generally of the opinion that in its primitive condition the amphibian heart was three-chambered and capable of accommodating two separate bloodstreams. Any other conditions which are encountered, such as those mentioned above, may be considered to be specializations or regressions.

A special coronary circulation, supplying and draining the muscular wall of the heart itself, is apparently lacking in many amphibians. In the frog, the conus arteriosus receives a branch bearing oxygenated blood from the carotid artery. This vessel breaks up on the conus into a network of small vessels which recombine to form two small veins entering the systemic (left innominate) and hepatic portal (anterior abdominal) veins, respectively. These vessels are called the *vena bulbi anterior* and *vena bulbi posterior.* All in all, despite the lack of even a partial ventricular septum, it seems that the amphibian heart is better equipped structurally and functionally than that of dipnoans to bring about a separation of oxygenated and unoxygenated blood streams.

REPTILES. The reptiles are the first group of chordates to become truly terrestrial. Except for certain aquatic turtles, which at times utilize cloacal respiration, lungs are the only respiratory organs. In this group, then, as well as in birds and mammals, an efficient pulmonary circulation is a necessity. With its development further changes have occurred in the structure of the heart.

Although a large sinus venosus is present in certain reptiles (turtles), in many it has been greatly reduced. Much of it may be incorporated within the wall of the right atrium. Valves which are present where veins enter the right atrium represent vestiges of the sinus venosus. A complete interatrial septum separates the oxygenated blood entering the left atrium from the unoxygenated blood in the right.

All reptiles have a three-chambered heart except crocodiles and alligators, in which it is four-chambered. Even in the three-chambered heart, however, the ventricle is partially divided by an incomplete *interventricular septum* which extends forward from the apex toward the center. The conus arteriosus no longer exists as such. Its distal portion as well as the ventral aorta has split into three main trunks (Fig. 12.26), each of which has a single row of semilunar valves at its base. One trunk is the *pulmonary trunk,* sometimes

called the *pulmonary aorta,* which gives off two pulmonary arteries going to the lungs. The pulmonary aorta leaves the right side of the ventricle. The two remaining *systemic trunks* are called the *left* and *right aortae,* respectively. The left aorta leads from the right side of the ventricle and crosses to the left side; the right aorta leads from the left side of the ventricle and crosses to the right side. An aperture, the *foramen Panizzae,* is located at the point where right and left aortae are in close contact and cross each other, so that their cavities are in communication. Even though the ventricle in most reptiles is only partially divided, the presence of an incomplete interventricular septum separates the two bloodstreams passing through the heart more effectively than is the case in amphibians. Recent studies (F. N. White, 1968, *Amer. Zoologist,* **8:**211), in which radiopaque media have been used, reveal a more complete separation of the two bloodstreams than one would expect from anatomical observations alone. The left aortic arch, because of ventricular-pressure differences, receives much more oxygenated blood than had formerly been suspected. Access of oxygenated blood from the right ventricle is via the foramen Panizzae which plays a much more important role than had previously been believed. In the Crocodilia, in which a complete interventricular septum appears for the first time, a true four-chambered heart is found. The left aortic arch, despite its origin from the right ventricle, receives a good quantity of oxygenated blood via the foramen Panizzae, as described above. Apparently the blood in the pulmonary artery is not so affected since this vessel arises from an area which, because of pressure changes, becomes functionally independent. The right atrioventricular valve in crocodilians is largely composed of muscular tissue rather than connective tissue.

In reptiles the right aortic arch, carrying oxygenated blood from the left side of the ventricle, gives off a large *brachiocephalic artery* which distributes blood to the anterior part of the body. At its base arise small coronary arteries which pass to the walls of the heart itself. A coronary sinus returns blood to the right atrium.

BIRDS. A completely double circulation (Fig. 12.14) occurs in birds for the first time, since at no point is there any opportunity for oxygenated and unoxygenated blood to mix. The sinus venosus has disappeared, and three large veins, two precavae and one postcava, enter the right atrium directly (Fig. 12.40). Pulmonary veins return oxygenated blood from the lungs to the left atrium. The heart in birds is relatively much larger and more compact than in forms previously discussed. Both atria are thin-walled. The ventricles are completely separated as in crocodilians. The muscular wall of the left ventricle is much heavier than that of the right and is partially surrounded by the latter. Blood from the left ventricle is distributed over a considerable

distance to all parts of the body, and a propulsive force of abundant power is required for such distribution. The relatively thin-walled right ventricle forces blood only a short distance through pulmonary arteries to the lungs. A single muscular and very well-developed atrioventricular valve separates the right atrium from the right ventricle. Two valves, however, known together as the *bicuspid valve,* are present at the left atrioventricular aperture. Chordae tendineae, which are fastened to the atrioventricular valves, are anchored at their other ends to the lining of the ventricle by heavy muscular projections called *papillary muscles.* It will be recalled that all structures projecting into the cavity of the heart are covered with endocardium.

The main advance shown by the heart of the bird over the four-chambered crocodilian heart lies in the elimination of the left aorta. Only two vessels leave the heart in birds: a pulmonary trunk, or aorta, from the right ventricle, and a systemic aorta, corresponding to the right aorta of reptiles, from

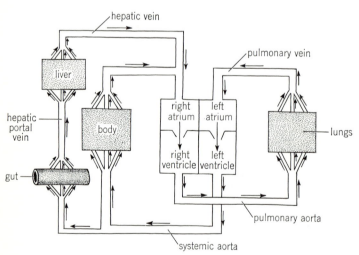

Fig. 12.14 Diagram of double type of circulatory system in a vertebrate having a four-chambered heart, ventral view.

Anatomy of the Chordates

the left. A single set of three semilunar valves is present at the base of each.

In birds the circulating blood passes from the left ventricle to all parts of the body. It is collected by the two precaval and single postcaval veins and enters the right atrium. From here it passes to the right ventricle, which then sends it to the lungs for aeration. Blood of high oxygen content is returned to the left atrium which sends it past the bicuspid valve into the left ventricle.

A well-developed coronary system is present in birds. Coronary arteries arise from the systemic aorta. A venous, coronary sinus enters the right atrium near the entrance of the postcava.

MAMMALS. The four-chambered mammalian heart (Fig. 12.15) is essentially similar to that of birds, the two sides of the heart being completely separated from each other by interatrial and interventricular septa. A thin area, the *fossa ovalis*, in the interatrial septum represents the position of an opening,

the *foramen ovale*, which is present during fetal life. A sinus venosus, present only during early embryonic development, is lacking in the fully formed mammalian heart, having been incorporated into the wall of the right atrium. Monotremes and the armadillo *Dasypus* are exceptions, a small and shallow chamber distinct from the right atrium being present, with semilunar valves at the point of junction. The sinus venosus is formed by the confluence of the precavae and postcava. In placental mammals the pulmonary veins open directly into the left atrium, but in monotremes and marsupials two pulmonary veins enter a common vestibule which in turn joins the left atrium.

When the atria are in a contracted state, a flaplike projection of each can be observed extending for a short distance over the ventricle. These are the *atrial appendages,* or *auricles*. Although the inner lining of the greater part of each atrium is smooth, that of the auricles is ridged by muscular bands,

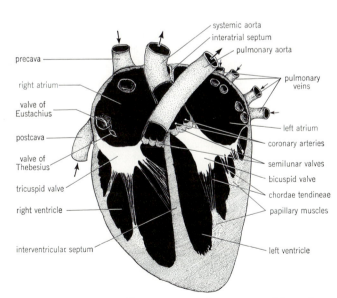

Fig. 12.15 Diagram showing internal structure of four-chambered mammalian heart, ventral view.

the *musculi pectinati*. No particular function has been ascribed to the atrial appendages. They actually represent the atria proper. The smooth area of the right atrium corresponds to the absorbed part of the sinus venosus; that of the left indicates where the proximal parts of the pulmonary veins have been incorporated in its walls.

Some variations occur among mammals in regard to the number of systemic veins entering the right atrium. A single postcava is present in all. In many species its opening into the right atrium is guarded by a *valve of Eustachius*. In bats and the mole a semilunar valve is present at the opening into the right atrium. It is believed to represent remnants of the sinuatrial valve of lower forms. There may be either a single precava, as in man and the cat, or two precavae, as in the rabbit, rat, and others. Between the superior and inferior vena cavae there is in certain species (man, wolf, deer, sea lion) a more or less marked prominence, the *tubercle of Lower*. This is apparently a vestige of a structure of importance in foetal circulation. Pulmonary veins, returning oxygenated blood from the lungs, enter the left atrium. Their number varies in different species.

As in birds, a *bicuspid,* or *mitral*, *valve,* consisting of two membranous flaps, prevents blood in the left ventricle from being regurgitated into the left atrium. Monotremes are exceptions in that the left atrioventricular valve is *tricuspid*. In most mammals a tricuspid valve, composed of three somewhat irregular flaps, is located between the right atrium and ventricle. Since a single valve is present in birds at this point, the presence of a tricuspid valve is a distinguishing feature of the mammalian heart. Chordae tendineae, attached to the irregular, ventricular surfaces of the bicuspid and tricuspid valves, are connected at their other ends either directly, to the inner walls of the ventricles and the interventricular septum, or indirectly to papillary muscles which are continuations of muscular ridges (*trabeculae carneae*) lining the inner surfaces of the ventricles. All are covered with endocardium. The right atrioventricular valve in monotremes is mainly vascular. In marsupials and placental mammals an occasional chorda tendinea is muscular, indicating that it is derived from the wall of the ventricle. Usually, however, the valve is composed of connective tissue alone.

In mammals, contrary to the condition in birds, there has been an elimination of the right aorta, so that only the left aorta persists. It arises from the left ventricle and distributes blood to all parts of the body. It is called the *systemic trunk*, or *aorta*. The *pulmonary trunk*, or *aorta*, from the right ventricle carries unoxygenated blood to the lungs. A single set of three semilunar valves is located at the bases of both pulmonary and systemic aortae.

A well-developed coronary system is present in the mammalian heart. Right and left coronary arteries arise from the systemic aorta as it leaves the left ventricle and just distal to the semilunar valves. In a number of mammals the ventricular septum is supplied by a special *septal artery* which arises from the left coronary artery near its origin. During diastole there is a rebound of blood in the aorta. The presence of semilunar valves prevents the blood from returning to the left ventricle. Some of the blood, however, is forced into the coronary arteries which distribute it to the tissues of the heart itself. Deoxygenated coronary blood is returned through several vessels which converge to enter the right atrium through the *coronary sinus* guarded by the *valve of Thebesius*. Other small openings, the *foramina of*

Thebesius, are the openings of small veins, the *venae cordis minimae*, which return some blood to the right atrium directly from the heart muscle.

One form of heart failure, known as *coronary occlusion*, is caused by a blood clot or some other fragment blocking a coronary vessel. Death often results when the myocardium fails to receive its normal supply of metabolic substances. In some cases the area of myocardium supplied by the blocked vessel becomes necrotic. Either it forms scar tissue or the heart may rupture at the affected point.

Just as the walls of blood vessels are supplied with blood by the vasa vasorum, and the walls of the heart by the coronary circulation, the tissues making up the framework of the lungs, bronchi, and even the walls of the pulmonary vessels themselves receive a supply of blood distinct from that inside the pulmonary arteries and veins. *Bronchial arteries* and *veins*, branches of certain systemic vessels close to the heart, function in this manner.

Furnishing support and preventing excessive dilatation of certain regions when contraction occurs in various parts of the heart are rings of dense fibrous connective tissue. These are especially well developed around the atrioventricular openings and in the regions from which the systemic and pulmonary trunks arise. They are also in continuity with the fibrous tissue in the interventricular septum. Since the free ends of cardiac muscle fibers may insert on this tissue, it is sometimes referred to as the "skeleton" of the heart.

In some mammals cartilaginous or bony tissue may actually be present in the heart. It appears in association with the fibrous rings. In the horse, for example, a plate of cartilage is usually present on the right side of the aortic ring. In older animals this may become calcified. In cattle it is common for two bones, the *ossa cordis*, to be present in the fibrous aortic ring. The one on the left is the smaller and is rather inconstant in appearance. The right one is generally about 1 in. long and roughly triangular in shape. These bones serve, respectively, as points of attachment for the left and right posterior cusps of the aortic valve.

The shape of the heart shows much variation among mammals but no adequate explanation has been offered to account for this. Generally speaking, there is some correlation between the shape of the heart and the shape of the thorax.

Evolution of the heart. Biologists have, for the most part, assumed that there has been progressive evolution through the vertebrate classes from the two-chambered heart (plus accessory chambers) of the Agnatha, Chondrichthyes, and actinopterygian Osteichthyes, pumping unoxygenated blood alone, to the four-chambered structure of crocodilians, birds, and mammals, which sends unoxygenated blood to the lungs and oxygenated blood to the rest of the body. There are those who question this interpretation and believe that there may have been a sharp dichotomy between the hearts of conventional fishes and those of tetrapods. The former handled one stream of blood and the latter two. According to this idea the double-type heart may first have appeared in crossopterygian fishes. Unfortunately these are known only from fossil remains except for the coelacanth, *Latimeria*, which is too specialized to reveal basic conditions. The various kinds of hearts found in dipnoans, amphibians, and most reptiles may represent nothing more than degenerate conditions and a regressive evolution, rather than intermediate stages of a progressive evolution from the two- to the four-chambered state, as is generally believed.

Rhythmicity of the heartbeat. The heart is capable of contracting rhythmically when removed from the body under experimental conditions. It would seem as though the impetus for contraction must arise within the heart itself. In the intact animal, however, the automatic beating of the heart is under the influence of nervous impulses. So-called cardiac centers in the medulla oblongata of the brain transmit impulses by means of efferent nerve fibers to the heart.

In the hearts of lower vertebrates, waves of contraction may be observed to pass from the sinus venosus to atrium to ventricle. The "pacemaker," which regulates or initiates the rhythmic beat of the heart, is a bundle of atypical muscle fibers located in the wall of the sinus venosus. It is called the *sinuatrial node.* In higher forms, in which the sinus venosus has become incorporated in the wall of the right atrium, the sinuatrial node lies embedded in the muscular atrial wall. In man it is actually in the wall of the superior vena cava, at the point where that vessel enters the right atrium. The muscle fibers of which the sinuatrial node is composed are smaller than ordinary atrial fibers and are specialized for conductivity rather than for contraction. They are embedded in a mass of collagenic connective-tissue fibers and have an unusually abundant capillary supply as well as being furnished with nerve fibers of both sympathetic and parasympathetic divisions of the autonomic nervous system. Another mass of atypical muscle fibers is the *atrioventricular* node which, in the four-chambered heart, is located in the lower part of the interatrial septum. It also acts as a pacemaker under experimental conditions when the sinuatrial node is destroyed

or prevented from functioning. There is no special pathway of fibers connecting the sinuatrial and atrioventricular nodes. Conducting impulses apparently are transmitted via normal cardiac muscle fibers located in the walls of the atria. From the atrioventricular node a bundle of peculiar muscle tissue, the *atrioventricular bundle*, or *bundle of His*, is distributed to the ventricular walls. Two main branches of this bundle course along the two sides of the interventricular septum, respectively, and then become continuous with other peculiarly modified muscle cells called *Purkinje fibers*. The latter are distributed to the papillary muscles and to the walls of the ventricles. The two nodes, the atrioventricular bundle, and the Purkinje fibers make up the conducting system which is responsible for the rhythmic sequence of the various phases of the heartbeat. It is of interest that the differences between these specialized structures and ordinary cardiac muscle become apparent rather early during embryonic development but do not reach their full development until some time after birth.

ARTERIAL SYSTEM

Although the arterial systems of various adult vertebrates appear to be different in arrangement, nevertheless a study of development reveals that all are built upon the same fundamental plan. The increasing complexity of the heart, from the simple two-chambered structure of lower forms to the four-chambered organ of crocodilians, birds, and mammals, is associated in part with the variations to be found in the blood-vascular system.

When the heart is forming in vertebrate embryos during development, a vessel, the *ventral aorta,* appears in the midline ventral to the pharynx. It soon establishes a connection with the conus arteriosus. At its anterior end the ventral aorta divides into two *aortic arches* which course dorsally in the mandibular region. Dorsal to the pharynx these are continued posteriorly, where they are known as the *paired dorsal aortae*. Additional pairs of aortic arches then appear, forming connections between ventral and dorsal aortae on each side. The aortic arches appear in sequence in an anterior-posterior direction, each coursing through the tissue between adjacent pharyngeal pouches. The typical number of aortic arches to form in vertebrates is six pairs (Fig. 12.16), although there are certain discrepancies among lower forms. The first aortic arch is known as the *mandibular aortic arch.* The second is the *hyoid aortic arch,* the remainder being referred to as the *third, fourth, fifth,* and *sixth aortic arches,* respectively. Each one lies anterior to the visceral cleft bearing the corresponding number. The paired dorsal aortae fuse pos-

teriorly to the pharyngeal region so that only a single dorsal aorta ultimately is present. It is continued into the tail region as the *caudal artery.* Various paired and unpaired vessels arise along the length of the dorsal aorta to supply all the structures of the body posterior to the pharyngeal region. Anterior continuations of the unpaired ventral aorta and the paired *radices* (singular, *radix*) of the dorsal aorta supply the head and anterior branchial regions. Although the method of branching of the dorsal aorta is fairly uniform throughout the vertebrate series, the aortic arches undergo rather profound modifications in different forms, the changes being rather similar in members of a given class. Blood, which is pumped anteriorly by the heart, passes through the ventral aorta to the aortic arches. These vessels then carry the blood to the paired dorsal aortae, from which it either goes anteriorly to the head or posteriorly to the single dorsal aorta which distributes it to the remainder of the body. Veins return blood to the sinus venosus or right atrium, as the case may be.

Changes in the aortic arches. AMPHIOXUS. The "heart" of amphioxus, sometimes called the *ventral aorta*, consists of nothing more than a single, contractile vessel lying ventral to the gill region. From the ventral aorta lateral branches, the *afferent branchial arteries,* are given off on both sides alternately. They extend up into the primary bars of the pharynx. In an adult specimen 60 or more pairs of afferent branchial arteries may be present. Each, as previously mentioned, bears a contractile enlargement, or *bulbillus,* at its base. The afferent branchial arteries give off small branches which connect with vessels in the secondary gill bars. The latter vessels have no direct connection with the ventral aorta, but those in both primary and secondary bars connect with the paired

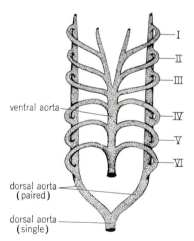

Fig. 12.16 Diagram illustrating the typical condition of aortic arches in vertebrate embryos, ventral view. Six pairs of arches connect dorsal and ventral aortae.

dorsal aortae by means of *efferent branchial arteries* (Fig. 12.17). The two aortae unite behind the pharynx to form a single median vessel which courses posteriorly. Oxygenation of the colorless blood takes place during its passage through the gill bars.

In amphioxus the aortic arches are much more numerous than in higher chordates. It is interesting to note that the vessels in the gill bars give rise to vascular plexuses which supply the excretory tubules before joining the dorsal aortae.

CYCLOSTOMES. The ventral aorta in cyclostomes is continued forward from the heart for a considerable distance. The number of aortic arches given off by the ventral aorta varies with the species and depends on the number of gill pouches. They are most numerous among the Myxinoidea, *Bdellostoma stouti* having 15 pairs. In the lamprey, *Petromyzon marinus,* which is representative of the group, 7 pairs of gill pouches are present. The ventral aorta leaves the heart as a single vessel which bifurcates at the level of the fourth gill pouch (Fig. 12.18*A*). Four afferent branchial arteries arise from each of the paired anterior extensions, and four pairs are given off by the unpaired portion. The first of the eight pairs of afferent branchial arteries on each side supplies the gill lamellae of the anterior hemibranch. The last furnishes blood to the most posterior hemibranch. Each of the remaining vessels arises at the level of an interbranchial septum and divides almost immediately to supply the lamellae on either side of the septum. Thus each holobranch is furnished with the two branches of an afferent branchial artery. Efferent branchial arteries, corresponding in position to the afferent vessels, collect blood from the gill lamellae (Fig. 12.18*B*). They join the single, median, dorsal aorta. The anterior end of this vessel is paired for a short distance, but the two portions come together again, thus forming what is referred to as the *cephalic circle.* From the circle arise arteries which supply the brain, eyes, tongue, and various parts of the head.

The salient features of the aortic arches in cyclostomes are (1) a reduction in number as compared with amphioxus, and (2) the manner in which they break up into interarterial capillaries between afferent and efferent arteries.

FISHES. Much variation is to be observed in the aortic arches of fishes. In general, there is a reduction in number within the superclass as the evolutionary scale is ascended. The greatest number occurs in certain primitive sharks in which the number is directly related to the number of gill pouches. In sharks of the genus *Heptanchus,* which

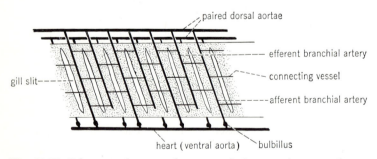

Fig. 12.17 Schematic diagram of portion of pharyngeal region of amphioxus, showing arrangement of aortic arches.

Anatomy of the Chordates

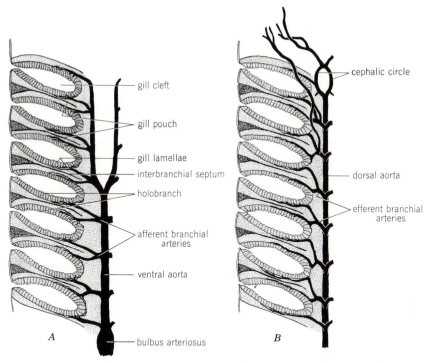

Fig. 12.18 Diagrams showing arrangement of arteries in gill region of lamprey, *Petromyzon marinus: A,* ventral aorta and afferent branchial arteries; *B,* dorsal aorta and efferent branchial arteries.

have seven pairs of gill clefts in addition to the spiracle, there are seven pairs of aortic arches in the adult. Six pairs are to be found in adult selachians of the genera *Hexanchus* and *Chlamydoselachus.* Since no hemibranch is present on the posterior wall of the last gill pouch, no arterial loop is present there.

Although in the rest of the fishes and in members of the higher classes the number of aortic arches is reduced or otherwise modified, practically all pass through a stage in embryonic development in which six pairs of aortic arches connect ventral and dorsal aortae (Fig. 12.16). Six, then, should be thought of as the primitive number of aortic arches for vertebrates, whose ancestors un-

doubtedly possessed a greater number. The anterior continuations of the paired dorsal aortae (*radices*) give rise to the *internal carotid arteries* supplying the brain. Those of the ventral aorta form the *external carotid arteries* which supply the jaws and face. In most fishes each aortic arch, except the first, or mandibular, consists of afferent and efferent branchial portions with an interarterial capillary network interposed. It is in the capillary network of the gill lamellae that aeration of the blood occurs. The afferent branchial arteries, as they leave the ventral aorta, course *through* the interbranchial septa, contrary to the condition in cyclostomes, described above. The blood collected by pretrematic and posttrematic branches of the efferent branchial arteries passes to the radices of the aorta. In *Protopterus* the third and fourth aortic arches pass directly to the dorsal aorta without interruption, no internal gills being associated with these vessels.

In sharks, other than those mentioned above, only five aortic arches persist, the first having been lost or modified. The most anterior afferent branchial artery on each side courses through the hyoid septum, supplying the hyoid hemibranch. It will be recalled that in elasmobranchs the *spiracle* represents the external opening of the modified first, or hyomandibular, cleft. Although five afferent branchial arteries are present on each side, there are but four pairs of efferent vessels (Fig. 12.19).

In teleost fishes and most others, only the last four pairs of aortic arches remain, numbers 1 and 2 having disappeared or having been reduced to small branches of the third (Fig. 12.20). In Chondrostei and Dipnoi a pulmonary artery arises from each sixth

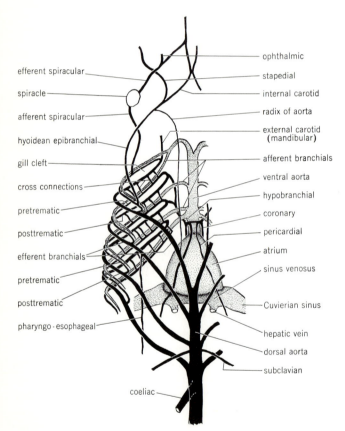

efferent spiracular
spiracle
afferent spiracular
hyoidean epibranchial
gill cleft
cross connections
pretrematic
posttrematic
efferent branchials
pretrematic
posttrematic
pharyngo-esophageal
coeliac

ophthalmic
stapedial
internal carotid
radix of aorta
external carotid (mandibular)
afferent branchials
ventral aorta
hypobranchial
coronary
pericardial
atrium
sinus venosus
Cuvierian sinus
hepatic vein
dorsal aorta
subclavian

Fig. 12.19 Diagram showing arteries in the left gill region of dogfish, *Squalus acanthias,* as seen from the dorsal side. The length of the arteries at the anterior end has been exaggerated for the sake of clarity. Capillary connections between afferent and efferent arteries have been omitted. (*Modified from Daniel.*)

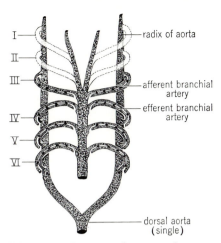

I — radix of aorta

II

III — afferent branchial artery

efferent branchial artery

IV

V

VI

dorsal aorta (single)

Fig. 12.20 Diagram of aortic arch region as found in most teleost fishes, ventral view. Arches I and II have degenerated. Each of the remaining arches is divided into afferent and efferent branchial arteries connected with each other by gill capillaries.

arch (or from the dorsal aorta) through which blood is carried to the swim bladder.

AMPHIBIANS. In amphibians and in the remaining vertebrate classes, there is a further reduction in the number of aortic arches together with a greater modification of the entire complex of vessels in the pharyngeal region. The aortic arches do not break up into afferent and efferent portions since in these higher forms internal gill lamellae do not develop. To be sure, amphibians possess external gill filaments, at least during early development, but these are not homologous with the internal gill lamellae of fishes, nor are they supplied with blood in the same manner (pages 219 and 564).

In anurans (Fig. 12.21), aortic arches 1, 2, and 5 disappear. The radix between arches 3 and 4 on each side gradually dwindles away. The anterior continuations of the ventral aorta become the external carotid arteries. The third arch, together with the an-

terior portion of the radix of the aorta on that side, becomes the internal carotid artery. A stapedial branch represents a remnant of the second aortic arch. The portion of the ventral aorta from which the internal and external carotid arteries arise becomes the common carotid. The fourth aortic arches persist to become the systemic arches, which unite posteriorly to form the dorsal aorta proper (Fig. 12.22). Arch 6 on each side sends a branch to the developing lung and to the skin, thus becoming the pulmocutaneous artery. The portion of arch 6 between the pulmonary artery and the radix is called the *ductus arteriosus,* or *duct of Botallus.* It disappears at the time of metamorphosis.

Slight differences from the above are to be found in urodeles (Fig. 12.23). In certain salamanders the fifth arch may persist in very much reduced form. Frequently the radix between aortic arches 3 and 4 fails to degenerate completely. The ductus arteriosus also persists in urodeles.

The external gills of larval urodele amphibians are supplied with vascular loops

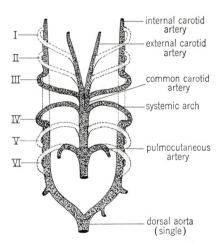

I — internal carotid artery

external carotid artery

II

III — common carotid artery

systemic arch

IV

V — pulmocutaneous artery

VI

dorsal aorta (single)

Fig. 12.21 Diagram showing modification of aortic arches as found in anuran amphibians, ventral view.

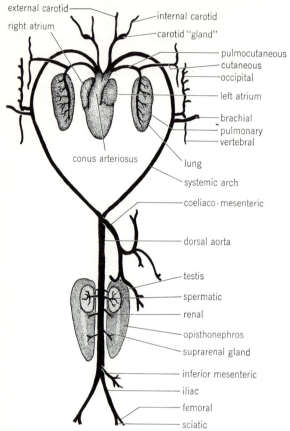

external carotid
right atrium
internal carotid
carotid "gland"
pulmocutaneous
cutaneous
occipital
left atrium
brachial
pulmonary
vertebral
conus arteriosus
lung
systemic arch
coeliaco-mesenteric
dorsal aorta
testis
spermatic
renal
opisthonephros
suprarenal gland
inferior mesenteric
iliac
femoral
sciatic

Fig. 12.22 Diagram of arterial system of adult frog, ventral view.

connected to aortic arches. The loops themselves are composed of afferent and efferent vessels connected by a capillary network. They lie lateral to the aortic arches. The latter are located at the bases of the gills, each serving as a gill bypass. Blood may pass directly through the aortic arches or through the adjacent gill loops. In urodele amphibians, at the time of metamorphosis the gills degenerate and the vascular loops atrophy. However, the main channel of each aortic arch concerned, which has maintained its integrity from the beginning, continues to persist.

Anuran larvae, after an early stage in which external gills like those of urodeles are present (Fig. 12.24), develop peculiar gills enclosed by opercular folds. The continuity of the aortic arches is then temporarily interrupted. Upon metamorphosis, however, the arches again become continuous structures.

Certain urodele amphibians, such as *Necturus,* are called *perennibranchiates* since they retain their gills throughout life and fail to undergo metamorphosis. Attempts to induce metamorphosis in *Necturus* under experimental conditions have resulted in fail-

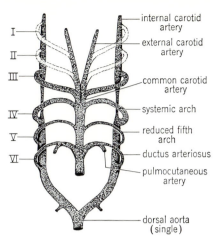

Fig. 12.23 Diagram showing modification of aortic arches as found in most urodele amphibians, ventral view.

internal carotid artery
external carotid artery
common carotid artery
systemic arch
reduced fifth arch
ductus arteriosus
pulmocutaneous artery
dorsal aorta (single)

ure. The reason for this has long puzzled zoologists. A morphological explanation has been suggested (F. H. Figge, 1930, *J. Exp. Zool.,* **56**:241), but its validity has been questioned. It seems that aortic arches 3, 4, and 5 persist in *Necturus* and that the ventral portion of arch 6 is missing. The dorsal portion of arch 6, representing the ductus arteriosus, is present, however, and connects the efferent portion of arch 5 to the pulmonary artery which supplies the lung (Fig. 12.25). In effect, then, the pulmonary artery comes off the fifth arch rather than the sixth as is usually the case. Blood going to a lung in *Necturus,* therefore, must first pass through a gill and has already been oxygenated. The pulmonary system, as a consequence, is of little value except in times of emergency when the oxygen content of the water may be depleted.

Experiments on larvae of the salamander *Ambystoma tigrinum* have been suggestive in this connection. In these larvae aortic arches 3, 4, 5, and 6 are present on each side although number 6 does not supply a gill.

It gives off a pulmonary artery to the developing lung but has a short connection, the ductus arteriosus, with the fifth efferent branchial artery. Animals in which the ventral portion of aortic arch 6 has been cut or ligated fail to metamorphose even after thyroid substance has been administered. In such cases the condition which normally occurs in *Necturus* has been simulated experimentally in *Ambystoma tigrinum.* Blood going to the lungs, therefore, has been oxygenated in the fifth aortic arch, and the lungs function very little, if at all. In normal specimens of *Ambystoma tigrinum,* at the time of metamorphosis the lungs begin to function as respiratory organs and the gills with their vessels are reduced. It has been suggested that in *Necturus* and in operated specimens of *Ambystoma tigrinum* it is impossible for the lungs to function because they receive only oxygenated blood. Therefore the gills are retained, and metamorphosis fails to occur. Other factors in *Necturus,* such as the absence of a spiral valve in the conus arteriosus, may also bear some relation to this peculiar state of affairs.

REPTILES. Just as in amphibians, reptiles retain aortic arches 3, 4, and 6. The fifth arch may also be retained in reduced form in certain lizards, and a remnant of the radix

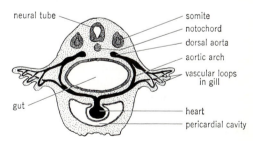

neural tube
somite
notochord
dorsal aorta
aortic arch
vascular loops in gill
gut
heart
pericardial cavity

Fig. 12.24 Diagrammatic cross section through gill region of early frog tadpole, showing relation of aortic arches to blood vessels in the external gills. (*After Maurer.*)

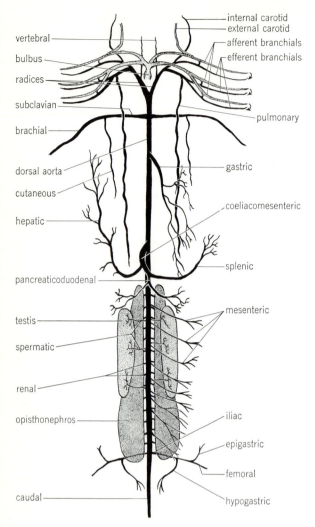

vertebral

bulbus

radices

subclavian

brachial

dorsal aorta

cutaneous

hepatic

pancreaticoduodenal

testis

spermatic

renal

opisthonephros

caudal

internal carotid

external carotid

afferent branchials

efferent branchials

pulmonary

gastric

coeliacomesenteric

splenic

mesenteric

iliac

epigastric

femoral

hypogastric

Fig. 12.25 Ventral view of arterial system of *Necturus*.

between arches 3 and 4 may persist on each side in certain snakes. In most reptiles, however, further modifications occur in the aortic arches. These consist chiefly of a splitting of the distal portion of the conus arteriosus and the ventral aorta into three vessels (Fig. 12.26). The fourth aortic arch on the left side establishes a separate connection with the right side of the partially divided ven-

tricle. It, together with a portion of the radix on the left side, becomes the *left arch of the aorta*. The sixth arch on each side gives off a pulmonary artery to the lung and in most cases loses its connection (ductus arteriosus) with the radix. It persists to a minor degree in some turtles and in *Sphenodon*. The two pulmonary arteries arise from a common trunk, the *pulmonary aorta,* from the right

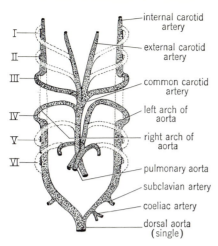

Fig. 12.26 Diagram showing modification of the aortic arches as found in reptiles, ventral view. The ventral aorta (truncus arteriosus) has split into three vessels: right and left systemic aortae and a pulmonary trunk, or aorta.

side of the ventricle. The remaining vessel derived from the truncus arteriosus connects to the left side of the ventricle and, as it courses forward, gives off the fourth aortic arch on the right side and finally divides into the two common carotid arteries with their external and internal branches. The right fourth aortic arch, together with a portion of the radix on that side, becomes the *right arch of the aorta* which joins the posterior continuation of the left arch to form the dorsal aorta proper. A coeliac artery supplying certain viscera arises from the left aorta near its junction with the right. The portion of the left aorta between the coeliac artery and the dorsal aorta proper is somewhat reduced since the major portion of the blood in the left aorta passes into the coeliac artery. Right and left subclavian arteries arise from the right and left aortae, respectively, in those reptiles having pectoral limbs.

Since the right aorta bears mostly oxy-genated blood and the left chiefly unoxygenated blood, mixing occurs in the dorsal aorta proper where these two vessels come together. Some mixing may also occur through the *foramen Panizzae* which joins the right and left aortae as they emerge from the heart. In crocodilians, in which two completely separated ventricles are present, mixing of blood occurs, as in other reptiles, at the foramen Panizzae and at the point where right and left aortae unite to form a single vessel. Mixing of oxygenated and unoxygenated blood seems to be associated with the poikilothermous mode of life.

BIRDS. The chief changes taking place in the aortic arches of birds correspond to those of reptiles. There is one important difference, however (Fig. 12.27). In birds the fourth arch and radix on the left side lose their connection with the dorsal aorta and finally disappear. The ventral aorta splits into two portions, a systemic aorta and a pulmonary aorta, or trunk. The systemic aorta is connected to the left ventricle, and the pulmonary aorta to the right.

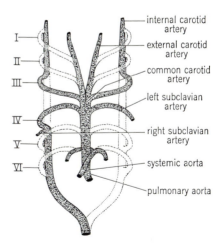

Fig. 12.27 Diagram showing modification of aortic arches as found in birds, ventral view.

The fourth aortic arch on the right side leaves the systemic aorta and, by means of the radix, leads to the main arterial channel, or dorsal aorta proper. The latter supplies the entire body with oxygenated blood. At its anterior end the systemic aorta gives rise to external and internal carotid arteries in the same manner as described for previous groups. The fourth aortic arch on the left may possibly contribute to the left subclavian artery, but this vessel is usually a branch of the third aortic arch. The right subclavian artery either comes off the right radix or off the third aortic arch on the right side.

The pulmonary aorta leading from the right ventricle gives off the pulmonary arteries, which are actually outgrowths of the sixth aortic arches. Until the time of hatching, there is a ductus arteriosus on the right side, representing the portion of aortic arch 6 between pulmonary artery and radix. This serves as a shunt from the right ventricle to the dorsal aorta at a time when the lungs are not functioning. It closes at the time of hatching, and all the blood in the right ventricle is then sent to the lungs for aeration. A cord of connective tissue on each side, the *ligamentum arteriosum,* or *ligamentum Botalli,* is all that remains of the former arterial shunt.

MAMMALS. The changes in the aortic arches of mammals are rather similar to those of birds except that the radix on the *right* side rather than the left loses its connection with the aorta (Fig. 12.28). The fourth aortic arch on the left side together with its radix, therefore, becomes the arch of the definitive aorta. The fourth arch on the right and a portion of the right radix become the right subclavian artery, and the left subclavian develops as an enlargement of one of the intersegmental arteries coming off the aorta in this region. There is much variation among mammals in the mannner in which the subclavian arteries arise. In mammalian embryos there is at first

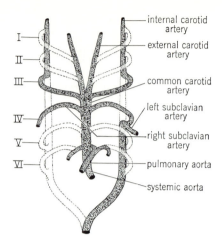

Fig. 12.28 Diagram showing modification of aortic arches as found in mammals, ventral view.

a ductus arteriosus on each side, but the one on the right persists only for a short time. The left one, which serves as a shunt between pulmonary and systemic aortae, persists until birth when it finally becomes occluded. A ligamentum arteriosum (ligamentum Botalli) of connective tissue is finally all that remains (Fig. 12.29).

In mammals, the actual presence of a fifth aortic arch during development has been questioned. However, remnants of a fifth arch have been observed in certain mammalian embryos, and there is little reason to doubt that mammals are similar to other vertebrates in passing through a primitive stage in which six aortic arches are present. In some mammals the dorsal part of the second aortic arch persists to become the stapedial artery with which the external carotid artery forms connections. Remnants of the first aortic arch may contribute to the mandibular artery.

The study of changes in the aortic arches of birds and mammals is one of the most striking illustrations of the law of biogenesis.

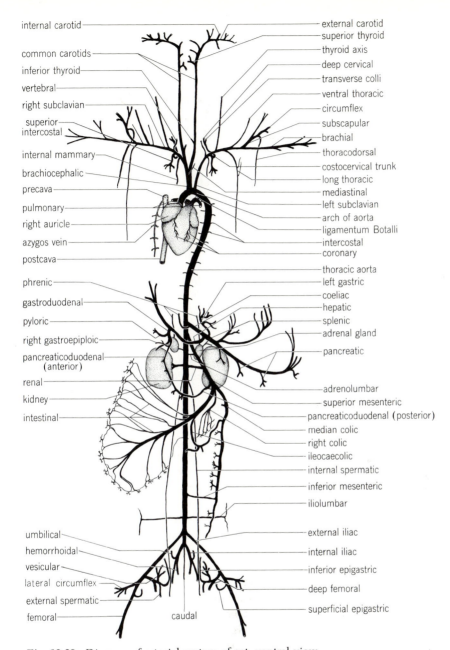

internal carotid
common carotids
inferior thyroid
vertebral
right subclavian
superior intercostal
internal mammary
brachiocephalic
precava
pulmonary
right auricle
azygos vein
postcava
phrenic
gastroduodenal
pyloric
right gastroepiploic
pancreaticoduodenal (anterior)
renal
kidney
intestinal
umbilical
hemorrhoidal
vesicular
lateral circumflex
external spermatic
femoral

external carotid
superior thyroid
thyroid axis
deep cervical
transverse colli
ventral thoracic
circumflex
subscapular
brachial
thoracodorsal
costocervical trunk
long thoracic
mediastinal
left subclavian
arch of aorta
ligamentum Botalli
intercostal
coronary
thoracic aorta
left gastric
coeliac
hepatic
splenic
adrenal gland
pancreatic
adrenolumbar
superior mesenteric
pancreaticoduodenal (posterior)
median colic
right colic
ileocaecolic
internal spermatic
inferior mesenteric
iliolumbar
external iliac
internal iliac
inferior epigastric
deep femoral
superficial epigastric

caudal

Fig. 12.29 Diagram of arterial system of cat, ventral view.

Remainder of the arterial system. The arteries supplying the head region are, for the most part, branches of the external and internal carotids. The internal carotid artery on each side enters the cranial cavity through the foramen lacerum. In some forms, branches of the fourth aortic arches or of the subclavian arteries also pass to the cephalic region. The occipitovertebral and vertebral arteries are examples. The aorta is continued posteriorly into the tail region where it becomes the *caudal artery.* Most of the vertebrate body is supplied with blood through branches of the dorsal aorta, which may for convenience be grouped under two divisions: somatic arteries, supplying the body proper, and visceral arteries, distributed to various portions of the digestive tract and associated structures.

SOMATIC ARTERIES. The arteries supplying the body proper are usually paired structures which clearly show evidences of metamerism or segmental arrangement. They supply portions of the body derived from the embryonic epimere, being distributed to the dorsal, or epaxial, musculature and vertebral column, where they are referred to as *parietal*, or *segmental*, *arteries*. Posterior branches pass through the intervertebral foramina into the neural canal to supply the spinal cord and its coverings. In higher forms, in which the body is divided into more or less definite regions, such terms as *intercostal, dorsolumbar,* and *sacral* are applied to these segmentally arranged vessels. Fusion of two or more segmental arteries may take place in various regions and obscure the fundamental metameric arrangement. The vessels going to the appendages, i.e., the *subclavians* to the pectoral appendages and the *iliacs* to the pelvic, may be composed of a union of several segmental arteries, the number of vessels concerned corresponding to the number of somites involved in the formation of the limb. During development a considerable shifting of vessels leading to the appendages occurs, and the end result shows little evidence of metameric origin.

VISCERAL ARTERIES. The arteries supplying the viscera are of two kinds, paired and unpaired. The paired arteries are segmentally arranged and supply portions derived from the embryonic mesomere, or nephrotome, from which the urogenital organs and their ducts arise. Although the mesomere itself is, for the most part, unsegmented, the arteries supplying its derivatives show pronounced evidences of metamerism, particularly in regard to the manner in which they supply the pronephros, opisthonephros, and mesonephros. Such terms as *renal, genital, ovarian, spermatic,* and *urogenital arteries* are applied to the paired visceral arteries. Renal and genital arteries are numerous in the lower vertebrates, but the number is markedly reduced in higher forms.

The unpaired visceral arteries supplying the spleen and the digestive tract and its derivatives are vessels which course through the dorsal mesentery of the gut. They branch profusely, the method of branching showing great variation even in members of the same species. There are usually three unpaired visceral arteries in vertebrates. The most anterior of these is the *coeliac artery,* supplying the anterior viscera, including stomach (*gastric*), spleen (*splenic* or *lienal*), pancreas (*pancreatic*), liver (*hepatic*), and duodenum (*duodenal*).

The second unpaired visceral artery is the *superior mesenteric*, which supplies the entire length of the small intestine with the exception of the pyloric end of the duodenum. This is taken care of by the coeliac artery. Branches of the superior mesenteric at its anterior end also supply a portion of the pancreas. The remainder of the vessel is distributed to the caecum and upper half of the large intestine.

The third unpaired artery is the *inferior mesenteric*, supplying the posterior part of the large intestine and rectum.

Variations from the above condition may be accounted for by fusions or separations. For example, in frogs and other amphibians, the coeliac and superior mesenteric arteries have united into a single coeliacomesenteric artery. In the dogfish, *Squalus acanthias*, an unpaired gastrosplenic artery arises directly from the aorta just posterior to the origin of the superior mesenteric to supply the spleen and posterior part of the stomach (Fig. 12.30). It actually represents a branch of the superior mesenteric which has become secondarily connected with the aorta.

Certain arteries play an important role during embryonic development and are, therefore, of unusual significance. The superior mesenteric artery, for example, is derived embryonically from a fusion of

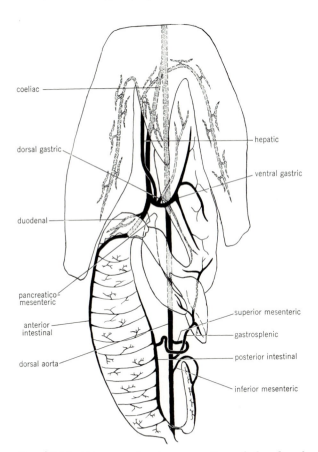

coeliac

dorsal gastric

duodenal

pancreatico-mesenteric

anterior intestinal

dorsal aorta

hepatic

ventral gastric

superior mesenteric

gastrosplenic

posterior intestinal

inferior mesenteric

Fig. 12.30 Diagram showing a portion of the dorsal aorta and its unpaired branches which supply the viscera of the dogfish, *Squalus acanthias*, ventral view.

paired *vitelline arteries* which originally branch off the aorta to supply the yolk sac (Fig. 12.31). A pair of *umbilical*, or *allantoic*, arteries, also present during development in higher forms, is lost at the time of hatching or birth. Their remnants are known as the *hypogastric*, or *internal iliac*, arteries, which branch to supply urinary bladder, rectum, tail, and thigh.

VENOUS SYSTEM

As in the case of the arterial system, a comparison of veins in the various vertebrate groups shows that they are arranged accord-ing to the same fundamental plan and that the variations encountered form a logical sequence as the vertebrate scale is ascended. In its development, the venous system of higher forms passes through certain stages common to the embryos of lower forms.

There is some difference in the formation of the earliest veins which appear in an embryo, depending upon whether or not a yolk sac is present. In forms without a yolk sac, a pair of *subintestinal* veins appears in the splanchnic mesoderm ventral to the gut. These veins soon fuse except in the region of the anus around which they form a loop, only to unite posteriorly where they continue

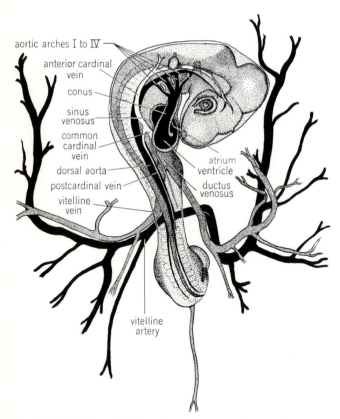

aortic arches I to IV
anterior cardinal vein
conus
sinus venosus
common cardinal vein
dorsal aorta
postcardinal vein
vitelline vein
atrium
ventricle
ductus venosus
vitelline artery

Fig. 12.31 72-hour chick embryo, showing the main parts of the circulatory system, dorsal view.

Anatomy of the Chordates

on into the tail as the *caudal* vein. In animals having a yolk sac, a pair of *vitelline (omphalomesenteric)* veins, draining the yolk sac (Fig. 12.31), joins the posterior end of the heart. Indeed, their fusion is primarily responsible for heart formation. The part of the heart into which the vitelline veins enter is destined to become the *sinus venosus*. The vitelline veins are concerned with obtaining nourishment used by the embryo during its development. Even though the yolk sac in mammals contains no yolk, vitelline veins (and arteries) are present. Each vitelline vein is joined posteriorly by a subintestinal vein arranged in a manner similar to that in forms in which a yolk sac is lacking.

Before long, two additional pairs of veins, *anterior cardinals* from the dorsal side of the head region and *posterior cardinals* from the posterior end of the body, establish connections with the fused vitelline veins. They join the sinus venosus on each side by a common vessel variously known as the *common cardinal, duct of Cuvier*, or *Cuvierian sinus*. The anterior cardinals are generally referred to as *jugular* veins which receive *internal* and *external* tributaries. In fishes and salamanders a pair of *inferior jugular* veins is usually present. These drain the ventrolateral portions of the head and join the common cardinal veins. *Polypterus* lacks inferior jugular veins. In many teleosts and in *Lepisosteus*, the two inferior jugular veins join each other to form a single trunk which enters the right common cardinal vein. The inferior jugulars apparently have no homologues among higher forms.

In lower vertebrates a pair of *ventral abdominal* or *lateral abdominal* veins, located in the ventral or lateral portions of the body wall, also enters the common cardinal veins. In amniotes these veins form secondary connections with vessels draining the allantois. Blood in the allantois or umbilical cord, as the case may be, therefore, courses through the abdominal veins in the body wall and enters the sinus venosus via the common cardinal veins. The term *allantoic,* or *umbilical*, veins is then applied to these vessels in amniote embryos. Although both left and right umbilical veins are present in reptiles, only the left persists in birds and mammals. When the liver develops, the umbilical veins lose their connection to the sinus venosus and course through the liver via a large channel, the *ductus venosus*, which in turn connects with the sinus venosus.

From the primitive venous complex just described (Fig. 12.32) are derived the prin-

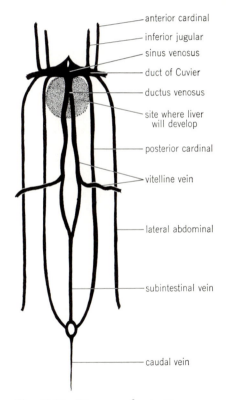

anterior cardinal

inferior jugular

sinus venosus

duct of Cuvier

ductus venosus

site where liver will develop

posterior cardinal

vitelline vein

lateral abdominal

subintestinal vein

caudal vein

Fig. 12.32 Diagram of primitive venous complex from which the venous systems of vertebrates are believed to have been derived, ventral view.

cipal veins of the body. Changes in the course of the blood as it returns to the heart account for the variations observed in venous vessels which, nevertheless, show a consistent arrangement within each vertebrate class. The vitelline and umbilical vessels are lost, as such, at birth or hatching, but remnants persist even in adult life.

The primitive veins so far described correspond, in the main, to the chief divisions of the arteries but are not generally so clear cut. The vitelline veins and the subintestinals give rise to the *unpaired visceral* veins, which later drain the digestive tract and its associated organs. They become located in the mesentery of the gut. *Paired lateral visceral* veins drain the urogenital organs derived from the embryonic mesomere. *Somatic* veins are generally paired structures which drain the body wall.

Hepatic portal circulation. All vertebrates have a hepatic portal system. A portal system, it will be recalled, is a system of veins that breaks up into a capillary network before the blood which courses through it is returned to the heart. It is the vitelline-subintestinal group of vessels which is concerned in the formation of the hepatic portal system.

In an embryo the liver develops as an outgrowth of the gut in the region just posterior to the heart. As the liver grows out it soon surrounds the vitelline (or subintestinal) veins. These veins break up into a network of small vessels (sinusoids) which ramify through the liver. Posterior to the liver several anastomoses, or junctures, form between the vitelline veins. The right vessel gradually dwindles away, leaving the left vitelline vein with its subintestinal tributary as the main vessel draining the digestive tract. This then becomes the hepatic portal vein, which collects blood from all portions of the alimentary canal as well as the spleen. It enters the liver, where it breaks up into sinusoids. The original connections of the vitelline veins to the sinus venosus become the *hepatic* veins which return the blood collected from the liver sinusoids to the heart.

Renal portal circulation. A renal portal vein is present in many vertebrates. It has its origin in venous vessels in the caudal region of the body and terminates in capillaries in the opisthonephros or mesonephros, as the case may be. Since there is so much variation in detailed arrangement of the renal portal veins in the various vertebrate classes, descriptions are given below under separate headings.

Hypophysio-portal circulation. A hypophysio-portal system of veins is associated with the pituitary gland. Although many variations occur among vertebrates, in general the hypophysio-portal vein begins in capillaries formed by arteries which pass through the pars tuberalis of the pituitary gland to reach the median eminence of the tuber cinereum of the hypothalamic portion of the brain and the neural stalk. The capillaries empty into venules which course back to the pars tuberalis, ultimately to enter the sinusoids of the anterior lobe of the pituitary gland. Thus this venous complex forms a true portal system, beginning as it does in capillaries in the brain and ending in a second set of capillaries in the anterior lobe. In man, monkey, sheep, goat, and rat, the anterior lobe has no arterial supply; it receives only portal venous blood. In the rabbit the anterior lobe, in addition to its portal venous supply, also receives a direct supply of arterial blood (P. M. Daniel and M. M. L. Prichard, 1960, *Anat. Record*, 136:180). Any hormones or neurosecretions formed in the hypothalamic portion of the brain undoubtedly reach the pars anterior through the hypophysio-portal pathway.

Venous systems of different chordates.
AMPHIOXUS. The veins of amphioxus represent the vertebrate venous system reduced to its simplest terms. The caudal vein from the tail is continued forward as the subintestinal vein. It also connects to the postcardinal on each side, so that blood can be returned from the tail via either channel. An anterior cardinal joins the postcardinal on each side, and they enter the posterior end of the heart (sinus venosus) by means of a common cardinal vein. The subintestinal vein breaks up in capillaries in the intestine. Another vein, beginning in capillaries in the intestine, leads to the hepatic caecum, where it breaks up into capillaries again. This vessel is a true hepatic portal vein. A contractile hepatic vein beginning in the hepatic caecum carries blood to the heart (Fig. 12.33). A pair of parietal veins draining the dorsal body wall and located dorsal to the gut furnishes additional somatic vessels which enter the posterior end of the heart.

CYCLOSTOMES. The venous system of the lamprey does not differ greatly from that of amphioxus. Two inferior jugular veins unite to form a single vessel which returns blood to the sinus venosus from the ventral, anterior region of the body. Paired anterior cardinals (jugulars) and posterior cardinals are present. The common cardinal veins appear to have fused, or else the left one has degenerated in the adult so that but a single common cardinal enters the sinus venosus on its dorsal side (Fig. 12.10). The caudal vein, which divides at the cloacal region, connects to the postcardinal veins which course on either side slightly ventrolateral to the dorsal aorta. They fuse as they approach the heart. Contractile *cardinal* and *caudal hearts* have been reported to be present in the hagfish, aiding in the return of venous blood to the heart.

The anterior part of the subintestinal vein remains as a small vessel which is found in the typhlosole of the intestine. Its posterior end has lost all connection with the caudal vein. The subintestinal vein becomes the hepatic portal vein, which also receives a branch from the head. A contractile *portal heart* is also present in the hepatic portal vein of cyclostomes. Blood from the liver is collected by a single hepatic vein which joins the sinus venosus. No renal portal system exists in cyclostomes. Segmentally arranged veins from the kidneys and gonad enter the postcardinal veins as do those from the body wall.

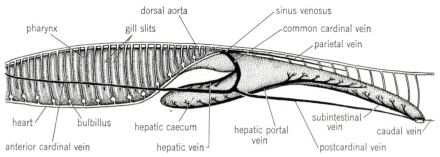

Fig. 12.33 Diagram showing the essential features of the circulatory system of amphioxus, lateral view.

FISHES. Many primitive features are retained by the venous systems of fishes, but important advances over the conditions in amphioxus and cyclostomes are to be found.

The sinus venosus receives a duct of Cuvier on each side (Fig. 12.34A). An anterior cardinal, or jugular, vein brings blood from the dorsal side of the head region to the duct of Cuvier. In most fishes a pair of inferior jugular veins from the ventrolateral part of the head also enters the common cardinal veins. These are lacking in *Polypterus* and are fused in *Lepisosteus*. The common cardinal on each side also receives a postcardinal vein from the posterior end of the body. Since fishes are the first vertebrates to possess paired appendages, veins bringing blood from the appendages first make their appearance in this group. Those from the pectoral appendages are called *subclavian* veins, and those from the pelvic are the *iliac* veins.

In young fishes the subclavian veins may enter the anterior ends of the postcardinals, but in adults a shifting has occurred so that the subclavian veins frequently enter the common cardinals directly. The iliac vein from each pelvic fin joins the *lateral abdominal* vein which courses through the body wall also to join the duct of Cuvier.

The posterior wall of the sinus venosus usually receives two hepatic veins of similar

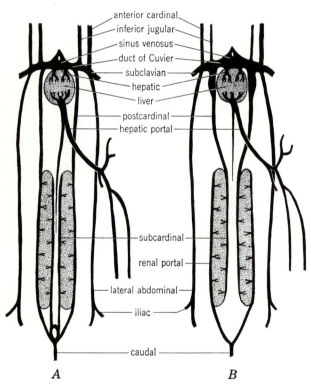

anterior cardinal
inferior jugular
sinus venosus
duct of Cuvier
subclavian
hepatic
liver
postcardinal
hepatic portal
subcardinal
renal portal
lateral abdominal
iliac
caudal

A　　　　　　　*B*

Fig. 12.34 Diagrams *A* and *B*, illustrating the changes over the primitive condition which occur in the venous systems of fishes, ventral view.

size which return blood from the liver to the heart. The hepatic portal vein of fishes is well developed (Fig. 12.35) and, as in other forms, is composed of the left vitelline vein and its subintestinal tributary. The subintestinal vein has, in most cases, lost its connection with the caudal vein.

The chief differences over the primitive condition to be found in the veins of fishes are concerned with changes which occur in

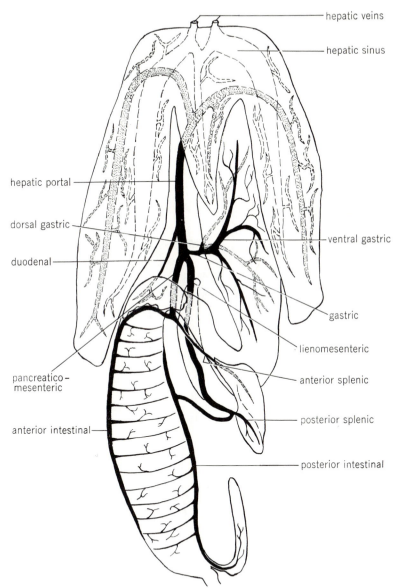

Fig. 12.35 Diagram of hepatic portal vein and its tributaries in *Squalus acanthias*, ventral view.

the posterior part of the body and which result in the development of the renal portal system.

With the development of the opisthonephros, the postcardinal veins grow backward, ultimately to unite with the caudal vein at the point where it divides to form a ring about the cloaca. The segmental veins in some elasmobranchs and teleosts bear semilunar valves which facilitate the return of blood to the heart. A new vessel or pair of vessels, the *subcardinal* veins, now develops in the center of the opisthonephros, and a connection with the caudal vein is established near the point where it is joined by the postcardinals. Small blood vessels, passing through the opisthonephros, connect the postcardinal and subcardinal veins. They are not associated with the glomeruli, which receive arterial blood. Blood in the caudal vein now has two alternative routes through which it may pass on its way to the heart. It may go through the postcardinals directly or through the subcardinals and renal veins to the postcardinals.

The next change which takes place involves an interruption in the course of each postcardinal vein, so that anterior and posterior portions are no longer continuous (Fig. 12.34B). The anterior ends of the subcardinals join the anterior portions of the postcardinals. The posterior ends of the subcardinals lose their direct connection with the caudal vein. Blood from the tail now can travel only by one route. It goes up the caudal vein into the original posterior portions of the postcardinals, which are now termed the *renal portal* veins. It then passes through the renal veins, which have assumed capillary dimensions, finally to enter the subcardinals. The blood then courses through the postcardinal veins, which join the duct of Cuvier. The postcardinals usually receive veins from the gonads en route to the heart.

The newly developed renal portal vein falls within the definition of a portal system, since it breaks up into a capillary network before the blood which it contains enters the heart. It is interesting to note that the blood passing through the renal veins, once the renal portal circulation is established, flows in a direction reverse to that occurring during an early stage of development.

The changes in the venous system described above have now reached the stage found in the elasmobranch fishes (Fig. 12.36). Interesting variations in the venous systems of other fishes are encountered, a few of which will be considered. Many of

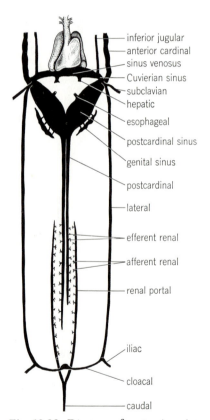

Fig. 12.36 Diagram of systemic veins and renal portal system of dogfish, *Squalus acanthias,* ventral view.

Anatomy of the Chordates

the differences involve only minor details, but certain major deviations are noteworthy.

In elasmobranchs, large cavernous *sinuses* occur in certain portions of the anterior and posterior cardinal veins.

In the catfish, *Ameiurus,* the caudal vein continues forward, after giving off the renal portal veins, and communicates with the hepatic portal vein. The connecting vessel undoubtedly represents the original connection of the subintestinal and caudal veins. In the eel, *Anguilla,* several vessels from each renal portal vein connect with the hepatic portal vein.

In such fishes as the tench and the cod, the renal portal system is imperfectly developed and some blood from the caudal vein may pass directly to the heart through a postcardinal vein which has not undergone the usual separation into anterior and posterior portions. In such cases there is an asymmetry in the development of the posterior veins. The tench and cod, like the catfish and eel, also possess connections between caudal and hepatic portal veins.

In teleost fishes the lateral abdominal veins, so conspicuous in elasmobranchs, are not present. In these fishes the subclavians enter the common cardinals and the iliacs join the postcardinals.

Veins from the swim bladder usually join the hepatic portal vein. This is a logical state of affairs, since the swim bladder is a derivative of the gut. However, in certain forms the connection is with the postcardinal veins instead. In *Polypterus* they join the hepatic veins directly. In the Dipnoi the pulmonary veins from the swim bladder enter the newly formed left atrium of the heart. In this group of fishes the double type of circulatory system appears for the first time.

It is in the true lungfishes that a connecting link is found which indicates how the venous system of amphibians may have arisen in evolution from the more simple and primitive arrangement found in other fishes. The veins of *Epiceratodus* show this transition most clearly.

In *Epiceratodus* a single, midventral, *anterior abdominal* vein, similar to that of amphibians, makes its appearance. The lateral abdominal veins have fused to form the anterior abdominal, which courses forward to enter the sinus venosus. In the meantime the iliac vein has formed a connection with the renal portal vein on each side (Fig. 12.37). The branch of the iliac which joins the anterior abdominal vein is called the *pelvic* vein.

The right postcardinal, including its posterior portion (originally the right subcardinal), becomes much larger than its counterpart on the left side. The connection with the caudal vein on both sides is retained. The larger vessel on the right side is now called the *postcaval* vein. It passes through the liver to open into the sinus venosus. The smaller left postcardinal passes over the liver and enters the left duct of Cuvier. The presence of the postcava in *Epiceratodus* clearly presages the appearance of the postcaval vein in amphibians, the anterior portion of which, however, has a different origin.

The venous system of *Protopterus*, except for the lack of an anterior abdominal vein, even more closely resembles that of amphibians than does the venous system of *Epiceratodus*.

AMPHIBIANS. The venous system in amphibians bears such a close resemblance to that of *Epiceratodus* that only salient differences need be mentioned. Chief among these is a change in the anterior connections of the anterior abdominal vein. Whereas in the lungfish the anterior abdominal opens directly into the sinus venosus, in amphibians it joins the hepatic portal vein (Figs. 12.38 and 12.39). Thus the renal and hepatic portal systems are brought into close association, since blood from the legs must pass through one or the other.

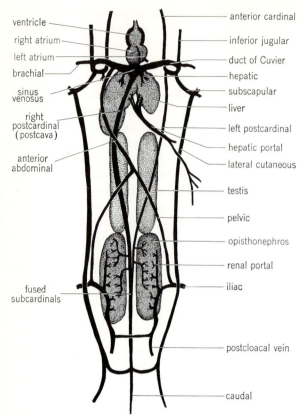

ventricle

right atrium

left atrium

brachial

sinus venosus

right postcardinal (postcava)

anterior abdominal

fused subcardinals

anterior cardinal

inferior jugular

duct of Cuvier

hepatic

subscapular

liver

left postcardinal

hepatic portal

lateral cutaneous

testis

pelvic

opisthonephros

renal portal

iliac

postcloacal vein

caudal

Fig. 12.37 Diagram of venous system of *Epiceratodus*, ventral view. (*Modified from B. Spencer.*)

Pulmonary veins from the lungs enter the left atrium as in the Dipnoi. In lungless salamanders, of course, pulmonary veins are absent and the left atrium is reduced in size.

The ducts of Cuvier which originally receive the subclavian, jugular, and postcardinal veins are further consolidated in amphibians and are now called the *precaval* veins which enter the sinus venosus on each side. The jugular vein now has *internal* and *external* tributaries. Since cutaneous respiration has developed to a high degree in amphibians, exceptionally large cutaneous veins are present which join the subclavians to enter the sinus venosus.

The postcava in amphibians differs from that of *Epiceratodus* in that its anterior portion is derived from the sinus venosus and portions of the original vitelline veins, rather than from the right postcardinal. The posterior part, however, like that of *Epiceratodus,* comes from the old subcardinals.

Although both urodeles and anurans are similar in the above respects, they exhibit certain differences in regard to the arrangement of the postcaval-postcardinal complex. In most urodeles and in a few anurans such as *Bombina* and *Ascaphus,* the anterior portion of each postcardinal persists in reduced form, connecting the middle portion of the postcava with the duct of Cuvier on each side. In adult anurans, however, the anterior portions of the postcardinals usually disappear and the postcava furnishes the only route through which blood from the kidneys and gonads can be returned to the heart.

REPTILES. The venous system of reptiles shows little change over the condition in amphibians. The large systemic veins entering the heart gradually shift to the right side, following the further partitioning of the heart into right and left sides.

Two precavae and one postcava enter the sinus venosus. The precavae are the original ducts of Cuvier which receive the jugular, subclavian, and postcardinal veins. The anterior portions of the postcardinals have degenerated into two small *vertebral* veins, which receive small segmental branches. In snakes the subclavians are lacking.

More blood from the posterior part of the body now courses through the anterior abdominal vein, which joins the hepatic portal vein anteriorly. The importance of the renal portal system is diminishing, and in some forms direct channels may even pass through

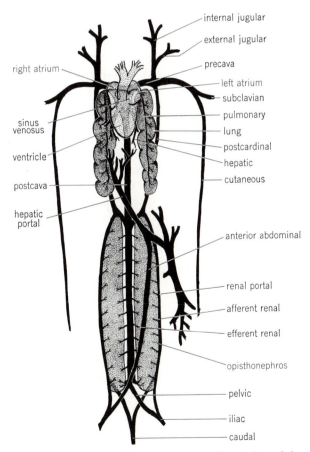

Fig. 12.38 Diagram of venous system of typical urodele amphibian, ventral view.

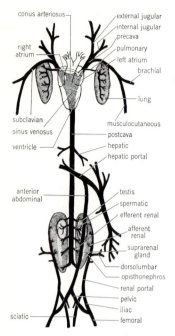

Labels on figure:
conus arteriosus
right atrium
subclavian
sinus venosus
ventricle
anterior abdominal
sciatic

external jugular
internal jugular
precava
pulmonary
left atrium
brachial
lung
musculocutaneous
postcava
hepatic
hepatic portal
testis
spermatic
efferent renal
afferent renal
suprarenal gland
dorsolumbar
opisthonephros
renal portal
pelvic
iliac
femoral

Fig. 12.39 Diagram of venous system of a typical adult anuran amphibian, ventral view.

the kidneys, connecting renal portal and post-caval veins. As in amphibians, blood in the hind limbs and tail reaches the heart either by the renal portal–kidney–postcaval route or by the pelvic-anterior abdominal–hepatic portal pathway.

Cutaneous respiration does not exist in reptiles, and the pulmonary circulation assumes greater importance. The pulmonary veins from the two lungs show discrepancies in size in various forms, depending upon the degree of asymmetrical development of these respiratory organs. One pulmonary vein may even be entirely absent in certain snakes in which the left lobe of the lung is not present.

The reptilian postcava, as in amphibians, is derived partly from the subcardinals and partly from the vitelline veins. It furnishes the main venous channel from the posterior

part of the body, since the postcardinals have practically disappeared.

BIRDS. The reptilian character of the venous system is clearly indicated in birds. The sinus venosus has been incorporated in the wall of the right atrium, so that caval veins enter the right atrium directly. There are two precavae and a single postcava. The precavae, as in reptiles, are formed by a confluence of subclavian and jugular veins. The original postcardinal connection is no longer in existence. Right and left jugular veins are joined, up in the neck, by an anastomosing vessel, and the right is larger than the left, which, in certain cases, is practically obliterated.

The postcava assumes an even greater importance in birds than in reptiles and is the chief pathway for the return of blood from the posterior part of the body. It is derived in the same manner as previously described for amphibians and reptiles. The posterior end of the postcava receives blood directly from the limbs (Fig. 12.40) via the "renal portal" veins (old posterior cardinals). The name "renal portal" is scarcely appropriate in birds, for although these vessels pass through the kidneys, the passage is, for the most part, direct and not interrupted by a capillary network. The hepatic veins of birds join the postcava as it nears the heart.

The caudal vein in birds is greatly reduced in correlation with the decrease in size of the tail. A vein, variously known as the *inferior mesenteric, coccygeomesenteric,* and *caudal mesenteric,* connects the caudal vein with the hepatic portal vein. Since this vessel connects the two "portal" systems, it may be homologous with the anterior abdominal vein of amphibians and reptiles, although this homology is questioned by some. Others consider a small *epigastric* vein of birds to be the homologue of the anterior abdominal vein of lower forms. This vessel carries blood

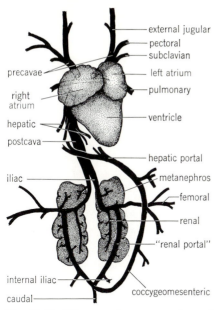

external jugular
pectoral
subclavian
left atrium
pulmonary
precavae
right atrium
ventricle
hepatic
postcava
hepatic portal
metanephros
iliac
femoral
renal
"renal portal"
internal iliac
coccygeomesenteric
caudal

Fig. 12.40 Diagram of venous system of bird, ventral view.

from the great omentum to one of the hepatic veins and is not located in the ventral body wall.

It should be recalled that the heart of birds is completely separated into right and left halves, the right half containing only unoxygenated blood and the left only oxygenated blood brought to the left atrium by the pulmonary veins. In birds the double type of circulation is fully developed.

MAMMALS. The shifting of the main venous channels to the right side is more clearly indicated in the class Mammalia than in others (Fig. 12.41). As in birds, pre- and postcaval veins enter the right atrium directly since the sinus venosus has been gathered into its walls. In some mammals two precavae are present, but in others, as in man, an anastomosis develops between the two through which blood from the left jugular and subclavian veins is shunted to the right side. The combined jugular and subclavian veins of the two sides are called the right and left innominate veins, respectively. The original basal portion of the left precava is modified to form the *coronary sinus*. In such cases, then, only a single precava is present. The internal jugular vein of mammals is the original anterior cardinal vein. It is usually smaller than its external jugular tributary.

In mammals a portion of the anterior end of the right postcardinal persists as the *azygos* vein. This vessel drains the intercostal muscles and enters the precava. It is homologous with the vertebral veins of reptiles. Considerable variation is to be found in the azygos vein in different mammals.

The greatest change in the mammalian venous system occurs in the postcava, which appears to have become considerably simplified. No trace of a renal portal system exists in the adult, and all blood from the posterior part of the body is collected by the postcava. The embryonic development of the mammalian postcava is complicated by the appearance of new vessels called *supracardinal* veins, which contribute to its formation. In other respects, however, its homologies with the postcavae of lower forms are clear.

The anterior abdominal vein has disappeared in mammals, being found only in the echidnas. The allantoic, or umbilical, veins of mammalian embryos, returning blood from the placenta, are homologous with the lateral abdominal veins of elasmobranch fishes and with the anterior abdominal vein of amphibians and reptiles. Usually only the left umbilical vein persists, passing through the liver as the *ductus venosus* to join the postcava before it enters the heart (Fig. 12.42). The umbilical vessels degenerate when they cease to function after birth.

The hepatic portal vein of mammals, now simply known as the portal vein, is in every

Circulatory System   583

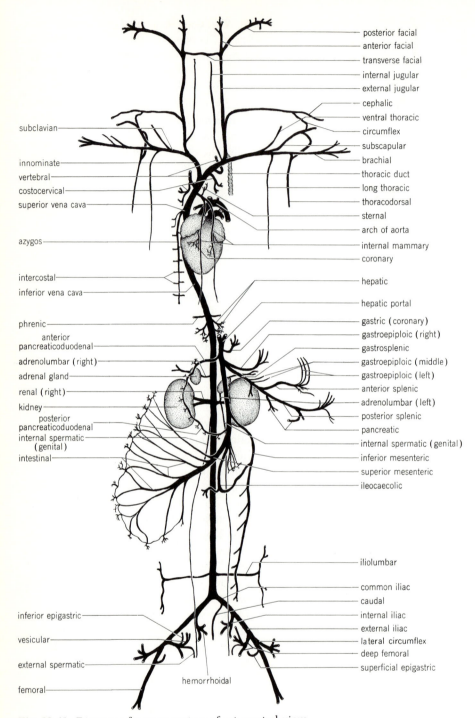

posterior facial
anterior facial
transverse facial
internal jugular
external jugular
cephalic
ventral thoracic
circumflex
subscapular
brachial
thoracic duct
long thoracic
thoracodorsal
sternal
arch of aorta
internal mammary
coronary
hepatic
hepatic portal
gastric (coronary)
gastroepiploic (right)
gastrosplenic
gastroepiploic (middle)
gastroepiploic (left)
anterior splenic
adrenolumbar (left)
posterior splenic
pancreatic
internal spermatic (genital)
inferior mesenteric
superior mesenteric
ileocaecolic
iliolumbar
common iliac
caudal
internal iliac
external iliac
lateral circumflex
deep femoral
superficial epigastric

subclavian
innominate
vertebral
costocervical
superior vena cava
azygos
intercostal
inferior vena cava
phrenic
anterior pancreaticoduodenal
adrenolumbar (right)
adrenal gland
renal (right)
kidney
posterior pancreaticoduodenal
internal spermatic (genital)
intestinal
inferior epigastric
vesicular
external spermatic
femoral
hemorrhoidal

Fig. 12.41 Diagram of venous system of cat, ventral view.

respect similar to the hepatic portal vein of lower forms.

Anomalies of the venous system are frequently encountered. They represent, for the most part, failure of primitive vessels to fuse or to shift positions, or even to degenerate in the orthodox manner. Actually, even in a single species, it is rare to find two individuals with identical venous systems. A few anomalies observed in man are the presence of two superior venae cavae; two postcaval trunks, one lying on either side of the aorta up to a point above the level of the renal veins; as many as seven renal veins, rather than the usual two, opening into the postcava; and failure of the external and internal iliac veins to unite to form a common stem, but opening directly into the postcava.

FETAL CIRCULATION

The allantoic, or umbilical, circulation is of great importance to the embryos of amniotes. In reptiles and birds, with their large-yolked eggs, the vitelline circulation is used in obtaining nutritive substances from the yolk. Although during early development the vitelline circulation is also used in respiration, it is soon superseded by the allantoic circulation, which, along with the developing allantois, has grown out to the periphery of the egg, beneath the porous shell. Here an exchange of oxygen and carbon dioxide takes place.

In mammals, the eggs of which contain little or no yolk, the vitelline circulation is comparatively of negligible importance. The

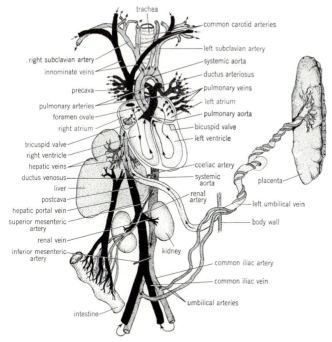

Fig. 12.42 Diagram showing arrangement of the fetal blood vessels of man, ventral view. Arrows indicate the direction of blood flow. Black represents unoxygenated blood; light stippling, oxygenated blood; intermediate stippling, mixed blood.

chief role of the vitelline arteries is the establishment of the superior mesenteric artery; that of the vitelline veins lies in the formation of the hepatic portal vein. The placenta usually develops in association with the allantois. Allantoic, or umbilical, arteries carry blood from the embryo to the placenta, and similarly named veins are employed in its return. The placenta is the organ in which exchange of gases, nutrient materials, and wastes takes place between the mother and her developing young.

A pair of umbilical arteries arises from the dorsal aorta and passes out the umbilical cord to the placenta (Fig. 12.42). At first a pair of umbilical veins returns oxygenated blood to the embryo. Later, in birds, the right artery degenerates, but in mammals both persist. In both birds and mammals the right umbilical vein degenerates and the left one alone continues to function. From the umbilical cord it courses through the body wall for a short distance and then enters the liver through a large channel, the *ductus venosus,* finally to join the postcava.

The lungs do not function during embryonic life, and the umbilical circulation is used for respiration. Conditions, therefore, are totally different during the developmental period from those existing after birth has taken place.

In a typical mammalian fetus, blood enters the right atrium through pre- and postcaval veins. A portion of it then goes to the right ventricle, past the tricuspid valve. Instead of being pumped to the nonfunctional lungs, however, much of this blood is shunted to the systemic aorta via the ductus arteriosus. It will be recalled that the ductus arteriosus is a portion of the sixth aortic arch which connects the pulmonary and systemic aortae during embryonic life. The rest of the blood in the right atrium passes through an opening in the interatrial septum (*foramen ovale*) into

the left atrium. Joined by a relatively small quantity coming from the pulmonary veins, it passes over the bicuspid valve into the left ventricle, from which it is pumped into the systemic aorta. This blood, together with that coming from the ductus arteriosus, is then distributed to all parts of the body, and a representative portion is sent out the umbilical arteries to the placenta.

From the above description it would seem as though a mixing of unoxygenated blood from pre- and postcaval veins, and oxygenated blood from the left umbilical vein, which joins the postcava, would occur in the right atrium. Although this matter has been the subject of much controversy and investigation for over a century, it now seems fairly clear that probably but little mixing occurs. The greater part of the blood from the postcava, carrying a large proportion of oxygenated blood from the left umbilical vein, crosses the dorsal part of the right atrium to enter the left atrium through the foramen ovale. A smaller amount passes into the right ventricle. The stream of unoxygenated blood from the precava(e) passes, for the most part, along the ventral portion of the right atrium into the right ventricle. Although different species vary somewhat in this respect, there is actually a moderately complete separation of oxygenated and unoxygenated blood in the fetal heart.

Changes occurring at birth. At the time of birth when the placenta is loosened from its moorings, the umbilical circulation ceases to function. The umbilical arteries no longer pulsate and gradually constrict. Blood in the umbilical vein drains from the placenta into the body. The stumps of the umbilical arteries become the *hypogastric,* or *internal iliac,* arteries which supply the posterior body wall and portions of the urogenital organs. The umbilical vein disappears, its former course being marked by a fibrous cord, the *ligamentum teres,* extending from the umbilicus (navel) to the undersurface of the liver. The remnant of the ductus venosus is found in the liver as a slender fibrous cord, the *ligamentum venosum.*

When the atrium is first partitioned, a membrane, the *septum primum,* grows down and divides it into left and right chambers. An opening, the *interatrial foramen,* appears in the septum primum, so that the cavities of the atria are in communication. Later a second septum, the *septum secundum,* semilunar in shape, grows down parallel to the septum primum on its right side. An oval aperture in the septum secundum is the *foramen ovale.*

With the cessation of the umbilical circulation, the amount of blood entering the heart through the postcava is materially decreased, since flow through the umbilical vein has stopped. At birth the lungs expand and pulmonary respiration begins. The ductus arteriosus is gradually occluded and becomes a fibrous rudiment, the *ligamentum arteriosum.* Blood from the right ventricle is no longer shunted to the systemic aorta but is sent to the lungs instead. The pulmonary veins now send a greatly increased amount of blood to the left atrium. The foramen ovale then closes. This is brought about by a decrease in pressure in the right atrium, which is now receiving less blood, and an increased pressure in the left atrium due to the increased flow through the pulmonary veins. The septum primum and septum secundum are pushed close together, causing the foramen ovale to be obliterated. A thin area, the *fossa ovalis,* in the interatrial septum indicates the former position of the foramen ovale.

Other factors may also play a part in the changes in circulation occurring at birth. Suffice it to say that after the above changes

have taken place, a perfect double circulation has been established.

Faulty development of the septum primum and septum secundum, or other abnormalities, may prevent the foramen ovale from closing properly. An open, or patent, foramen ovale is not necessarily fatal or even very harmful in human beings. In some cases, however, a mixture of oxygenated and unoxygenated blood takes place. As a result the child has a purplish color and is known as a "blue baby." In severe cases this will result in early death. Other anomalous conditions may result from improper closure of the ductus arteriosus or from a failure of the interventricular septum to develop properly.

LYMPHATIC SYSTEM

Origin. Lymphatic vessels originate during embryonic development considerably later than the vessels of the blood-vascular system and quite independently of them. The connections of the lymphatic vessels with veins are secondary developments. The first indication of the formation of lymphatic vessels is to be observed in the appearance of fluid-filled clefts or spaces which appear in the mesenchyme. The mesenchymal cells bordering the clefts become flattened and differentiate into the endothelial cells which come to line the lymphatic vessels. Various lymph spaces coalesce and branch in a complex manner, so that ultimately an elaborate system of anastomosing vessels is present throughout the entire body. The first lymphatic spaces to appear originate in close proximity to the larger veins. Enlargement of the lymphatic network in certain regions, or a union of several vessels, results in the formation of lymphatic sacs or sinuses. The larger lymphatic vessels usually course along with a vein and its corresponding artery. Smaller lymphatic vessels, however, unlike small veins, do not have a tendency to unite to form single vessels and several may be found close to a vein and its companion artery. The secondary connection of lymphatic vessels and veins, referred to above, usually, but not always, takes place with some of the large veins near the heart where blood pressure is at its lowest.

Valves of the lympathic system make their appearance considerably earlier than those of the venous system. The first ones to form are those in the larger lymphatic vessels in the vicinity of the heart.

Lymphatic nodes and nodules (page 559) are formed when connective-tissue elements condense about lymphatic plexuses associated with strands of mesenchymal tissue. Their development begins only after the primary vessels of the lymphatic system have been formed. Lymphatic nodules in man usually do not develop until after birth.

When the small lymphatic clefts and spaces first make their appearance, they are filled with tissue fluid. Some blood corpuscles may be present in these early vessels, probably coming from adjacent mesenchymal cells which are in the process of differentiating into blood-forming (hemopoietic) tissues.

Structure. The smallest subdivisions of the lymphatic system are called *lymph capillaries*. Although most of the tissue fluid is absorbed by capillaries of the blood-vascular system, much of it passes by diffusion through the endothelial walls of the lymph capillaries into the lymphatic vessels and is then referred to as *lymph*. Lymph capillaries end blindly at their free ends. Their bases unite to form large vessels which in turn combine to form vessels of still greater caliber. The largest vessels are those which, in most cases, open directly into the great veins near the heart. Although the lymphatic system is a closed system, it does not in itself form a complete circuit and in this respect differs from the blood-vascular system. Any colloidal material, which has somehow escaped from blood capillaries into the tissue fluid and which cannot reenter the blood capillaries, may diffuse through the endothelial walls of the small lymphatic vessels and thus ultimately be returned to the circulatory system.

Lymph capillaries are tubules of somewhat greater diameter than blood capillaries. Their caliber, however, is not uniform throughout (Fig. 12.43). No valves are present in the lymph capillaries. Their walls are extremely thin, consisting of a single layer of flat, squamous, endothelial cells. The blind ends of the tubules appear as small, distended knobs.

The larger vessels formed by the union of lymph capillaries have thicker walls and contain valves. The walls are covered with a thin layer composed of elastic and collagenous fibers as well as a few smooth-muscle cells. In still larger vessels, three layers, com-

Fig. 12.43 Lymph capillaries.

parable to those of small arteries and veins, make up the walls. As in blood vessels, they are called *tunica intima, tunica media,* and *tunica externa,* or *adventitia.* These layers are not so distinct as those in blood vessels.

The valves, which prevent backflow, are more numerous than venous valves and, when present, occur at close intervals. They are arranged in pairs and their structure is similar to that of the valves in veins, consisting of a connective-tissue core covered with endothelium.

The walls of the larger lymphatic vessels are supplied with tiny blood vessels similar to the vasa vasorum supplying the larger vessels of the blood-vascular system. The nerve supply to the lymphatics is abundant.

Before the lymph in the lymphatic vessels of mammals enters the blood-vascular system most of it filters through small, bean-shaped structures known as *lymph nodes.* These are usually absent in lower vertebrates. Lymph nodes vary considerably in size. The term lymph "gland" was formerly used in referring to these structures. Since they do not form secretions and are not true glands, the term has generally been discarded. Lymph nodes are masses of lymphatic tissue composed of a meshwork of collagenous and reticular connective-tissue fibers as well as numerous reticuloendothelial cells. They are encapsulated and enclose large numbers of lymphocytes as well as plasma cells and fixed macrophages. Lymphatic vessels ramify throughout the nodes, taking a rather circuitous course. The nodes are located here and there throughout the body along the lymphatic vessels. They are arranged in such a manner that lymph must filter through them on its way to the larger vessels. *Afferent lymphatic vessels* enter a lymph node at various points along its convex surface; *efferent lymphatic vessels* leave at the *hilus* where there is an indentation.

Valves in these vessels control the direction in which the lymphatic fluid travels.

Phagocytic cells in the nodes remove and even destroy particles of various kinds as well as bacteria which may be carried via the lymph to the nodes. Two main types of cells are formed within the lymph nodes: plasma cells and lymphocytes. Many lymphocytes from the nodes pass into the efferent vessels, ultimately to enter the bloodstream. The plasma cells are highly important in the production of antibodies (page 536).

Lymphatic nodules should not be confused with *lymph nodes.* The former are small, usually spherical masses composed largely of lymphocytes and plasma cells. They are not clearly circumscribed and have no capsules surrounding them. Lymphatic nodules may appear and disappear. Each consists of a central core and a peripheral zone, in which lymphocytic cells and plasma cells are densely packed. Numerous lymphatic nodules are located within a lymph node, but they also occur quite independently of nodes under wet epithelial surfaces in various parts of the body.

The small lymph capillaries of the intestine are called *lacteals.* Unlike the capillaries of the blood-vascular system, they are concerned with fat absorption.

Several factors are instrumental in propelling the slowly moving lymph through lymphatic vessels and nodes. These include: (1) muscular activity of various parts of the body, tending to squeeze the fluid along; (2) pulsations of neighboring arteries; (3) pressure built up in the smaller vessels by osmosis and absorption of tissue fluid; and (4) the action of pulsating lymph hearts.

Lymph hearts, when present, consist of enlargements in lymphatic vessels which have contractile walls. They are generally situated near the point where lymph enters the venous system. Valves are present which control the direction of flow. The rhythm of the beating lymph hearts bears no relation to the beat of the heart itself.

Not a great deal is known about the comparative anatomy of the lymphatic system. The small size and delicate structure of the vessels are barriers to detailed study. Furthermore, the great irregularity of the channels in the various classes of vertebrates precludes a study of evolutionary advance. Most studies have been carried out on mammals and birds. Less is known about conditions in the lower classes.

FISHES. Cyclostomes and elasmobranchs seem to lack well-defined lymphatic systems. Thin-walled sinuses opening into veins presumably serve a similar function. The lymphatic vessels of other fishes are extensively developed. Peripherally located channels extend into head, tail, and fins. Deeper channels follow the course of some of the larger veins.

Several connections occur between the lymphatic and venous systems in the posterior and middle as well as the anterior part of the body. Lymph hearts are usually not present, but some have been described in certain forms near the point of junction of lymphatic vessels and veins. The eel has a lymph heart in the tail. In the European catfish, *Silurus*, two caudal lymph hearts are present. Lymph nodes seem to be lacking in fishes.

AMPHIBIANS. In urodele amphibians, two main sets of lymphatic vessels are present. Superficial vessels, beneath the skin, carry lymph to cutaneous and postcardinal veins. Fourteen to twenty lymph hearts have been observed along their course in various forms. The deeper channels follow the dorsal aorta on each side and enter the subclavian veins.

Anurans are characterized by the presence of large lymph sacs, or spaces, beneath the skin. Most of the lymph flows toward the heart. Two pairs of lymph hearts are usually present in adult forms. In the frog the first pair lies behind the transverse processes of the third vertebra, pumping lymph into the vertebral vein. A posterior pair, located near the end of the urostyle, pumps lymph into the transverse iliac vein.

Lymph hearts are more numerous in larval and tadpole stages. It has been reported that over 200 lymph hearts are present in caecilians, located along the intersegmental veins beneath the skin.

REPTILES. A well-developed lymphatic system is present in reptiles. A large subvertebral trunk divides anteriorly to enter the precaval veins. In snakes the lymphatic vessels and sinuses are exceptionally large and numerous. A posterior pair of lymph hearts is to be found in many reptiles. They pump lymph into the iliac veins.

BIRDS. The lymphatic vessels of birds ultimately enter two thoracic lymph ducts which join the precaval veins. Transitory lymph hearts have been observed, during embryonic development, in the pelvic region. They are not usually present in adult birds. The *bursa Fabricii,* a lymphoid organ present in the young of most birds, is a cloacal derivative lined with epithelium of endodermal origin and having a cortical region derived in part from mesenchyme. It is important in the production of lymphocytes.

MAMMALS. Lymph hearts are altogether lacking in mammals. A main trunk, the *thoracic duct* (Fig. 12.41), drains all the lymphatic vessels of the posterior part of the body as well as those coming from the left side of the head, neck, and thoracic regions. It lies beneath the vertebral column and courses anteriorly to open into the left subclavian vein just past the latter's junction with the internal jugular vein. The lower end of the thoracic duct is expanded into a conspicuous enlargement, the *cisterna chyli.*

A so-called *right lymphatic duct* drains lymph from the right side of the head and neck, the right arm, and the right side of the thorax. It enters the *right* subclavian vein near the angle of junction with the right internal jugular vein.

It is through the lymph capillaries (lacteals) in the villi of the small intestine that fats enter the circulatory system. Since these vessels ultimately drain into the thoracic duct, the lymph in the thoracic duct after a fatty meal contains a much greater amount of fat than at other times. The emulsified fat, carried by the lymph, gives the latter a milky-white appearance. The term *chyle* is used to refer to this milky fluid. The chyle passes into the left subclavian vein, as was previously noted.

In mammals lymph nodes are particularly abundant in superficial regions of the head and in the neck, axillae, and groin. Within the body cavity great numbers lie in close association with the large blood vessels. They are exceptionally large and numerous in the mesentery of the intestine. In all these localities they serve to prevent invasion of the body by bacteria. The phagocytic action of the fixed macrophages, in particular, is primarily responsible for this phase of bodily protection. *Peyer's patches* are masses composed of numerous lymphatic nodules in the wall of the small intestine on the side opposite to that to which the mesentery is attached. They are especially numerous in the region of the ileum. In the horse a single patch of Peyer may be 2 in. long and ½ in. wide.

Other lymphatic organs. Among other structures considered to belong to the lymphatic system are certain organs of pharyngeal origin, previously discussed in that connection (pages 245 to 247). They include the *tonsils, adenoids,* and the *thymus gland.*

Tonsils and adenoids are masses of lymphatic tissue composed largely of coalesced lymphatic nodules lying close under the wet epithelial surface of the nasopharynx. Consisting primarily of lymphocytes and plasma cells, it would seem that their chief function is to form antibodies against antigenic substances that may possibly enter the tissues in this region. Filtering of tissue fluid and production of lymphocytes are among their other functions.

Many speculations have been advanced in the past concerning the function of the thymus gland. This has remained obscure, despite the great deal of investigation that has been carried out to determine the role of this glandular structure in the general body economy. Only rather recently has its relation to the lymphatic system, and the circulatory system in general, been elucidated.

The histological structure of the thymus tissue in lower forms is little known. In higher vertebrates it is divided by connective tissue into lobes which are further subdivided into numbers of lobules. Each lobule consists of a peripheral cortical area surrounding an inner, medullary portion. The medullary portions of adjacent lobules may be continuous with one another. The cortex is made up of small cells which seem to be identical in structure with small lymphocytes. Some authorities challenge their identity and have called them *small thymocytes.* Elongated *reticular cells* lie scattered among the smaller cells. The medulla, which is more vascular than the cortex, is made up of reticular cells, which are similar to those of the cortex but are more conspicuous. The lymphocytes are less numerous in the medulla. Small rounded bodies called *Hassall's corpuscles*, which are characteristic of the thymus gland, are also found in the medulla. The cells of these corpuscles are arranged in concentric circles. Hassall's corpuscles and the reticular cells are of epithelial origin. Their significance is not understood.

Perhaps the best-known fact concerning the thymus gland is that, although it is a comparatively large organ in the young animal or human being, it becomes relatively much smaller during adulthood. In man the cortical portions normally begin to involute at about four years of age, but the medullary regions do not begin to atrophy until the age of puberty. The gland reaches its greatest size between the eleventh and fifteenth years and then begins to decrease in size. Involution consists of a reduction of reticular cells and lymphocytes and a replacement by adipose tissue. This is referred to as "age involution" and is a normal occurrence. Age involution of the thymus, more than any other factor, has given rise to the belief that it is an endocrine organ in some way associated with growth. The general concept has been held in the past that the thymus stimulates growth and suppresses genital development and activity. Carefully controlled experiments have indicated that removal of the thymus has no effect upon the activity or size of the gonads. On the other hand, gonadectomy is known to delay involution of the thymus, especially in the male. Injection of sex hormones, either estrogenic or androgenic, is known to hasten involution. It is of interest that administration of the growth, or somatotrophic, hormone (STH) of the anterior lobe of the pituitary gland, or of thyroxine, from the thyroid gland, stimulates thymus growth.

Experiments have been reported which claim that thymus extracts injected into parent rats bring about precocious development and growth of the young and that, if such injections are continued over generations, there is a gradual acceleration in the time when certain physical manifestations first appear, e.g., opening of the eyes, eruption of teeth, appearance of fur, descent of the testes, etc. Efforts have been made to repeat these experiments but they have not been successful.

An enlarged thymus has been thought to be responsible for a condition in infants and young children referred to as *status thymicolymphaticus*. The sudden death of affected children has been attributed to interference of the thymus with normal respiration. In the young mammal, the thymus gland reacts to conditions of stress by rapid shrinkage in size. It is quite likely that, in cases of so-called status thymicolymphaticus, death has occurred so suddenly that there simply has not been enough time for stress atrophy to occur and so the large size of the thymus gland is retained. Thus, the very existence of such a condition as status thymicolymphaticus has been challenged.

It seems that during late prenatal and very early postnatal development the thymus gland probably furnishes the basic cells which are distributed to the lymphoid organs in other parts of the body such as the lymph nodes, lymph nodules, and the spleen. At these sites the cells multiply, forming large masses of lymphocytes as well as plasma cells (page 85). Many lymphocytes then later pass into the lymphatic stream and blood-vascular system where they play an important role in protecting the body against invading foreign tissues and bacteria, thus helping to resist infection. Lymphocytes, with their large nuclei and small amount of cytoplasm, are not generally phagocytic. Some lymphocytes, however, may possibly develop into monocytes which are phagocytic. The actual function of lymphocytes remains obscure, but it is quite possible that they are capable of carrying antibodies on their cell surfaces. Once the

Anatomy of the Chordates

thymus gland has done its work in furnishing the basic cells which are distributed to other lymphoid structures, it seems that it can be dispensed with, thus accounting for the fact that removal of the gland after a certain age causes no untoward effects and the normal process of age involution brings about no significant changes. It has been shown that if the minute thymus glands are removed from *newborn* mice, the animals at first grow and develop in the same manner as unoperated animals, but after a few weeks their lymph nodes and spleens fail to continue to develop normally and show signs of degeneration. Such animals seem to be unable to resist infection, and their ability to protect their bodies against foreign tissues is lost. Skin grafts from other strains of mice (J. F. A. P. Miller, 1961, *Lancet*, **2**:748), and even of rats, are not rejected as they would be in normal individuals. Death occurs at an early age. If removal of the thymus is delayed until after the young mice are two or three weeks old, then such disorders do not develop. Apparently the thymus has already played its role. It would seem probable that in other mammals, including man, the thymus has a similar function, but the length of time it takes for the gland to work would vary according to the species. It has been suggested that the presence of the thymus gland during early life may be important as the original site of plasma-cell production and thus in the establishment of normal immune reactions and in the production of antibodies later in life. It has been shown, however, that if thymus tissue, enclosed in a cell-tight millipore filter, is implanted into a neonatally thymectomized mouse, the animal's ability to form antibodies is restored. This would suggest that some blood-borne factor, rather than cells distributed to lymphoid organs, is responsible for inducing the latter to participate in immune reactions.

Further evidence for this hormonal concept of thymus function lies in the fact that in a pregnant neonatally thymectomized mouse immunological functions are restored, presumably by diffusion through the placenta of a hormonelike factor originating from the thymus glands of the young developing *in utero* (D. Osaka, 1965, *Science,* **147**:298).

In birds it seems that the bursa of Fabricius (page 192) is basically responsible for the formation of cells concerned with antibody production and for cellular reactions against invading bacteria, whereas cells from the thymus gland, after reaching the lymph nodes and spleen, are responsible for such a phenomenon as rejection of skin grafts. In mammals the thymus gland seems to be basically responsible for both these functions. More research needs to be done on the thymus gland and bursa of Fabricius before their functions can be unequivocally demonstrated.

It is possible that the thymus glands of mammals may ultimately be shown to be responsible for some, if not all, functions of the body having to do with immunological reactions.

Hemal nodes. Certain organs of the body closely resemble lymph nodes except that they are situated in the course of blood vessels rather than lymphatics. Blood, therefore, rather than lymph, filters through them. They are called *hemal nodes.* Hemal nodes are entirely devoid of lymphatic vessels. In pigs a special kind of *hemolymphatic node* has been described which has the properties of both lymphatic nodes and hemal nodes.

It is believed that hemal nodes function in a manner similar to that of the spleen and that they play a part in the destruction of old and worn-out red corpuscles. They may even have a role in erythrocyte formation. Their actual function, however, is not entirely clear. Most hemal nodes are very small. They are unusually abundant in retroperitoneal tissue near the points of origin of the superior mesenteric and renal arteries. Hemal nodes are most numerous in ruminating animals. Their occurrence in man is doubtful.

Spleen. The largest lymphoid organ in the body is the spleen. It first appears during embryonic development as a localized thickening in the dorsal mesogastrium, the mesentery supporting the stomach. When the stomach later swings over to the left, the spleen is also carried to the left side of the body. The organ itself, derived from mesenchymal cells, continues to grow and ultimately is connected to the stomach by a band, the gastrosplenic ligament, which is a portion of the original mesogastrium. Differentiation of the original mass of mesenchyme is complex and depends mainly upon the development and distribution of the blood vessels with which it is supplied.

The spleen is generally considered to be a hemolymphatic organ. It is interposed in the bloodstream rather than in lymphatic vessels. Blood, therefore, rather than lymph, filters through the spleen. Because of the peculiar arrangement of the blood vessels in the spleen, the blood comes in contact with phagocytes (macrophages) which engulf the fragments of disintegrating red corpuscles. The protein, hemoglobin, liberated by these old and worn-out cells, is the basis for manufacture of a substance known as *bilirubin.* This is carried by the splenic branches of the hepatic portal vein to the liver and excreted in the bile, imparting a yellowish-brown color to this secretion of the liver. It ultimately passes into the intestine and is responsible for the color of the feces. The iron which forms part of the hemoglobin molecule is also removed by the spleen and returned to the general circulation, which carries it to the bone marrow where it is used in synthesizing the hemoglobin of new erythrocytes which are formed there. Lymphocytes and plasma cells, probably originating in the thymus gland during early life, are formed later in large numbers in the spleen. The plasma cells are of special importance in the manufacture of antibodies. Thus the spleen serves as a mechanism of defense of the body against certain diseases. It is probable that most of the monocytes to be found in the blood are formed in the spleen, but their exact origin has not been convincingly demonstrated. Enlargement of the spleen is a feature of malaria and certain other diseases in which the organ may assume relatively enormous proportions.

The spleen also serves as a storehouse for erythrocytes. Large numbers may be housed by the spleen, which can return them to the blood-vascular system as requirements of the body demand. In addition to the usual capillary network between arteries and veins, there are in the spleen certain direct, sinus-like connections between the two. The presence of these sinuses may account for the blood-storing properties of the spleen. Contraction of the spleen occurs in response to the demand by the body for an increased supply of red blood corpuscles, which are thus forced into the bloodstream.

During embryonic development the spleen functions as a blood-forming (*hemopoietic*) organ since erythrocytes and lymphocytes are formed in its tissues. In mammals the production of erythrocytes by the spleen ceases in adult life but lymphocytes continue to be formed.

Study of the general microscopic structure of the spleen shows it to be composed of two general types of splenic tissue: *white pulp* and *red pulp.* The white pulp is actually made up of lymphatic nodules which produce the lymphocytes and plasma cells. Red pulp surrounds the nodules of white pulp. It is composed of a meshwork of reticular connective-tissue fibers in which great numbers of erythrocytes may be stored. The red pulp represents the filtering mechanism of the spleen, its appearance varying under different conditions. The white pulp is closely associated with branches of the splenic arteries; the red pulp is more closely connected to tributaries of the splenic veins.

Performing all the important functions that it does, it is surprising that the spleen is not essential to life. When the spleen is removed, most of its functions are carried on by other hemopoietic tissues in the body.

Among vertebrates, cyclostomes lack a spleen. Some peculiar tissue in the wall of the intestine has, however, been homologized with the spleen by certain authorities. Splenic tissue has been described in the stomach wall in *Protopterus*, which also lacks a discrete spleen but possesses a lymphoid mass partially surrounding the stomach. Red-cell formation seems to be limited to the submucosa of the stomach in lungfishes. In other fishes the spleen is usually large and distinct, but variable in shape in different species. It lies in close association with the stomach and is important in the formation of red corpuscles.

The spleens of higher forms, in general, vary but little from those of fishes in their location, usually being compact organs lying toward the left side of the body close to the stomach. Among amphibians it has been reported that the spleen of the newt is the only place in the body where erythrocytes are formed. In anurans the spleen first begins to be less significant in production of red corpuscles, and in the adults of higher forms this property has probably been lost altogether. The spleen continues to form lymphocytes, however.

Blood-forming (hemopoietic) tissues. A regular sequence of blood-cell formation occurs in various organs and regions of the body during development. In early embryonic life blood cells are first differentiated in the wall of the yolk sac, at least in cases where a yolk sac is present. Body mesenchyme then follows as a site of blood-cell formation. With the development of the liver, that organ takes over hemopoiesis for the time being. Then the spleen, thymus, and lymph nodes next play a part in blood-cell formation, followed finally by the bone marrow which, in addition to the lymph nodes, functions as the hemopoietic organ of the adult. It is in anuran amphibians that the bone marrow first becomes hemopoietic. In mammals the bone marrow is the chief hemopoietic organ.

A substance formed in the kidneys, termed *erythropoietin,* apparently stimulates the formation of erythrocytes even in early stages. Estrogens have an inhibiting effect upon erythropoiesis. It is not known whether estrogens act directly upon the bone marrow or upon production of erythropoietin (P. P. Dukes and E. Goldwasser, 1961, *Endocrinology,* **69**:21).

Lymphocytes originate in lymphatic tissue mostly by mitotic proliferation of previously existing lymphocytes. The granular leukocytes as well as erythrocytes, on the other hand, are formed in the bone marrow.

In some fishes and in certain amphibian larvae, the opisthonephros functions, at least for a time, as a blood-forming organ.

BLOOD

The composition of vertebrate blood has been discussed (pages 83 to 86). Several differences exist in the bloods of various vertebrates, a few of which will be considered. The chemical composition of blood plasma shows great variation, not only in different vertebrate classes but in the species within a class, or even a genus, as well. Dissimilarities are observed in the percentage of water and solids, including proteins and other organic constituents, salts, and other materials.

Among the formed elements, too, considerable differences exist.

The blood of amphioxus is practically colorless and contains but few red corpuscles. Apparently no leukocytes are present. Contrary to the condition in red-blooded invertebrates, in which the respiratory pigment hemoglobin is dissolved in the plasma, in vertebrates the hemoglobin is confined to the red corpuscles. The erythrocytes of vertebrates (Fig. 12.44), in all classes except mammals, are oval in shape and flattened. Each contains a conspicuous nucleus which causes the corpuscle to bulge in the center. Considerable variation in size occurs, the erythrocytes of the salamander, *Amphiuma,* being the largest of any animal. Among mammals, only the family Camelidae (camels and llamas) possesses oval corpuscles. Those of other mammals are usually round and biconcave. The mouse deer, *Tragulus javanicus,* has the smallest red corpuscles of all, averaging 1.5μ in diameter, and having a spherical configuration (K. L. Duke, 1963, *Anat. Record,* **147**:239). Those of man average 7.2μ in diameter. The erythrocytes of elephants are the largest found among mammals, but they do not exceed those of man by a great deal.

Mammals are unique in that the *erythroblasts,* cells which give rise to erythrocytes, extrude their nuclei before becoming mature erythrocytes. It is not surprising that the red cells should be relatively short-lived since they have lost their nuclei. It has been estimated that red blood cells in man live in the bloodstream anywhere from 100 to 120 days before they break down. In the dog, 124 days has been reported to be the maximum length of erythrocyte life.

About 10 million red cells are destroyed every second in the human being and, of course, a comparable number must be formed to replace them. Since the number of red corpuscles in man is about 5 million per cubic millimeter, the rate of destruction and formation is not so tremendous as it might seem at first glance. Persons living at high altitudes have unusually high red-blood-cell counts.

Leukocytes are much less numerous than erythrocytes. They retain their nuclei and may be classified into granular and nongranular types, each type having two or more subdivisions (pages 84 to 85). A sudden increase in their number often indicates the presence of an infection. Leukocytes in mammals are larger than erythrocytes, and their numbers are subject to much variation. An average healthy human being has approximately 7,000 per cubic millimeter of blood. The phagocytic action of certain leukocytes enables the body to combat

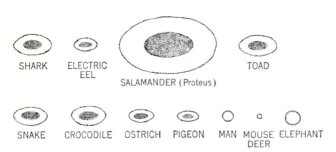

Fig. 12.44 Relative sizes of red corpuscles of representative vertebrates.

invasion by bacteria. A localized site of painful swelling and inflammation usually marks an area in which numbers of leukocytes have congregated in an effort to destroy bacteria. Usually the leukocytes destroy the bacteria, and the infected site is once more restored to normal. When pus forms, the bacteria have, at least temporarily, gained the upper hand, pus being composed of large numbers of dead leukocytes. Monocytes and granular neutrophils are the chief phagocytic cells of the body. They are not confined to blood vessels but, through their ameboid movements, are able to migrate throughout the entire body where they function as scavengers. They push between the flat endothelial cells of the capillaries (diapedesis) and in that manner are able to leave or enter the bloodstream. Most, if not all, phagocytic action takes place outside the bloodstream in the tissue spaces.

Blood platelets are formed elements found only in the blood of mammals. They are small, colorless, nonnucleated bodies which, in circulating blood, are oval in shape and biconvex, their length being approximately one-half to three-quarters the diameter of erythrocytes. When dried and stained they appear more rounded, with a portion which takes up certain stains, the remainder being rather transparent. It is generally believed that they are derived from detached fragments of the cytoplasm of certain giant cells, with very large oval-shaped or lobulated nuclei which are present in the bone marrow and known as *megakaryocytes*. Megakaryocytes are also found only in mammals. Platelets in man number between 250,000 and 500,000 per cubic millimeter of blood. It is very difficult to make an accurate platelet count since these tiny structures disintegrate when exposed to air or rough surfaces. Platelets live in the bloodstream only from 3 to 7 days. Worn-out platelets are probably phagocytosed in the same manner as degenerating erythrocytes. Blood platelets are lacking in vertebrates other than mammals. Instead, the members of the lower classes possess nucleated spindle cells, or thrombocytes, believed to have similar functions.

In order to understand the functions of blood platelets, it is important to have a rather clear conception of the phenomena involved in the clotting of blood. Several theories have, in the past, been advanced to explain what causes blood to coagulate, or clot, but the exact manner in which this occurs is still not known with certainty. Blood clotting is a complicated process, several phases of which have been recently elucidated.

A substance called *prothrombin* is present in the bloodstream but is normally inactive. Prothrombin cannot be formed if there is a deficiency of vitamin K. Prothrombin changes to an active enzyme, *thrombin*, when acted upon by another substance called *thromboplastin* in the presence of calcium ions. A small amount of thrombin is normally present in the circulating blood but is held in check by an inactivating substance called *antithrombin*. Thrombin then acts upon *fibrinogen*, one of the proteins in blood plasma, transforming it to *fibrin*, which is laid down at the site of an injury in the form of a fine meshwork of threads which are so closely interlaced that even erythrocytes cannot pass through the interstices. The mass of fibrin threads is known as a *clot*. It has been observed under the electron microscope that a thread of fibrin is composed of numerous filaments with alternating light and dark bands, so arranged that corresponding bands of adjacent filaments are in alignment. After a time the threads of fibrin contract and the clot itself becomes relatively solid and rather firm. Part of the blood plasma in the clotted area is squeezed out in the form of a clear fluid known as *blood serum*. The serum is like plasma except that, since fibrinogen has been removed, it is unable to clot.

At any rate, the source of thromboplastin, sometimes referred to as *thrombokinase*, which changes prothrombin to thrombin, has been the point of considerable contention. For a long time, it was believed that blood platelets liberated thromboplastin upon disintegration when exposed to air or rough surfaces, such as those caused by a cut or wound. It has been found, however, that other factors carried by the blood are also vitally concerned in the formation of thromboplastin. Furthermore, it has been demonstrated experimentally that extracts of platelets have little effect per se in causing blood to clot. The formation of thromboplastin is apparently a very complicated process involving the interaction of several factors. Blood clots are known to form under certain conditions when there is no injury of any kind. Under such circumstances, it is believed that thromboplastin is formed by some ingredient furnished by the blood platelets, together with two other substances normally found in the blood plasma and known, respectively, as *antihemophilic globulin* (AHG) and *plasma thromboplastin component* (PTC), or the *Christmas factor*, named for the individual in whom it was first discovered.

When a clot forms following a cut or abrasion, the following sequence of events probably occurs: PTC seems to become active when plasma comes in contact with with a rough surface. This, together with AHG, forms a combination which, in the presence of calcium ions, unites with an ingredient furnished by the blood platelets or by injured tissue cells to form thromboplastin. Platelets thus are not essential for the formation of this type of clot. It is highly probable that additional factors are involved

in the activation of thromboplastin since its action in converting prothrombin to thrombin is not uniform. It begins to act slowly, but subsequently its action becomes more and more rapid. In the condition known as *hemophilia,* in which a severe and even fatal hemorrhage may occur after a trivial injury, the blood fails to clot. Persons who are hemophiliacs lack the ability to form thromboplastin. True hemophilia is a hereditary condition which is transmitted as a sex-linked, recessive character. It appears only in males, who are termed "bleeders," and it is transmitted through females to their masculine offspring. It has become evident that true hemophiliacs lack the AHG factor, hence the name, antihemophilic globulin. Numerous cases, which formerly were believed to be examples of true hemophilia, are really examples of a very similar condition known as *Christmas disease*. This is known to be the result of a deficiency of PTC, the plasma thromboplastin component, or Christmas factor. The formation of PTC is also apparently under the control of a sex-linked, recessive gene. It is interesting that if the two types of blood from bleeders, one lacking AHG and the other lacking PTC, are mixed, normal clotting of blood will take place. Other types of "bleeding diseases" are known to be due to failure of thromboplastin to form properly. Some of these are hereditary but are not sex-linked as are those mentioned above.

The part played by blood platelets in the formation of thromboplastin is but one of their functions. In addition, they are highly important in causing threads of fibrin to contract and thus in causing the clot itself to become firm and solid. Still another function of blood platelets, not previously mentioned, is their ability to form *white thrombi*.

A *thrombus* is a clot or lump formed by an agglutination or coagulum of formed blood elements during life in any part of the blood-vascular or lymphatic systems. The term is applied only to clots or lumps which form *intravascularly*. There are two kinds of thrombi: red and white. Blood platelets are involved in the formation of both. Red thrombi can form only in blood that is still, in the sense that it is not circulating, and involve the conversion of fibrinogen to fibrin, as described above. White thrombi, on the other hand, are actually composed of masses of blood platelets alone. In an injured area, the platelets begin to stick to the walls of blood vessels through which blood is flowing. Increasing numbers adhere to one another, a phenomenon known as *agglutination*, and the mass gradually becomes larger. The white thrombus gradually obliterates the lumen of the vessel or vessels, preventing leakage and forming a coating over the injured tissue. When a clot forms over a cut or wound there is a tendency for platelets to agglutinate in a similar manner, but the coagulum formed by the fibrin meshwork supersedes it by far.

Administration of a substance called *heparin*, which can be extracted from the liver as well as from various other tissues in the body, delays or prevents clotting of blood.

It apparently has an effect which somehow nullifies the action of prothrombin or thrombin and thus prevents fibrinogen from being transformed into fibrin. Heparin also prevents blood platelets from agglutinating and forming white thrombi. Its use in persons prone to thrombosis has been of great medical importance in prolonging the lives of innumerable individuals.

Still another substance, known as *serotonin* (page 190), plays a role in the arrest of bleeding. It is produced by argentaffin cells located at the bases of the crypts of Lieberkühn in the intestinal mucosa, and is carried by, or absorbed by, the blood platelets. When a clot is formed, serotonin is liberated. It acts in two ways: as a local vasoconstrictor, and as an agent which tends to lower blood pressure and thus reduce the amount of bleeding to some extent. Serotonin, which is a derivative of the amino acid tryptophane, is somehow involved in nerve function. It has recently been found that the usually fatal type of food poisoning known as *botulism,* which affects the nerves, may be blocked by serotonin (R. D. Paine, Jr., 1963, *National Science Foundation,* 63–126).

SUMMARY

1. The circulatory system of higher metazoans serves to carry substances of physiological utility to all parts of the body and to remove from them the waste products of metabolism. Wastes are carried to appropriate organs for elimination.

2. The vertebrate circulatory system is of mesodermal origin and is composed of two systems of elaborately branched tubes which ramify throughout the body. They are (*a*) the blood-vascular system, composed of heart, arteries, arterioles, capillaries, venules, and veins, (*b*) the lymphatic system, made up of lymph capillaries, vessels, sinuses, nodes, and lymphoid organs of various types.

3. The blood-vascular system is subdivided into arterial and venous systems, the former carrying blood away from the heart and the latter returning it. One, two, or three portal systems of veins are present. Each breaks up into a capillary network before reaching the heart. Two of these systems are known as the hepatic

portal and renal portal systems, ending in liver and kidneys, respectively. The third is associated with the pituitary gland.

4. The walls of arteries, veins, and the large lymphatic vessels are composed of three layers, the tunica intima, tunica media, and tunica externa, or adventitia. The thickness and detailed structure of these three layers vary in the different types of vessels.

5. Both blood and lymphatic vessels first appear in embryonic development as spaces in the mesenchyme which subsequently grow and branch in a complex manner. If a yolk sac is present, the first vessels appear in its splanchnic mesoderm.

6. The heart is primarily a pulsating tube derived from a fusion of two vitelline veins. It becomes divided into chambers called atria and ventricles. In addition, two accessory chambers may be present, called the sinus venosus and conus arteriosus, respectively. Cyclostomes and fishes have two-chambered hearts, with one atrium and one ventricle; Dipnoi and amphibians with three-chambered hearts have two atria and one ventricle; the three-chambered heart of reptiles is like that of amphibians, but an incomplete partition appears in the ventricle. In crocodiles and alligators this becomes complete, forming a four-chambered heart with two atria and two ventricles. Birds and mammals have four-chambered hearts. Valves of various types are present which regulate the direction of blood flow.

7. Animals with two-chambered hearts have the single type of circulation in which only unoxygenated blood passes through the heart which pumps it to the gills for aeration. The double type of circulation exists in three-chambered and four-chambered hearts through which pass two streams of blood, one oxygenated and the other either unoxygenated or partly oxygenated. These two streams do not mix, at least to any appreciable extent.

8. Vessels which carry blood from the heart to the lungs and back comprise the pulmonary circulation. Unoxygenated blood or partially unoxygenated blood is pumped from the right ventricle (or right side of the ventricle) to the lungs. Oxygenated blood returns from the lungs to the left atrium. The remaining vessels make up the systemic circulation which sends oxygenated blood throughout the body and returns unoxygenated blood to the right atrium. A special coronary circulation supplies and drains the tissues of the heart itself.

9. The arterial systems of vertebrates are essentially similar, at least during developmental stages. Usually six pairs of aortic arch vessels extend from the ventral aorta to the dorsal side of the pharyngeal region where they fuse to form a vessel on each side. These two vessels combine posteriorly to form a single dorsal aorta. Changes in the aortic arches constitute the chief differences in the arterial systems of the separate vertebrate classes. A progressive reduction in the number of aortic arches occurs as the evolutionary scale is ascended. In cyclostomes and fishes which retain the greatest number of aortic arches, each arch is broken up into afferent and efferent portions connected by a capillary network. The gill lamellae through which the capillaries ramify serve as respiratory organs. In teleosts only the last four arches are retained. In remaining vertebrates arches 1, 2, and 5 disappear as does the radix on each side between arches 3 and 4. Arch 3 on each side, together with remnants of the ventral aorta and the anterior portions of the radices, becomes

the carotid complex, consisting of a basal common carotid artery with internal and external branches. The fourth pair of arches persists in amphibians and reptiles to become the systemic arches which unite to form the dorsal aorta. The left fourth arch in reptiles splits off the ventral aorta (truncus arteriosus) and establishes a connection with the newly forming right ventricle. In birds the left fourth aortic arch loses its connection with the dorsal aorta and degenerates. Mammals are similar to birds except that the right fourth aortic arch instead of the left degenerates in part. In Dipnoi and the higher classes which use lungs for respiration, pulmonary arteries grow out from the sixth aortic arches to the lungs. The connection of arch 6 to the aorta on each side functions during embryonic life as the ductus arteriosus. This is closed after birth. In reptiles, birds, and mammals the pulmonary arteries come from a separate pulmonary aorta formed by a splitting of the truncus arteriosus which connects with the right ventricle. The rest of the truncus, known as the systemic aorta, leaves the left ventricle.

10. The single dorsal aorta, as it courses posteriorly, gives off many paired somatic arteries to supply the body proper. Paired visceral vessels pass to the derivatives of the embryonic mesomere. Usually three unpaired visceral vessels, the coeliac, superior, and inferior mesenteric arteries, are distributed to the digestive tract and spleen.

11. The primitive venous system arises, in most cases, from several paired veins, all of which join the sinus venosus to enter the heart. These consist of (a) a pair of vitelline veins, continued posteriorly as the subintestinal veins; (b) paired anterior and posterior cardinals which, on each side, join to form a common cardinal vein, or duct of Cuvier, entering the sinus venosus; (c) lateral abdominal veins from the body wall which also enter the duct of Cuvier; and (d) subcardinal veins which appear in the center of the opisthonephros or mesonephros and which are joined by the caudal vein from the tail.

12. The vitelline and subintestinal veins give rise to the hepatic portal system of veins, which lose their connection with the sinus venosus and, instead, break up in sinusoids in the liver.

13. The anterior cardinal veins become the internal jugulars. In fishes and urodele amphibians a pair of inferior jugular veins also enters the sinus venosus, coming from the ventral head region. The latter vessels are without homologues in higher forms.

14. The caudal vein from the tail loses its connection with the subintestinal and subcardinal veins and splits in two at its anterior end, each portion connecting with the posterior part of the postcardinal. The postcardinals break up into two portions. The posterior parts, joined by the caudal vein, bring blood to the kidneys, and these become the renal portal veins. The anterior portion is joined by the subcardinals. In Dipnoi the right postcardinal is larger than the left and is called the postcava. In higher forms, however, the postcava is derived as an outgrowth of the vitelline veins which join the subcardinals. In such cases as these the postcardinals are reduced in size. In reptiles the postcardinal veins become the vertebral veins. The right postcardinal in mammals contributes to the formation of the azygos vein.

15. A subclavian vein from each forelimb enters the duct of Cuvier on each

side, and an iliac vein joins the posterior part of each lateral abdominal vein after it leaves the hind limb.

16. Beginning with the Dipnoi, the two lateral abdominal veins fuse to form an anterior abdominal vein, which then joins the hepatic portal vein near the liver. This vessel loses its importance in reptiles and birds and is absent in all mammals except the spiny anteater. The anterior abdominal vein connects the hepatic portal and renal portal systems. The umbilical, or allantoic, veins of amniote embryos are homologous with the lateral and anterior abdominal veins of lower forms.

17. The renal portal system also becomes reduced in reptiles and birds and is absent in mammals, the postcava furnishing the main venous drainage for the posterior part of the body.

18. The ducts of Cuvier, from amphibians on, become the precaval veins which in amphibians and reptiles enter the sinus venosus but in birds and mammals enter the right atrium directly, the sinus venosus having been incorporated in the wall of the right atrium. In several mammals the left precava loses its connection with the heart and a new vessel shunts blood from the left side of the anterior end of the body to the right precaval vein. The remnant of the left precava becomes the coronary sinus.

19. During fetal life, in amniotes, an umbilical, or allantoic, circulation exists which performs a role in respiration. The pulmonary circulation is inconsequential before birth, at which time it first begins to function and the umbilical circulation ceases. Umbilical arteries are branches of the dorsal aorta. The umbilical vein, formed primarily from the left umbilical vein, is homologous with the lateral and anterior abdominal veins of lower forms. An opening, the foramen ovale, in the interatrial septum, and the ductus arteriosus, connecting pulmonary and systemic aortae, make possible the peculiar routing of blood which is characteristic during fetal life. The umbilical vessels, foramen ovale, and ductus arteriosus do not exist as such after birth.

20. The lymphatic system collects tissue fluid from all parts of the body. After the fluid enters the lymphatic system, it is called lymph. The lymphatic vessels connect with the venous system at certain places, usually with the large veins near the heart. Much variation in detailed structure exists in the various groups of vertebrates. Some forms have lymph hearts which help to propel the lymph, but in others, muscular movements and the pressure of newly accumulated lymph cause the fluid to flow slowly through the lymphatic vessels. Lymph nodes, composed of lymphatic tissue and through which lymph slowly filters, interrupt the course of lymphatic vessels. Tonsils, adenoids, thymus glands, and Peyer's patches are lymphoid structures.

21. Many vertebrates have hemal nodes which interrupt the course of blood vessels in a manner similar to that of lymph nodes interrupting the course of lymphatic vessels.

22. The spleen is the largest lymphoid structure in the body. It functions as a hemolymphatic organ but, in addition, is able to store and discharge red corpuscles as the needs of the body dictate. The spleen lies in the mesentery of the gut near the stomach. It is absent in cyclostomes and *Protopterus*.

23. Blood-forming (hemopoietic) tissues include the splanchnic mesoderm of the yolk sac, body mesenchyme, liver, spleen, lymphoid organs, and bone marrow. Most of these function in this capacity only during embryonic life. In adult mammals only the lymphoid tissues and bone marrow ultimately have a hemopoietic function.

24. The composition of blood varies somewhat in different vertebrate groups. The main difference lies in the erythrocytes which are flattened, oval-shaped cells in all except mammals. Only camels and llamas, among mammals, have oval erythrocytes. The red corpuscles are nucleated except in mammals, in which the erythroblasts, which form erythrocytes, extrude their nuclei.

25. Leukocytes, formed in the bone marrow and lymphoid tissues, can wander about the body outside blood vessels. Certain kinds act as phagocytes, engulfing bacteria and particles of various kinds.

26. Blood platelets, possibly homologous with spindle cells of lower forms, are concerned with blood clotting in mammals.

| 13 |

NERVOUS
SYSTEM

Of all the cells in the body, those of which nervous tissue is composed have the properties of irritability and conductivity most highly developed. The detailed structure of nervous tissue has been discussed previously (Chap. 3). It will be recalled that all nervous tissues are of ectodermal origin. They trace their beginning, for the most part, from the embryonic neural tube and neural crests. In a number of cases, *placodes,* or thickenings of the superficial ectoderm in certain restricted areas, also give rise to nervous elements.

In addition to nerve cells, or *neurons,* of various types, there are present in the central nervous system (brain and spinal cord) other components referred to as *neuroglia.* These very numerous ectodermal cells are interspersed among the nervous elements, providing support and some degree of protection since true connective tissue of mesodermal origin is not present. It is coming to be recognized that they play a very important role in the synthesis of certain proteins which are vital elements in the proper functioning of such higher nervous functions as thought and memory (page 629). Several types of neuroglia cells are recognized. They generally consist of very much-branched cells, the processes of which form an interlacing network between neurons. Other nonnervous, supporting, *ependymal cells* of an epithelial nature line the cavities of brain and spinal cord. They are often ciliated.

Gray matter and white matter. Nerve-cell bodies, their dendrites, the supporting neuroglia cells, and the proximal unmyelinated portions of the axons have a grayish appearance and form the greater part of the *gray matter* of the brain and spinal cord. Groups of nerve-cell bodies are called *ganglia*. The term ganglion is generally used to refer to an aggregation of nerve-cell bodies forming a mass or enlargement *outside* the central nervous system. Gray ganglionic masses *within* the brain and spinal cord are usually designated as *nuclei,* or *nerve centers. White matter,* on the other hand, is composed chiefly of bundles of glistening white myelinated, or medullated, fibers. Such bundles, which it is possible to trace in prepared sections of brain and spinal cord, are known as *fiber tracts.* In some regions gray and white matter are intermingled to varying degrees, such an arrangement being known as a *reticular formation.*

Primary divisions. The nervous system is extremely complex, and all parts are structurally connected and functionally integrated. It is customary, nevertheless, to speak of two main divisions. These are (1) the *central nervous system,* composed of the *brain,* which lies within the cranial cavity of the skull, and the *spinal cord,* lying within the neural canal formed by the neural arches of the vertebrae, and (2) the *peripheral nervous system,* made up of nerves and ganglia which are connected to the brain and the spinal cord. Part of the peripheral nervous system

is composed of autonomic fibers which are distributed to those parts of the body under involuntary control. These, in turn, consist of sympathetic and parasympathetic components which function in an opposite manner in mediating various involuntary activities of the body. The autonomic portion of the peripheral nervous system is sometimes spoken of as a third division of the nervous system, called the *autonomic nervous system.* However, since autonomic fibers are anatomically associated with spinal nerves and certain cranial nerves, it is most logical to include them under the category of the peripheral nervous system. In the following pages we shall refer to the autonomic components as the *peripheral autonomic system.*

CENTRAL NERVOUS SYSTEM

Early development. The ectodermal neural plate from which the neural tube is formed consists at first of a single layer of epithelial cells. These usually proliferate as the neural folds are developing. By the time the folds have fused dorsally to form the neural tube, its walls are composed of several layers of cells. The original epithelial cells give rise to two types of daughter cells, *neuroblasts* and *spongioblasts.* Neuroblasts develop into typical neurons with their highly specialized properties of irritability and conductivity. Spongioblasts give rise to the nonnervous supporting ependymal cells and the various other neuroglial elements.

Soon after the neural tube has formed, it can readily be discerned that the cells of which it is composed become arranged in three rather definite layers: (1) an inner, epithelial *ependymal layer,* (2) an intermediate, nucleated *mantle layer,* and (3) an outer, fibrous *marginal layer,* free of nuclei (Fig. 13.1). It is from the neuroblasts in the mantle layer that the nerve cells forming the gray

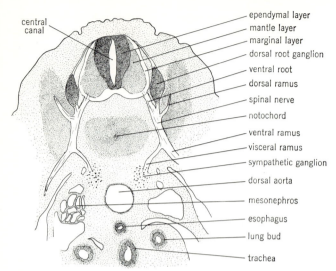

central
canal

ependymal layer
mantle layer
marginal layer
dorsal root ganglion
ventral root
dorsal ramus
spinal nerve
notochord
ventral ramus
visceral ramus
sympathetic ganglion
dorsal aorta
mesonephros
esophagus
lung bud
trachea

Fig. 13.1 Section through portion of 12-mm pig embryo, showing differentiation of neural tube and spinal nerves with roots and rami.

matter of brain and spinal cord develop. An axon and one or more dendrites grow out from each neuroblast in the development of a typical neuron. The axons of certain cells grow peripherally until they reach the marginal layer of the neural tube. They may then course anteriorly or posteriorly in that layer, ultimately to form the fiber tracts of brain and cord. The axons of other mantle-layer neurons penetrate through the marginal layer primarily in the ventrolateral angles of the neural tube. These become the *ventral,* or *motor, roots* of the spinal nerves and the motor components of certain cranial nerves. This segmental arrangement reflects the fundamental metamerism of the vertebrate body. Cilia develop from the ends of the ependymal cells in contact with the central canal. Although cilia are present during embryonic development, they rarely persist in adult life. The other ends of the ependymal cells send out long, branching processes which aid in support of the strictly nervous

elements and also contribute to the outer limiting membrane which comes to surround the neural tube. Ultimately these processes lose their cellular connections and the ependymal cells form the epithelial lining of the *central canal of the spinal cord* and of the *ventricles of the brain.*

Brain and spinal cord develop from the dorsal, hollow, neural tube itself. The anterior end of the neural tube, destined to form the brain, enlarges from almost the very beginning. Two constrictions appear in the enlarged portion so that three *primary brain vesicles* are rather clearly marked off. These are called the *forebrain (prosencephalon),* *midbrain (mesencephalon),* and *hindbrain (rhombencephalon),* respectively. The remaining portion of the neural tube, which fails to enlarge at the same rate as the brain, becomes the spinal cord, or *myelon.*

Subsequent changes in prosencephalon and rhombencephalon result in a further division of these two portions of the brain, but the mesencephalon undergoes no division. It gives rise to the *optic lobes*. Paired outgrowths from the anterior end of the prosencephalon give rise to the *telencephalon,* destined to form the *cerebral hemispheres;* the remainder constitutes the *diencephalon* (or *thalamencephalon*), which becomes the *'tween brain*. The anteroventral portion of the telencephalon later becomes modified to form the *olfactory lobes*, or

rhinencephalon. A dorsal projection of the *metencephalon*, or anterior end of the rhombencephalon, gives rise to the *cerebellum,* and the remainder of the hindbrain becomes the *myelencephalon,* or *medulla oblongata* (Figs. 13.2 and 13.3). The floor of the metencephalon in lower vertebrates does not differ from that of the myelencephalon. In higher forms, however, along with the more important role assumed by the cerebellum,

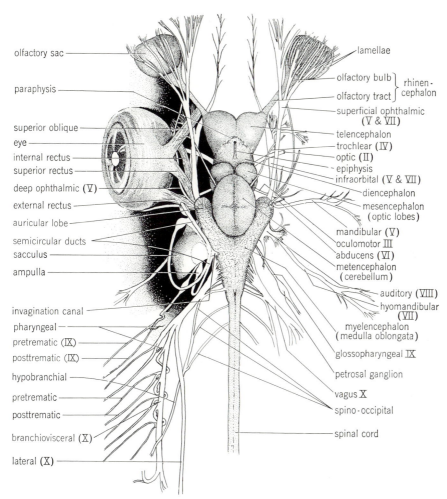

olfactory sac
paraphysis
superior oblique
eye
internal rectus
superior rectus
deep ophthalmic (V)
external rectus
auricular lobe
semicircular ducts
sacculus
ampulla
invagination canal
pharyngeal
pretrematic (IX)
posttrematic (IX)
hypobranchial
pretrematic
posttrematic
branchiovisceral (X)
lateral (X)

lamellae
olfactory bulb ⎫
olfactory tract ⎭ rhinencephalon
superficial ophthalmic (V & VII)
telencephalon
trochlear (IV)
optic (II)
epiphysis
infraorbital (V & VII)
diencephalon
mesencephalon (optic lobes)
mandibular (V)
oculomotor III
abducens (VI)
metencephalon (cerebellum)
auditory (VIII)
hyomandibular (VII)
myelencephalon (medulla oblongata)
glossopharyngeal IX
petrosal ganglion
vagus X
spino-occipital
spinal cord

Fig. 13.2 Dorsal view of brain, sense organs, and cranial nerves of *Squalus acanthias*.

Nervous System

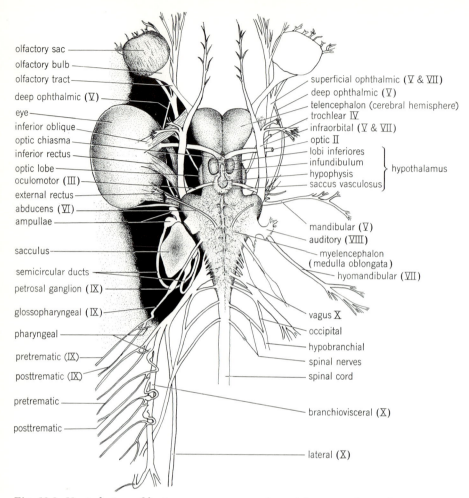

olfactory sac
olfactory bulb
olfactory tract
deep ophthalmic (V)
eye
inferior oblique
optic chiasma
inferior rectus
optic lobe
oculomotor (III)
external rectus
abducens (VI)
ampullae

sacculus

semicircular ducts
petrosal ganglion (IX)

glossopharyngeal (IX)

pharyngeal

pretrematic (IX)

posttrematic (IX)

pretrematic

posttrematic

superficial ophthalmic (V & VII)
deep ophthalmic (V)
telencephalon (cerebral hemisphere)
trochlear IV
infraorbital (V & VII)
optic II
lobi inferiores
infundibulum   } hypothalamus
hypophysis
saccus vasculosus

mandibular (V)
auditory (VIII)
myelencephalon
(medulla oblongata)
hyomandibular (VII)

vagus X
occipital
hypobranchial
spinal nerves
spinal cord

branchiovisceral (X)

lateral (X)

Fig. 13.3 Ventral view of brain, sense organs, and cranial nerves of *Squalus acanthias*.

the floor in this region becomes thickened by the development of fiber tracts, and in mammals forms the conspicuous *pons*. The posterior end of the myelencephalon is continuous with the spinal cord.

SUMMARY

I. PROSENCEPHALON (forebrain)
   A. Telencephalon
      1. Rhinencephalon (olfactory lobes)
      2. Cerebral hemispheres
   B. Diencephalon ('tween brain)

II. MESENCEPHALON (midbrain)

III. RHOMBENCEPHALON (hindbrain)

    A. Metencephalon (cerebellum, and pons if present)

    B. Myelencephalon (medulla oblongata)

IV. MYELON (spinal cord)

Cavities of Brain and Spinal Cord

With the differentiation of the neural tube into brain and spinal cord, its original cavity becomes modified to form the ventricles of the brain and the central canal of the cord (Fig. 13.4). When the telencephalon develops into the paired cerebral hemispheres, the cavity extends into those structures as the *lateral ventricles,* or *ventricles I* and *II.* The cavity of the diencephalon is then known as the *third ventricle.* Each lateral ventricle communicates with the third ventricle by means of an opening, the *interventricular foramen,* or *foramen of Monro.* In higher vertebrates a narrow canal extends posteriorly from the third ventricle through the mesencephalon and is referred to as the *cerebral aqueduct,* or *aqueduct of Sylvius.* In lower forms the cerebral aqueduct may be considerably expanded on either side and the term *optic ventricles,* or *mesocoele,* is then applied. Posteriorly the cerebral aqueduct communicates with the enlarged cavity of the rhombencephalon, or the *fourth ventricle.* In numerous lower forms the anterior part of the fourth ventricle becomes modified into a chamber, the *metacoele,* extending into the cerebellum. The portion of the fourth ventricle within the medulla oblongata is then referred to as the *myelocoele.* It is continued posteriorly as the *central canal* of the spinal cord.

As development of the brain continues, its walls become thickened with nervous tissue except in two regions where only the epithelial ependymal layer is retained. These regions include the anterior part of the roof of the diencephalon and the roof of the medulla oblongata. In each region a fusion of the roof with a highly vascular membrane, the *pia mater* (page 617), forms a *tela choroidea.* Folds from the tela choroidea project into the third and fourth ventricles and are known, respectively, as the *anterior* and *posterior choroid plexuses.* The ependymal layer lining the choroid plexus region is referred to as the *choroid plexus epithelium.*

Lymphlike cerebrospinal fluid is contained within the cavities of brain and spinal cord and circulates through the ventricles and central canal. It passes out of these cavities through openings, to be mentioned later, to bathe the entire central nervous system. The choroid plexuses are important in forming the cerebrospinal fluid.

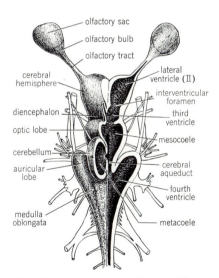

Fig. 13.4 Dorsal view of brain of dogfish with part of wall removed to show cavities of the brain.

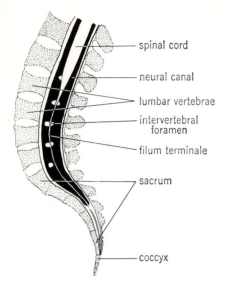

spinal cord

neural canal

lumbar vertebrae

intervertebral foramen

filum terminale

sacrum

coccyx

Fig. 13.5 Diagram of sagittal section of lower end of vertebral column of man, showing its relation to spinal cord and filum terminale.

The Spinal Cord

The portion of the neural tube which forms the spinal cord undergoes considerably less modification than that forming the brain. It generally assumes the shape of a more or less cylindrical, but slightly flattened, tube. It widens at the anterior end, where it is continuous with the medulla oblongata. The posterior end usually tapers down to a fine thread, the *filum terminale* (Fig. 13.5).

In cyclostomes and fishes, the spinal cord is of fairly uniform diameter throughout its length, but in most tetrapods two conspicuous swellings, or enlargements, occur at points where the nerves going to the limbs arise. Since, with the development of limbs, the nerve supply to these structures is increased, the enlargements are actually places in which greater numbers of nerve-cell bodies are situated than are present at other levels of the cord. The *cervical enlargement* is the

more anterior of the two. Here the white matter also shows an increase. It is the region where the large nerves supplying the forelimbs arise. The *lumbar enlargement,* near the posterior end of the spinal cord, marks the point of origin of nerves supplying the hind limbs. The difference in degree of development of forelimbs and hind limbs in various tetrapods accounts for the fact that in some forms the cervical enlargement exceeds the lumbar enlargement in size, whereas in others, the reverse condition obtains. In such limbless forms as snakes, neither enlargement is present. It is probable that in certain extinct dinosaurs, which possessed enormous limbs but had relatively little cranial capacity, the nervous tissue composing the cervical and lumbar enlargements exceeded that of the brain itself in size. Casts of the neural canal of the dinosaur *Stegosaurus* indicate that the lumbar enlargement may have been twelve times as large as the brain.

The spinal cord in higher forms (Fig. 13.6) has a conspicuous *ventral fissure* on its undersurface. A slight median depression, or *sulcus,* may be present on the dorsal side. From the sulcus a *dorsal septum* extends toward the interior of the cord. The dorsal septum and ventral fissure incompletely divide the spinal cord into symmetrical halves, connected across the middle by *commissures* of nervous tissue. The central canal, lined with a single layer of ependymal cells, is very small and lies in the center of the nervous mass connecting the two halves of the cord.

Length of Cord. At first, during embryonic development, the spinal cord and vertebral column are of approximately the same length. During subsequent development in many vertebrates, the growth of the spinal cord fails to keep pace with that of the

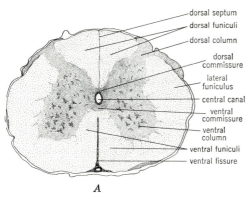

dorsal septum
dorsal funiculi
dorsal column
dorsal commissure
lateral funiculus
central canal
ventral commissure
ventral column
ventral funiculi
ventral fissure

A

B

Fig. 13.6 Cross sections of spinal cords of *A*, cat, and *B* frog, shown at same magnification. Gray matter is heavily stippled; white matter is lightly stippled.

vertebral column. As a result, the spinal cord of the adult may be considerably shorter than the backbone. In man it averages slightly less than 18 in. in length and reaches down only to the upper border of the second lumbar vertebra. The filum terminale, however, continues on to the first segment of the coccyx. In other vertebrates, the spinal cord may be relatively much longer than in man, but within the entire group there is a very definite tendency toward a reduction in length.

Gray and white matter of cord. In cross section the spinal cord is seen to be composed of gray and white matter, the former being almost completely surrounded by the latter (Fig. 13.6). The gray matter in amniotes is arranged somewhat in the form of a butterfly, or in the shape of the letter H. The portions corresponding to the upper bars of the H extend dorsally and are known as *dorsal columns.* The lower bars are the *ventral columns,* and the connecting bar, in which the central canal lies, forms the *dorsal* and *ventral gray commissures,* above and below the central canal, respectively. Considerable variation exists in the arrangement

of the gray matter within the spinal cords of different vertebrates, the H-shaped distribution being more typical of higher forms. The small nerve-cell bodies in the dorsal columns are, for the most part, those of *association neurons,* or *interneurons.* Their dendrites form synapses with the axons of sensory, or afferent, nerve fibers which enter the spinal cord via the dorsal roots of spinal nerves. The axons of the association neurons, which are of variable length, course up or down the spinal cord, cross to the opposite side, or pass ventrally on the same side, to form synapses with the dendrites of motor, or efferent, neurons, the very large cell bodies of which are located in the ventral columns. The axons of the motor neurons in higher forms typically emerge via the ventral roots of the spinal nerves (Fig. 13.18*B*), but in lower forms some may emerge through the dorsal roots as well. It is the cell bodies of these efferent motor neurons which are frequently destroyed in the disease known as infantile paralysis (anterior poliomyelitis). In such cases there results a paralysis of the muscles which would normally be supplied by the axons of these cells.

Neurologists have discovered that the

cells in the dorsal and ventral columns show a rather definite arrangement into areas, each of which is associated with a certain type of function. All functions carried on by the body can be resolved into two main categories, *somatic* and *visceral*. Somatic functions are those carried on by the skin and its derivatives, the voluntary musculature and skeletal structures. Visceral functions are those performed by the other organ systems of the body, i.e., digestive, respiratory, excretory, reproductive, endocrine, and circulatory systems. Somatic sensory fibers carry impulses from the somatic tissues to the central nervous system. They form synapses with cells in the *upper* portion of the dorsal columns. Visceral sensory fibers are those

carrying impulses from the visceral organs. They form synapses with cells in the *lower* portion of the dorsal columns. The motor neurons, the cell bodies of which are located in the ventral columns, are likewise of two types, somatic and visceral. The cell bodies of somatic motor neurons are located primarily in the *lower* portions of the ventral columns, whereas visceral motor neurons have their origin in the *upper* and *lateral* portions of the ventral columns. There are thus four areas in the gray matter on each side of the spinal cord, arranged in a dorsoventral sequence as follows: somatic sensory, visceral sensory, visceral motor, somatic motor (Fig. 13.7*A*). The visceral areas are, in general, smaller than the somatic

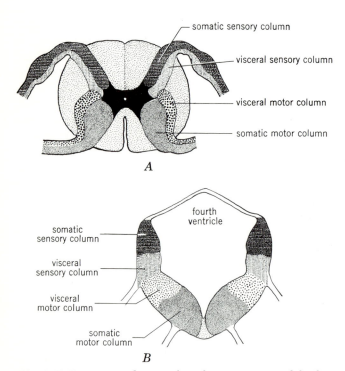

Fig. 13.7 Diagrams indicating the relative positions of the four columns of gray matter on each side of A, the spinal cord, and B, the medulla oblongata.

areas. A similar arrangement is present in the gray matter of the medulla oblongata (Fig. 13.7B).

The white matter of the cord is arranged in longitudinal columns called *funiculi* which lie outside the gray matter. The funiculi are in turn divided into *fasciculi*, or *fiber tracts*, composed of medullated fibers that carry impulses up and down the cord and to and from the brain. A *lateral funiculus* is present on each side between dorsal and ventral columns. A *dorsal funiculus* lies between the dorsal septum and the dorsal column on each side, and a ventral funiculus is located between the ventral fissure and the ventral column of gray matter. The two ventral funiculi are in communication through the *white commissure* which lies just below the ventral gray commissure. This serves as a bridge between the white matter of the two sides.

It is beyond the scope of this book to discuss the detailed arrangement of fiber tracts in which nerve fibers ascend and descend the spinal cord. Suffice it to say that those in the dorsal funiculi, for the most part, carry sensory nerve impulses *up* the cord and *to* the brain. Those in the ventral funiculi are primarily motor, carrying impulses *down* the cord and *from* the brain. The lateral funiculi carry both ascending (sensory) and descending (motor) fibers. In general, the lower vertebrates do not show so elaborate an organization or arrangement of the columns, funiculi, and fasciculi in the spinal cord as do higher forms.

Comparative anatomy of the spinal cord.
AMPHIOXUS. The spinal cord of amphioxus is of uniform diameter throughout except for a gradual tapering toward the end of the tail. It appears to be somewhat triangular in shape when viewed in cross section. The central canal is slit-shaped. There is no clear-cut distinction between gray matter and white matter, since in amphioxus medullated fibers have not as yet made their appearance. Only in a very general way does a cross section of the spinal cord of amphioxus bear a resemblance to that of the typical vertebrate. Nerve-cell bodies, some of exceptional size, lie in the central region in the vicinity of the central canal. The canal is not lined with a continuous layer of ependymal cells. Some of the giant cells even cross the slitlike central canal.

CYCLOSTOMES. A truly tubular spinal cord first appears in cyclostomes. The cord is considerably flattened and is ribbon-shaped (Fig. 8.27). The dorsal side is more or less convex, and the ventral side concave. Of fairly uniform diameter, the cord tapers gradually toward the posterior end. Gray and white matter are not sharply differentiated, the gray matter being in the form of a broad, longitudinal band surrounded on all sides by white matter, except where the roots of spinal nerves emerge. Dorsal septum and ventral fissure are lacking. In the lamprey a mass of tough, pigmented, fibrous tissue lies above the spinal cord, filling in the spaces between the incomplete neural processes and completing the enclosure of the neural canal dorsally and laterally. No other special function has been demonstrated for this peculiar fibrous tissue of the neural canal.

FISHES. A dorsal fissure is present in the spinal cords of most fishes, but a ventral fissure is absent. Gray matter is arranged in the form of a triangle, the apex of which points dorsally. Paired ventral columns of gray matter are present for the first time. In most fishes the cord tapers gradually to the end of the tail, but in certain teleosts there is a marked shortening and the spinal cord terminates at some distance from the end of the vertebral column. Thus the position of the posterior end is considerably anterior to the parts of the body which it supplies with nerves. The spinal cord of the marine sunfish, *Orthagoriscus*, has been reduced to almost ridiculous proportions. In this large fish, which may reach a length of 8 ft and weigh a ton or more, the spinal cord is less than ¾ in. long and is even shorter than the brain.

AMPHIBIANS. In salamanders the spinal cord extends to the posterior end of the vertebral column, but in frogs and toads it has been considerably shortened. The filum terminale in these forms extends into the urostyle. Cervical and lumbar enlargements appear for the first time, since these are the first forms in which the appendages have been modified into true limbs. A dorsal sulcus may be present, and a ventral fissure appears for the first time. Gray matter may appear in the form of an oval in cross section, but dorsal and ventral columns are usually present (Fig. 13.6B) and the beginning of the typical H-shaped arrangement found in higher forms is apparent. Fasciculi, or fiber tracts, in the white matter are more clearly indicated than in lower forms.

REPTILES. The reptilian spinal cord extends the entire length of the vertebral column. Cervical and lumbar enlargements are absent in snakes and limbless lizards but are present in other forms. They seem to be unduly conspicuous in turtles, however. The trunk muscles in turtles have been reduced, and their nerve supply is correspondingly less. Therefore, the portion of the spinal cord between the two enlargements is more slender than usual. The H-shaped arrangement of gray matter is typical of the condition found in other amniotes.

BIRDS. The spinal cord of birds extends the full length of the vertebral column, and a filum terminale is absent. The most conspicuous feature which distinguishes the spinal cord of birds from that of other vertebrates is the separation of the two halves of the cord on the dorsal side of the lumbar enlargement. An elliptical space, the *sinus rhomboidalis*, is present, filled with neuroglia cells of a gelatinous character. These make up the so-called *glycogen body*. The relative sizes of cervical and lumbar enlargements depend upon the degree of development of wings and legs. The ostrich is notable for the large size of its lumbar enlargement.

MAMMALS. In the duckbill platypus and a few rodents, the spinal cord extends to the sacral region of the vertebral column, but in other mammals it has been shortened. Even in mammals with well-developed tails, the cord fails to extend into the caudal region. Such tails are moved primarily by means of longitudinal muscles and ligaments which are innervated by nerves at their bases. In addition to the median dorsal sulcus, dorsolateral and ventrolateral sulci are frequently present. The ventral fissure is conspicuous. A marked discrepancy exists in the size of cervical and lumbar enlargements in bats, which have an unusually large cervical enlargement associated with the large wing muscles used in flight.

The Brain

With the evolution and development of bilateral symmetry in the animal kingdom and the necessity that one end of an animal precede the other in locomotion, the end first coming in contact with the environment developed a concentration of nervous tissue as well as sense organs. These structures better enabled animals to cope with their environment in such a manner that they could survive. The term *cephalization* is used in referring to such localization of structures and functions in the head region.

The brain of chordates, which is basically an enlargement of the anterior end of the neural tube, appears in its simplest form in amphioxus where it is but slightly larger in diameter than the spinal cord. In higher forms it becomes more complex as the evolutionary scale is ascended. *Latimeria* has the smallest brain in relation to the size of the animal of any living vertebrate. It lies almost entirely behind the hinge of the skull (page 443) and is much smaller than the braincase in which it lies, embedded in fat and cushioned on a large pair of blood sinuses. A word of caution should be interjected here about judging the size of the brain of fossil forms by the size of the braincase as determined by preparing cranial casts.

In the primitive condition the cell bodies of the neurons comprising the central nervous system are aggregated around the central canal or original cavity of the neural tube. This arrangement persists in the spinal cords of higher forms. In the brain region, however, a movement or migration of cells to peripheral areas has occurred so that gray matter and white matter differ in their spatial relations from the arrangement observed in the spinal cord. The degree of development of the various parts of the brain of vertebrates is correlated with their position in the evolutionary scale and with certain special requirements related to the particular environments in which they live. As the brain has developed it has gradually become the center of control over most body activities, aside from certain simple reflexes (page 657). Various nerve centers in the brain serve to integrate and correlate the impulses brought to it by afferent fibers from all parts of the body.

The three primary divisions of the brain previously mentioned—prosencephalon, mesencephalon, and rhombencephalon —make up what is often referred to as the *brain stem.* Each of these divisions may originally have developed in association with one of the three major sense organs, an arrangement which still persists in all vertebrates. Generally speaking, the sense of smell is related primarily to the prosencephalon, the sense of sight to the mesencephalon, and the sense of hearing to the rhombencephalon. Further developments from the brain stem such as the cerebral hemispheres, roof of the midbrain, and cerebellum appeared later as dorsal outgrowths into which nerve cells migrated, so that gray matter appears in the peripheral areas in these regions.

Such variations from the previously described ground plan of the brain as are to be found in the different classes of vertebrates are the result of certain mechanical factors which should be understood before details of brain structure are considered. These consist of (1) outpocketings or inpocketings (evaginations or invaginations) of various portions of the primitive brain vesicles, (2) folding or plaiting of the walls of different parts of the embryonic brain, thus bringing about an increased surface area of the folded region, (3) thickening of the walls in certain regions as a result of an increase in amount of nervous tissue, (4) failure of the thin-walled embryonic area to differentiate into nervous tissue, and (5) bending or folding of the brain along its longitudinal axis, such bends being called *flexures.*

Flexures. Only in amphioxus are "brain" and spinal cord arranged in a straight line. In vertebrates, however, owing to unequal rates of growth of various components, certain flexures occur during embryonic development which modify the original condition temporarily or permanently. Since the brain lengthens more rapidly than other head structures, the bending is apparently influenced by space limitations.

In all vertebrates a *cephalic,* or *cranial, flexure* occurs in the region of the mesencephalon in such a manner that the derivatives of the forebrain are bent downward at

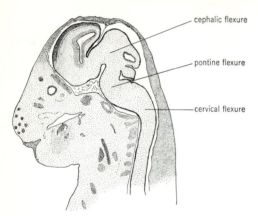

cephalic flexure

pontine flexure

cervical flexure

Fig. 13.8 Sagittal section through head of 18-day rat embryo, showing flexures of the brain.

right angles to the rest. This is more pronounced in some forms than in others and is least conspicuous in fishes. Later there is a tendency for the brain to straighten out once more, and only in birds and mammals, particularly in primates, is the cephalic flexure permanently retained. The second flexure to appear is the *cervical,* or *nuchal, flexure,* occurring near the junction of medulla oblongata and spinal cord. The bend is in the same direction as the cephalic flexure. The third, or *pontine, flexure* is found in the region of the metencephalon and is opposite in direction to the other two (Fig. 13.8). Cervical and pontine flexures practically never even appear in anamniotes but may be quite evident in the embryonic stages of amniotes. However, as development progresses, these two posterior flexures tend to straighten and either diminish or disappear. In birds and mammals the front part of the brain (cerebral hemispheres and diencephalon) comes to lie, at least partially, on top of, or over, the posterior portions.

Gray and white matter of brain. The posterior end of the medulla oblongata merges imperceptibly with the anterior end of the spinal cord. A point just anterior to the first pair of spinal nerves is generally considered to be the level at which brain and spinal cord are joined. The spinal cord lies posterior to the foramen magnum. The structure of the medulla oblongata at its posterior end differs but little from that of the spinal cord. In this region the gray matter is on the inside and the white matter surrounds it peripherally. Farther forward in the brain, gray and white matter lose their original relationship and are more or less irregularly intermingled in a reticular formation. In the cerebral hemispheres, roof of the midbrain (*optic tectum*), and cerebellum, however, the positions of these layers are reversed, so that the gray matter forms a sort of cap over the underlying white matter. Except for the outer layer, or cortex, of gray matter in the regions just mentioned, gray matter, surrounded by white matter, is still present in certain areas of the hindbrain, midbrain, and forebrain. This is roughly similar to that in the spinal cord but does not show the same sort of configuration. Even though there are no dorsal or ventral columns of gray matter in most of the brain, distinct areas, or nuclei, are still present which are functionally designated as sensory or motor, as the case may be. Some are related to certain cranial nerves in much the same manner as the dorsal and ventral columns of gray matter in the spinal cord are related to spinal nerves. There seems, however, to be a greater degree of complexity so far as their functions are concerned. The gray matter of the brain, like that of the spinal cord, consists, in general, of nerve-cell bodies with their dendrites and the proximal portions of their axons, as well as of neuroglia cells of various types. The white matter consists of tracts of myelinated fibers connecting various parts of the brain and of ascending and descending fibers carrying impulses to and from the spinal cord.

Myelencephalon. The lateral and ventral walls of the medulla oblongata thicken markedly, but the dorsal wall retains its epithelial character. The *pia mater,* a vascular membrane surrounding the brain, fuses with this thin layer of ependymal cells, the two together being called a *tela choroidea.* This membrane forms the roof of the large, triangular-shaped fourth ventricle of the brain. The fourth ventricle narrows posteriorly, where it is continuous with the central canal of the spinal cord. Vascular folds of the tela choroidea extend into the fourth ventricle to form the *posterior choroid plexus.* In lampreys the posterior choroid plexus forms an everted sac.

The thickened ventral and lateral walls of the medulla contain large white fiber tracts as well as several columns of gray matter. The gray matter is arranged in columns similar to those in the spinal cord, consisting of somatic sensory, visceral sensory, visceral motor, and somatic motor components (Fig. 13.7B). A longitudinal groove on the inner wall of each side marks the division between dorsal sensory and ventral motor portions. In higher vertebrates the integrity of the columns of gray matter is lost and a separation into various nuclei has occurred.

Cranial nerves V to XII form connections with the medulla. The points of origin of the purely somatic motor cranial nerves VI and XII lie in the most ventral or somatic motor column of gray matter of the medulla. The visceral motor components of the branchial cranial nerves, V, VII, IX, X, and XI, originate in the visceral motor column. The special sensory fibers of nerve VIII and the somatic sensory constituents of nerves V, VII, IX, and X enter the dorsal somatic sensory column. The visceral sensory components of nerves VII, IX, and X are associated with the visceral sensory column.

Cyclostomes, fishes, and amphibians possess only 10 pairs of cranial nerves. In these animals only cranial nerves V to X are associated with the myelencephalon.

Several fiber tracts in the medulla cross over (decussate) to the opposite side of the brain. This crossing occurs in the posterior region of a swollen part of the myelencephalon composed of paired *pyramids.*

The myelencephalon is sometimes referred to as the oldest part of the brain, since it is well developed in all vertebrates, even though other portions may be rudimentary or lacking. It contains important nerve centers which control such vital physiological processes as regulation of heartbeat, respiration, and metabolism. They represent portions of the visceral motor column.

In cyclostomes the medulla oblongata is the only part of the brain that can be said to be really well developed. In these forms, as well as in fishes and urodele amphibians, a pair of giant cells (Mauthner's cells) is located in the medulla. Their axons extend posteriorly to the muscles of the tail and are apparently concerned with execution of muscular contractions involved in swimming.

In some fishes a pair of prominent *vagal lobes* (Fig. 13.9) is present on the sides of the medulla oblongata. They represent enlargements of the visceral sensory area and are centers for the sense of taste, which is very well developed in these fishes. Vagal lobes are not to be found in adult amphibians or in higher vertebrates.

The dorsal anterior portion of the medulla oblongata is referred to as the *acousticolateralis area.* It is actually a development of the somatic sensory column and is continuous with the auricles of the cerebellum. This area, which in some forms is very prominent, contains nuclei associated with nerves from the lateral-line system and the inner ear. With the disappearance of the lateral-line system in terrestrial vertebrates,

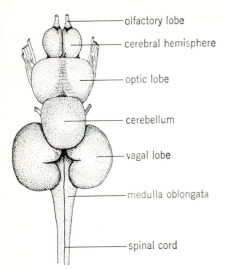

olfactory lobe

cerebral hemisphere

optic lobe

cerebellum

vagal lobe

medulla oblongata

spinal cord

Fig. 13.9 Dorsal view of brain of buffalofish, *Carpiodes tumidus*, showing conspicuous vagal lobes. (*After Herrick.*)

the area contains nuclei concerned only with the equilibratory and auditory functions of the ear.

Metencephalon. The dorsal part of the metencephalon, contrary to that of the myelencephalon with its thin epithelial covering, becomes the elevated and thickened cerebellum. The function of this portion of the brain is to coordinate the neuromuscular mechanism of the body. It is concerned with the proprioceptive muscle sense and with the sensory nerves connected with the equilibratory portion of the inner ear. One finds, as would be expected, that the cerebellum is much more highly developed in animals that are active, whether such activity occurs in water, on land, or in the air. In cyclostomes, sluggish fishes, and amphibians the cerebellum is poorly developed and consists of little more than a transverse shelf or ridge bordering the anterior edge of the tela choroidea of the myelencephalon. In such active

fishes as most elasmobranchs and certain teleosts, the cerebellum becomes a prominent structure (Fig. 13.2) extending dorsally and partially overlapping the mesencephalon anteriorly and the medulla oblongata posteriorly. Its cavity, the *metacoele,* or *cerebellar ventricle* (Fig. 13.4), connects below by a narrow aperture with the myelocoele near the point where the latter is joined by the cerebral aqueduct. A shallow, median, longitudinal depression on its dorsal surface, as well as a transverse groove, gives a lobed appearance to the cerebellum in certain elasmobranchs. The ventral portion of the metencephalon in the lower vertebrates is composed of heavy fiber tracts which merge with those of the medulla oblongata.

Prominent, irregular projections called the *auricular lobes,* or *restiform bodies,* continuous with the medulla oblongata, are actually parts of the metencephalon although some authorities consider them to belong to the medulla. They are situated at the anterior lateral angles of the medulla in certain fishes (Elasmobranchii, Chondrostei, Dipnoi, and a few others). Each auricle contains a cavity, or *auricular recess,* continuous with the fourth ventricle. The auricles are centers of equilibration. They serve to correlate muscular movements of the body in relation to impulses arising in the sensory nerves connected with the inner ears. The inner ears of these fishes are primarily organs of equilibration.

The cerebellum of reptiles is not so highly developed as that of elasmobranch fishes. It is rather simple in structure, in keeping with the sluggish life and poikilothermous nature of these animals. Swimming reptiles have a better-developed cerebellum than do others. The greatest advance in the reptilian cerebellum is seen in such higher forms as crocodilians, in which a pair of small, lateral, *floccular lobes* first appears near the ventral side of the cerebellum. These correspond to the auricular lobes, or restiform bodies, of the fishes mentioned above.

In birds, the cerebellum shows considerable advance over the reptilian condition. It consists of a prominent middle portion called the *vermis,* which is indistinctly divided into anterior, middle, and posterior locations. The vermis bears a number of transverse furrows. Lateral floccular lobes are much more highly developed than those of reptiles and lie on either side of the vermis. The cerebellum of birds extends quite far posteriorly and entirely covers the tela choroidea of the medulla oblongata. The *cerebellar cortex,* or gray matter of the cerebellum, covers the white matter which, in birds, shows a complexly branched arrangement called the *arbor vitae.* A small cavity, a projection of the fourth ventricle, extends into the cerebellum. In some birds there appears for the first time a sort of bridge of nerve fibers called the *pons* on the ventral side of the metencephalon and below the medulla oblongata. The fiber tracts of the pons connect the

cerebral cortex with the cerebellum. They decussate, crossing to the opposite side, as they pass to the cerebellum.

The mammalian cerebellum (Figs. 13.10 and 13.11) is more highly developed than that of other vertebrates. The middle portion, or vermis, is divided into anterior, middle, and posterior lobes. The anterior and posterior lobes are entirely median, but the middle lobe has bilateral extensions, the *cerebellar hemispheres.* The floccular lobes on either side of the posterior lobe of the vermis are much larger than those of reptiles and birds. The whole surface of the cerebellum is thrown into numerous folds, or *gyri,* separated from one another by deep grooves, or *sulci.* Branches of the arbor vitae extend into the cerebellar folds. The ventral part of the mammalian metencephalon is marked by a conspicuous *pons.* This is composed of a prominent mass of transverse nerve fibers, conspicuously arched. In its center is a median longitudinal groove, the *basilar sulcus,* where the basilar artery lies. In mammals, control of body movement has been

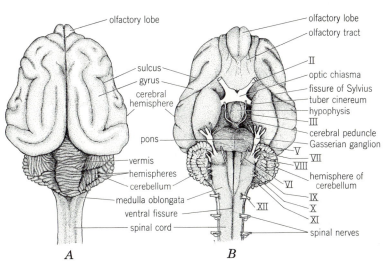

Fig. 13.10 Brain of cat from *A,* dorsal, and *B,* ventral, aspects.

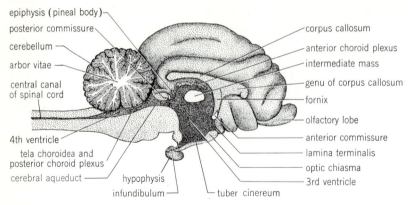

epiphysis (pineal body)
posterior commissure
cerebellum
arbor vitae
central canal of spinal cord
4th ventricle
tela choroidea and posterior choroid plexus
cerebral aqueduct
hypophysis
infundibulum
tuber cinereum
corpus callosum
anterior choroid plexus
intermediate mass
genu of corpus callosum
fornix
olfactory lobe
anterior commissure
lamina terminalis
optic chiasma
3rd ventricle

Fig. 13.11 Diagram of sagittal section of brain of cat. (*After Reighard and Jennings, "Anatomy of the Cat," Holt, Rinehart and Winston, Inc. By permission.*)

shifted to the cortex of the highly developed cerebral hemispheres. Motor fibers from the cerebral cortex pass to the cerebellum via the pons. The ventral portion of the pons consists mostly of nerve fibers interspersed with small amounts of gray matter; its dorsal portion is largely a reticular formation continuous with that of the myelencephalon and containing several large nuclei. In lower mammals the cerebellum shows a simpler form of organization than does that of higher forms. Advancement is shown primarily by an increase in number and complexity of the cerebellar folds.

The cortex of the cerebellum consists of three layers: (1) an outer, or *molecular, layer,* with few cells but numerous gray, sparsely medullated fibers; (2) a middle, or *intermediate, layer* in which lie large, complexly branched *Purkinje cells* (Fig. 13.12), characteristic of the minute structure of the cerebellar cortex; and (3) an internal *granular,* or *nuclear, layer* composed of numerous cell bodies of small nerve cells. The Purkinje cells are believed to play an important part in the process of correlating nerve impulses with equilibratory muscular movements.

Mesencephalon. The embryonic midbrain undergoes relatively less differentiation than other portions of the brain. It is marked off from the hindbrain very early in development by a conspicuous constriction, the *isthmus.* The floor and walls of the mesencephalon

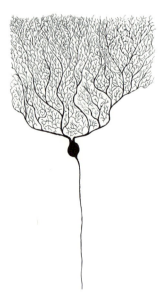

Fig. 13.12 Purkinje cell from cerebellar cortex.

Anatomy of the Chordates

are thick and composed of fiber tracts, the *cerebral peduncles,* connecting forebrain and hindbrain. The roof consists of a thick layer of gray matter, the *optic tectum.* Researches show that fibers from a particular quadrant of the retina of the eye grow preferentially toward specific areas of the tectum (G. R. DeLong and A. J. Coulombre, 1967, *Develop. Biol.,* **16:**513). The organizational factors controlling this specificity have not as yet been elucidated. In the optic tectum the nerve cells become organized in layers somewhat similar to those of the cerebellar cortex. In lower vertebrates two dorsal prominences develop in the roof of the mesencephalon. They are the *optic lobes,* or *corpora bigemina,* which serve as centers for the visual sense. In lower forms each optic lobe contains a large cavity, the *optic ventricle,* the two ventricles together forming the *mesocoele* (Fig. 13.4). In higher forms, however, the lobes are practically solid structures and only the narrow cerebral aqueduct passes through the center of the mesencephalon. A definite transverse fissure divides the optic lobes of snakes and mammals into four prominences, the *corpora quadrigemina.*

As the name implies, the optic lobes are primarily receptive centers for sensory nerve impulses coming from the eye. When corpora quadrigemina are present, as in mammals, the anterior pair, called the *superior colliculi,* contains receptive centers for the visual sense and the posterior pair, or *inferior colliculi,* serves to integrate auditory impulses. The superior colliculi in mammals have become less important as visual centers with the great development of the cerebral cortex, which seems to have taken over much of the integration and coordination of visual impulses.

At the anterior end of the roof of the mesencephalon is located the *posterior commissure,* consisting of fiber tracts which connect the two sides of the brain. Some of the fibers are probably derived from the diencephalon.

Nuclei for the oculomotor, or third cranial nerve, lie in the floor of the mesencephalon near the midventral line. They are actually portions of the somatic motor column of gray matter which is continued forward into the midbrain. Nuclei for the trochlear, or fourth, cranial nerve are more posteriorly located and lie near the junction of mesencephalon and metencephalon. The fibers of the trochlear nerve course through the lateral walls of this portion of the brain to emerge from the dorsal side. The oculomotor and trochlear nerves supply certain muscles which move the eyeball (pages 515 to 516).

The optic lobes are more conspicuous features of the brains of lower vertebrates than of higher forms. In birds, however, the lobes are unusually well developed in correlation with the highly organized visual-sensory system which is so necessary to these animals in flying and feeding.

It is noteworthy that a vascular choroid plexus is present in the middorsal portion of the mesencephalon in cyclostomes.

Diencephalon. The 'tween brain is bounded posteriorly by the posterior commissure, mentioned above, and anteriorly by a thin, membranous structure, the *lamina terminalis.* An *anterior commissure,* connecting the olfactory regions of the brain, is located in the lamina terminalis. The latter is usually considered to belong to the telencephalon. Its ventral termination is marked by a small recess just anterior to the optic chiasma. A fold, the *velum transversum,* marks the anterior end of the diencephalon on the dorsal side of the brain. It will be recalled that the third ventricle lies within the diencephalon. The optic vesicles, from which the sensory portions of the eyes develop (page 670), arise from the embryonic diencephalon. The original diencephalon undergoes considerable modification. It is composed of a dorsal *epithalamus*, a middle, very much thickened *thalamus*, and a thinner ventral portion, the *hypothalamus*. The thalamus in turn is divided into dorsal and ventral regions.

EPITHALAMUS. The anterior portion of the dorsal roof plate covering the third ventricle retains its original epithelial, ependymal character and, together with the pia mater, forms a tela choroidea similar to that covering the fourth ventricle. Vascular folds of the tela choroidea extend into the third ventricle forming the *anterior choroid plexus*. Portions of the plexus may pass through the interventricular foramina into the lateral ventricles of the cerebral hemispheres. During early development, a prominent irregular fold of the tela choroidea, the *paraphysis*, extends dorsally from its very anterior end, overlapping the cerebral hemispheres to some extent. The paraphysis is actually a part of the roof of the telencephalon since it lies anterior to the velum transversum. Its cavity, however, is considered to be a part of the third ventricle. The function of the paraphysis is unknown. It usually disappears in adult life. It has been demonstrated that in the salamander, *Ambystoma mexicanum*, the paraphysis produces glycogen which passes into the ventricular cerebrospinal fluid (A. Kappers, 1956, *Anat. Record*, **124**:144). In certain lower vertebrates large external folds of the anterior choroid plexus may be present in addition to the paraphysis.

Posterior to the tela choroidea the roof of the diencephalon consists largely of what is sometimes referred to as the *epiphysial apparatus*. This is primitively composed of two median, dorsal projections, an anterior *parapineal*, or *parietal*, *body* and a posterior pineal body, or *epiphysis*. In the most prim-itive condition these are separate outgrowths of the diencephalic roof, arranged in tandem. In some forms the parapineal body arises as an anterior outgrowth of the pineal body. It has long been suspected that these structures may originally have consisted of a bilateral pair which in the course of evolution became shifted so that one is now situated in front of the other. There is no very clear-cut evidence for this although in the Australian lamprey, *Geotria*, the two structures lie next to each other and are associated during embryonic development with the right and left habenular nuclei (page 624), respectively. Paired indentations observed in the skulls of some fossil fishes also give indirect evidence of the former existence of a bilateral pair of structures. Although both parapineal and pineal bodies are present in lampreys, some fishes, frogs, *Sphenodon*, and numerous lizards, only the pineal body has persisted in most fishes, urodeles, many reptiles, birds, and mammals. In some urodeles and birds a rudimentary parapineal body seems to be attached to the anterior end of the pineal body proper. The parapineal body, when well developed, forms a small, median, eyelike structure. In many lizards a small lens and retina are even present at the distal end of the parapineal stalk. The highest degree of development of this organ occurs in the peculiar "ancient" reptile, *Sphenodon punctatum*. An *interparietal foramen* in the skulls of these and certain other forms provides a passageway for the nerve connecting the median eye and brain. In no case is such an eye so well developed as to be comparable to the usual vertebrate eye. The *brow spot*, or *frontal organ*, commonly observed in the frog by students in elementary zoology courses, marks the location of the parapineal organ. It apparently is a vestige of what may at one time have been a third, median eye.

The pineal body, which lies posterior to the parapineal organ when the latter is present, is usually considered to represent the remnant of another eyelike structure which, together with the parapineal eye, presumably formed a second, but median, pair of visual organs. Only in lampreys, however, are *both* parapineal and pineal organs associated with such structures. The pineal body is present in all vertebrates and generally appears to be glandular in nature. A great many speculations, supported by experimental evidence of one kind or another, have been made in the past as to the possible function of the pineal body. Reports that a condition known as precocious puberty (macrogenitosomia praecox) is associated with tumors of the epiphysis or near the epiphysis gave rise to the belief that the organ was of an endocrine nature related in some way to the gonads. A tremendous amount of research has been directed toward elucidating such a relationship, but there has been, until fairly recently, scant evidence that the pineal body is actually an endocrine gland or that it belongs to the endocrine system. Numerous other lines of research have been followed in an attempt to determine a possible function of this structure, but conclusions have been vague. As a result, many investigators came to the conclusion that the pineal body, at least in higher forms, might be nothing more than an interesting vestigial structure of no functional significance and turned their interests in other directions. Even though the precise physiological role played by the organ is still rather uncertain, studies, using the electron microscope, indicate that it is a functional structure which in various vertebrates may exert its effect in numerous ways (D. E. Kelley, 1962, *Amer. Scientist*, **50**:597).

It seems that in lower vertebrates the parapineal and pineal organs contain photoreceptive elements, not necessarily organized in the form of eyelike organs, whereas in higher forms such elements are lacking and the pineal body has become a parenchymal mass of tissue in direct contrast to that of lower forms. The electron microscope has indicated that in the parapineal organs of lampreys, tadpoles, and lizards there are numbers of cells which are very similar to the rod and cone cells in the retinae of normal eyes. This finding, together with the results of experiments involving removal of the organs, exposure to or withdrawal of light, etc., in which altered behavioral patterns have been observed, indicates clearly and almost unmistakably that the organ is responsible for certain photoreceptive processes.

The practically solid mass of tissue making up the pineal body of higher forms is composed of parenchymal cells arranged in groups, together with neuroglial cells of one type or another. No cells of the photoreceptor type have been detected in birds or mammals. The nerve supply of the parenchymal type of pineal organ also differs somewhat from that of the photoreceptive type. A high degree of secretory activity is indicated, nevertheless, by electron microscopic studies which show that such cytoplasmic structures as mitochondria, other kinds of organelles, tubules, etc., undergo marked changes under various conditions. There is every indication that a high degree of physiological activity is going on in such cells and it is, therefore, difficult to consider these organs as inactive, nonfunctional vestiges.

It is quite possible that in the course of evolution there has been a change from the primitive photoreceptor type of organ, which in one way or another can translate photic stimuli into physiological controls of different types, to the type of secretory organ found in higher forms which, because of greater brain development and capacities, can carry out similar functions in response to stimuli affecting normal optic pathways.

Leghorn hens exposed to light have significantly larger pineal bodies and reproductive organs than do those receiving no light. Furthermore, their enzyme systems show a heightened activity (C. M. Winget, 1966, *Amer. Zoologist*, **6**:506). The effect of light on the reproductive cycles of rodents, as indicated in numerous experiments, may in some manner be mediated via the pineal body. Cytological and other changes detected in the pineal bodies of certain rodents following variations in exposure to light lend further credence to this idea. Pineal bodies of dark-exposed or blinded black rats secrete a substance which inhibits normal functioning of the endocrine system. Adrenal weights are reduced, and reproduction is inhibited. Primary effects are on the release of LH and on the growth of the pituitary gland (R. J. Reiter and R. J. Hester, 1966, *Amer. Zoologist*, **6**:313). Evidence has also been presented that the pineal body in the killifish, *Fundulus*, is light-sensitive (P. K. T. Pang, 1965, *Amer. Zoologist*, **5**:682). These recent findings imply that there may be a common embryonic origin of the photoreceptor types of cells, the parenchymal cells, and the neuroglial cells with which they are associated. A common origin of all, from embryonic ependymal cells, has been demonstrated.

It is of further interest that a tryptamine derivative, named *melatonin*, can be extracted from the pineal body. This brings about pigment concentration in melanophores of the frog and has been shown to inhibit the growth of ovaries and the occurrence of estrus in rats.

A pair of *habenular bodies* lies just in front of the pineal body in the roof of the diencephalon. These structures are connected by means of a delicate *habenular commissure.*

The epithalamic portion of the diencephalon is of relatively little significance as a nerve center.

THALAMUS. The thickened lateral portion of the diencephalon, known as the thalamus, which is divided into dorsal and ventral parts, contains important relay centers and consists largely of numbers of important nuclei of gray matter. They are integrating centers for impulses passing to and from the cerebral hemispheres. The dorsally located nuclei are concerned primarily with sensory impulses, and the ventral nuclei with motor impulses. Important sensory centers are the *lateral* and *medial geniculate bodies* which relay optic and auditory impulses, respectively. In reptiles the walls of the thalamus are thickened inwardly so as to meet in the center of the third ventricle. This condition is also found in mammals, the mass of gray matter connecting the two sides being termed the *intermediate mass,* or *soft commissure* (Fig. 13.11).

HYPOTHALAMUS. The ventral portion of the diencephalon is of great physiological importance, for both sympathetic and parasympathetic centers are located here (page 655). In other words, it is in the hypothalamus that the functions of the peripheral autonomic system are integrated with those of other nervous tissues of the body. The hypothalamus is the control center for such involuntary mechanisms as water, fat, and carbohydrate metabolism; temperature regulation; genital functions; and the rhythmicity of sleep. Certain peptides, called *release factors* (RF) have been extracted from the hypothalamus which, via the hypophysio-portal vein, stimulate the release of hormones by the anterior lobe of the pituitary gland (page 391). The close relationship between the nervous and endocrine systems, now frequently referred to as the *neuro-endocrine system,* should again be emphasized. In polyestrous species the rhythmic manifestation of reproductive phenomena in the adult female is characteristic. Males do not exhibit periodicity, being at all times in a state of sexual readiness. Since the hypothalamus indirectly brings about the release of gonadotrophic hormones from the anterior lobe of the pituitary gland, it is obvious that the hypothalamus of the female differs from that of the male. That differences in the hypothalami of the sexes are established early in life (in some cases even before birth) has been shown in experiments involving injection of small doses of gonadal hormones shortly before or after birth. Administration of small amounts of testosterone to rats or mice during the first few days after birth has demonstrated a profound effect upon developmental organization and later functioning of the hypothalamus in its control of gonadotrophin production by the anterior lobe of the pituitary gland. Ovarian hormones apparently have no such effect. Hypothalamic differentiation thus seems to be conditioned by the presence or absence of androgen during very early life. Genetic females injected neonatally with androgen upon maturity show wide deviations from the norm in respect to behavioral sexual patterns and rhythmic reproductive phenomena. Even urogenital deformities may result from hormonal imbalance during embryonic life before pertinent brain areas have fully matured. The hypothalamus is a conspicuous feature of the brains of fishes. It is less noticeable in amphibians, but in higher forms it becomes increasingly well developed.

The hypothalamus is composed of four main parts. The *optic chiasma,* or most anterior portion, is the point at which the two optic nerves cross as they pass from the eyes to the brain. Behind the chiasma is the *tuber cinereum* in which are located several nuclei thought to make up the parasympathetic center. Posterior to the tuber cinereum lies a pair of *mammillary bodies* containing nuclei for the integration of the olfactory sense in relation to other parts of the body. The *infundibulum* is an evagination of the hypothalamus at the level of the tuber cinereum. It grows ventrally to establish a connection with a diverticulum (*Rathke's pocket*) which arises from the roof of the embryonic mouth (stomodaeum). From the distal portion of the infundibulum is derived the *posterior lobe,* or *pars nervosa,* of the *pituitary gland.* Rathke's pocket, it will be recalled, gives rise to the anterior and intermediate lobes of that structure. A portion of the third ventricle frequently extends into the infundibulum.

In fishes and some urodele amphibians a pair of *inferior lobes* is present on either side at the base of the infundibulum. In these vertebrates the distal end of the infundibulum expands into a thin-walled, vascular sac, the *saccus vasculosus* (Fig. 13.3), of unknown function. It is most highly developed in deep-sea fishes. Inferior lobes and a saccus vasculosus are absent from the hypothalamic region in higher vertebrates.

Telencephalon. It is in the degree of development of the telencephalon that the greatest differences in the brains of vertebrates are to be found. In higher forms, the cerebral hemispheres, derived from the telencephalon, cover over the greater part of the remainder of the brain. The seat of consciousness lies in the cerebral hemispheres. Here are located the nerve centers controlling the activities which characterize the highly developed psychic life of man, such as intelligence, thought, and sensation.

Early in embryonic life two lateral swellings, or evaginations, appear at the anterior end of the prosencephalon. These grow anteriorly and dorsally. In some animals the two vesicles become fused secondarily along the median line. A longitudinal fissure between the two hemispheres is a conspicuous feature of most vertebrates but is only slightly indicated in fishes. The original anterior end of the neural tube remains practically unchanged in its position and becomes the lamina terminalis, mentioned above, which later marks the posterior border of the telencephalon on the ventral side of the brain. An *anterior commissure,* previously mentioned, connecting the olfactory regions of the two halves of the brain, is later located in the lamina terminalis.

At the anterior end of each hemisphere is an outgrowth called the *olfactory lobe,* or *rhinencephalon.* Projections of the lateral ventricles may or may not extend into the rhinencephalon. The olfactory lobes may come in contact with the posterior part of the olfactory apparatus. In some forms in which the nares are located at some distance from the brain, the olfactory lobes are drawn out to a considerable length. In such cases each lobe becomes subdivided into an anterior, expanded *olfactory bulb,* which borders upon the olfactory apparatus, and a narrow *olfactory stalk,* or *tract,* connecting the bulb with the remainder of the brain (Fig. 13.2). The comparative size of the olfactory lobes is closely correlated with the extent to which the sense of smell is developed.

The anterior paired telencephalic swellings, or cerebral hemispheres, enlarge to an increasing degree as the vertebrate scale is ascended and undergo the greatest change of any of the brain regions. In all vertebrates the floor of each hemisphere early differentiates into a thickened *corpus striatum,* a reticular formation, which receives its name from the fact that it has a striated appearance when viewed in microscopic section. The gray regions of the corpus striatum are often referred to as *basal nuclei.* The function of the corpus striatum is rather obscure. It is poorly developed in cyclostomes. The remainder of the hemisphere consists of a *pallium* which roofs over the lateral ventricle. It is the pallium that has become so highly developed and modified in the evolution of the higher groups of vertebrates (Fig. 13.14). In the lowest living vertebrates, the cyclostomes, each hemisphere is divided into an anterior olfactory bulb and a posterior olfactory lobe, sometimes called the *cerebral hemisphere.* This part of the brain is concerned mainly with receiving olfactory impulses from the olfactory apparatus and relaying them to the diencephalon.

The pallium is at first relatively thin walled, and the gray matter is present only on its inner walls adjacent to the ventricle. In teleost fishes a thin, nonnervous layer forms the roof of the ventricle (Fig. 13.13). Here the gray matter of the pallium has pushed laterally and downward toward the corpus striatum, which is very large and well developed, bulging upward into the lateral ventricle on each side. The cerebral

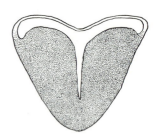

Fig. 13.13 Transverse section through cerebral hemispheres of teleost fish, showing thin-walled roof.

Anatomy of the Chordates

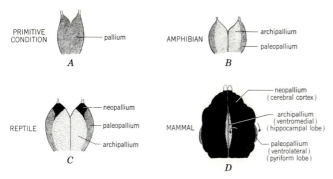

Fig. 13.14 Diagrams indicating evolutionary progress in development of pallium.

hemispheres of teleosts are usually small. The two halves are incompletely separated by a shallow median fissure. In elasmobranch fishes the cerebral hemispheres become more pronounced (Fig. 13.2). A shallow median groove on the dorsal side indicates a partial separation into two halves. The olfactory lobes in these fishes are large and drawn out into bulb and stalk, as previously described. In elasmobranchs and Dipnoi the pallium is fairly thick but remains chiefly related to the sense of smell. The gray matter is still undifferentiated and lines the cavities of the ventricles, white matter forming the outer surface. The corpus striatum is well developed. In fishes, in general, the telencephalon has not progressed beyond serving as an olfactory center. In certain species some cells migrate from the inner gray matter into the outer white region.

As the vertebrate scale is ascended there is an increasing tendency for nerve cells from the inner gray layer to migrate out into peripheral areas. In amphibians the arrangement differs but little from that of fishes except that the pallium is thicker and more cells from the gray matter have moved to peripheral positions. The pallium can now be divided into two general regions, a dorsal,

medial *archipallium* and a more lateral *paleopallium* (Fig. 13.14). The latter represents the olfactory region and lies above and adjacent to the corpus striatum. The archipallium is also still associated with olfactory nerve fibers. The cerebral hemispheres of amphibians are relatively larger than those of fishes. Olfactory lobes merge almost imperceptibly with the anterior ends of the cerebral hemispheres. It is questionable whether the telencephalon in amphibians functions in other than an olfactory capacity.

The first really marked change to be observed in the telencephalon of vertebrates occurs in reptiles. Most conspicuous is an increase in size of the cerebral hemispheres, which have grown backward to some extent and partially cover the diencephalon. The hemispheres are separated by a deep dorsal fissure. The corpus striatum is well developed but has assumed a more medial position. An increased amount of gray matter has migrated toward the peripheral surface. In certain reptiles, however, a new area, the *neopallium*, has appeared on the outer portion of each cerebral hemisphere at its anterodorsal end, between archipallial and paleopallial areas (Fig. 13.14). It is the growth and development of the neopallium that accounts for the large size of the cerebral hemispheres of mammals. In crocodilians, for the first time, nerve cells migrate into the neopallium and become arranged along its outer surface, thus forming a true cerebral cortex which serves as an association center. The surface of the cerebral hemispheres is smooth.

In birds, in correlation with the poorly developed sense of smell, the olfactory lobes are practically rudimentary. Archipallium and paleopallium are present as in reptiles but a neopallium seems to be lacking. There is, therefore, no cerebral cortex. The corpus striatum, or *hyperstriatum*, of birds is unusually large and thick. The cerebral hemispheres in birds are so large, because of the great size of the corpus striatum, that they push back against the cerebellum, thus covering over the diencephalon and optic lobes. Their surface is smooth, and the outer wall rather thin. The actions of birds, in most respects, are instinctive and follow rather stereotyped patterns. There is apparently little ability to reason or to learn. The basal nuclei of the prominently developed corpus striatum of birds are evidently the centers controlling their action patterns, some of which are highly complex.

It is in mammals, particularly in man, that the cerebral cortex reaches the height of its development. The neopallium has grown to a relatively enormous degree (Fig. 13.14), pushing the archipallial area medially and ventrally so that it becomes folded and lies in the ventromedial portion of the cerebral hemisphere, where it is known as the *hippocampal lobe*. This continues to serve as an olfactory center. Likewise, with the development of the neopallium, the paleopallium, originally in a lateral position, is pushed ventrally where it becomes the *pyriform lobe*, a portion of the olfactory lobe. The gray nerve-cell bodies in the neopallium becomes so numerous that they form a layer of gray

matter, the cerebral cortex, in the outer part of the cerebral hemispheres. This layer, however, even in man, where it is most highly developed, is only a few millimeters thick, the thickness varying somewhat in different portions of the cerebral hemispheres. The corpus striatum in mammals is less conspicuous than in some lower forms.

The cerebral cortex is actually composed of six layers of cells, the detailed structure and arrangement of which varies from region to region. As a result of the expansion of the cerebral hemispheres, the number of cells in the cortical layer increases accordingly. It has been estimated that the number of cells in the human cerebral cortex is between 9 and 10 billion in addition to around 100 billion neuroglia cells. The total mass of gray matter is relatively small, nevertheless.

In all vertebrates below mammals, the cerebral hemispheres are smooth. In many mammals, however, the surface becomes folded or convoluted so that ridges and depressions appear (Fig. 13.10). The folding is caused by the more rapid multiplication of cells in the cortex than occurs in the underlying structures. The ridges of the folds are called *gyri*, and the depressions *sulci*. Since the cortical gray matter follows the convolutions, there is a considerable increase in surface area and in the total amount of gray matter. The area of gray matter in the cerebral cortex of man is approximately 18 in. square. The extent to which the cerebral cortex is convoluted is not necessarily a criterion of intelligence, since some apes, with obviously highly developed mental faculties, have fewer convolutions than do certain other mammals with less-well-developed intellects. Large mammals generally have more convolutions than smaller species.

The presence of convolutions serves to divide the cerebral hemispheres into certain regions called *lobes,* among the most prominent of which are the temporal, occipital, frontal, and parietal lobes, named according to their position in relation to similarly named bones of the cranium. Each lobe is secondarily divided into numerous gyri. In the brains of primates the olfactory-lobe region shows a pronounced reduction, correlated with the poorly developed sense of smell of these animals.

Beginning with marsupials, a broad, white mass, the *corpus callosum,* or *anterior pallial commissure,* appears in mammals between the two hemispheres (Fig. 13.11). It is composed of a band of white medullated fibers which connect nearly all parts of the neopallial cortical regions of the two sides. Lying ventral to the corpus callosum is a similar band of white fibers, the *fornix,* which connects the hippocampal lobes with the hypothalamus.

The cerebral hemispheres in man are so large that they cover all the other portions of the brain. A description of their detailed structure is beyond the scope of this book, and special treatises on neurology should be consulted by the interested student. The numerous neurons in the cerebral cortex communicate with one another in an intricate manner by means of their processes. Recent researches indicate that the complex processes which govern behavior are determined by combinations of those nerve cells called *interneurons,* or *association neurons,* which are neither sensory nor motor. Coordinated activity is regulated by these cells acting in groups and chains. One large neuron of this type may have as many as 10,000 synaptic knobs, or points of contact with other neurons. Actually there is a small fluid-filled space between adjacent membranes but this is only from one hundred to a few hundred angstrom units wide. The very numerous surrounding neuroglia cells, suspended in a fluid bath, are most important elements. The fluid between neurons and neuroglia cells is a vital element in the electrochemical reactions between the two. It is now believed that stimulation of a neuron may cause some of the millions of ribonucleic acid (RNA) molecules within the neuron to bring about the manufacture of new proteins by adjacent neuroglia cells. The kinds of proteins and their molecular structure reflect what has been perceived and may be responsible in large part for what we refer to as memory. Only recently has there been any significant progress in identification of brain proteins. The whole subject is most complex, and with modern biological techniques it would seem that we are presently at the threshold of an era which will lead to a real understanding of the very intricate functioning of the nervous system.

Beneath the outer layer of gray matter is the *medulla of the cerebral hemisphere,* composed of white medullated fibers which either originate in cortical cells and carry outgoing impulses from the cortex or else come from other parts of the brain and bring impulses to the cortex. In addition, there are commissural fibers in the cerebral medulla which form complex connections between the two cerebral hemispheres.

LOCALIZATION OF FUNCTIONS. There is no evidence that certain portions of the cerebral hemispheres possess functions which correspond to various faculties of the mind. There is, for example, no particular area for mathematical reasoning, logic, honesty, etc. The practice of phrenology, which seeks to determine facts about the mind and character by studying the contours of the cranium, which fits so closely to the cerebral hemispheres. belongs in the realm of quackery. Much has been learned, however, about localization of various body functions in the cerebral cortex by study of persons who by accident have had portions of the cerebral hemispheres destroyed. The facts learned have been substantiated by experimental removal or electrical stimulation of areas in the cerebral hemispheres of laboratory animals. There is now a large body of evidence showing that there are definite areas in the cerebral cortex which mediate various functions of the body such as speech, sight, smell, movements of certain muscles, etc. Both sensory and motor areas are present. These areas are not clearly circumscribed, however. The different regions are so closely connected anatomically and physiologically that injury to one may bring about changes in others located in a distant portion of the brain. A rather large area of gray matter in the cerebral cortex of the frontal lobe of man, as well as some other areas, is not apparently concerned with motor or sensory functions. It seems that these areas have to do with those higher mental processes that characterize man, such as ability to learn, use of judgment, taking initiative, etc.

Decussation. The presence of a number of commissures in the brain and spinal cord has frequently been alluded to in the previous pages. They serve to connect similar regions in the two sides of the central nervous system and make bilateral integration possible. There are also fiber tracts in the brain which, in their course, cross over, or *decussate*, to the opposite side for no obvious reason.

It is a well-known fact that injury to the brain on one side of the body often results in paralysis of muscles on the opposite side. Impulses set up in certain *pyramidal cells* in the cerebral cortex travel down fibers leading to the nuclei of various cranial and spinal nerves. The fibers which course through the pyramids of the medulla oblongata, as previously mentioned, cross to the opposite side. In other words, injury anterior to the py-ramidal decussation will affect structures on the opposite side of the body. Damage to fiber tracts posterior to the decussation, in the main, affects only the same side of the body as that on which the damage occurred. Conversely, nerve impulses, entering the body, travel up the spinal cord to the brain, crossing over to the opposite side of the body en route.

Meninges. Both brain and spinal cord are surrounded by membranes, or *meninges* (singular, *meninx*), the complexity in arrangement of which increases according to advance in the evolutionary scale. These membranes protect and give some support to the central nervous system.

Cartilage and bone are covered with a tough, vascular membrane known as the *perichondrium* or *periosteum*, as the case may be. The cavities of the cranium, and the neural canal of the vertebral column, are thus lined with perichondrium or periosteum, depending upon whether the skeleton is composed primarily of cartilage or bone. In either case the lining membrane is called the *endorachis*. It is not considered to be a true meninx.

In cyclostomes and fishes (Fig. 13.15A) a single membrane, the *meninx primitiva*, forms a close union with brain and spinal cord, which it covers. Between the meninx primitiva and the endorachis is a space, the *perimeningeal* space, filled with mucoid and fatty tissue. Delicate strands of connective tissue cross the perimeningeal space, connecting the meninx with the endorachis.

Beginning with urodele amphibians (Fig. 13.15B), instead of a single meninx, two more or less distinct layers are present, an inner *pia-arachnoid layer* and an outer *dura mater*. The pia-arachnoid layer apparently has its origin, at least in part, from neural-crest cells, whereas the dura mater is com-

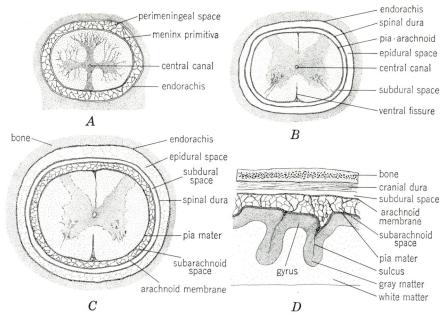

Fig. 13.15 Diagrams illustrating the relations of the meninges to the spinal cord and brain. A and B, cross sections through spinal cords of dogfish and salamander, respectively; C and D, sections of spinal cord and brain of mammal, respectively.

posed mostly of connective tissue. The pia-arachnoid is a very vascular membrane in intimate contact with the surface of brain and cord. Between pia-arachnoid and dura mater is the *subdural space;* the cavity between the dura mater and the endorachis is the *epidural space.* A small amount of fluid is present in the subdural space, but the epidural space is filled largely with alveolar tissue, fat, and numerous veins.

In mammals (Figs. 13.15C and D) the pia-arachnoid membrane differentiates further into two layers, an inner, very vascular *pia mater* and an outer, nonvascular *arachnoid membrane.* A network of fine trabeculae composed of collagenic and elastic fibers joins the arachnoid membrane and pia mater. A *subarachnoid space,* filled with cerebrospinal fluid, makes its appearance between the two membranes. In the brain

region the *cranial dura mater* fuses with the endorachis and the epidural space thus disappears. The *subdural space* between the dura and arachnoid membrane is shallow and contains only a minimum of fluid which is *not* cerebrospinal fluid. Melanocytes are usually present in the portion of the pia mater of mammals on the ventral side of the medulla oblongata.

About the spinal cord a fusion of *spinal dura* and endorachis does not occur, and the epidural space, which thus persists in this region, contains fatty and other connective tissues as well as a number of veins. The cranial dura is continuous with the spinal dura at the foramen magnum.

The cranial dura is a tough, dense membrane loosely attached to the cranium except at the sutures and the basal portion of the skull. The pia mater and arachnoid membrane, known collectively as the *leptomeninges,* are of a loose, delicate consistency. Those of the brain and cord are of similar structure and appearance.

Near the posterior end of the spinal cord the subdural space disappears and the spinal dura forms a close-fitting sheath around the filum terminale. At the points where cranial and spinal nerves emerge, the dura forms sheaths around the nerves and their roots (Fig. 13.16). The sheaths pass through the intervertebral foramina and are continuous with the periosteal covering of the bones. The dura is firmly attached to the margins of the foramen magnum.

The cerebrospinal fluid, which is present in the ventricles of the brain, the central canal of the spinal cord, and the subarachnoid space, and which is similar in composition to tissue fluid, circulates slowly through these various cavities and spaces. Elaborated, for the most part, by the anterior choroid plexus, the fluid passes into the lateral ventricles, through the interventricular foramina to the third ventricle, and thence through the cerebral aqueduct to the fourth ventricle. Three openings in the roof of the medulla oblongata—a single, median *foramen of Magendie* and two lateral *foramina of Luschka*—serve for the passage of

cerebrospinal fluid into the subarachnoid space. Thus the surfaces of brain and cord are constantly bathed by cerebrospinal fluid. The fluid is formed continuously and must be absorbed continuously in order to ensure an even flow and a constant intracranial pressure. Absorption is partially through lymphatic vessels, but most of it passes through small protrusions of the arachnoid membrane, the *arachnoid villi,* which project into the cavities of certain large venous sinuses in the dura mater. Only the thin mesothelial membranes of the arachnoid villi separate the cerebrospinal fluid in the subarachnoid space from the blood in the venous sinuses.

Certain modifications of the meninges are to be observed in mammals. The cranial dura sends a process, or fold, the *falx cerebri,* down into the fissure separating the two cerebral hemispheres. A similar process, the *tentorium,* pushes down between the cerebral hemispheres and cerebellum. In some forms, as in the cat, the tentorium becomes ossified and fused to the parietal bones. On the ventral side of the brain, the dura encapsulates the pituitary gland but also forms a fold, the *diaphragma sellae,* over the sella turcica, or depression in the dorsal face of the sphenoid bone, in which the pituitary gland lies. The stalk of the infundibulum, together with the pars tuberalis, if present, penetrates the diaphragma sellae.

PERIPHERAL NERVOUS SYSTEM

The nerves and ganglia which form connections with the central nervous system and which are distributed to all parts of the body comprise the peripheral nervous system. The autonomic portion of the peripheral nervous system is composed of those nerve fibers distributed to structures under involuntary control. Connection with the central nervous

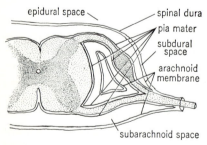

epidural space — spinal dura
— pia mater
subdural space
arachnoid membrane
subarachnoid space

Fig. 13.16 Diagram illustrating the relations of the meninges to the roots of a spinal nerve in a mammal.

system is mediated via spinal and cranial nerves. Spinal nerves, which form connections with the spinal cord, are paired, metameric structures. The paired cranial nerves, on the other hand, are connected with the brain, all but the first four being joined to the medulla oblongata. Metamerism is not so clearly indicated in the arrangement of the cranial nerves.

Spinal Nerves

Each spinal nerve connects to the spinal cord by means of two roots, *dorsal* and *ventral*. Except in amphioxus and the petromyzont cyclostomes, the two roots unite a short distance lateral to the cord to form the spinal nerve proper. The two roots have different embryonic origins.

Dorsal roots. The dorsal roots originate from neural crests,* which form in the crevice between neural tube and superficial ectoderm on each side (Fig. 3.10). A band of neural-crest cells is, therefore, at first present on either side of the spinal cord and extends in a longitudinal direction. At metameric intervals in each band, enlargements occur and the parts of the band between enlargements gradually disappear. Each neuroblast in these thickenings sends out two processes: (1) an axon, which grows toward the spinal cord and enters it in the region of the dorsal column; and (2) a dendrite, which grows peripherally to the skin, voluntary muscles, skeleton, or some visceral structure. The axon may pass up or down the cord for some distance. The dorsal root contains a rather conspicuous swelling, the *dorsal root ganglion,* which represents the location of the cell bodies of the sensory neurons present

---

* It may be recalled (page 435) that, at least in amphibians and birds, neural-crest cells seem to be responsible also for the formation of certain cartilages in the branchial region.

in the dorsal root. The dendritic processes of these cells have the histologic structure of axons but function as dendrites, nevertheless (page 97). Experiments have demonstrated that sensory nerve impulses travel through the spinal nerves *toward* the spinal cord and in doing so pass through the dorsal roots which, except in some of the lower forms, are strictly *sensory roots.* The sensory components of spinal nerves are, therefore, found in the dorsal roots and are spoken of as *afferent fibers.* The dorsal roots in certain lower vertebrates are not composed entirely of sensory fibers since some motor fibers (visceral efferent fibers) course through them.

When the axons of sensory nerve fibers enter the spinal cord, they form synapses with dendrites of nerve cells located in the gray matter of the cord. Sensory fibers are said to be either somatic or visceral. *Somatic sensory fibers* are those coming from the skin and its derivatives, voluntary muscles, and skeletal structures. They form synapses with cells in the somatic sensory column of gray matter. *Visceral sensory fibers* from visceral structures (page 612) terminate in the visceral sensory column of gray matter.

Ventral roots. The cell bodies of the neurons making up the ventral, or motor, roots lie in the gray matter of the ventral columns of the spinal cord. Their dendrites, therefore, lie within the cord itself, but their axons emerge, at metameric intervals, from the ventrolateral angles of the cord from which they pass to peripheral voluntary muscles or to autonomic ganglia. In motor fibers, impulses are conveyed *away from* the spinal cord. Motor fibers are therefore referred to as *efferent fibers.* *Somatic motor fibers* arise in the somatic motor column of gray matter in the spinal cord and are distributed to somatic structures. *Visceral motor fibers,* the cell bodies of which are located in the visceral motor column of gray matter, pass to autonomic ganglia where they form synapses with motor autonomic neurons.

In amphioxus, visceral motor fibers emerge from the spinal cord and pass out the *dorsal* roots, which are thus composed of both afferent and efferent fibers. This undoubtedly represents the primitive condition. Lampreys are much the same except that a few visceral motor fibers course through the ventral roots. In fishes and amphibians, visceral efferent fibers are found in both dorsal and ventral roots. Only in amniotes, therefore, is the dorsal root strictly sensory.

According to the above description the spinal nerve proper, formed by the union of dorsal and ventral roots, is composed partly of dendritic processes of sensory cells, located in the dorsal root ganglion, and partly of axons of efferent motor neurons, the cell bodies of which are situated in the ventral column of gray matter in the cord. The axons of the sensory cells are short, being confined to the proximal portion of the dorsal root. They pass to the somatic sensory or

visceral sensory columns of gray matter in the cord, as the case may be.

Rami. Not far from the point where dorsal and ventral roots unite to form the spinal nerve, three branches, or rami (Figs. 13.17 and 13.18), are usually given off. These include (1) a *dorsal ramus* supplying the skin and epaxial muscles of the dorsal part of the body, (2) a *ventral ramus* distributed to the hypaxial ventral and lateral regions, and (3) a *visceral ramus,* or *ramus communicans,* which courses medially and, in most cases, forms connections with one of the chain ganglia of the peripheral autonomic nervous system (Figs. 13.17 and 13.18). Both dorsal and ventral rami are composed of somatic sensory and somatic motor fibers. Some autonomic fibers going to the periphery are also included in these rami (page 635). A typical visceral ramus consists of two parts, a *white ramus* and a *gray ramus.* The white ramus is made up of visceral sensory and visceral motor fibers, mentioned above. Both visceral sensory and visceral motor fibers are medullated, giving this ramus a glistening white appearance. The axons of the visceral motor fibers are distributed to ganglia of the

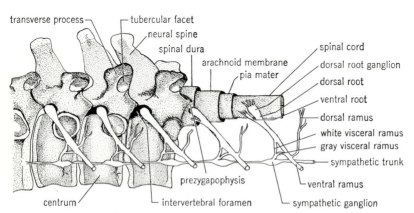

Fig. 13.17 Diagram showing relation of spinal nerves to spinal cord, meninges, vertebrae, and sympathetic trunk.

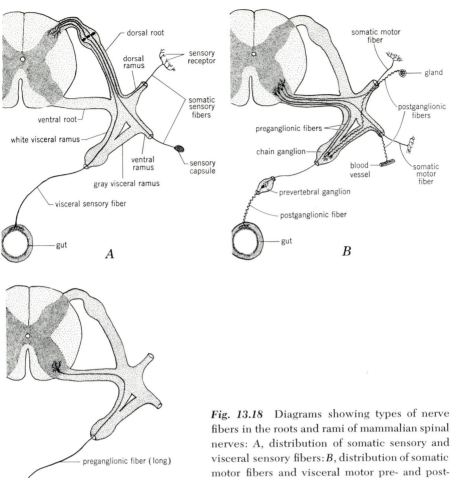

A

- dorsal root
- dorsal ramus
- sensory receptor
- somatic sensory fibers
- ventral root
- white visceral ramus
- ventral ramus
- sensory capsule
- gray visceral ramus
- visceral sensory fiber
- gut

B

- somatic motor fiber
- gland
- postganglionic fibers
- preganglionic fibers
- chain ganglion
- blood vessel
- somatic motor fiber
- prevertebral ganglion
- postganglionic fiber
- gut

C

- preganglionic fiber (long)
- postganglionic fiber (short)

Fig. 13.18 Diagrams showing types of nerve fibers in the roots and rami of mammalian spinal nerves: *A*, distribution of somatic sensory and visceral sensory fibers; *B*, distribution of somatic motor fibers and visceral motor pre- and postganglionic fibers of the sympathetic nervous system; *C*, distribution of pre- and postganglionic fibers of parasympathetic nervous system as found in certain sacral spinal nerves.

peripheral autonomic system. The fibers may go to the chain ganglia or pass through them to terminate in other peripheral ganglia. In either case they form synapses with non-medullated, or sparsely medullated, fibers *

* The electron microscope has demonstrated that so-called gray, nonmedullated fibers actually may have myelin sheaths surrounding them, but the quantity of myelin is minimal in comparison to that present around white, medullated fibers (page 97). The term *sparsely medullated* is often used in preference to *nonmedullated*.

of the peripheral autonomic system. The medullated visceral motor fibers are also referred to as *preganglionic fibers* since they terminate in autonomic ganglia. They form relays with nonmedullated *postganglionic fibers* which originate in the autonomic ganglia. The gray ramus communicans, composed of nonmedullated, gray, postganglionic fibers, runs parallel to the white ramus. The cell bodies of these fibers are located in chain ganglia. The fibers of the gray ramus join the spinal nerve and travel out either dorsal or ventral rami where they supply structures under involuntary control such as blood vessels, arrector pili muscles, and glands of the skin.

To summarize: dorsal and ventral rami carry medullated somatic sensory fibers, medullated somatic motor fibers, and sparsely medullated, or nonmedullated, postganglionic motor autonomic fibers; the white visceral ramus carries medullated visceral sensory and medullated preganglionic motor fibers; the gray visceral ramus carries only nonmedullated, or sparsely medullated, postganglionic autonomic motor fibers.

Plexuses. In the regions of the appendages, the ventral rami of certain spinal nerves are drawn out into the appendages. In such cases a more or less complicated network, or *plexus*, may be present, formed by the connection of branches of certain nerves with those of others (Fig. 13.19). In lower forms an anterior *cervicobrachial plexus* is present for each pectoral appendage and a posterior *lumbosacral plexus* for each hind limb. In higher vertebrates a further differentiation

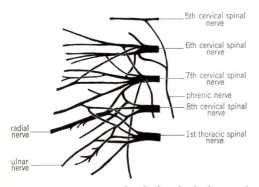

Fig. 13.19 Diagram of right brachial plexus of cat, ventral view. (*After Reighard and Jennings, "Anatomy of the Cat," Holt, Rinehart and Winston, Inc. By permission.*)

5th cervical spinal nerve
6th cervical spinal nerve
7th cervical spinal nerve
phrenic nerve
8th cervical spinal nerve
1st thoracic spinal nerve
radial nerve
ulnar nerve

takes place, so that separate cervical, brachial, lumbar, and sacral plexuses may be present. In mammals such large nerves as the *ulnar* and *median nerves* of the forelimbs and the *femoral* and *sciatic nerves* of the hind legs represent convergences of portions of the above-named plexuses.

Cauda equina. Since the spinal cord lies in the canal formed by the neural arches of vertebrae, the spinal nerves must pass out of the vertebral column in order to reach their destination. Dorsal and ventral roots, except in certain lower forms, lie within the neural canal. Openings, known as *intervertebral foramina,* are present between adjacent vertebrae, providing for the passage of spinal nerves. At first, during embryonic development, each spinal nerve passes through an intervertebral foramen located at the same level as the part of the cord from which the nerve arises (Fig. 13.20*A*). Later, however, growth of the cord fails to keep pace with that of the vertebral column. As a result, the spinal nerves, particularly those toward the posterior end, are drawn posteriorly so that they emerge through foramina which may be a considerable distance posterior to the point at which the spinal nerve leaves the cord (Fig. 13.20*B*). These nerves, therefore, must course for some distance through the neural canal, running obliquely or else parallel to the spinal cord, before they finally emerge. The more posterior the spinal nerve, the greater distance must it travel before reaching the foramen through which it passes. Because of the brushlike appearance of the posterior spinal nerves of higher forms and their fancied resemblance to a horse's tail, early anatomists gave to these nerves the name *cauda equina.*

Comparative anatomy of spinal nerves. AMPHIOXUS. The roots of the spinal nerves

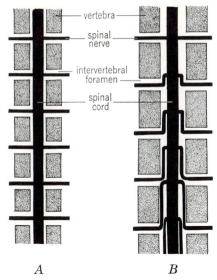

vertebra

spinal nerve

intervertebral foramen

spinal cord

A   B

Fig. 13.20 Diagram illustrating the formation of cauda equina: *A*, the spinal nerves emerge through the vertebral column at right angles to the spinal cord; *B*, because of the disproportionate growth of vertebrae and spinal cord, the nerves emerge at some distance posterior to their point of origin.

of amphioxus do not arise from the spinal cord in symmetrical pairs as do those of higher forms. Instead, those of one side alternate with those of the other in a manner similar to that of the myotomes, or muscle segments. Dorsal and ventral roots do not unite. Furthermore, they do not arise from the same level, the dorsal roots emerging opposite the myocommata and the ventral roots opposite the myotomes. The dorsal roots are composed of compactly arranged fibers, and dorsal root ganglia are lacking. The cell bodies of the sensory neurons in the dorsal roots lie within the spinal cord itself. The ventral roots are less compact, each arising from the cord by a number of loosely arranged bundles. The dorsal roots of the spinal nerves in amphioxus carry somatic

sensory, visceral sensory, and visceral motor fibers, the ventral roots being composed of somatic motor fibers alone. Each dorsal root branches into dorsal and ventral rami soon after leaving the cord. The ventral ramus gives off a visceral branch.

CYCLOSTOMES. As in amphioxus, the dorsal and ventral roots of the spinal nerves of cyclostomes alternate with each other as they arise from the cord, the dorsal root coming off somewhat posterior to the ventral root. In lampreys the roots do not join each other, but in hagfishes such a union is usual except in the caudal region. In either case each dorsal root bears a ganglion.

In the gill region of lampreys, fibers from adjacent dorsal roots unite as do fibers from adjacent ventral roots so as to form two nerves, one sensory and the other motor. These together form the *hypobranchial nerve*, which supplies the ventral part of the gill region. The motor components of this nerve presage the appearance of the hypoglossal nerve of amniotes.

Since paired appendages are lacking in cyclostomes, no spinal nerve plexuses are present.

FISHES. Dorsal and ventral roots unite in fishes, but the union occurs *outside* the vertebral column. In elasmobranchs the components form a looser union than is commonly the case. Also in elasmobranchs the roots do not arise from the cord in the same transverse plane, the dorsal roots coming off a short distance posterior to the ventral roots. A separate foramen is present for each root, that for the dorsal root being located in the intercalary plate and that for the ventral root in the neural arch proper (Fig. 10.14).

In some of the anterior spinal nerves, the dorsal roots have disappeared and only the ventral, motor roots of the original spinal nerves persist. They are called *spino-occipital nerves*. Fibers from these nerves contribute to the formation of a hypobranchial nerve (Figs. 13.2 and 13.3) similar to that described for cyclostomes except that it is composed of efferent fibers alone.

There is much variation in the complexity of the cervicobrachial and lumbosacral plexuses in fishes. The most complicated condition is found in skates, with their highly developed pectoral fins. As many as 25 spinal nerves may contribute to the formation of the cervicobrachial plexus in these fishes.

AMPHIBIANS. In amphibians, as in previous forms, the dorsal roots of the spinal nerves arise somewhat posterior to the level at which the ventral roots come off. The two roots unite as they pass through the intervertebral foramen. A swelling, the dorsal root ganglion, is located at the point of junction just outside the foramen. It is protected anteriorly by the transverse process of the vertebra and dorsally by the articular processes. From the spinal ganglion arise a dorsal ramus, in the form of a number of twigs, and a large ventral ramus. The visceral ramus comes off the ventral ramus.

In anurans, chalky masses known as *calcareous bodies,* apparently of lymphatic origin, surround the spinal ganglia. They are believed to serve as reserve supplies of calcium. In urodeles there are no calcareous bodies but the ganglia are surrounded by spongy, fatty tissue.

Between the occipital condyles and the first, or atlas, vertebra, the ventral root of a small *suboccipital nerve* emerges on each side. Some authorities consider it to be the first spinal nerve, but there is disagreement on this point. Usually the nerve emerging between the first and second vertebrae is called the first spinal nerve, at least in the anamniota. It also consists only of a ventral root, the dorsal root and ganglion having disappeared at the time of metamorphosis. The nerve is purely motor and may represent

the hypoglossal cranial nerve of higher forms. The remaining spinal nerves are typical. In urodeles the number varies with the number of segments in the body. At the time of metamorphosis in anurans, there is a reduction in number of spinal nerves, so that in most cases only 10 or 11 remain. Cervicobrachial and lumbosacral plexuses are present, the latter being more complex in anurans in accordance with the greater degree of development of the hind limbs. A conspicuous cauda equina is present in frogs and toads.

REPTILES. In the amniota there are 12 pairs of cranial nerves, in contrast to the condition in lower forms in which there are but 10. It is generally believed that the eleventh cranial nerve (spinal accessory) represents a separation of a visceral motor component of cranial nerve X (vagus) together with certain fibers contributed by spino-occipital nerves. The twelfth cranial nerve (hypoglossal) represents a union of two or three spino-occipital nerves which have been incorporated within the skull and, therefore, arise from the medulla oblongata. The hypoglossal is a motor nerve which lacks a dorsal root, as is the case with the spino-occipital nerves which contain only efferent fibers. The probable homology of the hypoglossal nerve with the hypobranchial nerve of lower forms has already been alluded to. With the appearance of the hypoglossal nerve in amniotes, spino-occipital nerves are no longer present.

The spinal nerves of reptiles show no peculiarities. As in other amniotes, dorsal roots contain only sensory fibers. Both visceral motor and somatic motor fibers are confined to the ventral roots. It is of interest that in certain snakes and limbless lizards a distinct, although poorly developed, lumbosacral plexus is present, indicating that these limbless forms arose in evolution from ancestors with limbs.

Anatomy of the Chordates

BIRDS. The arrangement of spinal nerves in birds is typical. In certain long-necked forms, the nerves making up the cervicobrachial plexus arise from the cord much farther posteriorly than is usually the case. The lumbosacral plexus may be composed of two or three distinct components, in which case the names *lumbar, sacral,* and *pudendal plexuses* are applied. The lumbar plexus supplies the thigh. The nerves of the sacral plexus unite to form the sciatic nerve passing through the thigh to the lower leg. The pudendal plexus sends branches to the cloacal and tail regions.

MAMMALS. The spinal nerves of mammals are named according to their relation to the vertebral column. There are, therefore, cervical, thoracic, lumbar, sacral, and caudal or coccygeal spinal nerves. Each is numbered according to the number of the vertebra which lies anterior to it except in the case of the cervical spinal nerves, the first of which emerges between the occipital bone and the atlas. In man there are 31 pairs of spinal nerves: 8 cervical, 12 thoracic, 5 lumbar, 5 sacral, and 1 caudal. The number varies a great deal in different mammals. The horse, for example, has the following numbers: 8 cervical, 18 thoracic, 6 lumbar, 5 sacral, and 5 caudal.

As a general rule, the somatic motor fibers of each spinal nerve supply a single myotome, or muscle segment. In each case the myotome supplied is the one adjacent to the spinal nerve during early embryonic development. No matter how the muscles derived from a certain myotome change or shift in position later on, the original nerve is maintained. This is the basis for determining muscle homologies which otherwise might be quite obscure. There are few exceptions to this.

The limb plexuses of mammals may be very complicated (Fig. 13.19). They are commonly divided into cervical, brachial, lumbar, and sacral divisions. The *phrenic nerve,* which supplies the diaphragm, arises from the cervical plexus (Fig. 13.19), indicating that at least some of the musculature of the diaphragm has been derived from myotomes in the cervical region. In the rat the phrenic nerve on each side arises from the fourth and fifth cervical nerves; in the cat cervical nerves 5 and 6 are involved; in man the phrenic nerve arises mainly from the fourth cervical nerve but also receives some branches from the third and fifth.

Cranial Nerves

The peripheral nerves which form connections with the brain are called *cranial nerves* (Figs. 13.2, 13.3, and 13.10*B*). They are much more specialized than spinal nerves and in many cases show little similarity to the latter in origin and distribution. Early human anatomists, not appreciating the functional characteristics of the various cranial nerves nor their homologies in lower vertebrates, assigned numbers to them in an anterior-posterior sequence, a system of classification which, in the light of modern research, has been shown to be very superficial and artificial. Nevertheless, the original terminology has persisted, making it rather difficult for the student to appreciate the complexity of the component parts of the various cranial nerves.

It will be recalled that dorsal and ventral roots of the spinal nerves in amphioxus and lampreys fail to join each other. It is believed that certain cranial nerves may originally have had a similar arrangement. If the two roots of a cranial nerve unite, they do so before emerging from the medulla oblongata. In some cases the dorsal roots and ganglia have apparently been lost, and in others the original ventral roots seem to have disappeared. For example, the *hypoglossal nerve,* or cranial nerve XII, when first formed, has two roots, the dorsal root bearing a ganglion; but the dorsal root and ganglion disappear during subsequent development. The hypoglossal nerve, as mentioned before, has been homologized with the hypobranchial nerve of lower forms. It is a purely motor nerve.

As in the case of spinal nerves, ganglia of the sensory portions of certain cranial nerves are located *on* the nerves, in this instance close to the brain. Some of the purely sensory nerves are special structures, not at all comparable in any way to the sensory components of spinal nerves.

It has already been mentioned that there are 10 pairs of cranial nerves in anamniotes and 12 in amniotes. Some are entirely sensory, composed of afferent fibers alone; others are purely motor. Still others are mixed nerves consisting of both motor and sensory fibers. The cranial nerves are usually listed as follows:*

0	Terminal	Sensory
I	Olfactory	Sensory
II	Optic	Sensory
III	Oculomotor	Motor
IV	Trochlear	Motor
V	Trigeminal	Mixed
VI	Abducens	Motor
VII	Facial	Mixed
VIII	Auditory (vestibulo-cochlear)	Sensory
IX	Glossopharyngeal	Mixed
X	Vagus	Mixed
XI	Spinal accessory	Motor
XII	Hypoglossal	Motor

In 1894 a new cranial nerve was discovered connecting to the anterior end of

*Medical students have long used the following non-sensical rhyme to help memorize the numerical sequence of the cranial nerves:
  "On old Olympus' towering tops
  A Finn and German viewed some hops."

the cerebral hemispheres. This nerve was first identified in *Protopterus* and has since been found in practically all vertebrates except cyclostomes and birds. It is lacking in man. In order to avoid the confusion which would result if the long-established nomenclature and symbols of the other cranial nerves were changed, the new nerve was called nerve 0, or the *terminal nerve.* Some authorities are of the opinion that nerve 0 is a somatic sensory, ganglionated remnant of the first of a series of branchial nerves (page 641). It apparently is unrelated to the olfactory nerve or olfactory sense.

We shall see in the following discussion that merely stating that a nerve is sensory, motor, or mixed has little meaning so far as functional attributes are concerned. Not all sensory nerves have similar origins, nor are the motor cranial nerves all made up of similar components. We must take into consideration the fact that, like spinal nerves, we have somatic sensory, visceral sensory, visceral motor, somatic motor, and autonomic fibers to consider. Furthermore, associated with certain cranial nerves of lower forms we have, in addition, sensory nerve fibers associated with the *lateralis,* or *lateral-line, system,* a system of sense organs and nerves, of importance in the aquatic mode of life, which has no counterpart in the higher vertebrate classes (page 723).

The sensory neurons of some of the cranial nerves originate, for the most part, from neural-crest cells in the head region. In other cranial nerves some of the sensory neurons may arise from thickenings in the superficial ectoderm which are known as *placodes.* The motor neurons, whether visceral or somatic, like those in the ventral roots of spinal nerves, arise within the neural tube itself. Postganglionic autonomic motor neurons are derived originally from neural-crest cells.

The acousticolateralis system. The lateral-line system consists of certain sense organs found in cyclostomes, other fishes, and aquatic urodele amphibians. The receptor organs for this system are called *neuromasts* and are composed of sensory cells and supporting cells. The neuromasts have definite connections with sensory branches of certain cranial nerves. The ganglia of these nerves, however, differ from other sensory nerves in that they arise in the embryo from placodes. The inner ear also arises from a placode in the same general vicinity, and its sensory cells rather closely resemble those making up the neuromasts of the lateralis system. The auditory, or acoustic, nerve and the acoustic ganglion, which lies close to the ear, are, however, derived from neural-crest cells rather than from a placode. Because of the similarity in origin of the inner ear proper to that of the sensory receptors, nerves, and ganglia of the lateralis system, it is customary to group these structures into what is called the *acousticolateralis system.* As development progresses, the ganglia of the lateralis system move inward and come to lie within the cranium close to the acoustic ganglion alongside of the medulla oblongata. The nerve fibers of the lateralis system become associated or fused with certain cranial nerves, namely, VII (facial), IX (glossopharyngeal), and X (vagus), of which they appear to be branches. Such fibers are entirely sensory and form connections with the somatic sensory column of gray matter near the anterior end of the medulla oblongata in the acousticolateralis area (page 617) close to the auricles, or restiform bodies, of the metencephalon. The lateralis nerves and their ganglia disappear in amniotes and are lost in amphibians at the time of metamorphosis. The nerve fibers of the lateralis nerves are classified as somatic sensory fibers.

The branchial nerves. Four of the cranial nerves are spoken of as branchial nerves, originally supplying the gill region with sensory and motor fibers. Each is primarily associated with a visceral arch. The branchial nerves include (1) the trigeminal nerve (V), which is the nerve of the mandibular arch; (2) the facial nerve (VII), supplying the hyoid arch; (3) the glossopharyngeal nerve (IX), associated with the third arch; and (4) the vagus nerve (X), which takes care of the remaining arches. Each branchial nerve has three main branches: pharyngeal, pretrematic, and posttrematic (Fig. 13.21). Occasionally a fourth, or dorsal, branch is present. The branchial muscles of the gill region are considered to be visceral muscles, despite the fact that they are voluntary and striated, because they are originally derived from the splanchnic mesoderm of the hypomere. Hence the motor nerves supplying them directly are composed of visceral motor fibers which are *not* preganglionic fibers. These are confined to the posttrematic branches of the branchial nerves. Visceral sensory fibers are found in all except the dorsal branch, the pretrematic and pharyngeal branches being made up only of such fibers. Somatic sensory fibers from the skin

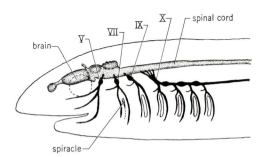

Fig. 13.21 Diagram showing relationship of the branchial cranial nerves (V, VII, IX, and X) to the visceral arches and gill slits. (*Modified from Johnston.*)

above the gill region are confined to the dorsal branch when this branch is present. The ganglia of the branchial nerves are to be found close to those of the lateralis nerves.

Special sensory nerves. Three of the cranial nerves are of a special nature. They are purely sensory nerves carrying impulses from the three major sense organs: nose, eye, and ear. These nerves are considered to be somatic sensory nerves although there is some dispute as to whether the olfactory nerve (I) is somatic sensory or visceral sensory. The cell bodies of the olfactory nerve are not derived from neural crests, but rather from a placode in the superficial ectoderm of the anterior head region. Those of the optic nerve (II) have a still different origin since they actually come from an outpocketing of the diencephalic region of the brain. The fibers of the auditory nerve (VIII), however, like the sensory fibers of the spinal nerves, originate from neural-crest cells.

Terminal nerve (0). Leaving the anterior, ventral portion of the cerebral hemisphere, although apparently originating in the diencephalon, the *terminal nerve* passes to the olfactory mucous membrane. It is a somatic sensory nerve and bears one or more ganglia. This nerve is best developed in elasmobranch fishes. In amphibians, reptiles, and mammals, it is associated with the vomeronasal, or Jacobson's, organ, which seems to be an accessory olfactory structure. The function of the terminal nerve is not clear. It has been suggested that it may represent the remains of an anterior branchial nerve which long ago lost its significance.

Olfactory nerve (I). The first cranial nerve is composed of nonmedullated, or sparsely medullated, fibers. They are outgrowths of neurosensory cells derived from a placode in the anterior head region. The olfactory nerve is considered by most authorities to be a special kind of somatic sensory nerve. The cell bodies of this nerve lie in the olfactory epithelium. Their axons grow toward the brain where they form synapses with neurons in the olfactory lobes. The terminal arborizations of several olfactory nerve fibers may synapse with the brushlike dendrites of a single cell in the olfactory bulb (page 626). The interlacing of these processes forms the so-called *olfactory glomeruli* of the bulb. No ganglia are present.

In those vertebrates in which the olfactory lobe is drawn out into bulb and tract, the olfactory tract is sometimes mistaken for the olfactory nerve. Actually the nerve proper usually consists of many separate fibers which are not gathered together in a sheath. In some forms, however, as in certain teleosts, amphibians, and lizards, the olfactory nerve is rather long and joins the olfactory lobe, which is not differentiated into a clearcut bulb and tract. In mammals as many as 20 separate nerve branches on each side pass through openings (olfactory foramina) in the cribriform plate of the ethmoid bone on their way to the brain.

That the single, median, olfactory organ found in cyclostomes is actually of bilateral origin is indicated by the presence of paired olfactory nerves passing from the olfactory mucous membrane through the olfactory capsule to the olfactory bulb of the brain.

A *vomeronasal* branch of the olfactory nerve is distributed to Jacobson's organ in those vertebrates possessing this structure. It originates in an *accessory olfactory bulb* which is a rather indistinct extension from the lateral side of the olfactory bulb proper.

Optic nerve (II). The second cranial nerve shows certain peculiarities, for an understanding of which the reader is referred to a description of the embryonic development of the eye (page 670). The optic nerve consists of a bundle of sensory nerve fibers, the cell bodies of which are located in the retina of the eye. The fibers grow down through the walls of the optic stalk, reaching the brain at the level of the diencephalon. The fibers of each optic nerve then usually cross beneath the diencephalon and pass to the lateral geniculate bodies or to the optic tectum in the roof of the optic lobes of the mesencephalon. Crossing is complete in all vertebrates except in those mammals having binocular vision. In these forms only about half the fibers in each optic nerve pass to the opposite side. Fibers from the nasal side of the retina are the ones which decussate, those from the temporal side of the retina going to the optic tectum on the same side of the brain as that from which they come. In mammals, impulses set up by optic stimuli are relayed from the optic tectum to the cerebral hemispheres. The point at which the nerves cross is called the *optic chiasma*. In cyclostomes the chiasma lies *within* the brain. The term *optic tract* is applied to the portion of the optic nerve which passes from the chiasma to the optic lobe. A *commissure* (*of Gudden*) is located at the posterior border of the chiasma, its fibers sometimes being confused with those of the optic nerves. Various optic chiasmas show unusual structural complications, a few of which are indicated in Fig. 13.22.

Actually the optic nerve should not be considered in the category of cranial nerves but rather as a fiber tract of brain, since it merely grows from one portion of the brain (optic cup) to another (mesencephalon). The optic nerve, however, is usually grouped in the category of special somatic sensory cranial nerves.

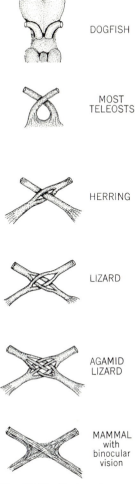

DOGFISH

MOST TELEOSTS

HERRING

LIZARD

AGAMID LIZARD

MAMMAL with binocular vision

Fig. 13.22 Various types of optic chiasmas. (*Those of herring and agamid lizard, after Wiedersheim, from Kingsley, "Comparative Anatomy of Vertebrates," The Blakiston Division, McGraw-Hill Book Company. By permission.*)

Oculomotor nerve (III). The eyeball is moved by six muscles, so arranged as to permit a maximum amount of rotary movement. These muscles are disposed in two groups consisting of two oblique and four rectus muscles, respectively, arranged as indicated in Fig. 11.13. Four of these muscles are derived from the myotome of the first, or premandibular, pro-otic somite (page 515) and are all supplied by the same nerve, the oculomotor. The somatic motor fibers, of which this nerve is chiefly composed, leave the ventral side of the mesencephalon and are distributed to the inferior oblique, superior rectus, inferior rectus, and internal rectus eye muscles. Some fibers also pass to the muscle (*levator palpebrae*) elevating the upper eyelid. Preganglionic autonomic fibers accompanying the somatic motor fibers are distributed to the ciliary ganglion. Postganglionic fibers then lead to the ciliary apparatus, which is concerned with accommodation, and to the sphincter muscles of the iris, which regulate the size of the pupil. A few somatic sensory proprioceptive fibers are borne by the oculomotor nerve.

In the lamprey the oculomotor nerve innervates only the superior rectus, internal rectus, and inferior oblique. The inferior rectus is supplied by the abducens nerve. For a possible explanation of this peculiar state of affairs the reader is referred to page 516.

Trochlear nerve (IV). The somatic motor trochlear nerve is the smallest of all the cranial nerves. Its nucleus lies near the floor of the mesencephalon. From here, fibers pass upward to the dorsal side where they decussate. The nerve leaves the dorsal side of the mesencephalon near its posterior end. It then enters the orbit to supply the superior oblique eye muscle derived from the mandibular, or second, pro-otic somite. The word

trochlear comes from a Greek term meaning pulley. The word is applicable because of the pulleylike arrangement of the superior oblique eye muscle in man. Like the oculomotor, a few proprioceptive sensory fibers are carried by the trochlear nerve.

Trigeminal nerve (V). The trigeminal nerve is a large nerve arising from the lateral side of the anterior end of the hindbrain at the level of the pons. Near its point of origin it bears a large *Gasserian,* or *semilunar, ganglion* which sends fibers to the somatic sensory column of gray matter in the medulla. The trigeminal is characteristically divided into three main branches, the *ophthalmic, maxillary,* and *mandibular nerves.* The first two of these branches bear somatic sensory fibers alone, but the mandibular branch is composed of both somatic sensory and visceral motor fibers. Visceral sensory and somatic motor fibers are lacking in the trigeminal nerve. The trigeminal is usually considered to be the first of the branchial nerves and is the nerve of the mandibular arch. It differs somewhat from the others in that its pretrematic (maxillary) and posttrematic (mandibular) branches bear somatic sensory rather than visceral sensory fibers.

In fishes the ophthalmic branch is composed of *superficial* and *deep (profundus)* portions, the latter possibly representing what originally may have been a separate branchial nerve. The superficial branch supplies the skin on the dorsal side of the head and snout with sensory fibers. A similarly named branch of the seventh nerve accompanies the superficial ophthalmic branch of the trigeminal, but it is distributed to portions of the lateral-line system in this region. The trigeminal does not supply any of the lateral-line organs. The deep ophthalmic, composed solely of fibers of the fifth nerve, courses separately behind the eyeball, which it supplies with some small ciliary nerves, and then joins the superficial ophthalmic and, along with that branch, is distributed to the skin in the dorsal and lateral regions of the snout. In higher vertebrates the two portions of the ophthalmic branch are not distinct and but a single ophthalmic nerve is present. It is the smallest of the three branches and supplies sensory fibers to the conjunctiva, cornea, iris, ciliary body, lacrimal gland, part of the mucous membrane of the nose, and the skin of the forehead, nose, and eyelids. The *ciliary ganglion* of the autonomic nervous system lies in close relationship to the ophthalmic branch.

The maxillary branch is the main nerve to the upper jaw, supplying the upper lip, side of the nose, lower eyelid, teeth of the upper jaw, and, when present, Jacobson's organ. The *sphenopalatine ganglion* of the autonomic nervous system lies close to the maxillary branch.

The somatic sensory portion of the mandibular branch is distributed to the lower lip and the teeth of the lower jaw. In mammals it also supplies the skin of the temporal region, external ear, and lower part of the face. A *lingual branch* goes to the mucous membrane of the anterior part of the tongue in mammals and certain reptiles. The visceral motor fibers of the mandibular branch go directly to the muscles used in chewing. *Otic* and *submaxillary ganglia,* belonging to the autonomic nervous system, are in close proximity to the mandibular nerve.

It should not be inferred from the above that the trigeminal nerve is part of the peripheral autonomic nervous system because of the close relationship of its branches to certain autonomic ganglia.

Abducens nerve (VI). The small abducens nerve arises from the ventral part of the medulla oblongata close to its anterior end,

courses anteriorly in an oblique direction, and is distributed to the external rectus eye muscle which has its origin from the third, or hyoid, pro-otic somite. Like the other nerves supplying the eye muscles, it is a somatic motor nerve which carries a few proprioceptive sensory fibers. In the lamprey, the inferior rectus muscle, in addition, is supplied by the abducens nerve (page 516). A branch of the abducens goes to the nictitating membrane if such a structure is present. In some vertebrates a separate muscle, the *retractor bulbi,* is used to draw the eyeball inward. It is a derivative of the external rectus and, when present, is supplied by the abducens nerve.

Facial nerve (VII). In lower aquatic vertebrates the large, somatic sensory components of the facial nerve are distributed to the lateral-line system. In fishes these include a *superficial ophthalmic branch*, accompanying the superficial ophthalmic branch of the trigeminal nerve on the dorsal side of the snout, and a *buccal branch,* which, with the maxillary branch of the trigeminal, courses along the floor of the orbit as the *infraorbital nerve.* The buccal and superficial ophthalmic branches supply the lateral-line organs on the snout. A third branch, the *hyomandibular,* represents the posttrematic branchial portion of the seventh nerve, which is the nerve of the hyoid arch. It contains visceral sensory, visceral motor, and somatic sensory fibers, the latter being lateralis components distributed to parts of the lateral-line system in the region of the hyoid arch. A *geniculate ganglion,* containing the cell bodies of the sensory neurons of the facial nerve, is located at the point where the hyomandibular branch leaves the medulla oblongata.

In elasmobranchs, the hyomandibular nerve lies just posterior to the spiracle. Near the geniculate ganglion it gives off a palatine branch composed of visceral sensory fibers alone. It represents the pharyngeal branch of this branchial nerve. The palatine branch passes to the roof of the mouth. A small twig from the palatine passes to the anterior side of the spiracle. This and other visceral sensory fibers, probably representing the pretrematic portion of the facial nerve, supply the taste buds and mucous membrane of the mouth. The visceral motor fibers of the hyomandibular innervate muscles of the hyoid region directly.

With the loss of the lateral-line system in terrestrial forms, the somatic sensory part of the facial nerve becomes relatively unimportant. Visceral sensory fibers in mammals supply the taste buds of the anterior two-thirds of the tongue. These fibers form part of the *chorda tympani,* a branch of the facial nerve which passes through the middle ear. After leaving the middle ear, its fibers course along with those of the lingual branch of the trigeminal nerve on their way to the tongue.

The visceral motor components of the facial nerve are more widely distributed. They pass to the muscles of the face, scalp, and external ear and to a few superficial neck muscles. Preganglionic visceral motor fibers, of the peripheral autonomic system, accompany the chorda tympani branch of the facial nerve and pass to the *submaxillary ganglion.* From this ganglion arise postganglionic fibers leading to the submaxillary and sublingual salivary glands. Other preganglionic fibers of the facial nerve pass to the *sphenopalatine ganglion,* where they relay with postganglionic fibers which are distributed, in turn, to the lacrimal gland and mucous membrane of the nose.

Auditory (vestibulo-cochlear) nerve (VIII). The purely somatic sensory auditory nerve bears an *acoustic ganglion* closely associated with the geniculate ganglion of the facial nerve. It arises from neural-crest cells, although possibly some cells may be contributed by the auditory placode from which the inner ear is derived. For a number of reasons (page 641), the inner ear is believed to have arisen in evolution in the same manner as the organs and nerves of the lateral-line system. The term *acoustico-lateralis system* is frequently used to designate the close relationship of the two. The auditory nerve comes from the lateral side of the medulla oblongata. Its ganglion is often inseparably fused with those of the trigeminal and facial nerves.

In higher forms, in which the inner ear serves both as an organ of hearing and of equilibration, the auditory nerve is divided into two main branches, a *vestibular branch,* which carries equilibratory impulses from the vestibular portion of the inner ear, and a *cochlear branch* for auditory impulses arising in the cochlea. In such cases the acoustic ganglion is divided into *vestibular* and *spiral (cochlear) portions.* In lower forms, in which the auditory function has not developed, the vestibular part of the nerve is the only portion that is present.

Strong fiber tracts in the brain pass from the eighth nerve to the cerebellum, where the center for equilibration is located. Impulses set up by auditory stimuli pass to the inferior colliculi of the optic tectum (roof of mesencephalon) and to the medial geniculate bodies in the roof of the diencephalon. In mammals, auditory impulses are relayed to centers in the cerebral hemispheres.

Glossopharyngeal nerve (IX). The ninth cranial nerve, arising from the medulla oblongata, is the nerve of the third visceral arch and the second gill pouch. Most typical of all the branchial nerves, in fishes its pretrematic, or hyoid, branch, made up of visceral sensory fibers, passes to the anterior side of the first *typical* gill slit.* Its post-trematic, or branchial, branch, composed of visceral sensory and visceral motor fibers, supplies the hemibranch on the posterior border of this gill slit. A small pharyngeal branch, containing visceral sensory fibers alone, goes to the pharynx, where it supplies taste buds and other receptors. In some lower aquatic vertebrates the glossopharyngeal sends a small dorsal somatic sensory branch to the anterior part of the lateral line proper. A prominent *petrosal ganglion* lies near the base of the glossopharyngeal nerve. It is situated close to the ganglia of the tenth cranial nerve and, in amphibians, actually fuses with them.

Two ganglia are present on the ninth nerve in higher forms. These are (1) a small, *superior ganglion,* close to the medulla, and (2) a *petrosal ganglion,* a short distance away. In these vertebrates some visceral motor components of the glossopharyngeal are distributed directly to muscles of the pharynx derived from the third visceral arch region. Preganglionic visceral autonomic motor fibers are also present, passing to the *otic ganglion.* Postganglionic fibers then relay impulses to the parotid salivary gland. In mammals the visceral sensory fibers of the ninth nerve pass to the taste buds of the posterior one-third of the tongue, to the mucous membrane of the pharynx, and to the palatine tonsils.

Vagus nerve (X). The vagus nerve is perhaps the most important of all the cranial nerves because of the preganglionic fibers of the parasympathetic portion of the peripheral autonomic nervous system which

it contains. Through these the vagus controls such vital activities as heartbeat, respiratory movements, and peristalsis. Each vagus nerve in man may contain more than 100,000 nerve fibers.

The tenth cranial nerve arises from the medulla oblongata by a number of roots. In lower aquatic vertebrates it bears a *lateralis ganglion* and a *jugular ganglion* near its base. A large, somatic sensory, lateralis component, the *lateral nerve,* branches off the vagus near its proximal end and extends the length of the body to nearly the end of the tail, lying underneath the lateral-line canal and located between the epaxial and hypaxial muscles of the body wall. The lateral nerve may be several feet long. It is not present in the amniota. In amphibians, with the exception of the perennibranchiates, it disappears at the time of metamorphosis. The lateralis ganglion is also lost.

In addition to the lateral nerve there is a large *branchiovisceral branch* with the jugular ganglion at its base. It is composed of visceral sensory and visceral motor fibers, together with a negligible number of somatic sensory fibers which are distributed to the skin in the area about the ear. Some sensory proprioceptive fibers are also present. Part of the branchiovisceral branch of the vagus serves as a branchial nerve supplying the visceral arches posterior to the third and the remaining gill pouches. In the gill region of fishes this branch gives off twigs, each of which in elasmobranchs (and lampreys) bears a small *epibranchial ganglion.* In teleosts the epibranchial ganglia fuse into a single mass located on the main trunk. Each twig in turn gives off a pharyngeal branch and then divides into pretrematic and post-trematic branches supplying the anterior and posterior hemibranchs, respectively, of each of the gill pouches posterior to that supplied by the glossopharyngeal nerve. As

---

* The spiracle, found in elasmobranchs and a few other fishes, actually represents the first gill pouch. It disappears in teleost fishes. The first gill pouch as found in teleosts, therefore, really represents the second pouch of elasmobranchs.

in the similarly named branches of the glossopharyngeal nerve, the pretrematic branches bear visceral sensory fibers alone, the posttrematic branches being composed of visceral sensory and visceral motor fibers distributed to the branchial muscles in this region. The pharyngeal branches pass to the pharynx, where they supply taste buds and other sensory receptors. Posterior to the gills the vagus courses caudally, giving off a branch to the heart, except in the hagfish, and then distributes its visceral sensory and visceral motor preganglionic fibers to the coelomic viscera, except those at the posterior end. The preganglionic autonomic motor fibers borne by the vagus terminate in small autonomic ganglia in, or close to, the walls of these organs, where they form synapses with very short postganglionic fibers (page 654).

With the disappearance of gills in amniotes and after metamorphosis in amphibians, the branchial branches of the vagus are lost for the most part. The visceral motor remnants of the posttrematic branches persist to supply directly certain muscles of the pharyngeal and laryngeal regions derived from this part of the gill area. Visceral sensory fibers pass to the taste buds and other receptors in the posterior pharyngeal region. In tetrapods preganglionic, visceral motor, vagal fibers supply the larynx, trachea, lungs, and abdominal viscera except those at the posterior end. In mammals a large *nodosal ganglion* is present at the base of the vagus a short distance distal to the jugular ganglion. It may possibly be homologous with the epibranchial ganglion of teleost fishes.

The fact that a *cranial* nerve supplies organs so far removed from the brain indicates that these structures must have migrated posteriorly from an originally more anterior position.

***Spinal accessory nerve* (XI).** An eleventh cranial nerve is found only in amniotes. It bears a very close relation to the vagus. The spinal accessory is apparently composed of a cranial portion, derived from certain posterior visceral motor fibers of the vagus, and a spinal portion, representing somatic motor fibers of some of the most anterior cervical spinal nerves (spino-occipital nerves). Fibers of the cranial portion accompany branches of the vagus to pharynx and larynx and should be considered to belong to the category of branchial nerves. Those of the spinal portion supply the sternocleidomastoid and trapezius muscles, which thus may be derivatives of myotomes. Their exact homologies are uncertain.

In lower vertebrates having but 10 cranial nerves in addition to the terminal nerve (0), the cranial portion of the spinal accessory is already indicated by a posterior branch of the vagus.

***Hypoglossal nerve* (XII).** The twelfth cranial nerve, also found only in amniotes, is a purely somatic motor nerve which arises from several roots and supplies the intrinsic muscles of the tongue, as well as a number of muscles below the tongue in the lower jaw. As has been previously suggested, the hypoglossal is represented in lower forms by the hypobranchial nerve, formed by the union of two or three spino-occipital nerves, which, in amniotes, has acquired cranial connections. The nerve has assumed greater importance with the appearance of muscles which move the tongue, and for this reason the twelfth nerve is best developed in mammals.

Although a hypoglossal nerve in living vertebrates appears only in amniotes, there is evidence from the fossil record that ancient crossopterygian fishes and amphibians possessed such a nerve. A separate foramen was present in the skulls of these forms which undoubtedly served for the passage of a hypoglossal nerve.

Autonomic Nerves

The autonomic portion of the peripheral nervous system is closely related anatomically to the spinal nerves and certain cranial nerves. Furthermore, nerve centers controlling the activity of autonomic neurons are located in the central nervous system. Consequently, all parts of the nervous system are so intimately bound together both structurally and functionally that none of its major constituents can be considered independently of the others.

The autonomic portion of the peripheral nervous system is composed of efferent neurons which send impulses to smooth muscles and glands in all parts of the body. It regulates the function of structures which are under involuntary control. The muscle of the heart, although striated, is not subject to voluntary control, and it, too, is innervated by autonomic fibers. The proper functioning of this part of the nervous system is necessary for regulating such activities as rate of heartbeat, respiratory movements, composition of body fluids, constancy of temperature, secretion of various glands, peristalsis, and other vital processes.

The complex connections of the autonomic neurons add to the difficulties embodied in an understanding of its structure. Most investigations have been carried out in mammals, and in man in particular. It will best serve our purpose first to discuss the autonomic system in man and then to make comparisons with other vertebrates.

It will be recalled that the cell bodies of the somatic motor neurons of cranial and spinal nerves are located in the ventral somatic motor column of gray matter of brain or spinal cord. Their medullated axons continue without interruption to the effector organ. A different pattern obtains in the autonomic system, in which efferent impulses must travel through *two* neurons, *preganglionic* and *postganglionic,* before they can bring about an effect (Fig. 13.18B and C).

The system is composed of these two types of neurons and a number of ganglia which serve as relay centers. The cell bodies of the preganglionic neurons are located in the visceral motor column of gray matter in the central nervous system. Their medullated fibers pass to outlying ganglia. These fibers make up the visceral motor components of certain cranial nerves and of spinal nerves. In the latter they are constituents of the white visceral rami. Postganglionic neurons have gray, sparsely medullated, or nonmedullated, axons. The cell bodies of these neurons, which are derived from neural-crest cells, lie in outlying ganglia which are often located at some distance from the central nervous system. It is here that the preganglionic fibers form synapses with the dendrites of postganglionic neurons. The axons of the postganglionic neurons are the ones which are distributed to smooth muscles, cardiac muscle, and glands in various parts of the body.

Although the peripheral autonomic system is essentially an efferent system, some authorities claim that certain visceral afferent fibers should be included. Such fibers pass from visceral organs up the dorsal roots of the spinal nerves to connect with the central nervous system.

Three large masses of nervous tissue, called ganglionated plexuses, are located near the vertebral column in the thoracic, abdominal, and pelvic regions. They are known as the *cardiac, coeliac,* or *solar,* and *hypogastric plexuses.* Each is composed of an intricate meshwork of nerves and ganglia containing visceral afferent fibers and visceral efferent fibers of the autonomic system. The

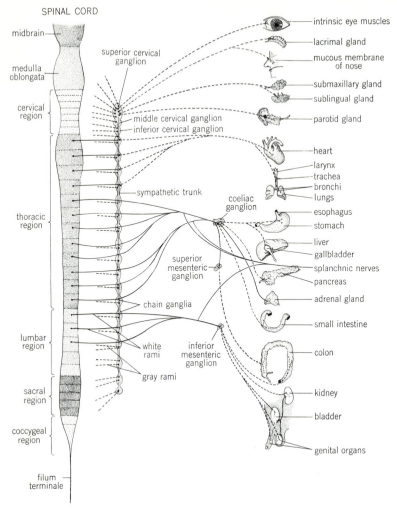

SPINAL CORD

midbrain

medulla oblongata

cervical region

superior cervical ganglion

middle cervical ganglion

inferior cervical ganglion

sympathetic trunk

coeliac ganglion

thoracic region

superior mesenteric ganglion

chain ganglia

lumbar region

white rami

inferior mesenteric ganglion

gray rami

sacral region

coccygeal region

filum terminale

intrinsic eye muscles

lacrimal gland

mucous membrane of nose

submaxillary gland

sublingual gland

parotid gland

heart

larynx

trachea

bronchi

lungs

esophagus

stomach

liver

gallbladder

splanchnic nerves

pancreas

adrenal gland

small intestine

colon

kidney

bladder

genital organs

Fig. 13.23 Unilateral diagram of essential parts of sympathetic nervous system of man. Preganglionic fibers are shown in solid lines; postganglionic fibers in dotted lines.

cardiac plexus is situated near the base of the heart. The large coeliac, or solar, plexus is located at the upper level of the lumbar region and surrounds the roots of the coeliac, superior mesenteric, and inferior mesenteric arteries. A heavy blow stimulating the solar plexus affects almost all parts of the body. The hypogastric plexus, in the region where the sacrum joins the last lumbar vertebra, is composed of fibers distributed to the pelvic viscera.

The autonomic portion of the peripheral nervous system is divided into two main parts, the sympathetic and parasympathetic systems (Figs. 13.23 and 13.24). The postganglionic fibers of both are, in most cases,

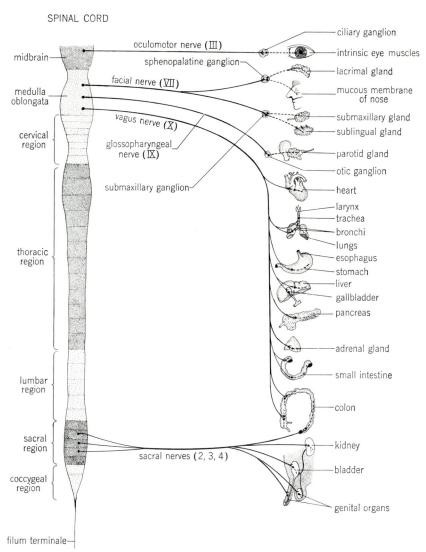

SPINAL CORD

midbrain — | oculomotor nerve (Ⅲ) — ciliary ganglion
intrinsic eye muscles
sphenopalatine ganglion — lacrimal gland
medulla oblongata — facial nerve (Ⅶ) — mucous membrane of nose
cervical region — vagus nerve (Ⅹ) — submaxillary gland
sublingual gland
glossopharyngeal nerve (Ⅸ) — parotid gland
otic ganglion
submaxillary ganglion — heart
larynx
trachea
bronchi
lungs
esophagus
stomach
liver
gallbladder
pancreas
thoracic region
adrenal gland
small intestine
lumbar region
colon
sacral region — sacral nerves (2, 3, 4) — kidney
bladder
coccygeal region
genital organs
filum terminale —

Fig. 13.24 Unilateral diagram of essential parts of parasympathetic nervous system of man. Preganglionic fibers are shown in solid lines; the very short post-ganglionic fibers in dotted lines.

distributed to all the involuntary structures of the body. The two systems, in general, work antagonistically. Whereas the one system (sympathetic) functions in such a manner as to strengthen an animal's defenses against adverse conditions by an expenditure of energy, the other (parasympathetic) is concerned with processes which tend to conserve and restore energy. Moreover, the manners in which these two systems bring about their effects differ for the most part. In both cases the postganglionic fibers give off chemical substances which, in the last analysis, bring about the effects. Postganglionic sympathetic fibers (except those going to the sweat glands and uterus, which cause vasodilatation) produce a hormone, formerly called *sympathin,* now known to be a catecholamine named *norepinephrine.* The same amine, in different concentration, along with epinephrine, is produced by the medulla of the adrenal gland. For this reason, postganglionic sympathetic fibers of this kind are usually spoken of as *adrenergic fibers.* Postganglionic parasympathetic fibers, and those postganglionic sympathetic fibers supplying the sweat glands and uterus, liberate a chemical called acetylcholine and are frequently referred to as *cholinergic fibers.* Preganglionic fibers of both sympathetic and parasympathetic nerves are also considered to be cholinergic since they liberate acetylcholine at the synapse (page 96). Epinephrine, or norepinephrine, and acetylcholine have opposite effects upon smooth muscles and glands. This further emphasizes the basic dual nature of the autonomic portion of the peripheral nervous system.

Sympathetic nervous system. (Fig. 13.23). Since the preganglionic fibers of the sympathetic nervous system connect to the spinal cord only in the thoracic and lumbar regions, the term *thoracolumbar outflow* is often applied to this portion of the autonomic system.

On either side of the ventral part of the vertebral column lies a long *sympathetic trunk,* extending from the foramen magnum to the coccyx. The trunk even enters the cranial cavity through the *carotid canal* in the temporal bone. At fairly regular intervals each sympathetic trunk bears enlargements known as the *chain ganglia* (Figs. 13.17 and 13.23). Groups of these ganglia at the anterior end of the sympathetic trunk have fused, forming three large masses, the *superior, middle,* and *inferior cervical ganglia.* The latter is located near the point of junction of the last cervical and first thoracic vertebrae. The chain ganglia are numbered according to the vertebrae opposite which they lie, but fusions may occur which obscure their segmental character. The inferior cervical ganglion on each side is frequently fused with the first thoracic ganglion, forming a *stellate ganglion.* In man there are cross communications between the lumbar portions of the sympathetic trunks.

The white visceral rami of all the thoracic spinal nerves and the first, second, and third lumbar are partly composed of preganglionic fibers which connect with the corresponding chain ganglia. They may terminate in the ganglion at the point where they enter or may send fibers up or down the sympathetic trunk for some distance. Other preganglionic fibers pass without synapses *through* ganglia of the sympathetic trunk and form connections with one or more of three large, *prevertebral ganglia* of the coeliac plexus, located in the abdominal region in front of the lumbar vertebrae. According to definition, prevertebral ganglia are sympathetic ganglia of the thorax and abdomen other that those of the sympathetic trunk. The prevertebral ganglia of the coeliac plexus include the *coeliac, superior mesenteric,* and *inferior mesenteric ganglia.* The nerves leading to these ganglia from the sympathetic trunk are called the *splanchnic nerves.*

From the chain ganglia and prevertebral ganglia arise sparsely medullated, or nonmedullated, postganglionic fibers which are distributed to the various structures in the body under involuntary control. A *gray visceral ramus* comes from each chain ganglion of the sympathetic trunk, whether or not a white visceral ramus is present. The postganglionic fibers of which the gray rami are composed are distributed via the dorsal and ventral rami of spinal nerves to the skin of the head, trunk, and limbs, where they supply smooth muscles in the walls of blood vessels, the arrector pili muscles, and skin glands. Other postganglionic fibers from the cervical ganglia at the anterior end of the sympathetic trunk pass to the intrinsic muscles of the eye, lacrimal glands, mucous membrane of the nose, palate, and mouth, and to the submaxillary, sublingual, and parotid salivary glands. From the inferior cervical ganglion and the first few chain ganglia in the thoracic region arise postganglionic fibers going to heart, larynx, trachea, bronchi, and lungs. The coeliac and superior mesenteric ganglia are the points of origin of postganglionic fibers which supply esophagus, stomach, small intestine, and the first part of the large intestine, as well as liver, pancreas, and blood vessels of the abdomen. The inferior mesenteric ganglion sends postganglionic fibers to the remainder of the large intestine and to the urogenital organs.

Preganglionic fibers pass *directly* to the medulla of the adrenal gland. This ectodermal, glandular structure, derived from neural-crest cells, is composed of specialized, or modified, sympathetic ganglionic cells which are homologous with postganglionic sympathetic neurons.

FUNCTION OF THE SYMPATHETIC SYSTEM. The action of the sympathetic neurons on the structures supplied by them has been the subject of many investigations. Among the reactions brought about by stimulation of the sympathetic system are (1) constriction of cutaneous blood vessels, causing pallor; (2) contraction of arrector pili muscles,

causing gooseflesh and making the hair stand erect; (3) secretion of sweat glands (cold sweat); (4) dilatation of the pupil, permitting more light to enter the eyeball; (5) reduction in amount of saliva secreted so that but small quantities of a thick, mucinous secretion are given off; (6) acceleration of heartbeat; (7) dilatation of the bronchi; (8) relaxation, or inhibition, of the smooth muscles of the digestive tract, causing a temporary cessation of peristalsis; (9) relaxation of bladder musculature; (10) lengthening of the urethral musculature (page 269); (11) increase in blood sugar; (12) rise in blood pressure; (13) increase in number of red corpuscles in the bloodstream; and (14) decrease in clotting time of blood.

The above reactions, when considered together, are those usually associated with pain, fear, and anger. They serve the body in a beneficial capacity in times of danger or when it is in an "aggressive" state. Many individuals, under conditions of extreme fear, may void urine involuntarily. This would seem to be paradoxical in light of the above. In such cases, however, certain reflexes, referred to as *conditioned reflexes,* usually established in early life, may bring about a reversal of reactions which normally occur in the aggressive state. Punishing or hurting a child in an effort to train it in toilet habits may be basically responsible for such poor conditioning.

Injection of epinephrine, a hormone of the medulla of the adrenal gland, causes effects which appear to be identical with those brought about by stimulation of the nerves of the sympathetic system. The hormone presumably acts upon the structures supplied by sympathetic nerve endings. Under normal circumstances the adrenal medulla secretes only insignificant amounts of epinephrine into the bloodstream. In times of stress, however, significant amounts may be liberated and the action of the hormone reinforces that of the sympathetic system so that maximal efficiency results. This action of epinephrine has been called the "emergency mechanism" of the body, for during times of stress the body is thus better adapted for self-preservation.

The adrenal medulla and its secretion are not essential to life. Sympathetic effects may be secured after removal of the adrenal medulla. It has already been indicated that stimulation of sympathetic postganglionic fibers (with the exception of those supplying the sweat glands and uterus, which are cholinergic) induces them to secrete norepinephrine. Epinephrine and norepinephrine are relatively stable compounds (catecholamines) which, circulating in the bloodstream, will simultaneously stimulate *all* structures (except sweat glands and uterus) supplied by the sympathetic system. Although these substances act rapidly, the effect lasts for only a short time.

Other tissues, known as *chromaffin tissues,* located in various parts of the body, also secrete norepinephrine. They, like the adrenal medulla, are actually modifications of sympathetic ganglionic cells.

Parasympathetic nervous system. The term *craniosacral outflow* is frequently used to designate the complex of preganglionic fibers of the parasympathetic nervous system (Fig. 13.24). The term is appropriate, since only certain cranial and sacral spinal nerves are involved.

Four cranial nerves are composed, at least in part, of preganglionic parasympathetic fibers. They are the oculomotor (III), facial (VII), glossopharyngeal (IX), and vagus (X) nerves. The ganglia in which they terminate are situated close to, or in, the organs supplied by this system. Hence the preganglionic fibers are rather long, and

the postganglionic fibers are very short.

The preganglionic parasympathetic components of the oculomotor nerve terminate in a *ciliary ganglion,* about the size of the head of a pin, situated in the back part of the orbit. Here they synapse with postganglionic fibers which pass to the eyeball, supplying the sphincter muscle of the iris as well as the ciliary muscles.

The facial nerve contains preganglionic fibers which pass either to the *sphenopalatine* or to the *submaxillary ganglion* in the head region. In the sphenopalatine ganglion, located close to the sphenopalatine foramen, certain fibers of the facial nerve synapse with postganglionic neurons, the fibers of which pass to the lacrimal gland, mucous membrane of the nose, palate, and upper part of the pharynx. Certain fibers of the chorda tympani branch of the facial nerve are preganglionic fibers which pass to the small submaxillary ganglion situated just above the submaxillary salivary gland. Short postganglionic fibers arising in the submaxillary ganglion are distributed to the submaxillary and sublingual glands.

An *otic ganglion,* located close to the foramen ovale of the sphenoid bone, receives preganglionic fibers of the glossopharyngeal nerve. Postganglionic fibers pass to the parotid gland and mucous membrane of the mouth.

The preganglionic fibers of the vagus nerve, like those of the other three cranial nerves forming part of the parasympathetic system, terminate in ganglia, but the ganglia are very small and are located in the walls of the structures which they supply. It is believed that those of the esophagus, stomach, small intestine, and upper part of the large intestine connect with the *plexuses of Auerbach* and *Meissner.* From these plexuses postganglionic fibers are distributed to the smooth muscles and glands of the alimentary tract. Other parasympathetic fibers of the vagus innervate the heart, larynx, trachea, bronchi, lungs, blood vessels of abdomen, liver, gallbladder, and pancreas.

The part of the parasympathetic system known as the *sacral outflow* is composed of efferent fibers which course through the white visceral rami of the second, third, and fourth sacral nerves, which together form the *pelvic nerve.* These preganglionic fibers, like those of the vagus, pass directly to the structures which they supply, to terminate in ganglia in, or very near, the organs. The pelvic nerve supplies the lower part of the large intestine, kidneys, bladder, and reproductive organs. Postganglionic fibers within these organs are relatively short.

FUNCTION OF THE PARASYMPATHETIC SYSTEM. Stimulation of the various components of the parasympathetic system brings about effects which are, in general. opposite to those secured by stimulating the sympathetic nerves supplying the same organs. Among these reactions are (1) dilatation of blood vessels (except the coronary vessels of the heart); (2) constriction of the pupil; (3) increase in salivary and gastric secretion; (4) constriction of bronchi; (5) contraction of walls of the digestive tract, bringing about peristalsis and other types of contractions: (6) contraction of bladder musculature; (7) shortening of the urethral musculature (page 269); and (8) dilation of blood vessels of the external genital organs.

The above reactions, when considered as a group, are those associated with comfortable or pleasurable sensations in the body and those which conserve energy. They form the mechanism of normal functions when the body is in a "receptive" state.

It has already been mentioned that the parasympathetic system brings about its effects by the liberation of a chemical substance, *acetylcholine,* at the terminal nerve endings of the postganglionic fibers. Acetylcholine may also be produced in other parts of the body besides the parasympathetic nerve endings (page 96). It is an unstable chemical substance, quickly rendered inactive in the body by an enzyme, *acetylcholinesterase,* the action of which is measured in thousandths of a second. Hence its effect is limited to the immediate region in which it is produced. It cannot affect parts of the body at any distance from the point of secretion or liberation. For example, when a savory odor induces the parasympathetic system to bring about muscular contractions of the stomach, it does not at the same time cause a desire to void, nor does it cause the pupils to constrict. The changes in the body associated with sexual stimulation, such as erection and glandular secretion, are the result of stimuli affecting the parasympathetic fibers of the sacral outflow which are distributed to the genital organs.

Hypothalamus. It has been amply demonstrated that the hypothalamus is a region of great importance in connection with the proper functioning of the autonomic nervous system. It is here that the nerve centers controlling the sympathetic and parasympathetic systems are located. Neurosecretory cells * supplying the posterior lobe of the pituitary are also connected with this portion of the diencephalon. It is in the hypothalamus that the autonomic system forms connections with all the other nervous tissues of the body and is thus influenced by every external and internal change. The possible role of the hypothalamus as an endocrine organ has

* Neurosecretory cells (page 96) are nerve cells in which secretory material formed in the cell bodies passes down the axons and is discharged into the circulatory system at the terminal end bulbs of the axons.

already been mentioned (pages 391 and 624).

The stable nature of the internal environment of the body, referred to as *homeostasis*, is controlled by hypothalamic nuclei. Such phenomena as constancy in amount of blood sugar, the part played by the sweat glands in regulating body temperature, composition of body fluids, and rate of heartbeat under different conditions are regulated and integrated in the hypothalamus.

Lesions of the hypothalamus are beleived to be associated with such abnormal conditions as *diabetes insipidus* and *adiposogenital dystrophy*.

Autonomic system of lower chordates. The general scheme of arrangement of the autonomic system in tetrapods is similar to that of man. The lower vertebrates show a progressive complexity of the autonomic system as the evolutionary scale is ascended.

If a peripheral autonomic system is actually present in amphioxus, it is apparently represented by the parasympathetic system alone. It will be recalled that in amphioxus dorsal and ventral spinal nerve roots remain separate. Visceral motor neurons course through the segmentally arranged dorsal spinal nerves directly to the visceral organs, where they form synapses with cells making up a nervous plexus. No intervening ganglia or relay-type connections are discernible. In cyclostomes segmental sympathetic ganglia are present, but these are not connected, nor are sympathetic trunks to be found. In lampreys, as in amphioxus, the dorsal and ventral spinal nerves do not unite. Preganglionic sympathetic fibers emerge from the spinal cord through the dorsal spinal nerves. In the hagfishes, in which the roots do unite, these fibers apparently pass through the ventral roots. A parasympathetic system appears for the first time in lampreys since visceral motor branches of

the vagus nerve (X) terminate in the walls of the heart and gut. The hagfish heart apparently lacks vagal innervation. The autonomic system is much better developed in elasmobranchs. Sympathetic and parasympathetic components are clearly demarcated. The sympathetic elements are confined to the abdominal region, there being no cranial sympathetic nerve fibers. It seems that no sympathetic nerve fibers innervate the hearts of fishes. Nevertheless, it has been shown that catecholamines cause an increase in the rate and force of the heartbeat in elasmobranchs and teleosts but not in the hagfish. It seems that regulation of the rate of heartbeat in fishes lacks the precision of that found in tetrapods. The consecutive sympathetic ganglia are sometimes joined by lateral connectives, but, in general, the sympathetic strands dorsal to the body cavity have a rather diffuse arrangement. All sympathetic ganglia connect with spinal nerves by white visceral rami, and each ganglion is closely associated with a small mass of chromaffin tissue. Parasympathetic fibers of the vagus pass to esophagus and stomach but do not seem to go to the intestine or urogenital organs. Parasympathetic fibers are also present in the oculomotor (III), facial (VII), and glossopharyngeal (IX) nerves. In teleosts, the autonomic system shows a still greater advance and considerable similarity to that of tetrapods. The sympathetic ganglia on each side are connected by a longitudinal trunk, forming a chain. This extends from the first spinal nerve to the region of the first hemal arch. The two trunks come together and fuse into a single strand between the kidneys, but diverge again posteriorly, becoming paired at the level of the first hemal arch. Parasympathetic components seem to be confined to the oculomotor and vagus nerves. The Dipnoi also have a pair of longitudinal sympathetic trunks, but segmental swellings, or ganglia, are inconspicuous or absent, although they are joined by metameric white visceral rami. The oculomotor and vagus nerves apparently contain parasympathetic fibers. In urodele amphibians the condition is similar to that of teleosts except that the sympathetic system is divided into cephalic, cervical, abdominal, and caudal regions. White visceral rami form connections in all but the cephalic region. Only cranial parasympathetic fibers are present, and these seem to be confined to the oculomotor and vagus nerves. In anurans, for the first time, a sacral outflow appears, in addition to the usual cranial parasympathetic fibers. The autonomic system in caecilians has not been extensively studied. A typical sympathetic system is present, but little is known about the parasympathetic system. For further information on the autonomic nervous system of lower chordates, the interested student should consult a paper by J. A. C. Nicol which gives a comprehensive description and discussion of this system (1952, Biol. Revs. Cambridge Phil. Soc., **27**:1–49).

REFLEXES

The term *reflex action* refers to an immediate involuntary response to a sensory stimulus. The so-called "knee jerk" is a familiar example of reflex action. If the knees are crossed and the patellar ligament is tapped, the lower leg jerks forward involuntarily because of sudden contraction of the *quadriceps femoris* muscle. In this case impulses set up in sensory fibers in the patellar ligament travel from the knee to the spinal cord, where they form synapses with motor fibers leading to the quadriceps femoris muscle. The pathway through which such impulses travel is known as a *reflex arc* (Fig. 13.25).

Reflex arcs make up the *functional units* of the nervous system. A reflex arc consists of two or more neurons together with a nonnervous component, called the *effector,* which in most cases is either a muscle cell or a glandular cell. A *receptor* element is present which receives stimuli and transmits impulses thus set up in the neuron. Receptors are sensory neurons, the dendritic ends of which are frequently associated with special accessory structures which facilitate the reception of stimuli. The terminal arborization of the axon of the receptor makes direct or indirect connections with the dendrites of a motor neuron within brain or spinal cord. The axons of the motor neuron terminate in the effector, the action being mediated by means of a neurohumor, in all probability. The simplest type of reflex arc consists of a sensory neuron synapsing directly with a motor neuron which leads to an effector.

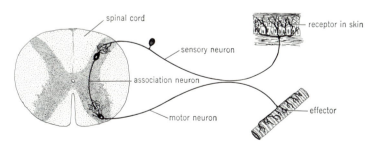

Fig. 13.25 Diagram showing arrangement of usual type of reflex arc.

In most cases one or more *intermediate neurons* are present between receptor and effector neurons. Through these, impulses are transmitted from sensory neurons to other neurons in brain or cord. Intermediate neurons on the same side of the central nervous system are called *association neurons* (Fig. 13.25). If they cross to the opposite side, they are known as *commissural neurons*. The presence of intermediate, association, or commissural fibers determines whether a reflex will occur on the same side or on the opposite side to that in which the sensory impulse travels.

Reflexes are commonly thought of as taking place only in the central nervous system, but they occur in the outlying ganglia of the autonomic nervous system as well.

Two types of reflexes, unconditioned and conditioned, are recognized. *Unconditioned reflexes* are those which are present at birth and which do not depend upon previous experience. They are characteristic of the species and are inherited as such. For example, the mere sight of food has no effect upon a newborn puppy, but placing food in the mouth will bring about a copious secretion of saliva. Later the sight or smell of food even at some distance may cause saliva to flow freely. Previous experience has established a *conditioned reflex*. The past association of the sight or odor of food with the secretion of the salivary glands has brought it about.

Much of the behavior of animals and human beings is due to conditioned reflexes, which often show a high degree of complexity. The more often certain reflexes occur, the more firmly do they become established and the more difficult are they to overcome or break.

The building up of proper reflexes is of utmost importance in the training of children. Among the earliest autonomic reflexes that take place in a child are those concerned with emptying the rectum and bladder. These are unconditioned reflexes. Reflex activity consists of contraction of the muscular walls in response to tension of the organ which stimulates sensory receptors. Evacuation of rectum and bladder results in a feeling of comfort. In these examples, the effectors, or muscles involved in the reflexes, are those supplied by parasympathetic nerve fibers. It may well be that the early association of comfort with contractions of these hollow organs and relaxation of the sphincters accounts for the feeling of pleasure with which the parasympathetic system is associated in later life. On the other hand, the discomfort felt by the child when the sphincters do not relax and when the organs are distended may be associated with feelings of anger, pain, and fear, with which the sympathetic system is later associated.

It is quite possible to reverse the action of these mechanisms by the establishment of the wrong kinds of conditioned reflexes during the early training period of a child. It has been already pointed out that unwise procedures, such as spanking or hurting a child in an effort to train it in toilet habits, must be considered in persons who later in life demonstrate abnormal reactions to certain stimuli. Poor conditioning of this sort may well explain why some adults have a tendency to urinate or defecate involuntarily when anxious or frightened, whereas the usual tendency is to do so during periods of uncontrollable laughter. The aim of modern education and habit training is to establish the association of desirable activities with parasympathetic reactions and undesirable activities with sympathetic responses.

SUMMARY

1. Almost all the nervous system is derived from the neural tube, neural crests, and ectodermal placodes, structures that appear very early in embryonic development.

2. The cells making up nervous tissue have the properties of irritability and conductivity more highly developed than do those of other tissues in the body. They are all derived from ectoderm.

3. Nerve fibers are, in general, of two types: medullated and very sparsely medullated (sometimes called nonmedullated), depending upon the thickness and appearance of their myelin sheaths. Medullated fibers are glistening white in appearance; sparsely medullated, or nonmedullated, fibers are gray.

4. Bundles of nerve fibers are called fiber tracts or nerves, depending upon their location. Fiber tracts are found in the brain and spinal cord; nerves are peripheral structures. Aggregations of nerve-cell bodies are referred to as nuclei or ganglia, again depending upon where they are located, nuclei being confined to the brain or spinal cord. A ganglion is an enlargement in the course of a nerve.

5. The nervous system is divided into two main parts: (1) the central nervous system, composed of brain and spinal cord, and (2) the peripheral nervous system, made up of cranial and spinal nerves. An important part of the peripheral nervous system, and often referred to as the autonomic system, consists of portions of certain cranial and spinal nerves, as well as numbers of outlying ganglia, connected with those structures of the body under involuntary control.

6. The central nervous system is composed of gray matter and white matter. Gray matter consists of nerve-cell bodies and sparsely medullated fibers, together with nonnervous ependymal and neuroglia cells. White matter is made up of bundles of medullated fibers.

7. Early in development the anterior end of the neural tube enlarges to form the brain. Two constrictions occur, forming three primary vesicles: the forebrain, or prosencephalon; midbrain, or mesencephalon; and hindbrain, or rhombencephalon. The first and third become secondarily divided, but the midbrain undergoes no division. The remainder of the neural tube becomes the spinal cord.

8. The prosencephalon differentiates into two regions: the telencephalon, consisting of olfactory lobes (rhinencephalon) and cerebral hemispheres, and the diencephalon, or 'tween brain. The rhombencephalon gives rise to an anterior metencephalon, the dorsal part of which forms the cerebellum, and a posterior myelencephalon, or medulla oblongata.

9. The original cavity of the neural tube forms the ventricles of the brain and the central canal of the spinal cord. A pair of lateral ventricles (I and II) is present in the cerebral hemispheres. These open into the third ventricle, located in the diencephalon, through the interventricular foramina, or foramina of Monro. The fourth ventricle lies in the medulla oblongata. It is connected with the third ventricle by the cerebral aqueduct. Expansions of the aqueduct in the optic lobes of lower

forms make up the mesocoele. A cavity, the metacoele, may extend from the fourth ventricle into the cerebellum. These cavities are filled with cerebrospinal fluid.

10. The spinal cord in tetrapods bears two enlargements at the levels where nerves going to the limbs arise. These are the cervical and lumbar enlargements opposite the forelimbs and hind limbs, respectively. In most forms the posterior end of the spinal cord tapers down to a fine thread, the filum terminale. There is a tendency toward a shortening of the spinal cord, and it rarely extends throughout the length of the vertebral column. In the spinal cord, gray matter is centrally disposed and almost completely surrounded by white matter. The gray matter is arranged in the form of a pair of dorsal, or sensory, columns and a pair of ventral, or motor, columns. The dorsal column has upper somatic sensory and lower visceral sensory regions. The ventral column, in turn, is composed of upper visceral motor and lower somatic motor regions. Sensory neurons, the cell bodies of which lie outside the central nervous system, form connections with cells in the dorsal column. The cell bodies of motor neurons lie within the ventral column of gray matter.

11. Because of the unequal rate of growth of various components during embryonic development, certain bends or flexures occur. There are three of these: (1) a cephalic flexure between forebrain and midbrain, (2) a cervical flexure near the junction of medulla oblongata and spinal cord, and (3) a pontine flexure in the region of the metencephalon. Later the brain tends to straighten out and the flexures diminish or disappear, but in birds and mammals the cephalic flexure is retained. Thus in these forms the front part of the brain comes to lie on top of the posterior portions.

12. It is in the development of the telencephalon that the greatest changes appear in the brains of vertebrates. In lower forms, the dorsal part, or pallium, is thin-walled, the ventral portion becoming the thickened corpus striatum. An outgrowth of the anterior portion of each cerebral hemisphere becomes the olfactory lobe, or rhinencephalon, concerned with the sense of smell. In higher forms the cerebral hemispheres have become the more significant part of the telencephalon and the olfactory lobes are of secondary importance. As the vertebrate scale is ascended there is an increasing tendency for nerve cells from the inner layer of gray matter in the pallium to migrate out to the periphery. A new area, the neopallium, first appears in reptiles in the outer, anterodorsal part of each cerebral hemisphere. The growth and development of the neopallium account for the large size of the cerebral hemispheres in mammals. Its outer surface, into which nerve cells have migrated, forms a gray layer, the cerebral cortex. The cerebral hemispheres are usually smooth, but in most mammals folds, called gyri, are present, separated by depressions, known as sulci. These increase the surface area of the cerebral cortex. In mammals, a white mass, the corpus callosum, contains nerve fibers connecting the cortical areas of the two cerebral hemispheres. The well-developed cerebral cortex of man is responsible for those phenomena which characterize the high degree of his psychic development.

13. The thickened lateral portion of the diencephalon on each side makes up the

thalamus. This portion contains important relay centers and consists largely of numbers of nuclei of gray matter. The floor, or hypothalamus, contains centers which integrate the activities of the peripheral autonomic nervous system with those of other nervous tissues. A ventral evagination of the hypothalamus forms the infundibulum, the distal portion of which gives rise to the posterior lobe of the pituitary gland. The anterior part of the roof of the diencephalon remains epithelial and, together with the pia mater, forms a tela choroidea. Vascular folds of the tela choroidea become the anterior choroid plexus where cerebrospinal fluid is liberated. A prominent fold at the anterior end, present during early development, is actually a portion of the telencephalon. It is called the paraphysis. The epithalamus, or roof of the diencephalon, posterior to the tela choroidea, consists largely of the epiphysial apparatus. In numerous forms this gives rise to parapineal, or parietal, and pineal outgrowths probably associated originally with additional eyes. In higher forms only the pineal body persists.

14. The floor and walls of the mesencephalon are thick and composed of fiber tracts connecting forebrain and hindbrain. The roof consists of a thick layer of gray matter, the optic tectum. Two prominences are present in the roof of the mesencephalon of lower forms. They are the corpora bigemina, or optic lobes, which serve as visual centers. In snakes and mammals there are four prominences, the corpora quadrigemina. The anterior pair serves as visual centers, and the posterior pair as auditory centers.

15. The dorsal part of the metencephalon becomes the cerebellum, the function of which is to coordinate the neuromuscular mechanism of the body. It is poorly developed in sluggish forms but is well developed in the more active vertebrates. In a few birds and in mammals there appears on the ventral side of the metencephalon a sort of bridge of nerve fibers, the pons. These fibers connect the cerebral hemispheres with the cerebellum in those forms in which certain areas of the cerebral cortex have assumed control of body movements.

16. The myelencephalon forms the medulla oblongata. The lateral and ventral walls are markedly thickened, but the dorsal wall retains its epithelial character. Together with the pia mater it forms a tela choroidea and posterior choroid plexus. The ventral and lateral walls contain large white fiber tracts as well as columns of gray matter similar to those of the spinal cord. Certain nuclei of the visceral motor column serve as centers for such vital functions as heartbeat and respiration. The anterodorsal portion of the medulla oblongata, referred to as the acousticolateralis area, is a development of the somatic sensory column. It contains nuclei associated with nerves from the lateral-line system, if present, and from the inner ear. Cranial nerves V to X (or XII) form connections with the medulla. Paired pyramids in the posterior portion mark the region where several fiber tracts cross (decussate) to the opposite side of the brain.

17. Brain and spinal cord are surrounded and protected by membranes called meninges. A single meninx primitiva is present in fishes. Two membranes, an inner pia-arachnoid membrane and an outer dura mater, exist in amphibians, reptiles,

and birds. In mammals, however, a further differentiation of the pia-arachnoid membrane occurs, forming an inner pia mater and an outer arachnoid membrane. The space between them is filled with cerebrospinal fluid.

18. The peripheral nervous system, made up of cranial and spinal nerves, consists of several kinds of fibers, each mediating a different type of function. Somatic sensory fibers, from peripheral parts of the body, lead to the somatic sensory column of gray matter in the central nervous system. Somatic motor fibers lead directly from the somatic motor column of gray matter to the periphery, where they supply voluntary muscles, except those in the gill region. Visceral sensory fibers bring impulses from the viscera to the visceral sensory column. Visceral motor fibers of certain cranial nerves pass from the visceral motor column directly to gill muscles, at least in aquatic forms. Preganglionic visceral motor fibers go to outlying ganglia where they synapse with autonomic neurons, the gray, sparsely medullated, or non-medullated, fibers of which go to smooth muscles and glands in the viscera and skin.

19. Spinal nerves are paired and segmentally arranged. They arise from the spinal cord by two roots, a dorsal root bearing a ganglion and a ventral root which lacks a ganglion. In higher forms these roots lie within the spinal canal, and the spinal nerves, formed by their union, pass out of the vertebral column through intervertebral foramina. Most spinal nerves give off three branches, or rami, a short distance beyond the point where they emerge. Dorsal and ventral rami are usually present, but a white visceral ramus is lacking in some. In the regions of the limbs, cross connections between certain spinal nerves form plexuses.

20. The cranial nerves, arising from the brain, number 10 in anamniotes and 12 in amniotes. The twelfth nerve and part of the eleventh nerve represent spinal nerves which have been taken over by the brain. Some cranial nerves are purely sensory (I, II, and VIII), and others are purely motor (III, IV, VI, XI, and XII). The remainder (V, VII, IX, and X) are mixed sensory and motor, but even these differ from spinal nerves in that if their roots unite they do so before emerging from the medulla oblongata. In lower vertebrates, nerves VII, IX, and X have branches related to the lateral-line system, a system peculiar to the aquatic mode of life. These branches disappear in terrestrial forms. Cranial nerves III, VII, IX, and X bear preganglionic visceral motor fibers which make up part of the parasympathetic nervous system. Another cranial nerve called the terminal nerve (0), discovered in 1894, is a sensory nerve bearing one or more ganglia and passing to the olfactory mucous membrane from the cerebral hemispheres. Its function is not clear.

21. The olfactory nerve (I) is a special somatic sensory nerve passing from the olfactory epithelium to the olfactory lobe of the brain.

22. The optic nerve (II) is not actually a nerve, but rather a fiber tract of the brain, passing from the retina of the eye to the optic lobes. The optic nerves of the two sides cross beneath the diencephalon to form the optic chiasma. In mammals having binocular vision, decussation is not complete, only half the fibers crossing to the opposite side.

23. The oculomotor nerve (III) is a somatic motor nerve arising from the mid-

brain and supplying the superior, inferior, and internal rectus muscles, the inferior oblique eye muscle, and the muscle elevating the upper eyelid. It also contains preganglionic autonomic fibers which go to the ciliary ganglion. Postganglionic fibers then lead to the iris and to the muscles of accommodation.

24. The trochlear nerve (IV), coming from the posterodorsal side of the midbrain, is a somatic motor nerve which supplies the superior oblique eye muscle.

25. The trigeminal nerve (V) is the nerve of the mandibular arch. It arises from the medulla oblongata and has three branches; ophthalmic, maxillary, and mandibular. The first two contain somatic sensory fibers; the last consists of both somatic sensory and visceral motor fibers. In fishes the ophthalmic is composed of superficial and deep (profundus) portions. In higher forms the two portions are not distinct and a single ophthalmic nerve is present. It supplies somatic sensory fibers to the conjunctiva, cornea, iris, ciliary body, lacrimal gland, mucous membrane of the nose, and the skin of the forehead, nose, and eyelids. The maxillary branch is the sensory nerve to the upper jaw. The mandibular branch sends somatic sensory fibers to the lower jaw and visceral motor fibers directly to the muscles used in chewing. The trigeminal nerve bears a Gasserian ganglion.

26. The abducens nerve (VI) is a somatic motor nerve, arising from the ventral part of the medulla and supplying the external rectus eye muscle, the nictitating membrane and retractor bulbi muscle, if present. In lampreys it also supplies the inferior rectus.

27. The facial nerve (VII), arising from the medulla, is a mixed nerve supplying the second visceral, or hyoid, arch. In lower aquatic forms it has large somatic sensory components supplying structures of the lateral-line system. In mammals its visceral sensory portion supplies the taste buds of the anterior two-thirds of the tongue. The major part of the nerve is visceral motor and is distributed directly to the muscles of the face. Preganglionic visceral motor fibers of the facial pass either to the submaxillary or to the sphenopalatine ganglion. From the submaxillary ganglion arise postganglionic fibers leading to the submaxillary and sublingual salivary glands; those from the sphenopalatine ganglion go to the lacrimal gland and mucous membrane of the nose. The facial nerve bears a geniculate ganglion at its base.

28. The auditory (vestibulo-cochlear) nerve (VIII) is purely somatic sensory. It connects with the medulla oblongata and carries impulses from the inner ear to the brain. The auditory nerve supplies the organs of equilibrium and those for sound reception.

29. The glossopharyngeal nerve (IX), arising from the medulla, carries visceral sensory and visceral motor fibers and in fishes sends some somatic sensory fibers to the lateral-line system. It is the nerve of the third visceral arch and the second gill pouch. In lower forms, it bears a petrosal ganglion near its base. In higher forms, two ganglia, superior and petrosal, are present. In higher vertebrates, visceral motor fibers pass directly to the muscles of the larynx derived from the third gill arch. In mammals, visceral sensory fibers pass to taste buds on the posterior one-third of

the tongue and mucous membrane of the pharynx. Preganglionic autonomic fibers go to the otic ganglion, synapsing with postganglionic fibers which supply the parotid gland.

30. The vagus nerve (X), coming from the medulla by several roots, is a mixed nerve. In lower aquatic forms, a somatic sensory branch supplies the lateral line proper. Visceral sensory and visceral motor fibers are distributed to the remaining visceral arches and to the pharynx. Visceral sensory fibers and visceral motor autonomic preganglionic fibers also pass to the heart (except in the hagfish), lungs (in higher forms), and all parts of the digestive system except those at the posterior end. Two ganglia, jugular and nodosal, are borne by the vagus nerve in higher forms.

31. The spinal accessory nerve (XI), arising from the medulla in amniotes, bears visceral motor fibers, some of which accompany the vagus to larynx and pharynx. Other somatic motor fibers pass to neck muscles which are possibly derived from myotomes. The nerve apparently is made up of a branch of the vagus which has become separated from that nerve, as well as of fibers from certain anterior cervical spinal nerves.

32. The hypoglossal nerve (XII), also arising from the medulla in amniotes, is a somatic motor nerve carrying fibers to the intrinsic muscles of the tongue and to certain muscles of the lower jaw. It actually represents the motor portions of two or three anterior spinal nerves which have been taken over by the brain.

33. The sympathetic components of the peripheral autonomic nervous system in man consist of medullated, visceral motor, preganglionic fibers leading from the spinal cord to outlying ganglia and of sparsely medullated postganglionic fibers leading from outlying ganglia to all muscles and glands of the body under involuntary control. Preganglionic fibers course through the visceral rami of all the thoracic and the first, second, and third lumbar spinal nerves. They are referred to as the thoracolumbar outflow. On either side of the vertebral column a sympathetic trunk courses longitudinally. Each trunk bears metameric chain ganglia with which the visceral rami of the above-mentioned spinal nerves connect. Other sympathetic ganglia—coeliac, superior mesenteric, and inferior mesenteric—lie among the viscera and are known collectively as the prevertebral ganglia.

34. The parasympathetic portion of the autonomic nervous system in man is composed of portions of cranial nerves III, VII, IX, and X and of the visceral rami of sacral nerves 2, 3, and 4. These consist of long preganglionic fibers, frequently referred to as the craniosacral outflow. They terminate in ganglia close to, or in, the organs which they supply. Postganglionic fibers are therefore very short. Four ganglia in the head region are associated with cranial nerves III, VII, and IX. They are called the ciliary (nerve III), submaxillary, sphenopalatine (nerve VII), and otic (nerve IX) ganglia, respectively. The ganglia associated with the vagus (X) lie within the walls of the structures which this nerve supplies. In the digestive tract connections are formed with the plexuses of Auerbach and Meissner.

35. The two parts of the autonomic system work antagonistically, thus serving to control the functions of all involuntary mechanisms in the body.

36. In other tetrapods the structure of the autonomic portion of the nervous

system is, in general, similar to that of man, but in lower forms it is less complicated. Sympathetic ganglia in cyclostomes are not connected. In elasmobranchs delicate nerve strands connect certain sympathetic ganglia, but true sympathetic trunks are lacking. In teleosts the sympathetic ganglia are connected for the first time by a longitudinal trunk. A parasympathetic system may first appear in amphioxus. In fishes and amphibians the cranial outflow seems to be confined to the oculomotor and vagus nerves. A sacral outflow first appears in anuran amphibians.

| 14 |

RECEPTOR
ORGANS

Closely associated with the nervous system are certain sensory receptor organs. They are so constructed as to be capable of responding to various stimuli by setting up impulses which are in turn transmitted by nerve fibers to the central nervous system. In the higher centers of the brain such impulses are interpreted as *sensations*. The thalamus serves as such a center in most vertebrates, but a well-developed cerebral cortex, if present, may be the seat of further sensory integration. The receptor organs themselves do not perceive anything but merely serve as a means of access to the nervous system.

In lower forms of animal life the entire integument may possess a sensory function. Sensory receptors, in such forms, are of a generalized character but are invariably derived from ectoderm. This is what one would expect, since the ectoderm is primitively the outer layer of the body and most likely to be affected by environmental change. Even though most vertebrate sense organs have been modified greatly in the course of evolution into structures of considerable complexity, nevertheless the actual receptor cells, with rare exceptions, are in each case derived from ectoderm.

Most of what is known about sense organs has been gathered from studies of man. In other animals little can be determined concerning their sensations. Investigators are therefore confined, for the most part, to studies of muscular and glandular responses to stimuli. Behavioral responses are usually analyzed in terms of human reactions under similar conditions. Much caution must be exercised, however, in interpreting the results of behavior experiments. Nevertheless, similarity in structure of the sense organs of man and other vertebrates leads us to assume that similar functions and even sensations are experienced by all.

The receptor organs with which we are most familiar are those for the senses of sight, hearing, smell, taste, and touch. However, numerous other sense organs are also present in the body. They include receptors for such senses as pain, heat, cold, equilibrium, hunger, thirst, fatigue, sex, and muscle position. Most of the sense organs are stimulated by environmental disturbances of various kinds. They are therefore spoken of as *external sense organs*. The receptors for the external sense organs are called *exteroceptors. Internal sense organs,* on the other hand, are affected by stimuli originating within the body itself. Certain internal sensory receptors are located in tendons, joints, skeletal muscles, heart, and other areas. The structures in which they lie are not in direct contact with the environment but may, nevertheless, be affected by factors of environmental origin. Receptors located in the walls of the digestive organs are known as *interoceptors*. They are responsible for such internal sensations as hunger, thirst, nausea, and others. No specific receptor organs have been discovered for some of these senses, the nerve endings themselves probably serving as receptors. Yet it would seem obvious that some sort of dif-

ferentiation must exist at the ends of dendrites of the nerves in question. Some authorities list over 30 different senses. The actual number enumerated by a particular author depends largely upon the manner in which the various senses are classified.

Terminology. The different kinds of receptors are named according to the nature of the stimulus by which they are affected. The following list includes the names of those more commonly known:

EXTERNAL SENSES	Exteroceptors
Sight	Photoreceptors
Hearing	Phonoreceptors
Smell	Olfactoreceptors
Taste	Gustatoreceptors
Touch (transient contact)	Tangoreceptors
Pressure (sustained contact)	Tangoreceptors
Temperature	Thermoreceptors
Heat	Caloreceptors
Cold	Frigidoreceptors
Pain	Algesireceptors
Currents of water	Rheoreceptors
INTERNAL SENSES	
Muscle position	Proprioceptors
Equilibrium	Statoreceptors
Hunger, thirst, etc.	Interoceptors
Pain	Algesireceptors

The term *teloceptor* is sometimes used to distinguish those exteroceptors which are able to detect stimuli coming from a distance. The olfactory, auditory, and visual receptors are examples. Caloreceptors might even be included under this category.

Each sense organ functions only within circumscribed limits. The limits, however, vary in different species of animals. Sounds of high frequency, for example, which can be detected by some mammals may be inaudible to man. The visible portion of the spectrum, produced when ordinary light passes through a prism, is in man confined to red on one hand and violet on the other. It has been demonstrated that certain insects can perceive ultraviolet light, which is invisible to man. The olfactory sense of the dog is clearly superior to the olfactory sense of man.

In some cases the application of a different kind of stimulus to an exteroceptor will affect it in a manner similar to that produced by the usual type of stimulus. Thus, mechanical or electrical stimulation of the retina of the eye, even if performed in total darkness, gives a sensation of light. Similarly, mechanical stimulation of the tympanic membrane gives a sensation of sound. When a certain chemical called menthol is applied to the skin, it produces an impression of cold.

The nerve fibers which convey impulses from receptor organs to the central nervous system are called sensory fibers. In the previous chapter it was emphasized that these may be grouped in two categories: somatic sensory and visceral sensory nerves. Somatic sensory nerves are those which are associated with exteroceptors, proprioceptors, and statoreceptors. Visceral sensory nerves, from the visceral organs, are associated with interoceptors.

SENSE OF SIGHT

Eyes are complicated photoreceptors, but not all photoreceptors are eyes. In certain protozoans a restricted area of the cell may serve as a photoreceptor. The eyespot of *Euglena* is an example. Scattered photosensory cells are frequently present in the integument of invertebrates. In some forms, aggregations of such sensory cells may form "eyes" of various types. Lateral-line organs in the skin of lampreys are said to be light-sensitive.

The eyes of vertebrates are highly specialized structures which have no true homo-logues in animals belonging to other phyla. They appear for the first time in cyclostomes, their evolutionary history being obscure. Reference has already been made to the presence of photoreceptive cells in one or two median eyelike structures, the parietal, or parapineal, organ, and the pineal body, or epiphysis. The reader is referred to pages 622 to 624 in Chap. 13, Nervous System, for a review of their structure and possible functions.

Photoreceptors of Protochordates

The *anterior pigment spot* of amphioxus, frequently referred to as an *eyespot*, has been shown to lack any sensory function. The so-called *infundibular organ*, composed of flagellated, ependymal cells, the long processes of which extend into the cerebral vesicle, is thought by some to function as a crude sort of eye which can detect shadows cast when light strikes the anterior pigment spot. Certain photosensory, ganglionic *cells of Joseph* are present on the dorsal side of the head region of amphioxus. Other photosensory cells, partially surrounded by pigment capsules, are located along the ventral side of the spinal cord.

The larval ascidian has a single eye with a retina and lens. It lies within a dilatation of the anterior end of the neural tube, the *cerebral vesicle*. The eye is situated slightly off the midline and is apparently the persistent member of a pair of structures, the other degenerating very early during development and being represented only by a small vestigial mass of tissue. Several reasons have been advanced for not homologizing the ascidian eye with those of vertebrates.

Vertebrate Eye

As an introduction to the discussion of the vertebrate eye, we shall first consider that of man, since its structure and function are better understood than are those of other vertebrates. The human eyeball (Fig. 14.1) is almost spherical in shape. It is composed of three coats, or layers, of which only the outer one is complete. The others are modified near the exposed portion of the eyeball.

The outermost coat protects and gives shape to the eyeball. It is called the *fibrous tunic.* This thick, tough layer is divided into two regions: a transparent exposed portion, the *cornea,* and a larger, opaque, posterior section, the *sclera,* or *sclerotic coat,* the anterior part of which is commonly referred to as the "white" of the eye. The muscles which move the eyeball are attached to the sclera. A thin membrane, the *conjunctiva,* which is continuous with the inner surface of the eyelids, covers and is fused to the outer surface of the cornea, constituting the *epithelium* of the latter. The cornea is practically circular in outline. It is somewhat thicker than the sclera and bulges out in front of it. The curved surface of the cornea is of importance in focusing light rays, thus sup-plementing the lens in this respect. Irregularities in the shape of the cornea are often responsible for the condition known as *astigmatism,* an optical defect causing imperfect images or indistinctness of vision.

The middle coat is called the *uvea.* It is divided into three regions: (1) a vascular, pigmented *choroid coat,* closely adherent to the sclera; (2) a portion which forms part of the thickened *ciliary body,* composed of *ciliary muscles* and *ciliary processes,* located anteriorly near the region where sclera and cornea join; and (3) a portion which forms part of the *iris,* a thin, circular disc, located at the anterior end of the ciliary body where the uveal coat turns sharply inward and away from the fibrous tunic. An opening in the center of the iris is spoken of as the *pupil* of the eye.

The innermost coat of the eyeball is the *retina.* It is actually made up of two layers: (1) an outer, thin, and relatively unimportant, nonsensory, *pigment layer* in contact with the entire uveal coat; and (2) a thicker, *sen-*

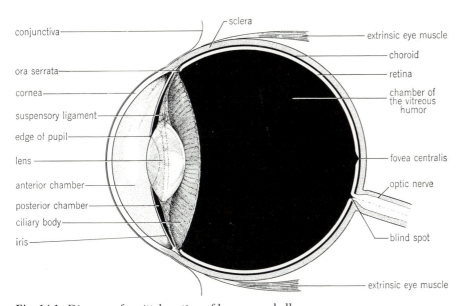

Fig. 14.1 Diagram of sagittal section of human eyeball.

sory, or *nervous, layer,* present only in the regions where the choroid coat is present. The nervous layer is the part of the eye that is sensitive to light. The pigment layer contributes to the ciliary body and iris and terminates at the rim of the pupil. The light-sensitive part of the retina ends abruptly at the ciliary body, its irregular, serrated margin being referred to as the *ora serrata,* or *ora terminalis.*

A biconvex *lens* lies immediately behind the iris, pushing the latter forward so as to cause a slight bulge. The lens has a clear, glassy appearance during life, but in preserved specimens it is frequently opaque. Its posterior side is more convex than its anterior surface. A thin *lens capsule* invests the lens closely. From the ciliary body arise numerous radially arranged fibers (Fig. 14.10) which are attached to the lens capsule. They form the *suspensory ligament (zonula of Zinn)* which keeps the lens in place. The shape of the lens is not fixed. *Accommodation,* or the adjustment of the eye which enables it to focus on objects at various distances, is accomplished by changing the shape of the lens (page 678).

The cavity of the eye in front of the lens is partially divided by the iris into *anterior* and *posterior chambers.* A clear watery fluid, the *aqueous humor,* fills these cavities, which are continuous with each other at the pupil. Aqueous humor is practically the same in consistency as tissue fluid and, except for its extremely low protein content, is similar to blood serum. The large cavity of the eye in back of the lens is called the *vitreal cavity,* or *chamber of the vitreous humor.* It contains a transparent semigelatinous substance, the *vitreous body,* or *vitreous humor,* apparently secreted by the retina during development of the eye.

The optic nerve leaves the eye from behind, piercing the retina, choroid, and scle-rotic coats. The point at which it emerges from the retina is the *blind spot.*

When light enters the eye, it passes first through the thin conjunctival epithelium covering the cornea, then through the cornea proper, aqueous humor, pupil, lens, and vitreous humor, finally to impinge upon photoreceptors in the retina.

Early development. The various structures which, in their aggregate, form the eye arise from three rudiments: (1) the retina and optic nerve are derivatives of the diencephalon; (2) the lens differentiates from the superficial ectoderm of the head; (3) the other portions come from neighboring mesenchyme.

The first indication of eye formation (Fig. 14.2) is to be observed in the appearance of a pair of lateral expansions, the *optic vesicles,* arising from the forebrain in the region which is later to become the diencephalon. The distal end of each optic vesicle soon enlarges and comes in contact with the superficial ectoderm of the head, but the proximal portion remains relatively unchanged. It is referred to as the *optic stalk.* The distal portion of the swollen vesicle next becomes indented in such a manner as to form a temporarily imperfect (page 672), double-walled *optic cup,* which is to become the retina. Pigment granules appear in the cells forming the outer layer of the optic cup which now is called the *pigment layer of the retina.* The inner layer gives rise to the *sensory,* or *nervous, portion of the retina.* The two layers later fuse.

In the meantime a thickening of the superficial ectoderm occurs at the point where the optic cup established contact with it. This is the *lens placode,* which gives rise to the lens of the eye. It invaginates at the same time as the optic cup is forming and soon becomes a closed vesicle which comes to lie

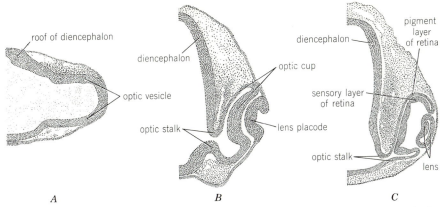

roof of diencephalon

diencephalon

optic vesicle

optic stalk

A

diencephalon

optic cup

lens placode

optic stalk

B

pigment layer of retina

diencephalon

sensory layer of retina

optic stalk

lens

C

Fig. 14.2 Successive stages *A, B,* and *C,* in the development of optic cups in chick embryo: *A,* 33-hour embryo; *B,* 48-hour embryo; *C,* 72-hour embryo. (*Drawn by R. Speigle.*)

within the optic cup, almost filling the latter during the early stages of development (Fig. 14.3). As the optic cup increases in size, the lens fails to grow at the same pace. It ultimately comes to lie at the mouth of the optic cup. The outer wall of the lens vesicle retains its epithelial character, but the inner wall thickens, its cells giving rise to *lens fibers.*

Experiments have shown that in some animals the developing optic vesicle actually induces the superficial ectoderm in this region to undergo the changes necessary for lens formation. When, for example, optic vesicles of certain amphibian embryos are transplanted beneath the skin on the side of the body, a lens is induced to form from integumentary ectoderm which otherwise would

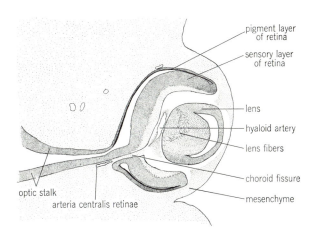

pigment layer of retina

sensory layer of retina

lens

hyaloid artery

lens fibers

choroid fissure

mesenchyme

optic stalk

arteria centralis retinae

Fig. 14.3 Sagittal section through optic cup of 12-mm pig embryo.

never give rise to such a structure. Furthermore, if an optic cup is prevented from developing at the normal site, no lens will develop. This induction phenomenon is not exhibited by all forms.

The indentation which formed the optic cup is continued along the ventral side of the cup as well as along the ventral border of the optic stalk. This forms the *choroid fissure,* which appears as a gap, extending from the optic cup toward the brain (Fig. 14.4). For a time the embryonic choroid fissure remains open, furnishing a pathway for blood vessels (*hyaloid artery, hyaloid vein*) which are of particular importance in connection with the development of the lens. Before long, however, the edges of the choroid fissure come together and fuse, thus forming a small and narrow tube within the already tubular optic stalk. Later the hyaloid vessels atrophy within the optic cup proper, but the vessels within the optic stalk are retained as the *central artery* and *vein* of the retina, supplying the retina proper. A lymphatic channel, the *hyaloid canal,* coursing through the vitreous humor from blind spot to lens, marks the former location of the hyaloid artery. The axons of nerve cells which develop in the sensory layer of the retina later converge toward the optic stalk and grow through the wall, or substance, of the optic stalk to the brain, thus forming the optic nerve (II). The optic canal, representing a portion of the original cavity of the brain, persists for a time but is gradually obliterated. The optic nerve, therefore, is actually the transformed optic stalk through which numerous nerve fibers pass to the brain. The central artery and vein of the retina actually course *through* the optic nerve. The optic nerve thus differs from other cranial nerves in that it is a composite of nerve fibers which grow from one part of the brain to another. It is, therefore, properly termed a *fiber tract,* rather than a peripheral nerve. In any event, it is usually classified under the category of special somatic sensory nerves.

The inner, or sensory, layer of the retina near the rim of the optic cup remains thin, whereas the remainder thickens as its cells become differentiated to form the truly nervous, or sensory, portion. The irregular line which marks the boundary of the two areas is the *ora serrata.* The thin portion of the inner layer fuses with the pigment layer overlying it, and these two, together with a part of the uveal coat, form the ciliary body and iris.

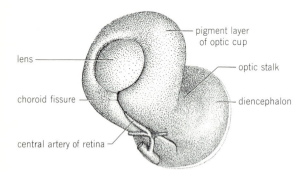

Fig. 14.4 Developing optic cup showing choroid fissure through which the central artery of the retina courses. (*Drawn by R. Speigle.*)

The mesenchyme surrounding the original optic cup differentiates into the uveal and fibrous coats of the eyeball. The cornea is derived from mesenchyme which grows in between the optic cup and superficial ectoderm with which it fuses, the latter forming the epithelial layer of the cornea. The transparency of the cornea is the result of inductive influences of both lens and retina affecting the overlying skin.

Fibrous tunic. In numerous lower vertebrates both sclera and cornea may be transparent. The opacity of the sclera is a secondary phenomenon. The tissue of which the sclera is composed consists of interlacing, tough, inelastic, tendinous, connective-tissue fibers, so arranged as to provide equal strength in all directions, thus keeping the eyeball constant in shape and causing it to be highly resistant to external or internal pressures. The sclera is practically noncellular and is not furnished directly with blood vessels. In many ray-finned fishes, certain reptiles, and birds, a ring of bony plates, the *scleral ossicles,* or *sclerotic bones,* lies embedded in the portion of the sclera near the cornea. These provide further protection to the eye. Scleral ossicles appeared early in evolutionary history since they have been identified in many fossil "ancestral" vertebrates. In the region of the cornea the connective-tissue fibers are arranged parallel to each other. Cells are present in small numbers, connecting with each other by long, thin, protoplasmic strands by means of which nutrient materials and wastes may be transferred from blood vessels located at the margins of the cornea. The cornea itself is devoid of blood vessels which, of course, would interfere with the transmission of light. It is possible that the cornea is nourished to some extent by the tears, or secretions of the lacrimal glands, which contain nutrient materials. Lacking blood vessels and lying in an exposed position, together with the fact that there is evaporation from its surface, the cornea has a temperature somewhat lower than that of the body. Sensory receptors for pain are numerous in the corneal epithelium, but other sensory receptors are apparently lacking. One noteworthy feature of the cornea is the rapidity with which it is repaired after injury. The cornea must be kept moist at all times.

Uvea. CHOROID COAT. The portion of the uvea which is adherent to the sclerotic coat consists of a thin, soft, dark brown, vascular membrane. Large and medium-sized arteries and veins, as well as numerous capillaries, form a rather rich plexus of blood vessels throughout the choroid coat. The pigment serves to keep light from entering the eyeball at random and also prevents internal reflections from occurring.

CILIARY BODY. At the point where the choroid portion of the uvea merges with the ciliary body, the uveal membrane becomes thicker, less vascular, and less heavily pigmented. Numerous involuntary *ciliary muscle fibers* are present in this layer. The inner face of the ciliary body is characterized by numerous radiating folds, the *ciliary processes.* These are formed partly from the uvea and partly from the nonnervous border of the retina mentioned above, lying beyond the ora serrata.

IRIS. The iris is composed of an outer layer, which is part of the uvea, and an inner layer derived from the nonnervous portion of the retina. In the latter region it is the *inner* of the two retinal layers (that nearest the lens) which is heavily pigmented, a condition just opposite to that obtaining in the remainder of the retina. The outer part of the retinal portion of the iris, as well as the uveal portion, contains relatively little pigment. The dark pigment in the inner portion of the iris is the only pigment present in individuals with blue eyes. This pigment as seen through the colorless tissue covering it, because of a physical effect known as the Tyndall-blue phenomenon, involving interference of light rays, gives the impression that the iris is blue. In brown and black eyes additional pigment is present in the outer, or uveal, region. The pink eye of the albino is devoid of all pigment, even in the retinal portion. It owes its color to blood in the numerous vessels of the iris. Although the albino condition is frequently encountered among vertebrates, true albinism rarely occurs in man. The ultimate color of the iris is not necessarily evident at birth. The iris of a newborn child, even in a Negro, is blue. Changes may occur later with the development of additional pigment in the uveal or retinal layers of the iris.

The iris contains two groups of involuntary, *ectodermal* muscle fibers called *sphincter* and *dilator muscles*, respectively, derived from the iridial portion of the retina. The muscle fibers of the iris are of the smooth type in amphibians and mammals, whereas those of reptiles and birds are striated. In the alligator an admixture of both kinds of fibers is present. The sphincter-muscle fibers are arranged in concentric bundles about the margin of the pupil. Contraction of these elements results in a diminution in size of the pupil. The dilator muscles are arranged in a radial manner, their contraction causing a dilatation of the pupil. The involuntary contraction of these two sets of muscle fibers regulates the amount of light that enters the eyeball under varying conditions of illumination. The dilator muscles are innervated by postganglionic fibers of the sympathetic nervous system which arise in the upper thoracic segments of the spinal cord and emerge from the superior cervical ganglion (Fig. 13.23).

They accompany the internal carotid artery into the cranial cavity through the foramen lacerum, finally entering the trunk of the nasociliary branch of the trigeminal nerve and terminating in the iris. The sphincter muscles of the iris are innervated by postganglionic fibers of the parasympathetic system, the cell bodies of which are located in the ciliary ganglion.

Retina. The pigment layer of the retina is closely applied to the uveal layer throughout its entire extent. The retina consists of three regions: (1) the *optic region*, the thickened portion in contact with the choroid coat, sensitive to light and terminating at the ora serrata; (2) the *ciliary region,* which together with the ciliary portion of the uvea forms the ciliary body; and (3) the *iridial portion,* contributing to the structure of the iris. It is in the iris that the uvea and nonnervous portion of the retina are most closely associated. We are chiefly concerned with the optic region, since it is here that the photoreceptors of the vertebrate eye are situated.

The term *optical axis* is used to refer to an imaginary line passing through the center of the cornea, pupil, and lens to the posterior pole of the eye. A small area of the optic portion of the retina in direct line with the optical axis is the *area centralis*, often called the *macula lutea* because of the yellow pigment which it contains. In the center of the macula lutea is a shallow depression, the *fovea centralis,* which is the area of most acute vision. The remainder of the retina in a fresh specimen is practically transparent.

The structure of the optic portion of the retina is highly complex. It consists of seven layers of nervous elements and two supporting membranes, the detailed discussion of which is beyond the scope of this book. Neurosensory photoreceptors called *rods* and *cones* are located in the *outermost* part of the retina, i.e., that which is farthest from the source of light and nearest the choroid coat. Rods and cones differ as to shape, the rods being slender and filamentous and the cones short and thick (Fig. 14.5). Electron microscopic studies of rods and their development indicate that they may possibly represent atypical cilia or flagellalike structures. Rods are believed to distinguish different degrees of light and darkness, whereas cones, in man and other higher primates, are chiefly concerned with both visual acuity and color vision. In the human eye, rods far outnumber the cones, the proportion being about 20 to 1. From the rod and cone cells, slender nerve fibers extend toward the vitreous humor, forming synapses with dendrites of *bipolar neurons*. The axons of the bipolar cells synapse in turn with the dendrites of neurons forming the *ganglionic layer* of the retina. This is the layer nearest the vitreous humor. The axons of the ganglionic layer course *over* the inner surface of the retina next to the vitreous humor. They converge at a single point a short distance from the fovea centralis, bend sharply, and then run parallel to each other to form the optic nerve. The nerve than passes through the retina, choroid, and sclerotic coats on its way to the brain. At the point where the fibers converge to form the optic nerve, neurosensory receptors are lacking. This area, therefore, is called the *blind spot,* since its presence is responsible for a gap in the visual field.

MACULA LUTEA AND FOVEA CENTRALIS. A true macula lutea, or yellow spot, is found only in man and some other primates. In other vertebrates the term *area centralis* is more appropriate. In the macula lutea rod cells have been practically eliminated and only cone cells are present. Furthermore, the cone cells in this region are longer, more slender, and more numerous than those in other parts of the retina. As a result of the increase in number of cones, the number of bipolar and ganglionic cells is increased in this area since here a single cone cell usually synapses with a bipolar cell and a single bipolar cell connects with a ganglionic cell. Each cone cell thus has its own line of ac-

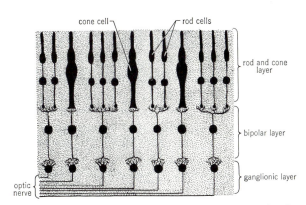

Fig. 14.5 Simplified diagram showing relation of rod and cone cells to bipolar and ganglionic layers of the retina. Note "convergence" of several rod cells onto a single bipolar cell. Bipolar cells may converge in a similar manner onto a single ganglionic cell.

cess to the brain and is individually more important than if several such cells connected to a single bipolar cell. The latter arrangement is typical of rod cells.

A depressed fovea centralis is present in man and numerous other vertebrates but does not occur in all forms. Birds have the best-developed foveas of all. In the foveal region retinal blood vessels, which would interfere with visual acuity, have been eliminated. According to G. L. Walls (1942, "The Vertebrate Eye," Cranbrook Institute of Science), the fovea is an effective magnifying device. Because of its shape and the manner in which light rays are refracted in the fovea, images which come to a focus on the fovea are magnified to a considerable extent and thus perceived more clearly than those in other portions of the retina. In the fovea the cone cells are arranged obliquely rather than perpendicularly. This also may be a factor in bringing about greater visual acuity. In order to see things more distinctly, we are constantly changing the position of the eyeball so as to focus the image of the object we wish to observe upon the fovea.

RHODOPSIN. The outer segment of each rod cell contains a reddish pigment called *rhodopsin,* or *visual purple.* Its chemical structure is not yet completely known. Vitamin A seems to be an important constituent. The presence of rhodopsin is necessary for any rod vision in dim light. In bright light it disappears or is reduced so that the threshold for vision is raised. In going from bright light into darkness or semidarkness it again makes its appearance, apparently being built up again, or regenerated, by the rod cells. This decreases the threshold for vision since much less light is required to stimulate the rods. Several minutes must elapse, however, before the eyes become adapted to such changed conditions. During this interval the concentration of rhodopsin is restored to a higher level. When emerging from darkness into bright light, we may be temporarily blinded until the rhodopsin has been depleted sufficiently to enable us to see normally. *Destruction* of rhodopsin, therefore, is important in *light* adaptation, whereas its *formation* is necessary in *dark* adaptation. Rhodopsin is not present in cone cells, which require a greater degree of illumination in order to function.

It has been determined that sufficient quantities of vitamin A must be available in order that the body can synthesize rhodopsin. The importance of supplying adequate amounts of vitamin A in the diet was emphasized in World War II, in which nocturnal activities of various kinds played so important a part.

If a fresh mammalian eye which has been kept for a time in total darkness is exposed to a bright object and then fixed immediately in a solution of alum, a photographic image of the object can be observed on the retina upon dissection. The formation of the image is brought about by bleaching or destruction of the rhodopsin in the rods upon which the bright portion of the image fell, the surrounding parts retaining their dark coloration. The fact that rhodopsin fades upon exposure to light is responsible for this phenomenon.

In such lower vertebrates as lampreys, fresh-water fishes, and amphibian larvae, the pigment in rod cells is somewhat different and more reddish in color and is referred to as *porphyropsin.* Cone cells also have been claimed by some to contain pigments of two types: *cyanopsin,* of a bluish color, and *iodopsin,* of a violet tinge. There is some doubt concerning this.

INVERSION OF THE RETINA. In the vertebrate eye light must pass *through* the various *layers* of the transparent retina before it can stimulate the sensory rod and cone cells. Impulses thus set up must then pass back *through* the *cells* comprising these layers to reach the optic nerve. Such an arrangement is certainly not what one would expect, nor can it be considered an efficient plan. This peculiar state of affairs is encountered only in the vertebrates which are said to possess an "inverted" eye.

Several theories have been advanced which seek to explain the inversion of the retina and its evolutionary origin, but none seems wholly adequate. The most satisfactory explanation for the origin of the photoreceptive elements of the retina and their inverted position was advanced by Studnicka in 1912 and later elaborated by Walls (1942). According to their idea, rods and cones are actually modified ependymal cells which line the cavity of the neural tube. The position of the sensory cells in the optic cup and optic stalk can be traced through the configurations of cup and stalk and are found to be continuous with the ependymal layer of the neural tube proper. Ependymal cells, at least during embryonic stages, are frequently flagellated, the flagella assisting in the circulation of cerebrospinal fluid in the central canal of the cord and the ventricles of the brain. If the flagella of ependymal cells are photosensitive, as indeed they are in the infundibular organ of amphioxus, then according to Walls' conception, the receptive elements of the rod and cone cells

are homologous with the flagella of ependymal cells and, therefore, lie in a reversed position in the fully formed vertebrate retina (Fig. 14.6). The fact that the receptive portions of rod and cone cells differentiate during development, in a manner strikingly similar to that of flagella, lends further support to this view.

Many authorities have sought to explain the evolutionary appearance of the vertebrate lens which in some vertebrates develops under the organizing influence of the optic cup. No really satisfactory explanation has been offered which accounts for all the factors involved.

Accommodation. Stimulation of the photoreceptive rods and cones may produce several sensations: light, various colors, form, perspective, and motion. In order to perceive objects and patterns in detail, it is essential that a picture of the object or pattern to be perceived fall upon the retina in perfect focus so that the photoreceptive rods and cones are stimulated in the correct manner. Practically all parts of the eye play a part in bringing this about, but the curved surfaces of the cornea and lens are of primary importance. The cornea is responsible for placing the image on the retina; the lens makes minor adjustments in sharpening the focus.

The eye must adapt itself in order to bring

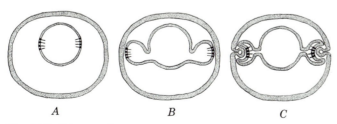

Fig. 14.6 Diagram illustrating Studnicka's and Walls' theory of retinal inversion.

objects at various distances into focus. This is known as *accommodation*. Problems of accommodation seem to have been solved in several different ways by the various groups of vertebrates in the course of evolution. In the higher vertebrates (amniotes) accommodation is brought about by changing the shape of the lens. In some, as in birds and possibly in certain reptiles, changing the degree of curvature of the cornea is employed as well. To appreciate the necessity of altering the shape of the lens in focusing, it is advisable at this point to review certain fundamental physical principles dealing with light and its passage through transparent media.

Refraction. Light, which is responsible for the various aspects of the sensation of sight, travels in straight lines through transparent media at incredible speeds. Whenever a ray of light passes obliquely into a transparent medium which differs in density from the one in which it is traveling (e.g., from air through water or glass), it is bent, or refracted (Fig. 14.7). In going from a rarefied medium into a more dense medium the ray is said to be bent *toward* the normal or perpendicular. Conversely, in passing from a dense to a more rarefied medium it is bent

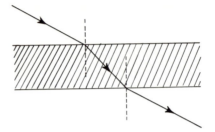

Fig. 14.8 Refraction of light in passing through a medium with two plane, parallel surfaces.

away from the perpendicular. If the second medium is surrounded by the first, as in the case of a pane of glass surrounded by air, the light ray is bent twice, once when it enters the glass and again when it leaves it (Fig. 14.8). When it passes through two such plane, parallel surfaces, it is found, upon emerging, to be displaced laterally to some extent but it proceeds in a direction parallel to the original. Rays which strike the surface between two media perpendicularly are not refracted. The more oblique the angle at which they strike, the greater is the angle of refraction.

When light passes through a lens or a refracting medium in which at least one of the surfaces is curved, its behavior is somewhat different from that just described for plane surfaces. In convex lenses, light rays are bent to such a degree that they meet at a certain point, or are said to come to a focus.

Image formation by lenses. A biconvex lens such as that found in the eye is a transparent refracting medium bounded by two spherical surfaces facing each other. In such lenses there is a point called the *optical center* so situated that any ray of light passing through it undergoes no angular deviation. A biconvex lens possesses what is known as a *principal axis*. This is a line

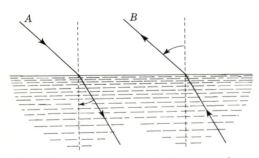

Fig. 14.7 Refraction of light: *A*, passing from rarefied to a more dense medium; *B*, passing from dense to a more rarefied medium.

passing through the center of the lens. Diagonal lines which intersect the principal axis at the optical center are referred to as *secondary axes* (Fig. 14.9). Rays of light traveling along the principal or secondary axes pass through the lens without changing their direction, but those passing through other parts are refracted. The farther a ray is removed from the principal axis, the greater will be its angle of refraction.

Light rays coming from objects at a distance of 20 ft or more may be considered as parallel, at least so far as the eye is concerned. When such parallel rays pass through a biconvex lens, all except those on the principal axis are first refracted and then meet at a point called the *principal focus*. The distance on the principal axis between the center of the lens and this point is known as the *focal distance* of the lens. The focal distance of a biconvex lens depends upon its degree of curvature as well as upon its refractive properties. A small, *inverted* image of the distant object appears in perfect focus beyond the focal point.

Light rays coming from near objects (nearer than 20 ft but farther away from the lens than the principal focal distance) are divergent rather than parallel. Such rays impinge upon the lens and are refracted, forming a small, *inverted* image beyond the

principal focus. The image distance of a near object varies with the distance of the object from the lens. The location of the image can be computed with little difficulty (Fig. 14.9). Let *AB* be an object in front of a biconvex lens but at a distance greater than its focal length. From point *A* one ray and only one, *AC*, will travel parallel to the principal axis. This will be refracted and pass through the principal focus *F*. The image of point *A* will be found somewhere on the line *CF*. In addition, from point *A* one ray and only one will impinge upon the lens so as to pass through its optical center *O*, and this ray will pass through the lens without undergoing any angular deviation. The image of point *A* will be found somewhere on the line *AO*. Actually the image of point *A* will be found where lines *CF* and *AO* intersect at *a*.

In the same manner the ray *BD* is parallel to the principal axis. It is refracted and passes through the principal focus *F*. Another ray, *BO*, travels along a secondary axis through the optical center. The point of intersection of lines *DF* and *BO* is at *b*, and it is here that the image of *B* will fall.

The line *AB* is made up of innumerable points, each of which gives off rays independently. For each point in the object there will be a corresponding point in the image *ab*. Because each image point appears on the

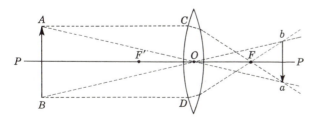

Fig. 14.9 Diagram illustrating image formation by lens: *AB*, object; *ab*, inverted image; *O*, optical center; *PP*, principal axis; *Aa*, *Bb*, secondary axes; *F*, principal focus; *OF*, focal distance.

opposite side of the principal axis from that of its corresponding object point, the image is inverted.

The image points for distant objects, e.g., those anywhere from 20 ft to *infinity* away from the lens, differ so little in position (no more than a fraction of a millimeter) that they are actually the same for all practical purposes. In the perfectly normal eye, the image point for distant objects falls upon the retina.

If an object is nearer the lens than its principal focal distance ($F'O$), its divergent rays, after being refracted, are still divergent and a *virtual, enlarged image* is formed. This is of no practical importance so far as the eye is concerned.

It should be understood from the above account that if a lens is rigid, as is true of all lenses made of glass, the image points for distant objects coincide. Each near object, however, has its own image point, depending upon its distance from the lens. Since the retina of the eye is the screen upon which both near and far objects form images, it is apparent that a rigid lens in the eye would be unsuitable. In such cases an image would fall exactly upon the retina only when the object was at a certain fixed distance away. Only at this distance would the image be sharply defined.

Image formation by the lens of the eye. In the lower vertebrates the shape of the lens remains constant but it may be moved forward or backward, as the case may be, so that light rays will focus on the retina. A remarkable characteristic of the amniote eye, however, is that it has the power of accommodating itself to objects at various distances by changing the convexity of the lens. Altering the shape of the lens changes its focal distance so that images of objects at various distances can be brought to a focus on the retina. The only variable lenses in existence are those in the eyes of amniotes.

If the eye follows a moving object, the lens changes its curvature, always in such a manner as to form a sharply focused image upon the retina. As the object approaches the eye the lens becomes more convex. Conversely, as an object moves away from the eye the lens tends to flatten out. Thus when our eyes are focused upon distant objects the lens is flattened and said to be at rest. Focusing upon near objects requires some effort in order to increase the curvature of the lens. The ciliary body is the structure in the eye which is responsible for accommodation in mammals.

Ciliary body. If a mammalian eye is cut into anterior and posterior halves and the anterior half viewed from its cut surface, the ciliary body appears as a concave bowl with the transparent crystalline lens in its center (Fig. 14.10). Its outer edge is bounded by the ora serrata. Thus the ciliary body extends from the ora serrata to a point near the circumference of the lens. It is attached to the lens capsule by numerous fine fibers which form the *suspensory ligament.*

The ciliary body, as previously mentioned, is derived from both the uveal coat and the nonnervous portion, or border, of the retina beyond the ora serrata. The greater part, however, is of uveal origin and is thus continuous with the choroid coat. It is made up of three main parts: (1) an outer rim, the *orbicularis ciliaris,* directly continuous with the choroid; (2) a large number of radiating folds, the *ciliary processes,* to which the fibers of the suspensory ligament are attached on the inner face; and (3) the involuntary *ciliary muscles,* on the outer face, lying between uvea and sclera at the point where the latter passes into the cornea and the

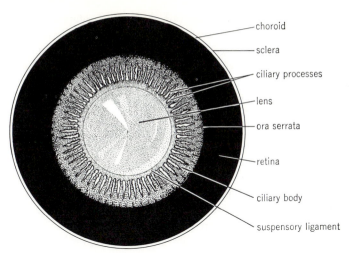

choroid

sclera

ciliary processes

lens

ora serrata

retina

ciliary body

suspensory ligament

Fig. 14.10 Anterior section of mammalian eye as viewed from behind after cutting the eyeball into anterior and posterior halves.

former gives rise to the ciliary body. Some of the ciliary muscle fibers are arranged circularly, but most are meridional or radial.

The capsule enclosing the lens is an elastic membrane which varies in thickness, being thinner on the posterior than on the anterior surface. The posterior face of the lens, therefore, bulges out farther and possesses a greater degree of curvature than does the anterior face. Any pull on the lens capsule would tend to flatten the elastic lens. When the eye is at rest, i.e., focused for distant objects, the ciliary muscles are relaxed. Under such conditions the pull upon the suspensory ligament is greatest and the lens is flattened. The ciliary muscles are controlled by sympathetic and parasympathetic nerve fibers, the preganglionic fibers of the latter being associated with the oculomotor nerve. When the muscles contract, as they do when the eye focuses for near objects (those within the critical distance of 20 ft), they tend to pull the choroid coat forward. This relieves the tension on the suspensory

ligament and lens capsule, and the lens tends to bulge and become more rounded. The greater part of the bulging occurs in the center of the anterior face. Pressure of the vitreous humor against the posterior surface prevents the latter from changing its shape to more than a slight degree. At any rate, the increase in curvature of the anterior face of the lens increases its refracting power, so that the image of the object is thrown into focus on the retina rather than at a point posterior to it.

Images which appear on the retina are inverted in the same manner as those formed by rigid lenses (Fig. 14.9). Actually we see them as though they were erect. Nerve centers for the sense of sight, located in the thalamus and cerebral cortex, are responsible for the fact that we see things right side up instead of upside down. This is a secondary phenomenon learned by experience.

The various transparent media of the eye play an important part in the refraction of light rays. They include the cornea, aqueous humor, lens, and the vitreous body. These media are of different densities, and each has its own refractive index. Actually the greatest amount of refraction occurs in the cornea, which is even more important than the lens in this respect. Since in most vertebrates the curved surface of the cornea is in a fixed position, it is of little importance in accommodation as compared with the variable lens.

In man, with increasing age, there is a tendency for the lens to lose water, to become more dense, and to lose some of its elasticity as a result. Customarily, additional lenses, in the form of eyeglasses, are used to compensate for the changes which occur.

Accessory structures of the eye. Numerous structures associated with the eye play an important part in the efficient functioning of this receptor organ. Of greatest significance are the muscles which move the eyeball, the eyelids, eyebrows, and the various parts of the lacrimal apparatus.

MUSCLES. Six broad, strap-shaped muscles are inserted on the sclerotic coat and are so arranged that, when they contract, they cause the eyeball to rotate for a variable distance, depending upon the degree of contraction. These extrinsic muscles are arranged in two groups (Fig. 11.13). The first group, known as the *rectus muscles,* is composed of four elements named the superior rectus, inferior rectus, internal (medial) rectus, and external (lateral) rectus, respectively. Each derives its name from the region of the eyeball on which it inserts, the points of insertion being rather evenly spaced around the equatorial region of the eyeball. All four muscles converge to their point of origin in the posteromedial portion of the orbit, near where the oculomotor nerve emerges. The second, or *oblique, group of muscles* is made up of two elements, the superior oblique and inferior oblique. They insert on the sclera very near the insertions

of the superior rectus and inferior rectus, respectively, but their points of origin are at the anteromedial wall of the orbit. These extrinsic muscles are innervated by three motor cranial nerves, the oculomotor, trochlear, and abducens. The oculomotor supplies four muscles: superior rectus, inferior rectus, internal rectus, and inferior oblique. The superior oblique is innervated by the trochlear nerve, and the external rectus by the abducens. The common origin of the four muscles supplied by the oculomotor, from a single pro-otic myotome, is discussed on page 516.

EYELIDS. Two transverse folds of skin, the upper and lower eyelids, lie in front of the eye, serving to protect it from injury. Each eyelid is lined on the inside with conjunctiva, continuous with that reflected over the surface of the cornea. A reinforcing band of dense connective tissue in each eyelid is called the *tarsal plate.* The opening between the lids is spoken of as the *palpebral fissure.* In numerous mammals (dogs, cats, rats, mice, etc.) the eyelids close and fuse during development only to open again several days after birth. Mice with open eyelids at birth have been reported by several laboratories. This condition is apparently due to some hereditary factor and is referred to as an *open-eyelid mutation.* Both eyelids are movable in most vertebrates, but the upper lid in man and some other forms has a greater range of movement than the lower. A muscle, the *levator palpebrae superior,* functions in elevating the upper eyelid. It is a derivative of the superior rectus and is innervated by the oculomotor nerve. Another muscle, the *orbicularis oculi,* is present in both eyelids of mammals. It is a flattened sphincter muscle and a derivative of the facial musculature.

In many vertebrates a third eyelid, or *nictitating membrane,* is present. This lies beneath the other two and passes from the inner angle out over the surface of the eye. It is usually transparent. In most mammals the nictitating membrane is reduced. In man a small fold, the *plica semilunaris*, in the inner corner of the eye, is believed by some to be a homologue of the nictitating membrane. If a nictitating membrane is present, it is supplied by the abducens nerve. The so-called "nictitating membrane" of amphibians is not homologous with those of higher vertebrates (page 688).

In man the edges of each eyelid are provided with three or four rows of hairs called *eyelashes* which gave added protection to the eyes. Modified sweat glands, the *glands of Moll*, open into the follicles of the eyelashes as do small sebaceous *glands of Zeis*. Infection of either type of gland results in a *sty*. On the margins of the eyelids just inside the lashes are the openings of the sebaceous *Meibomian*, or *tarsal*, *glands*, not associated with hairs (page 118).

EYEBROWS. In mammals, thickened areas of the integument may be situated above the upper edge of the orbit and provided with numerous thick, stiff hairs. These furnish added protection to the eyes.

LACRIMAL APPARATUS. The lacrimal apparatus of each eye consists of a *lacrimal gland*, which secretes a watery fluid, and a system of small canals by means of which the fluid is conveyed from the medial corner of the eye to the nasal passage.

In mammals the lacrimal gland lies beneath the lateral portion of the upper eyelid where several small ducts penetrate the conjunctiva. The tears, or *lacrimal fluid*, moisten the surface of the eyeball and provide nourishment for the nonvascular cornea. They are secreted continuously, even when the eyelids are closed, passing to the inner corner of the eye. A small opening, the *lacrimal punctum*, is present on the margin of each eyelid near its median border. Each punctum opens into

a small *canaliculus*. These, after coursing a short distance, unite to form the *lacrimal duct*, which in turn opens into the nasal passage, passing through an opening between the maxillary and lacrimal bones en route. Tears pass through the puncta, canaliculi, and lacrimal duct into the nasal passage. The constant flow of tears over the surface of the eyeball serves to moisten the cornea and to keep it clean by washing away dirt and other foreign particles which might otherwise accumulate. When lacrimal fluid is secreted in excess, the tears cannot drain away fast enough through the lacrimal duct. In such cases they break through the "dam" formed by the oily secretion of the Meibomian glands and overflow onto the lower eyelids and cheeks.

Occasionally, accessory lacrimal glands are present in mammals beneath the lower eyelid. This condition is reminiscent of that encountered in amphibians and reptiles, in which such glands lie under the lower lid. In these forms the lower eyelid is the more movable of the two.

MISCELLANEOUS STRUCTURES. *Optic pedicel*. In most elasmobranch fishes a cartilaginous rod, the *optic pedicel* (Fig. 11.13), connects the eyeball to the skull. It is situated in such a manner as to be surrounded by the converging rectus muscles. The eyeball end is frequently expanded and is loosely attached to the sclera. The optic pedicel gives support to the eyeball. Higher fishes do not have an optic pedicel. A ligamentous *tenaculum* is frequently present, however, helping to keep the eyeball in place. The tenaculum is not considered to be a homologue of the optic pedicel since in certain fishes (rays) both structures are present. These supporting elements have no counterparts in terrestrial vertebrates.

Harder's gland. A sebaceous gland known as the Harderian, or Harder's, gland is associated with the eye in many amphibians, reptiles, and birds, as well as in some mammals. It generally lies behind the eyeball and secretes an oily fluid which serves to lubricate the nictitating membrane. In snakes the secretion of Harder's gland contributes substantially to the saliva, thus facilitating the lubrication of captured prey prior to swallowing.

Muscles. In some vertebrates, in addition to the usual rectus and oblique muscles, other muscles may be present. These include (1) a *retractor bulbi*, derived from the external rectus and used to draw the eyeball into its socket and (2) a *levator bulbi,* which raises the eyeball back to its original position.

Scleral ossicles. In birds and reptiles, with the exception of snakes and crocodilians, the exposed portion of the sclera, adjacent to the cornea, contains a circle of thin, overlapping bones, the scleral ossicles, or sclerotic bones. They appear, however, to cover this portion of the sclera. The number varies in different forms, some birds having as many as 18 separate bones. They arise during embryonic development as dermal bones which sink down into the sclera. The scleral ossicles afford protection to the eye and are of some importance in accommodation in the forms in which they occur. Some modern fishes possess sclerotic bones, but no more than two are present. Four are known to have occurred in some primitive fossil forms. In some species they are fused to form a single ring. It is not clear whether these are actually homologues of the numerous small ossicles of reptiles and birds. A circlet of small *circumorbital* skull bones surrounds the eye in many fishes. These are sometimes homologized with the sclerotic bones of reptiles and birds, but in some fishes both types of bones are present, thus raising doubts concerning such homol-ogies. Numerous so-called "scleral" ossicles were present in extinct labyrinthodont amphibians, usually in the form of an incomplete dorsal ring. It is possible that they were located in the upper eyelid and in this way offered protection to the eye. Their homologies are also questionable. Scleral ossicles are not present in mammals.

Comparative anatomy of the vertebrate eye. The eyes of all vertebrates are constructed on the same fundamental plan, but many variations are to be found. In some forms the eyes are primitive; in others they are degenerate and functionless. The most important differences in the highly complex eyes of most forms are to be found in methods of accommodation, degree of development of the retina, and the shape of the pupil.

The differences in distribution of rods and cones among the various taxonomic groups, and even between closely related forms within a single group, seem to indicate that no evolutionary sequences are involved. The presence or absence of one or the other of these two types of sensory receptors appears to be more closely related to habit and habitat, as indicated in the following pages.

CYCLOSTOMES. In members of the class Cyclostomata two conditions are to be found. In hagfishes the eyes are degenerate and functionless, whereas those of adult lampreys are excellent visual organs showing no signs of degeneracy.

The hagfish eye is extremely small, being scarcely more than a millimeter in diameter. It lies entirely beneath the skin and is insensitive to light. The optic nerve is degenerate. Other nerves, as well as muscles and lens, are lacking. Cornea, sclera, and choroid are not differentiated, and pigment is absent. The retina is folded so as to fill entirely what would ordinarily be considered to be the chamber of the vitreous humor. A lens placode begins to form in the embryo, but it soon flattens out and disappears.

The eye of the lamprey, although primitive, is nevertheless a well-developed structure. It offers no clue as to the possible evolutionary origin of the vertebrate eye. From the time that the ammocoetes larva burrows into the mud, until metamorphosis, the eyes are undeveloped and functionless. At the time of metamorphosis the eye develops markedly and the skin covering it becomes transparent. Among the important features are a flattening of the outer surface of the eyeball, a thin membranous sclera composed of dense connective tissue, a thin cornea *not* fused to the skin, lack of suspensory ligament and ciliary apparatus, fixed size of the pupil, and a permanently spherical lens, held in place by the vitreous humor. Eyelids are lacking. In the lamprey the abducens nerve supplies both the external and inferior rectus muscles (page 516), whereas in other forms the latter is controlled by the oculomotor nerve which also supplies the superior rectus, internal rectus, and inferior oblique. Accommodation is brought about by flattening the cornea, which then pushes the lens *toward* the retina. This is accomplished by contraction of a special *corneal muscle*. Although both rod and cone cells are present in most forms, the rods far outnumber the cones. Both rods and cones are cone-shaped but differ in certain other respects. Walls (1942) is of the opinion that cones are phylogenetically older than rods and that the rods of lampreys are not related to the rods found in other vertebrates. Whether or not lampreys have color vision has not been established.

A cross section of the optic nerve reveals its primitive character. A core of ependymal cells is present, bearing the same relation to the fibers of the optic nerve as the ependymal cells of the neural tube bear to the brain wall. This relationship emphasizes the fact that the retina is actually an evagination of the brain.

ELASMOBRANCHS. The eyes of the cartilaginous fishes are exceptionally large, those of the deep-sea-dwelling holocephalians being largest of all in relation to body size.

Eyelids consist of mere folds of skin which have but little power of movement. In some species, however, the lower lid is movable. The presence of an optic pedicel is characteristic, although lacking in a few forms. The eyeball is usually elliptical in shape with a short anteroposterior axis. Conspicuous features include the presence of cartilage in the sclera, the lack of intrinsic muscles in the ciliary body, and a large, permanently spherical lens which pushes the iris far forward so that it almost touches the cornea (Fig. 14.11). The pupil is round in most forms but is slitlike in a few species. Muscles in the iris are poorly developed and are altogether lacking in deep-sea forms. The presence of a core of ependymal cells in the optic nerves of a few elasmobranchs is indicative of a relationship to the condition found in the lamprey.

In most elasmobranchs the surface of the choroid coat adjacent to the retina contains a layer of light-reflecting crystals composed of a chemical substance known as *guanin*. The layer is called the *tapetum lucidum*. Light rays striking the tapetum lucidum are reflected back, causing the eyes to "shine." This is, of course, most noticeable at night, and the presence of a tapetum in the many higher forms in which it occurs is generally restricted to those animals of nocturnal habit. Visibility is thought to be enhanced under such conditions since light rays, instead of being absorbed by the pigment in

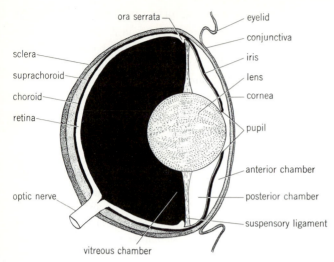

ora serrata
eyelid
conjunctiva
sclera
iris
suprachoroid
lens
choroid
cornea
retina
pupil
anterior chamber
optic nerve
posterior chamber
suspensory ligament
vitreous chamber

Fig. 14.11 Diagrammatic sagittal section through eye of dogfish shark.

the choroid coat, are reflected back to the sensitive retinal cells, thus conserving the light rays and making them more effective. The tapetum lucidum of elasmobranchs is exceptionally well developed. A thickened vascular portion of the choroid coat, called the *suprachoroid layer,* lies between the choroid and sclerotic coats on the medial aspect of the eyeball. It is present only in those elasmobranchs having an optic pedicel.

Cones are absent from the elasmobranch retina except in a few species. It is probable that color vision is lacking. Here, for the first time, an *area centralis* may be present, somewhat comparable to the macula lutea of primates except that it is devoid of pigment. No fovea is to be found.

The suspensory ligament which holds the elasmobranch lens in place consists of a washer-shaped gelatinous membrane. The peripheral portion is fastened over the inner surface of the ciliary body and its center border to a restricted, equatorially placed band on the surface of the lens. A portion of the ciliary body in the middorsal region

extends downward and helps to support the lens. The ventral side of the lens is supported by a papillalike projection of the ciliary body, which is attached anteriorly to the iris. A small muscle, the *protractor lentis,* of ectodermal origin, is located on the lens side of the papilla. It is so arranged that contraction of its fibers causes the lens to swing *forward,* somewhat in the manner of a pendulum. The protractor lentis thus serves as a muscle of accommodation, moving the lens forward, rather than changing its shape, when focusing for near objects. The distance through which the lens can move is restricted by the shallowness of the space between iris and cornea, thus limiting accommodatory efforts to a rather narrow range.

OTHER FISHES. In *Latimeria* cones are extremely rare but rods are abundant. The eyes "shine" at night with a yellow-green color like a cat's, but no really unusual features are to be found in the eyes of this ancient fish. A great many interesting variations are to be observed in the eyes of fishes. Only certain general features will be considered here. Typically there is a flattening of the anterior surface of the eyeball. Cartilage is present in the sclera, and in some forms bony plates may replace or supplement the cartilaginous elements. The cornea is composed of four layers, a *dermal layer* being interposed between conjunctival and scleral layers and a fibrous *autochthonous layer* lying inside the scleral portion. A tapetum lucidum is usually present in the choroid coat. The outer portion of the uvea in fishes is surrounded by a layer, the *argentea,* composed of silvery guanin crystals. This layer covers the dark pigment of the uvea and, by reflecting light, aids in rendering the almost transparent young fish inconspicuous. The argentea loses its significance during later life. A silvery *retinal tapetum* is present in a number of fishes. The light-reflecting crystals are located in the pigment layer of the retina. The visual cells of the retina consist of exceptionally long rods and two kinds of cones, single and "twin." Among fishes the latter are to be found only in teleosts. Color vision seems to be widespread among teleosts. An area centralis is present in the retinae of many bony fishes. A number of teleosts possess a fovea. Ciliary and functional iris muscles are lacking. Accommodation in the eyes of fishes is usually accomplished by changing the position of the spherical lens. In most cases a sickle-shaped *falciform process* is present in the form of a ridge of the choroid coat on the floor of the eyeball, projecting through the choroid fissure which, in teleosts, remains open. Beginning at the blind spot, its distal end extends forward and upward into the chamber of the vitreous humor. This end is attached to the lower border of the lens by a structure called the *campanula Halleri,* or *retractor lentis.* Small muscle fibers of ectodermal (retinal) origin located in the retractor lentis pull upon the lens (which is suspended from above, like a pendulum), drawing it *backward.*

Deep-sea fishes which live in a realm of total darkness have relatively enormous eyes. Most of these forms possess luminescent photophores which furnish the only source of light that may affect their eyes. Recognition of enemies, of prey, of members of the same species, and of members of the opposite sex is thus possible in these abyssal regions. The eyes have undergone modifications, all of which are conducive to increasing their sensitivity to light. Pupil and lens are extraordinarily large. Rod cells are extremely long and numerous, whereas cone cells are almost altogether lacking. In many forms the eyeball has assumed a tubular shape, its greatest diameter being in the region where the lens is situated.

Surprisingly enough, relatively few deep-sea fishes are blind. In some species, however, the eyes are degenerate. In others they are covered with a thick, opaque layer of skin. Only one fish (*Ipnops murrayi*) is known in which the eyes have disappeared altogether. This is the only vertebrate in which such a condition is known to exist. Better-known blind fishes, such as those found in certain caves, have degenerate, microscopic, functionless eyes. There has been much speculation as to the reason for the degenerate eye condition in so many cave-dwelling forms. The consensus is that the fish did not become blind because they lived in darkness but rather that those with degenerate eyes found refuge in caves and became adapted to such an environment. In such fishes, eyes begin to develop during embryonic life but they fail to differentiate further after a certain developmental stage is reached.

In flatfishes, as in the flounder and halibut, both eyes in an adult are on the same side of the head. These fishes lie on the bottom of the ocean. The underside may be either the right or left, depending upon the species. Early in life the eye on the side on which the animal will lie migrates over or even through the head, bringing about the odd condition seen only in these fishes.

The eyes of certain teleosts are adapted for vision in both air and water. Those of the mudskippers, *Periophthalmus* and *Boleophthalmus*, are set in the ends of retractable stalks. The so-called four-eyed fish, *Anableps*, of Central America, has two pupils in each eye. A horizontal partition, which appears in the iris during larval life, divides the original pupil in two, the upper one being the larger. The upper pupil is usually out of water and is used in air vision. Two corneal and two retinal areas are also present. Another fish, *Dialommus*, also appears to have two pupils in each eye, but these are actually circular, transparent areas in the otherwise pigmented, opaque cornea (Fig. 14.12). The use to which these areas are put is not understood.

The capture off the Oregon coast of a

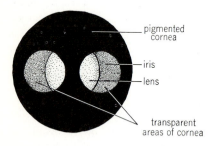

Fig. 14.12 Front view of eye of the blenny, *Dialommus*, showing two transparent areas in the opaque, pigmented cornea. (*After Breder and Gresser, from Walls, "The Vertebrate Eye." By permission of the author.*)

deep-sea fish with four eyes was reported in 1965. The fish, about 18 in. long, had two well-developed eyes with a secondary pair situated beneath them, enabling the fish to see light and movement of objects beneath it.

AMPHIBIANS. Among amphibians are to be found strictly aquatic types which never leave the water, as well as terrestrial forms which return to the water only for breeding purposes. Most of the latter, however, spend their larval lives in ponds or streams, but a few make use of tiny temporary pools formed when rainwater collects between leaves. It is probable that no amphibians possess both air and water vision comparable to that of the peculiar fishes mentioned above.

Aquatic amphibians have no problem to contend with in keeping their corneas moist, but terrestrial forms have been forced to make adjustments to keep the surface of the eye from drying in order that it may function efficiently. In such forms movable eyelids, moistened with glandular secretions, appear for the first time. The eyelids consist of a thick upper lid and a thinner, more movable lower lid. The upper border of the lower lid is transparent and is moved upward in closing the eye. This portion is particularly well developed in anurans, in which it is frequently referred to as the "nictitating membrane" although it is probably not homologous with the true nictitating membranes of other forms. When the eye is open, this transparent structure is folded inside the remaining portion of the lower lid. Closure of the eye is not accomplished merely by moving the eyelids but is brought about by retracting the entire eye within the orbit by means of the *retractor bulbi* muscle. This has its origin from the undersurface of the parasphenoid bone and its insertion on both upper and lower surfaces of the eyeball. It is believed to represent a bifurcation of the external rectus. Contraction of the retractor bulbi causes the nictitating membrane to be pulled upward. Conversely, protrusion of the eyeball is effected by means of the *levator bulbi,* which lies ventral to the orbit, together with contraction of a small muscle, the *depressor membrane nictitans,* which pulls the lower lid downward. The levator bulbi originates from the frontoparietal, parasphenoid, pterygoid, sphenethmoid, and palatine bones. It inserts on the superior border of the upper jaw.

The elements responsible, in terrestrial forms, for keeping the cornea moist first appear in land salamanders and consist, in these more primitive amphibians, of a row of glands in the lining of the lower lid. A group of these glands, located at the inner angle of the eye, may be the forerunner of Harder's gland. Another group in the outer angle of the eyelid may have given rise to the lacrimal glands of higher forms.

The eyes of anurans are best developed among amphibians. A prominent Harderian gland develops during metamorphosis. It is located at the inner angle of the eye and is pear-shaped. The gland is bound together by a capsule of connective tissue and is packed about the eyeball in the orbit. No true lacrimal gland is present, but a small "lacrimal" duct opens into the nasal passage behind and below the eye. Its outer end connects by a number of tubules to the outer angle of the eye. The duct is lined with ciliated columnar epithelium. The eyeball is almost spherical. Its anterior chamber is rather deep, the increased depth being brought about by the greater curvature of the cornea and the deeper position of the lens. Cartilage appears in the sclera during or after metamorphosis.

One of the most interesting features of the anuran eye is the ciliary body, which is more complex than in any of the lower vertebrates.

It is triangular in cross section, with the base of the triangle attached to the sclera. Numerous folds join the posterior surface of the iris. From the ciliary body come many fibers which are joined to the lens capsule, thus forming a suspensory ligament. These fibers lie entirely in the vitreous humor. Two small muscles of mesodermal origin, one dorsal and the other ventral, are responsible for accommodation. Each courses from the cornea to the inner part of the ciliary body. When these *protractor lentis* muscles contract, they draw the lens more closely toward the cornea. In doing so, the outer face of the lens may become slightly flattened.

No argentea is present in the outer layer of the choroid coat in amphibians, nor is a true tapetum lucidum present, dispite the fact that the eyes of certain anurans "shine."

The photoreceptor cells in the retina consist of four types: red rods, green rods, and single and double cones. Red rods are more numerous than green rods, the two differing considerably in shape. Red rods contain the reddish pigment rhodopsin (*visual purple*), whereas green rods are reported to contain a greenish substance sometimes called *visual green.* The latter may function in the same manner as visual purple. It has not been demonstrated that amphibians have any color vision whatever. Frogs possess an area centralis, but no fovea is present.

The beautiful colors of the iris observed in many anurans are noteworthy. The shape of the pupil shows many variations and is frequently used as an aid in classification. Vertical, horizontal, oval, and elliptical-shaped pupils are to be found, and in some forms the opening may even have an irregular, rhomboid shape.

The eyes of urodele amphibians are, on the whole, smaller and less complex than those of anurans. The lids are absent in permanently aquatic forms and are poorly developed in others. A nictitating membrane never develops beyond the rudimentary stage. The *retractor bulbi* muscle is present and is used in pulling the eye downward in its orbit. Such action aids in swallowing by compressing the mouth cavity. The beginning of differentiation of lacrimal and Harderian glands is discernible in certain members of the group. In the more terrestrial urodeles the eyes, in some respects, approach the degree of development found in anurans. In some permanently aquatic forms, such as *Necturus,* the cornea is fused to the skin, and the eyes, therefore, are capable of little, if any, movement. Generally speaking, the urodele lens is exceptionally large, the anterior chamber is shallow, cartilage is present in the sclera, and only a ventral *protractor lentis* muscle is present. The entire ciliary body is less highly developed than in anurans. The visual cells in the urodele retina include both red and green rods and single and double cones. Green rods, however, are not uniformly present. The red rods of *Necturus* are the largest rod cells that are known.

The eyes of some of the blind, cave-dwelling salamanders indicate that arrested development has occurred. Others show signs of secondary degeneration. Eye muscles and lens may be lacking, and retina and vitreal chamber are modified or degenerate.

The tropical, limbless caecilians, considered by many to be blind, actually have very small, almost microscopic eyes which can be utilized to some extent. The eyeball, which is only a fraction of a millimeter in diameter, lies beneath a transparent area of skin to which it may be fused. The lens protrudes through the pupil. No mechanism for accommodation is present. Rod cells alone are to be found in the surprisingly well-developed retina. A large Harderian gland almost completely fills the orbit. Its secretion, instead of affecting the eye, is utilized in lubricating the *tentacle*, a structure with highly developed olfactory and tactile properties. A tentacle is present on each side. It is a soft, rounded structure, located between the eye and the external naris. It has properties similar to those of erectile tissue and may be protruded by becoming turgid with blood. Retraction is brought about by contraction of a well-developed muscle. The tentacles are usually protruded during locomotion. Each tentacle lies in a pitlike depression and is supplied with a nerve at its base. The secretion of the Harderian gland is poured into the depression, apparently lubricating the tentacular apparatus and keeping it clean. It has been suggested that the structure has olfactory or gustatory as well as tactile properties and may be a derivative of the nasolacrimal duct.

Reptiles. The eyes of reptiles, with the exception of snakes, show a marked similarity in structure. In these animals, which have adopted a terrestrial mode of life, the chief changes to be observed over amphibians are found in perfections of structures first developed by the latter when making their transition from water to land. Among these features in the eyes of reptiles other than snakes and a few others are the increasingly movable eyelids, the lower lid still being the larger and more movable of the two. A true, transparent nictitating membrane is present, lying between the other eyelids and the eyeball. This membrane is lubricated by the secretion of a well-developed Harderian gland in the inner angle of the palpebral fissure. Except in *Sphenodon,* chameleons, and snakes, a lacrimal gland in the outer angle of the eye makes its appearance. A lacrimal duct drains the secretions of lacrimal and Harderian glands into the nasal passage. Turtles lack a nasolacrimal duct. One of the most important differences in the reptilian eye is the modification of the ciliary apparatus. The ciliary body is composed of a highly developed ring of padlike processes which are collectively referred to as the *annular pad.* The ciliary processes extend medially and are in contact with the peripheral portion of the lens. By a peripheral squeezing action the lens and cornea are altered in shape, thus increasing the curvature of the cornea as well as that of the outer surface of the lens. Snakes, again, are exceptions, the shape of the lens being fixed. The scleral ossicles, when present, are so arranged as to form a rigid, concave area, the function of which seems to be to resist the intraocular pressure which squeezes the lens. Of all vertebrates, the lens of the turtle is most flexible. A *conical papilla* extends from the blind spot into the vitreal chamber in most reptiles, but this is lacking in *Sphenodon* and is poorly developed in crocodilians. It is composed, for the most part, of ectodermal neuroglia tissue enclosing numerous small blood vessels which nourish the inner part of the retina by diffusion of substances through the vitreous humor. A relative increase in the number of cones is apparent in reptilian eyes, and in one turtle, *Amyda,* rods seem to be lacking. A central area for more acute vision is present in the retinae of most reptiles. A conspicuous fovea appears in *Amyda, Sphenodon,* and several snakes and lizards. In some chameleons it may be even larger than that of man. Color vision is believed to exist in turtles and lizards but is of doubtful occurrence in crocodilians and snakes.

The eyes of snakes present several features which cause them to differ markedly from those of other reptiles. They lack scleral ossicles and some of the usual eye muscles. Another chief difference lies in the lack of eyelids, the skin over the eye forming a transparent window, or "spectacle," which

is responsible for the fixed stare characteristic of these animals. In some of the limbless lizards and in a group of small lizards, the geckos, eyelids are also lacking. In snakes true lacrimal glands are absent. Instead of lacrimal fluid the oily secretion of Harder's gland flows in the space between cornea and spectacle. It drains through a duct into nose and mouth where it is added to the salivary secretions, helping to lubricate the captured prey preparatory to swallowing. In snakes a fine plexus of blood vessels ramifies over the retinal surface, thus providing nourishment to that structure as well as supplying the conical papilla. This structure in snakes is entirely of mesodermal origin and therefore is not homologous with the similarly named structure in the eyes of other reptiles which has a neuroglial framework of ectodermal origin. For these reasons, as well as others, some authorities are of the opinion that the snake eye has arisen along different evolutionary lines from those followed by other reptiles.

BIRDS. Just as is true of other anatomical features, the eyes of birds are strikingly uniform in structure. The black skimmer is the only bird in which the pupil is in the form of a vertical slit. There is but little deviation in essential features from the condition found in reptiles. The eyeball of the bird is very large in proportion to body size, a fact undoubtedly correlated with aerial life and the necessity for seeing with precision over considerable distances. The actual size of the bird's eye is usually unsuspected, since only the cornea is visible externally. A medium-sized hawk may have an eyeball as large as that of man.

The eyeball is never spherical, but is partly concave. The concave area is pronounced to varying degrees and is caused by the presence of a ring of scleral ossicles which surrounds the eyeball in the region of the ciliary apparatus (page 684). The portion in front of the ossicles is more or less conical in shape, whereas the posterior part is expanded and somewhat flattened. Flat, globular, and tubular shapes are recognized, depending upon how far back on the eyeball the concave area is located (Fig. 14.13). A supporting cup of hyaline cartilage is present in the sclera. In passerine birds and in woodpeckers an additional ring of sclerotic bones is present in the region where the optic nerve leaves the eyeball.

A highly developed nictitating membrane is present in birds. It is well lubricated and cleans the lining of the other lids as well as the corneal surface. In many birds the nictitating membrane moves independently of the upper and lower lids, which remain open during waking hours without blinking. During flight the nictitating membrane is said to be closed, offering protection to the delicate cornea. Harderian glands and lacrimal glands are well developed and situated in the usual positions. Accommodation in the bird's eye is controlled by the ciliary apparatus, as in reptiles. It involves deformation of the cor-

nea and increasing the curvature of the anterior surface of the lens by a squeezing action. The annular pad with its numerous processes is greatly developed in birds.

Rods and two kinds of cones, single and double, are present in the retina of the bird. Cones predominate in birds of diurnal habit, and rods in nocturnal forms. A conspicuous depressed fovea centralis is present in the area centralis. Color vision seems to be widespread in this class of vertebrates. In certain birds two "central" areas, each with its fovea, are present; one is central and the other temporal in position. This helps to explain the great visual acuity of many birds and is an important factor in flight (page 693).

Perhaps the most interesting feature of the bird's eye is the *pecten,* a serrated, fan-shaped structure, which extends into the vitreal cavity. The base of the pecten is attached to the blind spot and along the line formed by the fusion of the lips of the choroid fissure from which it is derived. Many variations in shape are to be noted. The best-developed pectens are found among diurnal birds, and the most poorly organized structures occur in

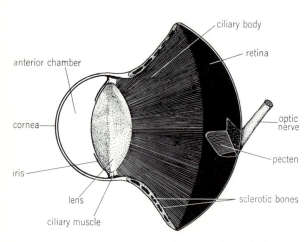

anterior chamber

cornea

iris

lens

ciliary muscle

ciliary body

retina

optic nerve

pecten

sclerotic bones

Fig. 14.13 Diagrammatic sagittal section through the eye of a bird of prey.

nocturnal forms. The function of the pecten has long been the subject of speculation. Even today no well-authenticated explanation has been forthcoming. Among the most plausible suggestions are (1) that the pecten is an erectile organ of accommodation which, by varying the intraocular pressure, can cause the lens to change its shape; (2) that a shadow of the pecten falls upon the retina, thus aiding in the perception of movement; (3) that the pecten is a supplemental nutritive device which supplies the retina with necessary substances by diffusion through the vitreous humor. If this last suggestion is accepted, it would indicate that the highly vascular pecten is an elaboration of the vascular conical papilla found in the eyes of numerous reptiles. The bird's eye lacks an arteria centralis retinae.

In many birds, as in numerous other vertebrates, the eyes are laterally located, each covering a different visual field. The term *uniocular vision* refers to this arrangement. Even in such animals, however, the two visual fields usually overlap to some extent. Birds of prey, such as owls, and, to a lesser degree, hawks and eagles, have *binocular vision*. Both eyes can be focused upon the same object, and the two visual fields overlap to a much greater degree than in animals having uniocular vision. Binocular vision is of advantage in judging distance and, therefore, in capturing prey. In uniocular vision a greater visual field is covered, enabling its possessor to detect danger coming from almost any direction. The additional area centralis and fovea in the eyes of numerous birds with uniocular vision are of great value during flight when clear vision straight ahead is of utmost importance.

MAMMALS. Although the eye of man, as previously described, is a more or less typical mammalian eye, there are, among members of the class Mammalia, certain deviations which are worthy of special mention. The upper eyelid in mammals is large and more movable than the lower. Eyelashes and Meibomian glands are usually present. Lacrimal glands in mammals are located under the upper lid at the outer angle. Harderian glands are lacking in most forms but are present in whales and certain semiaquatic forms, as well as in mice and shrews.

Prototheria. A well-developed nictitating membrane is present in *Ornithorhynchus* but altogether lacking in the echidnas. The eyes of these lower mammals show unmistakable reptilian affinities. There is little, if any, accommodation, however, and no trace of a conical papilla can be found. Cartilage is present in the sclera. The photoreceptors of the duckbill include rods and both single and double cones. In the echidnas, cones are lacking.

Metatheria. Among marsupials the nictitating membrane is reduced, but a Harderian gland is still present. No cartilage appears in the sclerotic coat, and with its elimination the eyeball has assumed a spherical shape. Ability to accommodate has not been definitely verified for the group, although the presence of a small ciliary muscle suggests that it is possible to change the shape of the lens. A vestigial conical papilla may occur in marsupials, but all traces of this structure are lost in higher forms.

Eutheria. In the large group of eutherian, or placental, mammals many variations are encountered but none is of primary importance. Numerous adaptations correlated with aquatic, terrestrial, or aerial life are to be found, as well as those fitting animals for nocturnal or diurnal existence. The eyes of moles are tiny, degenerate structures embedded in the skin. The smallest normal eyes are those of the shrews and certain bats, in which they are scarcely more than a millimeter in diameter. At the other extreme are the eyes of the blue whale, which measure between 4 and 5½ in. across, although much of the size is due to the presence of an exceptionally thick sclera. No cartilage is to be found in the eutherian sclera, however. A tapetum lucidum is present in many night-active forms. Hoofed mammals possess a *tapetum fibrosum.* This differs from the tapetum in elasmobranchs in that a portion of the choroid coat is composed of a tendinous type of connective tissue which glistens in a manner similar to a fresh tendon. Certain marsupials, elephants, and whales also have fibrous tapeta. Another type of tapetum, the *tapetum cellulosum,* is found in carnivores, seals, and lower primates. It is composed of several layers of cells which are filled with little rods or threads composed of some unknown organic material. It does not seem to be related to guanin. One or two other types of tapeta are recognized. The mechanism for accommodation is essentially similar to that described for man. There are many variations in shape of the mammalian pupil. The round pupil is most common. A vertical, slit-shaped pupil is characteristic of the cat family and certain others. The vertical slit, which assumes a rounded shape under nocturnal conditions, may be closed almost entirely in bright light. A wide, transversely arranged opening is met with among many ungulates, whales, and other mammals. A modification of the portion of the iris bordering the pupil and referred to as the *umbraculum* is encountered in the cony, gazelle, camel, and a few other species. This consists of an irregular, pigmented sort of fringe (Fig. 14.14) which is thought to protect the retina from excessive glare coming either from above or below. Well-developed nictitating membranes appear only sporadically among eutherian mammals. Among those possessing such membranes are the aardvark, horse,

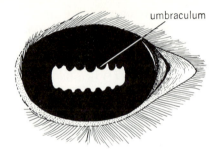

Fig. 14.14 Umbraculum of iris in the camel. (*After Lindsay Johnson, from Walls, "The Vertebrate Eye." By permission of the author.*)

scaly anteater, caribou, and panda. In carnivores it may be fairly well developed but cannot be drawn over the entire cornea. The retinae of most eutherians contain both rods and cones of the single type, but in members of the lowest orders cones are apparently lacking. This is true of most of the Edentata and Chiroptera as well as of certain shrews, hedgehogs, and rodents. Pure cone retinae are not found among mammals except in squirrels. The nocturnal flying squirrels, however, are reported to possess pure rod retinae (K. Tansley, 1960, *Anat. Record,* **136**:288). Capacity for color vision among mammals seems to be limited to the higher primates. Many mammals possess binocular vision, but this is best developed in man and other primates. Furthermore, because of the partial decussation of fibers in the optic nerves of mammals having binocular vision, the images observed by the two eyes are stereoscopically superimposed in the brain, giving third-dimensional effects not possible to obtain otherwise.

SENSES OF EQUILIBRIUM AND HEARING

The vertebrate ear, which is commonly thought of only as an organ of hearing, usually functions in a dual capacity, serving, at least in higher forms, as an organ of both hearing and equilibration. The equilibratory function of the ear is more fundamental. The structures responsible for this are to be found in all vertebrates with but little variation in different forms. The portion of the ear which is concerned with hearing first begins to differentiate in fishes and becomes more and more complex as the evolutionary scale is ascended.

When referring to the ear our attention is usually directed to the special apparatus found in higher vertebrates for reception of sound waves and their transmission to sensory receptors located in the head. This apparatus, as it occurs in man and other mammals, is composed of three portions referred to, respectively, as the *outer, middle,* and *inner ears.* The outer and middle ears are concerned only with receiving and transmitting sound waves. They are totally lacking in cyclostomes, fishes, salamanders, and some others. The inner ear, present in *all* vertebrates, is the portion in which the sensory receptors are located, whether it serves only for equilibration or for both equilibration and hearing.

The sensory receptors of the inner ear, described in the following pages, are composed of supporting cells and sensory cells with long, hairlike processes. The latter so nearly resemble the sensory neuromasts of the lateral-line system (page 723) that their close relationship is highly suggestive. This, together with the similarity in the embryonic origin of the inner ear proper and the ganglia, nerves, and sensory cells of the lateral-line organs, has made it customary to refer to them all together as the *acousticolateralis system.*

Inner Ear

The epithelium lining the inner ear is of ectodermal origin. It first arises in the embryo as a thickened placode in the superficial ectoderm of the head. Each of the paired placodes invaginates in the form of an *auditory pit,* which soon closes over to form a hollow *auditory vesicle,* or *otocyst.* It loses its connection with the superficial ectoderm (Fig. 14.15), lies close to the brain, and is entirely surrounded with mesenchyme.

Extensive changes occur as the otocyst develops further. When it has become fully

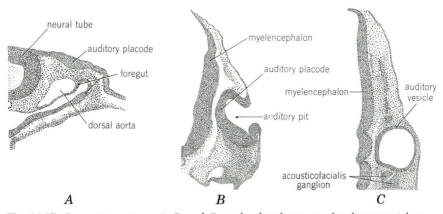

Fig. 14.15 Successive stages, A, B, and C, in the development of auditory vesicles in chick embryo; A, 33-hour embryo; B, 48-hour embryo; C, 72-hour embryo. (*Drawn by R. Speigle.*)

differentiated, its various divisions together with their supporting elements of fibrous connective tissue make up a delicate and complex structure, the *membranous labyrinth.* This consists of a closed series of tubes and sacs lined with epithelium of ectodermal origin.

Equilibrium. The typical membranous labyrinth of vertebrates consists of two chamberlike enlargements, an upper *utriculus* and a lower *sacculus.* These chambers are connected by a constricted area, the *sacculoutricular duct.* The degree of constriction in this region varies considerably in different forms. A narrow *endolymphatic duct,* a dorsomedial evagination of the otocyst, joins either the sacculus or the sacculoutricular duct. It terminates in a saclike swelling, the *endolymphatic sac,* or *sinus.* Three long narrow tubes, the *semicircular ducts,* connect at both ends with the utriculus (Fig. 14.16). They are arranged in three planes in such a manner that each is approximately at right angles to the planes of the other two. They are called the *external, anterior,* and *posterior semicircular ducts.* The external duct lies in a horizontal plane, the other two being vertical. At its lower end each duct bears an enlargement, or *ampulla.* The ampulla of the horizontal duct is situated anteriorly, close to that of the anterior duct. In a number of vertebrates the upper ends of the two vertical ducts are joined in a *crus commune* which connects with the upper portion of the utriculus. In lower forms, a slight projection of the ventral wall of the sacculus may be present. It is referred to as the *lagena.* The lagena is the forerunner of the auditory portion of the inner ears of crocodilians, birds, and mammals.

The membranous labyrinth is filled with fluid *endolymph,* the exact origin of which is not understood. Certain secretory *cells of Shambaugh* in the outer wall of the cochlear duct (page 701) are thought to keep the endolymph replenished. The viscosity of endolymph is from two to three times that of water. Almost completely surrounding the membranous labyrinth is the perilymphatic space, filled with fluid *perilymph.* The perilymph is actually cerebrospinal fluid, the perilymphatic space being in communication with the subarachnoid space. Surrounding the perilymphatic space is cartilage or bone, depending upon the species. In higher forms a *bony labyrinth* (Fig. 14.17), situated in the petrous region of the temporal bone, encloses the membranous labyrinth and follows all its configurations. The membranous labyrinth is attached to the bony labyrinth in certain places (page 701). Perilymph fills the spaces between the membranous and bony labyrinths. The *semicircular canals* are those portions of the *bony* labyrinth which surround the semicircular ducts.

The actual receptors for the sense of equilibrium consists of elevated patches of cells, the *cristae ampullares* (*cristae acousticae*) and the *maculae acousticae.* The former are located in the ampullae of the semi-

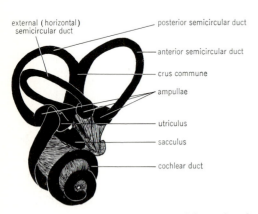

external (horizontal) semicircular duct
posterior semicircular duct
anterior semicircular duct
crus commune
ampullae
utriculus
sacculus
cochlear duct

Fig. 14.16 Right membranous labyrinth of rabbit. (*Modified from Retzius.*)

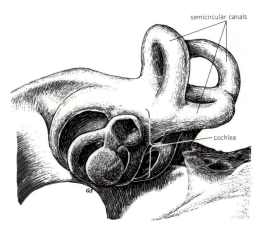

semicircular canals

cochlea

Fig. 14.17 Bony labyrinth of left ear of man exposed after a portion of the petrous region of the temporal bone has been removed. (*Drawn by G. Schwenk.*)

circular ducts. They are made up of *supporting cells* and sensory *hair cells* provided with long processes. The maculae, of which there are typically two, are also composed of supporting cells and sensory hair cells, the latter bearing short, bristly, hairlike processes. One, the *macula utriculi,* lies in the wall of the utriculus; the other, the *macula sacculi,* is located in the wall of the sacculus. It is possible that the latter may be more closely related to the sense of hearing than to equilibrium. In those vertebrates having a lagena, a third macula, the *macula lagenae,* lies at its base. Both cristae and maculae acousticae are supplied with fibers from the vestibular branch of the auditory nerve (VIII). A thickened mass forms in connection with each macula and covers its surface. This *otolithic membrane* is composed of gelatinous material into which groups of "hairs" from the hair cells penetrate. Each tuft is surrounded by a small space filled with endolymph. Small crystalline bodies called *otoconia* are deposited in the outer part of the otolithic membrane. They are composed

of a mixture of calcium carbonate and a protein. These deposits may be rather extensive, each forming a compact mass referred to as an *otolith.* Changes in position of the otoliths associated with the maculae, in response to gravitational forces, affect the hair cells and give information regarding the body when at rest (static sense) or of changes in velocity. It has been demonstrated that if guinea pigs are subjected to centrifugation strong enough to dislodge the otolithic membranes, the animals lose their static sense.

The supporting cells of each crista apparently secrete a gelatinous material which, in fixed preparations, appears in the form of a *cupula* into which the sensory hairlike processes extend. The cupula is believed by some to be merely an artifact and may not exist as such during life. It has been observed that when the cristae of fishes are being fixed for sectioning, the hairlike processes gradually shorten. Coagulated material appears in the spaces from which the "hairs" have withdrawn. This shrinks slowly and becomes the cupula.

Movements of the endolymph are believed to affect the cristae, stimulating the sensory hair cells, thus setting up impulses transmitted to the brain by branches of the auditory nerve. The cristae in the ampullae of the semicircular ducts are believed to be the receptors which give an awareness of movement (kinetic sense). They seem to be affected by rotational movements.

Comparative anatomy of the inner ear.

CYCLOSTOMES. The membranous labyrinth in cyclostomes is atypical and is considered by many to be degenerate. On the other hand, it may possibly represent a primitive condition, since there is evidence from the fossil remains of some of the oldest vertebrates that their membranous labyrinths differed but little from those of living cyclostomes. The suggestion (page 724) that the inner ear may have arisen in the course of evolution as a specialized portion of a primitive lateral-line system lends credence to the view that the membranous labyrinth in cyclostomes is primitive rather than degenerate (W. A. van Bergeijk, 1966, *Amer. Zoologist*, 6:372). In the hagfish, *Myxine*, only one semicircular duct is present (Fig. 14.18*B*). It lies in a verticle plane and bears an ampulla at either end. There is no distinction between sacculus and utriculus. A small endolymphatic duct is present with an enlarged sinus, or sac, at its dorsal end. The lamprey, *Petromyzon*, has two vertical semicircular ducts (Fig. 14.18*A*), each having an ampulla containing a crista ampullaris. An external, horizontal duct is lacking. A slight constriction is indicative of a separation into sacculus and utriculus. A *macula communis* lies at the bottom of the saccular region.

In all forms above the cyclostomes the typical three semicircular ducts are present without any basic modification.

FISHES. The membranous labyrinth of the elasmobranch fishes shows certain structural peculiarities (Fig. 14.19). The anterior vertical and the horizontal ducts join the utriculus dorsally. Their ampullae lie close together toward the anterior end. Where the posterior vertical duct lies in close proximity to the

sacculus, it is somewhat enlarged to form what is sometimes considered to be a second, or posterior, utriculus. The cavity of each "urticulus" connects near its base with that of the sacculus. A small projection from the ventral portion of the sacculus extends posteriorly for a short distance, forming the lagena. From the dorsal side of the sacculus in many elasmobranchs a small tube extends upward, opening on the surface of the head. It is frequently called the endolymphatic duct, although the term *invagination canal* is more appropriate. It is not homologous with the endolymphatic duct of higher forms. Fine sand grains, known as *otarena*, enter the cavity of the sacculus through the openings of the invagination canals and serve in the same manner as otoconia. They are clumped together in a large mass, the otolith. Sea water, rather than endolymph, probably fills the membranous labyrinth.

In other fishes the original invagination canal disappears and a new duct, the endolymphatic duct, forms as an outgrowth from the saccular region. Both utricular and saccular otoliths are present, the latter being more prominent and in some cases so large as almost to fill the saccular chamber. The utricular otolith is called the *lapillus;* that of the sacculus is the *sagitta.* It is possible that the saccular macula is stimulated by low-

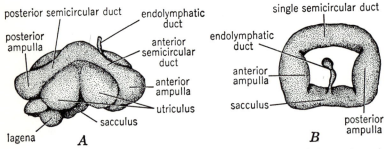

posterior semicircular duct   endolymphatic duct   single semicircular duct

posterior ampulla   anterior semicircular duct   endolymphatic duct

anterior ampulla   anterior ampulla

utriculus   sacculus

lagena   **A**   sacculus   **B**   posterior ampulla

Fig. 14.18 *A*, membranous labyrinth of lamprey (*modified from Krause*); *B*, membranous labyrinth of hagfish (*modified from Retzius*).

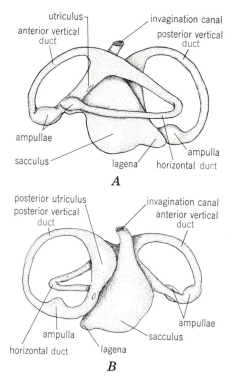

Fishes is not so completely enclosed in skeletal parts as are the ears of most vertebrates. Soft tissues provide a medium for transferring sound waves from the water to the sensory ear (J. M. Moulton, 1960, *Anat. Record*, **138**:371).

In teleost fishes of the order Cyprinoformes a chain of small bones, the *Weberian ossicles*, derived from the four most anterior vertebrae, is interposed between the swim bladder and the perilymphatic space, thus relating the function of the swim bladder with the equilibratory and possibly the auditory senses. The separate Weberian ossicles, proceeding from the ear to the swim bladder, are known, respectively, as the *scaphium*, *claustrum*, *intercalarium*, and *tripus*. Scaphium and claustrum are derived from the neural arch of the first vertebra, the claustrum in some cases being absent or nonfunctional. The intercalarium represents the neural arch of the second vertebra. It lies within a fibrous ligament connecting scaphium and tripus. The latter is the main element and is actually a modified rib of the third vertebra. A contribution from the fourth vertebra is sometimes used for support. In other fishes which lack Weberian ossicles, direct connections may exist between the swim bladder and the inner ears. This condition is found in a number of herringlike teleosts.

AMPHIBIANS. The membranous labyrinth in amphibians lies entirely within a closed, bony, *auditory*, or *otic, capsule*. The three semicircular ducts are arranged in the manner characteristic of all gnathostomes. A constriction partially separates utriculus and sacculus, the former being the larger of the two chambers and the portion to which the semicircular ducts connect (Fig. 14.20). A small endolymphatic duct extends dorsally from the sacculus and terminates in an expanded endolymphatic sac. In some frogs the sacculus is partially divided into a small upper

frequency vibrations and thus may exhibit the beginning of an auditory function. In teleosts an otolith, the *asteriscus*, is also present in association with the macula lagenae. In many fishes a small, sensory patch, the *macula neglecta*, is located in the wall of the utriculus. There is some suspicion that the macula neglecta of fishes may be concerned to some extent with sound reception. The macula lagenae in fishes, however, is apparently associated only with the sense of equilibrium. The cavity of the bony labyrinth connects with the cranial cavity, thus providing a means by which perilymph (cerebrospinal fluid) can reach and surround the membranous labyrinth. The ear of teleost

Fig. 14.19 Membranous labyrinth of dogfish shark: *A*, lateral view of left ear; *B*, medial view of left ear.

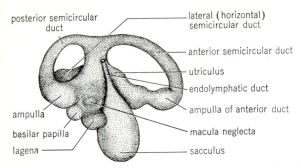

posterior semicircular
duct

lateral (horizontal)
semicircular duct

anterior semicircular duct

utriculus

endolymphatic duct

ampulla of anterior duct

macula neglecta

sacculus

ampulla

basilar papilla

lagena

Fig. 14.20 Medial view of membranous labyrinth of frog, *Rana esculenta. (Modified from Gaupp.)*

and a larger lower portion. A small lagena projects from the posterior portion of the sacculus. This is destined to give rise to the *cochlear duct* of higher forms. A sensory crista is located in each of the three ampullae. The utriculus is equipped with a macula utriculi and, in most amphibians, with an additional, smaller, sensory patch, the macula neglecta, which possibly has an auditory function. There are two of the latter in caecilians. A prominent macula sacculi is present in the wall of the sacculus, and a *macula lagenae* is located in the lagena. An additional sensory patch, the *papilla basilaris*, apparently derived from the macula lagenae, is present in the lagena in many amphibians. Others lack a basilar papilla but possess a small *papilla amphibiorum* in the wall of the sacculus near its junction with the utriculus. Urodele amphibians usually have both types of papillae. The papilla amphibiorum seems to function as a hearing device in amphibians but is of little or no significance as a forerunner of the auditory organs of amniotes. Because of certain structural peculiarities, the opinion is generally held that the papilla basilaris is of special significance in the phylogenetic development of the more elaborate hearing organs of reptiles, birds, and mammals. The further development of the acoustic organs is the chief advance made by

the ears of amphibians over the condition in fishes, but the main function of the inner ear remains that of equilibration.

REPTILES, BIRDS, AND MAMMALS. The significant changes in the membranous labyrinth in reptiles, birds, and mammals involve further development and lengthening of the lagena and an elaboration of the basilar papilla. These structures are concerned with the sense of hearing. The equilibratory portion of the inner ear remains relatively unchanged. The macula lagenae persists in reptiles, birds, and prototherians but is lacking in higher mammals.

Hearing. As mentioned above, it is in fishes that the membranous labyrinth first becomes involved in hearing which, at best, must be of a rather low order and insensitive to the direction of sound. The macula sacculi, macula neglecta, and the macula lagenae have all been considered as possible phonoreceptors, but this has not been actually demonstrated. Sound vibrations in the water may directly affect skeletal structures in the head, which transmit them to the inner ear (page 699). In some fishes the compressible swim bladder may receive vibrations, which are transmitted directly to the ear or else indirectly by way of the *Weberian ossicles.*

In amphibians the papilla amphibiorum,

Anatomy of the Chordates

as previously mentioned, may function as a phonoreceptor, but this structure is apparently without homologues in higher forms. The papilla basilaris, derived from the macula lagenae and present in many amphibians, alone seems destined for further phylogenetic specialization.

In amphibians, for the first time, an elaboration of the perilymphatic system furnishes a means of conducting vibrations from the middle ear (page 705) to the phonoreceptors of the inner ear by means of a *perilymphatic duct.* A portion of this duct comes in contact with the utricular region of the membranous labyrinth but is usually not in close proximity either to the papilla amphibiorum or to the basilar papilla. The perilymphatic duct terminates in a *perilymphatic sac* which lies beneath the medulla oblongata. Vibrations in the perilymph are probably transmitted to the endolymph, indirectly affecting one or the other or even both kinds of papillae, as the case may be. In the African clawed toad, *Xenopus laevis,* the relation of the perilymphatic duct to the basilar papilla is more intimate.

The macula neglecta, of doubtful significance in regard to the sense of hearing in fishes and amphibians, is of no further importance in the elaboration of the hearing apparatus in reptiles, birds, and mammals. In most reptiles the perilymphatic duct begins to encircle the lagena. In doing so it comes in close contact with the membrane underlying the basilar papilla. Sound vibrations in the perilymph are thus capable of stimulating directly the sensory cells of the basilar papilla. The lagena of snakes is more primitive than that of lizards, and its degree of development seems to be related to habitat, regardless of taxonomic affinities. Among snakes, burrowing forms have the most elongated lagenae. In lizards, taxonomic relationships are indicated by anatomical

similarities of the cochlear duct (M. R. Miller, 1966, *Amer. Zoologist,* 6:429).

In crocodilians, birds, and mammals the lagena and the perilymphatic duct are drawn out, or lengthened, to form the *cochlea,* and the sensory basilar papilla is stretched out into the *organ of Corti,* which is the true receptor organ for the sense of hearing in these vertebrates. The cochlea begins to assume a spiral form in mammals (Fig. 14.21). The degree of coiling varies, reaching its epitome in the alpaca, in which it makes 5 distinct turns. In man only 2¾ turns are present. The extended and coiled lagena becomes the *cochlear duct* filled with endolymph. Portions of the extended perilymphatic duct lie above and below the cochlear duct. The cochlear duct is thus fastened to the bony labyrinth on either side but is free from it above and below except for delicate connective-tissue strands which cross the perilymphatic space and help to keep the delicate membranous labyrinth in place. Thus the *cochlea proper* is divided into three longitudinal spaces, or chambers, called the scalae (Fig. 14.22). The upper one is the *scala vestibuli* and the lower the *scala tympani.* These are filled with fluid perilymph. Between the two is the *scala media,* or *cochlear duct,* which represents a modification of the original lagena. It, of course, like other parts of the membranous labyrinth, is filled with endolymph. The three scalae together make up the *cochlea.* The floor of the scala media is called the *basilar membrane.* This separates the endolymph in the scala media from the perilymph in the scala tympani. The thin-walled sloping roof of the scala media is referred to as the *vestibular,* or *Reissner's, membrane.* It separates the endolymph in the scala media from the perilymph in the scala vestibuli. At the apex of the cochlea the scala vestibuli is continuous with the scala tympani. The

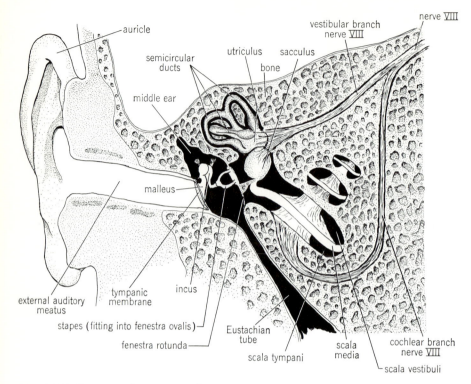

auricle

semicircular
ducts

middle ear

utriculus

bone

sacculus

vestibular branch
nerve VIII

nerve VIII

malleus

external auditory
meatus

tympanic
membrane

incus

stapes (fitting into fenestra ovalis)

fenestra rotunda

Eustachian
tube

scala tympani

scala
media

cochlear branch
nerve VIII

scala vestibuli

Fig. 14.21 Diagram of ear region of man.

point of junction, which is very small, is known as the *helicotrema*. The scala media comes to a point and ends blindly at the helicotrema.

The basilar membrane supports the complicated organ of Corti which consists of a ribbonlike band of neuroepithelial cells extending the length of the scala media. The proper functioning of the organ of Corti is highly dependent upon the organ receiving an adequate supply of oxygen. A detailed description of the organ of Corti is beyond the scope of this volume. Stated in general terms, however, it is composed primarily of *pillar cells* and *inner* and *outer hair cells,* the latter being supported from beneath by columnar *Deiter's cells.* A tunnellike space lies between the pillar cells. A thin, gelati-

nous, or jellylike, *tectorial membrane* hangs over and is in contact with the cilialike extensions of the hair cells. It is comparable to the cupula of a crista acoustica. It has been suggested that a reticular lamina actually forms the true boundary between the scala tympani and the scala media and that most of the organ of Corti may lie in the scala tympani with endolymph only in the scala media between Reissner's membrane and the reticular lamina (D. B. Webster, 1966, *Amer. Zoologist,* **6**:454). According to this concept, the "hairs" of the hair cells protrude through the reticular lamina, their distal portions making contact with the tectorial membrane. At the bases of the hair cells are fibers of the cochlear branch of the auditory nerve (VIII). They reach their destination via a portion of

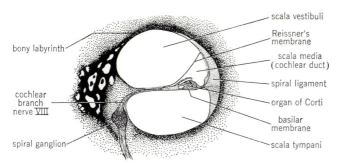

Fig. 14.22 Semidiagrammatic section through mammalian cochlea, showing the three scalae surrounded by the bony labyrinth.

the basilar membrane. The hair cells are very numerous and are arranged in a definite pattern from apex to base of the scala media. Their hairlike processes extend out into the endolymph where they are affected by vibrations in a manner to be described later (page 707). In the opinion of some, the tectorial membrane is the actual receptor for sound vibrations (M. W. Young, 1957, *Anat. Record*, **127**:391). It is originally formed as a cuticular secretion of the ectodermal cells of the auditory vesicle. It is believed that the hair cells at the base of the scala media respond to high tones, whereas those at the apex are affected by low tones. The organ of Corti in birds, although structurally different from the similarly named organ in mammals, appears to serve the same function.

Middle Ear

Urodele amphibians and some anurans lack a middle ear. This is believed to be a specialized or degenerate condition and not primitive. Beginning with the majority of anurans, however, a special mechanism is present in most vertebrates by means of which sound waves are transmitted to the auditory portion of the membranous labyrinth. The *middle ear,* or *tympanic cavity,* as

this portion of the ear is called, is concerned only with the auditory function of the ear and bears no relation to the sense of equilibrium. The middle ear is a modification of the first pharyngeal pouch. It traces its origin to the spiracular (hyomandibular) cleft which opens onto the surface of the head in the elasmobranchs and a few other fishes.

Formation. During development, in those forms in which a middle ear occurs, the first pharyngeal pouch elongates and comes in contact with the superficial ectoderm. The pouch fails to break through. A membrane, at first composed only of a layer of ectoderm and a layer of endoderm, thus separates the pouch from the outside. Later a small amount of mesodermal tissue pushes in between the two original layers. The membrane becomes the *tympanic membrane,* or *eardrum,* upon which sound waves impinge. The outer portion of this pharyngeal pouch becomes expanded to form the tympanic chamber. The portion nearest the pharynx becomes constricted so as to form a narrow *Eustachian tube,* which opens into the pharynx through a small aperture. The Eustachian tube provides a means by which atmospheric pressure on the two sides of the tympanic membrane can be equalized, thus permitting the membrane to vibrate freely. Both middle ear and Eustachian tube are lined with endoderm.

Structure. In most tetrapods two openings are present between the tympanic cavity and the cochlea. The upper one, or *fenestra ovalis* (*vestibular window*), connects with the scala vestibuli. A small bone, the *stapes,* generally serves as a plug which fits into the fenestra ovalis, much in the manner of a piston fitting into a cylinder, thus providing some freedom of movement. The lower opening, or *fenestra rotunda* (*cochlear window*), lies between the tympanic cavity and the scala tympani. A delicate membrane over the opening separates the two cavities.

In the majority of anurans a small rod-shaped bone, the *columella* (stapes), extends across the cavity of the middle ear from the center of the tympanic membrane directly to the fenestra ovalis. The rounded base of the columella, referred to as the *stapedial plate,* fits into the fenestra ovalis. In reptiles and birds the columella is usually divided into two parts, a basal portion, the *stapes proper* (sometimes called the *plectrum*), filling the fenestra ovalis, and an *extracolumella,* which is attached to the tympanic membrane. It should be recalled that the columella, or stapes, is derived from the hyomandibular cartilage of elasmobranch fishes along with the shift from the hyostylic to the autostylic method of jaw suspension (pages 434 to 452). The stapes frequently bears a foramen for the passage of an artery.

In mammals (Fig. 14.21) two additional bones, the *malleus* and *incus,* make their appearance between the tympanic membrane and the stapes. Malleus, incus, and stapes are referred to as *auditory ossicles.* The malleus, which is derived from the *articular* bone of lower forms, connects with the tympanic membrane at one end and with the incus at the other. The incus, which is homologous with the *quadrate* bone of lower vertebrates, lies between the malleus and stapes. The three auditory ossicles serve to transmit sound vibrations from tympanic membrane to perilymph in the scala vestibuli. They are arranged in such a manner as to exert some leverage. As a result, the force of the sound-wave vibrations is increased even though there is a lessening of the extent, or amplitude, of the vibrations. The auditory ossicles, despite their small size, are actually constructed in the manner of typical long bones. Epiphyses are lacking and full development is practically attained prior to birth. Articular cartilages are present at their ends. Marrow cavities are even present in the adult malleus and incus. Small ligaments are responsible for holding the ossicles together and for keeping them in place. A small *tensor tympani* muscle from the wall of the middle ear is attached to the malleus; a tiny *stapedius* muscle attaches to the stapes. The function of these muscles of the middle ear in relation to the transmission of sound vibrations is not clear. The former, by keeping the tympanic membrane tense, may play a role in the reception of high-frequency sound waves; the latter, by affecting the stapes, may protect the inner ear from injury when unusually loud noises set up violent vibrations. The surfaces of the auditory ossicles are covered with a thin layer of mucous membrane continuous with that lining the cavity of the middle ear. Thus, the ossicles do not actually lie in the middle-ear cavity. They bear the same relation to it as the peritoneum-covered intestine bears to the abdominal cavity.

Whether transmission of vibrations is accomplished by columella, extracolumella, and plectrum or by malleus, incus, and stapes, vibrations of the perilymph travel up the scala vestibuli to the helicotrema at the apex of the cochlea. They then pass down the scala tympani. A good share of the vibrations in the perilymph of the scala vestibuli is transmitted through the thickness of the cochlear duct to the perilymph of the scala tympani. The membrane over the fenestra rotunda moves in and out with corresponding movements of the incompressible perilymph, thus serving to damp the vibrations.

In mammals the chorda tympani branch of the facial nerve (VII) passes through the tympanic cavity but bears no functional relation to the ear.

Comparative anatomy of the middle ear.

FISHES. Until recently it was assumed that the middle ear and tympanic membrane first made their appearance in labyrinthodont amphibians. It is now evident that these structures may have originated in ancient rhipidistians. The Devonian fish *Eusthenopteron* is known to have possessed diverticula of the spiracle. It has been suggested that such a diverticulum may have been filled with air obtained via the spiracle and thus may have served as a middle ear. Its point of contact with an ectodermal ligamentous connection of the squamosal bone with certain bones of the parietal shield may have formed an effective two-layered tympanic membrane. The hyomandibular bone, which is homologous with the stapes (page 439), could readily have communicated sound waves from the tympanic membrane to the otic capsule (W. A. van Bergeijk, 1966, *Amer. Zoologist,* **6**:371).

Another point of view (K. S. Thomson, 1966, *Amer. Zoologist,* **6**:379) indicates that although the mechanism described above might have been effective in hearing when the animal was in water, migration to land would render it almost ineffective be-

cause of the small size, both of the tympanic membrane and the middle ear. Instead, it is suggested, the opercular bone of rhipidistians, with its direct connection to the hyomandibular bone, may be the forerunner of the tympanic membrane. When the ancestral fish migrated to land, the former operculum became a flexible structure capable of vibrating in response to airborne pressure waves and transmitting vibrations via the hyomandibular bone (stapes), surrounded by the spiracular diverticulum (middle ear), to the otic capsule.

AMPHIBIANS. No middle ear is present in caecilians, urodele amphibians, and a few anurans. Although tympanic membrane, tympanic cavity, and Eustachian tube are lacking, a columella is present. Often it is a vestigial structure. It lies within a ligamentous band, the *suspensor-stapedial ligament,* which extends from the fenestra ovalis to the ends of the squamosal and quadrate bones. In most amphibians, at the time of metamorphosis, another bony or cartilaginous element, the *operculum,* appears, fitting into the fenestra ovalis along with the columella. It is not represented in other vertebrates and appears to be a derivative of the wall of the otic capsule proper. In *Amphiuma* and *Necturus* columella and operculum are fused. The operculum is furnished with an *opercular muscle* which is attached to the suprascapula. In strictly aquatic forms it is believed that vibrations are transmitted to the ear via the jaws, suspensory apparatus, columella, fenestra ovalis, to the perilymph. In those terrestrial species lacking a middle ear the pathway is via the forelimbs, shoulder girdle, operculum, fenestra ovalis, to the perilymph. In either case the auditory sense must be of a rather low order and confined to the reception of deep vibrations.

Although most anurans have a middle

ear of the type previously described, there are some, i.e., certain burrowing toads, which lack a tympanic membrane and middle ear. In some forms, notably the tongueless toads, the two Eustachian tubes open by a single median aperture into the pharynx. In most species, however, two separate openings are present, and the tympanic membrane is flush with the surface of the head. In one frog from Thailand the tympanic membrane is located at the bottom of a pit corresponding to the external ear passage of mammals (page 707).

REPTILES. Middle ear, Eustachian tube, and tympanic membrane are altogether lacking in snakes, yet the sense of hearing appears to be acute in these reptiles. The extracolumella is attached at its outer end to the quadrate bone. Hence the path of sound vibrations is via the jaws, extracolumella, plectrum, fenestra ovalis, to perilymph. In turtles the tympanic membrane is thin and delicate in aquatic forms but thick and covered with skin in terrestrial species. The two Eustachian tubes in crocodilians open into the pharynx by a single median aperture.

In most reptiles the tympanic membrane is flush with the surface of the head, as in anuran amphibians. However, in certain lizards it lies at the base of an ear pit, which is a slightly depressed area on the surface of the head. It may be partially covered by a fold of integument. Similarly in crocodilians a movable integumentary fold covers and protects the depressed tympanic membrane.

BIRDS. Well-developed tympanic chambers are present in birds. The Eustachian tubes open by a single, median outlet into the pharynx. The tympanic membrane is located some distance from the surface of the head at the end of a narrow passage.

MAMMALS. The tympanic cavity in many mammals is surrounded by a bony structure, the *tympanic bulla,* which forms part of the temporal bone. In others it is in communication with a varying number of alveolar cavities, or spaces, in the mastoid region of the temporal bone. These are referred to as *mastoid air cells.* Middle-ear infections frequently involve infection of the mastoid air cells, a condition known as *mastoiditis.* The tympanic cavity connects with the pharynx by means of the Eustachian tube. In the true seals large sinuses connected to the middle ear combine to form a relatively large middle-ear cavity. This, together with the close proximity of distensible venous sinuses, helps to equalize pressure on the two sides of the tympanic membrane when the animal is feeding at great depths, with apparently little effect upon its sense of hearing. The extremely large tympanic cavity and tympanic bulla of gerbils, jerboas, and the kangaroo rat seem to aid the animals in some manner in avoiding capture by enemies (D. B. Webster, 1960, *Anat. Record,* **136**:299). It seems to serve as a sort of built-in sound amplifier. In the kangaroo rat the cochlea makes $4\frac{1}{4}$ turns. The Eustachian tubes of mammals open separately into the pharynx. The walls of the Eustachian tube are normally in close contact, thus virtually keeping the tube closed. They separate during the act of swallowing. The mammalian tympanic membrane is situated at the end of a fairly deep passageway called the *external auditory meatus.* In this respect it differs from most members of the lower classes. The external auditory meatus in man is 1 in. or more in length. It is of interest that in whales, air sacs which are diverticula of the Eustachian tube are associated with hearing and with regulating pressure on the tympanic membrane. In baleen whales the external auditory meatus is plugged by a waxy substance through at least part of its course (F. C. Fraser and P. E. Purves, 1960, *Proc. Roy. Soc. (London), Ser. B,* **152**:62).

External Ear

The outer portion of the external auditory meatus, or auditory canal, is supported by cartilage, but the remainder has bony walls. A true external ear, supported by elastic cartilage, which catches and directs sound waves into the auditory canal, is found only in mammals. Tufts of feathers, present in some birds near the opening of the auditory canal, may possibly function in a similar manner, but these do not constitute a *pinna,* or *auricle,* as the true external ear is called.

There seems to be a tendency to eliminate the external ear in aquatic mammals. Whales, sirenians, and certain seals lack a pinna. The external ear aperture of the dolphin is of pinhole size. During early development, however, it seems similar to that of other mammals. In other forms the opening of the external auditory meatus may be hidden in the pelage. Aquatic mammals must be able to close the ear passage so as to prevent the entrance of water. Various types of valvular arrangements are used by different aquatic mammals to effect such a closure. A degeneration of the pinna has also occurred in most burrowing mammals. Loose soil is kept out of the auditory canal by contraction of auricular muscles.

The pinna develops, for the most part, posterior to the opening of the external auditory meatus. Its size and structure show much variation among mammals. The deep cavity within the pinna is referred to as the *concha.* A pinna is lacking in the platypus, but the animal is nevertheless capable of changing the position of its external ear openings by the use of certain muscles. In some mammals, as in man, the pinna is capable of little, if any, movement. In others it is provided with well-developed voluntary muscles and can be moved in various directions so as to receive as many sound waves as possible from their source of origin. The large, flaplike external ears of elephants are probably not so effective in receiving sound waves as their size would indicate. In some species, as in the elephants and in rabbits in warm climates, the large pinna may serve primarily to dissipate heat rather than being useful as a hearing aid.

The external auditory meatus is lined with skin continuous with that covering the pinna. Tubular *ceruminous,* or *wax, glands,* considered to be modified sweat glands, as well as protective hairs, are integumentary derivatives found in the inner two-thirds of the auditory canal (page 121). Their secretion is believed to lubricate the lining of the canal and outer surface of the tympanic membrane. It also helps to keep out insects. Sometimes the cerumen, or ear wax, may be so abundant as to interfere with the reception of sound waves, in which case it must be removed.

Mechanism of Auditory Function

In most mammals, sound waves are caught by the pinna, concentrated by the concha, and directed into the external auditory meatus, finally to impinge upon the tympanic membrane. Vibrations of the tympanic membrane are transmitted to the malleus, and thence to incus and stapes. The stapes, situated in the fenestra ovalis, sets up vibrations in the perilymph of the scala vestibuli. They travel up to the helicotrema at the apex of the cochlea and then down the perilymph in the scala tympani, ultimately to be damped by the membrane covering the fenestra rotunda. Vibrations are also transmitted through the vestibular membrane to endolymph in the scala media and thence through the basilar membrane to the perilymph of the scala tympani.

There is some question as to whether the sensory receptors of the organ of Corti are stimulated by movements of the endolymph caused by vibration of the vestibular membrane or are affected directly by vibrations of the basilar membrane. The former seems more probable. In either case stimulation of the hairlike cells of the organ of Corti sets up impulses in the endings of the cochlear branch of the auditory nerve, which then travel to the brain where they are interpreted as sound.

OLFACTORY SENSE

The close association between the olfactory and respiratory organs of air-breathing vertebrates has been discussed previously. In most lower aquatic vertebrates with the gill type of respiration, the olfactory and respiratory mechanisms bear no relation to one another.

The *olfactoreceptors* consist of individual bipolar neurons derived from thickenings of the superficial ectoderm (olfactory placodes) in the anterior ventral portion of the head region. They are located in the *olfactory epithelium* which, in addition to olfactoreceptors, is composed of *supporting,* or *sustentacular, cells* and *basal cells.* The olfactory epithelium is of the pseudostratified type and has a thick lamina propria underlying it. The supporting cells have oval-shaped nuclei. In higher forms yellowish-brown pigment granules of unknown significance are present in the cytoplasm of these cells. They are responsible for the yellowish-brown color of the olfactory epithelium, which is quite in contrast to the pinkish hue of the respiratory epithelium. The function of the basal cells is not clear. It is of interest that olfacto-

receptors differ from other sensory-receptor cells in that their *cell bodies* are located in the periphery, or in the sense organ itself, and their axons make direct connection with the brain rather than depending upon other nerve cells and fibers to relay impulses. The relationship of the olfactory nerve to the brain is discussed in Chap. 13.

The cell bodies of the olfactoreceptors bear rounded nuclei. The dendritic end of each cell passes outward through the epithelium where it expands into a small *olfactory vesicle.* From 5 to 12 delicate *olfactory hairs,* each about 2 micra in length, extend outward giving in their aggregate a brushlike appearance. Tiny granules in the olfactory vesicle, each at the base of a hairlike process, are similar in appearance to the basal bodies seen in ciliated cells. The number of olfactory hairs varies with the species. They are kept moist by the secretion of branched, tubuloalveolar *glands of Bowman,* located in the lamina propria. The ducts of these glands pass through the olfactory epithelium. The secretion, in addition to being protective, serves as a moist medium in which minute amounts of volatile chemical substances are dissolved before stimulating the olfactoreceptors. In both chemical senses, taste and smell, it is necessary that substances be in solution before they can be detected. The continuous secretion of Bowman's glands provides for removal of old stimulating chemicals, thus rendering the olfactory hairs capable of responding to new stimuli.

Comparative anatomy of olfactory organs.

AMPHIOXUS. Near the pigment spot at the anterior end of amphioxus, but slightly to the left of the middorsal line, is a small *flagellated pit.* It has been suspected of possessing an olfactory function, but no nervous connections between it and the brain vesicle have been discovered, so that its actual function remains in doubt. The fact that the flagellated pit of amphioxus is unpaired is sometimes used as an argument in homologizing it with the median olfactory apparatus in cyclostomes. So-called *hair cells* in the velum and buccal cirri of amphioxus, supplied by sensory nerve fibers, may possibly be involved in olfaction.

CYCLOSTOMES. In the lamprey a single, median, nasal aperture opens onto the surface of the head. It leads, by means of a short passageway, to the blind *olfactory sac,* which undoubtedly represents a fusion of two since it is bilobed in the ammocoetes larva. In cyclostomes unmistakable evidence indicates a dual origin of the olfactory sac. Although the fossil remains of certain ancient ostracoderms indicate the existence of an unpaired olfactory apparatus rather similar to that of living cyclostomes, it is clear that others possessed paired, ventral nostrils. The olfactory sac is lined with numerous folds covered with olfactory epithelium. Paired olfactory nerves, which also indicate a bilateral origin of the olfactory apparatus, pass through a fibrocartilaginous *olfactory capsule* which surrounds the olfactory sac (Fig. 5.18) and which is situated immediately anterior to the brain. From the anterior edge of the olfactory sac a long tube extends ventrally and posteriorly, passing beneath the anterior tip of the notochord to the level of the second gill pouch where it ends blindly. This is the *nasopharyngeal pouch.* Between the ventral part of the diencephalon of the brain and the nasopharyngeal pouch lies the small pituitary gland. It has been suggested (W. J. Leach, 1951, *J. Morphol.,* **89:**271) that the nasopharyngeal pouch, which has no counterpart in higher vertebrates, may actually represent a portion of the embryonic stomodaeal cavity which has separated off from the main part of the stomodaeum. The peculiar position of the nasal aperture and its unusual relation to the

nasopharyngeal pouch, as compared with other vertebrates, are due to a shifting of these organs from their original ventral position during embryonic development (Fig. 14.23). Water is alternately forced in and out of the nasopharyngeal pouch and olfactory apparatus, the intake and outflow coinciding with respiratory movements in the gill region. In this manner water in contact with the olfactory mucous membrane is constantly replaced, enabling the animal to detect chemical substances in solution in the water. An electrical field produced by the lamprey in

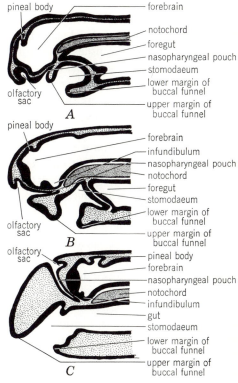

Fig. 14.23 Diagrams showing three stages, A, B, and C, in the development of the olfactory sac and nasopharyngeal pouch of the lamprey. Note how the nasal aperture has shifted to the top of the head. (*Based on Kuppfer and Dohrn.*)

the water surrounding the head aids the animal in locating its prey (H. Kleerekoper, 1958, *Anat. Record,* **132**:464). In the lamprey neither olfactory sac nor nasopharyngeal pouch has any connection with the pharynx.

In the hagfish the condition is somewhat similar to that found in the lamprey, but the single, median nostril is more anterior in position and the end of the nasopharyngeal pouch establishes a connection with the pharynx during late embryonic life. Since *Myxine* frequently lies buried in mud, it is possible for the respiratory current of water to pass to the pharynx and gills via the passage furnished by the nasopharyngeal pouch.

OTHER FISHES. Among other fishes the olfactory organs are paired. They may be nothing more than blind pits lined with olfactory epithelium. In the bony fishes the nostrils are usually situated on the dorsal aspect of the head at some distance from the mouth. In some, as in the perch, there are two nostrils on each side. The pits are so constructed as to have incurrent and excurrent openings which are sometimes quite far apart and connected to each other by a tube. Ciliary currents as well as movements of muscles of gills and jaws are responsible for the passage of water in and out the nostrils.

The nostrils of elasmobranchs are located on the ventral side of the snout. A flap of skin extends across the middle of each nostril, partially separating it into two divisions (Fig. 6.1). The flap is so arranged as to direct water in one side and out the other. In many elasmobranchs a prominent *oronasal groove* leads from each nostril to the corresponding corner of the mouth. The presence of an oronasal groove anticipates the condition first found in the Sarcopterygii, in which a true connection between the nostrils and the mouth is to be found. In the latter the olfactory epithelium, consisting of a few large folds, is located in the dorsal part of the nasal passage, which connects the internal and external nares.

Among fishes the olfactory epithelium is commonly thrown into numerous folds which greatly increase the epithelial surface. The folds are particularly well developed in elasmobranchs in which, of all the senses, the olfactory sense is probably of most importance to the animal. The fact that the folds are immersed in water accounts for their not collapsing or adhering to one another. The nasal sacs of typical fishes are always filled with water. Teleost fishes, in general, seem to have a poorly developed sense of smell, their well-developed sense of sight compensating to a high degree for this deficiency.

AMPHIBIANS. The short olfactory passage in amphibians usually arises as an oronasal groove which sinks into the upper lip and connects the external naris on each side with the mouth at a point just inside the upper jaw. In purely aquatic forms the nasal passage is lined with folds. Olfactory sense cells lie in the depressions between the folds, and ciliated epithelium covers the ridges. In larvae the cilia create a current which causes water to flow through the respiratory passage. During larval life *choanal valves,* consisting of simple folds of mucous membrane, regulate the direction in which the current of water flows. These are lost at metamorphosis, and new valves appear at the external ends of the nasal passages.

In terrestrial amphibians the olfactory epithelium is located in the upper medial portion of the nasal passages. Its surface is not folded to any extent. In some forms a shelflike fold from the lateral wall foreshadows the appearance of the conchae, or turbinal folds, which become highly developed in more advanced vertebrates. Glandular masses within the nasal passages serve to keep the olfactory epithelium moist. The secretion of the glands is of a mucous nature and covers the sensory epithelium. Olfactory hairs pass through this layer, so that their free ends are exposed but kept moist.

During inspiration the air passes toward the medial side of the nasal passage. Expired air passes along the lateral border, in which the olfactory epithelium is poorly developed or lacking.

A new structure, the *vomeronasal, or Jacobson's, organ* (page 208), first makes its appearance in amphibians. In frogs it arises as an anterior, ventromedial evagination of the nasal passage to which it ultimately connects by means of a short duct. In salamanders it is more lateral in position. Jacobson's organ is believed to be used in testing food substances held in the mouth and thus serves as a secondary olfactory area. The sensory epithelium of Jacobson's organ lacks glands of Bowman, whereas these glands are typically present in true olfactory epithelium. The organ is supplied with branches of the terminal (O), olfactory (I), and trigeminal (V) cranial nerves.

The olfactory organ in caecilians is particularly well developed in correlation with the poorly developed eyes and habit of burrowing underground. In these amphibians a peculiar structure called the *tentacle* is present on each side of the head between the eye and the external naris. Its base, which is lubricated by the secretion of the Harderian gland (page 690), is associated with the nasal passage and Jacobson's organ. The olfactory properties of the tentacles, which may be protruded and retracted, are quite independent of the respiratory mechanism. It is quite possible that caecilian tentacles are gustatory rather than olfactory in function. They also have tactile properties.

The sense of smell seems to play an important role in mating behavior among salamanders. Hedonic glands in the skin apparently secrete odoriferous substances which serve as sexual excitants.

REPTILES. With the development of palatal folds in most reptiles and of a complete secondary palate in crocodilians, the choanae come to open in the posterior portion of the roof of the mouth and the nasal passages become considerably elongated. A single concha is present in the lateral wall of each nasal passage. It consists of a projection from the maxillary bone covered with olfactory epithelium and brings about an increase in the olfactory surface. The concha is poorly developed in turtles but is quite large in crocodilians, in which it divides into three parts anteriorly. A so-called *pseudoconcha,* of questionable homology, lies posterior to the concha. The nasal passages of reptiles show, to an increasing extent, a tendency to become divided into anteroventral respiratory and posterodorsal olfactory regions.

Jacobson's organ is insignificant in turtles and crocodilians but is highly developed in *Sphenodon*, snakes, and lizards (Fig. 6.4). In these reptiles it has lost its connection with the nasal passage and opens directly into the mouth cavity, where it aids in distinguishing various types of food substances (page 208).

Those interested in details concerning evolution of nasal structure in the lower tetrapods should consult a recent paper by T. S. Parsons, 1967, *Amer. Zoologist*, 7:397–413.

BIRDS. The main advance to be observed in the nasal passages of birds is in the greater elaboration of the conchae. Three conchae are present in the lateral wall of each nasal passage. An anterior *vestibular concha* is located in the vestibular region. Smaller *middle* and *superior conchae* are located more posteriorly and are supported by *maxilloturbinate* and *nasoturbinate* bones, respectively. No olfactory epithelium is to be found covering the vestibular concha. Although the middle concha bears olfactory epithelium for a time after hatching, it soon disappears, so that only the superior concha is finally used to support the olfactory epithelium. Jacobson's organ in birds exists only as a transitory rudiment. Although the olfactory sense in birds, except in numerous marine species (albatross, shearwaters, and petrels), is rather poorly developed, there are a few which possess olfactory acuity. These include the kiwi (*Apteryx*), certain vultures of the New World, the oilbird of Trinidad and northern South America, and the honey guides of Africa and parts of southeast Asia.

MAMMALS. It is among mammals that the olfactory sense reaches the epitome of development. There are mammals, however, in which the sense of smell is poorly developed or altogether lost. Toothed whales, for example, lack a sense of smell, and in primates, seals, and whalebone whales it is but slightly developed.

It may be recalled that in mammals the nasal passages are composed of three more or less distinct regions. The first, or *vestibular region*, is lined with skin and leads to the mucous membrane inside. Hairs, sweat glands, and sebaceous glands are present. The remainder of each nasal passage includes both *respiratory* and *olfactory regions*. These comprise the greater portion of the nasal passage. In most mammals, extending from the lateral walls into these elongated regions are numerous projections, or folds, often of complicated structure. The folds may assume the shape of scrolls and greatly increase the surface of epithelium exposed in each nasal passage. The respiratory region comprises the greater portion of each nasal passage and occupies primarily the anterior and ventral areas. It is lined with pseudostratified, ciliated, columnar epithelium (respiratory epithelium). Here air is warmed and moistened. The innermost and upper portion is the olfactory region where the sensory olfactory epithelium is located.

The labyrinth of folds, or conchae, extending into the nasal passages is supported by cartilages or by turbinate bones. They are called the *ethmoturbinate, maxilloturbinate,* and *nasoturbinate,* since they arise from the ethmoid, maxillary, and nasal bones, respectively. Each turbinate bone may have secondary divisions. The epithelial covering of the maxilloturbinates is not sensory but is concerned with warming and moistening the air on its way to the lungs. The nasoturbinates, and particularly the ethmoturbinates, are the chief support for the olfactory epithelium. The more complicated they are, the keener the sense of smell. Man has three rather poorly developed conchae: superior, middle, and inferior. Turbinate bones, derived from the ethmoid, support the superior and middle conchae. An *inferior turbinate bone* supports the inferior concha. This is a separate bone which arises from the maxillary region during development. It articulates by means of sutures with the maxillary, ethmoid, palatine, and lacrimal bones. The turbinate bones are present during early development but disappear, for the most part. The nasoturbinate in man is practically rudimentary. The olfactory epithelium of the human being consists of a small area extending from the roof of the nasal cavity down both sides of the nasal septum for a short distance. It also covers part of the surface of the superior concha. The total surface area is about 250 sq mm on each side. The limited size of this sensory area accounts for the restricted olfactory sense of the human being.

Fibers of the olfactory nerve pass through small *olfactory foramina* in the cribriform plate of the ethmoid bone before establishing connection with the olfactory lobes of the brain.

Large air spaces called *sinuses* are located within certain bones of the skull of eutherian mammals. They are lined with an epithelium which is an extension of the nasal mucosa, but here the mucous glands are smaller and fewer in number. The mucosa is tightly adherent to the periosteal covering of the bones. The sinuses communicate with the nasal passages but are not concerned with the olfactory sense. Their function, if indeed they have any, is obscure. The presence of sinuses makes the skull lighter in weight and, at least in man, adds resonance to the voice. The principal sinuses are those in the frontal, ethmoid, sphenoid, and maxillary bones. The maxillary sinus (antrum of Highmore) is the largest of the sinuses. It is frequently the seat of infection.

Snouts or noses, commonly observed in mammals, have little if anything to do with the sense of smell. They are primarily tactile organs. There are a few species, however, in which both olfactory and tactile functions of the snout are highly developed. These include the platypus, echidnas, moles, pigs, shrews, and opossums. Such snouts are referred to as *chemotactile organs*. The snouts of several forms have become adapted for usages quite divorced from respiration and olfaction. Thus the trunk, or proboscis, of the elephant, representing the enormously elongated nose and upper lip, is used for many purposes. The external nares open at the end of the trunk. The snouts of some forms are used in digging and for sundry other purposes.

A well-developed, but small, Jacobson's organ is present in certain mammals belonging to the lower groups. In most, however, it appears only for a short time during early development (page 209).

The sense of smell is much more acute than the allied chemical sense of taste. Even man, with his relatively poorly developed olfactory organs, can detect certain substances in the air at extreme dilutions. Artificial musk and mercaptan may, for example, be distinguished at a dilution of one part to several billion parts of air.

Odors are caused by the emanation of volatile substances of molecular dimensions which reach the olfactoreceptors in the olfactory epithelium through air or water. The diffusion of volatile substances through water is slower than in air but is effective, nevertheless. Substances with the strongest odors are those which pass readily into a gaseous state, although not all gaseous substances have odors. It seems, furthermore, that the number of olfactoreceptors stimulated determines whether an odor is recognized as being weak or strong. Nonvolatile substances lack odors. Volatile substances must be dissolved in the moist surface of the olfactory epithelium before they are capable of stimulating the sensory receptors. It is believed by some that they may even dissolve in the olfactory hairs themselves before actually acting as stimuli.

The term *pheromones* is now used to designate a class of biologically active substances which are discharged externally and perceived by other individuals, initiating behavioral and even developmental changes in the latter. In the mouse, for example, a volatile substance (pheromone) emanating from the preputial glands of a male, is perceived by the olfactory epithelium of the female, transmitted via olfactory nerves, olfactory lobes, and cerebrum, to the hypothalamus and hence, via the hypophysio-portal-system pathway, to the anterior lobe of the pituitary gland. Release of gonadotrophic hormones may then affect the ovaries, thus effecting changes in estrous cycles, causing pseudopregnancies, or even terminating pregnancy in a newly impregnated animal. Secretion of pheromones by certain exocrine glands may be under endocrine control.

Attempts to classify odors have proved to be very difficult, and no truly satisfactory classification has been brought forth. The infinite variety of odors makes efforts at studying the mechanism of olfaction almost impossible of solution although several explanations have been proposed for such variations.

In general it is believed that olfactoreceptors may be of several different types, each type being specialized to receive only certain basic odors. Those for each type are segregated to some extent rather than being evenly distributed throughout the olfactory epithelium. The ability to detect such a great variety of odors may thus be due to the many combinations of receptors for basic odors capable of being stimulated.

Since olfactoreceptors are actually modified neurons with their nerve-cell bodies at the surface, they are more easily damaged or destroyed than are receptors for other senses. Once the cell body has been destroyed regeneration cannot occur. Hence, some deficiency in the sense of smell is common among human beings, particularly following infections, to which this region is very susceptible. In extreme cases the olfactoreceptors may be completely destroyed, and the sense of smell is then altogether lacking.

and tip (Fig. 14.24). The long axes of the cells extend from the base to the surface of the taste bud, reaching through the entire thickness of the epithelium. The number of taste cells in a taste bud is subject to much variation. Estimates in man range from 4 to 20. The supporting cells form the outer covering of the taste bud, the neuroepithelial cells lying in the interior. Some supporting cells, however, are interspersed among the taste cells. Each neuroepithelial cell bears a minute *taste hair* at its tip. The taste hairs project into a small, pitlike depression, the *taste pore*. Taste pores are found only in mammals.

Although it is rather difficult to differentiate between the chemical stimuli resulting in smell and those giving rise to the sensation of taste, it has been established that only four fundamental sensations of taste exist: bitter, sweet, salty, and sour or acid. Sometimes two additional tastes, alkaline and metallic, are added to the list, but there is no uniform agreement on this. Variations from the above are chiefly due to combinations of two or more of the fundamental tastes and to complications which arise from the part played by the olfactory nerve endings.

As with the other chemical sense, the sense of smell, substances must be in solution before they can be tasted. In mammals the

SENSE OF TASTE

The receptors for the sense of taste are spoken of as *gustatoreceptors*, or *taste buds*. The latter term is descriptive because of the resemblance of these structures to the unfolded leaf buds of plants. Taste buds are composed of two kinds of cells: *supporting cells* and *neuroepithelial cells*. The cellular elements are arranged in such a manner as to give the taste bud a sort of barrel shape, swollen in the middle and tapering at the base

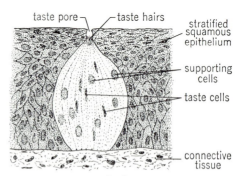

Fig. 14.24   A taste bud.

fluid in which the chemical is dissolved passes into the taste pore where it stimulates the taste hairs, setting up impulses in the neuroepithelial cells which are then transmitted to the nerve fibers associated with them. Although all taste buds have a similar appearance, it has not been determined whether a given taste bud responds to only one type of taste stimulus or to several types of stimuli.

The various taste sensations evoked by different chemicals have been the subject of careful study. Bitter taste is related chiefly to certain organic compounds, especially such alkaloids as morphine, quinine, and caffein. Glucosides, picric acid, and bile salts are additional organic substances which give a bitter taste. A few inorganic compounds such as calcium and ammonium salts induce the same sensation. Sweet taste is likewise evoked by a variety of organic and inorganic compounds. Among the former are sugars (glucose, maltose, sucrose, lactose, etc.), alcohols, saccharin, and chloroform. Lead acetate, which is sometimes called sugar of lead, is an inorganic compound giving a sweet taste. Salty taste is brought about chiefly by inorganic salts, especially the chlorides of sodium, potassium, magnesium, lithium, and ammonium. Some iodides, bromides, nitrates, and sulfates are also salty to the taste. Sour, or acid, taste is produced by acids and acid salts.

It is of interest that certain drugs have a specific action upon the various taste sensations, blocking some but leaving others to function normally. The drug cocaine completely blocks all taste sensations as well as those of touch and pain. When cocaine is applied, the various tastes are abolished in definite order, each apparently having its own threshold of response. Bitter disappears first, followed by sweet, salty, and sour in the order given.

Distribution of taste buds. In certain fishes the skin covering the entire body is abundantly supplied with taste buds. However, in most forms the gustatory receptors are confined to the oral region, being particularly numerous on the tongue. Here they are usually associated with certain small elevations of the surface epithelium, the *lingual papillae.* In mammals (Fig. 5.8) four types of papillae are recognized: filiform or conical, fungiform, foliate, and circumvallate.

Filiform papillae are small, conical elevations scattered over the upper surface of the anterior two-thirds of the tongue. They practically never have taste buds associated with them. In many forms, particularly in the cats, their free surface is highly cornified. This is of advantage to the animal in holding and in rasping minute particles of flesh from bones. *Fungiform papillae* are knoblike, rounded elevations, the free ends of which may bear a few taste buds (Fig. 14.25B). These papillae are few in number as compared with filiform papillae. *Foliate papillae* are leaflike structures running in three to eight parallel rows on either side of the tongue toward the base. They may be very much reduced in certain species, as in man. Taste buds are present along the sides of foliate papillae (Fig. 14.25C). The most prominent but fewest in number of all are the *circumvallate papillae.* These are rounded, somewhat elevated structures situated near the base of the tongue and arranged in the form of a V, with the apex pointing toward the esophagus. The *foramen caecum,* marking the point of origin of the thyroid gland, lies just posterior to the apex. Each circumvallate papilla is completely surrounded by a trenchlike depression. Taste buds are present in the epithelium which forms the walls of the trench (Fig. 14.25A). The circumvallate papillae and the taste buds associated with them are apparently of endodermal origin. This is an

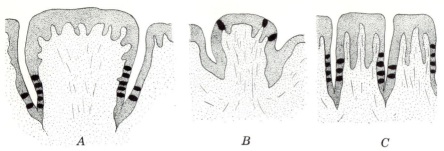

Fig. 14.25 Location of taste buds (shown in black) on *A*, circumvallate papillae; *B*, fungiform papilla; *C*, foliate papillae.

exception to the general concept that all sense organs and nervous structures are derived from ectoderm. The discrepancy would indicate that the potencies of germ layers, according to the germ-layer concept, are not so fixed as is generally supposed. *Von Ebner's glands* are small structures in the tongue, the ducts of which open into the bottom of the depression surrounding circumvallate papillae. They secrete an albuminous substance.

Comparative anatomy of gustatoreceptors.

AMPHIOXUS. On the velum and buccal cirri of amphioxus are groups of *hair cells* which may serve as gustatoreceptors or olfactoreceptors. They are supplied with branches of sensory nerves. Similar structures are not to be found over the general body surface. It is of interest that this primitive animal shows an aggregation of sensory receptors at the anterior end of the body.

CYCLOSTOMES. In adult cyclostomes, taste buds are present on the surface of the head in addition to those present in the pharyngeal region. In larval forms they are confined to the pharyngeal lining.

FISHES. In the elasmobranch fishes taste buds are associated with papillae located in the epithelium lining mouth and pharynx. Higher fishes show a greater diversity in taste-bud distribution. In some they are present on the surface of the head as well as in the mouth and on the pharyngeal epithelium. In such scavenger fishes as carp, suckers, and catfish, taste buds are present over the entire body surface, including fins and tail. It has been estimated that about 100,000 taste buds may be found in the integument of a single catfish. The barbels extending from the head are particularly well supplied. Taste has generally been considered as a close-up sense, used in determining whether food is palatable. It has recently been found (J. E. Bardach, J. H. Todd, and R. Crickmer, 1967, *Science,* **155**:1276) that in a blinded bullhead, for example, even after destruction of the olfactory apparatus, the sense of taste alone can enable the fish to detect chemical stimuli at 25 or more fish-lengths away, even when there are no water currents. A true gradient search can be enacted, using the sense of taste alone. Comparative stimuli affecting taste buds on the barbels and on right and left sides enable the fish to approach the source of the stimulus with precision, not merely by using random movements. It is of interest that the skin of the flat, ventral portion of the snout, or rostrum, of the sturgeon, *Acipenser*, lying anterior to the mouth, has a particularly abundant supply of taste buds.

Beginning with the Dipnoi in the evolutionary scale, the location of taste buds is limited to the lining of the mouth, pharynx, and tongue.

AMPHIBIANS. Taste buds of amphibians are present on the roof of the mouth, tongue, and lining of the jaws. In the frogs they occupy a position on the free surfaces of fungiform papillae. Only two kinds of papillae are present: filiform and fungiform. The latter are larger and fewer in number than the former, which do not have taste buds associated with them. Researches indicate that in tadpoles of the genus *Rana* a taste-perceiving apparatus is present even prior to the development of lingual papillae. A fine network of nerve fibers in premetamorphic papillae seems to serve the function of taste receptors (D. L. Hammerman and A. Goldfeld, 1966, *Amer. Zoologist*, 6:603). A distinction has been made between the responses of amphibians to olfactory and gustatory stimuli. Reaction to taste is indicated by snapping and swallowing movements; olfactory stimuli evoke head and body movements. It has been reported that newts are able to distinguish different tastes. In frogs apparently only salty and sour tastes are recognized. The possible gustatory function of the tentacles of caecilians has already been alluded to (page 690).

REPTILES. In the reptiles there is a more distinct localization of taste buds than is to be found in any preceding group. In lizards, snakes, and crocodilians they are practically confined to the pharyngeal region with few or none present on the tongue.

BIRDS. The horny tongues of most birds are lacking in taste buds, which are situated mainly in the lining of the mouth and pharynx. The actual distribution of taste buds, however, shows some correlation with the shape of the tongue. If the tongue is rather broad, the taste buds are located mainly on the lining of the upper beak or pharynx. If the tongue is long and narrow they are generally confined to the mucous lining of the mandible. The large, rather fleshy tongue of the parrot bears numerous taste buds.

MAMMALS. The number of taste buds in various species of mammals varies from a few hundred to over 10,000. The arrangement of the papillae on the mammalian tongue has already been referred to. Taste buds are also present on other parts of the mouth as well as in the pharynx. In man, for example, they are found on the mucous membrane covering the soft palate, the passage (fauces) between mouth and pharynx, posterior surface of the epiglottis, and vicinity of the arytenoid cartilages, and even on the vocal cords. Although taste buds on the tongue are almost always associated with papillae, those in other regions merely lie in the mucous membrane of the area in question.

In children the distribution of taste buds differs somewhat from that in adults, especially as regards the number at the tip of the tongue. During childhood many more are present in this area than in adult life. An interesting study on a newborn, male, Negro child indicates the presence of 2.583 taste buds in the oral and pharyngeal regions in addition to those on the tongue (E. Lalonde and J. Eglitis, 1961, *Anat. Record*, 139:310).

Variations in arrangement of the papillae in mammals are of some interest. Two rows of circumvallate papillae are found in monotremes, bats, rabbits, and certain others. Three rows occur in marsupials, some insectivores, and apes. Some monkeys have four rows. In man and certain carnivores the number has been reduced to a single row. Hyracoideans and guinea pigs lack circumvallate papillae. Foliate papillae are particularly well developed in rodents, but in many forms, as in man, they are almost rudimentary. In whales the gustatory sense is practically wanting.

Investigations have shown that each of the four tastes predominates in a different area of the surface of the tongue. Sweet prevails at the tip, bitter near the base, sour along the sides, and salty at both edges and the tip.

Nerve supply. Taste buds are innervated by branches of several cranial nerves, but there are no separate nerves for the sense of taste as there are for other senses. From subepithelial nerve plexuses in the vicinity of the taste buds, free nerve endings may penetrate into the taste buds themselves, terminating between the cells. Others may end in arborizations between adjacent taste buds. The manner in which the various types of nerve fibers function has not as yet been determined.

The cranial nerves for the sense of taste are the facial (VII), glossopharyngeal (IX), and vagus (X). In mammals a portion of the chorda tympani branch of the seventh nerve supplies taste buds in the anterior two-thirds of the tongue, those located in the fungiform papillae in particular. The glossopharyngeal furnishes fibers to the foliate and circumvallate papillae on the posterior third. The vagus is much less important in this respect, sending but a few fibers to the epiglottis and laryngeal region.

Visceral sensory branches of the ninth and tenth nerves supply the pharyngeal taste buds of fishes. Those in the mouth and on the surface of the head receive visceral sensory branches of the seventh nerve. In such forms as the catfish and buffalofish, with a general cutaneous distribution of gustatoreceptors, the visceral sensory area on either side of the medulla oblongata is expanded markedly. These prominences are spoken of as *vagal lobes* (Fig. 13.9).

The development as well as the function of the taste buds is dependent upon the nerves with which they are supplied. Experiments have shown that if the nerves of taste are cut, degeneration of the taste buds ensues. In the catfish they begin to degenerate shortly after the tenth day and have completely disappeared by the thirteenth day. After the nerve fibers have regenerated, the buds once more make their appearance. It has been suggested (J. M. D. Olmsted, 1920, *J. Comp. Neurol.,* **31**:465) that the organizing effect of the nerve fiber is mediated through hormonelike chemical substances given off by the fiber.

Development. Filiform and fungiform papillae first make their appearance as elevations of the tongue epithelium. Foliate papillae appear somewhat later as parallel folds along the lateral sides. Circumvallate papillae undergo the most complex development of all. Not only do slight elevations appear, but a circle of epithelial cells grows down into the mass of the tongue. Later this ring of cells splits to form the characteristic sulcus which surrounds the papilla. The glands of von Ebner are at first solid masses of cells which develop from the epithelium at the bottom of the sulcus. These masses later hollow out. Not long after the lingual papillae have formed, taste buds begin to appear. If Olmsted's conception is correct, the appearance of taste buds should coincide with the development of gustatory nerve fibers. Taste cells may be recognized as distinct from surrounding cells by their differential staining properties. Taste buds which are at first present on the crown of the circumvallate papillae later degenerate and disappear, so that only those in the walls of the surrounding trench persist. The taste pores do not appear until relatively late in development.

In such forms as elasmobranchs with pharyngeal taste buds, the latter seem to be of endodermal origin.

COMMON CHEMICAL SENSE

In the lower aquatic vertebrates the entire surface of the body is sensitive to mildly irritating chemicals. The moist integument of amphibians also comes under this category. In terrestrial forms, however, only those surfaces normally kept moist are sensitive to chemical irritants. The conjunctiva, lining of the mouth, nasal mucosa, lining of the anal canal, and genital apertures are all sensitive areas. Free nerve endings of spinal nerves and certain cranial nerves are the receptors concerned with receiving information regarding chemical irritants. The sense has no relation to the chemical senses of taste and smell. Ammonia fumes, for example, can be detected by the olfactory nerves but at the same time have a decidedly irritating effect on the nasal mucosa and conjunctiva.

OTHER SENSORY RECEPTORS

Cutaneous Receptors

In addition to the sense organs already referred to, other external sensory receptors affected by environmental stimuli are abundantly distributed in the skin over the entire surface of the body. The sensations which result from the stimulation of these receptors are referred to as *cutaneous sensations.* They include touch, heat, cold, and pain. All sensory nerve endings in the skin are external receptors of one kind or another. Some consist only of the free endings of nerve fibers. These probably serve as pain receptors and possibly for touch as well. Others are encapsulated nerve endings, and still others, in mammals at any rate, terminate in hair follicles. The skin of man contains a great number of extremely small sensory areas, each of which is responsive to one or more of the cutaneous sensations. Areas concerned with one sensation overlap others to some extent. The difficulty confronting investigators in assigning definite functions to the cutaneous receptors can be appreciated when one considers how much the sensory areas overlap and that interpretation depends so largely upon the degree of stimulation.

Until rather recently biologists were of the opinion that each type of cutaneous receptor subserved a particular function and was specifically concerned with one kind of sensation only. The whole question has been undergoing a reappraisal. Evidence has been presented which indicates that a given receptor may at different times respond to different kinds of stimuli. A given stimulus, furthermore, may affect more than one kind of receptor (D. R. Oppenheimer, E. Palmer, and G. Weddel, 1958, *J. Anat.,* **92**:321). Some even more recent investigators, on the other hand, tend to support the long-held concept that each type of receptor is concerned only with a specific kind of stimulus (H. L. Duthie and F. W. Gairns, 1960, *Brit. J. Surg.,* 47:585).

Sense of touch. Touch is the sense by which contact with objects gives evidence of certain of their qualities. The sensory endings for the sense of touch are called *tangoreceptors.* Branches of the dendritic processes of sensory nerve cells penetrating among the epithelial cells of the skin, cornea, and mucous membranes (Fig. 14.26*A*) are considered to be tangoreceptors. The nerve endings for the sense of pain (*algesireceptors*) also terminate freely in the skin. It is therefore very difficult to distinguish between these two types of receptors. In the skin of the pig's snout, nerve fibers which penetrate into the epidermis terminate in small expanded discs which are in contact with special epithelial cells (Fig. 14.26B). They are called *Merkel's tactile discs,* or *corpuscles,* and are believed to be tangoreceptors. Similar structures have been identified along the edge of the tongue and a few other areas in man. It is possible that they represent degenerate or rudimentary corpuscles of Meissner.

Several types of tangoreceptors consisting of encapsulated nerve endings have been recognized. They are located for the most part in the dermis of the skin or in the subepidermal connective tissue. Among these are *Meissner's corpuscles* (Fig. 14.26C), found in the skin of primates, particularly in areas which normally undergo friction. They are most abundant in the papillae of the corium of the dermis of the palms and soles. *Genital corpuscles* are special nerve endings in the skin of the external genitalia and of the nipples. *Grandry's corpuscles,* found on the beaks of certain birds, consist of two encapsulated tactile cells between which a nerve fiber terminates (Fig. 14.26D). *Herbst corpuscles* (Fig. 14.26E) are also found in the mouth parts of birds as well as in areas of the skin where feathers are few in number. They are composed of a capsule enclosing a double-rowed core of cells which surround the terminal nerve fiber. Rather large *Pacinian (Vater-Pacinian) corpuscles* (Fig. 14.26F) are to be found in the deep layers of the dermis as well as in such internally located parts as the tendons and periosteum. It is believed that these encapsulated nerve endings in the skin are stimulated by pressure. The sense of pressure is interpreted as being evoked by a stimulus greater in degree than that which causes the sensation of touch. It is not a true cutaneous sensation. *Corpuscles of*

Golgi and *Mazzoni,* found in the subcutaneous tissues of the fingers, are similar to Pacinian corpuscles but have thicker cores and thinner capsules.

Tactile nerve endings terminating in hair follicles are, of course, found only in mammals. The branching nerve endings are confined to the root of the hair and base of the follicle. Hairs are tactile organs with a fine degree of sensibility. When the hair is removed, hairy areas lose much of their sensitivity. Noteworthy among sensory hairs are the specialized vibrissae which are so well developed in mammals of nocturnal habit.

It is believed that touch or pressure causes a deformation of the surface of the skin with a consequent alteration in shape of the sensory endings. This is the important factor involved in the stimulation of tangoreceptors.

Temperature senses. Sensory thermoreceptors are confined for the most part to the skin. They are almost entirely lacking from the exposed surface of the eyeball. There is some disagreement concerning the specific end organs for temperature sensations. Some authorities believe that free nerve endings in the skin may serve as thermoreceptors, but it is more likely that specific encapsulated endings are responsible for receiving thermal stimuli.

COLD. The *end bulbs of Krause* (Fig. 14.26G) are usually considered to be *frigidoreceptors.* Each consists of a rounded meshwork of nerve terminations. A thin layer of connective tissue encloses the mass.

HEAT. The *end organs of Ruffini* (Fig. 14.26H) are generally believed to serve as *caloreceptors* for the sensation of warmth or heat. The endings for this sense lie somewhat deeper in the dermis than do those for cold and frequently are even located in

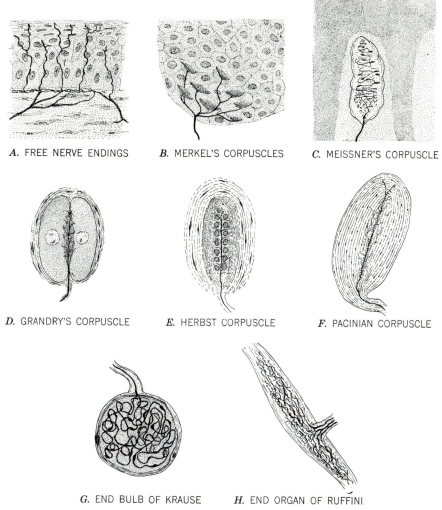

A. FREE NERVE ENDINGS   *B.* MERKEL'S CORPUSCLES   *C.* MEISSNER'S CORPUSCLE

D. GRANDRY'S CORPUSCLE   *E.* HERBST CORPUSCLE   *F.* PACINIAN CORPUSCLE

G. END BULB OF KRAUSE   *H.* END ORGAN OF RUFFINI

Fig. 14.26 Various types of cutaneous receptors found in vertebrates. Those depicted are *not* drawn to the same scale.

the subcutaneous tissue. The pit organs of the pit vipers are apparently heat receptors (page 36). Some boa constrictors and pythons have a number of very small pitlike structures along the margins of the jaws. Their function is believed to be similar to that of the pit organs of the pit vipers. Detection of warm-blooded prey from some distance by means of these organs may be of considerable survival value.

Sense of pain. *Algesireceptors,* which mediate pain sensations, consist of naked nerve endings lying in the deeper layers of the epidermis. They are also present in the cornea where tangoreceptors are lacking.

The terminal branches of a single neuron spread over a considerable distance or area and there is a high degree of overlapping. Algesireceptors are not specific in their response to any one kind of stimulus but react to several kinds of stimuli provided they are of sufficient intensity. Mechanical, thermal, and chemical stimuli all give warnings to the organism through pain receptors that the intensity of the stimulus is strong enough to induce injury.

Distribution of cutaneous receptors. Mapping out the sensory spots in the skin of the human being has been the subject of much careful study. This has been accomplished chiefly in reference to cold and warm spots by moving hot and cold metallic points over the skin. Tactile or pressure points as well as those for pain have been studied mainly by applying fine hairs of various diameters to the skin. Each of the cutaneous senses seems to have its own spots distributed through the skin. It has been estimated that there are in the entire cutaneous surface of man about 16,000 to 30,000 warm points, 150,000 to 250,000 cold points, 500,000 tactile or pressure points, and from 3 to 4 million pain points. In certain areas some types of points are much more abundant than in others. The lips and the tips of the fingers are particularly well supplied with receptors for all four of the cutaneous senses.

Little is definitely known concerning cutaneous sense organs in vertebrates other than man.

Internal Receptors

Proprioceptive sense. Receptors located in tendons, joints, and skeletal muscle give information to the central nervous system regarding the position and movements of various parts of the body. These *proprioceptors*

may consist of free nerve endings or of spindles and corpuscles of various kinds. *Pacinian corpuscles,* mentioned previously as being located in the deep layers of the corium, are also found in tendons, joints, and periosteum. Sensory *neuromuscular spindles* are located in striated muscle tissue, and *neurotendinous organs* are present in tendons but are to be found chiefly at the junction of muscle and tendon. The former are rather long, narrow structures consisting of small bundles of peculiar muscle fibers surrounded by a capsule of connective tissue. Several medullated and sparsely medullated, or nonmedullated, fibers may enter the spindle. Neurotendinous organs, sometimes called *Golgi organs,* are somewhat similar to neuromuscular spindles. Each is enclosed in a capsule containing several tendinous fasciculi. A single medullated nerve fiber usually enters the capsule and terminates in several nonmedullated, or sparsely medullated, branches. Some additional structures, closely resembling the end organs of Ruffini, have been identified in the synovial membrane lining joint cavities. Although their function is not clear, it would seem that they must in some way be concerned with the proprioceptive sense. Pacinian corpuscles, located in mesenteries, omenta, and visceral peritoneum, are believed to give information concerning stretching of mesenteries, distention of the colon, and the like.

Deep pain sensations. Free nerve endings are found lying between individual muscle fibers. Deep pain sensations are believed to arise from their stimulation.

Sense of equilibrium. Although, as mentioned on pages 696 to 697, a portion of the inner ear is chiefly concerned with the sense of equilibrium, other sensory organs may

Anatomy of the Chordates

also be involved. Proprioceptors and tangoreceptors, particularly those on the soles of the feet, are important in furnishing information to the central nervous system regarding the position of the body.

Other senses. Such sensations as appetite, thirst, hunger, and nausea, arising from within the body itself, may be mediated by so-called *interoceptors.* No specific endings or corpuscles have been described for these sensations. Algesireceptors, mediating pain sensations, are also present in internal serous membranes, mesenteries, and omenta.

Little is known concerning the sensations involved in what is commonly referred to as sexual desire. Certain *genital corpuscles,* present in the dermis of the external genitalia and of the nipples, are thought to be associated with sexual sensations.

LATERAL-LINE ORGANS

The lateral-line system of sense organs occurs only in cyclostomes, fishes, and aquatic amphibians. It is difficult to ascertain the function of these structures with accuracy. Reception of deep vibrations in the water and of stimuli caused by currents or movements of water, including those minor local currents produced by the animal itself, are among the functions ascribed to the lateral-line organs, which are sometimes referred to

as *rheoreceptors.* They may thus be functionally equivalent to proprioceptors since they furnish information regarding the position of the body in relation to the environment. It has been reported that in lampreys the lateral-line organs are sensitive to light.

In their most primitive condition the organs consist of sensory papillae called *neuromasts,* composed of groups of sensory cells surrounded by supporting cells. Each sensory cell has a hairlike process at its free end. Characteristically, the free ends of the hairlike processes are embedded in a gelatinous *cupula* in the same manner as described for the hairlike processes of the cristae acousticae of the inner ear (page 697). Stimulation is believed to be caused by a sliding movement of the cupula along the underlying cells, with a resulting displacement of, or pressure on, the bases of the hairlike processes. The cupula is apparently secreted by the cells of the neuromast. The neuromasts are usually arranged in rows or lines which follow the paths of nerves. In cyclostomes, certain fishes, aquatic amphibians, and amphibian larvae, the neuromasts are located on the surface of the skin, particularly in the head region. In other fishes they sink down into the skin in depressions or grooves, which in most forms become closed over to form tubes or canals (Fig. 14.27). In the Holocephali, however, the grooves remain open (Figs. 2.14 and

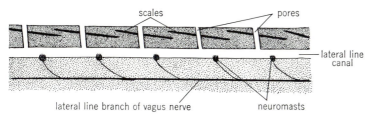

Fig. 14.27 Diagram showing relation of neuromasts to lateral-line canal and lateral branch of vagus nerve (X), as found in most fishes.

14.28). This also seems to have been true of numerous ancient amphibians, as indicated by their fossil remains. The closed tubes of most fishes open to the surface of the skin through minute pores which, in bony fishes, penetrate through the scales (Fig. 4.14). The canals are filled with a watery fluid. The canal openings, however, are usually covered with mucus. In embryonic and larval stages there are typically three longitudinal lateral-line canals, dorsal, ventral, and lateral, extending the length of the body. The dorsal and ventral canals disappear in the trunk region, so only the lateral canal persists. On the head, however, the original arrangement remains in the form of three main branches of the lateral-line canal proper (Fig. 14.28). The specific arrangement of the canals varies considerably in different fishes, but in general a large *supraorbital canal* courses forward above the eye and an *infraorbital canal*

lies below the eye. Other canals are given such names as *mandibular, hyomandibular,* and *opercular,* according to their position on the head. A *supratemporal,* or *occipital,* canal passes across the posterior part of the head, serving to connect the system on one side with that of the other. The canals are usually, but not necessarily, continuous with each other. Some authorities are of the opinion that the lateral-line system was originally derived from a network of superficial canals that covered the entire body (R. H. Denison, 1958, *Anat. Record,* **132**: 427). Since an inner ear consisting of semicircular ducts and canals is present in the fossil remains of the earliest known vertebrates, study of such fossils provides no clue as to the evolutionary origin of the inner ear. In these ancient forms a "pore-canal system" was of widespread occurrence and was obviously a primitive vertebrate character. It is quite possible that the

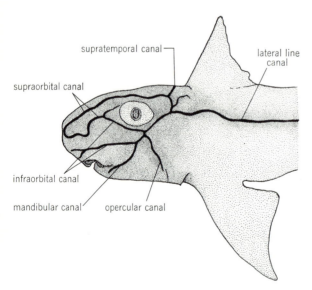

Fig. 14.28 Semidiagrammatic lateral view of head of *Chimaera monstrosa,* showing arrangement of grooves of lateral-line system in the head region.

semicircular canals of the inner ear evolved directly from the pore-canal system and that the lateral line canals were a later specialization of the same primitive sensory system (R. H. Denison, 1966, *Amer. Zoologist,* 6:369).

Of special interest is the fact that in the newt, *Diemictylus (Triturus) viridescens,* the lateral-line organs disappear when the animal takes to terrestrial existence, only to reappear when it later returns to water for breeding. Aquatic reptiles and mammals have no trace of a lateral-line system.

The lateral-line system is a somatic sensory system, supplied by branches of three cranial nerves, the seventh, ninth, and tenth. The seventh nerve is associated with the canals in the head region, the ninth with a restricted area at the base of the supratemporal canal, and the tenth with the lateral line proper. Even the portion of the lateral line near the end of the tail is supplied by the lateral branch of the vagus nerve, which in some cases may by many feet long. All three nerves bear sensory ganglia derived in part from thickened placodes in the skin. Because the inner ear arises from a similar placode and the sensory cells of the inner ear are similar in structure to neuromasts of the lateral-line system, the term *acousticolateralis system* is frequently used to embrace both systems. The two are believed to have a common evolutionary history.

Other peculiar sense organs found only in fishes are closely related to the lateral-line system and have a similar innervation. *Pit organs* consist of individual neuromasts which have sunken into small pits in the skin. They may be scattered about in no particular order or else occur in short rows. *Ampullae of Lorenzini,* found in elasmobranchs, certain other fishes, and even in some amphibians, may be present in large numbers in the skin of the head but rather deeply buried. Each consists of a lobulated, bulblike enlargement containing sensory cells supplied by the facial nerve, lying at the base of an elongated, fluid-filled tube or canal which connects to the surface by means of a small pore (Fig. 14.29). Certain studies indicate that the ampullae of Lorenzini, formerly considered to be thermoreceptors, are actually concerned with pressure changes. Other studies reveal their importance in the reception of electrical stimuli, enabling the animals to detect enemies and prey and to establish social patterns. A peculiar median *rostral organ* has been identified on the snout of *Latimeria.* It consists of a rostral sac with three pairs of openings to the outside. The sac is filled with a gelatinous substance and has sievelike connections with each tube leading to the exterior. Innervation seems to be from the superior ophthalmic branch of cranial nerve V. The function of this organ is unknown, but it invites comparison with the ampullae of Lorenzini of other fishes. Details of its innervation have not been described, so far as the author

Fig. 14.29 A group of sensory ampullae of Lorenzini.

is aware. *Vesicles of Savi,* found only on the ventral surface of the electric ray, *Torpedo,* are closed sacs separated from the rest of the epidermis. They function as receptors of mechanical stimuli in response to local pressures or to displacement of the skin.

The importance of the lateral-line organs in connection with the aquatic mode of life is indicated by their highly developed supply of branches of cranial nerves and by the fact that they have disappeared entirely in terrestrial forms.

SUMMARY

1. Receptor organs are structures capable of responding to certain types of stimuli by setting up impulses which are in turn transmitted by nerve fibers to the central nervous system. In the higher centers of the brain such impulses are interpreted as sensations. The actual receptor cells, with few exceptions, are of ectodermal origin.

2. Receptor organs are classified as external and internal sense organs. External sense organs, or exteroceptors, are those which are stimulated by environmental disturbances of various kinds. They include those for the senses of sight, hearing, smell, taste, touch, pressure, temperature, and pain. Internal sense organs, affected by stimuli arising within the body, include statoreceptors which, in response to gravitational forces, are concerned with equilibuium, proprioceptors for the sense of muscle position, and interoceptors for such internal sensations as hunger, thirst, and nausea. Internal receptors for deep pain sensations are also recognized. The lateral-line organs of lower aquatic vertebrates are exteroceptors.

3. The vertebrate eye has no true homologue among other phyla and appears for the first time in cyclostomes. Its evolutionary history is obscure. The retina, or sensory portion, develops as a cuplike outgrowth of the diencephalic region of the forebrain. The lens is derived from the superficial ectoderm which invaginates under the organizing influence of the optic cup in certain forms. Mesenchyme surrounding the optic cup differentiates into the uveal and fibrous coats of the eyeball.

In the vertebrate eye, light must pass through the various cellular layers of the retina before it can stimulate the sensory receptors, or rod and cone cells. Impulses thus set up must pass back through the cells comprising these layers to reach the optic nerve. The retina is thus said to be "inverted." Several theories have been advanced to explain the inversion of the retina.

Accommodation, or adjustment, made by the eye in order to bring objects at various distances into focus, is accomplished in different ways by various vertebrates. Lower forms accommodate by changing the position of the lens. Higher forms change the shape of the lens when accommodating.

The eye is moved by six broad strap-shaped muscles which are supplied by the purely motor third, fourth, and sixth cranial nerves. Other accessory structures are associated with eyes in different vertebrates, in each case serving to adapt the animal better to its own environmental needs.

4. The vertebrate ear, at least in higher forms, functions in a dual capacity as an organ of equilibrium and hearing. The equilibratory function is the more fundamental and primitive. The portion of the ear in which the sensory receptors are located is the inner ear, or membranous labyrinth. Its epithelial lining is derived from an invagination of the superficial ectoderm in the head region, known as the auditory placode. The inner ear contains endolymphatic fluid. The portion of the inner ear concerned with equilibrium is to be found in all vertebrates with but little variation in different forms. It consists typically of two chamberlike enlargements, an upper utriculus and a lower sacculus, the two being connected by a canal. An endolymphatic duct joins the sacculoutricular canal. Three long, narrow tubes, the semicircular ducts, arranged at right angles to each other, connect at both ends with the utriculus. Each semicircular duct bears a swelling, or ampulla, at its lower end. The sensory receptors for equilibrium are cristae ampullares (acousticae) and maculae acousticae. They are small, elevated patches of sensory cells located in the ampullae and in the utriculus and sacculus, respectively. An otolithic membrane containing crystalline bodies covers each macula, and a gelatinous cupula covers each crista. Movements of the endolymph affect the cupulae of the cristae, whereas changes in position of the otolithic membrane in response to the pull of gravity stimulate the maculae. In cyclostomes the membranous labyrinth is degenerate and only one or two semicircular ducts are present.

A slight projection of the ventral wall of the sacculus, the lagena, is usually present in lower forms. It evolves into the auditory portion of the ear in higher vertebrates. A macula is generally present in the wall of the lagena but is lacking in higher mammals. A basilar papilla, derived from the macula lagenae, first appears in amphibians and is the forerunner of the organ of Corti, the true sensory receptor for hearing in higher forms. In crocodilians and birds the lagena begins to elongate to form the cochlear duct, also filled with endolymph. In mammals the cochlear duct (scala media) assumes a spiral form and becomes more complicated. It contains the organ of Corti. Hairlike cells of the organ of Corti are stimulated by vibrations transmitted from the environment.

Beginning with anuran amphibians, a special mechanism is present in most vertebrates by means of which sound waves are transmitted to the auditory portion of the membranous labyrinth. The middle ear, or tympanic cavity, as this portion of the ear is called, is derived from the first pharyngeal pouch. A tympanic membrane, or eardrum, separates the middle ear from the outside. Small bones transmit vibrations from the tympanic membrane to fluid perilymph surrounding the membranous labyrinth. The perilymph in turn affects the endolymph within the membranous labyrinth. A Eustachian tube connects the cavity of the middle ear with the pharynx. It serves to equalize pressure on the two sides of the tympanic membrane. An outer, or external, ear is present only in mammals, although suggestions of such a structure appear in a few lower forms. It serves to catch and direct sound waves, which then impinge upon the tympanic membrane.

5. The receptors for the sense of smell consist of individual bipolar neurons derived from thickened areas in the superficial ectoderm. In aquatic vertebrates

these are confined to the olfactory epithelium lining the olfactory sacs. The latter are nothing more than blind invaginations of the outer epithelium. In the crossopterygian fishes, a connection between the nares and the mouth first appears. This passage is used in respiration primarily, the olfactory function being secondary. Further advances are to be found in a separation of the respiratory and olfactory regions from the mouth cavity by the development of a palate. Outgrowths from the lateral walls of the nasal passages in the form of folds and scrolls (conchae supported by turbinate bones) increase the respiratory and olfactory surfaces. Chemical substances must be in solution before they can stimulate the moist, mucus-covered olfactory epithelium.

6. Taste receptors are arranged in the form of taste buds, small structures composed of neuroepithelial cells and supporting cells. Substances to be tasted must be in solution in order to stimulate the sensory receptors. In some fishes gustatory receptors are distributed throughout the skin covering the entire body, but in most species they are confined to the oral and pharyngeal regions, being particularly numerous on the tongue. Here they are usually associated with small elevations of the epithelium called lingual papillae. The facial (VII) and glossopharyngeal (IX) cranial nerves are supplied with sensory fibers for the sense of taste. The vagus (X) nerve may also play a minor role in mediating taste sensations.

7. The entire surface of the body of lower aquatic vertebrates is sensitive to mildly irritating chemicals. In terrestrial forms, only those surfaces which are normally kept moist are sensitive to such stimuli. The receptors for this common chemical sense, which is in no way related to smell or taste, consist of free nerve endings of spinal nerves and certain cranial nerves.

8. Cutaneous receptors for the senses of touch, pressure, temperature, and pain consist of free nerve endings in certain cases and encapsulated nerve endings of various types in others. No specific nerves are concerned with the cutaneous senses.

9. Internal receptors for the proprioceptive sense consist of free nerve endings or spindles and corpuscles of different kinds. They are located in such places as muscles, tendons, joints, and periosteum. No specific receptors have been found which give rise to such sensations as hunger, thirst, and nausea. Internal pain receptors consist of free nerve endings.

10. Lateral-line organs of fishes and aquatic amphibians are related to the aquatic mode of life. They are believed to receive deep vibrations in the water and also to receive information concerning currents or movements of water, including those produced by the animal itself. The receptors consist of neuromasts made up of sensory cells and supporting cells. They are located either on the surface of the skin or inside grooves or canals in the skin. The lateral-line canal proper usually extends the length of the body but terminates on the head in several branches. The system is innervated by the seventh, ninth, and tenth cranial nerves. Because of the fact that the ganglia of these nerves arise in part from thickened placodes in the embryonic ectoderm similar to those giving rise to the inner ear, and because of the

similarity in structure of the sensory receptors of the ear and those of the lateral-line system, the two are often referred to together as the acousticolateralis system. Other structures belonging to the lateral-line system are the scattered pit organs found in fishes and the ampullae of Lorenzini and vesicles of Savi confined to the elasmobranchs.

SUMMARY:
CHARACTERISTICS
AND
ADVANCES

PHYLUM CHORDATA

Unique features. A notochord; a hollow dorsal nerve tube; gill slits connecting to the pharynx, or traces of them, present sometime during life.

Other features. Bilateral symmetry; cephalization; metamerism; true body cavity, or coelom; ventral-anterior direction of blood flow from heart. These features are also characteristic of numerous invertebrate animals.

Subphylum I. Hemichordata

Marine, burrowing forms, body wormlike and unsegmented with proboscis, collar, and trunk; numerous gill slits connecting pharynx with outside; a stiffened, forward extension of the gut into the proboscis probably *not* homologous with notochord; dorsal and ventral nerve strands, the dorsal one being larger and having a tubular structure only in the collar region; presence of giant nerve cells in anterior region suggests the beginning of a brain; dioecious; tornaria larva of certain forms closely resembling larval echinoderms. EXAMPLE: *Dolichoglossus (Saccoglossus) kowalevskii.*

Subphylum II. Cephalochordata

Marine; free-swimming; notochord well developed and extending entire length of body; no well-defined head; nerve cord slightly enlarged at anterior end, forming a cerebral vesicle; pharyngeal gill slits connecting indirectly with outside through an atrium and atriopore; metamerism very clearly indicated; dioecious; gonads arranged metamerically; hepatic caecum off ventral side of digestive tract homologous with the liver of higher forms; heart a ventral pulsating tube, sometimes said to be one-chambered; hepatic portal vein; metapleural folds suggest possible origin of paired appendages of vertebrates (doubtful); epidermis one cell layer thick; integumentary glands all of unicellular type; spinal nerves metameric, but those of one side alternate with those of the other; dorsal and ventral nerve roots not united; endostyle lying along midventral line of pharynx; no reproductive ducts. EXAMPLES: Amphioxus; *Branchiostoma.*

Subphylum III. Urochordata (Tunicata)

Marine; some free-swimming forms, but others becoming sessile after a free-swimming larval period; excurrent and incurrent siphons; pharynx large with numerous gill slits; notochord only in larva and then confined to tail region; nervous system of adult reduced to a small ganglion; hermaphroditic; endostyle in midventral part of pharynx perhaps forerunner of thyroid gland of higher forms; adneural gland around nerve ganglion possibly homologous with pituitary gland of higher forms; covering of body, or tunic, composed of tunicin (cellulose); heart

showing reversal of beat; mantle cavity present. EXAMPLES: Ascidians; *Molgula manhattensis, Botryllus, Doliolum, Salpa.*

Subphylum IV. Vertebrata (Craniata)

Anterior end of dorsal nerve cord enlarged to form a brain which is three-lobed during early development; protective and supporting endoskeleton; a cranium surrounding the brain, and a segmented spinal column composed of vertebrae enclosing the spinal cord; paired sense organs in head—olfactory, optic, and otic—are typical.

Superclass I. Pisces
Aquatic; in most cases respire throughout life by means of gills connected with pharyngeal gill slits.

Class I. Agnatha
Absence of jaws; absence of paired appendages even in embryo; a single median nostril representing a fusion of two; poorly developed cranium. In ostracoderms, known only through fossil remains, there were heavy dermal plates in the skin; some had an endoskeleton of a sort; some possessed lateral body lobes posterior to the head region, from which pectoral appendages may have evolved.

Order I. Cyclostomata. Living forms; body rounded, but tail laterally compressed; round suctorial mouth; median fins supported by cartilaginous fin rays; no scales in skin; skin soft with numerous unicellular mucous glands present; 6 to 14 pairs of gill pouches; poikilothermous; poorly developed brain not completely surrounded by cranium; vertebrae also poorly developed; protocercal tail; tongue specialized for rasping; epidermal horny teeth; persistent notochord; skeleton entirely cartilaginous; two-chambered heart; lateral-line system, but no true lateral line; no genital ducts; primitive in many respects, highly specialized in others. EXAMPLES: Lampreys and hagfishes.

Suborder I. Petromyzontia. Inhabiting rivers, lakes, and the sea; dioecious, ventral suctorial funnel at anterior end beset with horny teeth; seven pairs of gill pouches, each opening separately to the outside; ammocoetes larvae undergo metamorphosis; pharynx a blind pouch guarded by a velum; pancreas essentially lacking; opisthonephros; dorsal and ventral roots of spinal nerves not united; eyes primitive but well developed; only two semicircular ducts (the vertical ones) present in inner ear; nasal sac gives off nasopharyngeal pouch which ends blindly. EXAMPLES: Lampreys; *Petromyzon marinus, Entosphenus tridentatus.*

Suborder II. Myxinoidea. All marine; hermaphroditic; mouth nearly terminal and surrounded by four tentacles; no buccal funnel; few teeth; parasitic habit; 6 pairs of gill pouches in *Myxine* with a single opening to the exterior on each side; an asymmetrical esophageo-cutaneous duct on left side; 6 to 14 pairs of gill pouches in *Bdellostoma;* archinephros in embryo, pronephros and opisthonephros in adult; eyes degenerate and functionless; one semicircular duct in inner ear with an ampulla at either end; nasopharyngeal pouch connects with nasal sac in front and with pharynx behind; no metamorphosis. EXAMPLES: Hagfishes; *Myxine glutinosa, Myxine limosa, Bdellostoma.*

The remaining vertebrates typically have paired pectoral and pelvic appendages; true upper and lower jaws; paired nostrils; well-developed endoskeleton; closed cranium.

In those members of the superclass Pisces, higher in the evolutionary scale than members of the class Agnatha, paired appendages are present in the form of fins; median fins are also present; dermal scales of various types are in the skin; unicellular and multicellular mucous skin glands are abundant; they are dioecious; poikilothermous; have opisthonephros (persistent pronephros in a few); a two-chambered heart [a few (lungfishes) have a three-chambered heart]; 10 pairs of cranial nerves in addition to the terminal nerve (0); genital ducts.

Class II. Placodermi

Fossil forms only; most primitive of fishes; primitive jaws and paired fins; bony armor of dermal plates in skin; some with lunglike pharyngeal diverticula. EXAMPLE: *Bothriolepis.*

Class III. Chondrichthyes

Marine with few exceptions; cartilaginous skeleton; placoid scales; ventral, subterminal mouth; heterocercal tail; pelvic fins of male modified to form claspers used in copulation; no swim bladder; spiral valve in small intestine; lower jaw skeleton composed of Meckel's cartilage, upper jaw of palatopterygoquadrate bar; oviparous, ovoviviparous, and viviparous forms; eggs large with abundant yolk; meroblastic cleavage; variable number of aortic arches. EXAMPLES: Cartilaginous fishes.

Subclass I. Elasmobranchii

Five to seven pairs of gill slits opening separately to the outside; a pair of spiracles, representing the first pair of gill slits, usually opening on top of head; hyostylic method of jaw attachment; persistent notochord partially replaced by cartilaginous vertebrae; cloaca.

Order I. Selachii. Pectoral fins distinctly marked off from cylindrical body; gill slits lateral in position; tail used in locomotion. EXAMPLES: Sharks and dogfish; *Squalus acanthias, Sphyrna zygaena, Mustelus canis.*

Order II. Batoidea. Dorsoventrally flattened bodies; pectoral fins not sharply marked off from body; demarcation between body and tail distinct; gill slits ventral in position; spiracle large and well-developed; undulatory movements of pectoral fins used in locomotion. EXAMPLES: Skates (oviparous) and rays (ovoviviparous); *Pristis pectinatus, Torpedo, Raja, Manta.*

Subclass II. Holocephali
Persistent notochord; poorly developed vertebrae; operculum; open lateral-line canals; no spiracles; no cloaca. EXAMPLES: Chimaeras; *Chimaera monstrosa, Hydrolagus colliei.*

Class IV. Osteichthyes
Skeleton bony, at least to some degree; scales usually present of ctenoid, cycloid, or ganoid types; terminal mouth; operculum; swim bladder usually present; tail usually homocercal but sometimes heterocercal or diphycercal; Meckel's cartilage invested by bone; dermatocranium roofs over most of chondrocranium; four pairs of aortic arches; oviparous, ovoviviparous, and viviparous forms; no cloaca. EXAMPLES: Bony fishes.

Subclass I. Actinopterygii
All fins, paired and unpaired, supported by dermal skeletal fin rays. EXAMPLES: The ray-finned fishes.

Superorder I. Chondrostei. Fin rays of dorsal and anal fins exceeding in number the supporting skeletal elements; skeleton mainly cartilaginous; tail diphycercal or heterocercal; clavicles present; nostrils not connecting with mouth cavity; spiracles; spiral valve in small intestine. EXAMPLES:

So-called "ancient" fishes, or primitive ray-finned fishes; *Polypterus, Calamoichthys, Polyodon, Psephurus, Acipenser.*

Superorder II. Holostei. Dermal rays of dorsal and anal fins equal in number to internal radial skeletal elements; marine forms extinct; living species are fresh-water forms; internal skeleton partly cartilaginous; abbreviated heterocercal or homocercal tail; pelvic fins usually posteriorly located; no spiracles; no clavicles. EXAMPLES: The intermediate ray-finned fishes; *Amia,* the fresh-water dogfish; *Lepisosteus,* the gar.

Superorder III. Teleostei. Dominant fishes in the world today; homocercal tail; internal skeleton completely ossified; no spiracles; paired fins small; pectoral fins well up on sides of body; pelvic fins far forward in many; scales bony, thin, flexible, and rounded; well-developed jaws. EXAMPLES: Most familiar fishes; perch, sunfish, herring, etc.

Subclass II. Sarcopterygii
Clavicles present. EXAMPLES: Lobe-finned fishes and lungfishes.

Order I. Crossopterygii. Fins borne on fleshy, lobelike scaly stalks; skeleton of pectoral and pelvic fins with single point of attachment to girdles; premaxillary and maxillary bones present. EXAMPLES: The lobe-finned fishes.

Suborder I. Rhipidistia. Fossil forms only; of importance in evolution of tetrapod limb. EXAMPLE: Eusthenopteron.

Suborder II. Coelacanthini. Head short and deep; notochord extending far forward into head; ancient fishes, until 1938 known only by fossil remains; first dorsal fin in living forms is fan-shaped, rest are lobate; hypophyseal duct connects pituitary gland with mouth cavity; brain small and of simple construction; internal nares lacking in living species. EXAMPLE: *Latimeria chalumae.*

Order II. Dipnoi. Swim bladder basically bilobed and physostomous with ventral connection to esophagus and used as lung; internal as well as external nares in all; pulmonary circulation; three-chambered heart with double type of circulation; larval forms with external gills; sound production; diphycercal tail; cranium mostly cartilaginous; autostylic method of jaw attachment; spiral valve in intestine; premaxillary and maxillary bones lacking. EXAMPLES: The true lung-fishes; *Protopterus, Lepidosiren,* and *Epiceratodus.*

Superclass II. Tetrapoda
Paired appendages are limbs which are typically pentadactyl; cornified outer layer of epidermis; lungs; skeleton bony for the most part; sternum usually present; number of skull bones reduced; reduction in visceral skeleton.

Class I. Amphibia
Transition from aquatic to terrestrial life clearly indicated in this group; eggs laid in water or in moist situations and covered with gelatinous envelopes; metamorphosis; larvae with integumentary gills; respiration in adults (with few exceptions) by means of lungs and smooth, moist, vascular integument; buccopharyngeal respiration; poikilothermous; multicellular mucous and poison glands in skin; five pairs of pharyngeal pouches in embryo; lungs simple in structure, appearing at time of metamorphosis; opisthonephros; cloaca; reduction in number of bones in skull; skull flattened; two occipital condyles; three-chambered heart;

pentadactyl limbs usually modified; no nails or claws; 10 pairs of cranial nerves in addition to terminal nerve (0). EXAMPLES: Caecilians, salamanders, newts, frogs, and toads.

Numerous fossil forms of amphibians are known, the heads of which were covered almost solidly with bony dermal plates. They possessed more skull bones than modern amphibians and many had a ventral armor of overlapping bony plates (a few forms with dorsal bony plates in addition). Most familiar examples include the labyrinthodonts, lepospondyls, and phyllospondyls, varying primarily in the structure of their vertebral columns.

Order I. Anura. Absence of tail in adult; head and trunk fused; large, wide mouth; two pairs of well-developed limbs, the hind limbs being fitted for leaping and swimming; toes webbed; tympanic membrane flush with head; middle ear present for first time; true sound-producing organs; amplexus and external fertilization, therefore oviparous; larvae do not resemble parents; larvae with horny jaws in lieu of teeth; nine vertebrae plus urostyle; movable eyelids; lacrimal glands; well-developed sternum and limb girdles. EXAMPLES: Frogs and toads; *Rana pipiens, Bufo americanus.*

Order II. Urodela (Caudata). Body divided into head, trunk, and persistent tail regions; two pairs of weak limbs in most species; larvae closely resembling parents; internal fertilization (except *Cryptobranchus* and a few others); oviparous (except black salamander); larvae with true teeth in both upper and lower jaws; no middle ear; parietal muscles distinctly segmented; sternum poorly developed and primitive; lateral line lost at metamorphosis except in perennibranchiates.

EXAMPLES: Newts and salamanders; *Necturus maculosus. Diemictylus (Triturus) viridescens, Proteus.*

Order III. Apoda (Gymnophiona).
Snake-like bodies; no limbs or limb girdles; very short tail; anus almost at posterior end of body and ventral; in some, fishlike dermal scales embedded in skin; no gills or gill slits in adult; extremely small eyes embedded in skin; no eyelids; internal fertilization; male with protrusible copulatory organ; intestine not differentiated into large and small regions; compact skull. EXAMPLES: Caecilians; *Ichthyophis glutinosus, Typhlonectes.*

Class II. Reptilia
Many fossil forms. Living species with following characteristics: internal fertilization; oviparous and ovoviviparous forms; large-yolked eggs laid on land; three embryonic membranes, amnion, chorion, and allantois, appearing for the first time; poikilothermous; skin dry with dead corneal layer well developed; epidermal scales, some with bony plates underlying them; skin almost devoid of glands; limbs typically pentadactyl, terminating in claws (not in snakes or limbless lizards); respiration entirely by means of lungs except for cloacal respiration in aquatic chelonians; no metamorphosis; at least a partial separation of nasal and oral cavities; five pairs of pharyngeal pouches in embryo; vertebral column divided into cervical, thoracic, lumbar, sacral, and caudal regions (except snakes and limbless lizards); metanephros, one occipital condyle; three-chambered heart (except in crocodiles and alligators in which it is four-chambered); no lateral-line system; 12 pairs of cranial nerves in addition to terminal nerve (0).

Order I. Chelonia.
Living forms represent a very old group which has persisted with little change for 175 million years. Short wide bodies; no teeth, each jaw being covered by a horny scale; shell composed of plastron and carapace covering body and made of bone covered with large epidermal scales; poikilobaric respiration; single penis in male; oviparous; lumbar region of vertebral column lacking; thoracic vertebrae and ribs fused to carapace; solidity of skull bones well marked; limbs pentadactyl except in marine forms; terrestrial, fresh-water, and marine species; anus a longitudinal slit. EXAMPLES: Turtles, tortoises, and terrapins; *Chelydra serpentina, Chrysemys picta.*

Order II. Rhynchocephalia.
Well-developed parietal eye; no copulatory organs in male; anal opening a transverse slit; diapsid condition in regard to temporal fossae; amphicoelous vertebrae; lizardlike body form with scaly skin; gastralia; only one living species. EXAMPLE: *Sphenodon punctatum.*

Order III. Squamata.
Skin covered with horny epidermal scales; anal opening a transverse slit; paired eversible hemipenes in male used as copulatory organs; large palatal vacuities in roof of mouth; quadrate bone forms a movable union with squamosal; vertebrae usually procoelous; gastralia lacking. EXAMPLES: Lizards and snakes.

Suborder I. Sauria (Lacertilia).
Visible external earpits; tympanic membrane not at surface of head; movable eyelids and nictitating membrane; urinary bladder usually present; tail readily detachable; metachrosis well exemplified in most; usually two pairs of pentadactyl limbs terminating in claws; two halves of lower jaw firmly united and size of mouth opening restricted; well-developed protrusible tongue; air sacs in some; some with a well-developed parapineal eye; oviparous and ovoviviparous forms. EX-

AMPLES: Lizards, geckos, glass snakes, and skinks; *Heloderma horridum, Chameleon vulgaris.*

Suborder II. Serpentes (Ophidia). No limbs, but a few have vestiges of limb girdles; no sternum; no external ear openings or tympanic membranes; middle ear lacking; eyelids immovably fused and transparent; loose ligamentous attachment of jaws; ventral scales used in locomotion; left lung smaller than right and often absent altogether; hemipenes as copulatory organs; vertebral column divided into precaudal and caudal portions; tongue long, forked, protractile; no urinary bladder. EXAMPLES: Snakes; *Crotalus viridis, Naja tripudians.*

Order IV. Crocodilia (Loricata). Tail laterally compressed; two pairs of short legs; five toes on forefeet and four on hind feet; toes webbed; tympanic membrane exposed but protected by a fold of skin; teeth set in sockets in jawbones; oviparous; anal opening a longitudinal slit; single penis in male; four-chambered heart; first appearance of true cerebral cortex; adapted for amphibious life; diapsid condition in regard to temporal fossae; gastralia; dorsal and ventral scales reinforced with bony plates; tongue not protrusible; no urinary bladder. EXAMPLES: *Alligator mississippiensis, Crocodylus americanus,* caymans, gavials.

Class III. Aves
Feathers; legs and feet covered by horny skin or scales of reptilian type; light, hollow bones; loss of right ovary and oviduct except in certain birds of prey; very well-developed eyes; nictitating membrane; forelimbs modified to wings; homoiothermous; sexual dimorphism often highly developed; oviparous; large size of eggs; meroblastic cleavage; skin almost devoid of glands; one occipital condyle; four-chambered heart; claws at ends of digits of hind limbs; at least partial separation of nasal and oral cavities; four pairs of pharyngeal pouches in embryo; syrinx for sound production; air sacs from lungs; metanephros; fertilization internal; copulation by cloacal apposition except in a few forms with a single penis; rigidity of spinal column with posterior thoracic, lumbar, sacral, and proximal caudal regions firmly united; right aortic arch 4 persisting, left one disappearing; toes reduced to four or less; intratarsal joints in legs; flexible neck; no urinary bladder; 12 pairs of cranial nerves; absence of terminal nerve (0). EXAMPLES: Birds.

Subclass I. Archaeornithes
Fossil forms only; long, jointed tail with 18 to 20 separate caudal vertebrae; gastralia; three clawed digits on wings; thecodont method of tooth attachment; poorly developed sternum. EXAMPLE: *Archaeopteryx.*

Subclass II. Neornithes
Thirteen or fewer compressed caudal vertebrae; metacarpal and distal carpal bones fused; no free, clawed digits on wings (except young hoatzin); well-developed sternum; uncinate processes on ribs.

Superorder I. Odontognathae. Fossil forms only; teeth present in jaws and set in grooves or shallow sockets. EXAMPLES: *Hesperornis,* a flightless, swimming bird; *Ichthyornis,* a flying, toothed form.

Superorder II. Paleognathae (Ratitae). Flightless; wings poorly developed or rudimentary; no keel on sternum except in the tinamous; distal caudal vertebrae free; toothless; horny beaks, or bills. EXAMPLES: Walking or running birds; ostrich, rhea, kiwi, cassowary.

Superorder III. Neognathae. Keeled (carinate) sternum; wings well developed; five or six free caudal vertebrae terminating in a pygostyle; no teeth; horny beaks, or bills. EXAMPLES: Most modern birds; penguins use wings for swimming, not for flight.

Class IV. Mammalia
Hair present but sometimes scanty; homoiothermous; mammary glands; sweat glands and oil glands in skin with few exceptions; pentadactyl limbs often modified; claws, nails, or hoofs at ends of digits; movable lips and tongues in most forms; highly developed cerebral cortex; four-chambered heart; palate partially separates nasal and oral cavities; four pairs of pharyngeal pouches in embryo; metanephros; urinary bladder; single penis for copulation; seven cervical vertebrae (four exceptions); two occipital condyles; increased cranial capacity of skull; reduction in number of skull bones; long movable tail usually present; 12 pairs of cranial nerves in addition to terminal nerve (0); pinna usually present; lungs for respiration; diaphragm; nonnucleated red corpuscles except in embryo; left aortic arch 4 persisting, right one disappearing; blood platelets instead of spindle cells; testes located in scrotum in most cases; three auditory ossicles; dentition thecodont and usually heterodont.

Subclass I. Prototheria

Order I. Monotremata. Oviparous; eggs incubated outside body; cloaca present; no nipples or teats; oviducts distinct, undifferentiated, and opening separately into cloaca; temperature-regulating mechanism not highly integrated; rather poorly developed brain; separate coracoid and precoracoid bones; epipubic bones; no pinna; males with intra-abdominal testes; penis used only for sperm transport; monophyodont. EXAMPLES: Egg-laying mammals; *Ornithorhynchus anatinus, Tachyglossus.*

Subclass II. Theria
Mammary glands with nipples or teats; no cloaca (except pika); viviparous; eggs practically microscopic in size; testes usually in scrotum; penis for passage of both urine and spermatozoa; oviducts differentiated into Fallopian tubes, uterus, and vagina; pinna usually present; coracoid represented by a process on scapula. EXAMPLES: Marsupials and placental mammals.

Infraclass I. Metatheria

Order I. Marsupialia. Young born in immature condition, further development usually in marsupium; usually no placental attachment; epipubic bones usually present; paired uteri and vaginae; a third vaginal canal often present; mammary glands provided with nipples; no cloaca; testes in scrotum; brain having corpus callosum for first time. Most species confined to Australia and surrounding islands. EXAMPLES: Marsupials; opossum, *Didelphis virginiana;* kangaroo, *Macropus.*

Infraclass II. Eutheria (*Placentalia*)
Developing young nourished by means of an allantoic placenta; no marsupial pouch or epipubic bones; single vagina. EXAMPLES: The placental mammals.

Living eutherian, or placental, mammals are grouped in the following orders:

Order I. Insectivora. Small size; elongated snouts; prism-shaped molar teeth; pentadactyl; insectivorous diet usually. EXAMPLES: Moles, shrews, hedgehogs.

Order II. Dermoptera. Patagium; webbed feet; nocturnal; arboreal; herbivorous. EXAMPLE: *Galeopithecus,* the flying lemur.

Order III. Chiroptera. Pentadactyl forelimbs modified to form wings used in flight; keeled sternum; nocturnal; diet primarily insectivorous and herbivorous. EXAMPLE: Bats.

Order IV. Primates. Well-developed nervous system, particularly the cerebral portion; long pentadactyl limbs; eyes directed forward and completely encircled by bony orbits; thumb and big toe usually opposable; plantigrade foot posture; usually only one young at a birth. EXAMPLES: Lemurs, tarsiers, monkeys, apes, and man.

Suborder I. Lemuroidea. Long tails not prehensile; second digit of hind foot bearing claw, other digits bearing nails; thumb and hallux well developed; arboreal; crepuscular or nocturnal. EXAMPLE: Lemurs.

Suborder II. Tarsioidea. Elongated heel bone; small size; large protruding eyes and ears; second and third digits of hind foot bearing claws, nails on others; tails not prehensile; arboreal, nocturnal. EXAMPLE: Tarsiers.

Suborder III. Anthropoidea. Flattened or slightly rounded nails usually on all digits; arboreal or terrestrial; diurnal; tendency to upright posture; specialization of nervous system with highly developed, convoluted cerebral hemispheres. EXAMPLES: Monkeys, apes, and man.
SUPERFAMILY I. CEBOIDEA (PLATYRRHINII). Nostrils far apart and outwardly directed; tail may be prehensile; cheek pouches lacking. EXAMPLE: New World monkeys.

FAMILY I. HAPALIDAE (CALLITHRICIDAE). Tail not prehensile; claws on all digits except large toe, which bears a nail; small size. EXAMPLE: Marmosets.
FAMILY II. CEBIDAE. Long prehensile tails in some. EXAMPLES: Spider monkeys, howling monkeys, capuchins.

SUPERFAMILY II. CERCOPITHECOIDEA

FAMILY I. CERCOPITHECIDAE. Narrow septum between nostrils; nostrils directed downward; tails never prehensile; most have cheek pouches; ischial callosites often highly colored. EXAMPLES: Old World monkeys, macaques, baboons, mandrills.

SUPERFAMILY III. HOMINOIDEA. Lack tail and cheek pouches; narrow septum between nostrils; nostrils directed downward. EXAMPLES: Apes and man.

FAMILY I. PONGIDAE. Manlike, anthropoid apes; tendency to walk upright. EXAMPLES: Gorilla, orangutan, gibbon, and chimpanzee.

FAMILY II. HOMINIDAE. Great development of cerebral portion of brain; upright posture; generalized condition of skeleton. EXAMPLE: Man, *Homo sapiens.*

Order V. Edentata. Adults lacking teeth or having only poorly developed molar teeth. EXAMPLES: South American anteater, sloths, armadillos.

Order VI. Pholidota. Body covered with large, overlapping, horny scales with a few hairs interspersed; teeth lacking. EXAMPLE: Scaly anteater, *Manis,* the pangolin.

Order VII. Rodentia. Two incisor teeth on both upper and lower jaws, used for gnawing, grow throughout life; canine teeth lacking; diastema between incisors and premolars; foot posture plantigrade or approximately so; claws. EXAMPLES: Rats, mice, squirrels, beavers, gophers; capybara.

Order VIII. Lagomorpha. Four incisor teeth in upper jaw, one pair behind the other, grow continuously; canine teeth lacking; tails short and stubby. Pikas are the only members of the subclass Theria to have a cloaca in the adult. EXAMPLES: Rabbits, hares, pikas.

Order IX. Carnivora. Three pairs of small incisor teeth in both jaws; canine teeth well developed; flesh-eating mammals; poorly developed clavicles, sometimes absent; mobility of limbs; bipartite uterus; zonary type of placenta.

Suborder I. Fissipedia. Toes separated. EXAMPLES: Cats, dogs, weasels, martens.

Suborder II. Pinnipedia. Aquatic carnivores; toes webbed; appendages in form of flippers; males larger than females; short tail. EXAMPLES: Seals, walrus, sea lions.

Order X. Cetacea. Pectoral appendages webbed to form flippers; no claws; pelvic appendages lacking; tail (flukes) flattened dorsoventrally and notched; males with intra-abdominal testes; nasal openings on top of head; homodont dentition in toothed forms. EXAMPLES: Whales, dolphins, porpoises.

Order XI. Tubulidentata. Thick-set body; large, pointed ears; long snout; thick skin; scanty hair covering; few permanent teeth; those present lacking enamel; no incisors or canines; very long claws; plantigrade foot posture. EXAMPLE: Aardvark.

Order XII. Proboscidea. Great size; massive legs; nose and upper lip form proboscis, or trunk, with nostrils at tip; upper incisor teeth are tusks; thick skin with scant hairy coat; males with intra-abdominal testes. EXAMPLE: Elephants.

Order XIII. Hyracoidea. Somewhat resemble guinea pigs in size and shape; more closely related to ungulates; small ears; short tail; persistent growth of upper incisor teeth; four toes on forefeet; three toes on hind feet; second toe on hind feet bearing claw, the rest being hooflike nails; males with intra-abdominal testes; canine teeth lacking. EXAMPLE: Conies.

Order XIV. Sirenia. Herbivorous, aquatic forms; heavy dense bones; forelimbs in the form of flippers; hind limbs lacking; tail bearing horizontally arranged flukes which are not notched; no external ears; skin sparsely haired; males with intra-abdominal testes. EXAMPLES: Manatees and dugongs.

Order XV. Perissodactyla. Odd-toed, hoofed forms; unguligrade foot posture; axis of leg passes through middle toe; herbivorous; gallbladder lacking. EXAMPLES: Horses, donkeys, zebras, tapirs.

Order XVI. Artiodactyla. Even-toed, hoofed forms; unguligrade foot posture; axis of leg passing between third and fourth toes; herbivorous; true horns or antlers only in this group; incisors and canines of upper jaw usually absent (except in pigs). EXAMPLES: Cattle, pigs, hippopotami, camels, deer.

| INDEX |

Numbers in **boldface** indicate pages on which a term appears in an illustration. Words in *italic* refer to scientific names of animals. Structures coming under the following categories are not listed separately but are to be found under the headings Arteries, Bones, Cartilages, Muscles, Nerves, and Veins.

Amphibians:
chromatophores, 80, 106, 110
classification, 29, 734
"claws," 147
cloaca, 157, 188, 261
copulatory organs, 326
coronary circulation, 552
corpora lutea, 282, 372
cranial nerves, 617
dermal musculature, 527
dorsal mesentery, 155
ear, 699
early, 29
ecdysis, 341
eggs, 57
esophagus, 178
evolutionary origin, 26, 29, 30
eyes, 674, 688
fat bodies, 309
fertilization, 325
fins, 476
gall bladder, 194
gastralia, 470
gastrula, 61
gills, 219
girdles and limbs, 456
glands: adrenal, 351
   Harderian, 684
   integumentary, 113, 114, 116
   intermaxillary, 160, 161
   internasal, 160
   lacrimal, 683
   lingual, 161
   mouth, 160
   mucous, 113
   oral, 160
   parathyroid, 246, 346
   pituitary, 395
   poison, 113
   thymus, 246, 346
   thyroid, 339
   unicellular, 112, 116
gustatoreceptors, 69, 667, 714
Haversian systems, 89
heart, 544, 550, 552
hermaphroditism, 329
hypophysectomy in, 345, 363, 385, 387
hypothalamus, 625

Amphibians:
integument, 109, **110**, 113, 116, 527, 719
   glands of, 111
intestine, 188
islets of Langerhans, 361
Jacobson's organ, 208, 642, 711
kidney tubule, **256**
labyrinthodont, 33, 413, 470, 684, 705
larval, 29, 70, 111, 476
larynx, 229
lateral line organs, 641, 647, 723
lens formation, 671
limbs, 614
liver, 194
lower jaw, 440
lungs, 27, 221, 223, 235, 239
lymphatic system, 85, 590
mast cells, 80
meninges, 630
metamorphosis, 29, 648
modern, 449, 450, 470, 486
Müllerian duct, 288
nares, 204, 205, 710
nasal passages, 204
nephrostomes, 288
neural crests, 65, 435
neuromasts, 723
nictitating membrane, 683, 688
notochord, 64
occipital condyles, 463
olfactory nerve, 642
olfactory organs, 627, 710
opisthonephros, 255–257, **256**, 259
oral cavity, 159
ovaries, 278, **282**, 372
oviducts, 289, 291, 372
palate, 439
pancreas, 197
peritoneal funnels, 260
pharyngeal pouches, 70, 209, 219, 246, 346
primitive, 26
primordial germ cells, 276
pronephros, 255, 260, 288
reproductive ducts, 317
respiration, 203, 219, 239, 244, 551, 580

Amphibians:
ribs, 425, 470, 473
salientian, 31
scales, 125, 129, 403
secondary sex characteristics, 365
sex reversal, 329
skeleton, 29, 34
skull, 413, 450–454
spinal column, 423
spinal cord, 614
spinal nerves, 634, 638
spleen, 595
stegocephalian, 30
sternum, 473
stomach, 473
tadpoles (*see* Tadpoles)
tailed (*see* Urodeles)
tailless (*see* Anurans)
taste buds, 717
teeth, 165, 170
terminal nerve, 642
testes, 309
tongue, 162, 163, 512
toothless, 170
trachea, 233
trunk musculature, 510
urodele, 31
urogenital organs, 259–261
uterus, 291
venous system, 579
vertebrae, 419, **420**, 423
visceral arches, 449–454
Amphicoelous vertebrae, 421–423, **422**, **423**
Amphicraniostylic method of jaw attachment, 441, 464
Amphignathodon, 170
Amphigony, 275
Amphioxus:
anus, **15**
aortic arches, 559, **560**
atriopore, 14, **15**, 478
atrium, 15, **16**, 211, 251
autonomic nervous system, 656
blastula, 60
blood, 596
brain, 615
branchial bars, **16**
buccal cavity, 394
buccal cirri, 14, 15, 709, 716
characteristics, 731

Amphioxus:
circulatory system, 575
classification, 14, 731
cleavage stages, 58
coelom, 64
digestive tract, **15**
eggs, 57
endostyle, 15, **16**, 338
epipharyngeal groove, 16
excretory organs, 251, **252**,
560
fins, 14, 478
flagellated pit, 708
gastrula, 61, **62**
gill slits, 15, **16**
gills, 211
gonads, 275
gustatoreceptors, 716
head cavity, 394
heart, 546, 559
hepatic caecum, **15**, **16**, 194
infundibular organ, 677
integument, 108
integumentary glands, 111,
114
intestine, **15**
mesoderm formation, 63
metapleural folds, 14, **16**,
478
mouth, 158
nervous system, **15**, 65, 95
neurons, 95
notochord, **15**, **16**, 64, 414,
478
olfactoreceptors, 708
oral hood, 14
organ of Müller, 394
ovaries, 279, 371
pancreas, 197, 360
pharyngeal region, **15**, **560**
pigment spot, 668, 708
preoral pit, 394
respiration, 211
sperm, 314
spinal cord, 15, 16, 613, 615
spinal nerves, 633, 634, 636,
637, 639, 656
testes, **16**, 305, 364, 365
trunk musculature, 507
unicellular glands, 111
velar tentacles, **15**
velum, 15, 709, 716
venous system, 575

Amphioxus:
vestibule, 14
wheel organ, **15**, 394
Amphiplatyan vertebrae, 425,
429
Amphisbaenidae, 457
Amphistylic method of jaw
attachment, 438, **439**
Amphiuma, 220, 233, 473,
596, 705
Amplexus, 116, 292, 325, 326
Ampulla:
of ductus deferens, 322
of Lorenzini, **725**
of semicircular ducts, **696**
of testes, 365
of Vater, 192
Ampullary gland, 322
Amyda, 691
Amylase, 197
Anableps, 688
Anabolism, 250, 534
Anaerobic respiration, 202
Anal canal, 68
Anal fins, 291, 326, 476
Anal glands, 68, 122, 192
Anal membrane, 68
Anal plate, 68
Anamniota, 32, 253, 316, 616,
638, 640
Anapsid condition, 456
Anatomy, 2, 4
Ancient fishes, 733
Androgenic hormones, 82, 118,
312, 321, 324, 354, 357,
364–366, 370, 371, 625
Androgenic zone, adrenal
gland, 354
Androstenodione, 369
Androsterone, 369
Anemia, 185, 195, 355
Anger, 653, 658
Angiotensin, 268, 335
Angiotonin, 268
Angle, rib, 472, **473**
Angle gland, 161
Angler, 170, 361, 395
Angora hair, 142
Anguilla, 579
Angular process, 467
Animal pole, 57, 58, 60
Ankle, 484, 492, 496
Annular groove, **136**, **137**

Annular pad, 691, 692
Ant bear, 353, 428
Anteaters, 41, **42**, 48, **50**, 164,
172, 738
(*See also under* specific
headings)
Antebrachium, 484
Antelope, 54, 150, 328
blue, 286, 399
pronghorn, 149, **150**
South African, 121
Anterior chamber:
of eye, **669**, 670, 686
of swim bladder, 225
Anterior choroid plexus, 622,
632
Anterior lobe of pituitary (*see*
Pituitary gland)
Anterior palatine foramina,
465
Anterior palatine vacuities, 459
Anthropoidea, 47, 738
Antibodies, 85, 192, 536, 589,
591, 592, 594
Antidiuretic hormone (ADH),
267, 383
Antigenic substances, 591
Antihemophilic globulin, 598
Antilocapra, 150
Antithrombin, 597
Antithyroid drugs, 337
Antlers, 54, 149–**151**, 364, 368,
413, 466
Antrum:
of Highmore, 466, 713
ovarian follicle, **285**
Anurans:
adrenal glands, 261, 351
amplexus, 325
aortic arches, **563**–565
autonomic nervous system,
657
Bidder's canal, 318
bone marrow, 85, 410, 595
calcareous bodies, 638
characteristics, 30, 31, 734
classification, 30, 31, 734
clavicle, 486
cloaca, 188, 292
columella, 704
ear, 247, 703, 705
efferent ductules, 318
epaxial muscles, 511

Artiodactyla, 54, 149, 162, 268, 466, 739
Ascaphus, 580
Aschheim-Zondek test, 400
Ascidians, 731
  larval stages, 16, 668
Asexual reproduction, 274
Association neurons, 97, 611, 629, **657,** 658
Asteriscus, 699
Asthma, 243, 248, 359
Astigmatism, 669
Astrocytes, 95, 98
Astronethes, 115
Astroscopus, 204
Astylosternus, 244
Ateliotic dwarf, 388
Atlantal foramen, 428
Atlas vertebra (*see* Bones)
Atretic follicles, 278, 285, 373
Atrial appendages, 555
Atrichosis, 140
Atriopore, 14, **15,** 211, 279, 478
Atrioventricular bundle, 558
Atrioventricular node, 558
Atrioventricular valve, 547, **550,** 553, 554
Atrium:
  amphioxus, 15, 16, 211, 251
  heart, **543–545,** 547–555, **548, 549, 550, 554, 555**
Auditory canal, 86, 707
Auditory capsule (*see* Otic capsule)
Auditory function, 618
  mechanism of, 707
Auditory meatus:
  external, 455, 702
  internal, 455
Auditory organs, 700–708
Auditory ossicles (*see* Bones)
Auditory perception, 225
Auditory pit, 69, **695**
Auditory placode, 69, 646, **695**
Auditory vesicle, 69, **695**
Auricle:
  cerebellum, 617, 641
  ear, 707
  heart, **543,** 555, **569**
Auricular lobes, 618
Auricular recess, 618
Auricular surface, 431

Autochthonus layer, 687
Autonomic ganglia, 633–635, 645, 648
Autonomic nerve fibers, 540, 636, 644, 647–649
Autonomic nervous system (*see* Nervous system)
Autonomic neurons, 634, 649
Autosomes, **277**
Autostylic method of jaw attachment, 438, **439,** 704
Aves (*see* Birds)
Axial filament, 314
Axial musculature, 517
  (*See also* Muscles, axial)
Axial skeleton, 413–477
Axial thread, 314
Axilla, 108, 112, 120, 121
  hair, 142, 145
Axis cylinder (*see* Axon)
Axis vertebra (*see* Bones)
Axolotl, 220, 340
Axon, 93–96, **94,** 605, 606, 616, 633, 649, 657
  peripheral sensory, 97

B (beta) cells, 362, 389
Baboon, 47, 738
Baby teeth, 172
Backbone (*see* Vertebral column)
Baculum, 404
Badger, 307
Baldness, 144
Baleen, 51, 160, 706
Ball-and-socket joint, 411
Bandicoot, 43
Barbels, 244, 716
Barbicel, 135
Barbs, 133, **135–137**
Barbules, 134, **135**
Barracuda, **10,** 170
Barrow, 367
Basal bodies, 708
Basal cells, 708
Basal nuclei, 626, 628
Basalia (*see* Bones; Cartilages)
Basapophyses, 416, 422, 423, 470
Basement membrane, 73, 104, 110
  of capillary, 541

Basic lymphocytic cells, 592
Basicranial fenestra, 432, **433**
Basilar membrane, 701
Basilar papilla, **700**
Basilar plate, 432, **433,** 422
Basilar sulcus, 619
Basilingual plate, 454
Basipterygia, 476
Basitemporal plate, 462
Basophilic bodies, 93
Basophils, **84,** 343, 388, 410
Bat rays, 271
Batoidea, 22, 351, 733
Batrachoseps, 170
Bats, 41
  blood-sucking, 183
  cervical enlargement, 614
  characteristics, 46, 738
  circumvallate papillae, 717
  classification, 46, 738
  colic caecum, 189
  copulation, 315
  eye, 694
  forelimb, 6, **495**
  mandible, 467
  penis bone, 328
  reproduction, 315
  scent glands, 122
  spermatozoa, 315
  sternum, 46, 475
  teeth, 172
  uterus, 301
  wing, 5, 46, **495**
Bdellostoma, 20, 209, 213, 560, 732
Beaker cells, 111, 114
Beaks, 39, **41,** 111, 146, 158, 170, 462
Beards, 146
Bears, 50, 121, 138, 306, 307, 496
Beaver, 49, 138, 145, 173, 739
Belly of muscle, 504, **505**
Benzestrol, 370
Beta (B) cells, 362, 389
Beta-hydroxybutyric acid, 361
Bicarbonate, 190, 271
Bicipital ribs, 469, 472
Bicipital tubercle, **493**
Bicornuate uterus, **301, 302**
Bicuspid teeth, 173
Bicuspid valve, 554, 556, **585**

Bidder's canal, **260**, 318, 319
Bidder's organ, **261**, 283, 329, 372
Bilateral symmetry, 11
Bile, 195, 594
Bile capillaries, 193, 194
Bile duct, **177**, 192
Bile pigment, 196
Bile salts, 199
Bilirubin, 594
Bill, 111, 146, 462
   platypus, 41, 146, 159, 713
   (*See also* Beaks)
Binocular vision, 643, 693
Binomial nomenclature, 13
Biochemical mechanisms, 2
Biogenetic law, 5
Biology, 1, 2, 4
Bipartite uterus, 301, **302**
Bipedal locomotion, 37, 460, 491
Bipolar layer of retina, **675**
Bipolar neurons, **97**, **675**, 708
Birds:
   accommodation in, 678, 692
   adaptations, 39
   air sacs, 37, 236, 237, 241, 313
   allantoic circulation, 244
   allantois, 71, 244
   ancestry, 460
   androgen in female, 374
   antibody production, 593
   aortic arches, 6, **567**
   aquatic, 118, 132
   arteriovenous anastomoses, 542
   beak, 39, **41**, 111, 146, 158, 461, 462, 720
   bill, 111, 146, 461, 462
   bisexual potentialities, 329, 373
   bladder, urinary, 265
   blastula, 60
   blood pressure, 384
   bone marrow, 410
   bones, 37
   brain, 616
   bronchi, 236
   bursa Fabricii, 590, 593
   carotid body, 248
   cerebellum, 619
   cerebral cortex, 628

Birds:
   cerebral hemispheres, 616, 628
   cervicobrachial plexus, 639
   characteristics, 37, 736
   classification, 29, 37, 736
   claws, 147
   cleavage, 58
   cloaca, 157, 265
   colic caeca, 188
   columella, 704
   conchae, 711
   copulatory organs, 320
   corpus striatum, 628
   crop, 178
   dermal musculature, 527
   digits, 492
   dorsal mesentery, 155
   ductus deferens, 313, 320
   ear, 700, 706
   egg tooth, 172
   eggs, 57, 60, 295, 346
   elephant, 39
   esophagus, 178
   Eustachian tube, 706
   extinct, 38, 39
   eye, 37, 461, 674, 691, **692**
   feathers, 37, 110, 111, 132–138, 460, 527
   feet, 37–**40**, 110, 491
   fertilization, 39, 298
   flightless, 39
   fossil, 35, 36
   fovea centralis, 676
   gall bladder, 194
   gallinaceous, 118, 132, 147
   girdles and limbs, 490
   glands: adrenal, 352
     angle, 161
     Harderian, 684
     integumentary, 118
     mouth, 161
     nasal, 206
     parathyroid, 246, 346
     pituitary, 395
     sublingual, 161
     thymus, 246
     thyroid, 341
     uropygial, **118**
   gonads, 275
   gustatoreceptors, 717
   heart, 543–545, 554
   heat radiation, 527

Birds:
   hind limbs, 37, 491, **492**
   hyoid apparatus, 462
   hyperstriatum, 628
   integument, 110–111
   interclavicle, 486
   intermedin, 387
   intestine, 188
   islets of Langerhans, 361
   Jacobson's organ, 712
   kidneys, 261
   larynx, 231
   learning ability, 628
   legs, 37, 110, 539
   lips, 158
   liver, 194
   lower jaw, 462
   lumbosacral plexus, 639
   lungs, 37, 223
   lymphatic system, 536, 590
   mesonephros, 255, 261, 263
   metanephros, 264, 265
   modern, 737
   nasal passages, 206, 711
   neural crests, 65, 435
   nictitating membrane, 692
   occipital condyle, 426
   olfactory organs, 711
   optic lobes, 621
   orbit, 461
   origin of, **30**, **33**
   ovarian follicles, 283
   ovary, 37, 283, 295, 373
   oviducts, 37, 289, 295, 373
   palate, 159, 206, 439, 462
   pancreas, 197
   of paradise, 132
   parapineal body, 622
   passerine, 313, 320, 692
   penis, 327
   pharyngeal pouches, 221
   pharyngeal tonsils, 247
   pigment, 111
   pineal body, 622
   plumage, 374
   of prey, 146, 147, 163, 178, 283, 373, **692**
   primitive streak, **64**, 65
   primordial germ cells, 276
   progesterone in, 374
   ptyalin, 161
   raptorial, 295
   relaxin in, 374

Duct:
of Santorini, 196
of Wirsung, 192, 196
Ductless glands (*see* Endocrine glands)
Ductus aberrans, 263
Ductus arteriosus, 563, 566, 568, **585–587**
Ductus Botalli, 563
Ductus choledochus, 192
Ductus deferens, 76, 253, 259. 263, 276, **311**, 316
Ductus epididymidis, 253, 259, 316
Ductus venosus (*see* Veins)
Dugong, 53, 124, 175, 739
Duodenal mucosa, 190
Duodenal papilla, 192
Duodenohepatic omentum, 186
Duodenum, 156, **177**, 186, 188
Duplex uterus, **301**
Dura mater, 630
Dysmenorrhea, 380

Eagle, 693
Ear:
external, 36, 86, 112, 121, 695, **702**, 706, 707
inner, 69, 74, 455, 464, 617, 618, 646, 695, **702**
comparative anatomy of, 697–700
formation, 695
origin, 641
middle, 71, 76, 157, 203, 219, 247, 454, 695, **702**, 703
comparative anatomy of, 705
formation, 703
structure, 704
Ear pits, 35, 36, 706
Ear wax, 121
Eardrum (*see* Tympanic membrane)
Eccrine sweat glands, 113, 119
Ecdysis, 102, 110, 129, 131, 138, 341
Echidna(s), 41, 42, 172, 263, 299, 327, **328**, 583, 693, 713

Ecology, definition of, 2
Ectoderm, 61, **64**
proctodaeum, 68
stomodaeum, 68, 69
superficial, 68, 69, 670
Ectodermal derivatives, 68–70
muscles, 89
Edentata, 48, 138, 428, 738
Edentates, 119, 189, 413, 431
Eel, 19, 125, 218, **225**, 579, 590
Effectors, 96, 649, **657**
Efferent ductules, 259, **261**, 316
Efferent nerve fibers, 558, 611, 649, 655
visceral, 633
Egg cases, **290**
Egg cells, 275, 278
Egg tooth, 172, 175
Eggs, 41, 282, 346
abnormal, 297
of hen, **297**
size of, 57, 279
transport of, 304
Eggshell, 34, 294, 346
Ejaculatory duct, 322
Elasmobranch segments, 271
Elasmobranchii, 21, 618, 732
Elasmobranchs:
abdominal pores, 289
accommodation, 686
ampullae of Lorenzini, 725
aortic arches, 562
autonomic nervous system, 656
ceratohyal cartilage, 446
cerebellum, 618
cerebral hemispheres, 627
characteristics, 21–23, 732
chondrocranium, 433, **435**, 443
chromaffin bodies, 351
chromatophores, 109, 386
clasper, **325**, 483, 518
classification, 21, 732
coelom, 544
coloration, 109
conus arteriosus, 548
copulation, 324
coronary circulation, 549
corpus luteum, 280, 372
cranium, 436
dermal denticles, 412

Elasmobranchs:
ear, 698, **699**
egg cases, **290**
eggs, 280
esophagus, **177**, 178
eyes, 685, **686**
fertilization, 290
fin folds, 478
fin musculature, 517
gill slits, 21, **435**
gills, 214
glands: adrenal, 109
pituitary, 109
poison, 114
rectal, 270
thyroid, 339
head somites, 515
heart, **549**, 656
hyoid arch, 446
hyomandibular cartilage, 435
hypothalamus, 395
interrenal bodies, 350
intestine, 187
islets of Langerhans, 360
jaw attachment, **439**
kidney tubules, 271
male, **482**, 518
Müllerian ducts, 288, 289
muscles: appendicular, 517
branchial, 523
hypobranchial, 514
trunk, 517
myotomes, 517, 518
neurosecretory cells, 397
olfactory organs, 627, 710
opisthonephros, 258
optic pedicel, 683, 685
oronasal groove, 710
ostium tubae, **289**
ovaries, 280
oviducts, 288, 289
oviparous, 280, 290, 324
ovoviviparous, 280, 290, 324
pancreas, 197, 360
pectoral fins, 480, 517
pectoral girdle, 480
pelvic fins, 324, 482, 517, 518
pelvic girdle, 482
peripheral autonomic nervous system, 656
peritoneal funnels, 259

Ependymal cells, 35, 605, 610, 624, 668, 677, 685
Ependymal layer, 605, **606**
Ephedrine, 359
Epiblast, **61–64**
Epibranchial ganglia, 647
Epicardium, 544, 545
Epiceratodus:
  characteristics, **28**, 734
  classification, 27, 734
  gill clefts, 218
  heart, 548, 549
  liver, **580**
  opisthonephros, **580**
  pectoral fin, 481, **482**
  swim bladder, 222–**224**
  venous system, 579, **580**
Epidermal cells, **75**
Epidermal derivatives, 68, 413
Epidermal scales, 37, 68, 125, 129–139, 413, 470
Epidermal teeth, 125, 158, 162, **165**, 170, 172, **176**
Epidermis, 30, 35, 77, **102–105**, 109–111, 534
  regeneration, 145
Epididymis, 76, 259, 263, **311**, 316, 319, 320, 366
Epidural space, **631**
Epiglottis, 77, 86, 231, 232, **239**, 717
Epimere, 66, 68, 71, 502, **504**, 570
Epimyocardium, **545**
Epimysium, 91
Epinephrine (adrenalin), 109, 243, 350, 358, 359, 362, 652, 654
Epipharyngeal groove, **16**
Epiphysial apparatus, 622
Epiphysial foramen, 444
Epiphysial plate, 363, 389, **408**, 418
Epiphysis:
  bony, 368, 407, **408**, 418, 428
  brain, 622, 623, 668
Epiploic foramen, 186
Epipubis, 482, 489
Epispadias, 328
Episternum, 475, 492
Epistropheus (*see* Axon; Vertebrae)

Epithalamus, 622, 624
Epithelial bodies, 245
Epithelial cells, 74
Epithelial tissue, 73, 111
Epitheliomuscular cells, 89
Epithelium, 73
  functional classification, 78
Epitrichium, 139
Eponychium, 148
Epoöphoron, 263, 303
Equilibration, 226, 618, 646
Equilibratory organs, 222
Equilibrium, sense of, 618, 667, 694, 696–700, 722
Equine gonadotrophin, 399
Erectile tissue, 146, 207, 302, 304, 325, 327, 543, 690
Erection, 302, 327, 543, 655
Erepsin, 191, 198
Erinaceus, 45
Erythrinus, **224**
Erythroblasts, 596
Erythrocytes, 84, 202, **596**
  destruction of, 593
  formation, 85, 410, 593
  storage of, 594
  (*See also* Red corpuscles)
Erythrophores, 109
Erythropoiesis, 595
Erythropoietin, 268, 335, 410, 595
Esophageo-cutaneous duct, **213**
Esophagus, 70, 77, **157**, **177**, 179
  amphioxus, 15
  lamprey, **176**
  Molgula, 16, **17**
  muscle fibers in, 176
  nerve supply, 653
  papillae in, **178**
  Siren, **171**
Estivation, 28, 114
Estradiol, 284, 370, 371
Estriol, 381
Estrogens, 304, 307, 357, 370, 374, 375, 378, 379, 398, 493, 595
  testicular, 370
  tests for, 381
Estrone, 370, 381
Estrous cycle, 385
Estrus, 287

Ethmoid plate, 432, **433**
Ethmoid sinus, 466, 712
Ethynylestradiol 3-methyl ether, 379
Eunuch, 367
Eurycea bislineata:
  breeding season, 292, **319**, 365
  cloaca, **157**, 192, 294, **309**
  digestive system, **157**
  ductus deferens, 309, 319
  hyoid apparatus, 452
  mental gland, **309**
  oviduct, **292**
  sperm survival, 315
  spermatheca, **294**
  spermatophore, **294**, 315
  testes, **309**
  thyroid follicles, 339
  tongue, **163**, 452
  urogenital organs of male, **192**, **309**
  visceral arches, 452
Eustachian tube, 86, 219, 702, 703, 705, 706
  lining, 76, 704
  origin, 71, 157, 203, 247, 703
Eustachian valve, 556
Eusthenopteron, 26, 482, 483, **485**, 486, 705, 733
Euthacanthus macnicoli, **479**
Eutheria, 44, 123, 270, 328, 693, 737
Eutherians, 113, 301
Evocators, 67
Evolution:
  of amphibians, **30**
  of chordates, 12–55
  of heart, 557
  of man, 46, 47
  of marsupials, 44
  study of, 2
  of tetrapods, **30**
Ewe, 288
Excretion, 81, 259
Excretory ducts, 81
Excretory organs, 251, 502
Excretory system, 250–273
Excretory tubule, 251, **252**
Exocrine glands, 111, 196, 333, 335, 713
Exophthalmic goiter, 344

Heart:
chick, **65, 546**
comparative anatomy,
546–557
evolution of, 557
fetal, **585**
formation, **545, 546**
four-chambered, 543, 546,
553–**555**, 557
lining of, 73
location, 11
man, **547**
Molgula, **17**
muscle, 91
nerve supply, 648, 649, 653,
655, 667
origin, 72, 545
portal, 575
ruminants, 404
skeleton of, 557
three-chambered, 544, 546,
549, 552, 553
two-chambered, 543, 544,
546, 549, 557
Heat:
animal, 287, 534
production of, 250, 342,
361, 501, 534
radiation, 527, 542
receptors for, 720
sense of, 667, 719
Hedgehog, 45, 138, 528, 737
Hedonic glands, 117, 295, 711
Heel, 47, 495
Helicotrema, 702
Hellbender, 219
Heloderma, **36**, 161, 736
Hemal arch, 416, 418, 420,
422, 423, 468–470
Hemal nodes, 593, 594
Hemal processes, 422
Hemal spine, 325, 422, **423**,
477
Hemibranch, 211, 215, 647
hyoid, 442, 446
true, 215, 217
Hemichordata, 13, 730
Hemicyclaspis, **19**
Hemipenes, 320, **326**, 366
Hemogloblin, 84, 202, 594
in muscles, 202
Hemolymphatic node, 593
Hemolymphatic organs, 594

Hemophilia, 598
Hemopoietic organs, 594, 595
Hemopoietic tissues, 85, 410,
588, 595
Hen:
egg, **297**
reproductive organs, 295,
296
(*See also* Chicken)
Henle's layer, **141**
Hensen's node, 64
Heparin, 80, 599
Hepatic caecum, **15, 16**
Hepatic diverticulum, 192
Hepatic duct, 192–**193**
Hepatic portal circulation, 574
Hepatic portal system, 536,
574
(*See also* Veins, hepatic
portal)
Heptanchus, **217**, 218, 246,
560
Herbivores, 53, 189, 194
Herbst corpuscles, 720, **721**
Heredity, 4
Hermaphroditism, 277, 280,
308, 328–330
Hernia, 311, **312**
Heron, **40**
Herring, 329, 643, 733
Herring gull, 206
Hesperornis, 39, 172, 462, 736
Heterocercal tail, 21, 24, 25,
420, **477**
Heterocoelous vertebrae, **427**
Heterodont dentition, 168,
170–172, 464
Heterodontus, 168, 216, **290**
Heterophils, 410
Heterosexual twins, 330
Heterotopic skeletal elements,
404, 461
Hexanchus, 218, 561
Hexestrol, 370
HGF (hyperglycemic-
glycogenolytic factor),
362
Hibernating gland, 82
Hibernation, 82
Hilum (hilus) kidney, 265,
266
lung, 242

Hilum (hilus) kidney:
lymph node, 589
ovary, 287
Hind feet, 36, 38
toes on, 48, 52
Hind limbs, 431, **484**, 487, 488,
491
moles, 45
Necturus, **32**
nerves of, 610
veins of, 539
Hindbrain (rhombencepha-
lon), 69, 606
Hinge joint, 411, 615
Hinged teeth, 169
Hippocampal lobe, 628, 629
Hippopotamus, 54, 121, 183,
232, 739
Histamine, 80
Histidine, 268
Histiocytes, 80, 85
(*See also* Macrophages)
Histogenesis, 56–100
Histology, definition, 2, 72
Hoatzin, **147**, 491, 736
Hogs, 404
(*See also* Pig)
Hollow horn, **149**
Holoblastic cleavage, 58, 59
Holobranch, 211, 215
Holocephali, 215, 324, 420,
447, 733
Holocephalians:
clasping organs, 325
copulation, 325
eggs, 291
eyes, 685
jaw suspension, **439**
lateral-line organs, 723
skull, 447
Holocrine glands, 113
Holonephros, 252
(*See also* Archinephros)
Holosteans, 25, 26
Holostei, 25, 223, 723
Homeobaric respiration, 241
Homeostasis, 656
Hominidae, 48, 738
Hominoidea, 738
Homo sapiens, 47, 738
(*See also* Man)
Homocercal tail, 25, 420, **477**

Man:

  glands: parotid, 162
    pituitary, 574
    prostate, **322**
    salivary, 162
    suprarenal (*see* Adrenal
      gland)
    sweat, 119, **120**
    thymus, 593
    thyroid, 245, 336, 341,
      342, 347
  glans clitoridis, 302, **303**
  glans penis, **328**
  hair, 142, 143, 145
  heart formation, **547**
  hermaphroditism, 330
  ileum, **190**
  insulin, 363
  integument, 103, 107, 111
  intestine, 189
  islets of Langerhans, 361
  Jacobson's organ, 209
  kidney, 264, **266**, 267
  labia majora, 302, **303**, 330
  labia minora, 120, 301, **303**
  lanugo, 139, 140, 143
  larynx, 232, 368
  leukocytes, 596
  limb bud, **519**
  lips, 159
  liver, 194, 585
  lungs, 239, **242**
  Meckel's diverticulum, 189
  melanocytes, 106
  mental processes, 629
  mesonephros, 263
  muscles, **519**, **524**, 529
  nails, **148**
  nipples, 123
  nose, 206
  occipital condyles, 464
  olfactory sense, 668
  ovarian hormones, 286, 375
  ovulatory cycle, 287
  pancreas, **193**, 197, 361
  pancreatic ducts, 192, **193**,
    197
  parasympathetic nervous
    system, **651**
  pelvis, 376
  penis, **322**, **328**
  pigment, 144

Man:

  placenta, 398
  plica semilunaris, 683
  posture, 47
  precava, 556
  progesterone, 398
  pro-otic somites, **515**
  races, 120
  red corpuscles, **596**
  reproductive system:
    female, 303, 305
    male, **322**
  respiratory organs, **242**, 244
  ribs, **473**
  sacrum, 430, 431
  seminiferous tubules, 310
  sensory areas in skin, 719
  sinuatrial node, 558
  sinuses, 207, 208
  skeleton, 46, 47
  skull, 449, **463**
  spermatozoa, **314**, 315
  sphenoid bone, 464
  spinal cord, 611
  spinal nerves, 639
  sternum, 475, 476
  STH (somatotrophic hor-
    mone), 388
  stomach, **180**
  study of, 4
  supracondyloid foramen,
    494
  sympathetic nervous sys-
    tem, **650**
  taste buds, 714, 715, 717
  teeth, 168
  temperature control, 47
  testis, 310–312, **321**, **322**
  thumb, 46
  thyroid cartilage, 342
  tongue, 165
  tonsils, 247
  trachea, 235, 349
  tracheal cartilages, 235
  urethra, 76, **322**
  urogenital system: female,
    302, 303
    male, 322
  uterus, 301
  vagus nerve, 647
  vermiform appendix, 189,
    190
  vertebral column, 430, **610**

Man:

  vertebral formula, 431
  vertebrate ancestry, 415
  villi, 191
  vitamin B_{12} requirement,
    185, 195
  vocal cords, 233, 717
Manatee, 52, **53**, 124, 174, 428,
  429, 739
Mandible, **467**
  (*See also* Bones; Cartilages)
Mandibular arch, 164, 214,
  218, 435, 438, 445, 641,
  644
Mandibular canal, **724**
Mandibular fossa, 464
Mandibular somite, **515**
Mandibular symphysis, 457,
  460, 467
Mandrill, 143, 738
Manes, 145, 146
Manis, 48, 431, 738
Manta, 22, 733
Mantle, 16, **17**
Mantle layer, 605, **606**
Manubrium, **475**, 492
Manyplies (omasum), 182
Marbled salamander, **31**
Mare, 286, 287, 306, 375, 398
  pregnant, 399
  (*See also* Horse)
Marginal layer, 605, **606**
Marmoset, 47, 48, 738
Marrow (*see* Bone marrow)
Marrow cavity, 88, 406, **408**,
  409, 704
Marsupial bear, 43
Marsupial flying squirrel, 43
Marsupial mole, 43, 429
Marsupial pouch, 42, **43**, 45,
  300, 313, 528
Marsupial wolf, 43
Marsupialia, 42, 737
Marsupials:
  caecum, 189
  characteristics, 42, 737
  circumvallate papillae, 717
  classification, 42, 737
  cloaca, 270, 327
  corpus callosum, 629
  Cowper's glands, 327
  cranial cavity, 468
  delayed implantation, 307

Marsupials:
dental formula, 173
epipubic bones, 494
eye, 693
geographical distribution, 44
Jacobson's organ, 209
mesonephros, 263
Müllerian ducts, 299
nipples, 123
oviducts, 299
panniculus carnosus muscle, 528
penis, 313, 327, 328
placenta, 244, 398
pouch of, 42, 43, 45, 300, 313, 528
reproduction, 299
scrotum, 313, 327
skull, 468
spermatozoa, **314**
teeth, 172
urogenital canal, 299
vaginae, 299
Marten, 739
Masseteric fossa, 467
Mast cells, 80, 84, 350
Mastodon, 52
Mastoid air cells, 706
Mastoiditis, 706
Matrix:
intercellular, 78, 81, **86**, 167, 404
of nail, 148
(*See also* Intercellular substance)
Maturation of tissues, 390
Mauthner's cells, 617
Maxillary arch, 438, **440**
Maxillary sinus, 466, 713
Maxillary teeth, 171
Meatus:
auditory: external, **702**, 707
internal, 455
penis, 269, **322**
Meckel's cartilage (*see* Cartilages)
Meckel's diverticulum, 188
Median appendages, 476
Median eminence, 574
Median eye, 35, 622
Median fins, 14, 19, 21, 31, 476, 477, 510
Mediastinal septum, **242**

Mediastinal space, 242
Mediastinum, 242
Medulla:
adrenal gland (*see* Adrenal medulla)
cerebral hemispheres, 629
hair, **141**
kidney, 266
ovary, 278, 284
thymus gland, 591
Medulla oblongata, 558, **607**, 617
gray matter, **612**, 613, 617, 641
visceral sensory area, 718
Medullary bodies (*see* Chromaffin bodies)
Medullary plate (*see* Neural plate)
Medullary rays, 266
Medullated nerve fibers, **93**–95, 97, 540, 605, 613, 629, 636, 649
Megakaryocytes, 85, 409, 597
Meibomian glands, 113, 114, 118, 683
Meiolecithal eggs, 57–60
Meiosis, 278
Meissner's corpuscles, 720, **721**
Melanin, 81, 83, 106–108
Melanoblasts, 106, 108, 144
Melanocyte-stimulating hormone, 387
Melanocytes, 80, 106, 107, 144, 387, 631
dermal, 81
Melanophages, 106
Melanophore-stimulating hormone, 383, 386
Melanophores, 106, 109, 624
dermal, 110
Melatonin, 624
Membrane, 405
cell, 92
permeability, 96
Membrane bone, 132, 404–406, 412, 413, 433
Membranous cranium, 432
Membranous labyrinth, 69, 74, **696**
Membranous urethra, 323
Memory, 604, 629
Meninges, 630–632, 634

Meninx primitiva, 630
Meniscus, 411
Menopause, 377
Menstrual cycle, 121, 376, 377
Menstruation, 304, 354, 376, 377
Mental foramen, **467**
Mental gland, 112, 116, **117**, **309**, 365
Menthol, 668
Merganser, **41**
Merkel's corpuscles, 720, **721**
Meroblastic cleavage, **60**, 545
Merocrine glands, 113, 114, 119
Mesectoderm, 435
Mesencephalon (midbrain), 69, 606, 615, 620, 621, 643, 646
Mesenchyme, 66, 68, 78, 80, 89, 537
hemopoiesis in, 85, 595
Mesenteric ganglia, 650
inferior, 652
superior, 156, 652
Mesenteries, 72, 73, 75, **155**, 185
ventral, 545
Mesoblastic somites, 503
Mesobronchus, **236** ›
Mesocardium, 546
Mesocoele, 609, 621
Mesocolon, 155, 185
Mesodaeum, 154
Mesoderm, 11, 61, 63–**64**, 66
derivatives, 71–72, 78, 89, 537
differentiation, 66, 71, **502**
epimeric, 71
formation, 63
hypomeric, 71, 641
lateral, 474
mesomere, 71
parietal, 64
prechordal, 61
somatic layer, 63–66, 68, 102, 503, **504**, 518
splanchnic layer, 63–66, 70, 72, 503, **504**, 517, 545, 641
visceral layer, 64, 70
Mesodermal pouches, 63
Mesogastrium, 155, 594

Monkeys:
  stomach, 182
  vermiform appendix, 189
Monocytes, **84**, 85, 592, 594
Monoecious animals, 329
Monoiodotyrosine, 338
Monophyodont dentition, 168,
    170, 172
Monotremata, 41, 737
Monotremes:
  adrenal glands, 353
  appendicular skeleton, 492
  characteristics, 41, 737
  circumvallate papillae, 717
  classification, 41, 737
  cloaca, 157, 189, 267, **298**,
    327, 528
  Cowper's glands, 323
  egg tooth, 175
  eggs, 287, 298
  embryonic, 175
  epipubic bones, 43, 494
  interclavicle, 486
  Jacobson's organ, 209
  mammary glands, 112, 299
  mesonephros, 263
  Müllerian ducts, 298
  ovaries, 287, **298**
  oviducts, 298
  panniculus carnosus muscle,
    528
  pectoral girdle, 492
  penis, **327**
  peritoneal funnels, 263
  reproductive tract of female,
    298
  ribs, 472
  skull, 468
  stomach, **182**
  temperature control, 42
  temporal fossae, 456
  testes, 310
  thyroid cartilage, 231
  tricuspid valve, 556
  ureters, 269
  urogenital sinus, **298**
  uterus, 298
  vertebrae, 428
Mormyrus, 112, 529
Morphology:
  definition, 2
  and physiology, 4

Motor areas in cerebral hemi-
    spheres, 620
Motor end plate, 530
Motor nerve fibers, 633, 636
    (*See also* Somatic motor
      columns; Visceral
      motor columns)
Motor neurons, 611
Motor roots, spinal nerves,
    606, 633
Mouse:
  adrenal glands, 353, 354
  castration effects, 354
  classification, 49, 739
  clitoris, 269, 302
  corpora lutea, 286
  Harderian glands, 693
  hypothalamus, 625
  mesonephros, 263
  newborn, 307, 593
  ovarian capsule, 287, 299
  ovulatory cycle, 287
  parathyroid glands, 347
  pituitary gland, 399
  preputial glands, 324, 328,
    713
  relaxin in, 375
  salivary glands, 162
  scales on tail, 138
  secondary sex characters, 364
  sweat glands, 121
  thymus gland removal, 593
  urethra of female, 269
  uterus, 306
  young at birth, 307, 593
Mouse deer, **596**
Mouth, 154, 157–**176**, 653, 654
  angle of, 438
  glands of, 68, 160–162
  lining, 68, 77, 551
  primitive, 68
  roof of, 439, 457, 458
  round, 19
  subterminal, 21, 445
  terminal, 20, 23
Movable joints, 411
MSH (melanocyte-stimulating
    hormone), 383
Mucin, 81, 111, 114, 116, 162
Mucopolysaccharide, 185, 325
Mucosa, 155, 156
  nasal, 713
  vaginal, 304

Mucous cells, 111, 184
Mucous connective tissue, 81,
    82
Mucous glands, **21**, **110**, 113,
    114, 116
Mucous membrane, 73
  Fallopian tube, 304
  mouth, 646, 653, 654
  nose, 645, 646, 653
  olfactory, **176**
  palate, 653
  pharynx, 647, 654
  uterus, 304
Mucus, 73, 111, 114
Mud puppy (see *Necturus*)
Mudfish (see *Amia*)
Müllerian duct, 276, **288**, 299
  masculine homologues, 324
  vestigial, 288, 318, 320
  (*See also* Oviduct)
Müller's fibers, 95
Multicellular glands, 111–113
Multipolar neurons, 97
Mumps, 162
Murkat, 50
Muscle:
  attachment, 86, 88
  definition of, 90
Muscle buds, **517**
Muscle cells, 503
Muscle contractility, 501
Muscle fibers, 501, 508
  ectodermal, 68, 674
  involuntary, 90, 674
  smooth, 90, 505, 674
  striated, 90, 674
  voluntary, 90
Muscle glycogen, 361
Muscle position, sense of, 667
Muscle segment, 71, 639
  (*See also* Myotomes)
Muscles:
  abductor, 505, 519
  acromiodeltoid, 524
  acromiotrapezius, 524
  adductor, 505, 519
    femoris, 523
    longus, 523
    mandibulae, 524, 525,
      530
  anconeus, 521
  appendicular, 504, 510, 514,
    515

Muscles:
  as metameric structures, 11
  mimetic, 529
  multifidus spinae, 513
  mylohyoid, 526
  nerve supply, 507
  oblique, 510, 644
    external, 511–513
    inferior, **516**, 644, 682
    internal, 511–513
    superficial and deep, 511
    superior, **516**, 644, 682
  obturator: externus, 523
    internus, 523
  omohyoid, 513, 515
  opercular, 705
  orbicularis: ciliaris, 680
    oculi, 529, 682
    oris, 529
  orbital, 344
  origin, 504, 505
  panniculus carnosus, 528
  papillary, 554, 556
  parietal, 504, 506
  patagial, 528
  pectoral, 46, 91
  pectoral girdle and forelimb,
    519
  pectoralis, 519, 520, 528
    major, 520
    minor, 520
  pectoriscapularis, 520
  peroneus, 511
  pharyngeal, 216, 526, 648
  platysma, 103, 526, 529
  precoracohumeralis, 521
  pronator, 505, 521
  protractor lentis, 686, 689,
    690
  pterygoid, 526
  puboischiofemoralis, 522,
    523
  puboischiotibialis, 521–523
  pubotibialis, 522
  pyriformis, 522
  quadratus femoris, 523
  quadratus labii inferiores,
    529
  quadratus lumborum, 513
  quadriceps femoris, 523, 657
  rectus: abdominis, 509–513,
    517, 519
    cervicis, 514

Muscles:
  rectus: external, **516**, 645,
    682, 689
    femoris, 523
      anticus, **511**
    inferior, **516**, 644, 682
    internal, **516**, 644, 682
    internus minor, **511**
    lateral, 682
    medial, 682
    superior, **516**, 644, 682
  red, 91
  retractor: bulbi, 516, 645,
    684, 688, 690
    lentis, 687
  rhomboideus, 520
    capitis, 520
    major, 520
    minor, 520
    occipitalis, 520
  rotator, 505
  sacrospinalis, 513
  sartorius, 523
  scalenus, 512–514
  semimembranosus, 523
  semispinalis, 512
  semitendinosus, 523
  serratus, 512–514
    anterior, 519, 520
  smooth, 68, 89–91, 109,
    120, 503, 504, 527
  somatic, 504
  sphincter, 505, 658
    anus, internal, 527
    of Boyden, 193
    cloacal, 528
    colli, 528, 529
    iris, 654, 674
    marsupial pouch, 528
    mouth, 505
    of Oddi, 193, 195
    pyloric, 527
  spinalis dorsi, 512, 513
  spinotrapezius, **524**
  stapedius, 704
  sternalis, 528
  sternocleidomastoid, 520,
    526, 648
  sternohyoid, 513, 514
  sternomastoid, 526
  sternothyroid, 513, 514
  striated, 89–**91**, 156, 501,
    504, 506

Muscles:
  subarcuales, **515**
  subclavius, 520
  supinator, 505, 521
  supracoracoid, 521
  supraspinatus, 521
  temporal, 526
  tensor fasciae latae, 524
  tensor tympani, 526, 704
  tenuissimus, 523
  teres major, 520
  teres minor, 521
  thyrohyoid, 513, 514
  tibialis anticus, 511
  tongue, 91, 512, 515, 648
    intrinsic, 515, 648
  transverse abdominal, 513
  transverse spinalis system,
    512
  transverse thoracic, 514
  transversus, 511, 512, 514,
    520
  trapezius, 521, 525, 526,
    648
  triangularis sterni, 513
  triceps, 521
    brachii, **511**
    femoris, **511**
  trunk, 507–514
  vastus: externus, **511**
    intermedius, 523
    lateralis, 523
    medialis, 523
  visceral, 510, 641
  voluntary, 72, 89, 90, 156,
    501, 504, 506, 633
  white, 91
  xiphihumeralis, **524**
  zygomatic, 529
Muscular activity, 530
Muscular energy, 507
Muscular media cells, 267, 538
Muscular system, 501–533
Muscular tissue, 73, 89–92,
  501
Muscularis mucosae, **156**
Musculi pectinati, 556
Musk, 117
Musk glands, 117
Muskellunge, 71
Muskrat, 138, 145
Mustelus, 291, 733

Occipital segment, 436
Occipital vertebrae, 432, **433**
Octapeptides, 383, 386
Odobenidae, 310
Odontoblasts, 166, 167
Odontognathae, 39, 736
Odontoid process, 426, 429
Odors, 714
OIH (ovulation-inducing hormone), 383
Oil glands, 113, 114, 118
Olecranon fossa, 494
Olecranon process, 494
Olfactoreceptors, 667, 708, 713, 714
Olfactory bulb, 626, 627, 642
  accessory, 642
Olfactory capsule, **176**, 642, 709
  (*See also* Nasal capsule)
Olfactory centers, 627, 628
Olfactory epithelium, 97, 204, 206, 207, 642, 708–710, 712, **713**
Olfactory foramina, 437, 465, 642, 712
Olfactory glomeruli, 642
Olfactory hairs, 708, 710, 713
Olfactory lobes (rhinencephalon), 607, 626–628, 642, 712
Olfactory mucous membrane, **176**, 642, 713
Olfactory organs, 203, 626, 642, 708
  comparative anatomy, 708–714
Olfactory pits, 69, 204, 208
Olfactory placodes, 69, 204, 642, 708
Olfactory plate, 69
Olfactory sac, 158, **176**, **709**, 710
Olfactory sense, 465, 625, 640, 708–714
Olfactory stalk, 626, 627
Olfactory stimuli, 717
Olfactory tract, 626, 642
Olfactory vesicles, 708
Oligodendroglia, 95, 98
Omasum, 182
Omental bursa, 185
Omentum, 72, 73, **155**

Omosternum, **474**
Ontogeny, 6
Onychodactylus, 147
Oöcytes, 278, 282
Oögenesis, 278, 345
Oögonia, 278
Opercular aperture, 215, 218–220
Opercular canal, **724**
Opercular chamber, 220
Opercular fold, 564
Opercular gills, 217
Operculum:
  anuran larvae, **219**
  fishes, 23, 115, **215**, 442, 447
  middle ear, 705
Ophidia (*see* Serpentes)
Ophidians (*see* Snakes)
Opisthocoelous vertebrae, **422**
Opisthocomus critatus, **147**, 491
Opisthonephric tubules, **256**
Opisthonephros, 253, 255, **256**, 309, 317, 595
  comparative anatomy of, 258–261
  versus pronephros, 256
Opossum, 42, 43, 287, 299, 306, **328**, 353, 713, 737
  embryo, 300, 464
Opossum rat, 42
Optic canal, 672
Optic capsule, 432, **433**
Optic chiasma, 625, **643**
Optic cup, 69, 643, 670–**672**, 677
Optic foramen, 461
Optic lobes, **607**, 621, 643
  (*See also* Mesencephalon)
Optic organs, 668
Optic pedicel, **516**, 683, 685
Optic stalk, 643, 670–672
Optic tectum, 616, 621, 643, 646
Optic tract, 643
Optic ventricles, 609, 621
Optic vesicles, 69, 621, 642, 670, **671**
  as organizer, 67, 671
Optical axis, 674
Optical center, 678, 679
Ora serrata, **669**, 670, 672, 680, **681**, **686**

Ora terminalis, 670
Oral cavity, 159–175
  glands of, 68, 160–**162**
  (*See also* Mouth)
Oral contraception, 379, 380
Oral hood, 14, 158
Oral membrane, 68, 157
Oral plate, 68
Oral valve, **215**
Orangutan, 47, 495, 738
Orbit, 444, 453, 455, 465
Orbital fissure, 464
Orbital fossa, 457
Orbital glands, 162
Orbital process, **446**
Orbital shelf, **445**
Order, definition, 12
  (*See also under* specific names)
Organ:
  of Corti, 701–**703**, 708
  of Müller, 15, 394
Organ systems, 98
Organizers, 67, 677
Organs, 57, 64, 66
  definition, 98
  (*See also under* specific names)
Ornithorhynchus, 41, 693, 737
  (*See also* Platypus)
Oronasal groove, 204, 710
Orthagoriscus, 614
Orycteropus, 51
Os cornu, 149
Os penis, 328, 404
Os priapi, 404
Os uteri, 299
Osmotic pressure, 541
Osmotic regulation, 397
Ossa cordis, 557
Ossification centers, 407, 436
Osteichthyes, 23, 24, 27, 446, 557, 733
Osteitis fibrosa, 349
Osteoblast, 406
Osteoclasts, 409
Osteocytes, 87, 88, 167, 406
Ostium tubae, 285, 287, 288, 295, **296**, **298**
Ostracoderm theory, 479
Ostracoderms, 18, **19**, 21, 125–127, 166, 404, 709

Serotonin (5-hydroxy-trypta-
mine), 80, 190, 335, 350,
599
Serous fluid, 73
Serous glands, 114
Serous layer, 155
Serous membrane, 73, 185, 304
Serpentes, 36, 457, 736
Serranus, 329
Sertoli cells, 307, 370, 392
Serum (*see* Blood serum)
Sesamoid bones, 404, 491, 494,
495, 505
Sex:
determination of, 276, 277
sense of, 667
Sex cells, 307
Sex chromosomes, 277
Sex glands, 78, 118
Sex hormones, 195
female, 284
Sex-linked characters, 144, 598
Sex reversal, 277, 329, 372
Sexual activity, 117
Sexual allurements, 192
Sexual cycles (*see* Reproduc-
tive cycles)
Sexual desire, 723
Sexual dimorphism, 175, 364
Sexual patterns, 625
Sexual reproduction, 56, 274,
275
Sexual sensations, 723
Sexual skin, monkeys, 375, 377
Seymouria, 33
Shaft:
feather, 133, 134, **136**
hair, **141**
rib, 472, **473**
Sharks, 21, 22, 127, 325, 372,
439, 733
aortic arches, 560, 562
erythrocytes, **596**
lack of gall bladder, 194
respiration, **215**
teeth, 169
(*See also* Elasmobranchs)
Sheath of Schwann (neuro-
lemma), 93, **94**
Sheath cells, **94**, 95
Sheep, 54
glans penis, 328
hair, 141

Sheep:
horns, 149
ovulatory cycle, 277, 287
pancreas, 197
pancreatic ducts, 197
parathyroid glands, 347
pituitary gland, 574
placenta, 347, 398
sweat glands on muzzle, 121
teeth, 174
uterus, 301
vertebral formula, 431
Shell:
egg, **297**, 346
turtle, 34
Shell gland, 290, 294–**296**
Shell membranes, **297**
Shrew, 41, 162, 285, 322, **328**,
693, 713, 737
elephant, 285
European, **45**
Sight, sense of, 615, 630, 667,
668–694
Silurus, 590
Simple epithelium, 73–76
Simplex uterus, **301**, **303**
Single cones, 687, 689, 690
Sinuatrial aperture, 548, **550**
Sinuatrial node, 558
Sinuatrial valves, 548, 556
Sinus rhomboidalis, 614
Sinus venosus, **543**, 544, 547–
550, 555, **572**, **573**
Sinuses, 207, 466, 712
blood, 354
genital, **578**
hepatic, **577**
kidney, 266
postcardinal, **578**, 579
venous, 632
(*See also* Veins)
Sinusoids, 542
adrenal gland, 354, 542
carotid body, 248
liver, 536, 574
Siphon, 16, **17**, 325
Siphon sac, 325
Siren, 170, **171**, 219, 233, 486
dwarf, 314
Sirenia, 52, 428, 739
Sirenians:
ears, 707
flippers, 495

Sirenians:
hair, lack of, 145
lungs, 239
mammary glands, 124
pubis and ischium remnants,
493
salivary glands, 162
sternum, 475, 476
sweat glands, lack of, 119
teeth, 166, 172
trachea, 234
vertebrae, 428
Sirenidae, 294
Skates, **22**, 23, 115
cervicobrachial plexus, 638
characteristics, 733
classification, 21, 733
gill slits, 214
pectoral girdle, 480
pterygopodial glands, 325
reproduction, 325
spiracles, 216
Skeletal muscle, 89–91, 368,
501, 503, 667
Skeletal musculature, 503
Skeletal system, 403–500
Skeletogenous septum, 422,
468, 479
(*See also* Lateral septum)
Skeleton, 86, 403
chicken, **427**
effect of androgens on, 368
growth of, 345
of heart, 557
origin, 72
sex differences, 368
vertebrate, 8
(*See also under* specific
headings)
Skin (*see* Integument)
Skin glands (*see* Integumentary glands)
Skin grafts, 593
Skin joints, 103
Skink, 736
Skull, 18, 25, 29, 413, 712
comparative anatomy, 431
development, 432–443
evolution, 449
Skullcap, 405
Skunk, 50, 122
Sleep, 624

Sturgeon:
snout, 716
spiracle, 217
swim bladder, 226
taste buds, 716
teeth, lack of, 170
vertebral column, 420
Sty, 119, 683
Styloid process, 467, 493
Stylomastoid foramen, **465**
Subarachnoid space, 631, 632, 696
Subclass, definition, 13
Subcutaneous tissue, 79, 102, 103
Subdural space, **631**
Sublingual gland, 161, 172, 646, 653, 654
Submandibular duct, 161
Submandibular gland, 161
Submaxillary duct, 161
Submaxillary ganglion, 645, 646, 654
Submaxillary gland, 114, 161, 646, 652, 654
Submucosa, 155, **156**, 595
Subneural gland, 394
Suboccipital nerve, 638
Subpharyngeal gland, 338
Subphylum, definition, 13
Subpodocytic space, 268, **269**
Subterminal mouth, 21
Subunguis, 147, **148**
Subvertebral trunk, 590
Succus entericus, 191
Sucker (fish), **128**, 716
Sucking, 159
Sucrase, 191
Sudoriparous glands, 118
Sugars, 267, 507
in blood, 361–364
pentose, 3
Sulcus:
brain, 619, 628
spinal cord, 610, 614
Sulcus unguis, 148
Sulfonamides, 337
Summation, 96
Sunburn, 104, 107, 108
Sunfish, marine, 614, 733
Superclass, definition, 13
Superior colliculi, 621
Superior ganglion, 647

Superior lobe, lung, 239
Superior notch, **467**
Supernumerary mammary glands, nipples, and teats, 124
Superorder, definition, 13
Suprachoroid layer, **686**
Supracondyloid foramen, 494
Supraoptic nucleus, 383
Supraorbital canal, **724**
Supraorbital crest, 445
Supraorbital groove, 206
Suprarenal bodies, 350
(*See also* Adrenal gland)
Suprarenal glands (*see* Adrenal gland)
Suprarenal tissue, 69
Supraspinous fossa, **493**
Suprasternal notch, 341
Supratemporal arcade, 456–459, 461
Supratemporal canal, **724**
Supratemporal fossa, 456, **457**, 459
Suspensor-stapedial ligament, 705
Suspensory ligament, lens, **669**, 670, 680, **681**, **686**, 689
Sustentacular cells, 708
Sutural ligament, 411
Sutures, 411
Swallowing:
act of, 232, 341, 457, 706
of air, 203
initiation of, 176, 203
Swan, 132, 233, 298, 327, 426, 475
Sweat, 119
cold, 120, 653
color, 121
Sweat glands, 68, 89, **104**, 106, 112, 117–122, 653, 656
absence of, 119
apocrine, 113, 114, 119–122
dog, 164
ducts of, 105, 112
eccrine, 113, 119
Sweetbreads, 196, 246
Swim bladder, 27, 28, 203
absence of, 21, 204
bilobed, 226
blood supply, 227

Swim bladder:
comparative anatomy, 222–226
functions, 27, 219, 226, 700
phylogenetic origin, 221
physoclistous, 27, 221
physostomous, 27, 221, 239
Symbranchiates, 215
Sympathetic center, 624
Sympathetic ganglion, 351, 606, 653, 654, 656
(*See also under* specific headings)
Sympathetic nerve fibers, 542, 681
Sympathetic nervous system (*see* Nervous system)
Sympathetic trunk, **634**, 652, 656
Sympathin, 96, 652
Symphysis:
mandibular, 514
pubic, 374, 411, 483, 489, 490, 493, **494**
Synancia, 114
Synangium, **550**, 551
Synapse, 95, 96, 652
Synapsid condition, 456, 465
Synaptic knobs, 629
Synarthrosis, 411
Syncitium, 71, 74, 92
Synephrine, 359
Synergism, 345, 371, 389
Synovial capsule, 411
Synovial cavity, 411, **412**
Synovial fluid, 411
Synovial layer, 411
Synovial membrane, 411
Synsacrum, 427, 428, 472, 490
Syrinx, 231, 234
Systematic biology, 4
Systemic arch, **550**, 563
Systemic circulation, 551
Systemic division, 535
Systemic trunk, 553, 556
Systole, 551
Systolic blood pressure, 538

T filaments, 506
Tachyglossus, 41, **42**, 737
(*See also* Echidna)
Tactile organs, 719

Tonus, 505, 506
Tooth germ, 166
Tooth plates, 449
Tori, 139, 147, 148
Tornaria larva, 14
Torpedo, 22, 125, 517, 529, 530, 726, 733
Tortoise shell, 131
Tortoises, 32, 34, 146, 454, 735
  embryonic stages, **7**
Torus (*see* Tori)
Toucan, 134
Touch, sense of, 667, 719
Trabeculae:
  bony, 407
  carneae, 556
  cranii, 432
  liver, 123
Trachea, 71, 229, 233–235, 653, 655
  Jackass penguin, 233, **234**
  lining of, 75
  swans and cranes, 233, 475
Tracheal gills, **10**
Tracheal tubes, 9
Tragulus, **596**
Transitional epithelium, 77
Transitional layer, epidermis, **75, 102**, 103
Transverse processes, 416, 423, 429, 468, 469, 510
  accessory, 430
Triangular ligament, 195
Trichechus, 428
Triconodont shape (teeth), 175
Tricuspid valve, 556, **585**
Trifluormethyl-nitrophenol (TFM), 20
Triiodothyronine, 337, 338, 343, 349
Triiodothyropropionic acid, 345
Triton, 329
Tritubercular position, 175
Trituration, 181
Triturus, **294**, 393, 725, 735
Trochanter, **410**
Trochanteric (digital) fossa, **410, 493**
Tropomyosin, 506
True chambers, heart, 543
True cortex, adrenal gland, 354

True hermaphroditism, 329, 330
True jaws, 20
True ribs, 132, 468, 472
True teeth, 165, 166, 170
Truncus arteriosus, **518**, 544, 547, 567
Trunk:
  aortic, **550**
  arterial, 544, 553
  of body, 504
  Dolichoglossus, 14
  elephant, 52, 206, 713
  Necturus, **32**
  pulmonary, 554
  subvertebral (lymphatic), 590
  sympathetic, **634**, 652, 656
  systemic (arterial), 553
Trunk musculature, 504, 507
Trunk vertebrae, 420, 423, **425**
Trypsin, 197
Trypsinogen, 197
Tryptamine, 624
Tryptophane, 599
TSH (thyroid-stimulating hormone; thyrothrophic hormone), 341, 343, 344
Tuatara (see *Sphenodon punctatum*)
Tuber cinereum, 382, 574, 625
Tubercle:
  gizzard, 181
  of Lower, 556
  rib, 429
  teeth, 175
Tubercular facet, 429, 430, 469
Tubercular head, rib (*see* Tuberculum)
Tuberculosis, 243
Tuberculum, rib, 416, 469
Tuberculum impar, 163, 164
Tuberculum prelinguale, 205, 240
Tuberosity, **493**
Tubular glands, 112, **113**, 118, 119
Tubulidentata, 51, 739
Tufts:
  formed by feathers, 707
  formed by hairs, 143, 145
Tumors, 369
Tunic, 16, **17**

Tunica adventitia, 589
Tunica albuginea, 310
Tunica externa, **538**, 539, 589
Tunica interna, **538**, 539
Tunica intima, **538**, 589
Tunica media, 267, **538**, 539, 589
Tunica vaginalis, 310, **311**
Tunicata (*see* Urochordata)
Tunicates, 16, 394
Tunicin, 16
Turbinal folds, 710
Turkey, 56, 118, 370
Turtles, 34
  absence of teeth, 170
  anus, 34
  beak, 34, 146, 158, 170, 454
  carapace, 34, 426, 470
  characteristics, 34, 735
  classification, 32, 735
  cloacal respiration, 553
  concha, 711
  dermal skeleton, 8, 131, 403, 413
  ductus arteriosus, 566
  ecdysis, 131
  egg tooth, 172
  eggs, 295
  endoskeleton, 131
  epaxial muscles, 512
  epipubis, 494
  esophagus, **178**
  eye, 690
  gill rudiments, 221
  glands: adrenal, 352
    musk, 117
    oral, 161
    oviducal, 294
    parathyroid, 346
    thyroid, 341
  heart, 553
  hyoid apparatus, 456
  hypaxial muscles, 512
  Jacobson's organ, 711
  lack of neutrophils, 84
  limbs, 489
  lips, 158
  lungs, 236
  marine, 34
  mechanism of respiration, 241, 244
  origin, 33
  ovaries, 283

Vertebral section, rib, 471, 472
Vertebrarterial canal, 429, 470
    (*See also* Foramen
        transversarium)
Vertebrata, 13, 18, 731
Vertebrate plan, 5
Vertebrates:
    ancestry, 4, 71, 270, 673
    appendages, 477
    characteristics, 4
    ear, 694
    embryos of, 6, **7**
    erythrocytes, 596
    evolution, 413
    eye, 668
    integument, **107**
    intestines, 186
    limbs and girdles, 484
    neural tube formation, 65
    ribs, 469
    skeleton, 8, 86
    teeth, 404
    terrestrial, 30
    tongues, 162
Vesicles of Savi, 726
Vesicular connective tissue, 8,
    9, 83, 414
Vessels of vessels, 539
Vestibular concha, 711
Vestibular ganglion, 646
Vestibular glands, 303
Vestibular membrane, 701
Vestibular region, 206, 646,
    712
Vestibular window (*see*
    Fenestra ovalis)
Vestibule:
    amphioxus, 14, 15, 158, 159
    genital tract, 301
    lamprey, 372
    larynx, 232
    mouth, 158
    vagina, 301
Vibrissae, 145, 720
Villi:
    chorionic, 399
    intestinal, 89, **156**, 186,
        188, 191, 591
    uterine, 216, 290
Vipers, 36, 171, 458
Virilism, 354
Virility, 145

Visceral arches, 210, 214, 445,
    452, 463, 468, 503, 523,
    530, **641**
    attachment of teeth to, 168
    fishes, 214, 435, 449
    nerve supply, 641
    support of tongue, 164
Visceral clefts (*see* Gill clefts)
Visceral functions, 612
Visceral furrow, 209
Visceral layer, **254**, 311
Visceral motor columns, **612**,
    617, 633, 649
Visceral musculature, 510,
    526, 641
Visceral nerve fibers:
    motor, 523, 634–636, 641,
        644–649
    sensory, 612, 633–636,
        641, 645–648
Visceral peritoneum, **155**, **156**,
    179, 304, 544
Visceral pleura, 239, **242**,
    243, 544
Visceral pouch (*see* Gill
    pouches)
Visceral ramus, **606**, 634–636,
    638, 652, 653, 655, 656
Visceral sensory columns, **612**,
    633
Visceral skeleton, 210, 413,
    431, 432, 436, 443, 445
    muscles of, 523
Visual green, 689
Visual purple, 676, 689
Vitamins, 411
    A, 77, 195, 676
    B, 182, 370
    B$_{12}$, 185, 195
    D, 103, 349
    E, 312
    K, 597
Vitelline circulation, 585
Vitelline membrane, 296, **297**
Vitreal cavity, 670
Vitreous body, 670, 681
Vitreous humor, 670, 681, 689
Vitrodentine, 127, 166, 167
Viviparous animals, 324
Vocal cords, 31, **230**, 231, 717
    false, 232, **239**
    true, 82, 232, **239**
Vocal folds, 232

Vocal pouch, 231
Vocal sacs, 230, 293
Voice, 233
    modification, 159
    resonance, 208
Voice box (*see* Larynx)
Voluntary muscles, 72, 89, 90,
    156, 501, 504
Voluntary musculature, 72,
    526, 612
    (*See also* Voluntary
        muscles)
Vomerine teeth, 170, **171**, 453
Vomeronasal (Jacobson's)
    organ, **208**, 642, 711
Von Ebner's glands, 716, 718
Von Recklinghausen's disease,
    349
Vulva, 301, 372

Walking goby, 244
Walrus, 50, 174, 328, 739
Warbler, **41**
Warm-blooded animals
    (*see* Homoiothermous
        animals)
Warm points, 722
Warts, 116
Wastes, 83, 250, 257, 261,
    265, 270, 271
Water cells, 183
Water dog (see *Necturus*)
Watson-Crick model, DNA, 3
Wattles, 366
Wax-producing (ceruminous)
    glands, 112, 121, 170
Weasel, 50, 122, 144, 739
Weberian ossicles, 225, 699,
    700
Webs:
    appendages, 51
    feet, 30, 41, 45, 132
    toes, 36
    wings, 46, 495
Whalebone (baleen), 51,
    160, 706
Whales, 41
    baleen, 51, 160, 706
    blubber, 51
    blue, 41, **51**, 286, 694
    bony plates in skin, 138, 413
    bottlenosed, 51

Whales:
  bronchi, 235
  characteristics, 50, 51, 739
  chevron bones, 431
  classification, 50, 739
  colic caecum, 189
  ear, 706, 707
  Eustachian tubes, 706
  eye, 694
  fins, 390
  flippers, 5, 495
  flukes, 51
  forelimb, 6
  gall bladder, lack of, 194
  gray, 51
  gustatory sense, lack of, 718
  hair, 50, 139, 145
  Harderian glands, 693
  humpbacked, 51
  lips, 159
  lungs, 50, 239
  mammary glands, 50, 124
  median fins, 476
  nares, 51, 466
  nipples, 124
  os penis, 328
  palate, 465
  pelvic girdle, 51, 493
  pituitary gland, 397
  pubis and ischium,
      remnants of, 493
  ribs, 472
  right, 51
  salivary glands, 162
  skin glands, 119
  skull, 466
  sperm, 51
  spout, 51
  sternum, 475, 476
  stomach, 183
  sulfur-bottom, 51
  teeth, 172, 464
  tongue, 163
  toothed, 51, 168, 172, 712
  trachea, 234
  uterus, 301
  vertebrae, 429, 430

Whales:
  whalebone, 51, 172, 466,
      712
  (See also Cetacea)
Wharton's duct, 161
Wharton's jelly, 82
Wheel organ, 15, 158, 394
Whey protein, 184
Whiskers (see Vibrissae)
White of eye, 669
White commissure, 613
White corpuscles (see
      Leukocytes)
White matter, 605
  brain, 95, 97, 615, 616
  cerebellum, 619
  peripheral nerves, 97
  spinal cord, 95, 97, 611, 613
White pulp, 594
White ramus, 634–636, 649,
      653, 655, 656
White thrombus, 598
Whorls (hair), 143
Windpipe (see Trachea)
Wing feathers, 134, 491
Wing membrane, 495
Wings, 33
  bat, 5, 46
  birds, 33, 37, 46, 490
  flying dragon, 491
  pterosaur, 33
  rudimentary, 39
Wishbone, 179, 486, 490
"Witches'" milk, 124
Wolf, marsupial, 43
Wolffian body, 262
Wolffian duct, 253, 262, 316
Wombat, 43
Woodpecker, 40, 41, 163, 463,
      692
Wool, 141
Wrist, 484, 491, 496

X chromosome, 277
X zone, adrenal gland, 354
Xanthophores, 109

Xenopus laevis, 129, 147, 244,
      701
Xiphisternal cornua, 475
Xiphisternum, 474, 475
Xiphoid process, 475

Y chromosome, 277
Yellow body (see Corpus
      luteum)
Yellow bone marrow, 410
Yellow elastic fibers (see
      Elastic fibers)
Yellow spot, retina, 674, 675
Yolk, 34, 57, 58, 60, 62, 154,
      279, 297
Yolk plug, 61, 62
Yolk sac, 71, 154, 397, 572
  blood cell formation by,
      537, 595
  endoderm, 276
  respiration, 244
Yolk-sac placenta, 244, 397,
      398
Yolk stalk, 82, 154, 188

Z band, 506
Zaglossus, 41
Zebra, 53, 739
Zinc, 190
Zoarces, 255
Zona fasciculata, 353, 357
Zona glomerulosa, 353, 357
Zona pellucida, 284
Zona reticularis, 353
Zonation of adrenal gland,
      353
Zonula of Zinn, 670
Zoology, definition, 1
Zygantra, 426
Zygapophysis, 423
  (See also Postzygapophysis;
      Prezygapophysis)
Zygomatic arch, 459, 465
Zygomatic process, 465, 466
Zygophenes, 426
Zygote, 57, 275
Zymogen cells, 184